Computational Neuroscience
Trends in Research, 1997

Computational Neuroscience
Trends in Research, 1997

Edited by

James M. Bower
California Institute of Technology
Pasadena, California

Plenum Press • New York and London

Library of Congress Cataloging in Publication Data

Computational neuroscience: trends in research, 1997 / edited by James M. Bower.
   p.   cm.
   "Proceedings of the Annual Computational Neuroscience Conference, held July 14–17, 1996, in Boston, Massachusetts"—T.p. verso.
   Includes bibliographical references and indexes.
   ISBN 0-306-45699-0
   1. Nervous system—Computer simulation—Congresses. 2. Nervous system—Mathematical models—Congresses. 3. Neural networks (Neurobiology)—Congresses. I. Bower, James M. II. Computational Neuroscience Conference (1996: Boston, Mass.)
   [DNLM: 1. Neurosciences—congresses. 2. Neurons—congresses. 3. Computer Simulation—congresses. 4. Models, Neurological—congresses. 5. Neuronal Plasticity—congresses. 6. Neural Networks (Computer)—congresses. WL 100 C738 1997]
QP357.5.C64 1997
612.8′01′13—dc21
DNLM/DLC   97-23178
for Library of Congress   CIP

QP
357
.5
.C64
1997

Proceedings of the Annual Computational Neuroscience Conference,
held July 14 – 17, 1996, in Boston, Massachusetts

ISBN 0-306-45699-0

© 1997 Plenum Press, New York
A Division of Plenum Publishing Corporation
233 Spring Street, New York, N. Y. 10013

http://www.plenum.com

10 9 8 7 6 5 4 3 2 1

Printed in the United States of America

PREFACE

This volume includes papers presented at the Fifth Annual Computational Neuroscience meeting (CNS*96) held in Boston, Massachusetts, July 14 - 17, 1996. This collection includes 148 of the 234 papers presented at the meeting. Acceptance for meeting presentation was based on the peer review of preliminary papers originally submitted in May of 1996. The papers in this volume represent final versions of this work submitted in January of 1997.

As represented by this volume, computational neuroscience continues to expand in quality, size and breadth of focus as increasing numbers of neuroscientists are taking a computational approach to understanding nervous system function. Defining computational neuroscience as the exploration of how brains compute, it is clear that there is almost no subject or area of modern neuroscience research that is not appropriate for computational studies. The CNS meetings as well as this volume reflect this scope and diversity.

In order to emphasize the interrelated nature of computational neuroscience research, the papers in this volume are grouped into five different levels of investigation and analysis: subcellular, cellular, network, systems, and methodology. The papers found in each category represent research using a wide range of experimental preparations, analysis techniques, and technical approaches. In my view the focus of computational neuroscience on computational problems across numerous particular preparations and systems is one of the strengths of our field. As neuroscientific research as a whole continues to become more and more specialized, this breadth of interest and attention is not only refreshing, but I believe increasingly necessary. In this regard it is particularly encouraging to see the growing number of computational studies being conducted at the cellular and molecular levels. Perhaps no where else in neuroscience is the risk of getting lost in the trees and separated from overall brain function as great.

Of course, ultimately, any separation of brain studies into categories will be defeating. It is already clear, for example, that subcellular relationships can infleunce nervous system function even at the highest levels. Accordingly, it is easy to predict that computational studies initiated at one level of investigation will inevitably lead to other levels as well. Perhaps for this reason some of the most animated discussions at the CNS meeting each year involve cross-level analysis. The opportunity to interact with neurobiologists with such broad interests is one of the best features of the CNS meetings.

Jim Bower

REVIEWERS FOR CNS*96

The papers presented in this volume were first submitted in preliminary form in February of 1996. Each submitted paper was peer reviewed prior to its acceptance at the meeting under the supervision of the program committee. The meeting organizers are particularly thankful for the efforts of the reviewers in assuring acceptance of the highest quality papers.

CNS*96 PROGRAM COMMITTEE

- Jim Bower (California Institute of Technology)
- John Miller (University of California, Berkeley)
- Charlie Anderson (Washington University)
- Axel Borst (Max-Planck Institute, Tuebingen, Germany)
- Nancy Kopell (Boston University)
- Christiane Linster (Harvard University)
- Mark Nelson (University of Illinois, Urbana)
- Maureen Rush (California State University, Bakersfield)
- Karen Sigvardt (University of California, Davis)
- Philip Ulinski (University of Chicago)

CNS*96 REVIEWERS

Larry F. Abbott, Brandeis University; Charles H. Anderson, Washington University School of Medicine; Pierre Baldi, California Institute of Technology; Alexander Borst, Max-Planck-Society; Ron Calabrese, Emory University; Catherine Carr, University of Maryland; Erik De Schutter, Born Bunge Foundation; Bard G. Ermentrout, University of Pittsburgh; Michael Hasselmo, Harvard University; William R. Holmes, Ohio University; Gwen Jacobs, University of California at Berkeley; Leslie Kay, California Institute of Technology; Michel Kerszberg, Institute Pasteur; Nancy Kopell, Boston University; Zhaoping Li, Hong Kong University of Science and Technology; Christiane Linster, Harvard University; Bill Lytton, University of Wisconsin; Bartlett W. Mel, University of Southern California; John Miller, University of California at Berkeley; Kenneth D. Miller, University of California at San Francisco; Mark E. Nelson, University of Illinois; Bruno A. Olshausen, Massachusetts Institute of Technology; Mike Paulin, University of California at Los Angeles; Klaus Pawelzik, Universität Frankfurt; John Rinzel, MRB/NIDDK;

Maureen E. Rush, California State University, Bakersfield; Idan Segev, Hebrew University of Jerusalem; Shibab Shamma, University of Maryland; Gordon Shepherd, Yale University School of Medicine; Karen A. Sigvardt, University of California at Davis; Brian Smith, Ohio State University; Nelson Spruston, Northwestern University; Michael Stiber, Hong Kong University of Science and Technology; Greg Stuart, Australian National University; Philip S. Ulinski, University of Chicago; Gene V. Wallenstein, Harvard University; Charles Wilson, University of Tennessee; and Tony Zador, Salk Institute

SUPPORTING AGENCIES

National Institute of Mental Health and National Science Foundation

CONTENTS

Section III: Network

Section IV: Systems

Section V: Methodology

SECTION I

SUBCELLULAR

1

ACTIVITY-DEPENDENT REGULATION OF INHIBITION IN VISUAL CORTICAL CULTURES

Andrew DeWan, Lana C. Rutherford, and Gina G. Turrigiano*

Department of Biology and Center for Complex Systems
Brandeis University
Waltham, Massachusetts

Maintaining the correct balance of inhibition and excitation is extremely important for normal cortical function. Too little inhibition can lead to epileptiform activity, whereas too much inhibition can severely depress cortical responsiveness. This suggests that the balance of inhibition and excitation in cortical circuits should be tightly regulated. In visual cortex, activity has been shown to affect expression of the inhibitory neurotransmitter GABA in a manner consistent with a role in balancing excitation and inhibition; blocking activity in one eye leads to a down-regulation of GABA in the corresponding ocular dominance columns (Hendry and Jones, 1986). These data suggest that the level of activity is acting through some feedback signal to locally adjust the strength of cortical inhibition, although the mechanism by which this occurs remains unclear. Here we use a culture system to explore the role of activity in the control of cortical inhibition. We have found that blocking activity in culture leads to a reversible decrease in the number of neurons immunopositive for GABA, and that this decrease can be prevented by the coapplication of brain-derived neurotrophic factor (BDNF). These data suggest that activity levels can continuously adjust cortical inhibition in a bi-directional manner through a BDNF-dependent mechanism.

Primary visual cortex (area 17) was removed from postnatal (P4-P6) Long-Evans rat pups, dissociated using a papain dissociation procedure, plated onto collagen-coated glass-bottomed culture dishes at a density of approximately 200,000 cells/35mm dish, and maintained in serum-containing medium at 37 ° C in a 5% CO2 incubator. Cultures begin to show signs of synaptic activity after 3–4 days in vitro, and over the next few days develop waves of spontaneous firing. After 5–13 days in vitro cultures were fixed for 10 minutes in 4% paraformaldehyde and processed for double-label indirect immunofluorescence using a rabbit polyclonal GABA antiserum, and a mouse monoclonal MAP2 antibody as a neuronal marker. GABA immunoreactivity was visualized using a rhodamine-conjugated anti-rabbit antibody, and MAP2 immunoreactivity was visualized using a fluorescein-conjugated anti-mouse antibody. Cultures were examined and counted using a Nikon Diphot inverted microscope equipped with both DIC and fluorescence optics. Cell counts were performed on each culture by first counting all GABA-positive neurons in two independent strips through the center of the dish, then recounting the strips for the total number of MAP2 positive cells. The percent-

Computational Neuroscience
edited by Bower, Plenum Press, New York, 1997

age of neurons in each culture that were GABA-positive was then calculated. All numbers are expressed as mean ± SEM for the number of cultures indicated. Only cultures with more than 100 neurons/strip were included.

Blockade of neuronal activity for 48 hours after 5–13 days in vitro resulted in a decrease in the percentage of GABA-positive neurons in visual cortical cultures (Fig. 1A). To block action potential generation, the sodium channel blocker TTX was added to the cultures at a concentration of 0.1 uM for 2 days prior to fixation. TTX was refreshed after 24 hrs, and control experiments verified that spike generation remained completely blocked by this manipulation after 48 hrs. TTX treatment reduced the percentage of GABA-positive neurons from 29.6 ± 1.0 to 19.2 ± 0.9 (N=11). This represents a decrease to 64.0 ± 3.0% of control values. This reduction was statistically significant ($p < 0.01$, student's t test). The total number of neurons in these cultures was not reduced by incubation with TTX; control cultures had 184 ± 37 neurons/strip, while TTX treated cultures had 178 ± 27 neurons/strip, indicating that neuronal survival was not affected by treatment with TTX for this period of time. In addition, none of the other manipulations reported below produced a significant change in total neuronal survival.The effects of activity blockade and of neurotrophins on the percentage of GABA-positive neurons in cortical cultures. A. Cultures were treated with 0.1 μM TTX, either alone (TTX) or in the presence of 25 ng/ml BDNF (TTX + BDNF), 50 ng/ml NGF (TTX + NGF), or 25 ng/ml NT3 (TTX + NT3), for two days. * = significantly different from control, $p < 0.01$. B. The effects of different doses of BDNF on the ability of TTX to reduce the percentage of GABA-positive neurons was determined. TTX (0.1 μM) was applied in the presence of the indicated concentration of BDNF for two days, and the percentage of GABA-positive neurons determined. For each condition in A and B, the ratio of GABA-positive to GABA-negative neurons was determined, and these values are expressed as a percent of the values obtained for control cultures (control = 100 %, indicated by dashed line).

The neurotrophins, including BDNF, are a class of factors that have been shown to affect a diverse set of neuronal properties, including survival, outgrowth, and synaptic strengths. The expression of BDNF in intact cortex and in cortical cultures has been shown to be activity-dependent; high activity levels lead to increased BDNF expression, and vice versa (Isackson et al., 1991; Ghosh et al., 1994; Castren et al., 1992). In addition, BDNF has been shown to increase GABA expression in striatal neurons. These observations suggested to us that BDNF secretion might be the signal linking changing activity levels to the expression of GABA in visual cortical cultures. To explore this possibility we incubated cultures in TTX (0.1 uM) + BDNF (25 ng/ml). When TTX was applied in the presence of exogenous BDNF (N = 8), there was no reduction in the percentage of GABA-positive neurons (Fig. 1A).

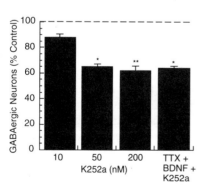

Figure 1. The effects of activity blockade and of neurotrophins on the percentage of GABA-positive neurons in cortical cultures. A. Cultures were treated with 0.1 μM TTX, either alone (TTX) or in the presence of 25 ng/ml BDNF (TTX + BDNF), 50 ng/ml NGF (TTX + NGF), or 25 ng/ml NT3 (TTX + NT3), for two days. * = significantly different from control, $p < 0.01$. B. The effects of different doses of BDNF on the ability of TTX to reduce the percentage of GABA-positive neurons was determined. TTX (0.1 μM) was applied in the presence of the indicated concentration of BDNF for two days, and the percentage of GABA-positive neurons determined. For each condition in A and B, the ratio of GABA-positive to GABA-negative neurons was determined, and these values are expressed as a percent of the values obtained for control cultures (control = 100 %, indicated by dashed line).

BDNF + TTX was significantly different from TTX alone (p < 0.01), and was not significantly different from control. BDNF was effective at preventing the effects of TTX at concentrations as low as 1 ng/ml (Fig. 1B). When BDNF was applied under control conditions, it produced no significant increase in the percentage of GABA positive neurons, suggesting that with activity intact endogenous BDNF levels are saturating for this effect. Neither NGF nor NT-3, two other neurotrophic factors present in cortex, were capable of blocking the effects of TTX application on GABA expression (Fig. 1A). These data indicate that BDNF can prevent the activity-dependent decrease in GABA-positive neurons in cortical cultures, and suggests that activity-dependent regulation of BDNF production or secretion may be the mechanism by which activity regulates GABA levels.

The above data suggest that activity blockade is reducing the percentage of GABA-positive neurons in culture by reducing the production of BDNF. This raised the question of whether blocking the action of endogenous neurotrophins might influence the expression of GABA in these cortical cultures. The neurotrophins act through the Trk tyrosine kinase receptors TrkA (NGF), TrkB (BDNF), and TrkC (NT3). This family of receptors is blocked by the compound K252a, which prevents autophosphorylation of the tyrosine kinase domain of the receptors. At concentrations of 200 nM or less, K252a is a remarkably specific inhibitor of Trk receptors, and blocks TrkA, TrkB, and TrkC with approximately equal efficacy, while leaving other tyrosine kinase signaling pathways, as well as protein kinase C pathways, intact.

To test whether endogenous neurotrophin signaling influenced GABA expression, cultures were incubated for two days with various concentrations of K252a. Concentrations of K252a as low as 10 nM produced a decrease in the percentage of GABA-positive neurons (n = 3), and at concentrations of 50 nM and above, K252a significantly reduced the percentage of GABA-positive neurons to 65.3 ± 1.8 and 62.1 ± 3.4% of control, respectively (Figure 2). The reduction produced by 50 nM (n = 3) and 200 nM (n = 9) K252a was comparable to that produced by blockade of activity with TTX (see Figure 1A). In the presence of 200 nM K252a, BDNF did not prevent the TTX-induced reduction in the percentage of GABA-positive neurons (Figure 2, TTX + BDNF + K252a, N = 3), indicating that K252a was effectively blocking Trk receptor signaling at this concentration. K252a treatment had no influence on neuronal survival at any of the concentrations tested. These data suggest that endogenous neurotrophin signaling through Trk receptors is regulating GABA expression in cortical interneurons. These data do not allow us to distinguish between signaling through the different Trk receptors, but given that neither NT3 nor NGF have any effect on GABA expression in this system, the most likely possibility is that this effect is mediated by endogenous release of BDNF.

To determine whether the effects of activity blockade on GABA expression were reversible, we first treated cultures for 2 days with TTX as described above, then washed out the TTX through several exchanges of medium. Dishes were then incubated for 2 days in either control medium, or medium supplemented with 25 ng/ml BDNF. Control experiments indicated that 48 hrs after wash of TTX spike generation was normal. Wash out of TTX partially restored GABA levels, to 82.7 ± 3.0% of control values (wash significantly different from TTX alone, p < 0.05). Wash + BDNF completely reversed the reduction in GABA levels produced by TTX, to 106.0 ± 8.7% of control values. These data suggest that activity is reversibly decreasing GABA expression by interneurons rather than selectively decreasing interneuron survival. The alternative possibility is that a new population of neurons is being induced to express GABA, but this is unlikely as no neurogenesis is occurring in these postnatal cultures.

The data we have presented suggest that activity can continuously and bi-directionally adjust the level of GABA expression in cortical cultures. This in turn suggests that ac-

A. DeWan *et al.*

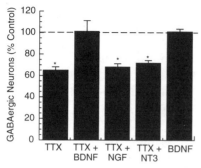

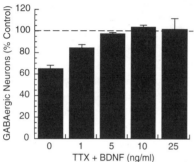

Figure 2. The effects of Trk receptor blockade on the percentage of GABA-positive neurons. K252a, a blocker of Trk receptor signaling, was applied for two days at the indicated concentration (10, 50, or 200 nM), and the percentage of GABA-positive neurons determined. The effect of BDNF (25 ng/ml) and TTX (0.1 μM) in the presence of K252a (200nM) for two days was also determined (TTX+BDNF+K252a). Numbers are expressed as a percentage of the value obtained for control cultures (control = 100%, indicated by dashed line). * = significantly different from control, $p < 0.05$; ** = significantly different from control, $p < 0.001$.

tivity is adjusting the level of functional inhibition. Consistent with this, we have found a decrease in the frequency and amplitude of inhibitory synaptic currents onto cortical pyramidal neurons following TTX treatment (Rutherford and Turrigiano, unpublished results). In addition, this treatment produces hyperexcitability in cortical cultures when washed out, as would be expected for manipulations that reduced inhibition (Rutherford et al., 1996). Electrophysiological recordings from visual cortex of dark reared rats also show an increase in spontaneous activity relative to control animals, which correlates with a decrease in the number of GABA-positive neurons (Benevento et al., 1995). Our data suggest that this activity-dependent regulation of circuit excitability is mediated through the activity-dependent regulation of BDNF. This in turn suggests that an important function for this neurotrophin is in the control of cortical excitability.

REFERENCES

Hendry S.H.C., and Jones, E.G. (1986). Reduction in number of immunostained GABAergic neurones in deprived-eye dominance columns of monkey area 17. Nature *320*:750–753

Isackson, P.J., Huntsman, M.M., Murray K.D., and Gall C.M. (1991). BDNF mRNA expression is increased in adult rat forebrain after limbic seizures: temporal patterns of induction distinct from NGF. Neuron *6*:937–948

Ghosh, A., Carnahan, J., Greenberg, M.E. (1994). Requirement for BDNF in activity-dependent survival of cortical neurons. Science *263*:1618–1623

Castrén E., Zafra F., Thoenen, H., and Lindholm D. (1992). Light regulates expression of brain-derived neurotrophic factor mRNA in rat visual cortex. PNAS *89*:9444–9448

Benevento, L.A., Bakkum, B.W., and Cohen, R.S. (1995). Gamma-aminobutyric acid and somatostatin immunoreactivity in the visual cortex of normal and dark-reared rats. Brain Research *689*:172–182

Rutherford, C.L, DeWan, A., and Turrigiano G.G. (1996) Role of BDNF in the activity-dependent regulation of cortical firing rates and GABA expression. Soc. for Neurosci. Abstr. 22:1017

2

THE DUAL ROLE OF CALCIUM IN SYNAPTIC PLASTICITY AT THE MOTOR ENDPLATE

Samuel R. H. Joseph, Volker Steuber, and David J. Willshaw

Centre for Neural Systems
Centre for Cognitive Science
Edinburgh University
Edinburgh EH8 9LW, Scotland, UK
E-mail: {sam,volkers,david}@cns.ed.ac.uk

ABSTRACT

Focal blockade of postsynaptic acetylcholine receptors (AChRs) in a small region of the neuromuscular junction may cause long-term synapse elimination at that site. Blockade of the whole junction does not cause synapse loss, indicating that it is the contrast in postsynaptic activity between the blocked and unblocked regions which causes withdrawal of the synaptic terminals. This phenomenon can be explained by the dual role of calcium, both in controlling AChR gene transcription and influencing AChR aggregation. A computational model is provided and the stability of the solutions is confirmed by theoretical analysis and computer simulation.

1. INTRODUCTION

Development of the nervous system involves an initial overproduction of synaptic contacts, followed by the removal of redundant connections. Although the mechanism responsible for the establishment of the final connectivity is one of the major issues of developmental neurobiology, it is still poorly understood. This is partly due to the difficulties posed by monitoring synaptic connections in the central nervous system over long time periods. The best studied synaptic system to date is the neuromuscular junction (NMJ) in the peripheral nervous system. As in many other parts of the nervous system, the NMJ achieves a final state of single innervation (each muscle fiber being connected to only one motor axon) after an initial stage of polyinnervation (Fig. 1). Electrical activity of the innervating axons is implicated in the withdrawal of the redundant connections during neonatal life. However, the exact role of activity in synapse elimination is complex and controversial.

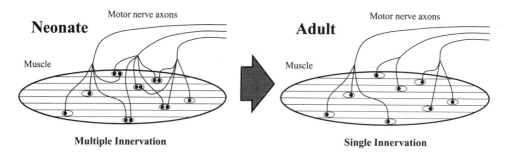

Figure 1. Developmental transition of Neuromuscular Junction.

Recent experimental results (Balice-Gordon and Lichtman, 1994) indicate that a threshold amount of postsynaptic activation is required to eliminate neighboring inactive synapses. Focal blockade of postsynaptic AChRs by application of α-bungarotoxin in a small region of the neuromuscular junction results in long-lasting synapse elimination at that site, while blockade of the whole junction does not cause synapse elimination. Furthermore, a small active area is neither eliminated by, nor able to eliminate a larger inactive region. This strongly implies that it is the difference in the activity between regions which causes withdrawal, but what mechanism underlies this process?

2. EXISTING MODELS OF SYNAPTIC PLASTICITY AT THE MOTOR ENDPLATE

Various computational models attempt to explain the effect of electrical activity in synapse elimination. Kerszberg and Changeux (1993) assume that the AChR gene transcription depends on the presence of a limiting amount of morphogen, eg a transcription factor of the Myo-D family. The production of the morphogen in the muscle fiber nuclei is inhibited by global postsynaptic activity and enhanced by local anterograde factors (such as CGRP) which are released by presynaptically active terminals. Calcium is postulated as mediating between electrical activity and AChR gene repression. Thus an active terminal will increase the AChR productivity of the nuclei associated with it and repress AChR expression at other, inactive, terminals (Fig. 2).

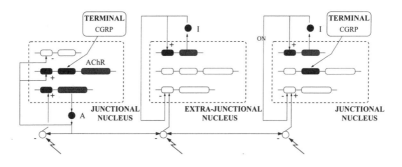

Figure 2. Kerszberg-Changeux model.

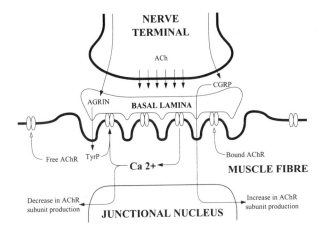

Figure 3. Schematic of molecular mechanisms at neuromuscular synapse.

Joseph and Willshaw (1995) extended the dual constraint model (DCM) (Rasmussen and Willshaw, 1993) by proposing that terminals compete for both the protein agrin, synthesized by motor neurons and transported down their axons, and AChRs, which are free to move laterally along the muscle fibers. The key to this competitive process is that as a terminal gets larger (agrin binding to the muscle fiber attracts AChRs) the increase in receptors leads to higher levels of depolarization caused by synaptic activity. This causes the calcium concentration around the area of the terminal to be raised, increasing terminal stability since calcium is required by agrin to assist in receptor accumulation.

Both this model and the Kerszberg and Changeux' model are able to reproduce the achievement of single innervation. However, neither of them can explain the effect of focal blockade observed in Balice-Gordon and Lichtman's experiments.

3. METHODS

Balice-Gordon and Lichtman's results can be explained by including calcium downregulation of AChR gene transcription in the DCM. Figure 3 displays the mechanisms operating at the NMJ that are incorporated in the Focal Blockade Model (FBM).

The FBM is described by two first order differential equations for the variation of the number of free receptors (R_f) in the muscle and the number of bound receptors (R_b) in a single terminal over time.

$$\frac{dR_f}{dt} = \alpha R_b + \beta R_b^{-2/3} - \gamma R_b R_f - \delta R_f \tag{1}$$

$$\frac{dR_b}{dt} = \gamma R_f R_b - \alpha R_b \tag{2}$$

where α, β, γ and δ are all rate constants. In equation 1 the first term represents the bound receptors that diffuse away from the terminal, leading to an increase in the number of free receptors, and is proportional to the size of the terminal, which itself is represented by the number of bound receptors (R_b). The second term represents the production of free receptors in the muscle fiber nuclei and is inversely proportional to the calcium concentration in the muscle fiber which is in turn proportional to the surface area of the terminal ($R_b^{2/3}$). The third term represents the agrin induced binding of free receptors into aggregates at the terminal. This term is proportional not only to the number of free receptors (the more free receptors the greater the chance of intersection with the terminal), but also to the diameter of the terminal ($R_b^{1/3}$) (the larger the terminal profile, the greater the chance of receptors intersecting with it) and the calcium concentration around the terminal ($R_b^{2/3}$) (the more calcium the greater the chance of aggregate formation). The fourth term represents the destruction/internalization of free AChRs by the muscle fiber and is simply proportional to the number of free receptors.

The first and second terms on the right hand side of equation 2 correspond, respectively, to third and first terms of equation 1. The FBM reproduces the standard situation with no blocked receptors with the equations as above. The blockade of the whole junction is represented by the equations:

$$\frac{dR_f}{dt} = [\alpha R_b + \beta] - R_f[\gamma R_b^{1/3} + \delta] \tag{3}$$

$$\frac{dR_b}{dt} = \gamma R_f R_b^{1/3} - \alpha R_b \tag{4}$$

In this case the calcium concentration is not affected by terminal size, so neither the production of AChRs nor the formation of bound AChR aggregates is proportional to the surface area of the terminal. The focal blockade of a small portion of the junction is represented by a similar pair of equations:

$$\frac{dR_f}{dt} = [\alpha R_b + \beta R_b^{-2/3}] - R_f[\gamma R_b^{1/3} + \delta] \tag{5}$$

$$\frac{dR_b}{dt} = \gamma R_f R_b^{1/3} - \alpha R_b \tag{6}$$

In this focal blockade case the stabilizing effect of calcium on the blocked region of the terminal is lost, but not on the production of AChRs by the muscle fiber nuclei.

4. RESULTS

A perturbation analysis performed on all three pairs of equations shows that the terminal is stable in the completely blocked and unblocked states, while the blocked area in a focal blockade is unstable. Specifically the analysis yields three inequalities which represent the stability of each system:

$$\text{Unblocked} \qquad R_f < \frac{3\beta}{2\delta R_b^{2/3}} + \frac{\alpha}{\gamma} \tag{7}$$

$$\text{Blocked} \qquad R_f < \frac{\alpha R_b^{2/3}}{3\gamma} \tag{8}$$

$$\text{Partial} \qquad R_f < \frac{\alpha R_b^{2/3}}{3\gamma} - \frac{\beta}{2\delta R_b^{2/3}} \tag{9}$$

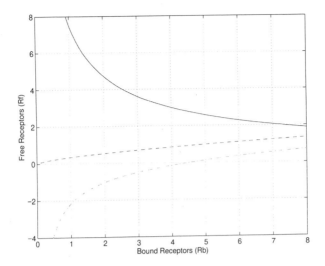

Figure 4. Stability relations. Unblocked junction (solid), blocked junction (dashed) and partial block (dot/dash).

In the relevant range of terminal size, while the blocked area of the terminal is small enough, the maximum level of free receptors to ensure stability is far lower than those of both the entirely blocked and unblocked cases (Fig. 4).

In the limit of the focal blockade case, as the size of the blocked area approaches that of the whole terminal, it becomes more stable again. What this implies is that in the partially blocked case the stabilizing effect of the calcium is missing in the affected region, while in the totally blocked case this is compensated for by the lack of negative effects of calcium downregulation of AChR production. One potential flaw in the analysis is that AChR production does not saturate with reducing calcium concentration. However, quantitative simulation of the FBM shows that the partial block case is less stable with or without this saturation limit. The simulation results (over 100 000 timesteps of $\Delta t = 0.0001$, rate constants set to unity) are shown in figure 5. The simulation would be strengthened by the inclusion of rate constants derived from experimental procedures, however this data was not available in full at the time of going to press.

5. SUMMARY

The combination of two antagonistic calcium dependent effects in the focal blockade model of the neuromuscular junction manages to explain further the role of activity in neural development. Understanding the interaction between neural activity and the establishment of neural connectivity is not only a major issue in developmental neurobiology, it could also be beneficial for the construction of artificial neural networks.

REFERENCES

[1] Balice-Gordon R. J. & Lichtman J. W. (1994) Long-term synapse loss induced by focal blockade of postsynaptic receptors. *Nature* **372**:519-524.

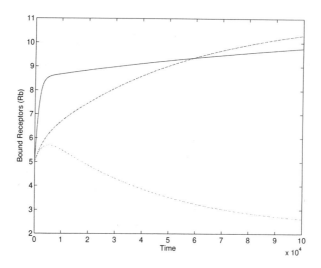

Figure 5. Simulation results. Unblocked junction (solid), blocked junction (dashed) and partial block (dot/dash).

[2] Joseph S. R. H. & Willshaw D. J. (1996) The role of activity in synaptic competition at the neuromuscular junction. *To be published in NIPS 8.*

[3] Kerszberg M. & Changeux J.- P. (1993) A model for motor endplate morphogenesis: Diffusible morphogens, transmembrane signaling, and compartmentalized gene expression. *Neural Computation* **5**:341-358.

[4] Rasmussen C. E. & Willshaw D. J. (1993) Presynaptic and postsynatic competition in models for the development of neuromuscular connections. *Biol. Cybern.* **68**:409-419.

[5] Wallace B. G. (1988) Regulation of agrin-induced acetylcholine receptor aggregation by Ca^{++} and phorbol ester. *Journal of Cell Biol.* **107**:267-278.

3

NEURONAL EXOCYTOSIS EXHIBITS FRACTAL BEHAVIOR

Steven B. Lowen,[1] Sydney S. Cash,[2] Mu-ming Poo,[3] and Malvin C. Teich[4]

[1]Boston University
Department of Electrical and Computer Engineering
44 Cummington St., Boston, MA
E-mail: `lowen@bu.edu`
[2]Columbia University
Department of Biological Sciences
1002 Fairchild Building
New York, NY
E-mail: `ssc8@columbia.edu`
[3]University of California at San Diego
Department of Biology
La Jolla, CA
E-mail: `mpoo@jeeves.ucsd.edu`
[4]Boston University
Departments of Electrical and Computer Engineering and Biomedical Engineering
44 Cummington St.
Boston, MA
E-mail: `teich@bu.edu`

ABSTRACT

The time sequence of exocytic events in both neurons and non-neuronal cells exhibits fractal (self-similar) properties, as evidenced by a number of statistical measures. Such fractal activity occurs in neurotransmitter secretion at Xenopus neuromuscular junctions and rat hippocampal synapses in culture, and in the exocytosis of exogenously supplied neurotransmitter from cultured Xenopus myocytes and rat fibroblasts. The magnitude of the fluctuations of the rate of exocytic events about the mean decreases slowly as the rate is computed over longer and longer time periods, the periodogram decreases in power-law fashion with frequency, and the Allan factor (relative variance of the number of exocytic events) increases as a power-law function of the counting time. These features are hallmarks of self-similar behavior. Their

Computational Neuroscience
edited by Bower, Plenum Press, New York, 1997

description requires models that exhibit long-range, power-law-decaying correlation (memory) in event occurrences. We have developed a physiologically plausible model that accords with all of the statistical measures that we have examined: the fractal lognormal-noise-driven doubly stochastic Poisson process (FLNDP). In particular, we show that the experimental rate function is well modeled by fractal lognormal noise (FLN). The appearance of behavior with fractal characteristics at synapses, as well as in systems comprising collections of synapses, indicates that such behavior is an inherent property of neuronal signaling.

INTRODUCTION

Communication in the nervous system is mediated by action-potential-initiated exocytosis of multiple vesicular packets (quanta) of neurotransmitter (Katz, 1966). Even in the absence of such action potentials, however, many neurons spontaneously release individual packets of neurotransmitter (Fatt and Katz, 1952). On arrival at the postsynaptic membrane, the neurotransmitter (ACh) molecules induce elementary endplate currents (EECs), which take the form of nonstationary two- (or multi-) state on-off sequences (Sakmann, 1992). Current flows when the ACh channel is open (i.e., when its two binding sites are occupied by agonist), and ceases when the channel is closed. A postsynaptic miniature endplate current (MEPC) comprises some 1000 EECs (Sakmann, 1992). It was shown by Del Castillo and Katz (1954) that superpositions of MEPC-like events comprise the postsynaptic endplate currents elicited by nerve impulses.

It has generally been assumed that the sequence of MEPCs forms a memoryless stochastic process (Fatt and Katz, 1952). However, Rotshenker and Rahamimoff (1970) discovered that exocytosis in the frog neuromuscular junction can exhibit correlation (memory) over a period of seconds, provided that extracellular Ca^{2+} levels are elevated above their normal values.

We have studied the statistical properties of exocytic events over a far larger range of time scales than previously examined (Lowen et al, 1997). MEPCs were recorded from innervated myocytes in Xenopus nerve-muscle cocultures and from rat hippocampal neurons in cell culture. MEPCs from non-neuronal preparations were also examined: the quantal secretion of ACh from isolated myocytes (autoreception) and from rat fibroblasts, both exogenously loaded with ACh (Dan and Poo, 1992; Girod et al., 1995).

We have directed our attention toward those statistical measures that reveal the presence of memory. Our analysis reveals that the time sequences of the MEPCs, and therefore of the underlying exocytic events, exhibit memory that decays away slowly, as a power-law function of time, in both neuronal and non-neuronal cells. This long-duration correlation is present over the entire range of time scales investigated, which stretches to thousands of seconds. The occurrence of a MEPC therefore makes it more likely that another MEPC will occur at some time thereafter. The analysis of long MEPC data sets reveals that the rate of events is consistent with a fractal process, exhibiting fluctuations over multiple time scales. Fractals are objects which possess a form of self-similarity: parts of the whole can be made to fit to the whole by shifting and stretching. The hallmark of fractal behavior is power-law dependence in one or more statistical measures, over a substantial range of the time (or frequency) scales at which the measurement is conducted (Lowen and Teich, 1995).

STATISTICAL MEASURES

Perhaps the simplest measure of a sequence of neuronal activity is its rate: the number of events registered per unit time. For vesicular release events, even this straightforward measure has fractal properties; the fluctuations of the rate do not decrease appreciably even when a very long counting time is used to compute it. This behavior derives from correlations in the sequence of interevent intervals, as confirmed by the observation that the fractal properties of the rate estimate are destroyed by shuffling (randomly reordering) the intervals. This operation removes the correlations among the intervals while exactly preserving their relative frequencies.

Another measure sensitive to fractal behavior is the Allan factor (AF) (Teich et al., 1996), defined as the ratio of the Allan variance of the event count to the mean. The Allan variance, in turn, represents the average variation in the difference between adjacent counts (Allan, 1966). To compute the Allan factor at a specified counting time T, the data record is first divided into adjacent counting windows of duration T, and the number of events falling within each window recorded. The difference between the counts in each window and the following window is then computed; the mean square of this quantity forms the Allan variance. Dividing this quantity by twice the average number of counts in each window yields the Allan factor. In general the Allan factor A varies with the counting time T. For fractal point processes, including all exocytosis recordings of sufficient length that we have analyzed, A rises as a power-law function of T for large counting times T (Lowen et al, 1997). Again, shuffling the intervals destroys this effect, rendering the Allan factor essentially constant with counting time.

The periodogram (PG), an estimator of the power spectral density, also reveals the presence of fractal activity. Much as for continuous-time processes, the power spectral density reveals how power is concentrated in various frequency bands. For low frequencies f (corresponding to long time scales T), the PG also varies in a fractional power-law fashion with frequency for the same data recordings, although with a negative exponent (Lowen et al, 1997). Finally, the PG computed from a shuffled version of the vesicular activity, in contrast, does not vary in power-law fashion, lending further credence to the notion that it is the ordering of the intervals, rather than their relative sizes, which is particularly responsible for the fractal aspects of the vesicular activity.

MODEL

A model of exocytic activity which successfully fits both the apparent fractal behavior seen over long time scales and interevent statistics such as the interevent-interval histogram (IIH) is the fractal-lognormal-noise driven Poisson point process (FLNDP), which we develop as follows.

Verveen and Derksen (1968) showed that the voltage of an excitable-tissue membrane at rest exhibits fractal ($1/f$-type) fluctuations with a Gaussian amplitude distribution, which they traced to fluctuating K^+-ion concentrations. Voltage-gated Ca^{2+}-ion channel openings are responsible for vesicular exocytosis (Zucker, 1993). For a fixed membrane voltage near the resting potential, calcium flow is negligible. Occasionally, however, random thermally induced channel openings will occur, often leading to spontaneous exocytic events for nearby vesicles. Such spontaneous behavior is almost completely memoryless, and is therefore well modeled by a homogeneous Poisson process (HPP) (Cox and Lewis, 1966), with a fixed rate λ given by the Arrhenius equation (Berry et al., 1980): $\lambda = \mathcal{A}\exp(-E_A/\mathcal{R}\mathcal{T})$. Here the rate

λ is the number of vesicular release events expected to occur in a unit time interval, $\mathcal{A}$ is a rate constant, E_A is the constant activation energy associated with the ion-channel opening, $\mathcal{R}$ is the thermodynamic gas constant, and $\mathcal{T}$ is the absolute temperature. (That only a fixed proportion of the channel-opening events leads to exocytosis can be incorporated into the value of $\mathcal{A}$.) Living cells generate non-zero voltages across their membranes which modify the rate, leading to $\lambda = \mathcal{A} \exp[-(E_A - zFV)/\mathcal{RT}]$, with z the valence of the charge involved in the channel opening, F the Faraday constant (coulombs/mole), and V the membrane voltage. Different fixed membrane voltages lead to spontaneous exocytic patterns which differ only in their average rates; all are HPPs. These rates are exponential functions of the membrane voltage, as prescribed by this equation.

However, the membrane voltage V is not fixed, but rather varies randomly in time, exhibiting Gaussian fluctuations which appear fractal; it is therefore well modeled by fractal Gaussian noise, or FGN [or a modified form of fractal Brownian motion, depending on the exponent α_S of the $1/f^{\alpha_S}$ noise $V(t)$]. Thus the rate λ of the Poisson process will also vary in time. Since the rate is the exponential transform of the voltage, which is described by FGN, the rate will behave as fractal lognormal noise (FLN). The resulting openings are therefore characterized by a *doubly stochastic* (rather than homogeneous) Poisson process with a rate that is FLN. Finally, then, the calcium-flow events, and therefore the exocytic events, are described by the FLNDP model. Unlike a homogeneous Poisson process, the FLNDP is not memoryless. Rather, the fluctuating membrane voltage imparts fractal correlations to the exocytic events so that the observation of a short (long) interevent time, for example, signifies a locally high (low) rate $\lambda(t)$, which in turn indicates that the next interevent time is also likely to be short (long).

Analytical predictions and computer simulations based on this three-parameter model were compared with the exocytic-event data for a variety of statistical measures. In particular, agreement with the data was excellent over all time scales for the AF, PG, and IIH (Lowen et al, 1997), with the same simulations employed for all three measures. Moreover, the AF and PG calculated from shuffled FLNDP simulations are in excellent accord with those of the shuffled data. We conclude that, aside from its physiological plausibility, the FLNDP provides an excellent mathematical model for characterizing the sequences of MEPCs observed in our experiments (Lowen et al, 1997).

LOGNORMAL RATE

In addition to serving as an excellent model for the the *point process* of exocytic events, the FLNDP process developed above also models the *rate* $\lambda(t)$ of event generation, which we now proceed to examine. Estimates of the rate were collected from a representative neuromuscular junction used for collecting AF and PG statistics (Lowen et al, 1997) by counting the number of exocytic events occurring in windows of $T = 1, 3, 10, 30, 100$, and 300 seconds. Each count was then divided by the window duration, to obtain a rate estimate, which was then expressed as a survivor function. The rate was relatively constant even over the $T = 300$ sec window, so that our estimate is indeed of the rate. Some fluctuation does exist, however, which serves to negatively bias the variance of the rate estimate for these longer windows. In contrast, the shorter counting times suffer from a lack of resolution, since the counting process necessarily yields an integral number of counts; this count is often zero for the shortest counting times. We plot rate estimates obtained for a range of counting times in Fig. 1 (solid curves), where the lowest curves on the left correspond to the shortest counting

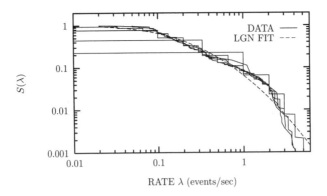

Figure 1. Doubly logarithmic plot of the estimated survivor function $S(\lambda)$ of the rate vs. the rate λ (probability that the rate is larger than the abscissa value) for spontaneous vesicular release obtained from a Xenopus neuromuscular junction (solid curves). The sequence of events from which this plot has been constructed has a duration $L = 8164$ sec and contains $N = 2644$ interevent intervals. Counting times were $T = 1, 3, 10, 30, 100$, and 300 seconds, increasing from the bottom up at the left edge of the figure. Also shown is the theoretical lognormal fit (dashed curve), which agrees with the experimental rate functions.

times. These rate estimates compare favorably with the theoretical lognormal rate function (dashed curve).

It will be of interest to conduct measurements of both the presynaptic membrane voltage and the exocytic process, for which the FLNDP model makes the following predictions. Given the parameters of $V(t)$, the rate $\lambda(t)$ can be shown to have a mean, variance, and autocorrelation function given by

$$E[\lambda] = \mathcal{A}\exp(\mu + \sigma^2/2) \tag{1}$$
$$\mathrm{Var}[\lambda] = \mathcal{A}^2 \exp(2\mu + 2\sigma^2)$$
$$R_\lambda(\tau) = E^2[\lambda(t)]\exp\left\{(zF/\mathcal{R}\mathcal{T})^2\left[R_V(\tau) - E^2[V]\right]\right\},$$

respectively, where we have defined

$$\mu = (E_A - zFE[V])/\mathcal{R}\mathcal{T} \tag{2}$$
$$\sigma = zF\sqrt{\mathrm{Var}[V]}/\mathcal{R}\mathcal{T},$$

and $R_V(\tau)$ is the autocorrelation function of the membrane voltage $V(t)$. Furthermore, if the rate $\lambda(t)$ [or equivalently the voltage $V(t)$] exhibits fluctuations which are slow in comparison with the average rate of channel openings $E[\lambda]$, then the moments and probability density function for the times t between openings are, respectively, given by

$$E[t^n] = n!\mathcal{A}^{-n}\exp\left[-n\mu + (n^2 - 2n)\sigma^2/2\right] \tag{3}$$
$$p(t) = \pi^{-1/2}\mathcal{A}\exp\left(\mu + 3\sigma^2/2\right)$$
$$\times \int_{-\infty}^{\infty}\exp\left[-x^2 - t\exp\left(\mu + 2\sigma^2 + \sqrt{2}\sigma x\right)\right]dx. \tag{4}$$

REFERENCES

[1] Allan DW (1966) Statistics of atomic frequency standards. Proc IEEE 54:221-230.

[2] Berry RS, Rice SA, Ross J (1980) Physical Chemistry. New York: Wiley, pp. 1146 ff.

[3] Cox DR, Lewis PAW (1966) The Statistical Analysis of Series of Events. London: Methuen.

[4] Dan Y, Poo M-m (1992) Quantal transmitter secretion from myocytes loaded with acetylcholine. Nature 359:733-736.

[5] Del Castillo J, Katz B (1954) Quantal components of the end-plate potential. J Physiol (London) 124:560-573.

[6] Fatt P, Katz B (1952) Spontaneous subthreshold activity at motor nerve endings. J Physiol (London) 117:109-128.

[7] Girod R, Popov S, Alder J, Zheng JQ, Lohof A, Poo M-m (1995) Spontaneous quantal transmitter secretion from myocytes and fibroblasts: comparison with neuronal secretion. J. Neurosci 15:2826-2838.

[8] Katz B (1966) Nerve, Muscle, and Synapse. New York: McGraw-Hill.

[9] Lowen SB, Teich MC (1995) Estimation and simulation of fractal stochastic point processes. Fractals 3:183-210.

[10] Lowen SB, Cash SS, Poo M-m, Teich MC (1997) Quantal neurotransmitter secretion exhibits fractal behavior. (in preparation).

[11] Rotshenker S, Rahamimoff R (1970) Neuromuscular synapse: stochastic properties of spontaneous release of transmitter. Science 170:648-649.

[12] Sakmann B (1992) Elementary steps in synaptic transmission revealed by currents through single ion channels. Science 256:503-512.

[13] Teich MC, Heneghan C, Lowen SB, Turcott RG (1996) Estimating the fractal exponent of point processes in biological systems using wavelet- and Fourier-transform methods. In: Wavelets in Medicine and Biology (Aldroubi A, Unser M, eds). Boca Raton, FL: CRC Press, ch. 14, pp. 383-412.

[14] Verveen AA, Derksen HE (1968) Fluctuation phenomena in nerve membrane. Proc IEEE 56:906-916.

[15] Zucker RS (1993) Calcium and transmitter release. J Physiol (Paris) 87:25-36.

4

STRENGTH AND TIMING OF GRADED SYNAPTIC TRANSMISSION DEPEND ON FREQUENCY AND SHAPE OF THE PRESYNAPTIC WAVEFORM

Farzan Nadim, Yair Manor, L. F. Abbott, and Eve Marder

Brandeis University
Waltham, Massachusetts

INTRODUCTION

Several classes of motor neurons within the stomatogastric ganglion of the spiny lobster *Panulirus interruptus* participate in the generation of the pyloric rhythm. Graded synaptic transmission among these motor neurons is an essential component in the production of the triphasic pyloric motor pattern[1]. Previous work characterized graded synaptic transmission[2], but in a "static context" where the time dependent dynamics were ignored.

In the behaving animal, the pyloric motor pattern is subject to changes in frequency and phase relationships[3]. The aim of the present work is to characterize graded synaptic transmission in such a dynamic environment.

The pyloric rhythm is driven by a pacemaker ensemble that includes two pyloric dilator (PD) neurons. The synaptic connection from the lateral pyloric (LP) neuron to the PD neurons provides the sole feedback from the rest of the pyloric network to the pacemaker ensemble. We studied this synapse, because of its importance in the regulation of phase and frequency of the pyloric rhythm.

METHODS

Standard procedures were used to isolate the stomatogastric nervous system and to identify the neurons[4]. In normal saline, 10^{-7} M TTX was applied to block action potential mediated transmission, leaving only graded transmission. Graded synaptic transmission was measured with two electrodes in the presynaptic LP neuron and one electrode in each postsynaptic PD neuron. Using two electrode voltage clamp, the LP neuron was driven with square pulses and realistic waveforms. Realistic waveforms were generated with the following procedure: the trajectories of the rhythmic membrane potential waveforms of several LP neurons were recorded in normal saline. Each recorded trace was divided into single cycles. Because we were interested in studying the graded component of synaptic transmission, these cycles were averaged and filtered to eliminate the action potentials, and the resulting "unitary" waveform was stored in the computer. The unitary waveform, scaled to various amplitudes and frequencies, was then used to voltage-clamp LP.

The PD membrane potential was recorded in current clamp. Tonic current injection was used to maintain these cells at depolarized levels (-40 mV to -20 mV) where the IPSP was large in amplitude. All recordings were done at room temperature (21–22°C).

RESULTS

Application of square pulses to the LP cell produced graded inhibitory postsynaptic potentials (GIPSPs) in the two PD neurons. These GIPSPs had a transient and a persistent component (Figure 1A). The transient component peaked approximately 50 to 100 ms after the onset of the pulse. Both the peak (open triangle) and the persistent (filled circle) components were sigmoidal functions of the presynaptic potential. In Figure 1C, we show the input/output relationships of the two components normalized with respect to their maximal values (in response to a presynaptic pulse to +10 mV). As the presynaptic potential was increased, the amplitude of the persistent component increased more gradually than that of the peak.

Square pulses are the simplest waveforms to generate, but not necessarily the best to use to characterize the temporal dynamics of graded synapses. During normal oscillations, the membrane potential of pyloric neurons shows smooth envelopes of slow-wave depolarizations. This motivated us to use realistic waveforms to voltage clamp the presynaptic cell. The postsynaptic response to a realistic waveform is shown in Figure 1B. Note that the amplitude of the GIPSP shows depression in time. As we increased the amplitude of the presynaptic waveform, the amplitude of the GIPSP gradually increased to a saturated level. The amplitude of the response to the realistic waveform was between the peak and persistent amplitudes of the response to a square pulse of the same amplitude (not shown). Moreover, the normalized input/output curve of the realistic waveform (grey square) fell between the normalized input/output curves of the peak and persistent components of the square pulse (Figure 1C).

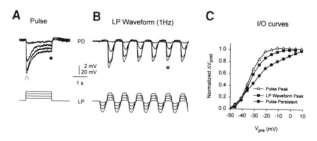

Figure 1. (A) The LP neuron was held at -50 mV and depolarizing voltage steps of various amplitudes (10–40 mV) in LP result in GIPSPs in PD that varied in shape and amplitude. The GIPSP had a transient (open triangle) and a steady-state (filled circle) component. (B) Synaptic response of the PD cell to periodic application of a 1 Hz realistic waveform (grey square; amplitude 10 to 40 mV) in LP (from a holding potential of -50 mV) is graded and shows depression. C, Input/Output curves for the peak and persistent components of the response to a square pulse, and to realistic waveforms.

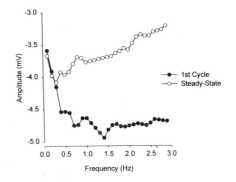

Figure 2. The amplitude of the GIPSP depends on the frequency of the injected waveform. LP was voltage clamped at frequencies ranging from 0.1 to 3 Hz. The amplitude of the 1st GIPSP (filled circle) and the steady-state GIPSP (open circle; average of the last 4 cycles) are plotted against frequency.

Because the pyloric rhythm operates at a wide range of different frequencies, we next asked whether the amplitude of the GIPSP depends on the frequency of the presynaptic waveform. In Figure 2, the LP neuron was voltage-clamped at frequencies ranging from 0.1 to 3 Hz. At each frequency, the waveform was applied until the amplitude of the GIPSPs recorded in the PD remained approximately constant (steady-state). The amplitude of the first GIPSP (filled circles) was independent of frequency except at frequencies below 0.6 Hz. The smaller amplitudes at low frequencies were most probably due to the slow rise of the presynaptic waveform. The amplitude of the steady-state GIPSPs (open circles) also increased at low frequencies. However, at higher frequencies it *decayed* as frequency increased. This decay was not due to cable attenuation, because such an effect would have been reflected in the amplitude of the first GIPSP as well. This suggests that the extent of synaptic depression is frequency-dependent.

In addition to its strength, we found that the time-to-peak of the GIPSP also depended on the frequency of the presynaptic waveform. To elucidate this effect, we defined Δt as the time difference between pre- and postsynaptic peaks (Figure 3A). We then calculated the phase of the GIPSP as the ratio of Δt to the period. At low frequencies, the peak of the GIPSP was phase-advanced ($\Delta t > 0$); it was phase-delayed ($\Delta t < 0$) at high frequencies (Figure 3B). At around 1 Hz, the presynaptic and postsynaptic peaks were aligned ($\Delta t = 0$).

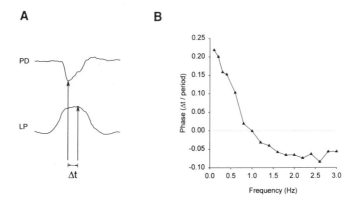

Figure 3. Phase relationship between the LP waveform and the PD response. (A) The difference between the peak time of the GIPSP and the peak time of the LP waveform is defined as Δt. (B) Phase (Δt normalized by period) is plotted against frequency. At frequencies below ~1 Hz, the GIPSP is phase-advanced with respect to the LP waveform. At frequencies above ~1 Hz it is phase-delayed.

SUMMARY

Traditionally, graded synaptic transmission has been measured by injecting square pulses in the presynaptic cell. We were interested to see how the graded synapses respond to more realistic waveforms. In response to square pulses in LP, the GIPSP in PD has a transient and a persistent component. These components depend differently on presynaptic potential. The response to a realistic waveform was a mixture of the two components seen with square pulses, both in amplitude and its dependence on the presynaptic potential.

The graded synapse between LP and PD shows temporal depression when LP is stimulated repeatedly. This depression can be seen with periodic application of a realistic waveform, and its extent can be quantitatively measured as the difference between the response to the first cycle and the steady-state response. Using such a protocol, we quantified the depression and recovery from depression as a function of presynaptic frequency. At frequencies above 0.5 Hz, the response of the first cycle was approximately constant whereas the steady-state response decayed with frequency. This occurred mainly because the *recovery* from synaptic depression was smaller as frequency increased. At frequencies below 0.5 Hz, however, both the first cycle and the steady-state responses increased with frequency. We suggest that when the presynaptic potential rises slowly (at low frequencies), there is less transmitter release, probably due to inactivation of presynaptic calcium channels. As a result, we found that the strength of the synapse was maximal around 0.5 Hz.

Because the LP to PD synapse provides the sole feedback to the pacemaker, the strength and timing of this synapse is important in determining the frequency and phasing of the elements of the pyloric rhythm. In addition to the effect of frequency on strength, we observed a clear effect of frequency on the relative time-to-peak of the synaptic response. At frequencies below 1 Hz, the GIPSP was phase advanced with respect to the presynaptic waveform. At frequencies above 1 Hz, it was phase-delayed. The 0-phase at ~1 Hz suggest a mechanism for adjusting the pyloric frequency.

In summary we stress that to capture the true biological functionality of neural circuits, synaptic strength should be treated as a dynamic variable, and not merely as a parameter, as often done in modeling.

ACKNOWLEDGMENTS

This research was supported by NS17813, MH46742, the Sloan Center for Theoretical Neurobiology at Brandeis University and the W.M. Keck Foundation.

REFERENCES

1. K. Graubard, J.A. Raper and D.K. Hartline, 1983, Graded synaptic transmission between identified spiking neurons, *J. Neurophysiol.* **50**:508–521.
2. B.R. Johnson, J.H. Peck and R.M. Harris-Warrick, 1995, Distributed amines modulation of graded chemical transmission in the pyloric network of the lobster stomatogastric ganglion, *J. Neurophysiol.* **74**:437–452.
3. E. Rezer and M. Moulins, 1983, Expression of the crustacean pyloric pattern generator in the intact animal, *J. Comp. Physiol.* **A153**:17–28.
4. A.I. Selverston, D.F. King, D.F. Russell and J.P. Miller, 1976, The stomatogastric nervous system: structure and function of a small neural network, *Prog. Neurobiol.* **7**:215–290.

5

AUTOPHOSPHORYLATION VERSUS DEPHOSPHORYLATION OF Ca^{2+}/CALMODULIN-DEPENDENT PROTEIN KINASE II

Switching Characteristics and Implication for the Induction of LTP

Hiroshi Okamoto[*] and Kazuhisa Ichikawa

Foundation Res. Lab.
Fuji Xerox Co., Ltd.
430 Sakai, Nakai-machi, Ashigarakami-gun, Kanagawa 259–01, Japan

1. INTRODUCTION

The hypothesis that usage-dependent change in the transmission efficiency of a synapse is a cellular representation of memory formation in the brain has been a foundation of modern neuroscience. Long-term potentiation (LTP) has been studied, most extensively in rat hippocampus, as an experimental model for this hypothesis. Considerable efforts have been made for elucidation of molecular mechanisms of LTP. Although molecular-signal transductions responsible for LTP might be very complicated and the entire picture of them has not yet been established, evidence has been accumulating strongly suggesting that Ca^{2+}/calmodulin-dependent protein kinase II (CaMKII) plays a crucial role in LTP [1]: CaMKII is highly concentrated in the postsynaptic region, which makes this enzyme a good target of Ca^{2+} influx through NMDA-receptor/channels known to be necessary for the induction of LTP; specific inhibitors of CaMKII prevent LTP; postsynaptic injection of a constitutively active form of CaMKII results in LTP-like enhancement of synaptic transmission [2]; and mice lacking αCaMKII gene are deficient in LTP. Therefore, tracing CaMKII activity in LTP will be incisive for understanding the main stream of molecular-signal transductions responsible for LTP. In the present study, we addressed this issue by theoretical investigation of an integrated model for postsynaptic biochemical-reaction networks involving CaMKII. Each plot of the networks was modelled on the basis of reports from *in vitro* experimental studies done for purified enzymes.

[*] okamoto@rfl.crl.fujixerox.co.jp

2. THEORY

CaMKII is an oligomeric enzyme composed of 10–12 almost identical subunits. There are three characteristic domains in the amino acid sequences of a subunit; a catalytic domain near the N-terminus, a regulatory domain in the central portion, and an association domain in the C-terminal half. The catalytic domain contains ATP-binding and substrate-binding sites and has potential kinase activity. The regulatory domain consists of two overlapping subdomains; an inhibitory domain and a Ca^{2+}/calmodulin complex (Ca^{2+}/CaM)-binding domain. In the absence of Ca^{2+}/CaM, the inhibitory domain interacts with the catalytic domain to block its activity (Fig. 1A). If Ca^{2+}/CaM binds to the Ca^{2+}/CaM-binding domain, however, the inhibitory effect is neutralised and, accordingly, the subunit becomes active (Ca^{2+}/CaM-dependently active, Fig. 1B). The association domain serves for oligomerisation of subunits. Remarkable properties of CaMKII can be observed in self-regulatory mechanisms by autophosphorylation. In the presence of Ca^{2+}/CaM, CaMKII rapidly autophosphorylates threonine-286/287 ($Thr^{286/287}$) located in the inhibitory domain of α/β subunit (Fig. 1C). The negative charge of phosphate at $Thr^{286/287}$ neutralises the interaction between the inhibitory and catalytic domains and, consequently, the enzyme thus autophosphorylated retains kinase activity even in the absence of Ca^{2+}/CaM (Ca^{2+}/CaM-independently active, Fig. 1D).

Autophosphorylation of CaMKII holoenzyme is an intra-molecular reaction. Therefore, it can be described by the following one-step processes:

$$K_n \xrightarrow{g(n)} K_{n+1} \qquad (n = 0, \cdots, N-1) \tag{1}$$

where K_n symbolises a CaMKII holoenzyme composed of n $Thr^{286/287}$-phosphorylated and $N - n$ $Thr^{286/287}$-dephosphorylated subunits. Recently, autophosphorylation of $Thr^{286/287}$ has been revealed as an inter-subunit reaction within a holoenzyme [3, 4]. It has also been shown that, preceding the inter-subunit reaction between a 'kinase' subunit and a 'substrate' subunit, Ca^{2+}/CaM must be bound to the substrate subunit [3, 5]. There are three kinds of 'kinase' subunits; $Thr^{286/287}$-phosphorylated subunit binding Ca^{2+}/CaM (CaMS*), $Thr^{286/287}$-phosphorylated subunit without Ca^{2+}/CaM (S*) and $Thr^{286/287}$-dephosphorylated subunit binding Ca^{2+}/CaM (CaMS). On the other hand, only CaMS can be 'substrate' subunit. Thereby, one can suppose three kinds of inter-subunit reaction processes for autophosphorylation of $Thr^{286/287}$: a) CaMS* phosphorylates $Thr^{286/287}$ of CaMS; b) S* phospho-

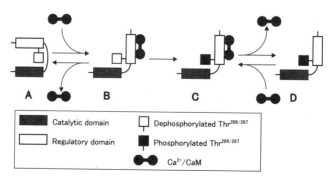

Figure 1. Regulatory model of CaMKII subunit.

rylates $Thr^{286/287}$ of CaMS; c) CaMS phosphorylates $Thr^{286/287}$ of another CaMS. The exact form of $g(n)$ is defined by statistical consideration, as follows:

$$g(n) = g_a(n) + g_h(n) + g_c(n)$$

$$g_a(n) = k_a \gamma^* \gamma n(N-n), \quad g_h(n) = k_h(1-\gamma^*)\gamma n(N-n), \quad g_c(n) = k_c \gamma^2 (N-n)(N-n-1) \quad (2)$$

where k_a, k_h and k_c are rate constants for reactions a), b) and c), respectively; γ^* is a probability that $Thr^{286/287}$-phosphorylated subunit binds Ca^{2+}/CaM and γ is a probability that $Thr^{286/287}$-dephosphorylated subunit binds Ca^{2+}/CaM. Note that $g(n)$ is a quadratic function of n because autophosphorylation of $Thr^{286/287}$ is an inter-subunit reaction. Approximating reactions among Ca^{2+}, calmodulin and CaMKII subunit by simple kinetics and assuming that a calmodulin molecule becomes active only when its four Ca^{2+}-binding sites are fully occupied, one has

$$\gamma^* = \frac{[Ca^{2+}/CaM]}{[Ca^{2+}/CaM] + K^*_1}, \quad \gamma = \frac{[Ca^{2+}/CaM]}{[Ca^{2+}/CaM] + K_1}, \quad [Ca^{2+}/CaM] = \frac{C_{CaM}[Ca^{2+}]^4}{[Ca^{2+}]^4 + K_2^{\ 4}} \quad (3)$$

where $[Ca^{2+}/CaM]$ and $[Ca^{2+}]$ are the concentrations of Ca^{2+}/CaM and Ca^{2+}, respectively; C_{CaM} is the total concentration of calmodulin; K^*_1 and K_1 are affinities of phosphorylated subunit for Ca^{2+}/CaM and dephosphorylated subunit for Ca^{2+}/CaM, respectively, and K_2 is an affinity of Ca^{2+}/CaM for Ca^{2+}.

It is thought of as a general principle that a phosphorylation process in cells accompanies a counter partner, a dephosphorylation process catalysed by phosphatase. As is usual, a lot of reports suggest that the degree of autophosphorylation of CaMKII in nerve cells is balanced by phosphatase activity. Thus, dephosphorylation of CaMKII should be included in our model, which will be described by the standard Michaelis-Menten scheme:

$$P + K_n \underset{k_{-3}}{\overset{nk_3}{\rightleftarrows}} PK_n \overset{k_4}{\longrightarrow} P + K_{n-1} \quad (n = 1, \cdots, N) \quad (4)$$

where P represents phosphatase. In the above, k_3 is multiplied by n because K_n has n phosphorylated sites to which the catalytic site of phosphatase is accessible.

To simplify analyses by reducing the number of variables and parameters, we postulate steady-state assumption for intermediate metabolites, PK_n ($n = 1,...,N$). Thereby, we have a set of equations describing the time evolution of the enzyme concentrations:

$$\frac{d[K_n]}{dt} = r(n+1)[K_{n+1}] + g(n-1)[K_{n-1}] - (r(n) + g(n))[K_n] \quad (n = 0, \cdots, N) \quad (5)$$

with

$$r(n) = \frac{nV_D}{K_D + \sum_{n=0}^{N} n[K_n]} \quad (6)$$

where $[K_n]$ is the concentration of K_n; K_D $(= (k_{-3} = k_{4)} / k_3)$ and V_D $(=k_4 C_P$ with C_P being the total concentration of phosphatase) are the Michaelis constant and maximal velocity for dephosphorylation of each subunit, respectively. Not all of equations (5) are independent because we have an identity, $\sum_{n=0}^{N} d[K_n] / dt = 0$, which provides the conservation law for CaMKII, $\sum_{n=0}^{N} [K_n] = C_K$, with C_K being the total concentration of the kinase.

3. RESULTS

To define the degree of autophosphorylation of the total amount of CaMKII in a postsynaptic spine, we introduced an order parameter, $\sum_{n=0}^{N} n[K_n]$, which represents the concentration of phosphorylated subunit. Since phosphorylated subunit is Ca^{2+}/CaM-independently active, the degree of Ca^{2+}/CaM-independent activity can also be measured by this quantity. The NMDA-receptor/channel activation during synaptic stimulation results in elevation of $[Ca^{2+}]$, and the amplitude of which will be a monotonic-increasing function of the stimulus intensity. For simplicity, $[Ca^{2+}]$ thus elevated during synaptic stimulation was approximated constant. Accordingly, we used $[Ca^{2+}]$ as a control parameter to characterise the stimulus intensity. We examined the time evolution of Ca^{2+}/CaM-independent activity of CaMKII measured by $\sum_{n=0}^{N} n[K_n]$ for various values of $[Ca^{2+}]$ by numerically solving equations (5). Throughout the examination, the parameter values were fixed as

$$N = 10, \quad k_a = k_b = k_c = 1s^{-1}, \quad K^*_1 = 6 \times 10^{-12} M, \quad K_1 = 3 \times 10^{-8} M, \quad K_2 = 10^{-5} M,$$

$$C_K = 10^{-6} M, \quad C_{CaM} = 5 \times 10^{-5} M, \quad K_D = 10^{-7} M, \quad V_D = 10^{-6} M s^{-1} \tag{7}$$

and the following initial condition was employed

$$[K_0] = C_K, \quad [K_n] = 0 \quad (n = 1, \cdots, N). \tag{8}$$

Results of the examination clearly show switching characteristics of the biochemical-reaction networks (Fig. 2): There is a threshold with respect to $[Ca^{2+}]$ between 0.85 µM and 0.86 µM (under the parameter values used in the present study); if $[Ca^{2+}]$ is above the threshold, autophosphorylation of CaMKII largely evolves, that is, Ca^{2+}/CaM-inde-

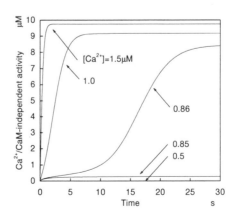

Figure 2. Relation between $[Ca^{2+}]$ and the time evolution of Ca^{2+}/CaM-independent activity of CaMKII.

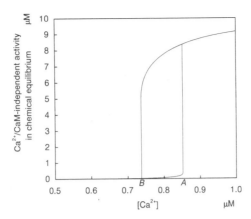

Figure 3. $[Ca^{2+}]$-dependence of Ca^{2+}/CaM-independent activity of CaMKII in chemical equilibrium.

pendent activity of CaMKII is switched on; if otherwise, Ca^{2+}/CaM-independent activity of CaMKII remains near zero (i.e. switched off).

These switching characteristics was also confirmed by numerically calculating $[Ca^{2+}]$-dependence of Ca^{2+}/CaM-independent activity of CaMKII in the chemical equilibrium. After the biochemical-reaction network for small $[Ca^{2+}]$ was settled in equilibrium, $[Ca^{2+}]$ was increased carefully so as not to disturb equilibrium; after $[Ca^{2+}]$ sufficiently exceeded the threshold, then, it was carefully decreased. Tracing Ca^{2+}/CaM-independent activity of the kinase along those increase and decrease in $[Ca^{2+}]$ yielded a hysteresis loop with a discreet jump at the threshold point (Fig. 3). These indicate that the biochemical-reaction networks have duplicated chemical equilibrium states for $[Ca^{2+}]$ between values indicated by A and B in Fig. 3. and they have a single chemical equilibrium state for $[Ca^{2+}]$ below A or above B. The appearance of the switching characteristics can therefore be explained as a discrete change in the dynamical structure of the biochemical-reaction networks at A, from bistable to monostable. These resemble first-order phase transition in statistical physics.

4. DISCUSSION

From the results obtained, we are lead to the following hypothesis for a role of CaMKII in the induction phase of LTP: Autophosphorylation versus dephosphorylation of CaMKII during synaptic stimulation transduces gradual information mediated by Ca^{2+} signalling via NMDA-receptor/channels into dichotomic information mediated by Ca^{2+}/CaM-independent activity of the enzyme. This dichotomic information encoding whether the stimulus intensity exceeds the threshold or not may determine the subsequent induction of LTP: If Ca^{2+}/CaM-independent activity of CaMKII is switched on, LTP may be induced; if it remains switched off, LTP cannot be elicited.

REFERENCES

1. For review for biochemical properties of CaMKII and involvement of the enzyme in LTP, see references in Okamoto, H. & Ichikawa, K. (1997), to appear in Cuthbertson, R., Holcombe, M. & Paton, R. (eds) *Computation in cellular and molecular biological systems*, World Scientific: Singapore.

2. Liedo, P.-M., Hjelmstad, G. O., Mukherji, S., Soderling, T. R. Malenka, R. C. & Nicoll, R. A. (1995), *Proc. Natl. Acad. Sci. USA* **92**, 11175–11179.
3. Hanson, P. I., Mayer, T., Stryer, L. & Schulman, H. (1994), *Neuron* **12**, 943–956.
4. Mukherji, S. & Soderling, T. R. (1994), *J. Biol. Chem.* **269**, 13744–13747.
5. Mukherji, S, Brickey, D. A. & Soderling (1994), T. R., *J. Biol. Chem.* **269**, 20733–20738.

6

A MODEL OF THE POSSIBLE ROLE OF GASEOUS NEUROMESSENGER NITRIC OXIDE IN SYNAPTIC POTENTIATION

Tao Wang and Nitish V. Thakor

Department of Biomedical Engineering
Johns Hopkins School of Medicine

1. ABSTRACT

The experimental studies show that NO can both enhance glutamate release from the presynapse and inhibit Ca^{++} influx in the postsynapse. In this paper, a theoretical model is presented to study these roles of NO in synaptic potentiation. In the model, NO is assumed to be produced in the postsynapse and diffuse into the extracellular space and then to the presynapse. Two kinds of excitatory receptor channels, NMDA and non-NMDA channels, are considered in the postsynaptic membrane. NO mediated glutamate release in the presynapse is expressed as the increase of the open channels in the postsynapse, and NO also decreases NMDA current since NMDA channel modulates Ca^{++} influx. Our simulations show that at low concentration of NO (0.5 uM), NO depresses the synaptic potentiation when a tetanus stimulation is delivered, but at high concentration (10uM), NO facilitates the synaptic potentiation. The membrane permeability, and the diffusion coefficient of NO are significant factors in determining the integrated function of NO in synaptic potentiation because they determine the concentration difference between the pre- and postsynaptic cells. Although accurate experimental measurements of all these parameters are not available, the model presented predicts that only at relatively high NO concentration levels, does NO facilitate production of LTP.

2. INTRODUCTION

Nitric oxide (NO), generated enzymatically from L-arginine by nitric oxide synthases (NOS), functions as a novel type of intercellular messenger molecule in many different tissues. NO mediates a great variety of phenomena, including endothelium-dependent vasorelaxation[1], macrophage-mediated cytotoxicity[2], inhibition of platelet adhesion and aggregation[2]. Especially, NO is a kind of neuronal messenger that carries out diverse signaling tasks in both the central and peripheral nervous systems[3,4] and as may play a role in synaptic plasticity.

As a neuronal messenger, once generated in a synapse (generally postsynapse), NO is able to diffuse from its site of production to the extracellular space and then to other synapses nearby to affect their synaptic transmission. The ways of NO mediating synaptic transmission are both pre- and postsynaptic. In one way, when NO diffuses into the presynapse, it regulates presynaptic transmitter release (especially increasing glutamate release.[5-7]), In the other way, NO also changes some ion flux in the postsynapse (especially decreasing Ca^{++} influx[8-10].). Based on these experimental results, This paper tries to build a model to investigate the integrated role of NO in synaptic transmission by studying the effects of NO in the induction of long-term potentiation (LTP).

3. MODEL DESCRIPTION

As described above, NO produced in the postsynapse decreases the postsynaptic Ca++ influx as well as diffuses into the presynapse to enhance the glutamate release. So there are three processes we should consider in the model: NO diffusing from the postsynapse to the presynapse, NO enhancing the glutamate release and NO inhibiting the Ca^{++} influx. Here, the mechanism of NO production is not considered and we assume that the amount of NO produced in the postsynapse (the source) stays constant. Figure 1 shows a schematic model assumed in this paper. NO generated in the spine head (postsynapse) diffuses into the presynaptic terminal to regulate glutamate release. Two kinds of glutamate receptor channels are considered here: NMDA channel (Ca^{++} influx) and non-NMDA (APMA and kainate).

3.1. NO Diffusion from the Postsynaptic Terminal

When modeling the diffusion process of NO, the consumption of NO during its diffusing should be taken into account because while diffusing, NO reacts with some biochemical substances, especially hemoglobin. This results in a very quick decay of NO and consequently the half-life of NO is very short (<10s) [11] . If we assume geometrical structure of the spine (shown in Fig.1) as a sphere with a radius of a, the concentration of NO in the spine is know as $Cs.$, the diffusion coefficient of NO in the tissue is D, the decay constant of NO is k (the relation k and the half-life $t_{1/2}$ can be expressed as $t_{1/2} = ln2/k$.), and the permeability of synaptic membranes to NO is p, the amount of NO diffusing from the postsynaptic cell to the presynaptic terminal can be expressed as:

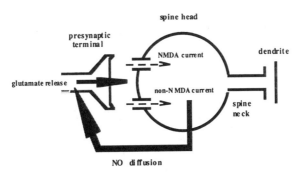

spine head

presynaptic terminal

NMDA current

dendrite

glutamate release

non-NMDA current

spine neck

NO diffusion

Figure 1. A schematic model of a synaptic connection. There are two synaptic channels in the spine: NMDA and non-NMDA, Only one neurotransmitter is considered to release from the presynaptic terminal: glutamate. NO is assumed to be generated in the spine and diffuses into the presynaptic terminal.

$$C_T = C_S \, p / \{ p + (D/a)[1 + a \, \text{sqrt} \, (k/D)] \}, \tag{1}$$

where, C_T is the concentration of NO in the presynaptic terminal.

3.2. NO and Synaptic Current

The two kinds of effects of NO on synaptic transmission (to enhance the release of glutamate and to inhibit Ca^{++} influx) finally result in the change of postsynaptic currents. On one hand, when glutamate release increase, the binding sites of its receptor channels should also increase, so both the NMDA and non-NMDA channel currents increase. On the other hand, NO inhibits Ca^{++} influx, and therefore NMDA current should decrease with the presence of NO. The former effect is due to the NO entering in the presynapse and the latter is caused by the NO staying in the postsynapse. Therefore, these two effects should relate to different concentrations of NO in different sites (pre- or postsynapse). So far, there haven't been enough experimental data to establish an accurate quantitative relationship between NO concentration and glutamate release. Actually, even measuring NO concentration *in situ* is also a difficult task. Here we try to formulate the relationships of NO and synaptic currents by making some simplifications.

3.2.1 Kinetics of NO Reacting with Its Target. The following kinetics of NO reacting with its target is assumed in this paper:

$$NO + R \longleftrightarrow NO*R \tag{2}$$

R is the target of NO in either pre- or postsynaptic site. The cellular response to the presence of NO is considered to be:

$$G = a[NOR], \; a \text{ is a constant.} \tag{3}$$

When at steady-state, G has a form of

$$G = a[R] [NO] / \{ K + [NO] \}, \tag{4}$$

where, K, a are constants. If we further assume $[R]$ retains as a constant during the reaction, G can be rewritten as:

$$G = c[R] [NO] / \{ K + [NO] \}, \; c \text{ is a constant.} \tag{5}$$

The above analysis is a simplified way to study the cellular responses to NO since the real chemical processes of NO reacting with its targets are too complicated to have been revealed up to now. Equation (5) is a saturation function which implies that the cellular responses to NO have limitations.

3.2.2. NO and NMDA and Non-NMDA Currents. As NO enhances glutamate release from the presynapse, the probability of one receptor channel binding to the transmitter increases. So the number of channels opened increases and the total conductance of the channels decreases. Therefore, the receptor channel currents can be written as:

$$I_{non\text{-}NMDA}(t) = (1 + r_e)g_{non\text{-}NMDA}(t)(V - E_n), \tag{6}$$

$$I_{NMDA}(t) = (1 + r_e)g_{NMDA}(t)m(V)(V - E_N). \tag{7}$$

where $g_y(t)$ is the channel conductance without the presence of NO and $Iy(t)$ is the total channel current ($y = non\text{-}NMDA, or NMDA$). V is the membrane potential at spine head and E_n, E_N are the channel equilibrium potentials, $m(V)$ is introduced to represent another mechanism of inhibition to NMDA receptor channel by extracellular Mg^{++} ions[12]. And r_e is a modification coefficient which has the same form of G. By considering that the enhancement of glutamate release relates to the presynaptic NO concentration:

$$r_e = fC_T/(K_T + C_T) \tag{8}$$

where C_T is the NO concentration in presynaptic terminal and f is a constant.

The another cellular response to NO is the inhibition of Ca++ influx, resulting in a decrease of NMDA current, so we modify the NMDA current as:

$$I_{NMDA}(t) = (1 + r_e - r_{i)}g_{NMDA}(t)(V - E_N). \tag{9}$$

In the above equation, the NO inhibition is represented by r_i which is defined in the same way as r_e:

$$r_i = hC_S/(K_S + C_S) \tag{10}$$

where Cs is the concentration of NO in postsynaptic space.

3.3. Equivalent Circuit

The following figure (Fig.2) shows the equivalent circuit of the model shown by Fig.1.

3.4. Long-Term Potentiation (LTP)

Long-term potentiation (LTP) is a long-lasting, activity-dependent increase in the efficacy of synaptic transmission. It can be induced by the delivery of tetanus stimulation to

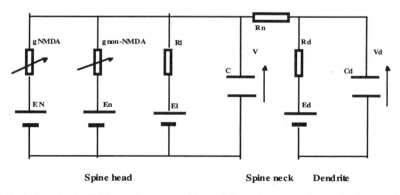

Figure 2. Equivalent circuit: V, Vd: membrane potentials, C, Cd: membrane capacitance, Rd, Rl : membrane resistance, Ed, El, membrane resting potentials.

afferent fibers and results in the increase in postsynaptic potential and intracellular Ca^{++} level[13]. So LTP can be expressed as the increase of postsynaptic potential amplitude and intracellular Ca^{++} level in the modeling study. To induce LTP, glutamate release and its receptor channel, NMDA channel, and Ca^{++} influx are necessary[13]. Here we will study what role NO plays in induction of LTP when a tetanus stimulation is delivered.

The temporal influx of Ca++ concentration in the postsynapse is modeled as [12]:

$$d[Ca^{++}]/dt = -K_a \left([Ca^{++}] - [Ca^{++}]_o \right) + K_b I_{NMDA}/V_h \qquad (11)$$

where, $[Ca^{++}]_o$ is the resting Ca^{++} level and V_h is the volume of the spine head. K_a and K_b are constants.

4. COMPUTER SIMULATION

4.1. Constants and Parameters Selection

Based on some experimental results[5-7,9,11,14,15], the parameters used in our simulations are shown in Table 1.

4.2. Results

By assuming a tetanus stimulation with 10 pulses and 100Hz, a potentiated synaptic potential can be induced (Fig 3(a)). If there is 0.5 uM NO being produced in the spine (the postsynapse), the synaptic potential decreases (Fig.3(a)), but it increases if the source concentration of NO is 10 uM (Fig.3(a)). By investigating Ca^{++} concentration, we find that 0.5uM NO also decreases Ca^{++} influx (Fig.3(b)), so in this case NO depresses the induction of LTP. But 10uM NO increases both the postsynaptic potential and Ca^{++} influx (Fig.3(a),(b)). From these results, we should conclude that high level of NO facilitates the induction of LTP.

For the pre- and postsynapse which are in the same synaptic connection, the permeability of the membrane to NO is essential in determining the NO diffusion between them, since the diffusion distance (synaptic cleft) is too small to be considered. Fig. 4 shows the effects of the permeability on potentiated synaptic behaviors when NO is present. Even with 10 uM NO, if the permeability is too small ($p=0.01$ um/ms), the postsynaptic poten-

Table 1. Parameters and constants

Geometrical Parameter:		NMDA Channel	
Radius of the spine head	$a_h=1um$	Peak conductance	$G_N=0.36nS$
Radius of the dendrite	$a_d=10um$	Peak Time	$t_N=4ms$
		Equilibrium Potential	$E_N=50mV$
		b=-55 mV	
NO Diffusion Parameters		non-NMDA Channel	
Diffusion coefficient	$D=3.8um2/ms$	Peak conductance	$G_n=0.36nS$
Half-life	$t_{1/2}=5s$	Peak Time	$t_n=4ms$
Permeability	$p=0.01, 0.1, 1$ um/ms	Equilibrium Potential	$E_n=50mV$
Membrane Properties:		Ca++ Parameters	
Membrane Capacitance	$C_m=2uF/cm2$	$[Ca++]o=0.1uM$	
Membrane Resistance	$R_m=25KW.cm2$	$Ka=0.2$	
Resting Potential	$E_r=-60mV$	$Kb=8$	
Spine Resistance	$R_n=0.025GW$		

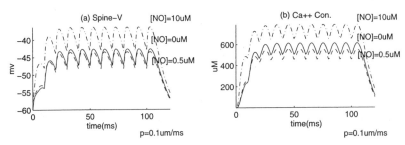

Figure 3. Simulation results with different NO source concentrations for a tetanus of 10 impulses at 100 Hz. (a) The potentials in spine head. (b) Ca^{++} influx. The NO concentration indicated in the figure is the concentration assumed in the spine head.

tial is less potentiated (Fig.4(a)). But if the permeability is large ($p=1um/ms$), NO facilitates the potentiation even its source concentration is small (Fig. 4(b)).

5. DISCUSSION AND CONCLUSION

The most important property of NO as a neuronal messenger is its mode of transmission — diffusion. While diffusing in the brain, NO can changes the synaptic efficacy by performing its functions on the presynapse (such as modulating neurotransmitter release) and postsynapse (such as regulating some ion channel currents). The results of our simulation in this project support this conclusion, and we have shown that the integrated role of NO in synaptic transmission depends on the amount of NO produced. If its level is low, NO acts to depress the synaptic potentiation in LTP, but it faculties LTP when its level is high [Fig. 3]. Also our simulations show the permeability of the cell membrane seems to play an important role in determining the functions of NO in synaptic transmission (potentiation or depression?) since the permeability value determine the amount of NO that releases from the postsynapse to the presynapse. It may be generally thought that NO is free permeable to the membrane since it is a gas. But based on the experimental data[15], the permeability is estimated as 0.1 to 1 um/ms, and this range of permeability values will cause significantly different results in our simulations [Fig.4].

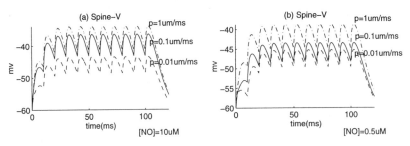

Figure 4. The effects of the membrane permeability to NO on the synaptic potentiation (a tetanus of 10 impulses at 100 Hz is used). (a) : The potentials in the spine head ([NO]=10uM), (b): The potentials in the spine head ([NO]=0.5uM).

6. REFERENCES

1. Palmer RMJ, Ferrige AG, et al. , Nitric oxide release accounts for the biological activity of endothelium-derived relaxing factor, *Nature* 327:524–526, 1987

2. Moncada S , The L-arginine: nitric oxide pathway *Acta Physiol. Scand.* 145:201–227, 1992

3. Schuman EM, Madison DV Nitric oxide and synaptic function , *Annu. Rev. Neurosci.* 17: 153–83, 1994

4. Garthwaite J. Charles, SL Chess-Williams R, Endothelium-derived relaxing factor release on activation of NMDA receptors suggests role as intracellular messenger in the brain, *Nature* 336:385–88, 1988.

5. Southam E and Garthwaite J , Comparative effects of some nitric oxide donors on cyclic GMP levels in rat cerebellarr slices , *Neurosci. Letters* 130:107–11, 1991

6. Lonart G, Wong J and Johnson KM , Nitric oxide induces neurotransmitter release from hippocampal slices, *European. J. of Pharmacology* 22:271–3, 1992

7. Lawrence AJ and Jarrott B, Nitric oxide increases interstitial excitatory amino acid release in the rat dorsomedial medulla oblongata , *Neurosci. Letters* 151:126–129 1993

8. Meffert MK , Premack BA and Schulman H Nitric oxide stimulates Ca2+ - independent synaptic vesicle release , *Neuron* 12:1235–44, 1994

9. Hoyt KR , Tang LH , et al. Nitric oxide modulates NMDA-induced increases in intracellular Ca++ in cultured rat forebrain neurons , *Brain Res.* 592:310–6, 1992.

10. Tanaka T , Endogenous nitric oxide inhibits NMDA- and kainates- responses by a negative feedback system in rat hippocampal neurons, *Brain Res.* 631:72–6, 1993

11. Meulemans A , Diffusion coefficients and half-lives of nitric oxide and N-nitroso-L-arginine in rat cortex *Neurosci. Letters* 171:89–93, 1994

12. Kitajima T and Hara K , A model of mechanisms of long-term potentiation in the hippocampus , *Biol. Cybern.* 64:33–9, 1990

13. Bliss TVP and Collingridge GL , A synaptic model of memory : Long- term potentiation in the hippocampus , *Nature* 361:31–9, 1993

14. Kelm M , Feelisch M, et al., Quantitative and kinetic characterization of nitric oxide and EDRF released from culture endothelial cells, *Biochem. Biophys. Res. Commun.* 154:236–44, 1988

15. Vanderkooi JM, Wright WW, et al., Nitric oxide diffusion coefficients in solutions, proteins and membranes determined by phosphorescence, *Biochimica et Biophysica Acta* . 1207: 249–54, 1994.

SECTION II

CELLULAR

7

AN INVESTIGATION OF TONIC VERSUS PHASIC FIRING BEHAVIOR OF MEDIAL VESTIBULAR NEURONS

Evyatar Av-Ron[1,2] and Pierre-Paul Vidal[3]

[1] Laboratoire de Biométrie
Institut National de la Recherche Agronomique
78026 Versailles Cedex, France
Tél: 33 1 30 83 33 67
E-mail: avron@bmve01.versailles.inra.fr
[2] B$_3$E — INSERM U444
ISARS, Faculté de Médecine
Saint-Antoine, 27 rue Chaligny
75571 Paris Cedex 12, France
[3] Laboratoire de Physiologie de la Perception et de l'Action
CNRS-Collège de France, UMR C 9950, 15 rue de l'Ecole de Médecine
75270 Paris Cedex 06, France

ABSTRACT

The vestibular-ocular and vestibulo-spinal network provides the ability to hold gaze fixed on an object during head movement. Within that network, the second-order neurons of the medial vestibular nucleus (MVNn) compute internal representations of head movement velocity in the horizontal plane. In vivo, these neurons can be classified as either tonic (type A) or phasic (type B), depending on their responses to head accelerations. In this study we have investigated to what extend the MVNn intrinsic membrane properties, could contribute to their dynamics. Biophysical models of the two categories were examined under ramp, step, sinusoidal and random depolarizing stimulations. Two factors were found major: the activation of the delayed potassium current and the rate of calcium flux.

1. INTRODUCTION

Stabilizing oculomotor and postular responses results from a complex multisensory integration. The neural network implementing gaze stabilization comprises amongst other cell types, the primary vestibular afferents, second-order (medial) vestibular interneurons (MVNn) and extraoculomotor and spinal motoneurons.

Our goal is to investigate to what extent the intrinsic membrane properties of the MVNn contribute to the specification of their dynamic responses during head movements. Two major classes of MVNn have been described in vitro. Type A MVNn were characterized by a broad action potential followed by a deep single afterhyperpolarization and a transient A-like rectification. Type B MVNn, in contrast, were distinguished by the presence of a thin action potential followed first by a fast, and then by a delayed and slower afterhyperpolarization. What could be the functional significance of this partition of the MVNn in two types? To investigate that question, we have modeled the type A and B MVNn membrane properties and investigated under various conditions the dynamic responses of these cell models.

2. NEURAL MODEL

The biophysical ion-conductance model (see Appendix), based on [1] consists of inward sodium (I_{Na}) and calcium (I_{Ca}) currents, outward potassium currents, delayed (I_K), transient (I_A) and calcium-dependent ($I_{K(Ca)}$) and a leak current (I_L). An ion current I_i was described by the product of three terms; the maximal conductance per unit area $\bar{g}_i$, the activation and inactivation variables, and the driving force ($V - V_i$). A simplified Hodgkin and Huxley [2] formalism was used whereby the steady state functions for activation and inactivation variables were either of the Boltzmann type for voltage-dependent processes or Michelis Menton kinetics for intracellular calcium concentration-dependent processes.

2.1. A-Cell Model

MVN A-type neurons in vitro were characterized as follows: spontaneous firing of 10-20Hz, wider action potential than B-type neurons, single deep AHP, and an A-like rectification when released from hyperpolarization [3,4]. These properties were modeled as follows: repetitive firing resulted from the delay in current I_K activation (position of N_∞), action potential width was altered by the rate of activation (slower τ_n), and delayed depolarization after hyperpolarization resulted from incorporating an A-current. The single deep AHP resulted from the contribution of the three potassium currents, though the calcium-dependent potassium current was most effective. Chemical blocker and ion substitution experiments were used to infer parameter values for the model. In calcium-free medium and the presence of TEA, A-type neurons continued to oscillate and did not develop a stable plateau potential. Experiments with TEA were simulated by reducing the value of $\bar{g}_K$. Viewed on a phase plane diagram, this caused the V-nullcline to shift to higher N-values (see e.g. Fig. 6b [5]). The model exhibited oscillations as long as its steady state point (SSP) was unstable, determined by the location of N_∞. Low-frequency firing was simulated by incorporating an I_A current [6] which also delayed the response after hyperpolarization [7,8]. The model exhibited a firing frequency of 61Hz with $\bar{g}_A = 0$, 20Hz with $\bar{g}_A = 4mS/cm^2$ (see Fig. 1), and 2Hz with $\bar{g}_A = 6.2mS/cm^2$. When released from hyperpolarization a delay of 100ms was seen for the first action potential. Finally, to describe the single deep AHP it was found that the calcium processes must be rapid, both in terms of calcium influx and intracellular calcium removal.

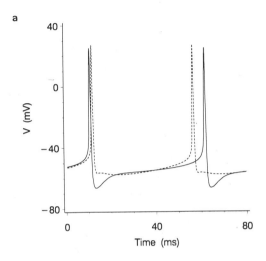

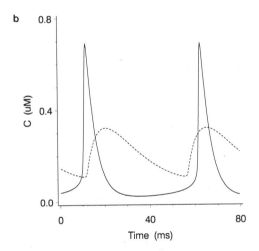

Figure 1. Behavior of type-A (*solid line*) and type-B (*dashed line*) models. **a** Membrane potential V. **b** Intracellular calcium concentration C. Model parameters: see Appendix.

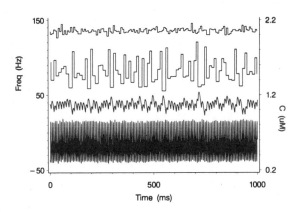

Figure 2. Random noise stimulation of MVNn models. Frequency diagrams of type-A (*top curve*) and type-B (*second curve*) models. Intracellular calcium concentration C of type-A (*fourth curve*) and type-B ($C+0.3$) (*third curve*) models. Model parameters: See Appendix. Peak noise stimulation: $20\mu A/cm^2$; using 'ran1' [10].

2.2. B-Cell Model

MVN B-type neurons in vitro were characterized as follows: spontaneous firing of 10-20Hz, narrower action potentials than A-type neurons, and both fast and delayed slow AHPs [3,4]. These properties were modeled by the following processes. The action potential duration was reduced by altering the current I_K. And the delayed AHP resulted from the calcium and calcium-dependent potassium currents.

When perfused with TEA, B-type neurons exhibited calcium plateau potentials [4,9] (i.e. stable SSP), different from A-type neurons which continued to oscillate (i.e. unstable SSP). This was realized in the model by shifting the position of the I_K activation function N_∞ to higher voltages. Hence, when simulating TEA experiments, a stable SSP was achieved for $\bar{g}_K$ values which for the A-type model caused an unstable SSP. The behavior of the B-type model is presented in Figure 1. The B-type model exhibited the three types of action potentials found experimentally (Fig. 2 [9]) by altering specific channel densities $\bar{g}_K$, $\bar{g}_A$ and $\bar{g}_{K(Ca)}$. By increasing I_K, the early AHP was more pronounced, while increasing $I_{K(Ca)}$ caused the delayed AHP to increase. Therefore, a natural variability of channel densities could bring about the different forms of action potentials observed experimentally.

3. DYNAMIC RESPONSES OF MVN NEURON MODELS

Examining the A and B MVN cell models under various depolarizing conditions and random noise stimulations showed that the tonic (type A) neuron model exhibited regular firing behavior, while the phasic (type B) model exhibited more irregular activity. This is an important point because the regularity of the resting discharge is a leading markers to bridge the gap between in vivo and in vitro preparations [11,12,13].

Indeed, in the guinea pig, a regular resting discharge is a common feature shared in vivo by the tonic second-order MVNn, and in vitro by the type A second-order MVNn in the whole brain. On the other hand, a more irregular discharge characterizes the phasic second-order MVNn in vivo, and the second-order type B MVNn recorded in the in vitro whole brain.

Based on our theoretical results (see Figure 2), the regular firing behavior exhibited by type A neurons is due to a fast calcium flux which causes a deep AHP and stabilizes the discharge rate of the neuron. And irregular firing activity is exhibited by type B neurons because of a slower calcium flux and hence weaker $I_{K(Ca)}$.

With step depolarizing stimulations, the tonic versus phasic firing behavior of the two models resulted mainly from the difference in the rate of calcium flux. The tonic (A) model showed faster calcium influx which brought about stable firing behavior within a relatively short transient period while the phasic (B) model behavior resulted from a slower rate of intracellular calcium accumulation so that the duration of transients was longer.

With ramp stimulations, the type A model showed a rapid increase in firing frequency for low input current values while the type B model exhibited a more linear current frequency (I F) response. What might be the role of such a difference in behavior? This result fits well with the in vivo data demonstrating a lower threshold for tonic neurons. One possibility would be that type A neurons respond in an all or none fashion to sensory inputs of low intensities while type B neurons code higher intensities of stimulation in a more graded manner. Finally, and very important to our point, for sinusoidal stimulations of various frequencies, the two models showed opposite behavior in terms of gain. With increasing stimulation frequencies, the A type model exhibited a decrease in gain while the B type model showed an increase in gain, see Figure 3. This difference in response was caused by the difference in calcium flux. For the type A model, the fast calcium flux caused the same peak frequency for both 2 and 5Hz stimulation. While the minimum firing frequency increased with 5Hz because the model could not slow down sufficiently during minimum sinusoidal input. On the other hand, the type B model achieved both higher peak and lower minimum firing frequencies with 5Hz. This was due to the difference in activation and inactivation of the currents I_{Ca} and $I_{K(Ca)}$. During sinusoidal peak stimulation, a stronger net depolarizing current occurred due to fast I_{Ca} activation versus slow $I_{K(Ca)}$ activation, the latter being dependent on intracellular calcium (C) accumulation. And during sinusoidal trough stimulation, a stronger net hyperpolarizing current occurred because of rapid I_{Ca} inactivation and slow intracellular calcium removal, thereby $I_{K(Ca)}$ was activated longer, which caused a longer interspike interval and hence a lower firing frequency.

Recently, the responses of MVN neurons stimulated with sinusoidal input were studied [14]. It was found that neurons with low background activity (e.g. 10Hz) showed a decrease in gain with increasing sinusoidal frequencies while other neurons with high background activity (e.g. > 30Hz) exhibited an increase in gain. This important experimental result fits well with our theoretical study which demonstrated opposite gain behavior for the two types of cells that were modeled.

4.   CONCLUSION

The characteristic behavior of type A and type B MVN neurons were modeled using a simple biophysical neural model. The type A and B models were then examined under four types of depolarizing stimulations random noise, step, ramp and sinusoidal inputs. It was found that the two models behave differently when challenged with input of various frequencies. The type A neurons appear to posses a set of conductances which render them more efficient to encode low frequency range head accelerations while the type B neurons would be more efficient at encoding the high frequency ones. The behavioral differences of type A and B neurons could correspond to tonic and phasic MVNn respectively. Therefore,

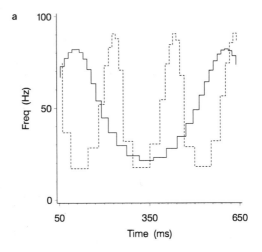

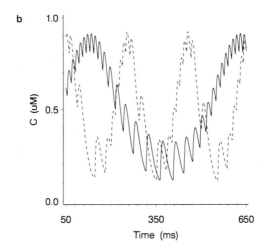

Figure 3. Sinusoidal stimulation of type-B model, 2Hz (*solid line*) and 5Hz (*dashed line*). **a** Frequency diagram. **b** Intracellular calcium concentration C. Model parameters: See Appendix, input $I = 0$ to $10\mu A/cm^2$.

our model supports the hypothesis that the MVNn dynamics are determined by their intrinsic membrane properties. We have shown that calcium related processes may play a key role in altering membrane conductances of neurons, both inward and outward currents, in bringing about the various responses observed experimentally. This theoretical investigation is a first step towards developing a network model for the vestibulo-ocular reflexes.

ACKNOWLEDGMENTS

We thank Sasha Babalian, Nicolas Vibert, Mauro Serafin, Michel Mühlethaler and Jean-François Vibert for helpful discussions. This work was supported by the International program of cooperation, PIC no. 292 and EA was partially supported by INSERM program *poste vert* and a post-doctoral position at INRA-Versailles.

5. APPENDIX

$$C_m dV/dt = I - I_{Na} - I_{Ca} - I_K - I_{K(Ca)} - I_A - I_L,$$

$$dC/dt = K_p(-I_{Ca}) - R \cdot C \quad dN/dt = [N_\infty(V) - N]/\tau_n(V),$$

$$dB/dt = [B_\infty(V) - B]/\tau_b, \quad \text{and} \quad dX/dt = [X_\infty(V) - X]/\tau_x.$$

$$I_{Na} = \bar{g}_{Na} \cdot M_\infty^3(V)(1-N)(V-V_{Na}), \quad I_A = \bar{g}_A \cdot A_\infty(V)B(V-V_K),$$

$$I_{Ca} = \bar{g}_{Ca} \cdot X^{xp}(K_c/[K_c+C])(V-V_{Ca}), \quad I_K = \bar{g}_K N^4(V-V_K),$$

$$I_{K(Ca)} = \bar{g}_{K(Ca)}(C^{cp}/[K_d^{cp}+C^{cp}])(V-V_K), \quad I_L = \bar{g}_L(V-V_L).$$

$$P_\infty(V) = \left(1 + exp[-2a^{(p)}(V - V_{1/2}^{(p)})]\right)^{-1} \quad \text{for} \quad P = \{A,B,M,N,X\}$$

$$\tau_n(V) = \left(\bar{\lambda}exp[a^{(n)}(V - V_{1/2}^{(n)})] + \bar{\lambda}exp[-a^{(n)}(V - V_{1/2}^{(n)})]\right)^{-1}$$

Default parameters: A-type model: $C_m = 1\mu F/cm^2$, $\bar{g}_{Na} = 10mS/cm^2$, $\bar{g}_K = 2mS/cm^2$, $\bar{g}_L = 0.1mS/cm^2$, $\bar{g}_{K(Ca)} = 1mS/cm^2$, $\bar{g}_{Ca} = 3mS/cm^2$, $\bar{g}_A = 4mS/cm^2$, $V_{Na} = 55mV$, $V_{Ca} = 124mV$, $V_K = -72mV$, $V_L = -50mV$, $V_{1/2}^{(m)} = -33mV$, $a^{(m)} = 0.055$, $V_{1/2}^{(w)} = -45mV$, $a^{(w)} = 0.055$, $\bar{\lambda} = 0.1$, $V_{1/2}^{(b)} = -70mV$, $a^{(b)} = -0.1$, $\tau_b = 10ms$, $V_{1/2}^{(x)} = -30mV$, $a^{(x)} = 0.08$, $\tau_x = 5ms$, $xp = 2$, $V_{1/2}^{(a)} = -40mV$, $a^{(a)} = 0.05$, $K_c = 0.1$, $K_p = 5$, $R = 25$, $K_d = 0.5$, $cp = 4$. B-type model: As above with $\bar{g}_K = 2.5mS/cm^2$, $\bar{g}_L = 0.3mS/cm^2$, $\bar{g}_{Ca} = 0.25mS/cm^2$, $\bar{g}_A = 1mS/cm^2$, $V_L = -40mV$, $V_{1/2}^{(w)} = -35mV$, $\bar{\lambda} = 0.2$, $\tau_x = 10ms$, $K_c = 1$, $K_p = 0.25$, $R = 0.05$, $cp = 1$.

REFERENCES

[1] Av-Ron E (1994) The role of a transient potassium current in a bursting neuron model. J Math Biol 33:71-87.

[2] Hodgkin AL, Huxley AF (1952) A quantitative description of membrane current and its application to conduction and excitation in nerve. J Physiol 117:500-544.

[3] Serafin M, de Waele C, Khateb A, Vidal PP, Mühlethaler M (1991a) Medial vestibular nucleus in the guinea-pig I. Intrinsic membrane properties in brainstem slices. Exp Brain Res 84:417-425.

[4] Serafin M, de Waele C, Khateb A, Vidal PP, Mühlethaler M (1991b) Medial vestibular nucleus in the guinea-pig II. Ionic basis of the intrinsic membrane properties in brainstem slices. Exp Brain Res 84:426-433.

[5] Av-Ron E, Parnas H, Segel LA (1991) A minimal biophysical model for an excitable and oscillatory neuron. Biol Cybern 65:487-500.

[6] Connor JA, Walter D, McKnown R (1977) Neural repetitive firing - modifications of the Hodgkin-Huxley axon suggested by experimental results from crustacean axons. Biophys J 18:81-102.

[7] Quadroni R, Knöpfel T (1994) Compartmental models of type A and type B guinea pig medial vestibular neurons. J Neurophsiol 72:1911-1924.

[8] Rush M, Rinzel J (1995) The potassium A-current, low firing rates and rebound excitation in Hodgkin-Huxely models. Bull Math Biol 57:899-929.

[9] Johnston AR, MacLeod NK, Dutia MB (1994) Ionic conductances contributing to spike repolarization and after-potentials in rat medial vestibular nucleus neurones. J Physiol 481:61-77.

[10] Press WH, Flannery BP, Teukolsky SA, Vetterlin WT (1988) Numerical recipes in C - The art of scientific computing. Cambridge University Press, Cambridge.

[11] Smith CE, Goldberg JM (1986) A stochastic afterhyperpolarization model of repetitive activity in vestibular afferents. Biol Cybern 54:41-51.

[12] Babalian A, Vibert N, Assie G, Serafin M, Mühlethaler M, Vidal, PP (1996) Central vestibular networks: functional characterization in the isolated, in vitro whole brain of guinea-pig. Neuroscience, submitted.

[13] Vidal PP, Babalian A, Vibert N, Serafin M, Mühlethaler M (1996) In vivo-in vitro correlation in the central vestibular system: a bridge too far ? In: New Directions in Vestibular Research, Higstein SM, Cohen B, Buttner Ennever JA (eds), Annals of the New York Academy of Sciences. In Press.

[14] du Lac S, Lisberger SG (1995) Cellular processing of temporal information in medial vestibular nucleus neurons. J Neurosci 15:8000-8010.

8

EFFECT OF POTASSIUM CONDUCTANCE CHARACTERISTICS ON PATTERN MATCHING IN A MODEL OF DENDRITIC SPINES

K. T. Blackwell,[1,2] T. P. Vogl,[1,2] and D. L. Alkon[3]

[1]George Mason University
Fairfax, Virginia 22030
[2]Environmental Research Institute of Michigan
1101 Wilson Blvd. Arlington, Virginia 22209
[3]Laboratory of Adaptive Systems
NINDS/NIH, Bldg 36, 4A21, Bethesda, Maryland 20892

1. ABSTRACT

Pattern matching is the ability to produce a stronger response to a previously learned pattern than to a novel pattern. Is it possible that the ability of mammals to recognize patterns is due to pattern matching by arrays of single neurons? Previous modeling studies have shown that plausible neuron models can match patterns of binary synaptic inputs. This study investigates the plausibility of analog pattern matching in a model of a dendrite with spines. Each dendritic spine includes a synaptic conductance and a calcium dependent potassium current whose properties depend on the previously learned value of synaptic conductance. The input to the model is a pattern of synaptic activation, and output of the model is the time integral of membrane potential (signal strength). Simulations show that signal strength is greatest when synaptic input equals the previously learned value, and is smaller when components of the synaptic input pattern are either smaller or larger than corresponding components of the previously learned pattern. The decrease in signal strength is proportional to the difference between input pattern and previously learned pattern. Pattern matching is robust to large changes in parameter values.

2. INTRODUCTION

Pattern matching is the ability to produce a stronger response to a previously learned pattern than to a novel pattern. The sensitivity and specificity with which animals can learn to recognize patterns has motivated the search for neuronal correlates of learning. In mammalian neurons, pairing weak stimulation of one pathway with strong stimulation of a separate but convergent pathway leads to strengthening of the synapses of the weakly stimulated pathway

(Bliss & Collingridge 1993). Associative conditioning (but not sham conditioning) results in the long term inactivation of the calcium dependent potassium conductance measured in hippocampal CA1 pyramidal cells (Disterhoft et al. 1986, Sanchez-Andres and Alkon 1991) and *Hermissenda crassicornis* (Alkon et al. 1985, Collin et al. 1988) as well as reduction of potassium currents in cerebellar Purkinje cells (Schreurs et al. 1996).

Is it possible the ability of mammals to recognize patterns is based on such changes in the biophysical properties of single neurons? Previous modeling studies have shown that plausible neuron models can recognize patterns of binary synaptic inputs. Siegel et al. (1994) investigated how changes in the properties of the calcium dependent potassium current allow for neurons to perform pattern recognition. Mel (1992) showed that the properties of the NMDA channel made neurons sensitive to the degree of spatial clustering of the activated synapses. In both of these studies, a compartmental model of a neuron is trained by repeated activation of one set of synapses, and tested by activating either the same or a different set of synapses. The neurons responded most strongly to the synaptic pattern used to train the neuron, demonstrating binary pattern matching.

The ability to discriminate among patterns that activate the same set of synapses, but with different intensities, allows for much richer information processing capabilities by the dendritic tree (Blackwell et al. 1995). This study investigates the plausibility of analog pattern matching in mammalian neurons by developing a biologically plausible model of a dendrite with dendritic spines.

3. METHODS

The model consists of a cluster of synaptic spines attached to a small area of dendritic membrane; each of the spines is modeled as an RC circuit (Fig 1a). The input to the model is the set of excitatory synaptic conductances located on the spine heads. Each synaptic conductance is approximated as an alpha function (Fig 1c) with peak value g_{smax}. Implicit in the model is an increase in spine intracellular calcium concentration proportional to g_{smax}.

Because long term inactivation of calcium dependent potassium conductances has been correlated with associative learning behavior, a calcium dependent potassium conductance, g_k, is included in each spine head (Fig 1b). g_k is non-inactivating:

$$g_k(t) = g_{kmax}m(t), \quad dm/dt = (m - m_{ss}) / \tau_m \tag{1}$$

The sensitivity of m_{ss} to calcium is modeled as a dependence on input synaptic conductance, $g_s(t)$ and the previously learned value of peak synaptic conductance, g_{s0} because the synaptic conductance is permeable to calcium:

$$m_{ss} = 1 / (1 + \exp(-15(g_s(t) - (g_{s0} + 0.2)))) \tag{2}$$

The output of the model is the integral of the potential (signal strength, Agmon-Snir 1995) across the somatic end of the dendritic branch, V_0.

4. RESULTS

The first set of experiments investigates the effect of parameters on pattern matching behavior in a model with a single dendritic spine. Figure 2 plots the signal strength versus

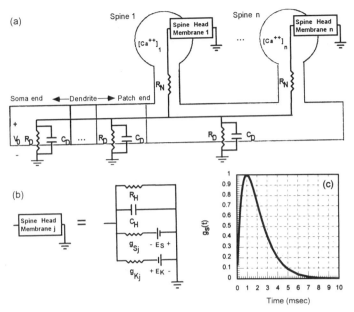

Figure 1. Circuit model of dendrite with dendritic spines. (a) dendrite is a passive cable. Spines are attached to distal end of cable. (b) The spine head membrane consists of a leak conductance, capacitance, synaptic conductance and calcium dependent potassium conductance. (c) The synaptic conductance is modeled as an alpha function: $g_s(t) = g_{smax} * e/\tau * \exp(-t/\tau)$. Intracellular calcium concentration is assumed proportional to g_{smax}.

synaptic input parameterized by the time constant (τ_m) of g_k, and the peak conductance (g_{kmax}). The heavy solid line in 2a and 2b corresponds to the "standard" parameters for τ_m and g_{kmax} used in all other simulations. Fig 2 demonstrates that the single spine model exhibits pattern matching because signal strength peaks for g_{smax} equal to g_{s0} for the standard parameters. These curves also illustrate that pattern matching is quite robust to changes in either of these parameters.

The second set of experiments investigates the sensitivity of the model to differences between the previously learned value and the synaptic input. Fig 3 illustrates pattern matching in a model with 8 spines in a single compartment. The previously learned value

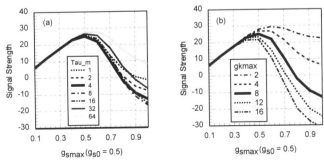

Figure 2. Signal strength peaks when g_{smax} equals g_{s0} in a single spine model, thus this model exhibits pattern matching. Pattern matching is relatively insensitive to changes in (a) the time constant of activation, τ_m, or (b) a change in g_{kmax}.

is 0.3 nA for all spines in 3a, and 0.7 nA for all spines in 3b. In both cases, signal strength decreases as the difference increases. Furthermore, for a given value of g_{s0}-g_{smax}, signal strength decreases with an increase in the number of spines whose input differs from the learned value by that value.

The third set of experiments investigates the sensitivity when the model learns a heterogeneous pattern of synaptic inputs. Fig 4a illustrates pattern matching in a model with 2 spines, which learned the pattern (0.3, 0.7). As in the previous cases, signal strength peaks when synaptic input equals the previously learned value. Fig 4b illustrates pattern matching in an 8 spine model which learned the pattern (0.1, 0.3, 0.3, 0.5, 0.5, 0.7, 0.7, 0.9). Each curve shows the signal strength when only a single component of the synaptic input pattern differs from the corresponding learned value. The symbols indicate the signal strength when the difference (g_{smax} - g_{s0}) equals 0.2. A difference of 0.2 results in a larger change from peak signal strength when g_{s0} is small than when g_{s0} is large. Furthermore, when g_{smax} is greater than g_{s0} signal strength is smaller than when g_{smax} is less than g_{s0}. This effect is also evident in Fig 3.

5. SUMMARY AND CONCLUSIONS

We showed that a group of spines in a single compartment responds more strongly to a previously learned pattern of synaptic inputs than to a novel pattern of synaptic inputs. Pattern matching behavior by this biologically plausible model suggests that neurons are able to discriminate input patterns that differ only in intensity distribution. The optimal parameters for pattern matching leads us to hypothesize the existence of a non-inactivating, fast activating calcium dependent potassium current in the spine head or on the dendritic branch. Evidence of voltage dependent potassium channels on distal dendrites (Wang et al. 1994, Veh et al. 1995), and voltage dependent calcium channels on dendritic spines (Mills et al. 1994) suggest the possibility that calcium dependent potassium channels may be found in the spine head, although calcium dependent potassium channels have

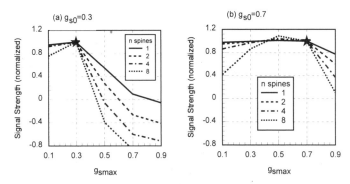

Figure 3. Sensitivity of pattern matching in an 8 spine model to differences between input pattern and learned pattern. Learned value of each spine is (a) 0.3 or (b) 0.7. The star marks the signal strength at which the input pattern equals the learned pattern. Each curve shows the decrease in signal strength when the value of *n* spines of the input pattern differs from that of the learned value, where *n* is indicated in the legend. The graph shows that the greater the difference between g_{smax} and g_{s0}, the smaller the signal strength. For a given difference, g_{smax} - g_{s0}, the larger is *n*, the smaller is signal strength.

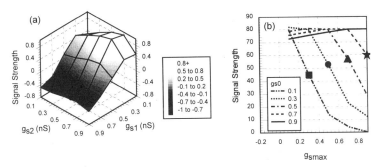

Figure 4. The effect of previously learned value on signal strength in (a) 2 spine model; g_{s0} = (0.3, 0.7), and (b) 8 spine model; g_{s0} = (0.1, 0.3, 0.3, 0.5, 0.5, 0.7, 0.7, 0.9). Symbols mark the signal strength for g_{s0} - g_{smax} = 0.2; comparison of symbols reveals that there is a larger change from peak signal when g_{s0} is small (e.g. 0.1) than when g_{s0} is large (e.g. 0.7).

not yet been found on dendritic spines. The use of a more accurate model of intracellular calcium concentration and potassium channel characteristics will permit investigations that elucidate the synaptic interactions that cause a change in properties of g_k steady state activation, m_{ss}.

6. REFERENCES

1. Agmon-Snir H. (1995) A novel theoretical approach to the analysis of dendritic transients. *Biophysical J.* **69**:1633–1656
2. Alkon D.L., Sakakibara M., Forman R., Harrigan J., Lederhendler I.I., Farley J. (1985) Reduction of two voltage-dependent K^+ currents mediates retention of a learned association. *Behavioral and Neural Biology* **44**:278–300
3. Blackwell K.T., Vogl T.P., Alkon D.L. (1995) Pattern recognition by a circuit model of dendritic spines. *Soc. Neurosci. Abstr.* **21**: 1225
4. Bliss T.V.P., Collingridge G.L. (1993) A synaptic model of memory: long-term potentiation in the hippocampus. *Nature* **361**: 31–39
5. Collin C.H., Ikeno J.F., Harrigan I., Lederhendler I.I. (1988) Sequential modification of membrane currents with classical conditioning *Biophysical J.* **54**: 955–960
6. Disterhoft J.F., Coulter D.A., Alkon D.L. (1986) Conditioning-specific membrane changes of rabbit hippocampal neurons measured *in vitro*. *Proc. Natl. Acad. Sci. USA* **83**: 2733–2737
7. Mel B. (1992) NMDA-based pattern discrimination in a modeled cortical neuron. *Neural Computation* **4**: 502–517
8. Mills L.R., Niesen C.E., So A.P., Carlen P.L., Spigelman I., Jones O.T. (1994) N-Type Ca^{++} channels are located on somata, dendrites, and a subpopulation of dendritic spines on live hippocampal pyramidal neurons. *J. Neuroscience* **14**: 6815–6824
9. Sanchez-Andres J.V., Alkon D.L. (1991) Voltage-clamp analysis of the effects of classical conditioning on the hippocampus. *J. Neurophysiology* **65**: 796–807
10. Schreurs B.G., Tomsic D., Gusev P., Alkon D.L. (1996) *J. Neurophysiol.* (In Press)
11. Siegel M., Marder E., Abbott L.F. (1994) Activity-dependent current distributions in model neurons. *Proc. Natl. Acad. Sci. (USA)* 91:11308–11312
12. Veh R.W., Lichtinghagen R., Sewing S., Wunder F., Grumbach I.M., Pongs O. (1995) Immunohistochemical localization of five members of the K_v1 channel subunits: contrasting subcellular locations and neuron-specific co-localizations in rat brain. *Eur. J. Neurosci* 7:2189–2205
13. Wang H., Kunkel D.D., Schwartzkroin P.A., Tempel B.L. (1994) Localization of Kv1.1 and Kv1.2, two K channel proteins, to synaptic terminals, somata and dendrites in the mouse brain. *J. Neuroscience* **14**: 4588–4599

9

CALCULATING FINELY-GRADED ORDINAL WEIGHTS FOR NEURAL CONNECTIONS FROM NEUROANATOMICAL DATA FROM DIFFERENT ANATOMICAL STUDIES

G. A. P. C. Burns,[1] M. A. O'Neill,[2] and M. P. Young[3]

[1]Neural Systems Group, Department of Psychology
University of Newcastle
Newcastle-Upon-Tyne, NE1–7RU, England
g.a.p.c.burns@ncl.ac.uk.
[2]Department of Engineering
University of Oxford
Oxford
mao@ermine.ox.ac.uk.
[3]Neural Systems Group, Department of Psychology
University of Newcastle
Newcastle-Upon-Tyne, NE1–7RU, England
malcolm@flash.ncl.ac.uk.

1. INTRODUCTION

Describing the gross organization of the brain at a systems level can be aided by the mathematical analysis of connections between brain regions (Young 1992, Young et al 1995, Scannell et al 1996). These analyses treat each brain region as a 'black box' and characterize its contribution to the system's organization according to its connections with other brain structures. The relative importance of any single connection in terms of the organization of the system may be influenced by the size or 'strength' of the connection in question which could be described in terms of the number of neurons that constitute it.

Most neuroanatomical papers offer qualitative descriptions of connection strength rather than quantitative measurements. In the rat, quantitative data does exist for a limited number of connections that have been particularly well studied (Martin 1986, Linden & Perry 1983), but does not exist for the vast majority of connection reports. For example, the number of retinal ganglion cells that project to the superior colliculus is of the order 10^5 (Linden & Perry 1983), and the weakest retinal efferents may only involve a few fibers, such as the retinal projection to the inferior colliculus (see fig 5 in Itaya & Van Hoesing 1982). In addition to this, the sensitivity of different neuroanatomical methods varies over at least two orders of magnitude (Wan et al 1982, Trojanowski et al 1982, Ter Horst et al 1984). The density of label produced in any neuroanatomical experiment is also dependent on the concentration and volume of tracer injected (Behazdi et al 1990).

The analysis presented here describes how neuroanatomical connection data can be analyzed objectively to give probabilistic ordinal connection weights between brain structures.

2. METHODS

Attempting to extract semi-quantitative connection weights from the neuroanatomical literature is a complicated by the following problems: the data are defined at the ordinal level of measurement; the nomenclature is disparate and non-standardized; different papers use a variety of techniques and approaches and interpret the structure of the brain according to different parcellation schemes.

We used data from the Neurobase neuroanatomical connectivity database, which is a literature based database that uses Microsoft Access and Excel to store, manipulate and present neuroanatomical tract tracing data. It presently describes 14732 connection reports taken from 1044 papers. It is designed to be open-ended and allow for systematic subsequent expansion and it provides a tool that could eventually incorporate the whole literature (Burns & Young 1996).

Each connection report in Neurobase contains the authors' description of the injection site and labeling pattern, and codes denoting the paper, experimental animal, experimental method, location of the soma and terminals labeled by the experiment, and the data-collator's interpretation of the density of labeling produced in the experiment.

The basic strategy of the approach adopted in this paper relies on the supposition that it is possible to compare tracer labeling patterns of connections to infer which connection is stronger, but only in certain circumstances: if two connections were labeled in the *same* experiment and one labeling pattern was significantly denser then it is possible to rank those connections relative to each other. If two connections were labeled in different experiments but with the same technique and one connection was very much more densely labeled, one could say that it was *probable* that one connection was denser than the other. If enough comparisons were made, it will be possible to rank the connections in order of connection strength in a simple one dimensional hierarchy. The two classes of comparisons are henceforth described as 'strict' and 'tentative' respectively. The difference in labeling density had to be much more pronounced for tentative comparisons to be accepted. If two separate comparisons of the same type contradicted each other then they were both dropped and strict comparisons were given higher priority than tentative comparisons.

The mboltzmann classifier evaluated the data by computing random hierarchies of connections and then shuffling them to minimize a cost function that was equal to $2(V_s)$ + V_t, where V_s is equal to the number of strict comparisons that were violated in any given connection strength hierarchy and V_t was equal to the number of violations of tentative comparisons. A full description of the mboltzmann classifier is described elsewhere (Hiletag et al 1996, Hilgetag et al, this volume).

3. RESULTS

We analyzed the output connectivity of the retina in the rat involving data taken from 23 separate papers yielding 821 comparisons between 87 connections. Mboltzmann found 4852 connection strength hierarchies where the number of levels varied between 21 and 32 and none of the comparisons had been violated. The most frequent number of levels was 26 with 1064 hierarchies. Probability density functions describing the connection strengths of are shown in figure 1 below. In subsequent analyses this information was interpreted by calculating the mean and variance of each probability density function providing both a derived value for the connection strength (referred to here as the 'ordinal probabilistic connection strength' or OPCS) and a measure of how well defined this value was, (i.e. the 'sharpness' of the probability density function's curve).

4. DISCUSSION

In previous studies of brain connectivity, little or no explicit attempt has been made to represent connection strength quantitatively. Scannell et al 1996 categorized connections into one of four classifications: 'dense', 'moderate', 'sparse' or 'non-existent' based on his interpretation of the literature and assigned ordinal weights of 3, 2, 1 and 0, respectively, to the connections. In other studies of neural connectivity, connections were not differentiated on the basis of their strength but on whether or not they were reciprocal (Young 1992, Young et al 1994) or on their cortical laminar pattern of origin and termination (Felleman & Van Essen 1991, Hilgetag et al 1996).

The mathematical discipline of Graph theory involves the analysis of networks made up of nodes and edges; such a formulation appears ideal for the analysis of systems level

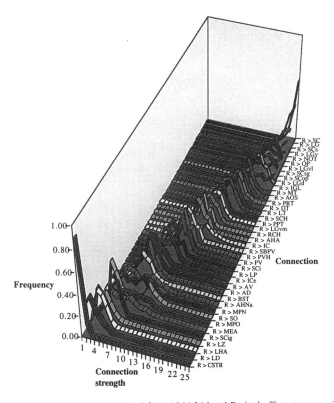

Figure 1. Probability density functions calculated from 1064 26 level Retinal efferent connection strength hierarchies using the MBOLTZMANN classifier.

Table 1. The meaning of abbreviations used to denote neuroanatomical structures in the rat brain

	Abbreviations and areas names		
AD	Anterodorsal nucleus of the thalamus	AHA	Anterior hypothalamic area
AHNa	Anterior hypothalamic nucleus	AOS	Accessory optic system
AV	Anteroventral nucleus of the thalamus	BST	Bed nucleus of the stria terminalis
CSTR	Corpus Striatum.	DT	Dorsal terminal nucleus of the accessory optic tract
IC	Inferior colliculus	ICe	Inferior colliculus, external nucleus
IGL	Intergeniculate leaflet	LD	Lateral dorsal nucleus of the thalamus
LG	Lateral geniculate complex	LGd	Lateral geniculate complex, dorsal part
LGv	Ventral part of the lateral geniculate complex	LGvl	Ventrolateral part of the lateral geniculate complex
LGvm	Ventromedial part of the lateral geniculate complex	LHA	Lateral hypothalamic area
LP	Lateral posterior nucleus of the thalamus	LT	Lateral terminal nucleus of the accessory optic tract
LZ	Lateral zone of the hypothalamus	MEA	Medial nucleus of the amygdala
MPN	Medial preoptic nucleus	MPO	Medial preoptic area
MT	Medial terminal nucleus of the accessory optic tract	NOT	Nucleus of the optic tract
OP	Olivary pretectal nucleus	PPT	Posterior pretectal nucleus
PRT	Pretectal region	PV	Periventricular nuclei of the hypothalamus
PVH	Paraventricular nucleus of the hypothalamus	R	Retina
RCH	Retrochiasmic area	SBPV	Subparaventricular zone
SC	Superior colliculus	SCH	Suprachiasmatic nucleus
SCi	Superior colliculus, intermediate layers	SCig	Superior colliculus, intermediate gray layer
SCop	Superior colliculus, optic layer	SCs	Superior colliculus, superficial layers
SCsg	Superior colliculus, superficial gray layer	SO	Supraoptic nucleus

neuroanatomical connectivity data. The development of this approach has been hampered by the absence of clearly defined connection weights in the raw data. Despite being only a probabilistic, ordinal measure, our method may aid the development of graph theoretical approaches (e.g. Jouve et al 1996).

A great wealth of neuroanatomical data exists in the literature but cannot properly be used without analysis. This methodology is time consuming and computationally intensive (Hilgetag et al 1996). It does, however, describe the 'raw data' of large scale connectivity analyses in more detail than has been achieved before.

5. REFERENCES

Behzadi, G., P. Kalen, et al. (1990). "Afferents to the Median Raphe Nucleus of the Rat - Retrograde Cholera-Toxin and Wheat-Germ Conjugated Horseradish-Peroxidase Tracing, and Selective D-[H-3]Aspartate Labeling of Possible Excitatory Amino-Acid Inputs." *Neuroscience* 37(1): 77–100.

Burns, G. A. P. C. and M. P. Young (1996). *Neurobase: a neuroanatomical connection database and its use in providing a description of connections in the rat hippocampal system.* Brain Research Assocation Abstracts, Newcastle upon Tyne.

Felleman, D. J. and D. C. V. Essen (1991). "Distributed hierarchical processing in the primate cerebral cortex." *Cerebral Cortex* 1: 1–47.

Hilgetag, C. C., M. A. O'Neill, et al. (1996). "Indeterminate organization of the visual system." *Science* 271(5250): 776–777.

Itaya, S. K. and G. W. Vanhoesen (1982). "Retinal Innervation of the Inferior Colliculus in Rat and Monkey." *Brain Research* 233(1): 45–52.

Jouve et al (1996) "A mathematical approach to the connectivity between the visual areas of the macaque monkey", Cerebral Cortex (In press).

Linden, R. and V. H. Perry (1983). "Massive Retinotectal Projection in Rats." *Brain Research* **272**(1): 145–149.

Martin, P. R. (1986). "The Projection of Different Retinal Ganglion-Cell Classes to the Dorsal Lateral Geniculate-Nucleus in the Hooded Rat." *Experimental Brain Research* **62**(1): 77–88.

Scannell, J. W., M. P. Young, et al. (1995). "Analysis of connectivity in the cat cerebral cortex." *Journal of Neuroscience* **15**: 1463–1483.

Sefton, A. and B. Dreher (1995). Visual System. *The Rat Nervous System.* G. Paxinos, Academic Press: 833–898.

Terhorst, G. J., H. J. Groenewegen, et al. (1984). "Phaseolus-Vulgaris Leuko-Agglutinin Immunohistochemistry - a Comparison Between Autoradiographic and Lectin Tracing of Neuronal Efferents." *Brain Research* **307**(1–2): 379–383.

Trojanowski, J. Q., J. O. Gonatas, et al. (1982). "Horseradish-Peroxidase (HRP) Conjugates of Cholera Toxin and Lectins Are More Sensitive Retrogradely Transported Markers Than Free HRP." *Brain Research* **231**(1): 33–50.

Wan, X. C. S., J. Q. Trojanowski, et al. (1982). "Cholera Toxin and Wheat-Germ Agglutinin Conjugates As Neuroanatomical Probes - Their Uptake and Clearance, Transganglionic and Retrograde Transport and Sensitivity." *Brain Research* **243**(2): 215–224.

Young, M. P. (1992). "Objective Analysis of the topological organization of the primate cortical visual system." *Nature* **358**: 152–155.

Young, M. P., J. W. Scannell, et al. (1994). "Analysis of Connectivity: Neural systems in the cerebral cortex." *Reviews in the Neurosciences* **5**: 227–250.

Young, M. P., J. W. Scannell, et al. (1995). "Non-metric multidimensional scaling in the analysis of neuroanatomical connection data and the organization of the primate cortical visual system." *Philosophical Transactions of the Royal Society* **348**: 281–308.

10

MODELING THE PASSIVE PROPERTIES OF NONPYRAMIDAL NEURONS IN HIPPOCAMPAL AREA CA3

Raymond A. Chitwood, Brenda J. Claiborne, and David B. Jaffe

Division of Life Science
The University of Texas at San Antonio
6900 N. Loop 1604 West, San Antonio, Texas 78249
randy@glu.ls.utsa.edu
brenda@harlan.ls.utsa.edu
david@glu.ls.utsa.edu

ABSTRACT

We have been studying the passive membrane properties of nonpyramidal neurons in the CA3 region of the hippocampal formation. Using a combination of current-clamp recording, three-dimensional reconstruction, and compartmental modeling we examined the passive somatic membrane response and dendritic voltage attenuation in identified nonpyramidal, and therefore presumed inhibitory, neurons. The input resistance (R_N) of compartmental models closely fit measured values for R_N in more than 50% of cells studied. Membrane charging in these models, however, did not fit experimentally measured values. Finally, analysis of voltage attenuation in our neuron models suggests that voltage signals significantly attenuate from most dendritic locations to the soma.

INTRODUCTION

The hippocampus is important for certain aspects of learning and memory and the CA3 region is an integral component of the so-called "tri-synaptic loop". Pyramidal neuron projections within this area form an excitatory, recurrent network that may have an autoassociative function.[11] The CA3 region has strong local inhibition to prevent epileptiform behavior due to this positive-feedback circuitry. Inhibitory interneurons represent approximately 10% of the total cells in this area[1] and extensively innervate CA3 pyramidal cells [5] thus controlling the excitability of the network. Another possible role for inhibitory neurons is that they normalize the output of this region.[11] Finally, inhibition may be important for the modulation and timing of oscillations; oscillations are thought to play a critical role in synaptic plasticity[8] and the integration of distributed processes.[6]

The physiology and morphology of interneurons is generally quite distinct from that of pyramidal neurons of the hippocampus. Their dendritic arborization is significantly smaller than that of pyramidal neurons and is generally more stellate or "nonpyramidal". The active properties of these cells are also distinct from those of pyramidal neurons; their action potentials are shorter, and exhibit little spike frequency accommodation.[5]

To study the passive properties of nonpyramidal, interneurons of area CA3, we have employed three methods. First, whole-cell, current-clamp recordings were used to measure the somatic passive membrane response of identified nonpyramidal neurons. Second, biocytin labeling of these neurons allowed us to visualize the dendritic arbors and digitize them using a computer-controlled reconstruction system.[3] Finally, both the physiology and morphometric data were combined to construct compartmental models. These were then used to estimate dendritic voltage attenuation.

METHODS

Hippocampal slices (300μm) were prepared form 14–30 day old Sprague-Dawley rats. They were maintained in a holding chamber at room temperature in oxygenated (95% O_2/5% CO_2) artificial CSF (aCSF) containing (in mM): 124 NaCl, 2.5 KCl, 26 $NaHCO_3$, 2 Mg Cl, 2 $CaCl_2$, 1.25 $NaHPO_4$, and 10 dextrose. Whole-cell patch clamp recordings were made from visually identified nonpyramidal neurons in the CA3b region using IR/DIC video microscopy.[14] All experiments were performed at room temperature (~23 °C). Micropipettes contained (in mM): 120 K-Gluconate, 20 KCl, 0.1 EGTA, 2 $MgCl_2$, 2 Na_2ATP and 10 HEPES (pH=7.3) and 0.5% biocytin or Neurobiotin. Membrane responses to depolarizing and hyperpolarizing current injections were measured within the linear range of the membrane. V/I curves were constructed to measure the R_N of each cell. Time constants were determined by fitting a double exponential function to the average of at least 100 membrane transients in response to hyperpolarizing current injections.

Cells were filled with biocytin or Neurobiotin by passive diffusion during whole-cell recordings. Slices were fixed in 3% gluteraldehyde, incubated in an avidin-HRP solution and developed using the peroxidase substrate DAB.[10] Non-resected slices were mounted and digitized in three dimensions using a computer controlled reconstruction system.[3] Only slices having minimal background staining and high contrast were selected for reconstruction. An example of a reconstructed CA3 nonpyramidal neuron is illustrated in Figure 2 (inset).

Coordinates from the morphometric analyses were used to construct compartment models using NEURON.[7] Values for specific membrane capacitance (C_m) and intracellular resistance (R_i) were 1.0 μF/cm^2 and 200 Ω-cm respectively.[12] Specific membrane resistivity (R_m) was determined from the slowest fitted time constant (τ_0). Voltage attenuation was determined by comparing the simulated voltage in each compartment to the voltage seen at the soma for each of two cases: the dendrosomatic (voltage in) and the somatodendritic (voltage out) directions.[2]

RESULTS

Whole-cell recordings were made from nonpyramidal neurons in the CA3b region of the hippocampus. These recordings were made from cell bodies of nonpyramidal neurons distributed in all substrata (e.g. *s. radiatum, s. lacunosum-moleculare, s. lucidum*, etc.). The somatic R_N of CA3 nonpyramidal neurons was calculated from the slope within the linear range of the voltage-current relationship.[13] Mean R_N for these cells was 456 ± 204

MΩ (mean ± SD, n=25). Membrane charging in response to a hyperpolarizing current pulse, again within the linear range of the V/I relationship, was best fit by two exponentials; a slow component (τ_0) of 63 ± 26 ms and a fast component (τ_1) of 10 ± 5 ms.

Each of these neurons was also labeled with biocytin or neurobiotin. Cells were reconstructed in three dimensions as described in Methods. For the 25 neurons studied, mean dendritic length was 268 ± 41 μm and the mean surface area was 12848 ± 4888 μm. The mean number of compartments comprising the entire arborization of the reconstructed neurons was 1136 ± 340 with a mean compartmental length of 2.6 ± 2.1 μm.

Passive neuron models were constructed using the data from the three-dimensional reconstructions described above. For each cell, specific membrane resistivity (R_m) was calculated from the measured value of τ_0, where $\tau_0 = R_m C_m$. Calculated R_N in these model neurons (642 ± 266 MΩ) closely fit measured values for R_N in more than 50% of cells studied (Fig. 1). Our criteria was an arbitrarily chosen cutoff of a 20% difference in R_N. In all cases where R_N differed between the model and experimentally measured value, the R_N of the model was larger. Only those cells with less than a 20% difference in R_N were used for further analyses (n=15).

We also examined whether the passive neuron models could reproduce the fast component of membrane charging, τ_1, that reflects the redistribution of charge into the dendrites. In none of our models did the calculated value of τ_1 fit the measured response. Mean τ_1 was 5 ± 3 for all modeled cells (n=15). In all cases modeled τ_1 was at least 50% less than measured values. We then tested the hypothesis that τ_1 was dependent upon the chosen value of R_i. R_i was varied in the model between 100–1000 Ωcm. Even at the highest value of R_i, where τ_1 was longer, simulated τ_1 was less than experimentally measured values.

We next examined dendritic voltage attenuation in our model nonpyramidal neurons. Figure 2 illustrates a typical distribution of voltage attenuation for simulated 100 Hz voltage signals into and out from the soma. Signals applied at the soma, and measured in each compartment relative to the soma (voltage out), were attenuated with a roughly uniform distribution. Mean attenuation out was 43.5 ± 10.5% (n=15). In contrast, dendritically applied voltage signals (voltage in), where voltage signals were applied at dendritic compartments and compared to the voltage seen at the soma, was not uniform; significant attenuation (>50%) was observed in the majority of dendritic compartments. Mean dendrosomatic attenuation was 80.2 ± 10.4%.

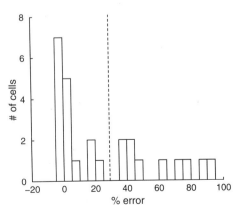

Figure 1. Histogram of modeled R_N error as compared to experimentally measured values. The dashed line represents the cutoff for an acceptable model. 15 of 25 cells fit R_N within 20%.

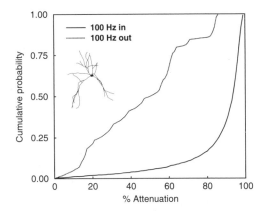

Figure 2. A cumulative probability histogram of attenuation for 100 Hz voltage signals in and out. A reconstructed *stratum radiatum* neuron from which this data was obtained is shown (inset).

Neurons are typically exposed to voltage changes over a wide range of frequencies. As illustrated in Figure 3, frequencies greater than 10 Hz had mean and median dendrosomatic attenuations for all compartments that were greater than 50%. In addition, asymmetric voltage-attenuation (voltage in versus voltage out) was seen at all frequencies. Note, however, attenuation of voltage in signals for all frequencies was greater than the mean. Voltage signals out from the soma, in contrast, had similar mean and median values consistent with a more normal or uniform distribution of voltage attenuation.

DISCUSSION

In this study we have used a combination of whole-cell recording, three dimensional reconstruction, and compartmental modeling to investigate the passive properties of hippocampal CA3 interneurons. The direct combination of experimental and modeling analysis of electrotonic structure is similar to the analysis previously applied to CA3 pyramidal neurons[10] as well as inhibitory neurons of area CA1.[15]

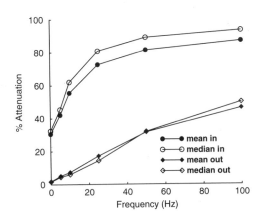

Figure 3. Comparison of median and mean voltage attenuation for different frequencies. Note the asymetrical distribution of voltage attenuations in and out.

We found that the measured R_N of nonpyramidal neurons were significantly higher than R_N of CA3 pyramidal neurons.[13] This is consistent with the smaller surface area and reduced dendritic arborization of nonpyramidal cells. In contrast, the slow membrane time constants (τ_0) for nonpyramidal neurons that we measured were not different than those for CA3 pyramidal neurons (~60 ms).

In over 50% of our neurons, the compartmental model successfully fit measured values for R_N. Those models that did not closely reproduce R_N in all cases had higher values. The most likely explanation for this difference was an underestimation of dendritic surface area. Underestimation of surface area could arise due to several factors including incomplete filling of dendrites or the inability to resolve deep neuronal processes in thick hippocampal slices. Another possibility for this difference could be a nonuniform distribution of R_m. Finally, a somatic shunt produced by the patch electrode, albeit smaller than that produced by sharp electrodes,[13] might also result in smaller measured R_N.[15] Both a nonuniform R_m and a somatic shunt might also account for our inability to fit the fast component of membrane charging to our model neurons.

Analysis of voltage attenuation indicates that for even modest signal frequencies (~10 Hz), there will be significant attenuation from most dendritic synaptic locations to the soma. Therefore, for a majority of synaptic locations there will be larger attenuation of synaptic potential propagation to the soma. This suggests that there may be a high degree of local dendritic processing in these cells in spite of their high R_N. Furthermore, nonpyramidal neurons, like pyramidal neurons, may have active dendrites. Amplification of electrotonically distant signals by active membrane could enhance the propagation of excitatory postsynaptic potentials from the synapse to the soma.[9] Finally, we found that nonpyramidal neurons, which are generally more stellate in morphology than pyramidal neurons, appear to have a unimodal distribution of voltage attenuation, both dendrosomatic and somatodendritic, across the dendritic tree. This is in contrast to the morphology of CA1 pyramidal neurons that have multimodal distribution of attenuation associated with lamellar distributions of dendrites.[4]

Our preliminary analysis of the morphology of nonpyramidal neurons suggests that there may be different classes of inhibitory neurons distributed throughout the substrata of area CA3. Based on the work presented here, we plan to further explorer the possibility that there may be distinct classes of nonpyramidal neurons in CA3, as defined physiologically, morphologically and electrotonically.

REFERENCES

1. Amaral, D.G., N. Ishizuka, and Claiborne, B.J. Neurons, numbers and the hippocampal network, in *Progress in Brain Research,* J. Storm-Mathisen, J. Zimmer, and O.P. Ottenson, Editor., Elsevier, Amsterdam, pp. 1–11., 1992.
2. Carnevale, N.T. and Johnston D., Electrophysiological characterization of remote chemical synapses. *J. Neurophys.* 47:606–621, 1982.
3. Claiborne, B.J., Use of computers for quantitative, three-dimensional analysis of dendritic trees, in *Methods in Neurosciences,* P.M. Conn, Editor., Academic Press, San Diego, pp. 315–330., 1992.
4. Claiborne, B.J., Zador, A.M., Mainen, Z.F., and Brown, T.H. Computational models of hippocampal neurons, in *Single Neuron Computation,* T. McKenna and J. Davis, Editor., Academic Press, Boston, pp. 61–79., 1992.
5. Freund, T.F. and Buzsaki, G. Interneurons of the hippocampus. *Hippocampus* 6:347–470, 1996.
6. Gray, C.M., Synchronous oscillations in neuronal systems: Mechanisms and functions. *J. Comput. Neurosci.* 1:11–38, 1994.
7. Hines, M., A program for simulation of nerve equations with branching geometries. *Int. J. Biomed. Comp.* 24:55–68, 1989.

8. Jensen, O., Idiart, M.A.P., and Lisman, J.E. Physiologically realistic formation of autoassociative memory in networks with theta/gamma oscillations: Role of fast NMDA channels. *Learning and Memory* 3:243–256, 1996.

9. Lipowksy, R., Gillessen, T., and Alzheimer, C., Dendritic Na^+ channels amplify EPSPs in hippocampal CA1 pyramidal cells. *J. Neurophys.* 76:1029–1038, 1996.

10. Major, G., Larkman, A.U., Jonas, P., Sakmann, B., and Jack, J.J.B., Detailed passive cable models of whole-cell recorded CA3 pyramidal neurons in rat hippocampal slices. *J. Neurosci.* 14:4613–4638, 1994.

11. McNaughton, B.L., Neuronal mechanisms for spatial computation and information storage, in *Neural connections, mental computation,* L. Nadel, *et al.* , Editor., MIT Press, Cambridge, pp. 285–350.,1989.

12. Spruston, N., Jaffe, D.B., Williams, S.H. and Johnston D., Voltage- and space-clamp errors associated with the measurement of electrotonically remote synaptic events. *Journal of Neurophysiology* 70:781–802, 1993.

13. Spruston, N. and Johnston D., Perforated patch-clamp analysis of the passive membrane properties of three classes of hippocampal neurons. . Journal of Neurophysiology 67:508–29, 1992.

14. Stuart, G.J., Dodt, H.U., and Sakmann, B., Patch-clamp recordings from the soma and dendrites of neurons in brain slices using infrared video microscopy. Pflugers. Arch. 423:511–518, 1993.

15. Thurbon, D., A. Field, and Redman, S.J. Electrotonic profiles of interneurons in stratum pyramidale of the CA1 region of rat hippocampus. J. Neurophys. 71:1948–1958, 1994.

11

AN ANALYSIS OF THE ADAPTIVE BEHAVIOR OF PIRIFORM CORTEX PYRAMIDAL CELLS

S. M. Crook[1] and G. B. Ermentrout[2]

[1]Mathematical Research Branch, NIDDK
National Institutes of Health
Bethesda, MD
[2]Department of Mathematics
University of Pittsburgh
Pittsburgh, PA

INTRODUCTION

Cortical pyramidal neurons respond to a depolarizing current pulse with a train of action potentials [4, 2]. These action potentials occur at a higher frequency during the initial stages of the current injection with a decreased firing rate or a cessation of firing at later stages of a sustained injection. This spike frequency adaptation can be mostly suppressed by local application of norepinephrine or acetylcholine [6, 7] which block a slow calcium-dependent potassium current.

Barkai and Hasselmo (1994) describe the adaptation characteristics of representative layer II pyramidal cells from piriform cortex, provide a measure for classifying these cells as showing either strong or weak adaptation, and develop a biophysical model to demonstrate the adaptation characteristics of both types of cells. In general, weakly adapting cells show repetitive action potential generation throughout the current injection period. In contrast, the strongly adapting cells often show a complete cessation of firing. The active conductances in their three-compartment model include a fast-activating voltage-dependent sodium current (I_{Na}) and a delayed rectifier potassium current (I_{K-DR}) which mediate the generation of simulated action potentials. They also include a noninactivating voltage-dependent potassium current (I_{K-M}), a calcium-dependent potassium current (I_{K-AHP}), a high-threshold voltage activated calcium current (I_{Ca}), and an additional voltage-dependent potassium current (I_{K-A}) similar to the currents in other pyramidal cell models [8, 10, 3]. The model parameters are adjusted to reflect characteristics of current clamp recordings of piriform cortex pyramidal cells where much of the focus concerns setting the maximal conductances of the different currents. In particular, the differences in the neuronal adaptation properties are represented by changing the maximal conductances of the I_{K-AHP}, I_{K-M}, and leak currents (I_L) where larger maximal conductances result in stronger adaptation.

Computational Neuroscience
edited by Bower, Plenum Press, New York, 1997

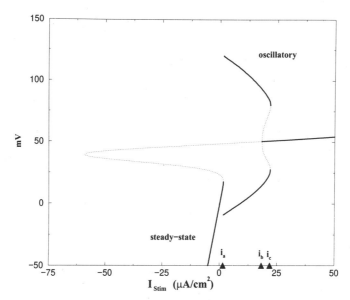

Figure 1. Bifurcation diagram depicting the behavior of the spiking model as current injection varies. Stable states are shown solid, and unstable states are shown dashed.

Vanier and Bower (1996) recently developed a similar five-compartment model for a weakly adapting pyramidal cell from layer II of piriform cortex with I_{Na}, I_{K-DR}, I_{K-M}, I_{K-AHP}, I_{Ca}, and I_L currents. The model was compared to actual current-clamp data recorded from a weakly adapting cell in a slice preparation of piriform cortex. A parameter search method was used to obtain the best match in terms of spike timing for current injections made at seven different levels of current where each stimulus above a threshold results in initial rapid firing adapting to a repetitive firing at constant frequency. The resulting model is consistent with the behavior of a weakly adapting cell but does not exhibit strong adaptation when the maximal conductances are increased for the I_{K-AHP}, I_{K-M}, and leak currents.

We use a simplified model of a pyramidal cell from piriform cortex to examine the underlying dynamics of the mechanisms that produce the adaptation in these models. Our analysis shows that the degree of adaptation is determined by the ionic conductance density of the adaptation currents and by the relative timing of the kinetics of this current and the rate of decay of intracellular calcium.

SPIKING BEHAVIOR

Consider a cortical oscillator where the dynamics of the cell can be represented by

$$C_M \frac{dV(t)}{dt} = -I_{Ion}(V, \vec{w}) + I_{Stim} \tag{1}$$

where $V(t)$ denotes the deviation of the membrane potential from some reference potential at time t ms, I_{Ion} is the sum of voltage and time dependent currents through the various ionic channel types, and $\vec{w}$ is the vector of auxiliary membrane variables such as intracellular calcium and the gating variables. The stimulus I_{Stim} represents the electrode current applied to the soma

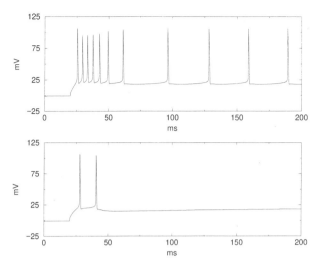

Figure 2. Voltage traces for $I_{Stim} = 5$ showing increased adaptation in the lower trace where the values of $\bar{g}_{K-AHP}$, $\bar{g}_{K-M}$, and $\bar{g}_{S-L}$ are increased.

divided by the total cell membrane area. We begin by examining the behavior in the case where the only currents present in the model are

$$I_{Na}(V, m, h) = \bar{g}_{Na} m^2 h (V - V_{Na}) \tag{2}$$

$$I_{K-DR}(V, n) = \bar{g}_{K-DR} n (V - V_K) \tag{3}$$

$$I_L(V) = \bar{g}_L (V - V_L). \tag{4}$$

The conductances and the kinetic equations for the gating variables for these currents and for those described in later sections are given in the appendix.

The bifurcation diagram shown in Figure 1 illustrates the behavior of the model as the level of current injection varies. Stable states are shown solid, and unstable states are shown dashed. The branch of steady states forms the S-shaped curve, and the oscillatory solutions are represented by the forked curve whose open end begins at $I_{Stim} = i_a$. The cell demonstrates rest behavior when provided with a sustained hyperpolarizing current or a depolarizing current below i_a. At this current level, corresponding to a saddle node bifurcation, we begin to see action potential generation. In this simple model, there is no adaptation in the repetitive firing regime; the frequency is constant for a given current injection. At $I_{Stim} = i_b$, a Hopf bifurcation occurs with an area of bistable behavior between the Hopf bifurcation point and the turning point where the oscillatory branch becomes unstable at $I_{Stim} = i_c$. These dynamics are identical to those of the Morris-LeCar model with a choice of parameters which provides an S-shaped steady state current-voltage relation for the model as described by Rinzel and Ermentrout (1992). This is an example of a Type I membrane which admits arbitrarily low frequency oscillations near the critical applied current due to the saddle node bifurcation.

ADAPTIVE BEHAVIOR

Now we supplement the spiking model with a high-threshold voltage activated calcium current and a calcium-dependent potassium current

$$I_{Ca}(V,s,r) = \bar{g}_{Ca}s^2r(V - V_{Ca}) \tag{5}$$

$$I_{K-AHP}(V,q) = \bar{g}_{K-AHP}q(V - V_K) \tag{6}$$

where the gating function q depends on a variable Ca representing the intracellular free calcium level. The calcium handling equation is

$$\frac{dCa}{dt} = -BI_{Ca} - Ca/\tau_{Ca} \tag{7}$$

where B determines the rate of increase of intracellular calcium due to the inward calcium channel current I_{Ca}. The second term represents the depletion of Ca as an exponential decay with time constant τ_{Ca}. As spiking begins, calcium enters the cell, activating the I_{K-AHP} current. When Ca is small, I_{K-AHP} has little effect, but as Ca increases, I_{K-AHP} also increases and counteracts the effect of the current injection. Since this process is slow, it is as if the influx of calcium causes the bifurcation diagram to slowly shift, resulting in a lower spike frequency or a steady-state. For a value of I_{Stim} which results in the generation of action potentials, the model demonstrates a progression from weak to strong adaptation as $\bar{g}_{K-AHP}$ is increased. The behavior is also dependent on the relationship between the rate constant of the I_{K-AHP} current and the rate of decay of intracellular calcium τ_{Ca}. If the rate constant for I_{K-AHP} is very small so that the gating variable changes quickly, then as intracellular calcium decays, the I_{K-AHP} current rapidly loses effect, allowing a return to spiking behavior. In this case, the model cannot achieve a steady state corresponding to strong adaptation when the $\bar{g}_{K-AHP}$ is increased; a slower rate of decay for intracellular calcium is also required. This is the case in the weak adaptation model developed by Vanier and Bower (1996).

Now that we understand the underlying dynamics, we supplement our model in order to better represent experimental data by adding a current

$$I_{K-M}(V_S,w) = \bar{g}_{K-M}w(V_S - V_K) \tag{8}$$

and creating an additional compartment representing the load provided by the dendrite. For our choice of cell dimensions and parameters, the coupling conductance between the two compartments is $g_c = 1.1\ mS/cm^2$. Figure 2 shows examples of the behavior of the full model for different values of $\bar{g}_{K-AHP}$, $\bar{g}_{K-M}$, and $\bar{g}_{S-L}$.

APPENDIX

The functions which determine the kinetic equations for the gating variables are listed below. We give most of the functions in the form $\alpha_y(u)$ and $\beta_y(u)$ where $y_\infty(u) = \alpha_y(u)/(\alpha_y(u) +$

$\beta_y(u))$ and $\tau_y(u) = 1/(\alpha_y(u) + \beta_y(u))$.

$$\alpha_m(V) = \frac{.32(30.1 - V)}{\exp(.25(30.1 - V)) - 1}$$

$$\beta_m(V) = \frac{.28(V - 57.1)}{\exp((V - 57.1)/5.0) - 1}$$

$$\alpha_h(V) = .128 \exp((34 - V)/18)$$

$$\beta_h(V) = \frac{4}{\exp((57 - V)/5) + 1}$$

$$\alpha_n(V) = \frac{.059(52.1 - V)}{\exp((52.1 - V)/5) - 1}$$

$$\beta_n(V) = .925 \exp(.925 - .025V)$$

$$\alpha_s(V) = \frac{.912}{\exp(-.072(V - 82)) + 1}$$

$$\beta_s(V) = \frac{.0114(V - 68.1)}{\exp((V - 68.1)/5) - 1}$$

$$\alpha_r(V) = 1.1 \min(.005, .005 \exp(-(V - 17)/20))$$

$$\beta_r(V) = 1.1(.005 - \alpha_r(V))$$

$$\alpha_q(Ca) = .000011Ca$$

$$\beta_q(Ca) = .0055$$

$$w_\infty(V) = \frac{1}{\exp(-(V - 42)/10) + 1}$$

$$\tau_w(V) = \frac{303.03}{3.3 \exp((V - 42)/20) + \exp(-(V - 42)/20)}$$

The maximal conductances in units of mS/cm^2 are $\bar{g}_{Na} = 220$, $\bar{g}_{K-DR} = 50$, $\bar{g}_{Ca} = 4$, $\bar{g}_{K-AHP} = 3$, $\bar{g}_{K-M} = 5$, $\bar{g}_{S-L} = 2$, and $\bar{g}_{D-L} = .05$. The reversal potentials in units of mV are $V_{Na} = 132$, $V_K = -13$, $V_L = 0$, and $V_{Ca} = 200$. The coupling parameter is $g_c = 1.1\ mS/cm^2$, and the current scaling parameter is $P = .05$. The capacitance is $C_M = .8\ \mu F/cm^2$. The calcium handling parameters are $B = 3.0$ and $\tau_{Ca} = .05$.

REFERENCES

[1] E. Barkai and M. E. Hasselmo. Modulation of the input/output function of rat piriform cortex pyramidal cells. *Journal of Neurophysiology*, 72:644–658, 1994.

[2] B. W. Connors, M. J. Gutnick, and D. A. Prince. Electrophysiological properties of neocortical neurons in vitro. *Journal of Neurophysiology*, 48:1302–1320, 1982.

[3] W. W. Lytton and T. J. Sejnowski. Simulations of cortical pyramidal neurons synchronized by inhibitory interneurons. *Journal of Neurophysiology*, 66:1059–1079, 1991.

[4] D. A. McCormick and D. A. Prince. Two types of muscarinic response to acetylcholine in mammalian cortical neurons. *Proceedings of the National Academy of Science*, 82:6344–6348, 1985.

[5] J. Rinzel and G. B. Ermentrout. Analysis of neural excitability. In C. Koch and I. Segev, editors, *Methods in Neuronal Modeling*, chapter 5, pages 135–169. The MIT Press, Cambridge, MA, 1992.

[6] S. M. Sherman and C. Koch. The control of retinogeniculate transmission in the mammalian lateral geniculate nucleus. *Experimental Brian Research*, 63:1–20, 1986.

[7] M. Steriade and R. R. Llinas. The functional states of the thalamus and the associated neuronal interplay. *Physiology Review*, 68:649–742, 1988.

[8] R. Traub, R. Wong, R. Miles, and H. Michelson. A model of a CA3 hippocampal pyramidal neuron incorporating voltage-clamp data on intrinsic conductances. *Journal of Neurophysiology*, 66:635–649, 1991.

[9] M. C. Vanier and J. M. Bower. A comparison of automated parameter-searching methods for neural models. In J. M. Bower, editor, *Computational Neuroscience: Trends in Research 1995*. Academic Press, 1996.

[10] M. Wilson and J. M. Bower. Cortical oscillations and temporal interactions in a computer simulation of piriform cortex. *Journal of Neurophysiology*, 67:981–995, 1992.

12

CHANGES IN RESPONSE PROFILES OF TASTE CELLS IN THE BRAIN STEM PARALLEL THOSE OBSERVED IN SOMATOSENSORY THALAMIC CELLS FOLLOWING ANESTHETIZATION OF PART OF THEIR RECEPTIVE FIELDS

Patricia M. Di Lorenzo, Christian H. Lemon, and Martin D. Kawamoto

Department of Psychology
Box 6000, SUNY at Binghamton
Binghamton, New York 13902-6000

Previous studies of electrophysiological responses to taste stimuli have revealed that taste cells at all levels of the nervous system are generally multisensitive, i.e. they respond to more than one of the four basic taste qualities (salty, sour, bitter and sweet). The study of the response profiles of taste cells, defined as their relative sensitivities across a variety of taste stimuli, and their organization has formed the basis for all major theories of taste coding. The careful examination of response profiles of taste cells, i.e. their extent and construction, is therefore of great import for our understanding of the operational principles of the gustatory system.

It can be argued that the response profile of a taste-responsive neuron is, in effect, it's receptive field. If one considers all the chemical stimuli that can be tasted as the stimulus domain, then the response profile is actually the receptive field of a taste cell, i.e. that portion of the stimulus domain to which the cell responds. If the response profiles are indeed analogous to receptive fields in other sensory systems, then they may show the same type of plasticity that has been reported in recent years under certain conditions in other sensory systems. To test this possibility, we modeled our experiment in the gustatory system after a study of somatosensory neurons recorded in the thalamus of the rat by Nicolelis et al.[10]. In that experiment, the application of a local anesthetic to the face induced an immediate and reversible reorganization of the receptive fields of thalamic neurons, including unmasking of responses not present before the anesthetization procedure and enhancement or suppression of responses to stimulation of regions surrounding the anesthetized zone. We reasoned that the analogous experiment in the taste system would entail an examination of response profiles before and after the elimination of sensitivity to one portion of the receptive field, i.e. to one taste quality. Accordingly, in the present experiment we use adaptation as a tool for examining plasticity of the receptive fields of taste responsive neurons in the nucleus of the solitary tract (NTS), the first synapse in the central gustatory pathway. (It is important to emphasize that adaptation per se is not the focus of our experiment, but is employed only for its effects on taste responses.)

Electrophysiological responses to representatives of the four basic taste qualities were recorded from NTS cells in deeply anesthetized rats (urethane, 1.5 gm/kg, i.p.). Small groups (n=2 to 3) of units were recorded simultaneously through an array of etched tungsten microelectrodes spaced about 115 μ apart. Taste stimuli consisted of aqueous solutions of NaCl (.1 M), HCl (.01 M), quinine HCl (.01 M) and sucrose (.5 M). Initially, all taste stimuli were presented in individual trials. Each trial consisted of a 10 sec baseline, 10 sec stimulus presentation, 10 sec wait and 20 sec distilled water rinse. Trials were spaced at least two min apart. A target stimulus was then chosen and applied to the tongue repeatedly without intervening water rinses until the NTS response was eliminated. A test stimulus selected from the remaining three taste stimuli was then applied to the tongue followed by a water rinse. This adaptation-test sequence was repeated until each of the tastants served as both target and test stimulus. At the end of the these tests, distilled water was bathed over the tongue for at least 3 min and the initial sequence of taste stimuli presentations (including rinse) was repeated.

Response profiles were recorded from 29 neurons in the NTS of 11 rats. In three animals there was one channel, in five animals there were two channels, in two animals there were three channels, and in one animal there were four channels that contained taste-responsive cells recorded simultaneously. In six animals, more than one taste-responsive cell could be isolated from the same channel. In all other cases, each cell was recorded from a separate channel.

Adaptation to all target stimuli affected at least half of the responses to all test stimuli, regardless of the response profile of the recorded cell. For all tastants, the most frequently observed effect of adaptation was attenuation of the responses to the non-adapting stimuli (cross-adaptation); however enhanced responses were sometimes seen, especially following adaptation to sucrose. Importantly, changes in taste responses following adaptation uncovered evidence of central inhibitory processes related to neural processing of tastants. These observations included a) adaptation to some stimuli occasionally produced inhibitory responses to a subset of the test stimuli, b) adaptation to some stimuli enhanced responses to a subset of the other tastants or produced responses where none existed previously, and c) adaptation to some stimuli which produced no response during baseline measures affected responses to other test stimuli. In each case where these effects were noted, cells that did not show these effects were recorded simultaneously from the same animal.

From these observations, it is possible to suggest that some taste-responsive cells in the NTS may receive information about (and potentially respond to) *every* taste stimulus, but that the intranuclear circuitry, perhaps especially the inhibitory interconnections, provides input that normally masks these responses to a greater or lesser extent. This arrangement would provide a mechanism with which the system might regulate its sensitivity to one or more tastants. Such changes have been reported following NaCl deprivation[1, 2], changes in corticofugal input[3, 5, 8] or with variations in the level of ovarian hormones[6,7].

Results of these experiments show that taste cells in the NTS exhibit the same type of evidence of reorganization of receptive fields following adaptation of taste stimuli that occurs in thalamic somatosensory neurons following anesthetization of part of their receptive field[10]. How these changes might affect the across neuron code for taste is being investigated with the aid of a mathematical model of the brainstem gustatory nuclei[4,9].

Supported by a grant from the Whitehall Foundation to PMD.

REFERENCES

1. Contreras, R.J. (1977) Changes in gustatory nerve discharges with sodium deficiency: A single unit analysis. *Brain. Res. 121*: 373–378.
2. Contreras, R.J. and Frank, M.E. (1979) Sodium deprivation alters neural responses to gustatory stimuli. *J. Gen. Physiol. 73*: 569–594.
3. Di Lorenzo, P.M. (1990) Corticofugal influence on taste responses in the parabrachial pons of the rat. *Brain Res. 530*:73–84.
4. Di Lorenzo, P.M. and Grasso, F.W. Representation of stimulus quality and intensity in a network model of brainstem gustatory neural coding. *Soc. Neurosci. Abstr. 20(2)*: 979, 1994.
5. Di Lorenzo, P.M. and Monroe, S. (1995) Corticofugal influence on taste responses in the nucleus of the solitary tract in the rat. *J. Neurophysiol.*, 74(1), 258–272.
6. Di Lorenzo, P.M. and Monroe, S. (1989) Taste responses in the parabrachial pons of male, female and pregnant rats. *Brain Res. Bull.*, *23*, 219–227.
7. Di Lorenzo, P.M. and Monroe, S. (1990) Taste responses in the parabrachial pons of ovariectomized rats. *Brain Res. Bull.*, *25*, 741–748.
8. Di Lorenzo, P.M. and Monroe, S. (1992) Corticofugal input to taste-responsive units in the parabrachial pons. *Brain Res. Bull.*, *29*, 925–930.
9. Grasso, F.W. and Di Lorenzo, P.M. Information processing in the gustatory nuclei of the brain stem: Gustatory Unit Stimulus Selective Topographical Organizer (GUSSTO), a network model. Paper presented at the 16th Annual Meeting of the Association for Chemoreception Sciences, Sarasota, Florida, 1994.
10. Nicolelis, M.A.L., Lin, R.C.S., Woodward, D.J. and Chapin, J. K. (1993) Induction of immediate spatiotemporal changes in thalamic networks by peripheral block of ascending cutaneous information. *Nature 361*: 533–536.

13

ROBUST PARAMETER SELECTION FOR COMPARTMENTAL MODELS OF NEURONS USING EVOLUTIONARY ALGORITHMS

Rogene M. Eichler West[1][*] and George L. Wilcox[2][†]

[1]Theoretical Neurobiology
Born Bunge Foundation
Universitaire Instelling Antwerpen–UIA
Universiteitsplein 1, B2610 Antwerp, Belgium
rogene@bbf.uia.ac.be
[2]3-249 Millard Hall
Department of Pharmacology
University of Minnesota
Minneapolis, Minnesota 55455
george@med.umn.edu

1. INTRODUCTION

Many modeling efforts have met skepticism among experimental neuroscientists largely because of a perception that system parameters are selected arbitrarily. In particular, a lack of high resolution data describing the spatial distribution of voltage-gated and/or calcium-dependent channels has rendered compartmental models suspect of representing singular solutions. On the other hand, a robust manifold of solutions may exist in nature and account for the behavioral repertoire of the biological system. We demonstrate the use of Evolutionary Algorithms (EAs) for the robust fitting of nonlinear channel distributions to experimental observations using the model and kinetics of Traub et al (1991). These results represent a successful method for optimizing parameters in high dimensional (> 100) nonlinear systems.

---

[*] Previous Affiliations: Minnesota Supercomputer Institute and the Graduate Program in Neuroscience, University of Minnesota. Current Affiliations: Universitaire Instelling Antwerpen and California Institute of Technology.

[†] Minnesota Supercomputer Institute, the Graduate Program in Neuroscience, and the Department of Pharmacology, University of Minnesota.

2. HOW DO EAS WORK?

Evolutionary algorithms are high dimensional parameter optimization strategies inspired by the principles of natural selection (Holland 1975, Forrest 1993). EAs define a parameter space as a string of values in an array, much like biological chromosomes encode proteins by lengths of nucleic acids. In the spirit of traditional Darwinian selection, the EA applies principles of natural selection (e.g., survival of the fittest, mutation, and recombination) to evolve the overall fitness of a *population* of parameter sets such that each individual achieves a desired set of behaviors. The values in the parameter strings are initially chosen at random. Each member of the population is evaluated for *fitness* and assigned a *fitness score* based on the degree of correspondence to an optimal behavior. Population members with high fitness have a higher probability of producing offspring. Generations are produced and evaluated until a fixed number of generations have passed or until a maximum fitness (if known) is achieved.

3. HOW DID WE DESIGN OUR EA?

3.1. Problem Representation

Each parameter string contained 114 real-valued parameters representing the 6 channel conductances in each of the 19 compartments of the Traub equivalent cylinder model. The parameters were ordered with like-channel types physically adjacent in the string. For example (using two channel conductance parameters):

$$Na_1 Na_2 ... Na_N ... KDR_1 KDR_2 ... KDR_N$$

where Na and KDR represent the sodium and potassium delayed rectifier conductance parameters, and the subscripts refer to compartment number (N total compartments).

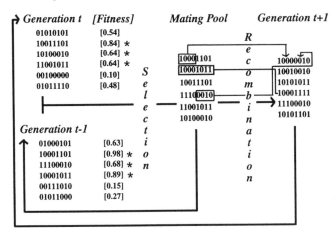

Figure 1. Generational flow diagram for one possible example of a population-based EA. Elite selection permits parameter strings from generations t and t-1 with sufficiently high fitness (designated by *) to join the mating pool. Recombination operators (described in text) are applied to members of the mating pool to produce generation t+1. The mating pool (which has already been evaluated for fitness) becomes the new generation t'-1. Generation t+1 becomes the new generation t' and is evaluated for fitness. The cycle continues until a goal fitness is achieved or until a predetermined number of generations has been evaluated.

3.2. Initialization

The initial population size was 1000 strings. The initial parameters were selected through random number generation and scaling. First, 114 real-valued random numbers are drawn between 0 and 1. Second, a random number is selected between 0.5 and 2.0 to scale the ratio between excitatory (inward) and inhibitory (outward) conductances. Third, a random number between 0 and 2,500 is selected to scale the total conductance between 0 and 2,500 mS/cm^2. We included recombination operators to perform scaling mutations such that the parameter sets of subsequent generations were not restricted to these initial bounds.

3.3. Simulation

A custom-written simulator using the Crank-Nicolson method was used for numerical integration using a fixed time step of 50 microseconds. State variables were saved every 250 microseconds both for graphical display and for postprocessing by the fitness filter. The accuracy of the simulator was verified using the Rallpack set of standards (Bhalla et al 1992).

3.4. Fitness

Simulated time series were assigned a score from 0.0 to 1.0. Time series with a high correspondence to experimentally observed behaviors were rewarded with higher scores. The criteria were an extension of Traub's with additional characteristics identified from the experiments of Wong and Prince (1978, 1981), Wong et al (1979), Regehr and Tank (1992), and Hablitz and Johnston (1981). The test stimulus was a 0.3 nA depolarizing current injection into the soma compartment. The criteria are described in Table 1.

Table 1.

Elimination of steady-state and inappropriate periodic solutions:
(All conditions must be satisfied before points may be accumulated under "Somaticburst behavior" or "Spatial criteria".)

1. Nonspiking solutions are rewarded for proximity to spike threshold;
2. Spikes or bursts are produced;
3. Spikes or bursts are produced beyond the initial transient;
4. Soma compartment initiates spikes or bursts by generating most of the driving current.

Select for somatic burst behavior:
(Partial credit assigned for proximity to "correct" solution)

5. Burst width between 50 and 75 milliseconds;
6. Individual spikes of appropriate width and height;
7. Peaks per burst between 3 and 8;
8. Post-burst afterhyperpotential 5 to 10 mV below resting;
9. Burst frequency between 0.8 and 1.0 Hz;
10. Least-squares fit to experimental CA3 burst waveform (criterion r > 0.8).

Select for spatial criteria:
(Partial credit assigned for proximity to "correct" solution)

11. Somatic peak height is greater than dendritic peak heights;
12. Calcium current is greatest in mid-apical compartments.

3.5. Selection

Population members from both the current and previous generation were admitted to the mating pool if their individual fitness value exceeded an "extinction threshold". The number of population members remained bounded between 500 and 1000 through adaptation of the extinction threshold.

3.6. Recombination

Two parent parameter strings were randomly selected from the mating pool for crossover operations. One parameter string was randomly selected from the mating pool for mutation operations. a, *Crossover 1 (2-pt, compartment ordering)*. Two points along the string are randomly selected. The parent strings contribute alternate regions to the offspring string. b, *Crossover 2 (2-pt, parameter ordering)*. The procedure is the same as in (a), except the parameters were first reordered such the parameters from the same compartment were adjacent. c, *Single Parameter Mutate.* Two random numbers were selected. The first number determined which of the parameters would be operated on. The second number was scaled to the range appropriate to that parameter and replaced the original value of the selected parameter. d, *Parameter-type Scale Mutation.* Two random numbers were selected. The first number determined which of the parameter types would be operated on. The second number was a scaling factor that ranged from 0.0 to 2.0. All conductances of the selected channel type were multiplied by the scaling factor. e, *Scale All Mutation.* Random number selection (a coin flip with probability 0.5) determined whether all values in a string were scaled by constant multiplicative factor 0.9 or 1.1.

All operators had the same initial probability of being chosen. Subsequent probabilities were based on the relative success of that operator. A success was indicated when a chromosome produced by a given operator scored a fitness value above the extinction threshold. The minimum probability of selection was fixed at 1%.

3.7. Termination

Termination criteria in the EA literature are somewhat arbitrary because the statistical nature of the search prohibits knowing *when* the population will evolve to a given fitness value. Our arbitrary termination criterion was satisfied after 200 CPU hours had elapsed on each of eight dedicated Silicon Graphics Power Challenge R8000 90 MHz processors.

4. RESULTS

The EA evolved a population of high fitness parameter sets from random initial conditions. Even after many generations, the EA produced a broad distribution of fitness scores suggesting that the algorithm had not overconverged. The best individual parameter string yielded a fitness of 0.92. Traub's parameter set yielded a fitness of 0.85. In contrast, random parameter solutions achieved a fitness greater than 0.75 in only 3 of 11,500 random number simulations. The most fit solution yielded an average burst width of 43 ms, an average postburst afterhyperpotential (AHP) of 1.8 mV below the resting potential, least-squares fit to the experimental burst waveform of 0.92, and bursts occurring at 1.0 Hz. Neither the EA-produced parameter sets nor Traub's parameter set produced a sufficiently large AHP, suggesting that additional channel kinetic equations may be necessary to achieve a higher correspondence to experimentally observed behavior. Figure 2 demonstrates the somatic time series for the highest fit parameter set.

The range of parameter distributions from high fitness individuals is qualitatively, and in many cases quantitatively, similar to those published by Traub. The optimal parameter values are small for some channels types (e.g., potassium A, < 3 mS/cm^2; potassium AHP, generally < 5 mS/cm^2) compared to the approximate range of all values explored (0 < x < 65). The large range of calcium-dependent potassium conductance parameters across all compartments has important functional implications for plasticity and signal integration: activity-dependent modulations of the these channel conductances may make little difference in overall fitness, but could significantly alter local electrical properties. The potassium delayed rectifier parameters have a distribution similar to the spatial expression of a delayed rectifier channel polypeptide, Kv2.1 (Maletic-Savatic et al 1995). In agreement with experimental estimates, we found spatial distributions for the sodium conductance parameters that appear to decrease exponentially along the length of the dendrites (Spruston et al 1995). The calcium channel distribution is difficult to interpret because Traub's kinetics describe a single high threshold channel that has subsequently been determined to consist of multiple distinct channel types. However, the distribution is consistent with somatodendritic staining patterns of N-type calcium channels (Mills et al 1994, Elliott et al 1995). We obtained similar results in three EA experiments starting from different initial random populations.

5. CONCLUSIONS

This study has applied a new method for parameter optimization that promises to dramatically improve the robustness of neuronal simulations. The results of our proof-of-concept experiment yield improved correspondence between simulated and experimental behaviors relative to that published previously for this model. Furthermore, the method should find general applicability in a broad range of simulation endeavors.

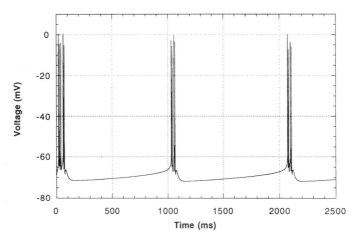

Figure 2. Highest fitness time series produced by EA parameter optimization. The voltage time series from the soma compartment displays 2500 milliseconds of simulated neuron behavior.

6. ACKNOWLEDGMENTS

This work was supported in part by a grant from the Minnesota Supercomputer Institute and Cray Research. The authors gratefully acknowledge generous contributions of supercomputing resources provided by the Laboratory for Computational Sciences & Engineering and the Minnesota Supercomputer Institute at the University of Minnesota.

7. REFERENCES

Bhalla, U. S., Bilitch, D. H., & Bower, J. M. TINS 15(11), 453–458 (1992).

Elliott, E.M., Malouf, A.T., & Catterall, W.A. J. Neurosci. 15(10), 6433–6444 (1995).

Forrest, S. Science 261, 872–878 (1993).

Hablitz, J. J. & Johnston, D. Cell. and Molec. Neurobio. 1(4), 325–334 (1981).

Holland, J. H. Adaptation in Natural and Artificial Systems. (Ann Arbor. The University of Michigan. 1975).

Maletic-Savatic, M., Lenn, N.J., Trimmer, J.S. J. Neurosci. 15(5), 3840–3851 (1995).

Mills, L.R., Niesen, C.E., So, A.P., Carlen, P.L., Spigelman, I.., & Jones, O.T. J. Neurosci. 14(11), 6815–6824 (1994).

Regehr, W. G. & Tank, D. W. J. Neurosci. 12(11), 4202–4223 (1992).

Spruston, N., Schiller, Y., Stuart, G., Sakmann, B. Science 268,297–300 (1995).

Traub, R. D., Wong, R. K. S., Miles, R., & Michelson, H. J. Neurophysiol. 66, 635–650 (1991).

Wong, R. K. S. & Prince, D. A. Brain Res. 159, 385–390 (1978).

Wong RKS, Prince DA (1981) J. Neurophysiol. 45:86–97.

Wong, R. K. S., Prince, D. A. & Basbaum, A. I. Proc. Natl. Acad. Sci. 76(2), 986–990 (1979).

14

HYBRID ANALYSES OF NEURONAL SPIKE TRAIN DATA FOR PRE- AND POST-CROSS INTERVALS IN RELATION TO INTERSPIKE INTERVAL DIFFERENCES

Michelle A. Fitzurka[1] and David C. Tam[2]

[1]Department of Physics
Catholic University of America
Washington DC 20064
fitzurka@stars.gsfc.nasa.gov
[2]Department of Biological Sciences and Center for Network Neuroscience
University of North Texas, Denton, Texas 76203
dtam@unt.edu

ABSTRACT

Two new hybrid spike train analysis methods called (1) pre-Cross-Interval/ Interspike-Interval-Difference (pre-CI/ISID) and (2) Interspike-Interval-Difference/ post-Cross-Interval (ISID/post-CI) phase plane analyses are introduced. They examine the dependency relationship between (1) the pre-cross-interval (pre-CI) and the interspike interval difference (ISID), and (2) the ISID and the post-cross-interval (post-CI) defined at a given reference spike. This allows for inferences to be made about how ISIDs in a spike train are related to the last cross interval and the next cross interval with respect to a compared spike train. Both methods were applied to simulated spike trains to display the capabilities of this new technique. The co-varying relationship between the pre-CI and post-CI with respect to the ISID can be revealed as clusters of bands and points in these phase plots.

1. INTRODUCTION

Single-unit analysis (which explores the temporal relationships within single spike trains)[4,7,8] and multi-unit analysis (which examines the spatial and temporal relationships across multiple spike trains)[5,6,9,10] have been studied extensively. Recently, several new single-unit interval statistics have been introduced. *Joint Interspike Interval Difference (JISID)* scatter plots were designed to characterize the relationship between adjacent *interspike interval differences (ISIDs)* [3] while *First and Second Order* phase plots were used to compare the *interspike intervals (ISIs)* and their first order *(ISI/ISID)* and second order *(ISI/ISI2D)* differences[1,2].

In this paper, we introduce two new hybrid spike train analysis methods which combine single unit interval measures (in this case, the *ISID*) with multi-unit interval measures (in this case, the *pre-CI* and *post-CI*). These are called (1) the *pre-Cross-Interval/Interspike-Interval-Difference (pre-CI/ISID)* and (2) the *Interspike-Interval-Difference/post-Cross-Interval (ISID/post-CI)* phase plane analyses. In this manner, the effects of the preceding cross interval on an interspike interval trend and the effects of that trend on the subsequent cross interval can be explored in the *pre-CI/ISID* and *ISID/post-CI* plots, respectively. The interspike interval trend, defined by the *ISID*, can be either increasing, decreasing or constant.

2. METHODS

If spike trains $r(t)$ and $c(')$ are taken to be the reference and compared spike trains respectively (see Fig. 1), they can be defined as:

$$r(t) = \sum_{n=1}^{n=N} \delta(t - t_n) ,$$

(1)

with N indicating the total number of spikes in the reference spike train, t_n denoting the time of occurrence of the n-th spike in the reference spike train, and $\delta(t)$ denoting the delta function, and as:

$$c(t') = \sum_{m=1}^{m=M} \delta(t' - t'_m) ,$$

(2)

with M indicating the total number of spikes in the compared spike train, and t'_m denoting the time of occurrence of the m-th spike in the reference spike train.

Then, relative to the n-th reference spike in the reference spike train, the n-th *interspike interval (ISI)*, τ_n and the $(n-1)$-th *ISI*, τ_{n-1} are defined as follows:

$$\tau_n = t_n - t_{n-1}$$

(3)

and

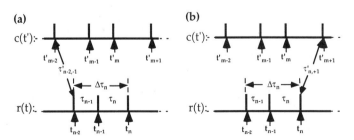

Figure 1. Segments of the reference, $r(t)$, and compared, $c(t')$, spike trains showing **(a)** the n-th *ISID* ($\Delta\tau_n$) in relation to the *pre-CI* ($\tau'_{n-2,-1}$) with respect to the $(n-2)$-th reference spike and **(b)** the n-th *ISID* ($\Delta\tau_n$) in relation to the *post-CI* ($\tau'_{n,+1}$) with respect to the n-th reference spike.

$$\tau_{n-1} = t_{n-1} - t_{n-2}$$ (4)

which satisfies the condition:

$$t'_m < t_n < t'_{m+1}$$

while the n-th *ISI difference (ISID)*, $\Delta\tau_n$ is:

$$\Delta\tau_n = (\tau_n - \tau_{n-1}) = (t_n - t_{n-1}) - (t_{n-1} - t_{n-2}) = t_n - 2t_{n-1} + t_{n-2} \ .$$ (5)

A *cross interval (CI)* is defined as the time between the occurrence of a spike in the reference spike train and the occurrence of the last (*pre-CI*) or next (*post-CI*) spike in the compared spike train. The *post-CI* ($\tau'_{n,+1}$) relative to the n-th reference spike is (see Fig. 1b):

$$\tau'_{n,+1} = t'_{m+1} - t_n$$ (6)

while the *pre-CI* ($\tau'_{n-2,-1}$) relative to the $(n-2)$-th reference spike is (Fig. 1a):

$$\tau'_{n-k,-1} = t'_{m-k} - t_{n-2} \ .$$ (7)

which satisfies the condition:

$$t'_{m-k} < t_n < t'_{m-k+1} \ .$$

From these defined variables then, a *pre-CI/ISID* phase plot can be constructed to compare the *pre-CI* with respect to the *ISID* ($\tau'_{n-2,-1}$ vs. $\Delta\tau_n$), while an *ISID/post-CI* phase plot would show the *ISID* with respect to the *post-CI* ($\tau'_{n,+1}$ vs. $\Delta\tau_n$). In this way, the relationship between the local trend in the reference train, the *interspike interval difference*, can be investigated with respect to the *cross interval* that preceded it as well as the *cross interval* that succeeded it to see if any dependency relationships exist.

Local trends, *ISIDs*, may be *positive* (i.e., $\Delta\tau_n > 0$ implying increasing *ISIs*, $\tau_n > \tau_{n-1}$), *negative* (i.e., $\Delta\tau_n < 0$ implying decreasing *ISIs*, $\tau_n < \tau_{n-1}$), or *constant* (i.e., $\Delta\tau_n = 0$ implying equal adjacent *ISIs*, $\tau_n = \tau_{n-1}$). These two constructions can be combined into one composite plot by plotting the *ISID* on the y-axis, the *pre-CI* on the $-x$-axis (left half of the plot), and the *post-CI* on the $+x$-axis (right half of the plot). This construction would be called the composite *pre-CI/ISID/post-CI* phase plot and would allow for comparison between the left and right halves of the graph to reveal the relationship of the *pre-CI* and the *post-CI* with respect to the *ISID* respectively. In other words, whether a given *CI* leads to a specific local trend (*ISID*) and whether that trend leads to a specific *CI* will be exposed as clusters or patterns on the left and right halves respectively.

3. RESULTS

3.1. Gaussian Example

We simulated a periodic Gaussian spike train (spike train A) with a 250 msec mean *interspike interval* and a 25 msec Gaussian variance. Then, a dependent spike train (spike train B) was generated relative to this Gaussian spike train with each spike in the original spike train inducing a spike in the dependent spike train with a probability of 0.75 at a 3 msec latency with 0.3 msec variance.

Figure 2 shows the composite *pre-CI/ISID/post-CI* phase plot for the coupling relationship between spike trains A and B, using spike train A and spike train B as the reference spike train in Figs. 2a and 2b, respectively. Distinct clusters can be seen in each of the plots indicating the fundamental periodic nature of the spike trains. Also, in Fig. 2a, the clusters lie in a horizontal band about the *x*-axis indicating that the *ISIDs* are fairly small independent of the *CIs*, while in Fig. 2b the clusters lie in vertical bands close to the *y*-axis indicating that the *CIs* are fairly constant independent of the *ISIDs*. Note that the similarity between the points in the left and right halves of the plot suggests a similar coupling relationship between *pre-CI* and *post-CI* with respect to the *ISID*.

3.2. Burst Firing Example

To simulate burst firing, we create a synthetic spike train (spike train A) in which the *ISIs* mimic a burst of approximately equally spaced action potentials (25 msec, with a 2.5 msec Gaussian variance). This is defined as the *intraburst period* and is followed by a pause of no activity referred to as the *interburst period* (125 msec, with a 12.5 msec Gaussian variance). The *interburst* and *intraburst periods* are repeatedly alternated to mimic burst firing. A dependent spike train (spike train B) is generated based on this

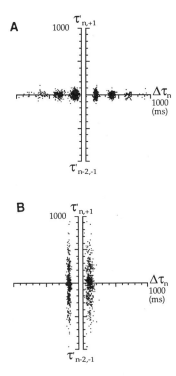

Figure 2. (a) The composite *pre-CI/ISID/post-CI* phase plot for this Gaussian example with spike train A taken as the reference spike train. (b) The composite *pre-CI/ISID/post-CI* phase plot for this Gaussian example with spike train B taken as the reference spike train.

bursting spike train by each spike in spike train A inducing a spike in spike train B with a probability of 0.75 at a 3 msec latency with 0.3 msec variance.

Figure 3 shows the composite *pre-CI/ISID/post-CI* phase plot for the coupling relationship between spike trains A and B, using spike train A and spike train B as the reference spike train in Figs. 3a and 3b, respectively. As in the Gaussian example case, these coupling relationships are not reciprocal with respect to the two neurons in that Fig. 3a and 3b show significantly different features.

In Fig. 3a, the *pre-CI/ISID* (left half) and *ISID/post-CI* (right half) plots are fairly symmetric whereas in Fig. 3b such symmetry is not seen. Symmetry between the left and right halves of the plots implies similar firing before and after a given *ISID* trend.

4. DISCUSSION

With this new analysis, the relationship between a local trend in a given spike train (as reflected in the *ISID*) and the cross intervals preceding and succeeding it (the *pre-CI* and *post-CI*, respectively) are revealed. These relationships are captured by the distribution of points in the left and right halves of the composite plot (i.e., *pre-CI* vs. *ISID* and *post-CI* vs. *ISID*). In this way, the similarity in the conditions leading to and from the local trend can be represented graphically.

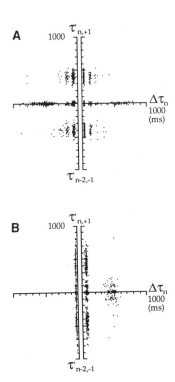

Figure 3. (a) The composite *pre-CI/ISID/post-CI* phase plot for this bursting example with spike train A taken as the reference spike train. (b) The composite *pre-CI/ISID/post-CI* phase plot for this bursting example with spike train B taken as the reference spike train.

Both the dependent and independent spike trains can be taken as the reference spike train with respect to the other. This then allows for the identification of reciprocity in the coupling relationships between the *cross intervals* and the associated firing trend (*ISID*).

ACKNOWLEDGMENTS

This research was supported in part by ONR grant number N00014–94–1–0686 to DCT.

REFERENCES

1. Fitzurka, M. A. and Tam, D. C. (1996a) First order interspike interval difference phase plane analysis of neuronal spike train data. In: *Computational Neuroscience.* (J. M. Bower, ed.) Academic Press Inc.: San Diego, CA, pp. 429–434.
2. Fitzurka, M. A. and Tam, D. C. (1996b) Second order interspike interval difference phase plane analysis of neuronal spike train data. In: *Computational Neuroscience.* (J. M. Bower, ed.) Academic Press Inc.: San Diego, CA, pp. 435–440.
3. Fitzurka, M. A. and Tam, D. C. (1995) A New Spike Train Analysis Technique for Detecting Trends in the Firing Patterns of Neurons. In: *The Neurobiology of Computation* (J. M. Bower, ed.) Kluwer Academic Publishers, Boston, MA, pp. 73–78.
4. Perkel, D. H., Gerstein, G. L. and Moore, G. P. (1967a) Neuronal spike trains and stochastic point process. I. The single spike train. *Biophysical Journal.* 7: 391–418.
5. Perkel, D. H., Gerstein, G. L. and Moore, G. P. (1967b) Neuronal spike trains and stochastic point process. II. Simultaneous spike trains. *Biophysical Journal,* 7: 419–440.
6. Perkel, D. H., Gerstein, G. L., Smith, M. S. and Tatton, W. G. (1975) Nerve-impulse patterns: A quantitative display technique for three neurons. *Brain Research,* 100: 271–296.
7. Rodieck, R. W., Kiang, N. Y.-S. and Gerstein, G. L. (1962) Some quantitative methods for the study of spontaneous activity of single neurons. *Biophysical Journal.,* 2: 351–368.
8. Smith, C. E. (1992) A heuristic approach to stochastic models of single neurons. In: *Single Neuron Computation* (T. McKenna, J. Davis and S. Zornetzer, eds.) Academic Press, San Diego, CA. pp. 561–588.
9. Tam, D. C. (1993) A multi-neuronal vectorial phase-space analysis for detecting dynamical interactions in firing patterns of biological neural networks. In: *Computational Neural Systems.* (F. H. Eeckman and J. M. Bower, eds.) Kluwer Academic Publishers, Norwell, MA. pp. 49–53.
10. Tam, D. C., Ebner, T. J., and Knox, C. K. (1988) Cross-interval histogram and cross-interspike interval histogram correlation analysis of simultaneously recorded multiple spike train data. *Journal of Neuroscience Methods,* 23: 23–33.

15

LONG-TERM POTENTIATION: EFFECTS ON SYNAPTIC CODING*

Ricci Ieong[†] and Michael Stiber[‡]

Department of Computer Science
The Hong Kong University of Science and Technology
Clear Water Bay, Kowloon
Hong Kong
E-mail: ricci@cs.ust.hk
E-mail: stiber@cs.ust.hk

1. INTRODUCTION

Learning, in the context of individual neurons, is represented in both physiological and Artificial Neural Network (ANN) models by change of synaptic strength. Many of the changes in neural responses as a result of the learning process may be explainable by simple consideration of synaptic strength alteration. For instance, sensitization and habituation of vertebrate and invertebrate neurons are usually interpreted in terms of incrementation and decrementation of synaptic strength [1]. However, there are still many unexplained cases [2]. In the mammalian hippocampus, neurons do not receive one stimulus, but a number of different stimuli from a variety of different locations. So simply looking at the increment and decrement of synaptic strength is insufficient to explain what has been modified by learning.

Neural responses include input and output rate and their *pattern*. By "pattern", we mean the ordered sequence of interspike intervals and cross-intervals.

To study the effect of learning, we use an accepted physiological model: Long-Term Potentiation (LTP), first observed in excitatory synapses of rabbit hippocampal neurons [3]. When a high frequency afferent tetanus is applied to these neurons, postsynaptic potentials are enhanced. Other than the magnitude of postsynaptic potential enhancement, LTP also involves persistent maintenance of this enhancement for minutes or even hours [3]. LTP has also been

*This work was supported by the Hong Kong Research Grants Council (HKUST187/93E, HKUST527/94M, HKUST668/95E). ODEPACK software, developed at Lawrence Livermore Laboratories, and libf77 routines, copyright 1990 & 1991 by AT&T Bell Laboratories, were used. Poster presentation preferred. Categories: Modeling and simulation; theory and analysis. Themes: Learning and memory; excitable membranes and synaptic mechanisms.
†To whom correspondence should be addressed.
‡Current Address: Department of Molecular and Cell Biology, 315 Life Sciences Addition, University of California, Berkeley, CA

recorded in invertebrate neurons, for instance in the crayfish opener-excitor neuromuscular synapse.

Similar effects have also been observed in the *inhibitory* synapse of the goldfish, *Carasius auratus* Mauther cell [4]. Thus, though originally investigated in excitatory preparations, there is no *a priori* reason to expect LTP to be an exclusively excitatory phenomenon. One might also consider that cerebellar Purkinje cells, involved in motor learning, are also connected through inhibitory synapses.

LTP aftereffects have been primarily studied in terms of induction as a function of stimulus rate. When a high rate stimulus is applied, LTP duration and PSP strength increase are recorded as the modification of the neural response as a result of learning. However, we can also consider LTP in the context of presynaptic spike temporal pattern, rather than just their rate. There is evidence that this is a better description of LTP dependence on the presynaptic discharge than is presynaptic rate [5]. Tsukada and collaborators showed that the increase in PSP size recorded strongly depends on the correlations among presynaptic interspike intervals [5].

In this report, we consider not only presynaptic pattern, but postsynaptic pattern, too, to describe LTP induction and its effects on neural behavior — the patterns of spikes it produces in response to identical presynaptic spike trains before and after "learning". >From our simulation results, we observe that the proportion of synchronization between the neural responses increases after learning, with shorter presynaptic interspike intervals producing a greater proportion of synchronization.

2. METHODS

The model used is of the crayfish slowly adapting stretch receptor organ (SAO), the recognized prototype of an inhibitory synapse [6]. This model was chosen because of the level of familiarity with its (and the living preparation's) dynamics in response to stationary [7] and nonstationary [8] inputs, and its convenient representation of synaptic membrane variables. It consists of a single presynaptic inhibitory fiber (IF) and a postsynaptic neuron [9]. Trains of input spikes are presented to the SAO through the IF via their common inhibitory synapse. When the SAO is unperturbed, it fires like a pacemaker with all Action Potentials (APs) separated by a nearly invariant natural interval, N. When inputs are applied, a more complex timing pattern may be produced, either *aperiodic* or *periodic*.

In our model, synaptic strength corresponds to the *maximum permeability* $\bar{P}_{syn}$, of the cell membrane to a charge carrier ion, which is chloride (Cl^-):

$$I_{syn} = A\bar{P}_{syn} \frac{V_m F}{RT} \frac{[Cl^-]_0 - [Cl^-]_i \exp\left(-FV_m/RT\right)}{1 - \exp\left(-FV_m/RT\right)} \times \sum_{k=1}^{n} \left(e^{(s_k-t)/\tau_+} - e^{(s_k-t)/\tau_-}\right) \quad (1)$$

Though the synapse in our model is an inhibitory synapse, we assume the potentiation is independent of the carrier ion, so modification of $\bar{P}_{syn}$ depends solely on LTP controlling parameters and PSP interspike intervals.

As shown in Fig. 1, we record the time of occurrence of each pre- or post- synaptic spike (s_k or t_i, respectively, $i, k = 0, 1, 2 \ldots$) thus assimilating both trains to point processes. Based on these times, certain interspike intervals are calculated, including postsynaptic, $T_i = t_i - t_{i-1}$, and presynaptic, $I_k = s_k - s_{k-1}$. Cross intervals or *phases*, $\phi_i = t_i - s_*$, were also computed from an AP back to the most recent preceding PSP.

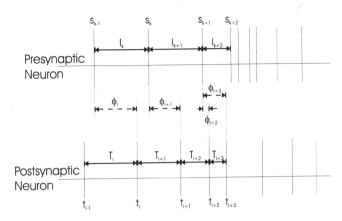

Figure 1. The presynaptic stimulus, s_k, the postsynaptic response, t_i, the phase, ϕ_i, the presynaptic interval, I_k and the postsynaptic interval, T_i.

Pacemaker inputs, with $I_k = I$ (typically normalized as N/I), were applied to the model SAO as the simplest stationary presynaptic pattern, and to duplicate the most common LTP experiments.

We incorporated a general mathematical model of LTP into the SAO model, rather than one specific to pyramidal neurons with uniquely mammalian messengers [5]. The modifiable component of permeability, $\bar{P}_{syn,m}$ in the model after the kth PSP is:

$$\bar{P}_{syn,m}(I_k) = \left(1 - \frac{1}{\gamma}\right)\bar{P}_{syn,m}(I_{k-1}) + \frac{A\bar{P}_{syn,f}}{\gamma}$$
$$\times \left(e^{-\alpha(I_k + I_{k-1} + \ldots + I_1)} + e^{-\alpha(I_k + I_{k-1} + \ldots + I_2)} + \cdots + e^{-\alpha I_k}\right) \qquad (2)$$

where γ is the LTP growth rate, A is the LTP glutamate transmitter amount, α is the LTP decay time constant and $\bar{P}_{syn,f}$ is the fixed component of the maximum permeability $\bar{P}_{syn} = \bar{P}_{syn,f} + \bar{P}_{syn,m}$. It causes the maximum synaptic permeability of our simulation to vary along time. These changes are compared with the unpotentiated value to compute the magnification ratio,

$$M = \frac{\bar{P}_{syn,m}}{\bar{P}_{syn,f}} \qquad (3)$$

Magnification ratio plots were produced by recording the magnification ratio at presynaptic spike arrival times for multiple simulations with different N/I. Additionally, *bifurcation diagrams* are used to show the change in neural response due to LTP. These plot a measure of neural *behavior*, here ϕ_i, verses normalized presynaptic rate N/I. These behaviors include *locked* and *non-locked* behaviors. $p{:}q$ locked behaviors are stationary, periodic behaviors where q postsynaptic spikes appear for every p presynaptic spikes in a repeating sequence of T_i and ϕ_i. Non-locked behaviors include a variety of aperiodic behaviors. In bifurcation diagrams, locked behaviors produce ranges of N/I with few, discrete categories of ϕ_i, while non-locked ones produce a vertically dense distribution of points within a range of ϕ_i.

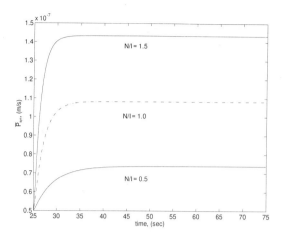

Figure 2. Magnification Ratio versus simulation time for $N/I = 0.5, 1.0, 1.5$.

3. RESULTS

Magnification ratios obtained from the LTP simulation, shown in Fig. 2, mimic those found in living preparations [3, 10, 5]. They increase continuously from their unpotentiated values to their asymptotic values which depend directly on presynaptic rate for pacemaker input. This has also been reported in other experiments [10].

This increase in synaptic permeability modifies the global behavior of the neuron. From the bifurcation diagrams in Fig. 3, one can see the increase in locked responses as a result of LTP. Without LTP enhancement, 62% of the neural responses were locked. However, with LTP, 84% of input rates in the identical range produced locking.

Additionally, the widths of the input rate domains within which certain locking ratios occurred increased with LTP. Although not all the locking domains increased in width, the major ones, such as 1:1, 2:3 and 3:2, widened dramatically. Responses were still ordered along N/I in the same sequence with and without LTP.

Locked responses in the high rate region ($N/I > 1.0$, $p{:}q > 1{:}1$) increased more rapidly than the low rate region ($N/I < 1.0$, $p{:}q < 1{:}1$). For low rates, locking domains increased by 6.1%; for high rates, 81.3% of the responses were locked with LTP — a 34.7% increases.

4. CONCLUSIONS

Application of a general model of LTP causes a time-varying increase in maximum synaptic permeability in response to pacemaker input, with a concomitant increase in the dynamical postsynaptic aftereffects of IPSPs. Because the potentiating effect of PSPs decay over time, shorter interspike intervals result in greater accumulated potentiation [5]. Therefore, maximum permeability increases as interspike interval decreases.

LTP enhancement clearly changes the neuron's behavior. The abundance of locked responses increases. The lower rate boundary of the 1:1 locked response has been shifted towards the low rate region after "learning". However, the changes were not uniformly distributed along the inhibitor rate scale, being concentrated in the high rate region. Higher input rates produce greater potentiation, hence higher maximum permeability. So more higher

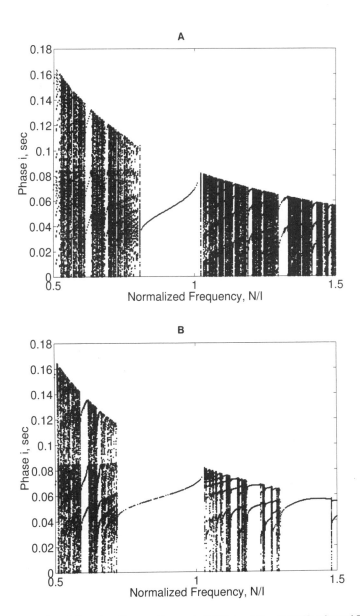

Figure 3. Comparison of two bifurcation diagrams for constant PSP simulation and LTP enhanced PSP simulation. (A) Constant PSP simulation with $\bar{P}_{syn} = 5 \times 10^{-6}$ cm/s. (B) LTP enhanced PSP simulation with $\bar{P}_{syn,f} = 5 \times 10^{-6}$ cm/s, LTP glutamate transmitter amount, $A = 0.533$, LTP decay time constant, $\alpha = 1/223$ msec^{-1} and LTP growth rate, $\gamma = 25$.

locked responses ($p{:}q > 1{:}1$) reside in the identical simulation range ($N/I > 1.0$) after LTP enhancement.

We can summarize that the neuron synchronizes more readily with the input stimulus with LTP. Shorter input intervals produce more periodic neural responses. Therefore, the responses and the dynamics of the neuron depend on the effects of LTP. How each individual LTP parameter affects this synchronization is still an issue to be determined.

REFERENCES

[1] J. Martinez, Jr. and R. Kesner, *Learning and Memory: A Biological View*. Florida: Academic Press, 1986.

[2] T. Berger, G. Chauvet, and R. Sclabassi, "A biologically based model of functional properties of the hippocampus," *Neural Networks*, vol. 7, no. 6/7, pp. 1031–1064, 1994.

[3] T. Bliss and T. Lomo, "Long-lasting potentiation of synaptic transmission in the dentate area of the anaesthetized rabbit following stimulation of the perforant path," *The Journal of Physiology*, vol. 232, pp. 331–356, 1973.

[4] S. Charpier, Y. Oda, and H. Korn, "Long-term enhancement of inhibitory synaptic transmission in the control nervous system," in *Long Term Potentiation 2* (M. Baudry and J. Davis, eds.), pp. 151–168, MIT PRESS, 1995.

[5] S. Shinomoto, M. Crair, M. Tsukada, and T. Aihara, "The stimulus dependent induction of long-term potentiation in ca1 area of the hippocampus ii: Mathematical model," tech. rep., Department of Information-Communication Engineering, Tamagawa University, Machida, 194 Japan, 1992.

[6] J. Segundo, E. Altshuler, M. Stiber, and A. Garfinkel, "Periodic inhibition of living pacemaker neurons: I. locked, intermittent, messy, and hopping behaviors," *Int. J. Bifurcation and Chaos*, vol. 1, pp. 549–81, Sept. 1991.

[7] M. Stiber and J. P. Segundo, "Dynamics of synaptic transfer in living and simulatied neurons," in *Proc. ICNN-93*, (San Francisco), pp. 75–80, Mar. 1993.

[8] M. Stiber and R. Ieong, "Hysteresis and asymmetric sensitivity to change in pacemaker responses to inhibitory input transients," in *Brain Processes, Theories and Models. W.S. McCulloch: 25 Years in Memoriam* (R. Moreno-Díaz and J. Mira-Mira, eds.), (Grand Canary, Spain), pp. 513–22, Nov. 1995.

[9] M. Stiber and J. Segundo, "Learning in neural models with complex dynamics," in *Proc. IJCNN-93*, (Nagoya, Japan), pp. 405–8, 1993.

[10] T. Bliss and G. Collingridge, "A synaptic model of memory: Long-term potentiation in the hippocampus," *Nature*, vol. 361, pp. 31–39, Jan. 1993.

16

MEASURING THE INFORMATION EXPRESSED BY NEURAL DISCHARGE PATTERNS

Don H. Johnson*

Computer and Information Technology Institute
Department of Electrical and Computer Engineering
Rice University, MS 366
Houston, TX
E-mail: dhj@rice.edu

ABSTRACT

Various measures have been used to assess how well single neurons represent information. Modeling discharge patterns as stochastic point processes, we determine how well certain measure accomplish this task. We show that the information theoretic measure—capacity—can do a poor job. The mean-squared error measure more accurately describes the fidelity to which sensory signals can be extracted. Calculation of fundamental bounds on mean-squared error show that time-varying signals must have bandwidths orders of magnitude less than the average discharge rate (under a Poisson model) if accurate signal representations are to result. This result indicates that neural ensembles must be considered to understand information encoding by neurons.

1. INTRODUCTION

In sensory systems especially, and in neural information processing systems more generally, we are ultimately concerned with how a neuron's discharge pattern represents information and what fidelity that representation provides. From the point process viewpoint, information-bearing signals or indicators are expressed by the intensity's temporal variations. Two measures can be used to assess the expressiveness: the information capacity and the mean-squared estimation error. Investigators have attempted to analyze [12] and measure the capacity of a single neuron's discharges [8]. The second measure is more specific, and characterizes how well an optimal signal processing system can extract temporal extrinsic intensity variations beyond those induced by spike generation mechanisms [1, 4, 6, 7]. The theoretical understanding of

*Supported by grants from the National Science Foundation and from the National Institute of Mental Health

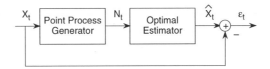

Figure 1. The stimulus-related signal X_t is encoded by a point process N_t. To quantify the quality of this encoding, envision an optimal system that estimates—decodes—the stimulus based solely on observing the point process. Denote its estimate by $\widehat{X}_t$ and the estimation error by ε_t. Because the estimator cannot create information, the capacity of the combined encoding-decoding system cannot exceed the capacity of the encoding system considered alone.

optimal signal processing system can extract temporal extrinsic intensity variations beyond those induced by spike generation mechanisms [1, 4, 6, 7]. The theoretical understanding of both measures in the point process case is immature, with the Poisson process being the only one for which broad results exist. In the Poisson case, the intrinsic discharge properties are the simplest, but, because of refractory effects, this model can only approximate neural discharge patterns.

To understand either measure, we model a point process's intensity as a stochastic process. Mathematically, let X_t be a stationary process that describes extrinsic (stimulus-related) information; the intensity of the point process is related to X_t in a causal way: $\mu(t; N_t, \mathbf{w}_t) = F[X_s, s \leq t; N_t, \mathbf{w}_t]$. We must assume that the functional $F[\cdot]$ is known, with the notation meant express the presence of extrinsic and intrinsic (neuron-dependent) components. The intensity's dependence at time t on the input's past expresses the possible presence of filtering (linear or nonlinear) of the signal. This relationship between a stimulus signal and a point process is shown in Fig. 1. The capacity-based measure attempts to quantify the stimulus encoding by comparing the statistical properties of X_t and N_t. Mean-squared error captures how well the stimulus can be optimally estimated from the point process. One interesting approach combining the two measures has been taken by Bialek (1991); he and his co-workers measured the capacity of the system having X_t as input and the estimation error $\varepsilon(t)$ as the output. Since information cannot be created, the capacity of the first system must exceed or equal Bialek's capacity. In this paper, we quantify the first system's capacity and learn how limited this measure of encoding can be. We then derive the mean-squared estimation error for an interesting special case, which we then use to quantify neural encoding.

2.   INFORMATION CAPACITY

The information capacity (or simply the capacity) of a point process is the maximal mutual information between the intensity $\mu(t; N_t, \mathbf{w}_t)$ and the sequence of events expressed by the counting process N_t over the interval $[0, T]$, with the maximum obtained by searching over all possible intensities that meet certain constraints [2, pp. 180–186]. Typical constraints are a maximal rate $\lambda_{\max}$, a minimal rate $\lambda_{\min}$, an average rate $\bar{\lambda}$, and a bandwidth limit on the extrinsic signal. Letting $\mathcal{L}$ represent constraint set, the capacity is abstractly expressed by

$$C = \lim_{T \to \infty} \max_{X_t} \max_{\mu(t; N_t, \mathbf{w}_t) \in \mathcal{L}} \frac{1}{T} I_T[\mu(t; N_t, \mathbf{w}_t); N_t].$$

Mutual information is a similarity measure: The more statistically dependent N_t is on X_t, the greater the mutual information. The maximization over the extrinsic input means that capacity depends solely on intrinsic properties: the form of the encoding and the dependence structure

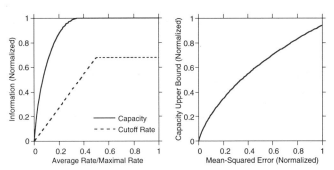

Figure 2. The relation between information measures and rate or mean-squared error are depicted. In each case, the minimal rate λ_{min} equals zero, and the vertical scales have been normalized by λ_{max}/e, the maximal capacity of a point process. In the left panel, capacity and cutoff rate for Poisson processes are plotted against $\bar{\lambda}/\lambda_{max}$. The relation between a normalized upper bound on capacity and the mean-squared error (normalized by its maximum) is shown in the right panel. In addition to $\lambda_{min} = 0$, $\bar{\lambda} = \lambda_{max}/2$ for this calculation, and the intensity was that of a random telegraph wave. This latter plot indicates that capacity increases even as the mean-squared error becomes worse.

of the point process. Maximal similarity is achieved for a specific extrinsic signal, which, in empirical terms, is not easy to find experimentally and can be "non-physical". When we impose rate constraints without restricting the signal's bandwidth, the capacity of *any* point process having an intensity description corresponds to that of a Poisson process, the point process having no intrinsic properties, and is given by [3, 5]

$$
C = \begin{cases}
\lambda_{min} \left[\dfrac{1}{e} \left(\dfrac{\lambda_{max}}{\lambda_{min}} \right)^{\lambda_{max}/(\lambda_{max}-\lambda_{min})} \right. \\
\qquad \left. - \ln \left(\dfrac{\lambda_{max}}{\lambda_{min}} \right)^{\lambda_{max}/(\lambda_{max}-\lambda_{min})} \right], & \bar{\lambda} - \lambda_{min} > \lambda^{\circ} \\[2ex]
(\bar{\lambda} - \lambda_{min}) \ln \left(\dfrac{\lambda_{max} - \lambda_{min}}{\bar{\lambda} - \lambda_{min}} \right), & \bar{\lambda} - \lambda_{min} < \lambda^{\circ}
\end{cases}
$$

where capacity has units of nats/s and

$$
\lambda^{\circ} = \lambda_{min} \left[\frac{1}{e} \left(\frac{\lambda_{max}}{\lambda_{min}} \right)^{\lambda_{max}/(\lambda_{max}-\lambda_{min})} \right].
$$

The capacity achieving extrinsic signal is a random telegraph wave, which flips between maximal and minimal rates, and has infinite bandwidth. In the important special case where $\lambda_{min} = 0$, $\lambda^{\circ} = \lambda_{max}/e$ and the capacity expressions simplify greatly.

$$
C = \begin{cases}
\lambda_{max}/e \text{ nats/s} \quad (0.53\lambda_{max} \text{ bits/s}), & \bar{\lambda} > \lambda_{max}/e \\
\bar{\lambda} \ln (\lambda_{max}/\bar{\lambda}), & \bar{\lambda} < \lambda_{max}/e
\end{cases}
$$

In most cases, the average rate is less than the maximal rate divided by e; consequently, the second expression best applies to neural discharge patterns.

The definition of capacity does not directly express a communications model and does not necessarily provide an accurate measure of how well information is expressed by a point process. The fundamental theorems of information theory only suggest that capacity bounds the ability of an information encoder to reliably send information through a system. For a given system, achieving this bound may be impossible, and it would thus overestimate what can be achieved. An alternate measure, the cutoff rate R_0, more directly quantifies information transmission capabilities. The conceptual communications model underlying its calculation is that one of a set of q discrete symbols is transmitted every T seconds (the symbol interval) using a symbol-specific intensity waveform. The cutoff rate is smaller than the capacity, and it provides a more accurate view of the information bearing capability than the capacity. Using results from [3, 11] derived for a Poisson process, when $\lambda_{\min} = 0$, we have that $R_0 = \bar{\lambda}/2$ when $\bar{\lambda} < \lambda_{\max}/2$, and $R_0 = \lambda_{\max}/4$ otherwise (Fig. 2).

3. MINIMUM MEAN-SQUARED ERROR

Employing the minimum mean-squared error measure amounts to finding the optimal system that extracts the estimate of X_t from the point process, and has the smallest mean-squared error ε^2. The optimal estimator (the conditional mean of X_t given the event sequence) is nonlinearly related to the point process and is difficult to find, even in the case of a Poisson process [10, Section 6.5.2]. Instead of explicitly finding the estimator, lower bounds on the minimum mean-squared error are known [10, Section 6.5.4]. Simpler expressions for the error and explicit formulation of the estimator become possible if we consider the Poisson process and restrict the estimator to be a linear filter (i.e., a Wiener filter). Here, the intensity is proportional to a positive-valued extrinsic signal X_t. Calculation of the mean-squared error reveals that it depends on the power spectrum of the input X_t, in terms of both the bandwidth and the shape of the power spectrum. When the input has a first-order, lowpass power spectrum having bandwidth B, the mean-squared estimation error normalized by the maximal error is given by

$$\tilde{\varepsilon}^2 = \frac{\bar{\lambda} B}{\mathcal{V}[\lambda]} \left(\sqrt{1 + \frac{2\mathcal{V}[\lambda]}{\bar{\lambda} B}} - 1 \right).$$

Thus, the normalized error depends only on $B/(\mathcal{V}[\lambda]/\bar{\lambda})$, the ratio between the bandwidth and the intensity's variance-to-average ratio. As shown in Fig. 3 for the case of Poisson process having a random telegraph wave intensity, this normalized error increases as the bandwidth/rate ratio increases, nearly reaching the saturation value when the bandwidth equals the average rate (about 75% of $\varepsilon_{\max}^2$). Furthermore, note that accurate estimates, arbitrarily defined to occur when the normalized squared error is less than 10%, occurs when the intensity's bandwidth is less than a small fraction (about 0.01) of the average rate. Consequently, sensory information can be extracted reliably from a single discharge patterns only when the signal's temporal variations are much slower than the average discharge rate.

4. RELATIONSHIPS BETWEEN CAPACITY AND MEAN-SQUARED ERROR

To gain a more realistic evaluation of capacity, Shamai and Lapidoth [9] found a relationship between a upper bound on capacity and the mean-squared estimation error. This relation depends in a complex way on the maximal and minimal intensity constraints, but the

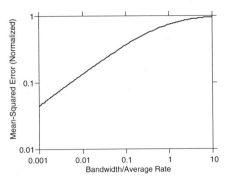

Figure 3. The normalized mean-squared error produced by a Wiener filter is plotted against the ratio of bandwidth and the average rate. In this example, the intensity equaled a symmetric random telegraph wave having $\lambda_{\min} = 0$, $\bar{\lambda} = \lambda_{\max}/2$, and $\mathcal{V}[\lambda(t)] = \bar{\lambda}^2$. For all bounded intensities having a first-order power spectrum, this curve serves to lower bound the mean-squared error.

general form of this relation can be exemplified by the special case $\bar{\lambda} = \lambda_{\max}/2$ and $\lambda_{\min} = 0$, wherein (Fig. 2)

$$\widetilde{C} \leq \frac{e}{2} \frac{\tilde{\varepsilon}^2}{2 - \tilde{\varepsilon}^2} \ln \frac{2}{\tilde{\varepsilon}^2}$$

with $\widetilde{C}$ denoting the capacity normalized by its maximal value $\lambda_{\max}/e$. Somewhat surprisingly, the capacity is small when the mean-squared estimation error is small, and large when the errors are worse.* Thus, capacity does *not* assess the information bearing capabilities of a neuron's output; large capacities can only be achieved when the neuron's extrinsic intensity component *cannot* be well estimated by any system. This surprising result can be traced to the definition of capacity for a point process: This definition (and many other cases as well) does not correspond to a communication problem. A closer statistical association between the extrinsic signal and the point process does not necessarily imply increasing ease of extracting information. The cutoff rate portrays more accurately than the capacity the ability to extract information represented by a point process, but does not accurately portray how transient, time-varying signals so prevalent in sensory systems can be estimated from discharge patterns.

5. SUMMARY

We have shown that capacity calculations, made either analytically or empirically, do not reflect how accurate analog information is represented by neural discharge patterns. For the moment, mean-squared error, though more difficult to determine, gives a more accurate picture. Finding it in simple examples leaves the conclusion that high discharge rates are necessary to accurately represent time-varying information. Sufficiently high rates are not found in neural discharges. However, simple pooling of statistically independent patterns creates high discharge rates. It would seem that some measure of ensemble activity is not only interesting, but necessary for understanding how sensory systems work.

The importance of ensemble modeling can be demonstrated from our results and structural considerations. Take the collection of auditory-nerve fibers that innervate a common

---

*A similar derivation for the additive white Gaussian noise channel yields an exact relationship between capacity and mean-squared estimation error that has essentially the same form.

frequency region. Estimates suggest a few hundred nerve fibers respond to stimuli filtered similarly in the cochlea. Our mean-squared error results indicate that fidelity in estimating an intensity having a 300 Hz bandwidth (the approximate bandwidth of each fiber's cochlear filter) requires an average response rate of 10^5 spikes/s. To produce this rate by simple pooling of individual responses, each of which has an average rate of 100 spikes/s, would require 1,000 nerve fibers. Thus, the auditory pathway would need to perform close-to-optimal processing *if* the system were to reconstruct the acoustic signal waveform, a demanding task. If, on the other hand, the auditory system were a (complicated) feature extractor, focusing on signal amplitude, fundamental frequency, etc., much lower combined discharge rates would be needed. It may well be that the latter information extraction model applies to the auditory system.

REFERENCES

[1] W. Bialek, F. Rieke, R. R. de Ruyter van Steveninck, and D. Warland. Reading a neural code. *Science*, 252:1852–1856, 1991.

[2] P. Brémaud. *Point Processes and Queues*. Springer-Verlag, New York, 1981.

[3] M. H. A. Davis. Capacity and cutoff rate for Poisson-type channels. *IEEE Trans. Info. Th.*, IT-26:710–715, 1980.

[4] D. H. Johnson and A. Swami. The transmission of signals by auditory-nerve fiber discharge patterns. *J. Acoust. Soc. Am.*, 74:493–501, 1983.

[5] Y. M. Kabanov. The capacity of a channel of the Poisson type. *Theory Prob. and Applications*, 23:143–147, 1978.

[6] K. E. Mark and M. I. Miller. Bayesian model selection and minimum description length estimation of auditory-nerve discharge rates. *J. Acoust. Soc. Am.*, 91:989–1002, 1992.

[7] M. I. Miller. Algorithms for removing recovery-related distortion from auditory-nerve discharge patterns. *J. Acoust. Soc. Am.*, 77:1452–1464, 1985. Also see Erratum. *J. Acoust. Soc. Am.*, 79: 570, 1986.

[8] F. Rieke, D. Warland, and W. Bialek. Coding efficiency and information rates in sensory neurons. *Europhysics Letters*, 22:151–156, 1993.

[9] S. Shamai (Shitz) and A. Lapidoth. Bounds on the capacity of a spectrally constrained Poisson channel. *IEEE Trans. Info. Th.*, 39:19–29, 1993.

[10] D. L. Snyder. *Random Point Processes*. John Wiley and Sons, Inc., New York, 1975.

[11] D. L. Snyder and I. B. Rhodes. Some implications for the cutoff-rate criterion for coded direct-detection optical communication systems. *IEEE Trans. Info. Th.*, IT–26:327–338, 1980.

[12] R. B. Stein. The information capacity of nerve cells using a frequency code. *Biophysical J.*, 7:67–82, 1967.

17

A CONTINUOUS TIME MODEL OF SYNAPTIC PLASTICITY IN THE CEREBELLAR CORTEX

Garrett T. Kenyon

Department of Neurobiology and Anatomy
University of Texas Medical School
Houston, 6431 Fannin, Houston, Texas 77030

1. INTRODUCTION

Patients with cerebellar damage exhibit a variety of motor deficits[1], and both human and animal studies indicate that cerebellar lesions disrupt several forms of motor learning[2]. A number of competing hypotheses regarding the nature of cerebellar involvement in motor function have been proposed[3-8]. One particularly influential class of models, based on the original proposals of Marr[9], and shortly thereafter amended by Albus[10], asserts that motor memories are stored in the cerebellar cortex at synapses from parallel fibers onto Purkinje cells (pf*Pkj). This class of models can be characterized by the following four hypotheses. 1) Climbing fiber inputs to the cerebellum, which originate from the inferior olive, are topographically organized by motor function. Each climbing fiber conveys a precise motor instruction to a group of target Purkinje cells, which in turn project to cells in the cerebellar nuclei capable of generating the instructed movement via projections to brain stem motor nuclei. 2) Mossy fiber inputs to the cerebellum, which originate from the pontine and other brain stem nuclei, are activated by virtually every sensory modality, as well as by direct inputs from the cerebral cortex. Granule cells, whose axons give rise to parallel fibers, receive a highly divergent input from mossy fiber afferents. The precise pattern of parallel fibers active at any given moment provides a sparse distributed representation of the mossy fiber input. 3) For any given Purkinje cell, each climbing fiber input induces a cell wide signal that causes the weights of any coactive pf*Pkj synapses to be depressed. 4) By repeatedly pairing a particular movement context with the appropriate pattern of climbing fiber activity, the occurrence of that context alone becomes sufficient to initiate the instructed movement. In particular, the occurrence of a previously paired movement context leads to a reduction of Purkinje cell firing, thereby initiating the instructed movement by releasing the appropriate target neurons in the cerebellar nuclei from tonic Purkinje cell inhibition. In order to account for the reversibility of learned motor behavior, the above model is typically augmented by an additional hypothesis: 5) When pf*Pkj synapses are active in the absence of a climbing fiber input, their synaptic weights are potentiated[11,12]. Thus, repeated presentations of a previously established movement context in the absence of activity in those climbing fibers which instruct for that movement will eventually eliminate the motor response associated with that context.

There is considerable experimental support for Marr/Albus based models of cerebellar motor learning. Although climbing fiber dependent plasticity at pf*Pkj synapses was originally postulated on the basis of the unique anatomical arrangement of the parallel and climbing fiber inputs to Pukinje cells, several studies have subsequently reported that a long-term depression (LTD) of parallel fiber to Purkinje cell transmission can be induced by conjunctive stimulation of the parallel and climbing fiber pathways[13]. Further experiments using an *in vitro* cerebellar slice preparation have demonstrated that plasticity at pf*Pkj synapses is bidirectional; pf*Pkj synapses are depressed when activated in temporal proximity to a climbing fiber input, but exhibit a long-term potentiation (LTP) when activated in the absence of recent climbing fiber input[14]. These findings have since been extended by several laboratories, and plasticity at pf*Pkj synapses is increasingly well understood at both the physiological and the molecular levels[15,16]. Experimental data also supports the postulated motor topographic organization of the cerebellum and the inferior olive. Specifically, the cerebellar cortex appears to be organized in terms of parasagital microzones that receive climbing fiber input from localized regions of the inferior olive and in turn project to localized areas of the deep nuclei[17]. Although the role of the cerebellum in motor learning, and its importance as a site of plasticity, remains controversial[18,19], several learned motor behaviors appear to engage the cerebellum in the manner generally predicted by the Marr and Albus models[20–23].

Empirical studies of plasticity at pf*Pkj synapses demonstrate that it is pathway specific; the induction of plasticity in one parallel fiber pathway has no effect on an unstimulated control pathway[16]. This property ensures that any changes in pf*Pkj synaptic weights responsible for an adapted movement are restricted to the subset of synapses which represent the context in which that movement occurs. The requirement that plasticity at pf*Pkj synapses be activity dependent, however, carries with it the implication that the weights of pf*Pkj synapses will be constantly fluctuating in proportion to their background levels of activity. Furthermore, since each pf*Pkj synapse is likely to contribute to the representation of multiple movement contexts, a certain degree of background activity is inevitable. All pf*Pkj synapses to a given Purkinje cell will therefore tend to drift up or down, depending on the level of spontaneous climbing fiber input to that cell. In order for motor memories to remain stable for prolonged periods, it follows that the spontaneous level of climbing fiber activity must be carefully regulated to an intermediate, or equilibrium value, at which LTD and LTP are in balance at all pf*Pkj synapses simultaneously. Although this is assumed *a priori* in several cerebellar models[11,12], it is important to investigate the physiological conditions under which an equilibrium level of climbing fiber activity could be maintained. This is particularly true since the potential for instability appears to be an inherent feature of any reversible model of associative motor learning. In this paper, we therefore investigate the stability of pf*Pkj synapses using a continuous time model of cerebellar-olivary dynamics.

2. MODEL

The essential circuitry of the cerebellar-olivary system addressed in the present model is illustrated schematically in figure 1. If the weight of the 'ith' pf*Pkj synapse at time 't' is denoted by $w_i(t)$, then an expression for the firing activity of a representative

Purkinje cell, denoted by 'f', can be approximated as the sum of the responses of a passive RC circuit to the synaptic inputs from each parallel fiber,

$$\frac{df}{dt} = -\frac{f}{\tau} + \sum w_i(t_{ij}) \cdot \delta(t - t_{ij}) \,,$$

(1)

where τ is the time constant of the Purkinje cell, t_{ij} denotes the 'jth' firing of the 'ith' pf*Pkj synapse, and the sum is over all combinations of i and j.

Since Purkinje cells inhibit output neurons in the cerebellar nuclei, and the cerebellar nuclei in turn send inhibitory projections to the inferior olive[24], the net synaptic influence of Purkinje cells on climbing fiber activity is excitatory (figure 1). To facilitate the analysis, we will assume that this influence can be approximated by a linear relationship, such that climbing fiber activity is simply proportional to Pukinje cell activity, where for simplicity we take this proportionality factor to be unity.

We next assume that each climbing fiber input increments a cell wide signal (presumably related to the intracellular calcium concentration[25]), whose amplitude we denote by C. Letting K denote the increment produced by each climbing fiber input, the cell wide signal can then be described by an equation of the form,

$$\frac{dC}{dt} = -\Gamma C + K \cdot \sum \delta(t - t_k) \,,$$

(2)

such that between climbing fiber inputs the cell wide signal decays exponentially with a time constant Γ, and t_k denotes the arrival times of successive climbing fiber inputs.

To model synaptic plasticity, we assume that whenever a pf*Pkj synapse is active, the weight of that synapse changes by an amount proportional to the current value of the cell wide signal;

$$\frac{dw_i(t)}{dt} = -\varepsilon \sum (C - \theta) \cdot \delta(t - t_{ij}) \,,$$

(3)

where ε is a gain factor coupling changes in synaptic weight to the current value of C, and θ specifies the threshold signal level separating LTD from LTP. Note that synaptic weights remain constant when they are not active (i.e. there is no decay term in equation 3).

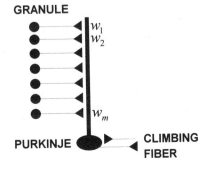

Figure 1. A schematic of the cerebellar-olivary circuitry addressed in the present model. Purkinje cells receive up to 200,000 excitatory synapses from parallel fibers, but from only a single climbing fiber. Since Purkinje cells inhibit neurons in the cerebellar nuclei, which in turn sends an inhibitory projection to the inferior olive, the net influence of Purkinje cells on climbing fibers is excitatory, as shown.

We now assume that pf*Pkj synapses are activated independently in a Poisson distributed manner, where the mean firing rate of the ith synapse is given by F_i. We likewise assume that the climbing fiber is activated in a Poisson, or memory independent manner, where the rate of climbing fiber input is given by f.

The above assumptions define a Markov process for the $m + 2$ dynamical variables, f, C, and $\{w_i\}$, where m is the number of pf*Pkj synapses. For simplicity, we will assume that the time constant of the Purkinje cell, τ, is much shorter that Γ^{-1}, so that f may be considered an instantaneous function of the $\{w_i\}$. We may then define the conditional probability that a representative Purkinje cell/climbing fiber is in the state (C, f), at time t, given the state (C', f') at time t'. When $t = t' + \Delta t$, it is possible to construct an explicit expression for this conditional probability by considering all possible events which can occur as the time interval Δt goes to zero. Proceeding in a standard fashion, we may then obtain a Fokker-Planck equation for the conditional probability from which explicit equations for the expectation values of C and f can be derived. Although length restrictions do not permit an explicit derivation of these steps to be presented here, such techniques have been previously described[26].

3. RESULTS

Defining the equilibrium level of climbing fiber activity as f_0, it is possible to show that on average

$$\frac{dC}{dt} = -\Gamma(C - \theta) + K(f - f_0) \, , \tag{4}$$

where the quantity f_0 is given by

$$f_0 = \Gamma\theta/K \, . \tag{5}$$

In equation 4, all dynamical quantities are to be interpreted as expectation values defined over an ensemble of identically prepared systems. For simplicity, we have used the same variable to denote the expectation value of the corresponding dynamical quantity. In the following, all dynamical quantities are to be interpreted as expectation values unless otherwise noted. Equation 4 shows that on average, the time derivative of C is given by the sum of a standard decay process plus a source term that is proportional to the instantaneous rate of climbing fiber input relative to its equilibrium value. This conforms to our intuition that the cell wide signal should increase when climbing fiber activity is high, and decrease when climbing fiber activity is low. Equation 5 further shows that the equilibrium level of climbing fiber activity is determined by three quantities; the decay rate, amplitude, and threshold of the cell wide signal produced by each climbing fiber input. The functional form of this dependence is also intuitively reasonable. As Γ becomes smaller, the cell wide signal will decay more slowly, and thus f_0 will be smaller as well, since fewer climbing fiber inputs would be necessary to maintain the same threshold signal level. Likewise, the larger the value of θ, the larger f_0 must be to maintain this signal level. A similar argument applies to the dependence on K, the larger the amplitude of each climbing fiber induced signal, the smaller f_0 must be to maintain the same threshold signal level.

Similarly, it is possible to obtain an expression for average value of the climbing fiber activity as well,

$$\frac{df}{dt} = -\varepsilon\left|\vec{F}\right|^2 (C - \theta) \, . \tag{6}$$

Equation 6 shows that on average, the negative rate of change in the instantaneous level of climbing fiber activity is proportional to the difference between the expected value of the cell wide signal and the threshold signal level. This again conforms to our intuition, since a high value of C will cause pf*Pkj synaptic weights to decrease, and thus reduce the expected value of f, while the opposite occurs for a low value of C.

It is clear that an equilibrium solution to equations 4 and 6 is given directly by $C = \theta$ and $f = f_0$. A simple argument is sufficient to show that the quantities C and f are always driven to their corresponding equilibrium values from any initial state. For instance, if the expected rate of climbing fiber activity is too high, ($f > f_0$), then according to equation 4 this will drive the value of the cell wide signal higher ($C > \theta$). According to equation 6, however, this in turn drives climbing fiber activity back down until equilibrium is restored. Physiologically, this is because elevated climbing fiber activity causes a net depression of pf*Pkj synaptic weights, which in turn reduces the amount of tonic Purkinje cell inhibition received by target neurons in the cerebellar nuclei. Since cerebellar output inhibits climbing fibers, cerebellar output continues to increase only until climbing fiber activity has been driven back down to an equilibrium level at which pf*Pkj synaptic weights remain constant. Thus, negative feedback from the cerebellum back to the inferior olive ensures that bidirectional plasticity at pf*Pkj synapses is self-regulating. Climbing fiber activity is always driven to an equilibrium level at which pf*Pkj synaptic weights remain constant, despite the continuous weight fluctuations produced by their background levels of activity.

Figure 2 shows the solutions to equations 4 and 6 following a step perturbation. We see that the cell wide signal rapidly returns to its equilibrium value, θ, with a time constant governed by Γ. Climbing fiber activity also returns to equilibrium, but with a longer time course that depends on the quantities $\{F_i\}$ and ε. This is as expected, since the rate at which climbing fiber activity returns to equilibrium should clearly depend on the background level of parallel fiber activity, as well as on the size of the individual changes in synaptic weight produced by the background levels of pf*Pkj input.

4. DISCUSSION

The present results show that an electrophysiologically characterized form of bidirectional plasticity at pf*Pkj synapses can be stable when incorporated into the cerebel-

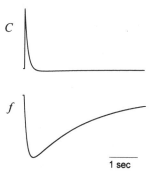

Figure 2. The average response to a step perturbation predicted by the continuous time model. The top trace shows the dynamics of the climbing fiber induced cell wide signal, C, as a function of time following a step perturbation intended to mimic the effects of a single climbing fiber input. The cell wide signal decays to a point slightly below its equilibrium level, with a time constant given by Γ, and then recovers with a much slower time course. The bottom trace shows the change in the expected value of the climbing fiber firing rate, f, following the same step perturbation. Initially, while the cell wide signal is above equilibrium, f decreases due to the induction of LTD. The reduced climbing fiber activity then causes the cell wide signal to fall below equilibrium, which in turn drives the climbing fiber activity back up due to the induction of LTP. The continuous time model is thus stable with respect to arbitrary perturbations.

lar-olivary system. Specifically, we have shown how the inhibition of climbing fibers by cerebellar output can dynamically maintain an equilibrium level of climbing fiber activity at which pf*Pkj synaptic weights remain approximately constant despite the continuous fluctuations produced by their background levels of activity. It was further argued that plasticity at pf*Pkj synapses must be activity dependent in order to ensure that motor learning is both associative and reversible. A finite level of background activity was also argued to be an inevitable component of any biologically realistic model, since the storage capacity of pf*Pkj synapses would be grossly compromised if each synapse contributed to only a single movement context. We therefore believe that the present results are somewhat general in their application, since they address issues that are likely to be confronted by any biologically realistic model of motor learning that is both associative and reversible.

Although the present results suggest that a bidirectional from of activity dependent plasticity at pf*Pkj synapses is not inconsistent with the stable storage of motor memories at these synapses, this conclusion only holds if each pf*Pkj synapses is stable at the same level of climbing fiber activity. In particular, equation 5 shows that the equilibrium level of climbing fiber activity is determined by three quantities: the decay rate, amplitude, and threshold of the cell wide signal produced by climbing fiber input. The simultaneous stability of all pf*Pkj synapses therefore requires that these quantities occur in the same ratio at every synapse. Otherwise, the same level of climbing fiber activity could not produce an equilibrium between LTD and LTP at all pf*Pkj synapses simultaneously. We must ask, therefore, whether the restrictions postulated by the present model could be reasonably implemented in a biological context?

First, it seems unlikely that the threshold signal level could be exactly the same at every pf*Pkj synapse. For this to be true, it would be necessary to assume that the chemical pathways mediating LTD and LTP at the individual pf*Pkj synapses all possessed highly uniform kinetics and relative concentrations. Furthermore, such uniformity must hold not only across widely separated cellular compartments, but between different Purkinje cells as well. Second, for the increment in the cell wide signal produced by each climbing fiber input to be the same at every pf*Pkj synapse, it would apparently be necessary to require that the influx and diffusion of calcium be highly uniform throughout the Purkinje cell dendrite, even down to the scale of individual spines. The restrictions required by the present model to ensure the stability of plasticity at all pf*Pkj synapses simultaneously therefore appear to be unreasonably severe to admit a plausible biological implementation.

The above arguments suggest that certain assumptions of the present model need to be modified. In particular, we briefly consider the consequences if the changes in individual pf*Pkj synaptic weights occured in a discrete manner, such as the turning on or off of a binary synapse. If this were the case, it appears that a biological implementation could be much more feasible, since it is simpler to assure uniformity across all pf*Pkj synapses when the underlying processes are discrete, rather than continuous. For example, we might assume that each climbing fiber input raises the cell wide signal above a critical threshold level that always completely shut off any concurrently active pf*Pkj synapses. Although to ensure stability it would still be necessary to assume that the decay of the cell wide signal resulted in the same temporal eligibility window for the induction of LTD at every pf*Pkj synapse, it is possible that this restriction alone would not be insurmountable in a biological context. For instance, each climbing fiber input could initiate a chemical cascade whose kinetics were highly uniform across all pf*Pkj synapses. Further study will be necessary to clarify this issue.

5. CONCLUSION

Our results indicate that bidirectional plasticity at pf*Pkj synapses can be stable when incorporated into the cerebellar-olivary system, despite the continuous fluctuations in synaptic strength produced by background activity. Our results therefore support the hypothesis that motor memories are stored at pf*Pkj synapses. Our results further suggests that pf*Pkj synaptic weights could be binary, in order to guarantee a uniform stability across all pf*Pkj synapses.

REFERENCES

1. S. Gilman, J.R. Bloedel, and R. Lechtenberg, 1981, Disorders of the Cerebellum, first ed. Contemporary Neurology Series, (F. Plum ed.), Volume 21, F. A. Davis Company, Philadelphia.
2. M. Ito, 1984, The cerebellum and neural control, Raven Press, New York.
3. J.C. Houk and S.P. Wise, 1995, Distributed modular architectures linking basal ganglia, cerebellum, and cerebral cortex: Their Role in planning and controlling action, *Cerebral Cortex* **2**:95–110.
4. J.-H. Gao, L.M. Parsons, J.M. Bower, J. Xiong, J. Li, and P.T. Fox, 1996, Cerebellum implicated in sensory acquisition and discrimination rather than motor control., *Science* **272**:545–47.
5. J.R. Bloedel, 1992, Functional heterogeneity with structural homogeneity: How does the cerebellum operate?, *Behav. Brain Sci.* **15**:666–678.
6. M. Ito, 1982, Cerebellar control of the vestibulo-ocular reflex - around the flocculus hypothesis., *Ann. Rev. Neurosci.* **12**:85–102.
7. R. Llinas and J.P. Welsh, 1993, On the cerebellum and motor learning., *Cur. Opin. Neurobiol.* **3**:958–968.
8. A. Pellionisz and R. Llinas, 1980, Tensorial approach to the geometry of brain function: Cerebellar coordination via a metric tensor, *Neurosci.* **5**:1125–1136.
9. D. Marr, 1969, A theory of cerebellar cortex, *J. Physiol.* **202**:437–470.
10. J.S. Albus, 1971, A theory of cerebellar function, *Math. Biosci.* **10**:25–61.
11. M. Fujita, 1982, Adaptive filter model of the cerebellum., *Biol. Cybern.* **45**:195–206.
12. T.J. Sejnowski, 1977, Storing covariance with nonlinearly interacting neurons, *J. Math. Biol.* **4**:303–321.
13. M. Ito and M. Kano, 1982, Long lasting depression of parallel fiber-Purkinje cell transmission induced by conjunctive stimulation of parallel fibers and climbing fibers in the cerebellar cortex, *Neurosci. Let.* **33**:253–258.
14. M. Sakurai, 1987, Synaptic modification of parallel fibre-Purkinje cell transmission in *in vitro* guinea-pig cerebellar slices, *J. Physiol. (Lond)* **394**:463–480.
15. D.J. Linden and J.A. Connor, 1993, Cellular mechanisms of long-term depression in the cerebellum., *Cur. Opin. Neurobiol.* **3**(3):401–6.
16. M. Ito, 1989, Long term depression, *Ann. Rev. Neurosci.* **12**:85–102.
17. J.C. Houk and A.R. Gibson, 1986, Sensorimotor processing through the cerebellum, in: *New Concepts in Cerebellar Neurobiology*, (J.S. King and J. Courville, Eds), pp. 387–416, Liss: New York.
18. J.P. Welsh and J.A. Harvey, 1989, Cerebellar lesions and the nictitating membrane reflex: performance deficits of the conditioned and unconditioned response., *J. Neurosci.* **9**(1):299–311.
19. T.M. Kelly, C.-C. Zuo, and J.R. Bloedel, 1990, Classical conditioning of the eyeblink reflex in the decerebrate-decerebellate rabbit., *Behav. Brain. Res.* **38**:7–18.
20. Gilbert and W.T. Thach, 1977, Purkinje cell activity during motor learning., *Brain Res.* **128**:309–328.
21. D.A. McCormick and R.F. Thompson, 1984, Neuronal responses of the rabbit cerebellum during acquisition and performance of a classically conditioned nictitating membrane-eyelid response., *J. Neurosci.* **4**(11):2811–2822.
22. N.E. Berthier and J.W. Moore, 1986, Cerebellar Purkinje cell activity related to the classically conditioned nictitating membrane response., *Exp. Brain Res.* **63**:341–350.
23. E. Watanabe, 1984, Neuronal events correlated with long term adaptation of the horizontal vistibulo-ocular reflex in the primate flocculus, *Brain Res.* **297**:169–174.
24. T.J. Ruigrok and J. Voogd, 1990, Cerebellar nucleo-olivary projections in the rat: an anterograde tracing study with Phaseolus vulgaris-leucoagglutinin (PHA-L)., *J. Comp. Neurol.* **298**(3):315–333.
25. C.F. Ekerot and O. Oscarsson, 1981, Prolonged depolarization elicited in Purkinje cell dendrites by climbing fiber inputs in the cat, *J. Physiol.* **318**:207–21.
26. G.T. Kenyon, R.D. Puff, and E.E. Fetz, 1992, A general diffusion model for analyzing the efficacy of synaptic input to threshold neurons, *Biol. Cybern.* **67**:133–141.

18

EFFECTS OF A DIFFUSING MESSENGER: LEARNING TEMPORAL CORRELATIONS

Bart Krekelberg and John G. Taylor

Centre for Neural Networks
King's College London
The Strand, London, UK
E-mail: bart@mth.kcl.ac.uk

1. INTRODUCTION

The neural messenger nitric oxide (NO) could radically alter our view of information processing in the central nervous system. Unlike ordinary neurotransmitters, NO is not confined to transporting messages across the synaptic cleft. Instead, due to its small size and long persistence in the extracellular fluid, it can convey information across considerable spatial ($170\mu m$) and temporal ($4-5s$) gaps [15]. In previous work we discussed the possible computational role of such a diffusing messenger in the development of cortical maps in the early stages of development [9]. Elsewhere, we briefly discussed the possible consequences of the temporal persistence of NO [8]; the current work is an extension of those results.

NO, a gas that is produced in response to post-synaptic depolarization has been shown to regulate long term potentiation (LTP) and depression (LTD) in a way that depends on the pre-synaptic firing rate. To wit, [16] found that LTP occurs in the presence of NO and a high pre-synaptic firing rate whereas LTD occurs if the pre-synaptic firing rate is low. The persistence of NO in tissue leads to the development of a memory trace: a high concentration of NO signals a recent activation of the neurons in that area. As NO influences the learning process through the gating of LTP and LTD, this memory trace can become an important factor in self-organizing development. In this paper we explore the consequences of this hypothesis in a simulation model. For specificity we concentrate on the self-organization in the upper-layers of primary visual cortex, although we believe that the process of learning with a memory trace has a more general validity.

Gilbert et al. [6] showed that pyramidal neurons in the upper layers of cat's V1 have excitatory lateral connections to neurons at a distance of up to 8mm. This has been corroborated in many laboratories and across species. Physiological data show that this connectivity is mainly between neurons with similar orientation selectivity [11]. Developmental studies in the cat show that the patchy distribution of the long range connectivity starts to develop in the second postnatal week [2] and is refined during weeks three to five. This onset of clustering coincides with the time at which cells in layer IV (i.e the input layer for layer II/III) are known to be visually responsive, orientation selective and roughly ordered in the orientation domain [1, 14].

We hypothesise that the iso-orientation connectivity develops as a results of temporal correlations in the natural visual environment. These temporal correlations are coded into the lateral weights due to the effect NO has on the learning process.

2. METHODS

We use a simple model to explore the role of a diffusing messenger in the refinement of long range lateral connections in layers II/III of V1 in the cat. A fixed afferent connectivity that implements the orientation selectivity is assumed, while the lateral connectivity is initially random but small over a large distance and purely excitatory. All cells produce nitric oxide in proportion to their depolarization. The NO is assumed to decay at a rate of $0.5s^{-1}$ and has a diffusion constant of $2600 \mu m^2/s^{-1}$ ([12]).

A small part of visual cortex is represented by 100 leaky-integrator neurons in a one-dimensional network. These neurons represent 4 hypercolumns. In this perfectly ordered network, neurons with similar orientation preference are 25 nodes away. A randomly chosen oriented stimulus is moved across the retina at a fixed speed. The next stimulus is put on the retina ISI seconds later. A synaptic strength is changed if the local NO concentration is above the NO threshold. The change is an increase if the pre-synaptic firing rate is high but a decrease if the pre-synaptic firing rate is low. Note that the post-synaptic depolarization is not directly relevant in this rule; this learning rule is not synapse-specific. The simulations are stopped when the total weight change averaged over ten different stimuli is less than 1 part in 1000.

3. RESULTS

Not too surprisingly, given the right choice of parameters, the network develops lateral connections that preferentially connect neurons with similar orientation preference. Although encouraging, this cannot be the reason for building an abstract model like this. A more important goal is to derive the relation between the various parameters in the model and the way in which they influence the lateral connections. To investigate this, a large number of simulations with different parameter values was run. The converged weights were analyzed along the lines of Weliky et al's study of ferrets' long-range connectivity [13].

First, a connectivity vector is built that represents the hypothesis of iso-orientation connectivity. Secondly, the correlation of the weights of each neuron with the hypothesized connectivity is determined. This is called the iso-correlation. Thirdly, we determine the frequency with which the iso-correlation exceeds the correlation with the connectivity of all the other neurons in the simulation. (For details, see [13, figure 9].) This index eliminates spurious correlations in the simulation that could, for instance, be due to the small network size. Figure 1 displays histograms of the frequency index for three simulation runs. In a network which is strongly iso-orientation connected, most cells will be better correlated with the hypothesized connectivity than with an arbitrary other cell's connectivity. A histogram of the iso-frequency index of such a network can therefore be recognized from its skewedness. Two tests were performed on the histograms. First, a χ^2-test was performed to determine whether

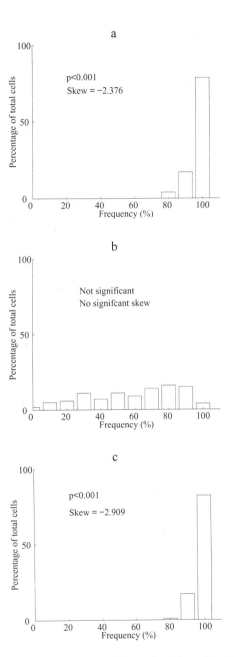

Figure 1. Histograms of the iso-frequency index in three different simulations of the NO learning rule. The statistical measures in the figures indicate whether a histogram is significantly different from flat (χ^2-test and skewedness). (**top**: ISI = 0.75s, NO threshold = 1.2, A flat distribution, **middle**: ISI = 0.7s, NO threshold= 3; A skewed distribution, **bottom**: ISI = 1.3s, NO threshold = 3.5; A skewed distribution.)

the distributions were significantly different from a flat distribution. Secondly, the skewedness of the distribution was determined. Skewedness was only considered to be significant if it exceeded two times the standard deviation of the distribution of the skewedness of a normal distribution. These measures of statistical significance are shown in figure 1. The simulation on the left of figure 1 has not led to a connectivity that is iso-connected in a statistically significant manner.

With the frequency histograms the iso-correlation indices have acquired a well defined meaning, as they allow us to determine whether a network's connections are significantly iso-oriented or not. With such a significance analysis underlying every correlation index from now on, we turn back to the dependence on the temporal parameters of the environment and the nitric oxide related parameters of the model.

Figure 2 shows the mean of the iso-correlation indices in a network as a function of the interstimulus interval. The different lines show different NO threshold parameters. The figure shows that there is a lower limit to the interstimulus interval that is required for significant iso-orientation connectivity to develop. This lower limit shifts downward with the NO threshold, allowing neurons with a high NO threshold to develop iso-connectivity for a larger set of stimulus environments. As hardly any data are available on the *temporal* statistics of the natural environment, this dependency is rather difficult to test.

Figure 2 shows the dependence on the NO threshold: for any choice of the interstimulus interval there is a region of NO thresholds that will lead to the development of iso-orientation connectivity. Outside this restricted domain, the iso-connectivity will be lost. This is a strong prediction because it does not depend on the interstimulus intervals which are unknown and difficult to control in experiments. Experimentally the predictions of this simulation experiment could be investigated by inhibiting the NO synthase in developing animals or by looking at the development of transgenic animals in which the gene for NOS has been "knocked out".

An unexpected result of the simulations is that the peak of the iso-connectivity curve shifts leftward with increasing ISI (figure 2; right). This implies that the slow aspects of the environment (ISI =low) are preferentially learned by neurons with a high NO threshold whereas the fast stimuli are only picked up by the low NO threshold neurons. This points to an intriguing relation between a neuron's NO threshold and the temporal information it will process. We hope to explore this relation in future work.

Our hypothesis is that the lateral connectivity in the model develops as a result of the *temporal* correlations in the visual environment. Until now, however, the input to the network has been correlated both temporally and spatially. This is due to the assumption of perfect ordering of the orientation map and is unwanted for two reasons. Firstly, even though layer IV neurons in cat's V1 show orientation selectivity before the development of the long-range connections [1], the ferret develops these two systems in synchrony [4]. If the model is to have any general validity, it should be able to cope with a non-ordered input layer. Secondly, perfect ordering is a far cry from any actual cortical orientation map: fractures and singularities are as prevalent as linear zones.

To investigate the influence of spatial correlations, a new parameter called *disorder*, is introduced. This parameter denotes the fraction of cells in the input layer that have a random orientation preference. The other cells' orientation preferences are left ordered. The dependence of the iso-correlation on the disorder parameter is shown for three different networks in figure 3. The simulations show that in the regimes where solid iso-connectivity would be formed (figure 3, middle and right), the influence of the disorder parameter is small. The connectivity that develops, even when the orientation map is random, is rarely significantly different from the connectivity in the ordered case. This supports the claim that the NO learning

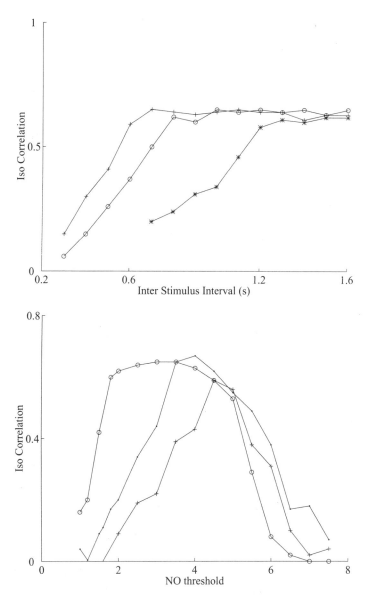

Figure 2. **top**: The iso-correlation as a function of the interstimulus interval for three choices of the NO threshold. +) NO threshold = 4 ∘) NO threshold = 3 ∗) NO threshold = 2 **bottom**:The mean iso-correlation index as a function of the NO threshold for three choices of the interstimulus interval. +) ISI =0.5 ·)ISI=0.75 ∘) ISI = 1.3

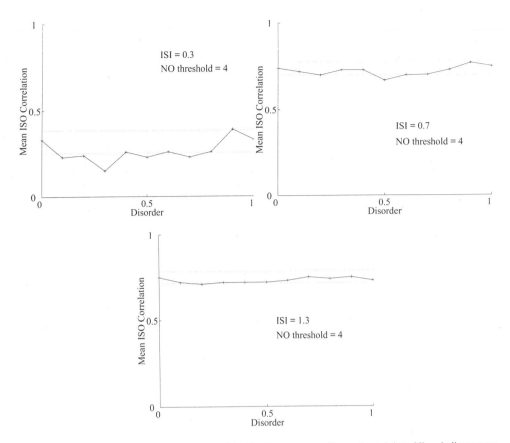

Figure 3. The iso-correlation index as a function of the disorder parameter. The horizontal dotted lines indicate mean iso-correlations that are different from the iso-correlation that develops in the ordered orientation map (disorder=0) ($p \leq 0.05$, Student's t-test). Only the left most diagram shows significant (but variable) effects of the disorder on the self-organization process.

rule codes the *temporal* correlations of a stimulus presentation process into the lateral weights of a network.

4.   DISCUSSION AND CONCLUSION

The basic mechanism in the current model is the memory trace set up by nitric oxide. Firstly, we have shown that this allows *temporal* correlations in the stimuli to be converted into spatial connection strengths. Secondly, the particular way in which the statistics of the lateral connections depend on the properties of the environment and the diffusing messenger have been derived as predictions in figure 2. Thirdly, an unexpected property has emerged that relates the NO threshold for learning in a neuron to the temporal information it will process.

In the real nervous system, the mechanism we propose may just be one of many operating in parallel. The direct way to determine the relevance of our model is to change the NO production in a developing cat and determine its long-range connections afterwards. Such an experiment has not yet been performed. The most obvious other mechanism that could account for some of the data on the long-range connections, is plain Hebbian learning driven by purely spatial correlations.

We can distinguish between the results of our temporal model and the plain Hebbian model by considering the statistics of the environment and the lateral connectivity. In a static model, connections between neurons with similar orientation preference, located at right angles to the orientation preference, would need quite peculiar statistics of the environment. To be specific, only if a large part of the stimuli consists of periodic oriented bars, would such connectivity develop in the static model. Secondly, the connectivity between neurons with like orientation preference *along* the axis of their orientation preference would require oriented bars covering at least 8 degrees of the visual field. Furthermore, no such connections could exist along the axis of orientation preference between cells that are end-stopped.

Data to decide between the alternative models are becoming available. Gilbert [7], for instance, observed that the preferred direction of connectivity is along an axis in cortex which is perpendicular to the orientation preference of the cells. As explained above, such an anisotropy would be hard to understand in a static model. In a dynamic model, on the other hand, this fits with the observation that orientation selective neurons respond most vigorously to oriented bars that move in a direction perpendicular to their orientation. Contrary to Gilbert's findings and more in line with a static model, recent anatomical data [4] show more long-range connectivity along the axis of the orientation preference of the cells. Without an accurate knowledge of the statistics of both the connectivity and the visual environment or a way to influence the latter in an experiment, a firm conclusion cannot be reached.

Martin and Marshall [10], proposed a model for the development of long-range lateral connections in which a memory trace is incorporated into the membrane dynamics. This limits the temporal extent of the memory to the temporal extent of a depolarization which is of the order of $30ms$. To connect neurons laterally over a cortical distance of over $4mm$ which corresponds to approximately $4°$ of visual field, the visual stimuli would have to move at a speed of $\frac{4}{30} °/ms \simeq 100 °/s$. This seems rather fast for the average stimulus in the kitten's visual environment. Stimuli used to elicit responses in recordings, usually move at speeds below $10 °/s$. The temporal scale at which NO operates (seconds) would seem to make it a better candidate for a memory trace that deals with natural stimuli.

Our model is similar to the model developed in [5]. There, an abstract memory trace was introduced without making the identification with a particular chemical process like NO. [5] showed that the development of lateral connections could lead to complex cells. The main difference with the current work is that here the long-range laterals are individually subthreshold. Furthermore, unlike [5], we have identified a process that could be responsible for the memory trace and explored the dependence on the temporal statistics of the environment and the chemical properties of the memory trace. Finally, by identifying the trace with NO, the temporal trace formation has become intricately linked to spatial organization [9]. An interesting extension along these lines would be to test the model not on randomly disordered input layers, but on orientation maps with a particular realistic ordering. The structures known to exist in layer IV's orientation preference map could be introduced and their influence on the lateral weights determined. We expect that the non-homogeneous properties of the orientation map will lead to specific anisotropies in the lateral connectivity. This would make for another interesting experimental prediction.

REFERENCES

[1] K. Albus and W. Wolf. *J. Physiol.*, 348:153–185, 1984.

[2] E. Callaway and L. Katz. *J. Neurosci.*, 10:1134–1153, 1990.

[3] J. C. Durack and L. C. Katz. *Cerebral Cortex*, 6:178–183, 1996.

[4] D. Fitzpatrick. *Cerebral Cortex*, 1996.

[5] P. Földiák. *Neural Comp.*, 3:194–200, 1991.

[6] C. D. Gilbert and T. N. Wiesel. *Nature*, 280:120–125, 1979.

[7] C. D. Gilbert and T. N. Wiesel. *J. Neurosci.*, 9:2432–2442, 1989.

[8] B. Krekelberg and J. G. Taylor. *Neural Networks World*, 6:185–189, 1996.

[9] B. Krekelberg and J. G. Taylor. *Network*, 8:1–16, 1997.

[10] K. E. Martin and J. A. Marshall. *Neural Information Processing Systems*, 5:417–424, 1993.

[11] D. Y. Ts'o, C. D. Gilbert, and T. N. Wiesel. *J. Neurosci.*, 6:1160–1170, 1986.

[12] J. M. Vanderkooi et al. *Biochimica et Biophysica Acta*, 1207:249–254, 1994.

[13] M. Weliky, K. Kandler, D. Fitzpatrick, and L. C. Katz. *Neuron*, 15:541–552, 1996.

[14] T. Wiesel and D. Hubel. *J. Comp. Neurol.*, 158:307–318, 1974.

[15] J. Wood and J. Garthwaite. *Neuropharmacology*, 33:1235–1244, 1994.

[16] M. Zhuo, E. Kandel, and R. Hawkins. *Neuroreport*, 5:1033–1036, 1994.

19

INFORMATION PROCESSING IN A CEREBELLAR GRANULE CELL

Huo Lu,[1] F. W. Prior,[2] and L. J. Larson-Prior[1]

[1]Department of Neuroscience and Anatomy
[2]Department of Radiology
Pennsylvania State University
College of Medicine
M.S. Hershey Medical Center
Hershey, Pennsylvania 17033
hlu@neuro.hmc.psu.edu
prior@xray.hmc.psu.edu
ljlp@neuro.hmc.psu.edu

ABSTRACT

A fifteen compartment, biologically realistic model of a cerebellar granule cell (GC) was developed to examine the signal processing capabilities of this most numerous element in the cerebellar cortical circuit. The model explicitly includes compartments for the soma, axon hillock, proximal axon, dendrites and terminal bulbs. All synaptic inputs were transduced via activation of glutamate receptor subtypes located on the dendritic bulb compartments, and were systematically varied in their number and frequency. An intriguing morphological feature, in which axonal location is shifted from the soma to a dendrite, was specifically examined to determine its impact on granule cell output. The GC was shown to be electrotonically compact, resulting in a lack of biasing of output based on axonal location. Biasing of output could be driven by changes in the passive parameters of the model, but required an unrealistically large change in resistive coupling between dendritic and somal compartments. Thus, axonal location does not induce physiologically relevant phase shifts between synaptic inputs located on multiple dendritic bulbs, suggesting that the GC relies heavily upon temporal aspects of its input signals for integrative processing.

INTRODUCTION

The cerebellar cortical circuitry receives multi-modal sensory information which it integrates to organize appropriate outputs. Data enters the cerebellum from two major sources: the climbing fibers which arise almost exclusively from the inferior olivary nucleus in the brainstem, and mossy fibers (MFs) which take their origin from widely distributed sensory systems throughout the central nervous system (CNS). Information entering the cerebellar cortex via MFs is processed in the granule cell layer before being sent to Purkinje cells (PCs). As the sole output neuron of the intracortical circuit, the PC represents the final integrator of cortical information processing.

Cerebellar granule cells (GCs) exhibit a characteristic morphology consisting of a spherical soma (5–10 μm) from which 4–6 small dendrites arise. Each dendrite ends in a convoluted enlarged bulb (Mugnaini et al. 1974; Palay and Chan-Palay 1974) on which MFs synapse. Frequently, the GC axon originates not from the soma, but from one of the dendrites (Palay and Chan-Palay 1974). This feature may privilege sensory information to the axon-bearing dendrite, since a single MF synaptically contacts any given GC only once (Albus 1971).

A recent investigation of synaptic integration in dopaminergic neurons of the substantia nigra has led to the hypothesis that a dendritic site of axonal emergence privileges synaptic input to that dendrite, with action potential initiation shifted to dendritic sites. This hypothesis was strengthened by results obtained from a passive cable model which illustrated substantial differences in the electrotonic decay of excitatory postsynaptic potentials (EPSPs) generated from stimulus sites on distal versus proximal dendrites (Haüsser et al. 1995). The functional consequences of such privileging are significant, providing a mechanism by which synaptic input to a single dendrite can disproportionately influence neuronal output. In addition, the location of inhibitory input becomes a major factor in the ability of synaptic input to the privileged dendrite to control neuronal output.

The influence of axonal location on synaptic integration in cerebellar granule cells was investigated using a recently developed, biologically realistic model (Lu et al. 1995) which incorporated active conductances on both somal and axonal compartments while maintaining passive dendritic conduction.

METHODS

Based on experimentally defined morphology (Palay and Chan-Palay 1974; Mugnaini et al 1974) and physiology of cerebellar GCs (Cull-Candy et al., 1989; Silver et al 1992; Gabbiani et al 1994), a cerebellar GC model was constructed using the GENESIS simulator (v2.0.1, Bower and Beeman 1994). This compartmental model included a spherical somal compartment, four dendrites and an axon which was placed either on the somal or one of the dendritic compartments. The dendrites were modeled with three compartments; one proximal to the soma, one distal to the soma and a dendritic bulb on which all synaptic contacts were made (Fig. 1). Each compartment had a specific capacitance of $1 \mu F/cm^2$, with an axial resistance (RA) of $1 \frac{1}{2} \cdot m$ (Silver et al., 1992). Specific membrane resistances were defined as 19,450 $\frac{1}{2} \cdot cm^2$ for the somal compartment and 30,300 $\frac{1}{2} \cdot cm^2$ for all other compartments (Gabbiani et al 1994).

Six active conductances were included in the somal compartment : *gNa* and *gK(DR)*, which represent Hodgkin-Huxley fast sodium and delayed potassium conductances (Cull-Candy et al 1989; Gabbiani et al 1994); *gK(A)*, a transient potassium conductance (Bardoni and Belluzzi 1993; Cull-Candy et al 1989; Gabbiani et al 1994; Gorter et al 1995); *gH*, a mixed sodium/potassium conductance (Gabbiani et al 1994); *gCa(HVA)*, a high-voltage-activated calcium conductance (Gabbiani et al 1994); and *gK(C)*, a large-conductance Ca^{2+}-dependent K^+ (Fagni et al 1991). The voltage-dependent kinetics of *gK(C)* and *gK(DR)* were modeled as equal based on experimental data (Gabbiani et al., 1994; Yamada et al., 1989). The fast sodium and delayed potassium conductances, were also included in the initial segment of the axon.

Two synaptic conductances, N-methyl-D-aspartate (NMDA) and α-amino-3-hydroxy-5-methyl-4-isoxazolepropionic acid (AMPA) were implemented on each of the four dendritic bulb compartments. The biophysical parameters by which these synaptic conductances were modeled were derived from literature (D'Angelo et al 1990; Silver et al 1992; Gabbiani et al 1994).

RESULTS

Two morphological configurations for axon location were tested: one in which the axon arose from a proximal dendritic compartment and a second in which the axon arose from the somal compartment (Fig. 1) When the axon arose from the dendrite (Fig. 2A), stimulation of the axon-bearing dendrite resulted in an EPSP which reached the axon Is more rapidly (latency to peak = 43.5 ms) and with a slightly greater amplitude (peak amplitude = 4.08 mV) than EPSPs produced by stimulation of the Db most distal to the Is (latency to peak = 45 ms, peak amplitude = 4.03 mV). Under these conditions, the EPSP generated in the somal compartment was invariant regardless of the site of stimulation (Fig. 2B), exhibiting a latency to peak of 41 ms and a peak amplitude of 4.09 mV. The axon was then shifted to the somal compartment and the same two Dbs were stimulated. This protocol resulted in an EPSP of equal magnitude to that noted at the soma when the axon was located on a dendrite, differing only in its latency to peak (44 ms). These data indicate a degree of biasing towards inputs from an axon-bearing dendrite. However, the ratio of the peak amplitudes of EPSPs at the Is generated by identical stimuli applied at the Db of the axon-bearing dendrite and at the Db most distal to it was 1.01, indicating that each EPSP would contribute equally to the integrated signal at the Is. (Fig. 2A, B).

Action potentials generated either by stimulation of the Db of the dendrite carrying the axon or by stimulation of the Db most distal to it exhibited a difference in their latency to peak of 120 μs (Fig. 2C, D). This stands in contrast to the rather large shift in latency to peak of action potentials generated by a similar stimulation protocol in some dopaminer-

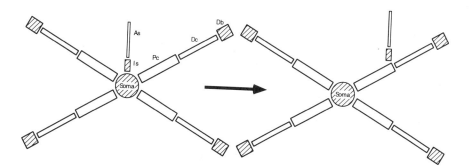

Figure 1. Schematic of the cerebellar granule cell model. Compartment dimensions were as follows: soma = 10 μm diameter, Pc and Dc = 40 μm length, Db = 8 μm length, Is = 10 μm length, As = 40 μm length. Shaded compartments were constructed with active conductances. Two configurations are indicated: one in which the axon emerged from the soma and one in which the axon emerged from an initial dendritic compartment. Db, dendritic bulb; Dc, distal compartment; Pc, proximal compartment; Is, initial segment; As, axon segment.

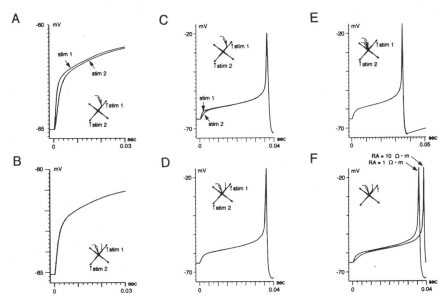

Figure 2. Effect of axon location on granule cell responses. With the axon emerging from a dendrite, EPSPs recorded from (**A**) the initial segment with stimulation of the Db of the axon bearing dendrite and from that Db most distal to the site of axon emergence show differences in both latency and amplitude. By comparison, EPSPs recorded from the somal compartment (**B**) under the same conditions exhibit no shifts in latency or amplitude. Action potentials recorded at either the dendritically arising axon Is (**C**) showed a slight difference in activation latency, but no difference in peak latency with stimulation at either the proximal (stim 1) or distal (stim 2) Db. (**D**) Action potentials recorded in the somal compartment resulted in simultaneous action potential initiation with activation at either stim 1 or stim 2. In cases where the axon originated from the somal compartment, action potentials of the same latency and amplitude were evoked by stimulation of any Db (**E**). However, changes in the resistive coupling between the axon bearing dendrite and the soma could drive shifts in action potential latency-to-peak at resistive values of [3] 10 Ω*m (**F**).

gic neurons from the substantia nigra (Haüsser et al. 1995). By progressively increasing axial resistance of the dendritic compartment that links the axon Is and the soma from 1 - 40 Ω*m it was possible to artificially generate significant shifts in the latency to peak of action potentials recorded from the soma and Is at values of [3] 10 ½·m (Fig. 2F). These data suggest that under physiologically realistic conditions, no significant shift in the arrival time of synaptically generated potentials is noted with shifts in axonal location from somal to dendritic sites at distances which are morphologically justified for the cerebellar granule cell.

Although minimal differences in arrival times were observed using a single pulse stimulation paradigm, these differences might be magnified under conditions of tonic stimulation. If action potentials originate first in a dendritically located axon and spread to the soma, the introduction of sufficient time delays by the intervening segment of dendrite could produce back propagating reverberations. Similarly, slight shifts in latency between dendritically located axons and the soma could produce interference patterns which might significantly modify the output frequency of the neuron. Therefore, the effect of tonic stimulation paradigms on granule cell output were tested.

A two second stimulus train ranging in frequency from 10 to 400 Hz was delivered to the model cell in two configurations: one in which the axon was located on a dendrite and another in which it arose from the soma (Fig. 3). The relation between GC firing rate and synaptic stimulation frequency was linear across the frequency range of 10–100 Hz, becoming increasingly non-linear at higher stimulation frequencies (Fig. 3A). Throughout the full range of stimulation frequencies; however, the response is essentially the same regardless of whether the axon arose from the dendrite or the soma, and whether the stimulus was delivered proximal or distal to the site of axon emergence (Fig. 3A, diagrams B2-4).

An increase in the resistive coupling between the Pc and the soma was introduced by changing RA from 1 to 10 Ω*m, and the GC firing response was then re-tested under four experimental configurations (Fig. 3B, C). One condition, in which the Db of a dendrite which carried the standard axial resistance was stimulated (Fig. 3C2), represents the control condition, with the axon arising from the somal compartment. A significant nonlinearity was introduced by the change in resistive coupling when the Db of the dendrite carrying that change was stimulated (Fig. 3B1, 3C1), with increasing divergence from the control response noted at stimulus frequencies >100 Hz. Although this condition did not introduce a significant phase shift between soma and somally located Is, it did introduce a

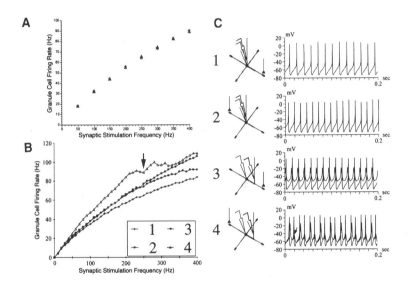

Figure 3. Effects of axon location on firing rate. In (**A**) GC firing rate is plotted as a function of synaptic stimulation of the dendritic bulb with the axon located either on the cell soma or a dendrite. Stimulation was delivered to the Db of the dendrite from which the axon arose (♦), and to a distal dendrite (X) with the response recorded at the Is. The same stimulus was also applied with the axon arising from the soma (Δ). (**B**) Effects of dendritic stimulation on GC firing rate after increasing axial resistance of the proximal dendritic compartment (RA = 10 ½*m). Each of the four experimental conditions plotted is illustrated in (**C**) for a stimulus frequency of 250 Hz (arrow), the value at which the greatest divergence of responses was noted between morphologic configurations. Comparison of the firing rate vs. synaptic inputs in the four experiments illustrated). (**C**) Four different morphological configurations are illustrated. In (1), the axon arose from the soma and the Db of the dendrite on which the Pc axial resistance was increased to 10 Ω*m was stimulated. The plot to the right shows that the response of both soma and Is were completely overlapping. (2) represents the control condition, in that the stimulated Db lay on a dendrite in which the axial resistance remained unaltered. Recordings from both Is and soma, with the axon at the soma, were again overlapping. Note that, in this case, the response train was phase shifted relative to that illustrated in (1). The axon was then placed on the Pc which carried an increased axial resistance and the Db proximal (3) and distal (4) to the recorded Is was stimulated. Under these conditions, there was a phase shift in the response between Is and soma (somal action potentials in both cases are represented by largest spike amplitudes). In addition, interference patterns were exhibited under condition (4, right panel, arrow).

phase shift with respect to the control response which is illustrated in the response plot (Fig. 3C1,2). When the axonal location was shifted to the dendrite carrying the increased resistive load, phase shifts between somal and Is compartments were clearly introduced (Fig. 3C3,4). In one condition, that in which the Db farthest from the Is was stimulated, the GC firing rate increased at stimulation frequencies >50 Hz (Fig. 3B4) and interference patterns were introduced (Fig. 3C4). While stimulation of the proximate Db also introduced phase shifts between somal and Is responses, no interference pattern was noted.

SUMMARY AND CONCLUSIONS

Several different neurons exhibit the morphological property of axonal emergence from a dendrite, including cortical pyramidal cells (Palay et al., 1968; Peters et al., 1968; Sloper and Powell, 1979) and dopamine neurons of the substantia nigral (Häusser et al., 1995). This morphological feature may have significant functional consequences, potentially privileging the input to the dendrite from which it arises. This suggestion has been recently investigated in nigral dopamine neurons, in which a biasing of activation was noted in cells having a dendritic origin for their axon (Haüsser et al., 1995).

Based on our simulation studies, axonal location in cerebellar granule cells does not appear to bias the synaptic inputs to the dendrite from which the axon arises. Several factors may act together to produce this result. First, in both experimental (D'Angelo et al., 1995) and these simulation studies, the granule cell has been shown to be electrotonically compact, with short, radially symmetric dendrites and a small soma (Palay and Chan-Palay, 1974). Second, a significant difference between cerebellar GCs and nigral dopamine neurons lies in the distance from the site of the axon initial segment to the soma, which can be as great as 240 μm for nigral cells but is usually 30–40 μm in GCs. Finally, the presence of active conductances on nigral dopamine cells was shown to affect signal attenuation in nigral dopamine neurons (Haüsser et al., 1995). In cerebellar GCs, regardless of axonal location, synaptic input to any Db activates the Is equally.

The production of dendritic privilege could be artificially induced by changing the passive properties of the dendritic compartment between the axon initial segment and soma. An increase of RA in this compartment from 1 ½*m to 10 ½*m clearly illustrates that synaptic input to that dendrite is privileged relative to inputs at non-axon bearing dendrites or the soma. This change in RA is functionally equivalent to increasing the dendritic length 10 times without changing RA. Currently, there is no physiological evidence to support the hypothesis that such differences in passive dendritic properties exist in GCs. Thus, axonal location does not appear to significantly impact signal processing in cerebellar GCs.

Information entering the cerebellar cortex via mossy fiber pathways may be represented by both phasic and maintained stimuli (Bloedel and Courville, 1981). The effect of axonal location, in which very small differences in signal propagation from dendrite to soma are introduced, may be more significant when the input signals are represented as tonic, maintained stimuli. Our simulation studies show that, while minor differences in signal propagation do occur when the axon arises from a dendrite, they do not significantly influence the GC output over the stimulation frequency range tested. Once again, changes in dendritic RA are capable of introducing significant changes in GC signal processing in those instances where the axon has a dendritic origin. Thus, the relatively short distance between the soma and an axon of dendritic origin in cerebellar GCs appears to negate any effect of that site of origin on GC signal processing.

ACKNOWLEDGMENTS

The authors gratefully acknowledge the financial support of the National Institutes of Health (NS 30759) and the National Science Foundation (IBN 9514844) to LLP.

REFERENCES

Albus JS (1971) A theory of cerebellar function. Mathematical Biosciences 10:25–61

Bardoni R and Belluzzi O (1993) Kinetic study and numerical reconstruction of A-type current in granule cells of rat cerebellar slices. J Neurophysiol 69:2222–2231

Bloedel JR and Courville J (1981) Cerebellar afferent systems. In: Brooks VB (ed), Handbook of Physiology, The Nervous System, pp. 735–830

Bower JM and Beeman D (1994) The book of GENESIS : exploring realistic neural models with the GEneral NEural SImulation System. TELOS, Speinger-Verlag, New York.

Cull-Candy SG, Marshall CG and Ogden D (1989) Voltage-activated membrane currents in rat cerebellar granule neurones. J Physiol 414:179–199

D'Angelo E, Rossi P and Gathwaite J (1990) Dual-component NMDA receptor currents at a single central synapse. Nature Lond 346:467–470.

D'Angelo E, De Filippi G, Rossi P and Taglietti V (1995) Synaptic excitation of individual rat cerebellar granule cells *in situ* : evidence for the role of NMDA receptors. J Physiol 484:397–413

Fagni L, Bossu JL and Bockaert J (1991) Activation of a large-conductance Ca^{2+}-dependent K^+ channel by stimulation of glutamate phosphoinositide-coupled receptors in cultured cerebellar granule cells. Eur J Neurosci 3:778–789

Gabbiani F, Midtgaard J and Knöpfel T (1994) Synaptic integration in a model of cerebellar granule cells. J Neurophysiol 72:999–1009

Gorter JA, Aronica E, Hack NJ and Balázs R (1995) Developement of voltage-activated potassium currents in cultured cerebellar granule neurons under defferent growth conditions. J Neurophysiol 74:298–306

Häusser M, Stuart G, Racca C and Sakmann B (1995) Axonal initiation and active dendritic propagation of action potentials in substantia nigra neurons. Neuron 15:637–647

Lu H, Prior FW and Larson-Prior LJ (1995) Signal transduction in a cerebellar granule cell: a modeling approach. Neurosci Abstr 21:916

Mugnaini E, Atluri RL and Houk JC (1974) Fine structure of granular layer in turtle cerebellum with emphasis on large glomeruli. J Neurophysiol 37:1–29

Palay SL, Sotelo C, Peters A and Orkand PM (1968) The axon hillock and the initial segment. J. Cell Biol. 38:193–201

Palay SL and Chan-Palay V (1974) Cerebellar cortex cytology and organization. Springer-Verlag, New York

Peters A, Proskauer CC and Kaiserman-Abramof IR (1968) The small pyramical neuron of the rat cerebral cortex. The axon hillock and initial segment. J. Cell Biol. 39:601–619

Silver RA, Traynelis SF and Cull-Candy GC (1992) Rapid-time-course miniature and evoked excitatory currents at cerebellar synapses in situ. Nature 355:163–166

Sloper JJ and Powell TPS (1979) A study of the axon initial segment and proximal axon of neurons in the primate motor and somatic sensory cortices. Phil. Trans. Roy. Soc. (Lond) B 285:173–197

Yamada WM, Koch C and Adams PR (1989) Multiple channels and calcium dynamics. In: Koch C and Segev I (eds) Methods in neuronal modeling from synapses to networks, pp 97–133

20

A MODEL FOR FAST ANALOG COMPUTATIONS WITH NOISY SPIKING NEURONS

Wolfgang Maass

Institute for Theoretical Computer Science
Technische Universitaet Graz
Klosterwiesgasse 32/2
A-8010 Graz, Austria
E-mail: maass@igi.tu-graz.ac.at

ABSTRACT

We show that networks of spiking neurons can simulate arbitrary feedforward sigmoidal neural nets in a way which has previously not been considered. This new approach is based on temporal coding by single spikes (respectively by the timing of synchronous firing in pools of neurons), rather than on the traditional interpretation of analog variables in terms of firing rates.

As a consequence we can show that networks of noisy spiking neurons are "universal approximators" in the sense that they can approximate with regard to temporal coding *any* given continuous function of several variables.

1. RESULTS

Traditionally one views the firing rate of a neuron as the representation of an analog variable in analog computations with spiking neurons. However with regard to fast cortical computations this view is inconsistent with experimental data. Thorpe and Imbert (Thorpe, 1989) have demonstrated that visual pattern analysis and pattern classification can be carried out by humans in just 100 msec, in spite of the fact that it involves a minimum of 10 synaptic stages from the retina to the temporal lobe. The same speed of visual processing has been measured by Rolls et al. in macaque monkeys. Furthermore they have shown that a single cortical area involved in visual processing can complete its computation in just 20-30 msec (Rolls, 1994a, 1994b). On the other hand the firing rates of neurons involved in these computations are usually below 100 Hz, and hence at least 20-30 msec would be needed just to sample the current firing rate of a neuron.

We explore in this article the theoretical possibilities of analog computations with noisy spiking neurons in *temporal coding*, where analog variables $x \in [0, \gamma]$ are encoded by the firing time $T - x$ of a spiking neuron (as in Hopfield, 1995), respectively by the median firing time of a pool of neurons in a more noise-robust implementation. It is well-known that EPSP's and IPSP's can *shift* the firing time of a neuron by a small amount. We show that this simple mechanism is in principle sufficient for computing in temporal coding arbitrarily complex bounded continuous multi-variable functions in just 20 msec.

The results reported in this article are rigorous theoretical results. We refer to (Maass, 1995) for all details that have to be omitted in this summary.

Assume that the membrane potential at the trigger zone of a neuron v is determined at time t by n postsynaptic potentials $h_1(t), \ldots, h_n(t)$, which were caused by the firing of presynaptic neurons $a_1, \ldots, a_n$ at times $T_0 - s_1, \ldots, T_0 - s_n$. We will focus on the initial segments of length $d + \triangle$ of these PSP's, during which they can be approximated well by a piecewise linear function

$$h_i(t) = \begin{cases} 0 & , \text{ if } \quad t - (T_0 - s_i) < d \\ w_i \cdot (t - (T_0 - s_i) - d) & , \text{ if } \quad d \le t - (T_0 - s_i) \le d + \triangle, \end{cases}$$

where $w_i \ge 0$ in the case of an EPSP and $w_i < 0$ in the case of an IPSP. Let t_v be the time when the sum of these PSP's reaches the firing threshold Θ and neuron v fires. Assume that the n PSP's are at time t_v within their linearly rising (EPSP) respectively linearly decreasing (IPSP) segments of length $\triangle$, i.e.

$$d \le t_v - (T_0 - s_i) \le d + \triangle \quad \text{for} \quad i = 1, \ldots, n \quad . \tag{1}$$

Then we have $\Theta = \sum_{i=1}^n h_i(t_v)$, and hence

$$t_v = \frac{\Theta}{\sum_{i=1}^n w_i} + T_0 + d - \frac{1}{\sum_{i=1}^n w_i} \cdot \underline{w} \cdot \underline{s} \quad . \tag{2}$$

Thus neuron v essentially outputs for inputs $s_1, \ldots, s_n \in [0, \gamma]$ the (normalized) inner product $\underline{w} \cdot \underline{s}$ in temporal coding. The constraint (1) is satisfied if $\sum_{i=1}^n w_i > 0$, $\sum_{i=1}^n w_i \cdot s_i \ge 0$, and Θ is chosen so that

$$\gamma \cdot \sum_{i=1}^n w_i \le \Theta \le (\triangle - \gamma) \cdot \sum_{i=1}^n w_i \quad . \tag{3}$$

The parameter γ has to be chosen $\le \triangle/2$ in a theoretically rigorous simulation. However we expect that a larger value of γ (around 5 msec) can actually be employed for computations in biological neural systems (see Maass, 1995). If we fix $s_1 \equiv 0$, then we can choose w_1 so that $\sum_{i=1}^n w_i = 1$ for *arbitrary* given $w_2, \ldots, w_n \in \mathbf{R}$, and hence compute in this way a linear function of $n - 1$ inputs $s_2, \ldots, s_n$ with *arbitrary* given weights $w_2, \ldots, w_n \in \mathbf{R}$.

If all weights w_i are multiplied with a factor $\lambda > 1$, this will increase the slope of the membrane potential $\sum_{i=1}^n h_i(t)$ at the time when it crosses the threshold. As a consequence the output of this neuron in temporal coding will become more noise-robust, both from the point of view of the common mathematical models for noise in spiking neurons (Gerstner, 1994, Maass, 1995, 1996), and from the point of view of experimental results (Mainen, 1995).

In an even more noisy setting when synapses and/or neurons fail with significant probability, one may replace each single neuron v in our construction by a pool P_v of neurons with approximately identical connection. In this case the mean firing time of those neurons in P_v that do fire plays the role of the firing time t_v of the single neuron v. One should note that in this generalized interpretation a network of spiking neurons can in principle perform analog computations in temporal coding with high reliability, but single-neuron recordings from a small number of neurons in this network would not necessarily yield repetitions of the same firing pattern for repetitions of the same computation.

So far we have shown that a spiking neuron can compute (for a certain range of its parameters) in temporal coding any *linear* function $\underline{s} \mapsto \underline{w} \cdot \underline{s}$. In particular it can detect to what degree a learned pattern $\underline{w}$ (which is stored in the efficacies of its synapses) is present in a stimulus $\underline{s}$.

We will now show that by using *two* layers of spiking neurons, one can in fact approximate in temporal coding *any* given bounded continuous function $F : [0, \gamma]^n \to [0, \gamma]^k$, with any desired degree of precision. It is well-known that any such function F can be approximated by a feedforward sigmoidal neural net with two layers. Furthermore it is known that the exact form of the activation function of the gates is irrelevant for this result (Leshno, 1993). Hence one can employ in particular the piecewise linear activation function π_γ with

$$\pi_\gamma(x) = \begin{cases} \gamma & \text{, if } x > \gamma \\ x & \text{, if } 0 \le x \le \gamma \\ 0 & \text{, if } x < 0 \end{cases} .$$

We show that one can simulate with spiking neurons in temporal coding a given sigmoidal gate G with activation function π_γ , i.e. a gate that outputs

$$G(s_1, \cdots, s_n) = \begin{cases} \gamma & \text{, if } \underline{w} \cdot \underline{s} > \gamma \\ \underline{w} \cdot \underline{s} & \text{, if } 0 \le \underline{w} \cdot \underline{s} \le \gamma \\ 0 & \text{, if } \underline{w} \cdot \underline{s} < 0 \end{cases}$$

for any $s_1, \ldots, s_n \in [0, \gamma]$. One adds auxiliary neurons to the preceding construction whose firing is time-locked with the onset of the stimulus. These can prevent a firing of neuron v before time $T - \gamma$ for $T := \Theta + T_0 + d$ (assume for simplicity that $\sum_{i=1}^n w_i = 1$), and they can force v to fire at the latest by time T (see Maass, 1995, for details).

With k layers of spiking neurons one can thus simulate in temporal coding with any desired precision any given k-layer net of sigmoidal neurons that employ the activation function π_γ . Hence in combination with the abovementioned results we have shown that with 2 layers of spiking neurons one can in principle approximate in temporal coding any given bounded continuous function with a computation time of ≤ 20 msec.

Finally we would like to point out that with biological neurons one can carry out very similar multi-layer computations *without* simulating explicitly the activation function π_γ (i.e. without forcing v to fire within a specific time window $[T - \gamma, T]$). The natural biological form of EPSP's and IPSP's has in general the effect that if v fires before $T - \gamma$, the resulting PSP in a neuron v' on the next layer is at the time when v' fires no longer within its initial *linearly* increasing (respectively decreasing) phase. As a result the "too large" value of the output of v in temporal coding is effectively *reduced* in size through the form of the PSP in v' . Dually, if v fires *after* the time window $[T - \gamma, T]$, the resulting PSP will in general still have its initial value 0 at the time when v' fires, which amounts to replacing the (theoretically) negative value of this output of v in temporal coding by the value 0 . In this way the *bounded* functional form of PSP's in v' achieves a similar effect as the application of a bounded activation function to the value $\underline{w} \cdot \underline{s}$ that is output by v in temporal coding.

Our new model for linear and nonlinear computations with spiking neurons also provides a new perspective on learning. For example in this model the Hebb rule would want to increase the efficacy w_i of the synapse between a presynaptic neuron a_i and v if the outputs of a_i and v in *temporal* coding are correlated, i.e. if v fires within a certain time window after a_i (and decrease the efficacy otherwise). Exactly this type of synaptic modulation has recently been found in pyramidal neurons (Markram, 1995).

2. REMARKS

1. In an alternative biological interpretation of our construction one can exploit finer details of dendritic integration in order to simulate *several* layers of a sigmoidal neural net with gates of fan-out 1 by a *single* spiking neuron. The idea is here to exploit boosting phenomena via voltage-dependent channels at branching points ("hot spots", see e.g. Mel, 1993, Sheperd, 1995) of the dendritic tree in order to simulate by a hot spot a sigmoidal gate in temporal coding (where the output corresponds to the "firing time" of such hot spot). In this interpretation the weight on the edge from this gate to a gate on the next level of the simulated sigmoidal neural net would correspond to the amplitude of the "action potentials" resulting from this hot spot, combined with the distance of this hot spot to the next "higher" hot spot in the dendritic tree (respectively to the soma). In this biological interpretation one can avoid the use of (unreliable) synapses for the simulation of internal connections in the sigmoidal neural net.

2. Our construction shows as a special case that a *linear* function of the form $\underline{s} \rightarrow \underline{w} \cdot \underline{s}$ for real-valued vectors $\underline{s} = \langle s_1, \ldots, s_n \rangle$ and $\underline{w} = \langle w_1, \ldots, w_n \rangle$ can be computed very efficiently (and very fast) by a spiking neuron v in temporal coding. In this case no auxiliary neurons are needed.

   A fast computation of linear functions is obviously relevant in many biological contexts, such as coordinate transformations between different frames of reference, or the analysis of a complex stimulus $\underline{s}$ in terms of many stored patterns $\underline{w}, \underline{w}', \ldots$.

   For example in an olfactory neural system (see e.g. Hopfield, 1991, 1995) the stimulus $\underline{s}$ may be thought of as a superposition of various stored basic odors $\underline{w}, \underline{w}', \ldots$. In this case the output $y = \underline{w} \cdot \underline{s}$ of neuron v in temporal coding may be interpreted as the amount by which the basic odor $\underline{w}$ (which is stored in the efficacies of the synapses of v) is present in the stimulus $\underline{s}$. Furthermore another neuron $\tilde{v}$ on the next layer might receive as its input $\underline{y} = \rangle y, y', \ldots, \langle$ from several such neurons $v, v', \ldots$, i.e. $\tilde{v}$ receives the "mixing proportions" $y = \underline{w} \cdot \underline{s}$, $y' = \underline{w}' \cdot \underline{s}$, for various stored basic odors $\underline{w}, \underline{w}', \ldots$ in temporal coding. This neuron $\tilde{v}$ on the second layer can then continue the pattern analysis by computing for this input $\underline{y}$ the inner product $\underline{W} \cdot \underline{y}$ with some stored "higher order pattern" $\underline{W}$ (e.g. the composition of basic odors that is characteristic for an individual animal) that is encoded in the efficacies of the synapses of neuron $\tilde{v}$. Such multi-layer pattern analysis is facilitated by the fact that the here considered neurons encode their output in the same way in which their input is encoded (in contrast to the approach in Hopfield, 1995).

   One also gets in this way a very fast implementation of Linskers network (Linsker, 1988) with spiking neurons in temporal coding.

REFERENCES

[1] Gerstner, W. and van Hemmen, J. L. (1994) *How to describe neuronal activity: spikes, rates, or assemblies?* Advances in Neural Information Processing Systems, vol. 6, Morgan Kaufmann (San Mateo) 463-470.

[2] Hopfield, J. J. (1991) *Olfactory computation and object perception.* Proc. Nat. Ac. Sci. USA, vol. 88, 6462-6466.

[3] Hopfield, J. J. (1995) *Pattern recognition computation using action potential timing for stimulus representations.* Nature, vol. 376, 33-36.

[4] Leshno, M., Lin, V. Y., Pinkus, A., and Schocken, S. (1993) *Multilayer feedforward networks with a nonpolynomial activation function can approximate any function.* Neural Networks, vol. 6, 861-867.

[5] Linsker, R. (1988) *Self-organization in a perceptual network.* Computer Magazine, vol. 21, 105-117.

[6] Maass, W. (1995) *Fast sigmoidal networks via noisy spiking neurons*, to appear in Neural Computation.

[7] Maass, W. (1996) *On the computational power of noisy spiking neurons.* Advances in Neural Information Processing Systems, vol. 8, MIT-Press (Cambridge), 211-217.

[8] Mainen, Z.F. and Sejnowski, T. J. (1995) *Reliability of spike timing in neocortical neurons.* Science, vol. 268, 1503-1506.

[9] Markram, H. and Sakmann, B. (1995) *Action potentials propagating back into dendrites trigger changes in efficacy of single-axon synapses between layer V pyramidal neurons.* Abstract 788.11 in Vol. 21 of the Proc. of the Conference of the Society for Neuroscience.

[10] Mel, B. W. (1993) *Synaptic integration in an excitable dendritic tree.* J. of Neurophysiology, vol. 70, 1086-1101.

[11] Rolls, E. T. (1994a) *Brain mechanisms for invariant visual recognition and learning.* Behavioural Processes, vol. 33, 113-138.

[12] Rolls, E. T. and Tovee, M. J. (1994b) *Processing speed in the cerebral cortex, and the neurophysiology of visual backward masking.* Proc. Roy. Soc. B., vol. 257, 9-15.

[13] Shepherd, G. M. (1994) *Neurobiology*, 3rd ed. Oxford University Press (New York).

[14] Thorpe, S. J. and Imbert, M. (1989) *Biological constraints on connectionist modelling.* In: Connectionism in Perspective, Pfeifer, R., Schreter, Z., Fogelman-Soulié, F., and Steels, L., eds., Elsevier (North-Holland).

21

STOCHASTIC MODELING OF THE PYRAMIDAL CELL MODULE

Bruce H. McCormick,[1] Glen T. Prusky,[2] and Sandeep Tewari[3]

[1] Scientific Visualization Laboratory
Texas A&M University
College Station, TX
E-mail: mccormick@cs.tamu.edu
[2] Department of Psychology
The University of Lethbridge
Lethbridge, AB, Canada
E-mail: pruskyg@hg.uleth.ca
[3] Scientific Visualization Laboratory
Texas A&M University
College Station, TX
E-mail: tewari@aic.lockheed.com

ABSTRACT

The neuronal environment of the pyramidal cell module (PCM) is morphologically modeled as perhaps the most important example of a space-filling neuronal structure. PCMs are fundamental information processing units of the monkey striate cerebral cortex. The PCM is modeled as a cylinder, 1600 μm long and 31 μm in diameter, which contains approximately 142 cells, the majority of which are pyramidal cells. Granular cells, spiny stellate cells, and other inhibitory nonpyramidal cells have a minority presence. The influence of peripheral PCMs is modeled by considering a larger cylinder around the PCM in question. We assume a hexagonal packing of the PCMs inside the larger cylinder, much like rods in a chilled water nuclear reactor. A PCM has a six-layer architecture, and the number and type of cells varies from one layer to the other. The PCM is then modeled as a stack of coaxial wafers of varying thickness. The dendrite morphology of the constituent cells of the PCM, as derived from digital neuron tracing, is used to stochastically estimate distribution functions for wafer-to-wafer interactions. Translational and rotational invariance is invoked to simplify the functional form of the critical distribution functions.

Computational Neuroscience
edited by Bower, Plenum Press, New York, 1997

1. INTRODUCTION

Although Nissl-stained preparations show that the cell bodies of neurons in the cerebral cortex are arranged in horizontal layers, the more interesting architectural issue from the point of view of function is how the neurons are organized in vertically oriented units [2]. There is strong evidence for the existence of vertically oriented neuronal columns called *pyramidal cell modules* [2]. The pyramidal cell modules are the fundamental information-processing units of the monkey striate cortex. Our objective is to model the cylindrical environment in which these pyramidal cell modules develop, and demonstrate the feasibility of using stochastic L-systems [4] and ray-casting growth strategies [1] to model the pyramidal cell module as an example of a dense space-filling neuronal structure. Because of its computational complexity a detailed morphological model of the pyramidal cell module has not been hitherto attempted.

2. METHODS

2.1. The pyramidal cell module

Apical dendrites of pyramidal cells are organized into vertically oriented groups or clusters. Counts show that there are some 1270 clusters of layer V apical dendrites per mm^2 in the tangential plane. Thus, the clusters of apical dendrites have a center-to-center spacing of 31μm. These clusters represent the axes of modules of pyramidal cells. Since it is known that there are 1.8-1.9×10^5 neurons beneath 1 mm^2 of cortical surface, the pyramidal cell module is modeled as a cylinder 1600 μm long and 31 μm in diameter which contains approximately 142 cells, the majority of which are pyramidal cells [2]. Granular cells, spiny stellate cells, and other inhibitory nonpyramidal cells have a minority presence.

A pyramidal cell module has the following neuronal constituents: Kernel cell dendrites, peripheral dendrites, and ascending/descending axons. (1) *Kernel cell dendrites* are from dendritic arbors whose soma are within the pyramidal cell module. There are approximately 142 such soma within the pyramidal cell module. (2) *Peripheral dendrites* are dendrites that do not originate from soma within the the pyramidal cell module under consideration but enter from peripheral pyramidal cell modules. (3)*Ascending/descending axons* are of two principal types, the thalamocortical connections and the collaterals. The thalamocortical connections enter from outside the cerebral cortex and generally terminate in layer IV. The collaterals largely arise from axons of the pyramidal cell module under consideration and from neighboring pyramidal cells. These axons typically branch and then grow vertically upwards.

2.2. Design of the PCM modeling environment

A morphological model of the pyramidal cell module (PCM) and its connections is developed below. The model first embeds the PCM in a large cylindrical array of surrounding PCMs. Then each PCM is decomposed into submodules called wafers. The neuronal constituents of these wafers are then defined, and their distributions within the submodules estimated. Finally, detailed morphological modeling of these neuronal constituents is undertaken.

A pyramidal cell module is influenced by other such modules around it, and the radius of influence of such peripheral modules can be determined. The influence of such peripheral PCMs is modeled by considering a larger cylinder around the PCM under consideration. We

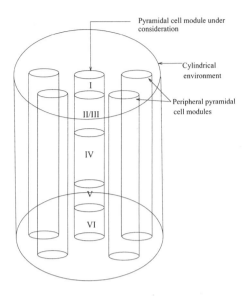

Figure 1. The modeling environment for a PCM [1, 3]

assume a hexagonal packing of the PCMs inside the larger cylinder, much like rods in a chilled water nuclear reactor. A diagrammatic representation of this modeling environment is shown in Figure 1 [1, 3]. In our model the pyramidal cell module under consideration is the *sink* which receives the neuronal constituents originating in the peripheral PCM which form the *sources*. The model described here can be used for modeling the information flow by reversing the designations *source* and *sink*.

The number of peripheral dendritic segments increases exponentially as the distance from the module's cylindrical axis increases and is assumed to drop to zero at a distance R_0. This is modeled by considering a large cylinder of radius R_0, called the "radius of influence" around the PCM under consideration (see Figure 2). The radius of influence R_0 depends upon the cell type.

2.3.   Decomposition into wafers

A pyramidal cell module (PCM) has a six-layered architecture, and the number of cells varies from one cortical layer to the other. This is an important consideration in modeling the PCM. Each cortical layer can be modeled as one or more wafers. The PCM is then a collection of wafers of varying thickness. The layered model of the distribution of cells in the PCM used in this work is based on the estimates made by Peters [2].

Figure 2 shows the modeling environment at the wafer level. Here, the index l refers to the PCM under consideration and the index k refers to a peripheral PCM. The sub-index i refers to individual wafers within the peripheral PCM and the sub-index j refers to individual wafers within the PCM under consideration.

2.4.   Wafer-to-wafer connections

Our model makes some simplifying assumptions about the dendritic segment distribution within a PCM. (1) The distribution function is invariant under translation in the tangential X-Y

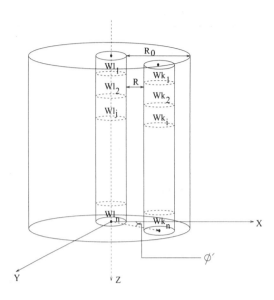

Figure 2. The modeling environment at the wafer level [1, 3]

plane, that is, the pyramidal cell module under consideration could be translated in the plane to an adjacent PCM and this would not affect the stochastic distribution functions. (2) The distribution function is also invariant under rotation in a plane, that is, the peripheral pyramidal cell modules can be rotated around the PCM under consideration and, if the fibers inside the PCM under consideration are appropriately rotated, then this again would not affect the distribution function. The above assumptions hold only within a common architectonic area of the cortex. The distributions could be different in different architectonic areas or on the boundaries between adjacent areas.

Our interest is in estimating the distribution of dendritic segments within a wafer (for example, wafer W_j in Figure 3). Such wafers will be referred to as the wafer-of-interest. There are two sources of the dendritic segments within the wafer-of-interest:

- Wafers in peripheral PCMs (for example, wafer W_i in Figure 3). Such wafers will be referred to as peripheral wafers.

- Ascending/descending dendrites from wafers within the same pyramidal cell module

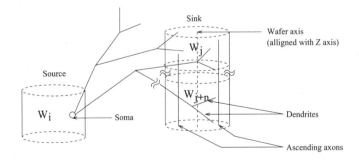

Figure 3. Interaction at the wafer level [1, 3]

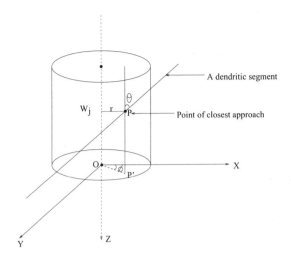

Figure 4. Spatial distribution of dendritic segments within a wafer [1]

(for example, wafer W_{j+n} in Figure 3).

2.5. Distribution Functions for Peripheral Dendrites

The length of the dendritic segment is ignored in this model. An important consideration here is that its point of closest approach **P** to the cylindrical axis is unique (Figure 4). The probability density function of the dendritic segments within a wafer is given by:

$$f(\mathbf{P}, \theta | R, \phi') d\mathbf{P} d\theta R d R d\phi'. \tag{1}$$

where $\mathbf{P} \equiv (r, \phi, Z)$ is the point of closest approach on the dendritic segment and is specified by cylindrical co-ordinates (r, ϕ) and the depth Z from the top of the PCM. Here θ is the angle the dendritic segment makes with the vertical Z-axis; R is the cylindrical radial distance of the peripheral PCM in which the peripheral wafer (and hence the soma of the peripheral dendrites arise) lies; and ϕ' is the azimuthal angle of the peripheral PCM containing the peripheral wafer. Substituting in Equation 1 we have:

$$f(\mathbf{P}, \theta | R, \phi') = f(r, \phi, Z, \theta | R, \phi'). \tag{2}$$

Because of rotational invariance we can say that:

$$f(r, \phi, Z, \theta | R, \phi') = f(r, \phi - \phi', Z, \theta | R) \tag{3}$$

which by symmetry is an even function of $\phi - \phi'$. Since we are considering interaction at the wafer level we can ignore Z^*. Equation 1 can now be simplified to:

$$f(r, \phi, \theta | R). \tag{4}$$

The closest-approach points **P** of the dendritic segments entering the wafer-of-interest at a particular azimuthal angle ϕ lie along a straight line (Figure 5). If R (i.e., the cylindrical

---

*More accurately, a wafer can be modeled by many sub-wafers thereby eliminating the dependence on Z.

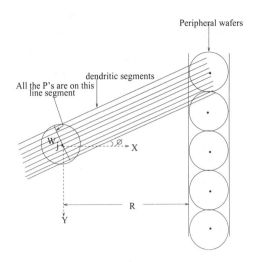

Figure 5. The uniform distribution in *r* of dendritic segments [1]

radial distance between the wafer-of-interest and the peripheral wafers) is large the distribution of such dendritic segments is uniform in R (the shortest distance between the wafer-axis and the dendritic segment). This follows because when R is large the peripheral wafers can be considered to be arranged in a straight row. Assuming that the kernel cells in these peripheral wafers are uniformly distributed, it can be easily seen that dendritic segments entering the wafer-of-interest at a particular azimuthal angle ϕ will be uniformly distributed in r.

In summary, the probability distribution function for dendritic segments from peripheral wafers reduces to $f(\phi, \theta | R)$.

REFERENCES

[1] S. Tewari, *Implicit Function Modeling of Neuron Morphology*, M.S. Thesis, Department of Computer Science, Texas A&M University, College Station, TX, July, 1995.

[2] Peters, A., "The organization of the primary visual cortex in the macaque," A. Peters and K.S. Rockland (Eds), *Cerebral Cortex*, vol. 10, Plenum Press, 1994.

[3] B. H. McCormick and S. Tewari, "Implicit Function Modeling of Neuron Morphology," in *Proc. Computational Neuroscience 1995 (CNS*95)*, Monterey, CA, July, 1995.

[4] B. H. McCormick and K. Mulchandani, "A Framework for Modeling Neuron Morphology," in *Proc. Computational Neuroscience 1995 (CNS*95)*, Monterey, CA, July, 1995.

22

HOW NEURONS MAY RESPOND TO TEMPORAL STRUCTURE IN THEIR INPUTS

Bartlett W. Mel,[1] Ernst Niebur,[2] David W. Croft[3]

[1] Department of Biomedical Engineering
University of Southern California
Los Angeles, California
E-mail: mel@quake.usc.edu
[2] Krieger Mind-Brain Institute and Department of Neuroscience
The Johns Hopkins University
Baltimore, MD
E-mail: niebur@jhu.edu
[3] Tanner Research Corporation
Pasadena, CA
E-mail: croft@tanner.com

1. INTRODUCTION

The way in which a neuron responds to temporal structure in its synaptic input stream is an issue of fundamental importance in the study of nervous system function. One historically significant notion that has resurfaced frequently within the field of neuroscience entails that spiking neurons act as "coincidence detectors", i.e. generating an output spike only when a sufficient set of inputs is activated quasi-synchronously. The biophysical justification for this notion is straightforward: any neuron whose time constant of integration is relatively short, and whose firing threshold is relatively high and sharp, would seem to qualify as a neuronal coincidence detector. Canonical exemplars of such cells have been found in the auditory brainstem specialized for detecting simultaneous arrival of action potentials from the left and right ears. [1] has argued for the utility of neuronal coincidence detection in his theory of "synfire chains", which would allow long sequences of neuronal activation patterns to be preserved in the cortex. Recent discoveries of both oscillations and short- and long-rage correlations among spike trains in cerebral cortical neurons [6] has rekindled interest in temporal structure in neuronal spike trains, has led to a variety of physical models for the genesis of either oscillatory, random, and/or synchronous spike trains (e.g. [10]), and has motivated a variety of models for the possible functional roles of temporal structure in neuronal spike trains, such as for selective visual attention [9, 8], or visual awareness [4].

Though a number of previous modeling efforts have examined the systems-level consequences of temporal structure in neuronal spike trains, relatively few modeling studies have
been carried out at the level of the biophysically-detailed single neuron [3, 7] to assess the
degree to which a "standard" neuron, presumably without anatomical or physiological specialization for high-precision temporal processing, responds differentially to various aspects
of the temporal structure of its synaptic input stream.

In two recent studies of model pyramidal cells [3, 7], it was demonstrated that synchronous activation of synapses could lead to a stronger cell responses (measured as mean
output firing frequency) than the same number of desynchronized inputs, but only when the
stimulus conditions were weak relative to firing threshold. Under stronger stimulus conditions, "overcrowding" of synaptic inputs within the refractory period led to suboptimal use
of the input events; temporal desynchronization was found to increase output rates in these
cases. These studies highlight a subtlety not entailed by the classical picture of a single-neuron
coincidence detector, which emphasizes the difficulty of bringing a neuron to threshold by a
single, coincident "blast" of synaptic input.

The present study extends earlier work in several ways. We (1) systematically study the
effects of both synchronicity and periodicity in input spike trains, (2) emphasize steady-state
rather than one-shot stimulus regimes, and (3) consider a wide range of biophysical conditions,
high and low stimulus frequencies, passive and active dendrites, large and small neurons, and
the effects of three different spiking mechanisms.

Several novel conclusions have resulted from this work that suggest much more complex
and subtle temporal processing properties than anticipated in classical notions of single neuron
function.

2. METHODS

We used NEURON to run (1) a 164-compartment cortical pyramidal cell with passive
or active dendrites, and (2) a single-compartment leaky integrate-and-fire neuron. In a typical
run, 100 randomly placed excitatory synapses (using the kinetic synapse model of [5]) were
activated at 100 Hz, and the cell's mean output firing rate was recorded over 500 ms. In different
runs, and over a range of peak excitatory synaptic conductances, the temporal structures of
input spike trains was varied along 2 continuous dimensions: synchronicity (S) and periodicity
(P). S=0 meant complete asynchrony and S=1 complete synchrony among trains; P=0 meant
Poisson and P=1 periodic trains (fig. 1).

3. RESULTS

Consistent with earlier work [3, 7], our steady state results showed that synchronizing
synaptic inputs could have non-trivial effects on output spike rates. For weak synapse conditions, temporal coincidences among input events were needed to boost the output firing rate
off the baseline (fig. 2A). For very strong inputs, output firing rates could be either boosted or
depressed (fig. 2AB). In particular, strong highly synchronized inputs could boost the output
rate by "pulling it up" to the input rate in a stimulus-locked fashion. Depression of output
spike rates could also occur for strong, highly synchronized inputs based on overcrowding of
input events during the spiking refractory period, as reported in [3, 7]. Finally, in large regions

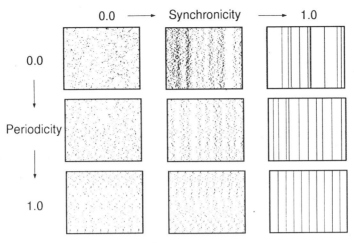

Figure 1. Examples of spike rasters delivered to neuron during an experimental run. Selected frames show the temporal pattern of input spikes as S and P vary from 0 to 1.

of the parameter space we explored, synchronization of inputs had remarkably little effect in either direction on output spike rates.

In contrast to the relatively modest and inconsistent effects of input synchronization on output firing rates, we found output rates increased more consistently and substantially with increases in spike train periodicity, i.e. as spike trains became individually more regular, *whether or not synchronized* (fig. 2CD). Thus, steady state firing rates could be substantially increased (e.g. nearly doubled from 55 to 100 Hz) by simply replacing 100 independent Poisson inputs with 100 randomly phase-shifted periodic inputs. A similar pattern of results was observed for both the full dendritic model, and the single-compartment soma.

3.1.   Modeling the Effects of S and P

To better understand the relations between our S and P parameters and output spike rates, we studied the effects of S and P on the mean and variability of the somatic voltage trace in the post-synaptic cell, a crude global parameterization of post-synaptic current injection that could be more intuitively related to output spike rates. We observed nearly opposite effects of S and P on these biophysical intermediaries, where increases in P led to substantially larger, smoother post-synaptic voltage traces, while increases in S led to marginally smaller, but much more variable voltage traces. These influences were largely explained by a simple RC-circuit model analyzed in response to pairs of conductance pulses of variable aspect ratio and inter-pulse separation (figs. 3 and 4).

In brief, we found that the key consequence of making input trains more regular (periodic), was the elimination of the very brief ISI's which predominate in random spike trains. As plotted in fig. 3, closely spaced conductance pulses led to reduced post-synaptic charge injection (and hence lower mean voltages), since the second pulse rides on the voltage tail of the first, and hence "suffers" a reduced driving force. A second, mechanistically different cost associated with short ISI's occurred as a result of synaptic receptor saturation, where two closely spaced release events produced significantly less integrated post-synaptic conductance than two pulses separated by an interval greater than the conductance decay time constant.

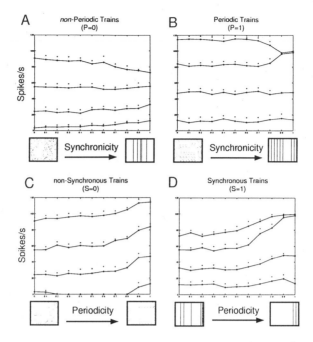

Figure 2. Experiment 1: Synchronicity effects for (A) P=0 and (B) P=1, and periodicity effects for (C) S=0 and (D) S=1.

Coupled with the observed increase in post-synaptic voltage mean, the elimination of short ISI's that occurred as spike trains were made increasingly periodic led to a reduction in post-synaptic voltage *variability*, as the large spurious voltage excursions associated with randomly timed inputs were replaced by regularly spaced voltage events of uniform size (fig. 4).

With respect to the consequences of *synchronizing* input trains, post-synaptic driving-force saturation accounted for reductions in post-synaptic current injection that varied in size from negligible (for weak synapse conditions) to moderate (for strong synapse conditions). The model in fig. 3C illustrates the suppressive nature of the effect. Finally, in contrast to the variability reduction seen with increasing spike train regularity, synchronizing input trains led to increases in post-synaptic voltage variability, as schematized in fig. 4.

In summary, the different effects of S and P on output spike rates could be explained in terms of their differential effects on post-synaptic voltage mean and variability. Since P acted primarily to boost steady-state levels of post-synaptic current injection, it was consistently associated with boosts in output spike rates. In contrast, the large increases in post-synaptic voltage *variability* produced by increases in input synchronization produced inconsistent changes in spike rates, including (i) modest boosts in weak synapse conditions, where larger fluctuations were needed to generate threshold crossings, and (ii) stimulus locking conditions in cases of strong synaptic inputs, which could either boost or depress output spike rates depending on parameters.

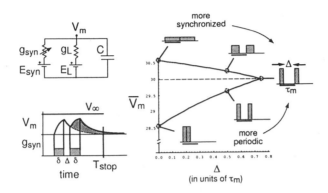

Figure 3. A single-compartment RC "soma" is presented with pairs of conductance pulses. Random spike trains contain many short ISI's, which are gradually eliminated as trains are made more regular. Lower curve in graph shows growth of mean somatic voltage V_{soma} (computed over the interval T) as the inter-pulse interval is increased. As multiple spike trains are synchronized, conductances are progressively "bunched up" in time. Upper curve shows predicted drop in V_{soma} with increasing synchronization.

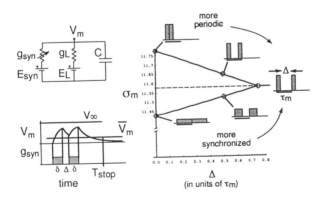

Figure 4. Simple analytical model as in fig. 3, but plot is of σ_{soma} as pulse shape and inter-pulse interval are varied. Upper curve in graph shows drop in voltage variability σ_{soma} (computed over the interval T) as the inter-pulse interval is increased. Lower curve shows model for growth in σ_{soma} as inputs are increasingly synchronized.

4. DISCUSSION

Our experiments currently have a number of limitations. First, only excitatory synapses were used; the "balancing" effects of inhibition have been shown elsewhere to increase a neuron's sensitivity to input variability [2]. Second, we have used a highly simplified model for synaptic physiology that assumes a constant, high release probability, complete saturation of receptors for every release event, and no paired-pulse (or higher order) inter-spike interactions that modulate release probability or magnitude at individual synapses. Different assumptions for any of these conditions could alter our results. Third, our simple, modified Hodgkin-Huxley model for spike generation lacks currents that produce spike-train adaptation and other dynamics often observed in cortical neuronal responses. Fourth, we have only considered the transduction behavior of an isolated neuron, free of the influences of the network in which it normally operates. Many of the interesting dynamical features of neuronal responses are likely due to network effects, including feedback excitation within a cortical column, and feedforward and feedback inhibition between and among neighboring columns.

Perhaps the most interesting conclusion of this work lies in the possibility that regulation of spike train statistics, in particular the transition from random to regular spike trains (as has been observed to occur in cerebral cortex [6]), could in effect be a novel, "covert" mechanism for manipulating the effectiveness of those neuronal outputs onto their post-synaptic targets—without altering mean firing rates.

REFERENCES

[1] Abeles, M. (1991). *Corticonics – Neural circuits of the cerebral cortex*. Cambridge University Press.

[2] Bell, A., Mainen, Z., Tsodyks, M., and Sejnowski, T. J. (1995). "Balancing of conductances may explain irregular cortical spiking" INC-9502, Institute for Neural Computation, UCSD, San Diego, CA.

[3] Bernander, O., Koch, C., and Douglas, R. J. (1994). "Amplification and Linearization of Distal Synaptic Input to Cortical Pyramidal Cells" *J. Neurophys.*, 72(6), 2743–2753.

[4] Crick, F., and Koch, C. (1990). "Towards a neurobiological theory of consciousness" *Seminars in the Neurosciences*, 2, 263–275.

[5] Destexhe, A., Mainen, Z., and Sejnowski, T. (1994). "An efficient method for computing synaptic conductances based on a kinetic model for receptor binding" *Neural Computation*, 6, 14–18.

[6] Gray, C., and Singer, W. (1989). "Stimulus-specific neuronal oscillations in orientation columns of cat visual cortex" *Proc. Nat. Acad. Sci., USA*, 86, 1698–1702.

[7] Murthy, V. N., and Fetz, E. E. (1994). "Effects of input synchrony on the firing rate of a 3-conductance cortical neuron model" *Neural Computation*, 6(6), 1111–1126.

[8] Niebur, E., and Koch, C. (1994). "A model for the neuronal implementation of selective visual attention based on temporal correlation among neurons" *Journal of Computational Neuroscience*, 1(1), 141–158.

[9] Niebur, E., Koch, C., and Rosin, C. (1993). "An oscillation-based model for the neural basis of attention" *Vision Research*, 33, 2789–2802.

[10] Usher, M., Stemmler, M., and Olami, Z. (1995). "Dynamic pattern formation leads to 1/f noise in neural populations" *Physical Review Letters*, 74(2), 326–329.

23

THE DENDRITIC ORIGINS OF FAST PREPOTENTIALS IN PYRAMIDAL CELLS

Mark M. Millonas and Philip S. Ulinski

James Franck Institute and Department of Organismal Biology and Anatomy
The University of Chicago
Chicago, IL

ABSTRACT

We used compartmental modeling of a pyramidal cell to examine the hypothesis that prepotentials originate in hot spots or localized regions of high sodium and potassium conductances in the dendritic tree. The full range of experimentally observed prepotentials could be reproduced using compartmental models by incorporating hot spots containing fast sodium conductances and delayed rectifier potassium conductances in densities that were high relative to conductance densities estimated in labs where prepotentials were not observed.

1. INTRODUCTION

The visual cortex of freshwater turtles contains pyramidal cells that show a regular spiking (Connors et al., 1982; Connors and Gutnick, 1990) pattern of large amplitude, overshooting action potentials following intrasomatic current injections (Connors and Kriegstein, 1986). An interesting feature of these cells is they also show small amplitude action potentials or "spikelets". Spikelets are abolished by tetrodotoxin, implicating sodium conductances in their generation. Collision experiments involving antidromic activation of pyramidal cells suggest that spikelets are generated at sites remote from the soma. Connors and Kriegstein (1986) noted the resemblance of spikelets to the fast prepotentials described in hippocampal pyramidal cells by Spencer and Kandel (1961) and followed them in hypothesizing that spikelets are generated in dendritic "hot spots" situated some distance from the soma. Evidence that action potentials can originate in the dendritic tree and propagate to the soma has been obtained for hippocampal pyramidal cells (Turner et al., 1989, 1991) and neocortical pyramidal cells in the visual (Hirsch et al., 1995), somatosensory (Amitai et al., 1993) and motor (Regehr et al., 1993) cortices. For a more complete discussion and list of references see Millonas and Ulinski (in press). It is now clearly established that dendritic action potentials can also propagate in a retrograde direction into the dendritic tree in hippocampal and neocortical pyramidal cells (Stuart and

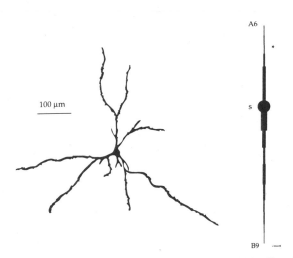

Figure 1. Compartmental model. A. Camera lucida drawing of the pyramidal cell used to construct the compartmental model. B. Reduced compartmental model obtained as described in the text. The sixth apical dendritic compartment (A6), the soma compartment (S) and the ninth basal dendritic compartment (B9) are labeled in this and following illustrations. Active conductances were in many cases incorporated in the third or fourth apical compartment as indicated by an asterisk.

Sakmann, 1994), but the circumstances under which retrograde versus orthograde propagation occurs and the significance of either occurrence for the integrative physiology of pyramidal cells remains ambiguous.

2.   MATERIALS AND METHODS

Simulations with the reduced compartmental model shown in Fig. 1 show that the full range of prepotentials observed in real turtle pyramidal cells can be generated by incorporating localized dendritic hot spots with relatively high densities of both fast sodium and delayed rectifier potassium conductances in the dendritic tree.   Fast sodium and delayed rectifier conductances were present on the soma (S) compartment for all simulations and included on specific dendritic compartments (*) for many simulations.   These active conductances were specified by Hodgkin-Huxley equations. Kinetic parameters were modified from the values used by Traub (1982) in his model of rat hippocampal pyramidal cells. The densities of the sodium and potassium conductances, $\bar{g}_{Na}$ and $\bar{g}_K$, in the soma compartment were constrained by altering the densities until the waveform of the first action potential in the model could be precisely superimposed on the first action potential recorded in the real pyramidal cell for depolarizing current pulses ranging from + 0.1 to +1.0 nA. Specific details will appear separately (Millonas and Ulinski, submitted).   No attempt was made to include the active conductances responsible for spike adaptation in pyramidal cells. The densities used in the soma compartment were $\bar{g}_{Na} = 370$ mS/cm^2 and $\bar{g}_K = 250$ mS/cm^2. Sodium and potassium conductances with the same kinetic parameters were incorporated into dendritic compartments in some simulations.   The conductance densities were varied between simulations in these cases.

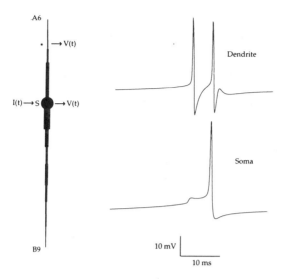

Figure 2. Simulated prepotential. Fast sodium and delayed rectifier conductances with densities of 370 mS/cm^2 and 250 mS/cm^2, respectively, were added to the fourth apical dendritic compartment . A 100 msec, +0.3 nA current pulse was injected into the soma compartment. Voltage responses in the fourth dendritic compartment (top) trace and soma compartment (bottom trace) were simulated. Notice that the somatic voltage response shows a prepotential that appears as a rounded shoulder prior to the generation of the large action potential. See text for detailed description of the simulations.

3. RESULTS

Fast prepotentials were obtained when active conductances were incorporated into specific dendritic compartments. Figure 2 shows the results obtained by incorporating both sodium and potassium conductances in apical compartment 4 of the model. Channel densities were $\bar{g}_{Na} = 370$ mS/cm^2 and $\bar{g}_K = 250$ mS/cm^2, the same as those in the soma compartment. The voltage transient in the soma compartment (bottom trace) consists of a low amplitude fast prepotential that is followed by a large amplitude action potential and, ultimately, a hyperpolarization. The same intrasomatic current pulse produced a more complex voltage transient in the dendritic compartment (top trace). It included two depolarizing events with relatively large amplitudes and two hyperpolarizing phases. The depolarizing events are regenerative action potentials that are dependent upon the fast sodium conductance in the dendritic compartment. The hyperpolarizing events are due to the delayed rectifier conductance in the dendritic compartment. Comparison of the two traces indicates that the first action potential in the dendritic compartment is initiated prior to the onset of the prepotential in the somatic compartment. Although prepotentials in the soma are triggered by events in the dendritic tree, they involve activation of both the somatic and dendritic spike generating mechanisms.

These simulations demonstrate that fast prepotentials are consistent with the existence of a dendritic hot spot containing both sodium and potassium conductances that is situated some distance from the soma. The relationship between the shapes of prepotentials and parameters in the compartmental model was next examined by systematically varying conductance density and position of the hot spot in the reduced compartmental model. Varying

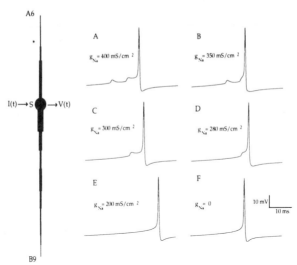

Figure 3. Effect of varying the density of sodium conductance in the hot spot. Simulations shown in this figure involved recording the somatic voltage response produced by intrasomatic current injections of +0.3 nA. The fourth apical dendritic compartment had a hot spot with a delayed rectifier potassium conductance that was held constant at 250 mS/cm². A - F. Somatic voltage responses produced with hot spot sodium conductances of 400 mS/cm² (A) through 0 mS/cm² (F). See the text for detailed description of the results.

the density of the fast sodium conductance in apical compartment 4 in the reduced compartmental model is typical of manipulations in other compartments in showing that the timing, amplitude and frequency of prepotentials in the somatic compartment are dependent upon the density of sodium conductance in the hot spot (Fig. 3). The potassium conductance was held constant at 250 mS/cm² in these simulations. Removing the sodium conductance from the hot spot results in a simple, large amplitude action potential in the somatic compartment (Fig. 3F). A sodium conductance of 280 mS/cm² results in a fast prepotential consisting of a shoulder that precedes the large amplitude action potential (Fig. 3D). Increasing the sodium conductance to 350 mS/cm² results in a low amplitude prepotential and a shoulder that grades into the large amplitude action potential (Fig. 3B). Further increasing the sodium conductance density produces a pair of prepotentials (Fig. 3A).

Varying the density of the potassium conductance in the hot spot demonstrates that prepotentials occur only within a restricted range of potassium conductance densities (Fig. 4). Sodium conductance was held constant at 350 mS/cm² in these simulations. Low (Fig. 4A) or high (Fig. 4F) potassium conductance densities in the hot spot resulted in large amplitude action potentials without prepotentials in the somatic compartment. However, potassium conductance densities of 200 - 300 mS/cm² produced distinct prepotentials in the somatic compartment with the exact density of the potassium conductance determining the timing of the prepotential relative to the large amplitude action potential (Fig. 4C,D).

4.   CONCLUSIONS

These simulations demonstrate that the range of prepotentials observed in turtle pyramidal cells could result from hot spots situated at different loci on the dendritic tree and containing different densities of fast sodium and potassium conductances. It is not necessary

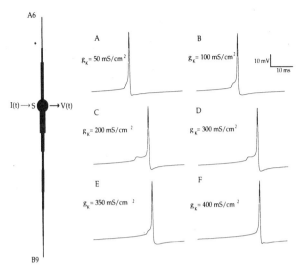

Figure 4. Effect of varying the density of the potassium conductance in the hotspot. Simulations shown in this figure involved recording the somatic voltage response produced by intrasomatic current injections of +0.3 nA. The fourth apical dendritic compartment had a hot spot with a fast sodium conductance that was held constant at 370 mS/cm^2. A - F. Somatic voltage responses produced with hot spot potassium conductances of 50 mS/cm^2 (A) through 400 mS/cm^2 (F). See the text for detailed description of the results.

to assume a wide range of conductance densities or locations since only small changes in the timing can produce the whole range of observed behaviors. However, it is necessary to assume a localized distribution of the conductances with densities on the order of those found in the soma. A detailed description of these results, and an analysis of the integrative implication of dendritic hot spots will appear shortly (Millonas and Ulinski, in press).*

REFERENCES

[1] Amitai, Y, Friedman, A, Connors, BW and Gutnick, MJ (1993) Regenerative activity in apical dendrites of pyramidal cells in neocortex. Cerebral Cortex 3: 26 - 38.

[2] Connors, BW, Gutnick MJ and Prince DA (1982) Electrophysiological propertes of neocortical neurons *in vitro*. Journal of Neurophysiology 48: 1302 - 1320.

[3] Connors BW, and Gutnick MJ (1990) Intrinsic firing patterns of diverse neocortical neurons. Trends Neuroscience 13: 99 - 104.

[4] Connors BW and Kriegstein AR (1986) Cellular physiology of the turtle visual cortex: distinctive properteis of pyramidal and stellate neurons. Journal of Neuroscience 6: 164 - 177.

[5] Hirsch JA and Gilbert CD (1991) Synaptic physiology of horizontal connections in the cat's visual cortex. Journal of Neuroscience 11: 1800 - 1809.

[6] Millonas MM and Ulinski PS, Dendritic origins and integrative implications of fast prepotential in pyramidal cells from turtle visual cortex (submitted).

[7] Regehr W, Kehoe, J, Ascher P and Armstrong C (1993) Synaptically triggered action potentials in dendrites. Neuron 11: 145 - 151.

[8] Spencer WA and Kandel ER (1961) Electrophysiology of hippocampal neurons IV. Fast prepotentials. Journal of Neurophysiology 24: 272 - 285.

[9] Stuart G and Sakmann B (1994) Active propagation of somatic action potentials into neocortical pyramidal cell dendrites. Nature 367: 69 - 72.

*Address Correspondence to: Dr. Mark Millonas, James Franck Institute, 5640 S. Ellis Ave., The University of Chicago, Chicago, IL 60637.

[10] Traub R (1982) Simulation of intrinsic bursting in CA3 hippocampal neurons. Neuroscience 7: 1233 - 1242.

[11] Turner RW, Meyers DER and Barker JL Localization of tetrodotoxin- sensitive field potentials of CA1 pyramidal cells in the rat hippocampus. (1989) Journal of Neurophsyiology 62: 1375.

[12] Turner RW, Meyers DER, Richardson TL and Barker JL (1991) The site for initiation of action potential discharge over the somatodendritic axis of rat chippocampal CA1 pyramidal neurons. Journal of Neuroscience 11: 2270 - 2280.

24

ENTRAINMENT OF A SPIKING NEURON MODEL TO TWO PERIODIC PULSE TRAINS

Hirofumi Nagashino and Yohsuke Kinouchi

Department of Electrical and Electronic Engineering
Faculty of Engineering
The University of Tokushima
Tokushima 770 Japan
E-mail: nagasino@ee.tokushima-u.ac.jp
E-mail: kinouchi@ee.tokushima-u.ac.jp

ABSTRACT

In multifrequency coordinated rhythmic movements in humans, the frequency of voluntary rhythm is entrained by periodic external force so that the frequency ratios of the voluntary rhythm to the external force are rational numbers in low order. We propose a conceptual model in the neural network level that generates such synchronization. In the model, a spiking neuron receives two impulse trains, whose frequencies are different, as the input from the central nervous system and the external force. The ratios of output frequency of the neuron to the external frequency are rational numbers in low order.

1. INTRODUCTION

The nervous system in vertebrates and invertebrates processes information employing impulse trains. Therefore it is important to elucidate the basic mechanisms of how the impulse trains are transformed through neuronal processing.

It has been reported that in multifrequency coordinated rhythmic movements in humans, the frequency of voluntary rhythm is entrained by periodic external force so that the frequency ratios of the voluntary rhythm to the external force are rational numbers in low order [1], which is below referred to as $n : m$ synchronization, where n and m are positive integers. In this behavior two input pulse trains are considered to be incorporated in the neural network level, from the central nervous system and as external signal. To clarify its underlying mechanism it is necessary to investigate the integration property of the two inputs in a neuron.

The characteristics of firing of a neuron has been investigated by physiological experiments and theoretical modeling when a periodic impulse train activates a neuron.

Computational Neuroscience
edited by Bower, Plenum Press, New York, 1997

It has been shown that in both experimental and theoretical approaches the average firing rate of the neuron behaves like Cantor functions of the intensity or interval of the input pulse [2]-[6]. The response characteristics of neuron models, however, has not been shown to two pulse input trains with different frequencies.

We propose a conceptual model in the neuronal network level that generates $n : m$ synchronization described above. In the model, a spiking neuron receives two periodic impulse trains, whose frequencies are different, as the input from the central nervous system and the external force, respectively.

The neuron model receives the input pulse train from the central nervous system and outputs a impulse train that gives the frequency of the voluntary rhythmic movement. It receives another pulse train as external signal that has the different frequency from that of the central input. The output of the neuron model ia affected by the external input. Over a relatively wide range of the period of the external input the model is entrained to it in such a way that the frequency ratios of the external input to the output of the neuron are rational numbers in low order.

2. A MODEL FOR MULTIFREQUENCY RHYTHM

A number of spiking neuron models have been propsed so far [7]-[9]. In the present work we formulate the neuron model in the following way to express the function of a neuron as simply as possible. Let u_k and v_k denote the membrane potential and the threshold of neuron k, respectively; The dynamics of u_k is expressed as

$$\tau_u \frac{du_k}{dt} = -u_k + E \sum_{i=0}^{\infty} \delta[t - iT_c] + P \sum_{i=0}^{\infty} \delta[t - iT_p], \tag{1}$$

where T_c and E are the period and intensity of the central input, T_p and P are the period and amplitude of the external force, τ_u is the time constant of the membrane potential. The output of the neuron y_k is expressed as

$$y_k = \delta[1(u_k - v_k) - 1], \tag{2}$$

where $\delta[\cdot]$ is Dirac's impulse function and $1(\cdot)$ is the unit step function. This equation means that an output impulse is elicited when the membrane potential exceeds the threshold.

The relative refractory period is modeled by the change of the threshold v_k driven by the output of the neuron y_k as

$$\tau_v \frac{dv_k}{dt} = -v_k + ry_k, \tag{3}$$

where τ_v is the time constant of the threshold and r ia a constant.

The output impulse is assumed to reset the central input, that is next central input impulse comes out T_c later than the moment of the output impulse.

Two neurons are incorporated in the model; one is associated with an extensor and the other with a flexor. For the former $k = e$ and for the latter $k = f$. The impulse trains for the extensor and those for the flexor have the phase difference of half the period in the central input as well as in the external force. Eq. (1) is rewritten as

$$\tau_u \frac{du_k}{dt} = -u_k + E \sum_{i=0}^{\infty} \delta[t - (i - \frac{j}{2})T_c] + P \sum_{i=0}^{\infty} \delta[t - (i - \frac{j}{2})T_p], \tag{4}$$

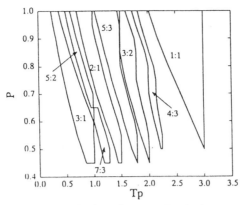

Figure 1. Regions of $n : m$ synchronization.

where $k = e, f$, $j = 0$ with $k = e$ and $j = 1$ with $k = f$.

The intensity of the impulses from the central input, E, and the constant for the threshold dynamics, r, are set to the values in which without external force the neuron fires once to every central input impulse. In the simulation the following constants were used; $\tau_u = 1$, $\tau_v = 2$, $E = 1$, $T_c = 3$, $r = 4$.

3. RESULTS

Characteristic simulation results are shown in the following. In the simulation, the frequency of the external input was made larger than the that of the central input as in the experiments in [1].

Fig. 1 shows the regions in the plane of the intensity P and the period T_P of the impulses of the external force where the frequency ratios of the external input to the output are rational numbers in relatively low order. When the intensity of the impulses of the external force P is too large, the neurons are entrained completely to the external force. When P is too small, they fire mostly by the central input. For the appropriate range of P, the neurons are entrained mostly in such a way that the ratios of the external frequency of the neuron to the output frequency are rational numbers in low order such as 1:1, 2:1, 3:1, 3:2 and so on.

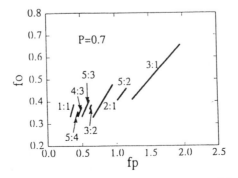

Figure 2. The relationship between the output frequency f_o and the frequency of the external input f_p, $P = 0.7$.

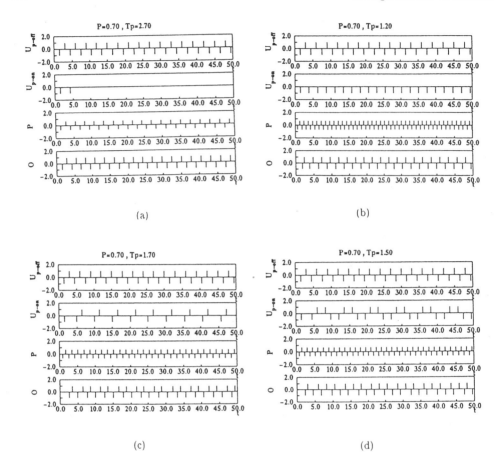

Figure 3. Simulation Results, $P = 0.7$. (a) $f_p : f_o = 1 : 1$, (b) $f_p : f_o = 2 : 1$, (c) $f_p : f_o = 3 : 2$, (d) $f_p : f_o = 5 : 3$.

Fig. 2 shows the relationship between the output frequency ratio f_o and the frequency of the external input f_p in which $n : m$ synchronization in low order occurs. It is observed that the output frequency tends to become higher and higher as the frequency of the external input increases.

Some runs of time series are shown in Fig. 3. First row U_{p-off} shows the central input in case that there is no external input. Second row U_{p-on} shows the one in case that the external input (third row P) is added to the neuron models and the central input is reset by output of the neuron models (fourth row O). In the first to fourth rows upper lines to the horizontal line denote the impulses that corresponds to the extensor and lower lines to the flexor. Fifth and sixth rows are the waveforms of the joint angles θ_1 and θ_2 respectively. It is assumed that the impulses of the external force are applied to the neurons at the moments when the joint angle θ_1 goes across the value of 0.

When $f_p : f_o = 1 : 1$, the output impulses are elicited only from the external input except some initial period. The central input is reset at every output impulse so that no central input impulses are applied. In the example illustrated here when $f_p : f_o = 2 : 1$, an output impulse is elicited corresponding to every two impulse of the external force for the extensor, whereas it is synchronized by the central input for the flexor, so that the phase difference between the outputs of the extensor side and flexor side is not equal to π. In the example of $f_p : f_o = 3 : 2$, the output impulse synchronized by the central input and that by the external input are alternately elicited for both the extensor and flexor, which does not give the output impulse train of regular intervals. Assymetric output for the extensor and flexor above is caused by the reset of the central input. In the example of $f_p : f_o = 5 : 3$, two output impulses synchronized by the central input and the subsequent output impulse synchronized by the external input are elicited repeatedly for both the extensor and flexor.

Let us notice the relationship between the external impulses and output impulses in the examples in Fig. 3. For $f_p : f_o = 3 : 2$, the sequence of the pair of the number of the external impulses and the number of the output impulses is (1:1) + (2:1). For $f_p : f_o = 5 : 3$, it is (1:1) + (1:1) + (2:1). In this way, the sequence of the response found in the simulation is the combination of (n:1) and ($n + 1$:1). Although such characteristics have been reported in the case of single pulse train input [2], the output pulse intervals in two pulse train input are closer to regular ones than in the single input.

4. CONCLUSIONS

We have shown that a spiking neuron model that receives two pulse trains with different frequencies from the central nervous system and as external input are entrained in such a way that the frequency ratios of the external input to the output of the neuron are rational numbers in low order. Such relationship is accomplished with more complicated phenomena caused by the mixed influence of two pulse trains than the case for a single pulse train. There are various types of synchronization, that only to the external force, only to the central input and complex entrainment to the external force and the central input.

Simulation was made of the case where some amount of noise input is added to the neuron and of the model in which the two neurons are coupled with mutual inhibition. Qualitatively similar characteristics are observed to those of the model described above.

ACKNOWLEDGMENTS

The authors wish to thank Prof. J. A. Scot Kelso and Dr. Gonzalo C. deGuzman for their invaluable discussion.

REFERENCES

[1] G. C. deGuzman and J. A. S. Kelso, Multifrequency behavioral patterns and the phase "attractive circle map," Biological Cybernetics, 64:485-495, 1991.

[2] J. Nagumo and S. Sato, On a response characteristic of a mathmatical neuron model, Kibernetik, 10:155-164, 1972.

[3] G. B. Ermentrout, n:m phase locking of weakly coupled oscillators, J. Math. Biol., 12:327-342, 1981.

[4] T. Kiemel and P. Holmes, A model for the periodic synaptic inhibition of a neuronal oscillator, IMA J. Math. Appl. Med. Biol., 4:145-169, 1987.

[5] T. Nomura, S. Sato, S. Doi, J. P. Segundo and M. D. Stiber, Global bifurcation structure of a Bonhoeffer van der Pol oscillator driven by periodic pulse trains, Biological Cybernetics, 72:55-67, 1994.

[6] S. Doi and S. Sato, The global bifurcation structure of the BVP neuronal model driven by periodic pulse trains, Mathematical Biosciences, 125:229-250, 1995.

[7] R. F. Reiss, A theory and simulation of rhythmic behavior due to reciprocal inhibition in small nerve nets, Proc. AFIPS Spring Joint Computer Conference, 171-194, 1962.

[8] J. P. Segundo, D. H. Perkel, H. Wyman, H. Hegstad and G. P. Moore, Input-output relations in computer simulated nerve cells, Kybernetik, 4:157-171, 1968.

[9] H. Nagashino, H. Tamura and T. Ushita, Existence and control of rhythmic activities in reciprocal inhibition neural networks, Trans. IECE Japan, E62:768-774, 1979.

25

DISCRIMINATION OF PHASE-CODED SPIKE TRAINS BY SILICON NEURONS WITH ARTIFICIAL DENDRITIC TREES

David P. M. Northmore and John G. Elias

Departments of Psychology and Electrical Engineering
University of Delaware
Newark, Delaware 19716

1. ABSTRACT

Artificial neurons with dendritic trees, modeled in VLSI, were used to evaluate large numbers of different patterns of synaptic connections onto dendrites. Random search yielded connections that produced well-differentiated responses to input spike patterns that differed only in phasing. Feedforward networks of up to 3 layers exhibited increasing variability that can be exploited for discrimination purposes.

2. INTRODUCTION

Parallel pathways in sensory systems typically encode data in both the spatial distribution of spiking activity and in the cross-fiber temporal patterning. Recently we have shown how such spatiotemporal spike patterns can be discriminated by networks of silicon neuromorphs with artificial dendritic trees [2,6]. An error-correcting delta rule was implemented by weight changes achieved by moving synapses along the branches of the dendritic tree. This approach, similar to that commonly used with artificial neural networks, successfully trained a network to discriminate spike inputs coded by spatial patterns and by average frequency, and to generate spike responses coded in similar terms. However, this "Perceptron approach" is biologically simplistic in that neurons in the brain also convey information in the temporal pattern of spike-firing (e.g. [7]). Here we explore one of many possible temporal codes [1], focusing on the decoding of information contained in the relative phasing of parallel spikes trains. Our aim is show how phase differences in the millisecond range can be discriminated in a biologically plausible way using the properties of passive dendritic trees, as implemented in our silicon neuromorphs.

3. HARDWARE AND METHODS

VLSI techniques (2 μm CMOS double-poly n-well process on a 2x2 mm MOSIS Tiny Chip format) were used to fabricate the silicon neuromorphs, each comprising an artificial dendritic tree (ADT) and an integrate-and-fire soma [2]. In the present work, all neuromorphs had a dendritic tree with four primary branches shown schematized in Fig. 1. A branch consisted of a series of 32 identical compartments, each with membrane capacitance, two programmable resistors (R_m and R_a) representing membrane and axial cytoplasmic resistances, and transistors emulating an excitatory and an inhibitory synapse. The excitatory synapse depolarizes the compartment, while the inhibitory synapse hyperpolarizes it. The synapses, when activated by a 50 ns pulse, generate impulse responses at the soma whose amplitude and latency-to-peak depend upon synapse proximity. In these neuromorphs, all synapse on-conductances are fixed and sufficiently high to drive the potential of the activated compartment almost to the supply rails. (In newer chips, the on-conductance of each synapse is programmable.) The impulse responses are summed at the soma by a leaky RC integrator, and when the result exceeds a threshold (V_{th}), an output spike is fired and the integrator is discharged. The value of V_{th} is set by the frequency ratio of spikes impinging on two special soma synapses, one of which ("upper") raises V_{th}, the other ("lower") reduces V_{th} [3]. A neuromorph's output spikes can be routed via high speed digital circuitry to an arbitrary set of synapses in a network with a timing accuracy of 0.1 msecs. Separate programmable delays can be imparted to the spikes travelling on each pathway, although the present results were all obtained with nominally zero conduction delays.

In this work, excitability of each neuromorph was regulated by a single, feedback connection to the upper soma synapse; all lower synapses were driven at 20 spikes/sec (see Fig. 1). With this arrangement, a sustained excitatory train applied to the dendrite elicits from the soma a transient ON excitation, and OFF inhibition, each lasting about 0.5 secs.

The input stimulus patterns consisted of three concurrent spike trains of constant frequency (250/sec) lasting 40 msecs. Different input patterns were constructed by phase shifting the 2nd and 3rd train with respect to the 1st by steps of 0,1,2 and 3 msecs. The sixteen input patterns shown in Fig. 2 represent a complete set of phase relations between the three trains. Thus, the patterns did not differ in frequency or spike number, but only in phase shifts, which were small compared to the dendritic "membrane" time constant (20–100 msecs) set by R_m.

4. RESULTS

Figure 1 represents a neuromorph with a discriminative set of synaptic connections to its four-branch dendritic tree. Three-train input patterns were simultaneously applied to

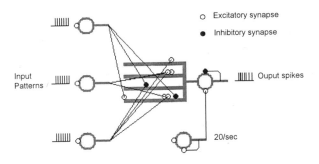

Figure 1. Input patterns, each consisting of three concurrent spike trains, are input to relay units that distribute the spikes to excitatory and inhibitory synapses on the 4-branch dendritic tree. Soma excitability is regulated by feedback to the upper soma synapse and a tonic 20/sec train to the lower soma synapse.

relay units that connect to excitatory and inhibitory synapses at the positions shown on the tree. The connections were found by random search in which each relay unit was allowed to make up to four synaptic connections anywhere on the tree. With this particular connection pattern, the 16 input trains stimulated the neuromorph to generate highly differentiated spike responses, as shown by the peristimulus time histograms in Fig. 2. These responses varied principally in the relative numbers of ON and OFF spikes, but there were also differences in latencies of both ON and OFF responses, as well as differences in histogram profile. To get a quick and simple indication of the discriminative ability of a set of connections, the variability of histogram shape across the 16 input patterns was quantified by a sum-of-squares measure, the "response variance".

As another approach to assessing connections, we measured the average information transmitted by each. The 16 stimuli were presented 20 times to generate a set of response histogram templates. The 16 stimuli were then presented anew, and each response was classified as one of the 16 inputs from its best matching template. From 80 repetitions of the stimulus, we obtained a joint probability distribution (P(S,R)), and a measure of average transmitted information in bits. A number of randomly generated sets of connections were screened by computing the response variances. Figure 3 (circles) shows results for 8 sets of connections onto a single neuromorph that spanned the range of observed response variances. The average information transmitted by each connection set was well correlated (0.98) with response variance, the latter being much quicker to calculate.

The connection set of Fig. 1 conveyed 2.1 bits of information about the stimuli and was among the best found, a success attributable to convergence of the different input trains onto the same dendritic loci (see Discussion).

To study how phase-coded inputs are discriminated by feedforward multilayer networks, the same 16 input patterns were input to networks with two or three layers of neuromorphs, each layer having three neuromorphs. As before, every neuromorph made up to four randomly chosen synapses on the dendrites of the next layer. In a study of 230 different randomized three-layer networks, the average response variances for neuromorphs in successive layers was 5.82, 9.64, and 12.04, suggesting a variance "amplification". In the most discriminating networks, neuromorph firing in the output layer would be highly differentiated, for example, giving an ON burst to only two or three of the 16 input patterns.

The squares in Fig. 3 show the relationship between the average transmitted information and the response variance of neuromorphs in a second layer. The connections onto the first layer were high-variance connections that were not changed; the connections onto

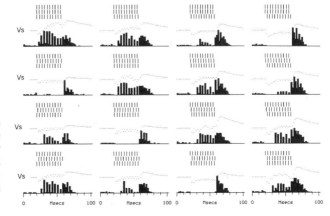

Figure 2. Sixteen panels each showing: phase-shifted input spike patterns, and responses generated by the connections of Fig. 1. Vs is soma voltage. Histograms show neuromorph output spikes accumulated over 50 trials.

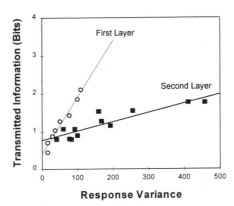

Figure 2. Average transmitted information as a function of response variance for neuromorphs in the first and second layers of feedforward networks with randomized connections.

the second layer were varied randomly, yielding responses with up to five times the variance of the first layer. Average transmitted information at the second layer output was lower than at the first, although still well correlated (0.90) with response variance.

5. DISCUSSION AND CONCLUSIONS

Where inputs differ in spatial and temporal components, as they did for example in a study of discrimination of spike trains representing speech sound waveforms, randomized connections to ADTs can provide partial separation of input patterns that are easily exploited by a single-layer perceptron [6]. In the present work, the input patterns were more of a challenge to discriminate as they differed only in temporal patterning, and then only in phase shifts that were considerably less than the membrane time constant. Small time delays between spikes originating in two different trains are decoded by the dendritic tree as follows. If two excitatory inputs arrive close together on the dendrite and within a few milliseconds of each other, they combine their effects sublinearly due to dendritic saturation [4,5]. However, if the same two excitatory inputs arrive on spatially separated sites, they sum their effects linearly. The degree of summation determines the amplitude of the postsynaptic potentials appearing at the soma, and hence the soma's firing rate. The temporal patterning of this firing depends on dendritic delays, and on the transient response of the soma due to the negative feedback to the soma threshold. The result is that subtle differences in the temporal patterning of the stimuli give rise in the neuromorph's spike response to substantial differences in latency, ON vs. OFF firing, and other temporal patterning features, as well as differences in spike number. Subsequent stages of dendritic processing tend to amplify these differences.

The information theoretic analysis indicated that response variance at any one layer correlates with information transmitted through that layer. However, the amplification of response variance with ascending layers occurs with a loss of average information in the output of individual neuromorphs. While part of the reason is noise accumulation, the average information measure is also diminished for neuromorph responses that exhibit high stimulus selectivity.

Dendritic nonlinearities and delays make it difficult to find algorithms for training neuromorphic, or indeed real neuronal networks to perform discrimination tasks. Although established algorithms such as the delta rule can be implemented with our neuromorphs [6], they neglect much of the power of dendritic processing. The present results, showing the amplification of response variance, suggest a biologically plausible method for developing discriminative networks by making random feedforward connections and eliminating those connections onto neurons that least vary their responses to stimulation. After a few such feedforward stages, the highly differentiated responses that emerge can readily be associated with a desired response using Hebbian or other learning principles.

6. ACKNOWLEDGMENTS

Supported by NSF Grants BCS-9315879, and BEF-9511674.

7. REFERENCES

1. Cariani, P. (1994) As if time really mattered: Temporal strategies for neural coding of sensory information. In *Origins: Brain and Self-organization*, Ed. K. Pribram. Lawrence Erlbaum, Hillsdale, NJ. Pp 208–252.
2. Elias, J.G. and D.P.M. Northmore (1995) Switched-capacitor neuromorphs with wide-range variable dynamics. IEEE Trans. Neur. Net. 6:1542–1548.
3. Elias, J.G., D.P.M. Northmore and Westerman, W. (1997) An analog memory circuit for spiking silicon neurons. Neural Computation, in press.
4. Ferster, D. and Jagadeesh, B. (1992) EPSP-IPSP interactions in cat visual cortex studied with in vivo whole-cell patch recording. J. Neurosci. 12:1262–1274.
5. Northmore, D.P.M. & J.G. Elias (1996) Spike train processing by a silicon neuromorph: The role of sublinear summation. Neural Computation, 8:1245–1265.
6. Northmore, D.P.M. & J.G. Elias (1997) Spatio-temporal spike pattern discrimination by networks of silicon neurons with artificial dendritic trees. In *Spatiotemporal Models in Biological and Artificial Systems*, edited by F.L. Silva et al., IOS Press, Amsterdam. Pp 136–143.
7. Optican, L.M. and Richmond, B.J. (1987) Temporal encoding of two-dimensional patterns by single units in primate inferior temporal cortex. III. Information theoretic analysis. J. Neurophysiol. 57: 162–178.

26

CHOLINERGIC MODULATION OF INHIBITORY SYNAPTIC TRANSMISSION IN THE PIRIFORM CORTEX

Madhvi Patil,[*] Christiane Linster, and Michael Hasselmo

Department of Psychology
Harvard University
33 Kirkland St. Cambridge, Massachusetts 02138
patil@katla.harvard.edu
linster@katla.harvard.edu
hasselmo@katla.harvard.edu

1. INTRODUCTION

The modulatory effects of acetylcholine in cortical structures have been previously studied in this laboratory and by others, using both experimental and computational tools (Hasselmo and Bower, 1992; Pitler and Alger, 1992). Cholinergic modulation in the piriform cortex shows laminar selectivity in suppression of excitatory synaptic transmission at the intrinsic (in layer Ib) but not afferent (in layer Ib) fiber synapses, and may set the dynamics for recall and storage of information (Hasselmo and Bower, 1992). Cholinergic modulation has also been shown to enhance long term potentiation of synaptic potentials (Hasselmo and Barkai, 1995). The present work addresses the effect of cholinergic modulation on primarily inhibitory, and also excitatory components of synaptic transmission using intracellular recording techniques in a brain slice preparation.

Our experimental results show laminar specificity in the suppression of inhibitory potentials. In order to study the possible cellular mechanisms underlying the observed results, we use a computational model of piriform cortex. Electrophysiological data from piriform cortex brain slices was stimulated using a network of 20 pyramidal cells, and 20 of each feedforward and feedback interneurons. With the help of the intracellular experimental data and the computational model, this work attempts to propose a mechanism underlying the cholinergic modulation of inhibitory transmission in the afferent and intrinsic layers of the piriform cortex, specially with regards to the feedback and feedforward GABAergic interneurons.

---

[*] To whom correspondance should be addressed

Computational Neuroscience
edited by Bower, Plenum Press, New York, 1997

2. EXPERIMENTAL BACKGROUND

·   Intracellular recordings were obtained from pyramidal neurons in layer II of the piriform cortex in the *in vitro* slice preparation. Slices 400 μm thick were obtained from adult female Sprague-Dawley rats (150–200 gm), using standard procedures as described in previous publications (Hasselmo and Bower 1990; 1992) in accordance with institutional guidelines.

Intrinsic Fiber Inhibitory Potentials

Recordings of inhibitory potentials obtained at different membrane potentials during stimulation of layer Ib are shown in Figure 1A. Depolarization of the neuron membrane potential allowed observation of the early and late components of the IPSPs, as can be seen in Figure 1A. Laminar differences in the components of the IPSPs were observed, with the early Cl⁻ component being more prominent during the stimulation of the association fiber layer (layer Ib) whereas the late component was observed during stimulation of both the layers. As shown in Figure 1D and F, inhibitory synaptic potentials elicited during stimulation of the afferent fiber layer (layer Ia) rarely evoked a prominent early inhibitory component.

Perfusion of the cholinergic agonist carbachol in the slice chamber caused a significant suppression of inhibitory synaptic potentials elicited by stimulation of association/intrinsic fibers in layer Ib, as shown in Figures 1B and C. As can be seen in the figure, carbachol caused suppression of both the early and late components of the inhibitory synaptic potential. The strong suppression of the fast IPSP component made quantitative measurements difficult. Change in the slow inhibitory synaptic potentials (n=14) revealed that layer Ib inhibitory synaptic potentials were significantly reduced by 79.4 ± 5.71 % (p< 0.0007). After washing the carbachol out of the slice chamber, the IPSP amplitudes recovered to about 60% of their control value. This partial recovery may be due to insufficient wash period in some slices, since wash takes over 45 minutes.

Afferent Fiber Inhibitory Potentials

In comparison, perfusion of carbachol had a much weaker effect on inhibitory synaptic potentials elicited by stimulation of afferent fibers in layer Ia, as shown in Figure 1D and E. Measurement of the change in IPSPs in a number of slices (n=16) revealed that IPSPs elicited by layer Ia stimulation were decreased on average by 18.5 ± 10.8 %. This effect was just statistically significant (p< 0.05). The mean effect of carbachol on inhibitory synaptic potentials elicited by stimulation in the two layers is summarized in Figure 2 and compared to previously reported (Patil and Hasselmo, 1996) effects on excitatory synapses in the two layers.

3. COMPUTATIONAL MODELING

The piriform cortex model is based on the piriform cortex circuitry shown in Figure 3, constructed using anatomical and electrophysiological data (Haberly, 1991). For the simulations described here, 20 of each pyramidal cells and feedforward and feedback interneurons were implemented. Pyramidal cells have three compartments, whereas interneurons are simulated as point neurons. The equations used for synaptic transmission,

A) EPSPs/IPSPs evoked by association fibre
stimulation at different membrane potentials

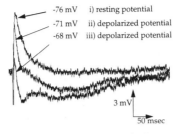

-76 mV   i) resting potential
-71 mV   ii) depolarized potential
-68 mV   iii) depolarized potential

3 mV

50 msec

Comparison of association layer (Ib) IPSPs
and EPSPs on exposure to carbachol

B   depolarized potential

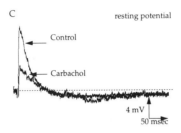

Carbachol

Control

C   resting potential

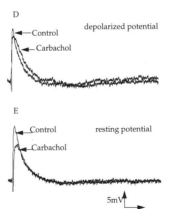

Control

Carbachol

4 mV

50 msec

Comparison of afferent layer (Ia) IPSPs
and EPSPs on exposure to carbachol

D   depolarized potential

Control

Carbachol

E   resting potential

Control

Carbachol

5mV

50 msec

Figure 1. Cholinergic modulation of inhibitory synaptic potentials. A: Intracellular recorded response of pyramidal cell to stimulation of layer Ib at different resting membrane potentials. B: Comparison of effect of carbachol on IPSPs evoked by stimulation of layer Ib at depolarized membrane potential and C: at resting membrane potential. D: Effect of carbachol on pyramidal cell response evoked by stimulation of layer Ia at depolarized potential and E: at resting membrane potential.

**%Decrease in layer Ia and Ib evoked
EPSPs and IPSPs**

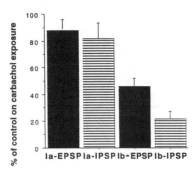

Figure 2. Histogram comparing the percent suppression of IPSP amplitude (shaded bars) and EPSP amplitudes in the intrinsic association fiber layer. (Patil and Hasselmo 1996).

reversal potentials and time courses of different types of synapses (AMPA, NMDA, GABA-A and GABA-B; Figure 3B) are taken from previous computational models of piriform cortex (Wilson and Bower, 1992). We use a simplified neuron model, were action potentials are triggered when the membrane potential at the soma reaches a threshold (8 mV above resting membrane potential). After an action potential, the membrane potential is reset to resting membrane potential. The two sets of interneurons (FF - feedforward and FB - feedback) are both believed to be GABAergic, with slow K^+ components (GABA receptors linked to K channels) towards the distal dendrites and the fast Cl^- components (GABA receptors linked to Cl channels) acting at the soma (Tseng and Haberly, 1988). However, more recent literature indicates that it is somewhat controversial whether FB and FF interneurons both activate fast and slow components. In the model, interneurons were initially computed with a varying probability for the distribution of K^+ and Cl^- components. They were then adjusted to produce IPSPs that resemble those obtained experimentally at different values of membrane depolarization.

Electrical stimulation in the model is simulated as a brief presynaptic (1ms) square input pulse to either afferent inputs (layer Ia) or intrinsic connections (layer Ib) (to both pyramidal cells and interneurons). In order to compare our results to the experimental data, where pyramidal cells were kept from spiking, spiking thresholds for pyramidal cells are set high, such that pyramidal cells don't spike for the given stimulus strength.

4. RESULTS

Distribution of Synaptic Channels

With the chosen parameter set, synaptic potentials obtained by the simulation of the model are comparable to the experimental data (Figure 4A and 4B). The results suggest the following distribution of the fast and slow inhibitory components: that fast inhibitory (Cl^-) components are mainly activated by FB interneurons (80%), whereas the slow (K^+) components are mainly activated by FF interneurons (80%). In addition, the simulations suggest a larger effect of the NMDA mediated slow excitatory component in layer Ib than in layer Ia.

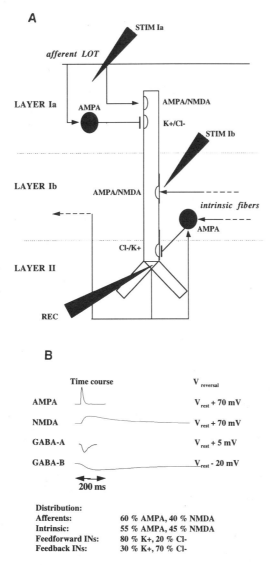

Figure 3. Schematic representation of the model architecture.

Cholinergic Modulation of Synaptic Potentials

The experimentally observed suppression of intrinsic but not afferent evoked EPSPs due to cholinergic modulation (Hasselmo and Bower, 1992 and present results) was first modeled by decreasing the synaptic transmission (by 30%) at excitatory synapses onto pyramidal cells in layer Ib. The results show a proportional decrease of EPSPs and a lesser degree of suppression of IPSPs evoked by layer Ib stimulation. However experimental data indicates that while EPSPs decrease by about 30%, the IPSPs decrease by about 90%. Hence a simple decrease in excitatory synaptic input to pyramidal cells cannot account for the decreased IPSP in layer Ib.

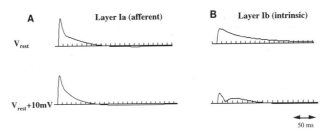

Figure 4. Intracellular responses of pyramidal cells in the model A: EPSPs and IPSPs evoked by layer Ia stimulation at two resting potential and at 10 mV depolarization B: EPSPs and IPSPs evoked by layer Ib stimulation.

As a second mechanism, we propose that acetylcholine could also presynaptically suppress glutamate onto both pyramidal cells and interneurons. Results of this simulation show a similar suppression of EPSPs and IPSPs in both layer Ia and layer Ib (Figure 5A). However, the experimental results show that IPSPs decrease almost twice as much as EPSPs in both layers.

These results suggest that suppression of GABA release from interneurons is implicated in the observed cholinergic effects. We have implemented two mechanisms leading to equivalent results: (1) presynaptic suppression of glutamate release onto interneurons (30%) and presynaptic suppression of GABA release (30%), or (2) no suppression of glutamate release but enhanced (90%) suppression of GABA release (Figure 5B).

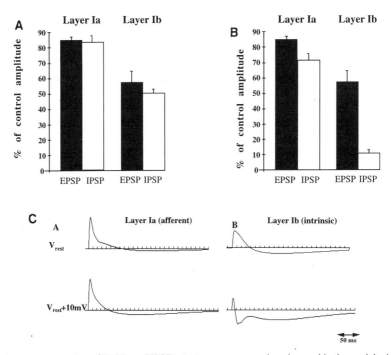

Figure 5. Average suppression of EPSPs and IPSPs. A: Average suppression observed in the model with presynaptic suppression of glutamate release onto pyramidal cells and interneurons (30%). B: Average suppression observed in the model using suppression of glutamate release (30%~) and suppression of GABA release (30%). C: Simulated intracellular recordings of cholinergic modulation in layer Ia and layer Ib.

5. SUMMARY

Intracellular recordings in piriform cortex pyramidal cells have shown that in both layer Ia and layer Ib, suppression of evoked IPSPs is larger than the corresponding suppression of evoked EPSPs. The observed suppression of IPSPs can not be explained only by decreased excitatory input to interneurons due to decreased pyramidal cell activation. Using a computational model of piriform cortex, we have proposed two possible mechanisms underlying this observation. The first mechanism proposes a suppression of glutamate release from pyramidal cells onto all excitatory synapses, and a suppression of GABA release of similar magnitude. Alternatively, presynaptic suppression of glutamate only onto pyramidal cells and suppression of GABA release could be proposed. Both mechanisms are in accordance with experimental data from both piriform cortex and hippocampus (Madison and Nicoll , 1988; Pitler and Alger , 1992 and present data). In order to support the predictions of the model, regarding the inactivation of the inputs to the interneurons, we need to study the FB interneurons, and understand the process (pre/post) involved in this suppressed synaptic transmission.

REFERENCES

Haberly L.B, "Olfactory Cortex", in Synaptic organization of the Brain, (1991) 317–345.

Hasselmo M.E. and J.M. Bower, "Cholinergic suppression specific to intrinsic not afferent fiber synapses in the rat piriform (olfactory) cortex", (1992) J. Neurophys 67(5):1222–1229.

Hasselmo M.E. and Barkai, "Cholinergic modulation of activity-dependent synaptic plasticity in the piriform cortex and associative memory function in a network biophysical simulation", (1995), J Neuroscience, 15(10):6592–6604.

Madison, D.V. and R.A. Nicoll, "Norepinephrine decreases synaptic inhibition in the rat hippocampus'" (1988), Brain Research, 442, 131–138.

Patil, M. and Hasselmo, M.E. Modulation of inhibitpry synaptic potentials in the piriform cortex (1996), submitted.

Pitler, T.A. and B. E. Alger, "Cholinergic excitation of GABAergic interneurons in the rat hippocampal slice", (1992) J Physiology, 450, 127–142.

Tseng, G-F and L.B. Haberly, "Characterization of synaptically mediated fast and slow inhibitory processes in piriform cortex in an in vitro slice preparation", (1988) J. Neurophysiology, 59(5), 1352–1376.

27

MEASURING INFORMATION IN FIRING RATES

Michael G. Paulin

Department of Zoology
University of Otago
Box 56, Dunedin, New Zealand
mpaulin@otago.ac.nz

1. INTRODUCTION

It might be thought that one spike carries one bit of information because it is an all-or-none event. However, spikes convey information not merely because they occur, but because of when they occur. The amount of information that can be transmited by one spike depends on how accurately its arrival time can be controlled by the sender and measured by the receiver, and can be arbitrarily large.

Neuroscientists often represent spike trains using rate histograms. From an information theory point of view it would seem desirable to have bins that are short compared to average interspike intervals in order to retain information about timing of individual spikes. On the other hand, from a signal processing point of view, this leads to worst-case aliasing artifacts (Paulin, 1992). Thus if we analyse spike trains in terms of spike counts in time intervals, we are forced into a choice between a representation that throws away details of spike timing and one that treats spike trains as discrete binary sequences. Rather than belabour the shortcomings of either choice, I will demonstrate a better one.

Aliasing errors in rate estimation can be avoided by using phaseless anti-aliasing filters followed by sampling. Any low-pass filter with approximately the right bandwidth gives a substantial improvement over a rate histogram (Paulin, 1992), but there are theoretical reasons for preferring a Gaussian filter. A Gaussian filter is one with a Gaussian gain characteristic in the frequency domain, which transforms spikes into Gaussian functions in the time domain. This estimator of firing rate may be called Gaussian local rate or GLR. Precise timing occurs in one limit, when the variance or smoothing parameter goes to zero. Mean firing rate occurs in the other limit, when this parameter goes to infinity.

GLR can be thought of as a procedure that smears spikes across time, or as one that introduces uncertainty about times at which spikes occur while retaining the notion that they occur at precise times. Since Gaussian distributions have minimum variance for a given entropy, GLR produces a representation of spikes that is in a least squares sense the most spikelike or temporally localized representation with a specified finite average capacity to convey information. Thus GLR satisfies the constraint that a mathematical model of a spike train should, like the real thing, convey finite amounts of information in terms of points along a one-dimensional continuum (Paulin, 1992; 1996).

The aim of this paper is to show, using simulations with a simple integrate-and-fire model neuron, how much information about a spike train can be captured by using GLR in comparison to a rate histogram. The model neuron used here is merely a method of generating spike trains containing information about a known stimulus. It is not intended to be a realistic representation of a neuron, but to illustrate how it is possible to encode signals in spike trains at information transmission rates that surpass not only the fallacious common sense notion that one spike has the capacity to transmit one bit, but also the slightly deeper but equally fallacious notion that in a discrete-time representation of a spike train containing less than one spike per sampling interval we are limited to one bit per sampling interval.

2. MODEL NEURON

Computations were carried out using MATLAB on a pentium-based machine. Spike trains were generated by passing band limited digital Gaussian noise into an integrate-and-fire mechanism. The simulation time step length was 0.001s (1ms). Lowpass filtering was via a second order digital butterworth filter. The integrator was initially charged with a uniform(0,1) random number, then the input signal was integrated until the threshold of 1 was reached. The threshold crossing time was determined by linear interpolation to specify a spike time. The integral was reduced by 1 and the procedure repeated. This is called integral pulse frequency modulation or IPFM. Note that spike times are specified at higher resolution than the simulation step size. A stimulus encoded by IPFM can be recovered without error by low-pass filtering when its bandwidth, in cycles per time unit, is somewhat lower than its mean, which corresponds here to the mean number of spikes per time unit in the response. This is obviously the case when the stimulus is constant. With stimulus bandwidths comparable to mean spike rates or higher the spike train spectrum contains stimulus-carrier interaction terms that preclude accurate decoding by such simple means (Paulin, 1992). Nonetheless, an IPFM spike train may contain stimulus-related information at frequencies above its mean rate.

3. MEASURING INFORMATION

The maximum average amount of information conveyed by a sequence of numbers with specified variance in the presence of Gaussian white noise is $I = \Sigma \log_2(1+SNR(f))/2$ bits per unit time, where the signal to noise ratio or SNR is the ratio of signal power to error power at the output and the sum is taken over all frequencies. This maximum is attained when the signal is Gaussian. If the signal is non-gaussian then this expression, the discrete Shannon formula, provides an upper bound. For example, if signal and noise are chosen independently from the same Gaussian distribution at each time step then SNR=1 at all frequencies and the discrete Shannon formula gives 1/2 bit per time step, or r/2 bits per second at sampling rate r.

GLR estimates and rate histograms of spike trains generated from filtered Gaussian white noise stimuli of different bandwidths were treated as estimates of the stimulus, and the reconstruction error was treated as noise. The bandwidth of GLR with smoothing parameter (variance) v is $1/\text{sqrt}(2\pi v)$ (Paulin, 1996), and the bandwidth of a rate histogram is its Nyquist frequency, 1/(2binwidth). The discrete Shannon formula was applied using the etfe function in MATLAB, which estimates power spectra from short data records. Tests with simulated Gaussian channels verified that this gives unbiased information capacity estimates from signals containing 4000 samples, or 4 seconds of data at 1KHz.

With IPFM the amount of information that can be transmitted per unit time interval scales with the background firing rate. This can be seen to be true by noting that we could simply define the simulation time steps to be micoseconds instead of milliseconds and thereby increase information transmission rates by three orders of magnitude. Of course the redefinition also decreases the nominal intervals between spikes by three orders of magnitude, so the amount of information per spike is invariant. Bits per unit time interval are not. For this reason results are presented in terms of bits per spike. This is a matter of representing the information capacity of a spike train in terms of an intrinsic time scale. It does not entail viewing single spikes as carriers of finite quanta of information.

Estimated information transmission rates are averages from 144 trials at each bandwidth. Amplitude distributions for stimuli, reconstructed waveforms and error waveforms were estimated from single trials by sampling at the relevant Nyquist rate in each case. The distributions were compared to Gaussian distributions with the same variance using a chi-squared goodness of fit test.

4. RESULTS

GLR stimulus reconstruction is extremely accurate when the stimulus bandwidth and reconstruction bandwidth are both lower than half the mean rate of the spike train. That is, when there are at least two spikes per cycle at the stimulus bandwidth and at the reconstruction bandwidth. However, in this case errors are non-Gaussian and the information transmission rate estimates are unreliable.

If stimulus bandwidths and reconstruction bandwidths are comparable to mean firing rate then GLR waveforms are Gaussian and so is the reconstruction error. Rate histogram reconstructions are not Gaussian but the reconstruction errors are (Figure 1). It follows that at reconstruction bandwidths comparable to mean firing rate, the discrete Shannon formula correctly estimates the amount of information about the stimulus captured by the GLR reconstruction, but overestimates the amount of information in a rate histogram.

When reconstruction bandwidth equals mean firing rate, GLR recovers an estimated 3.9 bits of information per spike, whereas the estimated upper bound on the information capacity of the rate histogram is 0.66 bits per spike. GLR information capacity is maximized at this point, while rate histogram information capacity falls continuously with increasing bandwidth.

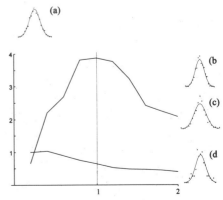

Figure 1. Estimated average amount of stimulus-related information recovered by GLR (upper trace) and rate histograms (lower trace) at various bandwidths (horizontal axis), normalised by mean firing rate to give bits per spike. Inset graphs show single-trial amplitude distributions at a reconstruction bandwidth corresponding to one spike per cycle: Stimulus (a), GLR reconstruction (b), GLR reconstruction error (c) and histogram reconstruction error (d).

5. SUMMARY AND CONCLUSION

The present study quantifies the difference in performance of GLR and rate histograms at a bandwidths corresponding to a single spike per cycle using information capacity as an absolute measure. It shows that GLR outperforms rate histograms by a factor of more than six at this bandwidth and that GLR information capacity is maximised there.

This result suggests that to reconstruct a stimulus from an IPFM spike train GLR bandwidth should equal mean firing rate. Although IPFM is clearly a 'rate' code, substantial amounts of information about the stimulus may be lost if the spike train is examined on a temporal scale that is unable to resolve individual spikes. Using GLR implies that the issue in spike coding is not timing versus rate, but how to define temporal locality in order to represent messages transmitted as points in time. The present results indicate that for deterministic integrate and fire neurons there is a natural timescale for representing spike trains defined by the mean interspike interval.

The computed estimates measure of the relative performance of two representations, GLR and rate histograms, in reconstructing a stimulus encoded by IPFM. For GLR we have an estimate of actual information transmission rates while for rate histograms we have an estimate of an upper bound. It must be noted that there may be more efficient ways to encode stimuli in spike trains than IPFM. Therefore the amount of stimulus-related information in a GLR or rate histogram representation of a spike train, even one generated by IPFM, may be more than can be determined by treating the representation as an estimate of the stimulus. Information in this context is a measure of how variability among responses to one stimulus is related to variability among responses over all stimuli, and this is not necessarily connected to how closely the representation resembles the stimulus. Thus it may be possible to capture more than 3.9 bits of information per spike in a GLR waveform, and more than 0.66 bits per spike in a rate histogram.

A GLR waveform can be reconstructed from samples taken at the Nyquist rate. It follows that the optimal GLR representation of an IPFM spike train, whose bandwidth is equal to the mean firing rate, can be captured by sampling at twice the mean firing rate. At this point the probability of finding a spike in a randomly chosen sampling interval is 0.5 and GLR captures nearly two bits per sampling interval.

The maximum entropy distribution of spike counts in bins is Poisson, corresponding to an exponential distibution of interspike intervals. With on average one spike per two bins the information capacity is one bit per bin (50–50 chance of an event per bin) or two bits per spike (one spike every two bins). This seems to be a strict upper bound on the information capacity of a spike train at this bandwidth. How can GLR capture nearly two bits per sample and four bits per spike at the same bandwidth?

The solution of the apparent paradox lies in the fact that GLR captures information about location of spikes within sampling intervals using continuous sample values, while rate histograms deal only with counts or presence-absence of spikes. Thus GLR has a much higher information capacity than a rate histogram with the same number of sample points. GLR not only does about six times better at the optimal bandwidth with a spike train generated by IPFM, it does nearly twice as well as the best that could be done using a rate histogram with a spike train generated by any means.

REFERENCES

Paulin, M.G. (1992) Digital filters for firing rate estimation. Biol. Cybern. 66: 525–531

Paulin, M.G. (1996) System Identification of Spiking Sensory Neurons using Realistically Constrained Nonlinear Time Series Models. In: Advances in Processing and Pattern Analysis of Biological Signals. I. Gath and G. Inbar (eds). Plenum, NY.

28

A MODEL OF POSITION-INVARIANT, OPTIC FLOW PATTERN-SELECTIVE CELLS

Robert I. Pitts,[1,2] V. Sundareswaran,[2] and Lucia M. Vaina[2]

[1] Department of Computer Science
Boston University
Boston, MA
[2] Brain and Vision Research Laboratory
Department of Biomedical Engineering
Boston University
Boston, MA
E-mail: rip@bu.edu
E-mail: sundar@bu.edu
E-mail: vaina@bu.edu

ABSTRACT

Two apparently inconsistent proposals for the functionality of MSTd cells exist, based either on optic flow patterns being represented by motion *components*[2-4], such as translation, expansion/contraction and rotation, or by a *continuum* of motion patterns that includes spirals[6]. A model, consisting of excitatory and inhibitory subunits, has been proposed[3] to support the component view. Here, we used a neural network to show that a model of this type can be selective to the continuum of patterns[6]. We extended this model by adding inhibitory connections between units to account for the reported[6] position-invariant characteristics of MSTd cells.

1. INTRODUCTION

Neural analysis of visual motion has been studied extensively in monkeys revealing a hierarchy of motion processing with areas that integrate local motion into complex motion patterns[10]. One area with many motion-sensitive neurons is the middle temporal area (MT), containing cells with moderate-sized receptive fields that respond to translational motion, responding most strongly to a *preferred direction*. In the dorsal section of the medial superior temporal area (MSTd), many neurons have been found with large receptive fields and that respond to complex motions. These neurons prefer large-field stimuli and respond to translation, expansion/contraction and rotation. Such motions are called *optic flow* because they are produced when three-dimensional motion is projected onto a two-dimensional surface, such as our retinas. Many MSTd cells respond well to just the directional component in complex motion[10], even in the absence of other cues.

2. ELECTROPHYSIOLOGICAL STUDIES

Duffy and Wurtz[2-4] have suggested that optic flow patterns are encoded by response to motion *components*: planar, circular and radial motion. Their physiological studies have found cells that respond to one, two or three of these motion components: single-component neurons responding to planar, radial or circular motion; double-component neurons responding to planocircular or planoradial motion; and triple-component neurons responding to all three.

Their experiments with large-[2] and small-field[3] patterns revealed overlapping regions of excitation and inhibition within the receptive fields of MSTd neurons. In addition, they found that single-component neurons had more overlap of excitatory/inhibitory regions, stronger inhibitory responses, and the most position-invariant responses, i.e., responses that change less as a stimulus is moved from place to place in the receptive field. This suggests that inhibition plays an important role in refining the response of MSTd neurons.

Later work by Duffy and Wurtz[4] showed that many MSTd neurons exhibit *relative position invariance*, i.e., these neurons respond preferentially to the same optic flow patterns, but the strength of the response varies systematically when the stimuli is shifted within the receptive field of the cell.

In contrast, Graziano *et al.*[6] have suggested that optic flow patterns are encoded by position-invariant responses to a *continuum* of patterns, which, in addition to circular and radial patterns, includes spiral patterns. They tested the responses of MSTd cells using these patterns and found cells preferring each pattern. Many of the cells found by Graziano *et al.* were position-invariant, single-component cells. Those that might be considered a double-component cell if tested with just circular and radial patterns, in fact, preferred intermediate spiral patterns.

3. NEURAL NETWORK MODELS

Neural network models of MSTd neurons that use MT-like responses as inputs have been proposed and implemented. The existence of projections from MT to MST is one justification for using MT-like responses as inputs. In addition, psychophysical experiments[8] suggest that the performance of a human in correctly detecting radial and circular patterns is limited by a preprocess that detects local motion.

Zhang *et al.*[12] used Hebbian learning to simulate units that decomposed optic flow patterns (with position invariance) *and* spiral-sensitive units (with approximate position invariance). In addition, they characterized the network weights that gave these responses. A model proposed by Lappe and Rauschecker[7] used a network to form a population coding of heading directions. It reproduces some of the performance properties of humans in determining heading and gives response profiles like triple-component neurons. Wang[11] used a simple competitive learning mechanism to produce single-, double- and triple-component units (with position invariance strongest for single-component units). Finally, Beardsley *et al.*[1] explored

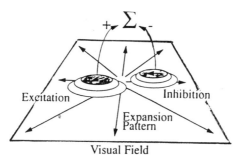

Figure 1. Duffy and Wurtz'[3] proposed model of a pattern-selective unit (see text).

optic flow selectivity in the hidden units of a backpropagation network. They trained the network to categorize inputs as expansion, contraction or rotations, and found hidden units consistent with both the component and continuum hypotheses.

We implemented a simple model proposed by Duffy and Wurtz[3] that uses an excitatory and an inhibitory subunit that each respond preferentially to specific local motions (Figure 1). We showed that this model can implement units that are selective to expansion/contraction, rotation and spirals. With this model, we focused on spiral patterns and developed an *extension* that gives position-invariant units.

4. METHODS

The patterns used in these experiments consist of expansion, contraction, rotation and spiral patterns, a subset of the same patterns used by Wang[11]. Each pattern is represented by a 5 × 5 grid consisting of local motions (Figure 2).

The inputs to our model are the responses of direction-selective units with narrow tuning to the four cardinal and four oblique directions. These input units connect to a pair of excitatory and inhibitory subunits that respond only to one direction of motion. The response of a subunit depends on the following parameters: its *center*; its *peak response at the center*, which falls off exponentially from the center; and the *extent* of its receptive field.

The output unit's response is the excitatory subunit's response minus the inhibitory subunit's response. This output unit is what gives a pattern-selective response. We alway define the *preferred pattern* as the pattern which elicits the strongest response when patterns are presented in the center of the receptive field and the *anti-preferred pattern* to be the pattern with opposite local motions. We did not normalize the response of our output units since

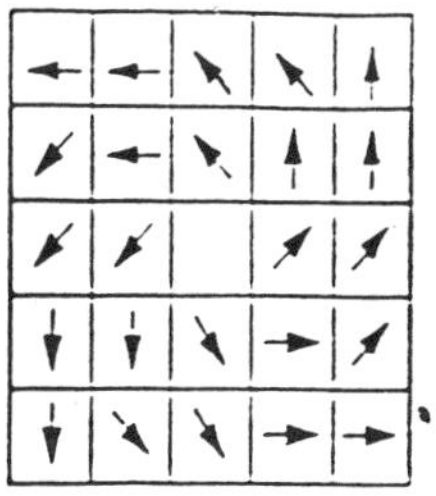

Figure 2. The counterclockwise outward spiral pattern.

there is evidence[10] that, for MSTd cells, it is the relative response of a neuron to two different patterns that matters.

Finally, our measure of position invariance is shown in Equation 1:

$$invariance = \sum_{\forall i} \min \left(\frac{resp_{pref,i} - resp_{anti,i}}{resp_{pref,center} - resp_{anti,center}}, 0 \right) \qquad (1)$$

where $resp_{pref,i}$ is the response of the unit to the preferred pattern when centered at location i, $resp_{anti,i}$ is the response to the antipreferred pattern, and the summation variable i runs through all the locations where patterns are centered.

This measure is based on a measure of invariance used by others[6]. It accumulates penalty when the response to the antipreferred pattern is higher than the response to the preferred pattern at some location.

5. RESULTS

We studied the selectivity of output units by varying the parameters of the model, i.e., those of the subunits. We varied only the preferred direction and center of subunits since empirical experiments showed that these parameters influenced the selectivity much more than the peak response and extent of the subunit. Various values for these parameters produced units selective to each of the eight optic flow patterns[9].

We concentrated on units that were selective to counterclockwise outward spiral motion. For these *spiral units*, the types of excitatory/inhibitory organizations present were expected: the excitatory subunit responds to a direction that is present in the preferred pattern and the inhibitory unit responds to a direction that is opposite to motion in the preferred pattern. The inhibitory subunit helps to refine the response, distinguishing between patterns with similar local motions. The two subunits are often centered at different locations.

6. POSITION INVARIANCE OF SPIRAL-SELECTIVE UNITS

We measured the position invariance of spiral units for patterns presented in four locations besides the center. Plotted in Figure 3a is the response of one of the units with the highest invariance measure. Even this unit with a good invariance measure does not respond maximally to the preferred pattern at all shifted positions.

The problem is that we sometimes get higher responses to patterns that are "close" to the preferred pattern—for example, counterclockwise rotation and expansion. To fix this problem, we needed to reduce the response to those patterns. Since a unit selective to the anti-preferred pattern (Figure 3b) gives a higher response to counterclockwise rotation and expansion, we added an extra level of opponency, including such a unit's output as an inhibitory contribution.

Thus, to get a position-invariant spiral unit, we created a new *combined unit* that took as inputs the response of the outward spiral unit minus the response of the inward spiral unit. The combined unit responds maximally to counterclockwise outward spiral motion at all 5 locations at which patterns are centered (Figure 3c).

In other experiments[9], we've constructed combined units that respond invariantly to many of the 25 locations. The improvement provided is significant because some combined units can achieve higher invariance from units with low or mediocre invariance. Furthermore,

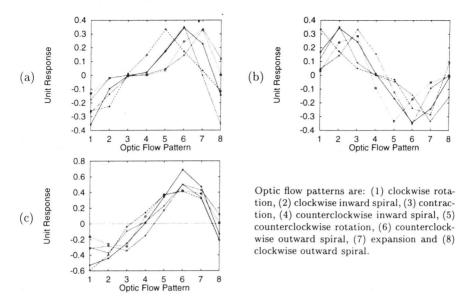

Optic flow patterns are: (1) clockwise rotation, (2) clockwise inward spiral, (3) contraction, (4) counterclockwise inward spiral, (5) counterclockwise rotation, (6) counterclockwise outward spiral, (7) expansion and (8) clockwise outward spiral.

Figure 3. Responses of spiral units with patterns presented at 5 locations. Units selective to: (a) counterclockwise outward spiral (position-dependent); (b) clockwise inward spiral (position-dependent); and (c) counterclockwise outward spiral (position-invariant).

we've testing combined units (with high position invariance) in the presence of noise, by introducing noise into the input patterns. Results[9] with four units showed a steady decline in the ability of the units to correctly respond to the preferred pattern as the amount of noise increased.

Like Duffy and Wurtz[4], our units produce graded responses depending on the location at which a pattern is centered. Duffy and Wurtz[4] observed response *shapes* falling in to one of three categories and found an overall preference for higher responses in the middle of the visual field. We examined the response shapes of our highly position-invariant combined units[9]. Although ours do not have the exact same response shapes, all give high responses in the middle of the visual field.

7. SUMMARY

Neurons in area MSTd of monkeys are selective to optic flow patterns. Duffy and Wurtz have proposed that MSTd cells respond to *components* of motion, i.e., expansion/contraction and rotation[2-4]. In contrast, Graziano *et al.* have proposed that they respond to a *continuum* of motion patterns, including spirals[6]. We implemented a model proposed by Duffy and Wurtz'[3] and created units selective to all of these patterns. By adding another layer of opponency, we've extended their model to produce more position-invariant units, like those of Graziano *et al.*[6]

We produced thousands of spiral-selective units and focused on them because spiral selectivity was a key difference between the results of Duffy and Wurtz[2-4] and Graziano *et al.*[6] (Duffy and Wurtz[5] have recently reported spiral-selective MSTd neurons.) Our units show graded responses depending on the centered location of a stimulus pattern, like neurons found by Duffy and Wurtz[4], although they do not show the exact same graded response *shapes*. In any case, we do not focus on the graded response as a means to extract *heading* information (using the center of motion) because recent work by Duffy and Wurtz[5] has shown that graded responses can be produced even when the center of motion is not shifted.

It is possible that some of our units are more selective for planar motions. Selectivity to planar patterns may be important because recent work by Duffy and Wurtz[5] shows a relationship between MSTd neurons with strong selectivity for planar motion and for optic flow patterns with strong *planar motion components*.

Other neural network models of MSTd neurons[7,11,12] fail to explain the regions of excitation and inhibition reported for MSTd neurons[3]. One model[7] has "opponent weights," but they do not relate them to *regions* of the receptive field. In contrast, this model, consisting of an excitatory and an inhibitory subunit, directly models this behavior. Because it does not "learn," this model does not suffer from "biased learning," i.e., it is difficult to say what it means to develop spirally-selective units when a large number of spirals are part of training. If it can be determined that a significant number of spirals are seen during the development of area MSTd, then it may be reasonable to present spirals during a neural network learning procedure. Certain types of motion do produce spiral-like optic flow patterns. Whether these patterns turn up enough in visual experience and whether they are significantly "like" the spiral patterns used by modelers remains in question.

We believe that this model is biologically plausible because it is simple, and opponency is found in other parts of the visual system. While we have shown that it is possible to construct units with position-invariant receptive fields, we have not suggested how this organization develops.

REFERENCES

[1] Beardsley, S. A., Vaina, L. M. and Poggio, T. The development of optic flow selectivity in MSTd neurons using backpropagation networks. *Soc. Neurosci. Abstr.* **22**, 1619 (1996).
[2] Duffy, C. J. and Wurtz, R. H. Sensitivity of MST neurons to optic flow stimuli. I. A continuum of response selectivity to large-field stimuli. *J. Neurophysiol.* **65**, 1329-1345 (1991).
[3] Duffy, C. J. and Wurtz, R. H. Sensitivity of MST neurons to optic flow stimuli. II. Mechanisms of response selectivity revealed by small-field stimuli. *J. Neurophysiol.* **65**, 1346-1359 (1991).
[4] Duffy, C. J. and Wurtz, R. H. Response of monkey MST neurons to optic flow stimuli with shifted centers of motion. *J. Neurosci.* **15**, 5192-5208 (1995).
[5] Duffy, C. J. and Wurtz, R. H. *Personal communication*.
[6] Graziano, M. S. A., Andersen, R. A., and Snowden, R. J. Tuning of MST neurons to spiral motion. *J. Neurosci.* **14**, 54-67 (1994).
[7] Lappe, M. and Rauschecker, J. P. A neural network for the processing of optic flow from ego-motion in man and higher mammals. *Neural Comp.* **5**, 374-391 (1993).
[8] Morrone, M. C., Burr, D. C., and Vaina, L. M. Two stages of visual processing for radial and circular motion. *Nature.* **376**, 507-509 (1995).
[9] Pitts, R. I. and Vaina, L. M. A computational model of MSTd neurons sensitive to optic flow patterns. (Submitted).
[10] Saito, H., Yukie, M., Tanaka, K., Hikosaka, K., Fukada, Y., and Iwai, E. Integration of direction signals of image motion in the superior temporal sulcus of the macaque monkey. *J. Neurosci.* **6**, 145-157 (1986).
[11] Wang, R. A simple competitive account of some response properties of visual neurons in area MSTd. *Neural Comp.* **7**, 290-306 (1995).
[12] Zhang, K., Sereno, M. I., and Sereno, M. E. Emergence of position-independent detectors of sense of rotation and dilation with hebbian learning: an analysis. *Neural Comp.* **5**, 597-612 (1993).

29

PYRAMIDAL CELL RESPONSE TO PATTERNS OF SYNAPTIC INPUT BELIEVED TO UNDERLIE GAMMA OSCILLATIONS

Alexander D. Protopapas and James M. Bower

Division of Biology, 216-76
California Institute of Technology
Pasadena, CA
E-mail: alexp@bbb.caltech.edu
E-mail: jbower@bbb.caltech.edu

1. INTRODUCTION

The fast oscillations induced in the piriform cortex in response to olfactory stimuli were first recorded some 50 years ago [1]. Since then, 30-80 Hz oscillations (also called gamma) have been found in many other cortical areas as well [4]. The ubiquity of stimulus-induced oscillatory activity in the brain suggests that the neural activity that underlies such oscillations might be critical to neural computation [2]. The question then becomes: How does one elucidate the neural activity that gives rise to these oscillations?

One approach is to try to deduce the synaptic currents that give rise to the oscillatory field potentials. If the brain area of interest is organized in a laminar fashion, and its anatomical structure is well understood, this can be done using a technique called current source density (CSD) analysis [9]. This technique allows one to locate in space and time the current sources and sinks that are responsible for generating field potentials. By correlating the known anatomical positions of different synaptic pathways to different current sources and sinks it is possible to determine, or at least suggest where in space and time different synaptic pathways are active during the course of the oscillations.

The layered architecture and distributed nature of synaptic connections in piriform cortex has made it especially amenable to the CSD approach. Using this method in combination with system modeling, Ketchum and Haberly [5, 6, 7] have been able to suggest a stereotyped spatio-temporal pattern of synaptic input that might underlie gamma oscillations.

In this study, we use the patterns of synaptic activity described by Ketchum and Haberly as input for a realistic single cell model of a piriform pyramidal cell to study single cell response to physiologically plausible patterns of synaptic input.

Computational Neuroscience
edited by Bower, Plenum Press, New York, 1997

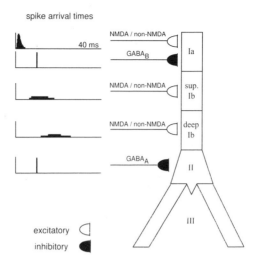

Figure 1. Synaptic input underlying gamma oscillations. Spike arrival times are shown as histograms on the left. Different pathways terminate on different portions of the apical dendrite labeled by anatomical layer (Ia, superficial Ib, etc.). Synaptic receptors mediating the responses of different pathways are shown above each "axon."

2.   METHODS

The model was constructed using the GENESIS neural simulator [3] and was composed of 1000 electrical compartments and eight different voltage-gated currents. The morphology of the model was based on a stained layer II pyramidal neuron. Passive and active properties were tuned to data collected from recordings in our laboratory.

In order to simulate background synaptic activity, we used data from a study that recorded spiking activity from neurons in the piriform cortex of awake-behaving rats in the absence of olfactory stimuli [8]. Background input was modeled as a Poisson process and the experimental data was used to tune the mean rates of excitatory and inhibitory inputs.

Figure 1 shows the sequence of synaptic events believed to underlie a single gamma oscillation. In this paper we refer to this full sequence of synaptic events as a "gamma pattern." Theta oscillations (7-12 Hz) often encompass multiple gamma oscillations. Consequently, we repeat gamma patterns at intervals of 25 ms to simulate the synaptic activity that may underlie activity during the course of a theta oscillation. We refer to this sequence as a "theta pattern."

3.   RESULTS

In order to study the effects of background synaptic input on pyramidal cell behavior, we plotted the power spectrum of the membrane potential of the cell during background stimulation. Interestingly, we found that despite the unpatterned nature of the background stimulation, the power spectrum showed a peak in the 10-20 Hz range; however, this peak was absent when voltage-gated currents in the soma were removed. This is shown in figure 2. This suggests that voltage-gated channels in the pyramidal cell may act to amplify fluctuations in the 10-20 Hz range (i.e. roughly theta).

The main thrust of this simulation study was to understand the consequences of the Ketchum-Haberly pattern of synaptic input for individual neurons. One issue we wished to

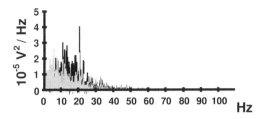

Figure 2. Power spectra of membrane potential in response to simulated background synaptic input. Gray trace shows power spectrum in the absence of voltage-gated channels in the model. Black trace shows spectrum in the presence of voltage-gated channels. Note amplification of frequency components in the 10-20 Hz range in the case where voltage-gated channels are present.

explore was the degree to which the inputs underlying a single gamma oscillation would effect the response of the cell during subsequent gamma oscillations. In order to address this issue, we repeated the Ketchum-Haberly pattern of synaptic input eight times to represent eight distinct gamma oscillations. Each repeat differed in that the magnitude of the synaptic input was scaled differently. Two simulation runs (of eight gamma oscillations) were performed in which all but the first gamma pattern were scaled identically. The results of the two simulation runs were then compared. If the neuron's response to a single gamma pattern was independent of other gamma patterns, then we would expect that the traces from the simulation runs would diverge for the duration of only a single gamma oscillation. On the other hand, a divergence that would last longer than a single gamma oscillation would indicate that the neuron's response to one gamma pattern will have a significant effect on subsequent inputs. The results of these simulation runs are shown in figure 3. As can be clearly seen, a single gamma pattern can have an effect on a neuron that can last up to 100 *ms*. The reason for this long lasting effect is due to a combination of factors arising from the time constants associated with channel kinetics (voltage-gated and synaptic) and the passive properties of the neuron.

An identical approach was used to determine the level of interaction between two theta patterns. In this case, two simulation runs were done, each with two theta patterns separated by varying amounts of time. The first theta patterns in the simulation runs were different but the second were identical. The level of interaction between the two inputs can then be determined by looking at the divergence between the two traces in response to the second theta pattern. We found that, when the two theta patterns were separated by 200 *ms* or more, there was

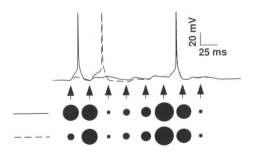

Figure 3. The effects of a single gamma pattern on the model's response to subsequent inputs. Traces show model's response to 8 gamma patterns. Arrows indicate the start of each pattern and filled circles indicate the scaling of each pattern. Stimulation used for both simulation runs are identical except that the first gamma pattern for the solid trace is much larger. Divergence of the two traces is indicative of the long lasting effects of a single gamma pattern.

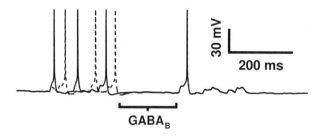

Figure 4. Two theta patterns separated by roughly 200 *ms*. The first theta patterns in the sequence are very different but the second are identical. If the theta patterns are separated by an adequate amount of time, the first theta pattern has no impact on the second. A biologically plausible mechanism for such a separation exists in the presence of a very slow $GABA_B$ inhibition.

negligible divergence as is shown in figure 4. Such a separation in the biological system might be accomplished by a slow $GABA_B$-mediated potassium inhibition as previously suggested by Wilson and Bower [10].

4. SUMMARY AND CONCLUSIONS

We constructed a single cell model to examine the implications of a stereotyped pattern of synaptic input believed to underlie the gamma oscillations seen in response to odor stimuli in the piriform cortex. As a first step in constructing a physiologically plausible pattern of synaptic input for our neuron, we simulated background activity based on experimental data from awake-behaving animals and found that the neuron amplified fluctuations in the membrane potential in the 10-20 *Hz* range. This result suggests that one function of the voltage-gated currents in this cell might be to tune the cell to the theta frequency which also happens to be the sniffing rate of the rat.

When looking at the model's response to multiple gamma patterns during the course of a theta cycle, we found the response to individual gamma patterns was not independent of previous gamma patterns. This suggests that the neuron integrates over multiple gamma patterns. In contrast, theta patterns appear to have little effect on each other provided they are separated by roughly 200 *ms* or more. A biologically plausible mechanism for such a separation is found in the slow $GABA_B$ inhibition which lasts several hundreds of milliseconds and may act to space out oscillations of the theta cycle [10]. One computational speculation that might be gleaned from these results is that the cell integrates over multiple gamma patterns, but over only a single theta pattern. In this sense one can imagine that from the perspective of the neuron theta patterns are computationally isolated from each other, but gamma patterns are not. Because theta patterns occur in sync with the animal's sniffing rate, the isolation of theta patterns may serve to prevent the corruption of an incoming olfactory signal from one sniff with that of another.

ACKNOWLEDGMENTS

We wish to thank Mark Domroese for providing us with an anatomically stained pyramidal cell for our single cell modeling work and John Miller and Gwen Jacobs for their generous help in digitizing the cell's anatomy. AP is supported by NSF BIR-9400878.

REFERENCES

[1] E. D. Adrian. Olfactory reactions in the brain of the hedgehog. *J. Physiol. (London)*, 100:459–473, 1942.

[2] J. M. Bower. Reverse engineering the nervous system: An *in vivo, in vitro*, and *in computo* approach to understanding the mammalian olfactory system. In S. F. Zornetzer, J. L. Davis, and C. Lau, editors, *An introduction to neural and electronic networks*, pages 3–28. Academic Press, New York, N. Y., second edition, 1995.

[3] J. M. Bower and D. Beeman. *The Book of GENESIS: Exploring Realistic Neural Models with the GEneral NEural SImulation System*. Springer-Verlag, New York, 1995.

[4] C. M. Gray. Synchronous oscillations in neuronal systems: mechanisms and functions. *J. Comput. Neurosci.*, 1:11–38, 1994.

[5] K. L. Ketchum and L. B. Haberly. Membrane currents evoked by afferent fiber stimulation in rat piriform cortex. I. Current source-density analysis. *J. Neurophysiol.*, 69(1):248–260, 1993.

[6] K. L. Ketchum and L. B. Haberly. Membrane currents evoked by afferent fiber stimulation in rat piriform cortex. II. Analysis with a system model. *J. Neurophysiol.*, 69(1):261–281, 1993.

[7] K. L. Ketchum and L. B. Haberly. Synaptic events that generate fast oscillations in piriform cortex. *J. Neurosci.*, 13(9):3980–3985, 1993.

[8] J. McCollum, J. Larson, T. Otto, F. Schottler, R. Granger, and G. Lynch. Short-latency single unit processing in olfactory cortex. *J. Cogn. Neurosci.*, 3(3):293–299, 1991.

[9] U Mitzdorf. Current source-density method and application in cat cerebral cortex: Investigation of evoked potentials and EEG phenomena. *Physiol. Rev.*, 65(1):37–100, 1985.

[10] M. Wilson and J. M. Bower. Cortical oscillations and temporal interactions in a computer simulation of piriform cortex. *J. Neurophysiol.*, 67:981–995, 1992.

30

A DETAILED MODEL OF SIGNAL TRANSMISSION IN EXCITABLE DENDRITES OF RAT NEOCORTICAL PYRAMIDAL NEURONS

Moshe Rapp, Yosef Yarom, and Idan Segev

Department of Neurobiology
Institute of Life Sciences
The Hebrew University
Jerusalem 91904, Israel

1. ABSTRACT

Simultaneous patch clamp recordings from the soma and dendrites of neocortical pyramidal neurons of young rats show that, independent of the synaptic input location, the sodium action potential (AP) always starts at the soma and is then carried along the axon, but also propagates backward decrementally into the dendritic tree[1]. This back-propagating AP is supported by a low density ($\bar{g}_{Na}$ = ~4 mS/cm^2) of Na$^+$ in the dendrites and soma membrane of the pyramidal neurons.

In the work described here we built a detailed biophysical model, based on a fully reconstructed layer V pyramidal cell. Data obtained from electrophysiological measurements were used to restrict the model. By investigating model parameters, which reproduced the findings of Stuart and Sakmann, we were able to address the following questions: 1. Under what conditions will the AP always be seen first at the soma and only later in the dendrites? 2. What is the degree of amplification of the back-propagating AP due to the Na$^+$ channels in the dendrites? 3. How resistant is the back propagating AP to background synaptic perturbation?

The model presented here can then be used to predict how the back-propagating AP will be seen in different locations in the dendrites and what will be its functional role.

2. INTRODUCTION

The question 'what are dendrites for?' has been at the core of single neuron research and neuroscience in general for many decades. It is now clear that dendrites do not serve only as simple transmission lines, but that they also participate in the computational tasks of neurons. Advanced recording and visualization techniques revealed that dendrites of most central neurons are equipped with excitable ionic channels. In most cases these ionic channels operate in the sub-threshold range, and are responsible for the non-linear input-output properties of dendrites.

Using excised patches from well-identified locations along the soma and apical dendritic trunk, Stuart and Sakmann[1] found that, in layer V cortical pyramidal neurons of young rats, the dendrites contain a low density (~4 mS/cm^2) of fast inactivating voltage dependent Na$^+$ channels. These channels are distributed homogeneously over the membrane of the soma and the apical dendrite. Simultaneous recordings from the soma and apical dendrite show that the AP always starts at the soma, then propagates actively, yet decrementally, backwards into the dendrite.

The purpose of our study was to build a minimal, experimentally-based, model of layer V pyramidal cells, that will reproduce the experimental findings of Stuart and Sakmann[1]. Investigating the behavior of such a model is a crucial step in understanding the functional role of the back-propagating AP and the information processing implemented by the dendrites of pyramidal neurons. For more detailed discussion see [2].

3. MODEL

Four physiologically characterized and anatomically reconstructed layer V pyramidal neurons from 14 day old rats were modeled. The results for one representative neuron is depicted in Figure 1, with similar results for the other three modeled cells. Dendritic

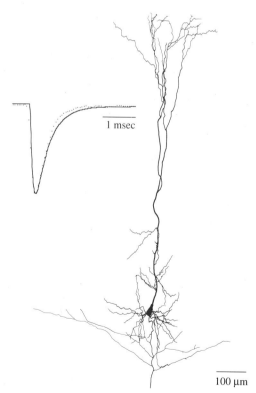

Figure 1. 3D reconstruction of a layer V pyramidal neuron from a 14 day old rat. In inset: The experimental fast inactivating Na$^+$ current as measured by Stuart and Sakmann[1] in excised patch (dotted line), compared with the simulated Na$^+$ current used in the present model.

spines were assumed to be passive and were incorporated globally into the modeled dendritic tree (see Rapp et. al.[3]), in agreement with the spine distribution described by Larkman[4]. Numerical computations were performed using NEURON[5]. 1202 compartments were used to represent the neuron shown in Figure 1. Based on the experimental data from this neuron, τ_0 was estimated to be 48 msec and C_m was assumed to be 0.8 $\mu F/cm^2$, implying membrane resistivity of 60,000 Ωcm^2. R_i was assumed to be 300 Ωcm in the dendrites[6] and 70 Ωcm in the axon[7]. These parameters yield a somatic input resistance of 122 $m\Omega$. Since the goal of the work was to understand the mechanisms underlying the initiation and propagation of the back-propagating AP, only the active currents that were directly involved in generating the observed behavior were incorporated into the model. Namely the fast-inactivating sodium current and the potassium delayed rectifier.

The kinetics of the excitable channels were modeled using Hodgkin and Huxley (H&H) formalism[8] at 20°C. To match experimental results[9] α's and β's for the dendritic and somatic Na^+ conductance were shifted by +10 mV. m was explicitly described by $1/(1 + \exp((V_{1/2} - V)/K))$, with $V_{1/2} = -29.5$ mV and K = 7.5 mV. To fit the time course of the Na^+ transient (Figure 1, inset), τ_h was increased by a factor of 1.6. In the axon, τ_h was further increased by a factor of 1.6 and the m curve was as above with $V_{1/2} = -39.5$ mV. For the K^+ conductance a single n-gate was used[10], with $\alpha_n = 0.006*(v-20)/(1 - \exp(-(v-20)/9))$; $\beta_n = -0.0006*(v-20)/(1 - \exp((v-20)/9))$. The ratio between the maximal sodium and potassium conductances was kept as in the original H&H model. $E_{Na} = +90$ mV and $E_K = -85$ mV; the resting potential was -70 mV. Fast excitatory (AMPA) and inhibitory ($GABA_A$ and $GABA_B$) inputs were each modeled as an 'α function': $g_{syn}(t) = g_{peak}(t/t_{peak})\exp(1-t/t_{peak})$ with $g_{peak} = 0.4$ nS, $t_{peak} = 0.3$ msec and associated reversal potential of 0 mV[11] for the AMPA input. For the $GABA_A$ input, $g_{peak} = 0.4$ nS at $t_{peak} = 0.2$ msec and for the $GABA_B$ input $g_{peak} = 0.3$ nS at $t_{peak} = 50$ msec. Only the shunting effect of the inhibitory synapses was considered in this study and, therefore, their corresponding reversal potential was not taken into account. The NMDA conductance input was modeled by: $g_{NMDA}(t) = (g_N(\exp(t/\tau_1) - \exp(t/\tau_2))/(1 + \eta[Mg+2] \exp(\gamma V_m))$ with $g_{NMDA} = 0.2$ nS; $\tau_1 = 80$ msec; $\tau_2 = 0.67$ msec; $\eta = 0.33/mM$; $[Mg+2] = 2$ mM, and $\gamma = 0.06/mV$ [12], with a reversal potential of 0 mV.

4. RESULTS

Since the goal of this work was to build a *minimal* model that will incorporate the experimental findings, the first attempt was to assume that the excitable channels are homogeneously distributed over the membrane of the neuron (dendrites+soma+axon). Figure 2 shows that an homogeneous model cannot reproduce the experimental observations. When the density of the voltage-dependent Na^+ channels is low ($\bar{g}_{NA} = 4$ mS/cm^2) current injection at the soma produces only a small regenerative response that does not propagate actively into the dendrites (Figure 2B). With higher, homogeneous, Na^+ channels density of 12 mS/cm^2, current injection at the soma initiates an AP at the soma (Figure 2C) which propagates (non-decrementally) into the apical dendrite. However, unlike the experimental results, under these conditions the dendrites are too excitable, and an AP is generated at the dendritic site in response to current input delivered to the dendrite (Figure 2D).

The conclusion is that in order to reproduce the experimental results, the voltage threshold at the axon (near the soma) must be lower than in the dendrites. Therefore, we assumed that the axon is equipped with a higher density of a different type of Na^+ channel[13], and that (compared to the dendrites) the activation and inactivation curves (m_ h_)

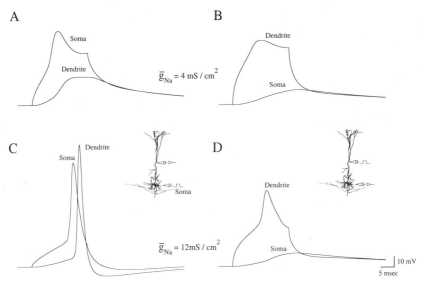

Figure 2. Homogeneous model cannot explain the experimental results. A. With a weakly excitable membrane ($\bar{g}_{NA}$ = 4 mS/cm^2) over the whole neuron surface and the measured kinetics of the Na$^+$ current, a full-blown AP cannot be initiated. In these conditions, a strong (1.3 nA, 40 msec) depolarizing pulse to the modeled soma produced only a small local regenerative response (upper trace). lower trace: The attenuated response at the apical dendrite (550 μm from the soma). B. Injection of a 0.6 nA, 40 msec, pulse at the dendritic site gave rise to only a small local regenerative response which attenuated significantly towards the soma. C. With uniform $\bar{g}_{NA}$ of 12 mS/cm^2, a soma pulse (300 pA, 40 msec) produced a brief, 90 mV somatic AP; this AP propagated, decrementally, backwards into the apical dendrite. D. A current pulse (500 pA, 40 msec) to the dendrite initiated a local response first in the dendrite which then attenuated to the soma.

are shifted by 10 mV to the left and that the inactivation kinetics are slower. Based on recent findings of Colbert and Johnston[14] the axon hillock was assumed to contain the same Na$^+$ channel density and type as the soma and dendrites, and the highly excitable axonal region starts only 50 μm away from the soma. We found that under these conditions, the experimental results can be reproduced with Na$^+$ channel density at the axon that is 40 times higher than that of the soma and dendrites. It should be noted that a lower threshold at the axon can also be achieved by using the same type of Na$^+$ channels as at the soma and dendrites. In this case the Na$^+$ channels density in the axon should be hundreds of times higher than in the soma and dendrites. The behavior of the model that agrees with the findings of Stuart and Sakmann is shown in Figure 3.

In panel A a current injection to the soma produces a rapid, 100 mV, AP that propagates decrementally into the apical dendrite. When the current stimulus is applied to the apical dendrite (550 μm from the soma) the AP still appears first at the soma and then it propagates backwards into the dendrite (Figure 3B). From Figure 3C it is clear that the AP is actually initiated in the axon and the weakly excitable dendrite amplifies this AP when it propagates along the dendrites. Indeed, when blocking the Na$^+$ channels the passively attenuated signal at the dendrite is 4 times smaller at the distance of 550 μm from the soma, compared to the excitable case (compare dotted line to continuous line). Figure 3D shows that in the excitable model the profile of the AP peak along the path from the soma to the apical dendrite agrees well with the experimental observations.

In Figure 4 we examined the effect of background synaptic shunt on the back-propagating AP. Synaptic shunt was modeled assuming that in *in vivo* conditions, 8000 excita-

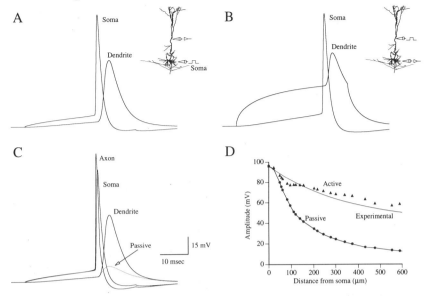

Figure 3. To reproduce the experimental results an excitable axon with a different type of Na$^+$ channels was assumed. $\bar{g}_{NA}$ in the axon was 140 mS/cm^2. In the dendrites, a linear decrement in $\bar{g}_{NA}$ (from 8 mS/cm^2 at the soma to 4 mS/cm^2 in the dendritic tuft) was used to best fit the experimental results. A. A brief, 100 mV, somatic AP is generated in response to a 150 pA, 40 msec pulse to the soma. This AP propagates decrementally into the apical dendrites (550 μm from the soma). B. A 300 pA, 40 msec, pulse to the apical dendrite initiated the AP at the axon first; this AP then propagated backward into the dendrite. C. The AP is initiated in the axon rather than in the soma (axonal trace recorded 50 μm from the soma). Stimulus parameters as in A. Comparison between the active and passive cases shows that the AP propagates actively back into the dendrite. In the passive case the somatic AP recorded in the active case was utilized as a voltage command to the soma (a "simulated" AP, see [1]). D. Amplitude of the back-propagating AP as a function of distance from the soma. Continuous line: the experimental results[1]; filled triangles: model result; filled circles: passive case.

tory inputs (AMPA+NMDA), 1000 GABA$_A$ and 1000 GABA$_B$ synaptic inputs are activated randomly, each at a frequency of either 1.5 or 3 Hz. The effect of this synaptic input was modeled by globally changing the effective R$_m$ (see [3]). Figure 4 shows that the back-propagating AP is quite resistant to background synaptic activity, yet, this network activity can modulate the AP's amplitude (and extend the invasion into the tree) in a graded manner.

5. SUMMARY AND CONCLUSIONS

Investigation of a detailed, biophysically-constrained, model of reconstructed pyramidal cells shows that: 1. In the pyramidal neurons, where the repolarizing conductances are relatively small, a sodium channel density as small as 30 times less than in the squid giant axon is sufficient to carry the AP actively in the apical tree. 2. The initiation of the AP first in the axon cannot be explained solely by morphological considerations; the axon must be more excitable than the soma and dendrites. The minimal Na$^+$ channel density in the axon that fully accounts for the experimental results is about 40-times that of the soma, assuming that $\bar{g}_{Na}$ in the axon hillock and the initial segment is the same as in the soma. 3. The weakly excitable apical dendrite enables the invasion of a significant portion of the somatic AP into most of the dendritic tree. 4. A backward propagating AP in weakly excitable dendrites is gradually modulated by background synaptic shunt.

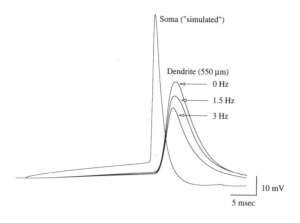

Figure 4. The back-propagating AP can be gradually modulated (rather that blocked) by the background synaptic activity. In this simulation, 8,000 excitatory inputs (AMPA + NMDA), 1,000 GABA$_A$ and 1,000 GABA$_B$ synaptic inputs were activated. In all cases the same "simulated" AP was imposed at the soma.

In conclusion, weak dendritic excitability, together with the boundary conditions imposed by dendritic morphology, enable the dendrites to function as a highly non-symmetrical transmission line. In the forward direction, full-blown APs that invade the whole tree are unlikely to occur and the dendrites can uninterruptedly integrate the many synaptic inputs they receive. In the backward direction, however, the dendrites become highly nonlinear devices, acting more like an axon, thereby enabling the soma to send a powerful global message ("I fired") to the whole dendritic tree.

REFERENCES

1. G. J. Stuart and B. Sakmann, "Active propagation of somatic action potentials into neocortical pyramidal cell dendrites" *Nature*, vol. **367**, pp. 69–72, 1994.
2. M. Rapp, Y. Yarom, and I. Segev, "Modeling back propagating action potential in weakly excitable dendrites of neocortical pyramidal cells" *PNAS*, vol. **93**, pp. 11985–11990, 1996.
3. M. Rapp, Y. Yarom, and I. Segev, "The impact of parallel fiber background activity on the cable properties of cerebellar Purkinje cells" *Neural Comp.*, vol. **4**, pp. 518–533, 1992.
4. A. U. Larkman, "Dendritic morphology of pyramidal neurones of the visual cortex of the rat. III. Spine distributions" *J. Comp. Neurol.*, vol. **306**, pp. 332–343, 1991.
5. M. Hines, "A program for simulation of nerve equations with branching geometries" *Int. J. Biomed. Comp.*, vol. **24**, pp. 55–68, 1989.
6. G. Major, A. Larkman, P. Jones, B. Sakmann, and J. Jack, "Detailed Passive Cable Models of Whole-Cell Recorded CA3 Pyramidal Neurons in Rat Hippocampal Slices" *j. Neurosci.*, vol. **14**, pp. 4613–4638, 1994.
7. R. Traub, "Model of synchronized population burts in electrically coupled interneurons containing active dendritic conductances" *J. of Computational Neuroscience*, vol. **2**, pp. 283–289, 1995.
8. A. L. Hodgkin and A. F. Huxley, "A quantitative description of membrane current and its application to conduction and excitation in nerve" *J. Physiol.*, vol. **117**, pp. 500–544, 1952.
9. O. Hamill, R. Huguenard, and D. Price, "Patch-clamp studies of voltage-gated currents in identified neurons of the rat cerebral cortex" *Cerebral cortex*, vol. **1**, pp. 48–61, 1991.
10. Z. Mainen, J. Jeorges, J. Huguenard, and T. Sejnowski, "A model of spike initiation in neocortical pyramidal neurons" *Neuron*, vol. **15**, pp. 1427–1439, 1995.
11. P. Stern, F. Edwards, and B. Sakmann, "Fast and slow components of unitary EPSCs on stellate cells elicited by focal stimulation in slices of rat visual cortex" *J. Physiol.*, vol. **449**, pp. 247–278, 1992.
12. C. E. Jahr and C. F. Stevens, "Calcium permeability of the *N*-methyl-D-aspartate receptor channel in hippocampal neurons in culture" *PNAS*, vol. **90**, pp. 11573–11577, 1993.
13. R. Westenbroek, D. Merrick, and W. Catterall, "Differential subcellular localization of the R$_i$ and R$_{ii}$ Na$^+$ channel subtypes in central neurons" *Neuron*, vol. **3**, pp. 695–704, 1989.
14. C. M. Colbert and D. Johnston, "Mechanisms of action potential initiation in soma, axon hillock and initial segment of pyramidal neurons" *Society of Neuroscience Abstracts*, vol. **21**, pp. 682.2, 1995.

31

INTER-ARRIVAL TIME SPIKE TRAIN ANALYSES FOR DETECTING SPATIAL AND TEMPORAL SUMMATION IN NEURONS

David C. Tam[1] and Michelle A. Fitzurka[2]

[1]Department of Biological Sciences and Center for Network Neuroscience
University of North Texas
Denton, Texas 76203
dtam@unt.edu
[2]Department of Physics
Catholic University of America
Washington DC 20064
fitzurka@stars.gsfc.nasa.gov

ABSTRACT

Two spike train analysis techniques are introduced to establish how the probability of firing in a neuron is dependent on the number of sequential spikes fired in another neuron (temporal summation) and in any other neuron (spatial summation). The *Joint Inter-Neuronal-Arrival-Time/Cross-Interval (J-INAT/CI) Probability Mass Function (PMF)* is used to deduce how a sequential (burst) firing pattern in any neuron is contributing to the generation of a spike in a reference neuron. Analogously, the *Joint Inter-Spike-Arrival-Time/Cross-Interval (J-ISAT/CI) PMF* is defined similar to the former PMF except that the contribution from only *one* other neuron is considered. These analyses can be used to establish the precise coupling relationship between the firing times of neurons both spatially and temporally.

1. INTRODUCTION

With advances in multi-neuronal recording technology, large amounts of simultaneously recorded spike trains can be obtained from biological neuronal networks. As a consequence, there has been a renewed interest in multi-unit spike train analysis to decipher the significance and contribution of synchronous and burst firings in a network[1,7]. Conventional auto-correlation[2,5] and cross-correlation[3,4] techniques have been used for these types of analyses traditionally, and recently newer techniques have been introduced[6,8]. However these methods usually provide an estimate of probability between the spike firings in neurons, they do not explicitly extract how many neurons or how many sequential spikes are involved in a particular correlation. Thus, it is of interest to reveal the conditional probability of spike firing based on the exact number of prior spikes fired in one neuron and in any other neuron so that the precise firing patterns contributing to the generation of a spike can be quantified. These analyses can be used to establish the precise coupling relationship between the firing times of neurons that are contributing to coordinated firing patterns in a network, typically characterized by synchronized (spatial patterns) firing and burst (temporal patterns) firing patterns, which are often observed in central pattern generating (CPG) networks.

Two spike train analysis techniques are introduced to establish how the probability of firing in a neuron is dependent on the number of sequential spikes fired in another neuron (temporally) and in any other neuron (spatially). The *Joint Inter-Neuronal-Arrival-Time/Cross-Interval (J-INAT/CI) Probability Mass Function (PMF)* is used to deduce how the sequential (burst) firing pattern in any neuron is contributing to the firing of a spike in a reference neuron. The *J-INAT/CI histogram* is constructed to estimate this joint probability between a sequence of near-synchronous spike firing in any number of neurons in the network and the cross-interval. Analogously, the *Joint Inter-Spike-Arrival-Time/Cross-Interval (J-ISAT/CI) PMF* is defined similar to the former *PMF* except that the contribution from *one* other neuron is considered instead of *any* other neuron. The *J-ISAT/CI histogram* is built to estimate the joint probability between a given number of sequential spike firings and the cross-interval.

2. METHODS

Let $r(t)$ and $c_1(t'_1)$ represent the reference and compared spike trains with total numbers of spikes, N and M_1, respectively, as follows:

$$r(t) = \sum_{n=1}^{n=N} \delta(t - t_n),$$

(1)

and

$$c_1(t'_1) = \sum_{m_1=1}^{m_1=M_1} \delta(t'_1 - t'_{m_1}),$$

(2)

where t_n denotes the time of occurrence of the n-th spike in train r, t_{m_1} denotes the time of occurrence of the m_1-th spike in train c_1, and $\delta(t)$ denotes a delta function.

The *first-order pre-cross-interval, (pre-CI)* $\tau'_{n,-1}$, relative to the n-th reference spike in train r is defined as:

$$\tau'_{n,-1} = t'_{m_1} - t_n$$

(3)

which satisfies the condition:

$$t'_{m_1} < t_n < t'_{m_1+1},$$

while the *second-order pre-CI*, $\tau'_{n,-2}$ relative to the same n-th reference spike in train r is defined as:

$$\tau'_{n,-2} = t'_{m_1-1} - t_n .$$

$$(4)$$

Similarly, the *k-th order pre-CI* can be defined as:

$$\tau'_{n,-k} = t'_{m_1-k+1} - t_n .$$

$$(5)$$

When the two spike trains fire coincidentally, the zeroth order *CI* is defined as:

$$\tau'_{n,0} = t'_{m_1} - t_n = 0$$

$$(6)$$

which satisfies the condition:

$$t'_{m_1} = t_n < t'_{m_1+1} .$$

Similarly, the cross-intervals between the reference spike train r and other compared spike trains, $c_2, c_3, \ldots$, can be defined. Let the compared spike trains c_2 and c_j be defined as:

$$c_2(t'_2) = \sum_{m_2=1}^{m_2=M_2} \delta(t'_2 - t'_{m_2})$$

$$(7)$$

and

$$c_j(t'_j) = \sum_{m_j=1}^{m_j=M_j} \delta(t'_j - t'_{m_j}) .$$

$$(8)$$

Then the *k-th order pre-CIs*, $\tau'_{n,m_2,-k}$ and $\tau'_{n,m_j,-k}$, between these spike trains c_2 and c_j and the reference spike train r can be defined as:

$$\tau'_{n,m_2,-k} = t'_{m_2-k+1} - t_n$$

$$(9)$$

and

$$\tau'_{n,m_j,-k} = t'_{m_j-k+1} - t_n$$

$$(10)$$

respectively. Since it is the duration of these intervals that is of interest, we consider only their magnitudes in the following discussion. Let the *probability density function (pdf)*, $f(tau')$, of the *k-th order pre-CI* between the compared train c_j and the n-th spike in the reference train r be defined as:

$$f(\tau'_{n,m_j,-k}) = \int_{-\infty}^{\infty} \delta(t - \tau'_{n,m_j,-k}) dt .$$

$$(11)$$

Then for the whole spike train, taking all spikes into account, the *pdf* becomes:

$$f(\tau'_{n.m_j,-k}) = \int_{-\infty}^{\infty} \delta(t - \tau'_{n.m_j,-k}) dt \qquad \forall n .$$

(12)

Also, let the *probability distribution function (PDF)*, $F(\tau')$, of the *k-th order pre-CI* between the compared train c_j and the reference train r be:

$$F(\tau'_{j,-k}) = \int_{-\infty}^{\tau'} f_{j,-k}(t) dt \qquad \forall n$$

(13)

and the *probability mass function (PMF)*, $P(\tau')$, of the *k-th order pre-CI* between the compared train c_j and the reference train r be:

$$P(\tau'_{j,-k}) = F(\tau'_{j,-k}) - F(\tau'_{j,-k^-}) = \int_{\tau'_{j,-k^-}}^{\tau'_{j,-k}} f_{j,-k}(t) dt \qquad \forall n$$

(14)

where $\tau'_{j,-k}$ denotes the cross-interval just greater than but not equal to $\tau'_{j,-k-1}$. This *k-th order pre-CI PMF* gives the probability of firing exactly k spikes in the train c_j prior to the firing of a spike in the reference train r.

To show the relationship between this probability and the preceding spike in another neuron leading to the firing of a spike in the reference spike train (i.e., the *first-order pre-CI*, $\tau'_{j,-1}$), the above *k-th order pre-CI PMF* can be modified to incorporate the *joint* probability function, the *k-th order Joint Inter-Spike-Arrival-Time/Cross-Interval (J-ISAT/CI)* *PMF* as follows:

$$P(\tau'_{j,-k}, \tau'_{j,-1}) = F(\tau'_{j,-k}, \tau'_{j,-1}) - F(\tau'_{j,-k^-}, \tau'_{j,-1}) = \int_{\tau'_{j,-k^-}}^{\tau'_{j,-k}} f_{j,-k,n}(t) dt .$$

(15)

This function will provide the joint probability between the firing of a spike in the reference neuron r in relation to the precisely k spikes that have fired prior to the last spike in the compared spike train c_j. This function can be established by constructing a two-dimensional histogram based on the spike counts to produce the *J-ISAT/CI histogram*. This histogram can be used to estimate, from experimentally recorded spike train data, the joint probability of exactly k spikes in the compared neuron contributing to the firing of a spike in the reference neuron.

Similarly, to show the relationship between how the sequential near-synchronous firing in any of the other compared neurons and the *first-order pre-CI*, $\tau'_{j,-1}$, eq. (14) can be modified to incorporate the joint probability function, the *J-INAT/CI PMF* as follows:

$$P(\tau'_{l,-k,n}, \tau'_{j,-1}) = F(\tau'_{l,-k,n}, \tau'_{j,-1}) - F(\tau'_{l,-k^-,n}, \tau'_{j,-1}) = \int_{\tau'_{l,-k^-,n}}^{\tau'_{l,-k,n}} f_{l,-k,n}(t) dt .$$

(16)

This function will provide the joint probability between the firing of a spike in the reference neuron r in relation to the exactly k spikes that have fired in any other neuron C_l within the network prior to the last spike in the compared spike train c_j. Similarly, the two-dimensional *J-INAT/CI histogram* can be constructed from experimentally recorded spike train data to estimate the *J-INAT/CI PMF* to reveal the time interval in which *k* spikes from any other recorded neurons in the network are needed to generate a spike in the reference neuron.

3. RESULTS

To illustrate how these analyses can be used to extract the exact number of sequentially fired spikes in other neurons in the network that contribute to the firing of a spike in a neuron, we simulated a series of spike trains with different firing characteristics. Figure 1 shows the *second-order J-ISAT/CI histogram* of a spike train that fired periodically (mean ISI = 5 msec) with a Gaussian variance, coupled with another periodic spike train (mean ISI = 5 msec).

It can be seen that, congruent with the theoretical prediction, the maximal joint probability for two and exactly two spikes that fired before the reference neuron has fired is at the time-lag of 10 msec (i.e., two periods) along the inter-arrival time axis. The bell-shaped probability profile shows the gradual increase in conditional probability of firing given that two successive spikes have fired prior to the firing of the reference neuron. Thus, this is an illustration of how temporal summation can be detected using this spike train analysis.

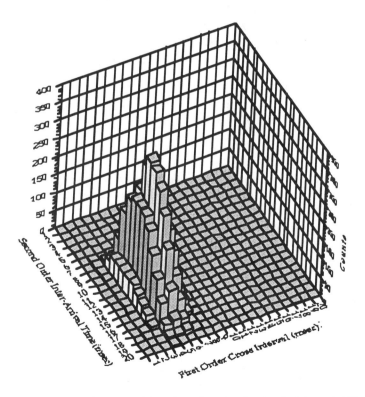

Figure 1. *Second-order J-ISAT/CI histogram* of Gaussian spike trains showing the joint probability reflecting the contribution of two and exactly two spikes in one compared neuron to the firing of reference neuron, representing *temporal summation.*

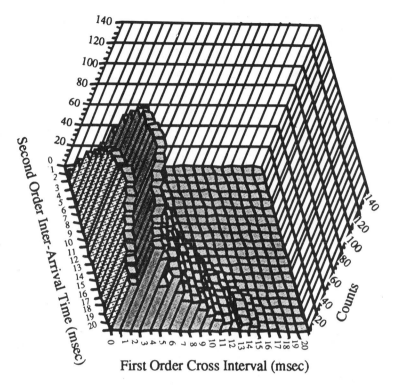

First Order Cross Interval (msec)

Figure 2. *Second-order J-ISAT/CI histogram* of Poisson spike trains showing the joint probability reflecting the contribution of two and exactly two spikes in any other compared neurons to the firing of reference neuron, representing *spatial summation*.

Figure 2 shows the *second-order J-INAT/CI histogram* of another simulation with a Poisson spike trains. It can be shown that the lopsided bell-shaped probability profile along the inter-arrival time axis is broader than the previous example due to the large variability contributed by multiple neurons used as the compared neurons. The negative exponential profile along the cross-interval axis is due to the Poisson firing characteristic of the reference neuron used in this example. This example shows the conditional probability of firing given that two spikes have fired prior to the firing of the reference neuron independent of which neuron those two spikes originated. This serves as an illustration of how spatial summation can be detected using this spike train analysis.

4. DISCUSSION

The two spike train analysis techniques introduced, *J-INAT/CI* and *J-ISAT/CI* show the temporal and spatial summation of preceding spikes leading to firing as designed. The plot in the first example illustrates the probability of next spike firing with respect to two and only two preceding spikes summed temporally, while the second plot shows the probability of next spike firing with respect to two and only two preceding spikes summed spatially. These methods allow for the quantification of the contribution of the *exact* number of spikes leading to the next spike firing both temporally or spatially rather than *all* of the preceding spikes as revealed by conventional cross-correlation methods.

ACKNOWLEDGMENTS

This research was supported in part by ONR grant number N00014-94-1-0686 to DCT.

REFERENCES

1. Aertsen, A. M. H. J., Gerstein, G. L, Habib, M. K. and Palm, G. (1989) Dynamics of neuronal firing correlation: modulation of "effective connnectivity." *Journal of Neurophysiology.* 61: 900–917.
2. Perkel, D. H., Gerstein, G. L. and Moore, G. P. (1967a) Neuronal spike trains and stochastic point process. I. The single spike train. *Biophysical Journal.* 7: 391–418.
3. Perkel, D. H., Gerstein, G. L. and Moore, G. P. (1967b) Neuronal spike trains and stochastic point process. II. Simultaneous spike trains. *Biophysical Journal,* 7: 419–440.
4. Perkel, D. H., Gerstein, G. L., Smith, M. S. and Tatton, W. G. (1975) Nerve-impulse patterns: A quantitative display technique for three neurons. *Brain Research,* 100: 271–296.
5. Rodieck, R. W., Kiang, N. Y.-S. and Gerstein, G. L. (1962) Some quantitative methods for the study of spontaneous activity of single neurons. *Biophysical Journal.,* 2: 351–368.
6. Smith, C. E. (1992) A heuristic approach to stochastic models of single neurons. In: *Single Neuron Computation* (T. McKenna, J. Davis and S. Zornetzer, eds.) Academic Press, San Diego, CA. pp. 561–588.
7. Tam, D. C. (1993) A multi-neuronal vectorial phase-space analysis for detecting dynamical interactions in firing patterns of biological neural networks. In: *Computational Neural Systems.* (F. H. Eeckman and J. M. Bower, eds.) Kluwer Academic Publishers, Norwell, MA. pp. 49–53.
8. Yang, X. and Shamma, S. (1990) Identification of connectivity in neural networks. *Biophysical Journal,* 57: 987–999.

32

INTEGRATE-AND-FIRE NEURONS MATCHED TO PHYSIOLOGICAL F-I CURVES YIELD HIGH INPUT SENSITIVITY AND WIDE DYNAMIC RANGE

Todd W. Troyer and Kenneth D. Miller

Keck Center for Integrative Neuroscience
University of California
San Francisco CA
E-mail: todd@phy.ucsf.edu, ken@phy.ucsf.edu

ABSTRACT

Simple integrate-and-fire neurons that accurately reproduce *in vitro* data from cortical regular spiking cells can display surprisingly sophisticated behavior. To reproduce *in vitro* f-I plots, voltage after spikes was reset to 5 mV below threshold, and simple spike rate adaptation was added. Small reset results in input sensitivity (high gain) on short time scales; adaptation leads to wide dynamic range over longer time scales. The model displays physiological ISI variability using either delta function or temporally realistic synaptic conductances. Cross correlation between pre- and post-synaptic spikes suggests that cortical neurons may be capable of transmitting information on the millisecond time scale.

1. INTRODUCTION

One line of investigation into the issue of neural coding has been to explore the information processing abilities of simple model neurons. Is the classical notion of spike generation consistent with what we know about information transmission in cortical neurons? By "classical notion" we mean viewing neurons as integrating many synaptic inputs until some voltage threshold and then producing a spike. Such a neuron spikes at a rate dependent upon the net synaptic current reaching the soma and thus can embody the standard notion of "rate coding." Are simple "integrate-and-fire" models also capable of more sophisticated "temporal coding" strategies? We have found that such simple model neurons exhibit surprisingly complex behavior when their parameters are fit to reproduce the input/output properties of cortical regular spiking cells.

Computational Neuroscience
edited by Bower, Plenum Press, New York, 1997

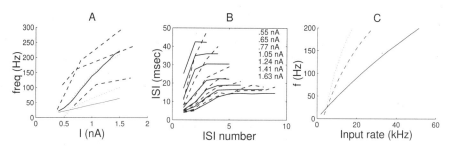

Figure 1. Model Parameters: *Solid lines: model; dashed lines: experiment (A,B).* RC parameters matched to slice recordings of regular spiking cells [3]: $V_{rest} = -74\ mV$; $1/g_{leak} = 40M\Omega$; $C = \tau g_{leak}$, $\tau = 20\ msec$; spike width $t_{spike} = 1.75\ msec$. $V_{thresh} = -54\ mV$ was set 20 mV above V_{rest}. Adaptation was modeled as a spike-triggered K conductance ($E_K = -90$), $g_K = 15\ nS$ at spike conclusion, decaying with $\tau_K = 40\ msec$. V_{reset}, τ_K, and g_K were fit to experimental f-I curves [3]. $V_{reset} = -59\ mV = V_{thresh} - 5\ mV$. Synaptic inputs initially were Poisson-distributed delta-function conductance changes. g_{syn} was taken from an exponential distribution with mean $\bar{g}_{syn}$; values $> 4\bar{g}_{syn}$ were then reset to this maximum value. For excitatory synapses, $V_{ex} = 0mV$, and $\bar{g}_{ex} = 3.4\ nS\ msec$, yielding mean EPSP size of .49 mV. For inhibitory synapses, $V_{in} = -70mV$, and $\bar{g}_{in} = 22.8\ nS\ msec$, yielding IPSPs twice as large as EPSPs (at V_{thresh}). Inhibition was quantified as the ratio R of the mean inhibitory current to the mean excitatory current at V_{thresh}. **A.** Firing frequency f (1/1st ISI) vs. current I for model (solid) and experiment (dashed, 3 cells). Lower line shows 1/7th ISI (after adaptation). Dotted lines show the model without adaptation for both high gain ($V_{reset} = -60\ mV$) and low gain ($V_{reset} = V_{rest}$) cases. **B.** Adaptation: ISI length vs. ISI number for model (solid) and experiment (dashed) at different levels of current injection (shown at right). Note: to match physiological current ranges with adaptation, V_{rest} was changed to $-60\ mV$. This shifts the f-I plot to the left but does not affect the slope (gain). **C.** Steady state output rate vs. excitatory input rate for high gain nonadapting (dotted), low gain nonadapting (dashed), and high gain adapting (solid) models ($R = .75$).

2. RESULTS

2.1. Matching Physiological f-I Curves

A rate encoding neuron is characterized by its input/output function, i.e. the transformation from presynaptic input rates to postsynaptic firing rate. While experimental estimates of synaptic input *in vivo* provide only weak constraints [4], the input/output behavior *in vitro* of a neuron can be characterized by an f-I curve: a plot of instantaneous spike frequency (f) vs. intracellularly injected current (I). By matching the *in vitro* data from McCormick et al. [3], we have constructed an integrate-and-fire model of a cortical "regular spiking" cell (fig. 2A,B; [8, 9]). After matching RC parameters and setting spike threshold, *the only free parameter in our basic model is the after-spike reset voltage* V_{reset}. Traditionally, $V_{reset} = V_{rest}$. However, the fast spiking conductances and those contributing to the resting potential are unrelated. Therefore, *we set* V_{reset} *to match physiologically observed f-I curves.* We also incorporated spike rate adaptation by adding a spike-triggered potassium conductance. Parameters for this conductance were also set by fitting experimental f-I curves.

2.2. Sensitivity and Dynamic Range

The neuron's sensitivity to changes in input current *in vitro*, the *neuronal gain*, is simply the slope of the f-I curve. By definition, a cell with high gain is sensitive to small changes in input current. However, input sensitivity implies narrow dynamic range: beyond a small range of superthreshold input currents, the output rate saturates. This problem is greatly reduced by incorporating spike rate adaptation into the model. Adaptation provides negative feedback

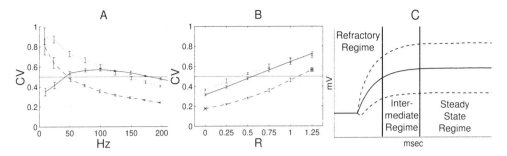

Figure 2. ISI Variability and Dynamic Range: A,B: Solid lines: full model; dashed lines: low gain nonadapting model ($V_{reset} = V_{rest}$); dotted lines: high gain nonadapting model. Variability assessed as CV of the ISI distribution (mean CV ± standard deviation for 10 simulations of 10 sec each). Thin line shows CV = .5 (rough lower bound for data in [6]). **A.** CV vs. firing rate ($R = .75$). **B.** CV vs. inhibition ratio R. Output firing rate = $100(\pm5)$ *Hz*. Approximations to an EPSP integrator [6] and balanced inhibition [4] models marked by 'x' and '*' respectively. **C.** Schematic showing three regimes of spike behavior (see text). Shown are the mean (solid) and ± 2 standard deviations of membrane voltage following a spike ($V = V_{reset}$ for initial $t_{spike} = 1.75$ msec; spiking thereafter turned off).

that reduces the gain to slowly varying inputs. In our model, spike repolarization determines the neuronal gain on the time scale of single interspike intervals (ISIs), whereas adaptation determines neuronal gain to slowly varying input currents. This "gain control" feature of adaptation applies both to current injection *in vitro* (lower line in fig. 1A) and synaptic input (fig. 1C).

While there is no direct way to quantify *in vivo* "sensitivity", an indirect measure is found in the variability in the distribution of interspike intervals. Softky and Koch [6] have shown that spike trains from cortical neurons display high variability, reporting values of the coefficient of variation (CV – the standard deviation divided by the mean) of the ISI distribution in the range .5 to 1. For an integrate-and-fire neuron needing N randomly occurring excitatory postsynaptic potentials (EPSPs) to reach threshold, one expects the CV to be approximately $1/\sqrt{N}$. For reasonable values of N (≈ 30), CV $\approx .18$, i.e. *an EPSP integrator yields low variability*. Shadlen and Newsome [4] pointed out that if large numbers of excitatory inputs are tightly balanced by inhibition, the membrane voltage follows a "random walk" and ISIs are quite variable. However, neither of these groups considered models that accurately reproduce *in vitro* f-I curves. Recently, we showed that a simple integrate-and-fire model matched to physiological f-I curves displays high ISI variability for a wide range of firing rates and levels of inhibition (Fig. 2A,B; [8, 9]).

2.3. An Intuitive Picture

After each spike, there is a transient period in which the cell is less likely to fire, the so-called "relative refractory period". Spike rates *in vitro* are determined by the rate at which the injected current overcomes this refractory tendency not to spike. When considering the input the cell receives *in vivo*, however, refractoriness only tells half the story. After the effects of the previous spike become negligible, the cell spikes when *fluctuations* in the synaptic input are sufficient to push the membrane voltage above threshold. Thus, in a simple intuitive picture for *in vivo* spiking, ISIs are divided into three regimes (fig. 2C; [1, 9]). In the initial *refractory regime*, the state of the neuron is dominated by the recovery from the previous spike. In this regime, $1/\sqrt{N}$ arguments are valid and spiking is regular. In the final, *steady state regime*, synaptic variation causes the cell to fluctuate about some steady state mean voltage

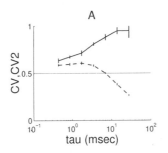
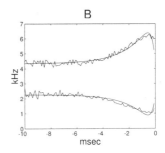

Figure 3. Effects of Finite Synaptic Time Course: A. *Solid lines: CV. Dashed lines: CV2.* Two variability measures vs. synaptic decay constant τ. The area under the conductance curve was fixed at 5 *nS msec* for excitatory and 33.6 *nS msec* for inhibitory synapses. To quantify "burstiness", we plot τ vs. both CV (solid) and an alternate measure of variability, CV2 (dashed) [2]. CV2 is the CV of consecutive pairs of ISIs, averaged over the whole ISI distribution. Since CV2 depends only on consecutive intervals, it is much less sensitive than CV to changes in short term firing rate (e.g. alternating low and high firing rates), while closely approximating CV for Poisson distributed spikes with constant rate. Total input was adjusted to ensure final firing rate of 100 ($\pm$5) Hz. Inhibition/excitation ratio $R = .75$. **B.** Average rate of presynaptic input before a spike (occurring at t=0). Upper trace: excitatory input. Lower trace: inhibitory input. Smooth line is fit with scaled and inverted trace of the postsynaptic conductance (rise time = .25 *msec*; decay = 1.75 *msec*). Mean output rate $\approx$ 25 *Hz*; $\approx$ 5000 spikes collected.

(necessarily subthreshold). Since threshold crossings in this regime result from random voltage fluctuations, spike statistics in this regime are Poisson – threshold crossings are equally likely to occur in any small interval. In the *intermediate regime*, the mean voltage is still significantly rising, yet typical voltage fluctuations can be sufficient to cross threshold. For neurons with high gain and hence weak refractoriness, steady state behavior predominates. Thus, in contrast to the common intuition, *we should expect cortical neurons to display high ISI variability.*

2.4. Neuronal Processing on Multiple Time Scales

To ensure robustness of these mechanisms, we varied the time course of synaptic transmission. Short synaptic time constants give results similar to delta function conductances. If there is only a single synaptic time constant, and it becomes long relative to the mean ISI, then spike trains come to display an unrealistic burstiness: correlations induced in the somatic input current yield long periods of subthreshold current (long ISIs) alternating with periods of superthreshold current (bursts of spikes with short ISIs) (fig. 3A). However, we do not see this burstiness with physiologically realistic mixtures of fast and slow synapses: with at least one fast synaptic time constant, e.g. AMPA and/or GABA$_A$, physiologically realistic CVs and distributions are robustly obtained.

In the steady state regime, the *timing* of individual spikes depends on the *fluctuations* in the synaptic input current, whereas the *probability* of spiking depends on the mean current. This can be seen from the cross-correlation between an output spike train and its presynaptic inputs (fig. 3B). As expected, before each spike a cell receives more excitatory and fewer inhibitory inputs [5, 7]. Surprisingly, the time course of this effect closely follows the postsynaptic conductance (1-2 msec), and is largely independent of the membrane time constant.

3. SUMMARY

By fitting the parameters of a very simple neural model to available *in vitro* data [3], we see that classical notions of neuronal integration are consistent with sophisticated information processing abilities. With parameters set to reproduce the high gain of cortical regular spiking cells, the neuron mainly operates in the steady state regime and is sensitive to the fluctuations in input current. Cells operating in this regime may be capable of spiking with precision on the millisecond time scale (fig. 3B), and could process information using a scheme radically different from the usual "rate code" ([1, 7], but see [5]). However, our simulations show that such cells can also function as "noisy rate encoders", with neuronal gain modulated in time to increase dynamic range. This work suggests that the different temporal filtering properties of (fast) AMPA and (slow) NMDA receptors in combination with differing neuronal gain on fast and slow time scales may enable networks of simple integrate-and-fire neurons to *simultaneously* process information on *multiple* time scales.

REFERENCES

[1] M. Abeles. *Corticonics: Neural Circuits of the Cerebral Cortex.* Cambridge University Press, Cambridge, 1991.

[2] G.W. Holt, W.R. Softky, C. Koch, and R.J. Douglas. A comparison of discharge variability *in vitro* and *in vivo* in cat visual cortical neurons. *Journal of Neurophysiology*, 75(5):1806–1814, 1996.

[3] D.A. McCormick, B.W. Connors, J.W. Lighthall, and D.A. Prince. Comparative electrophysiology of pyramidal and sparsely spiny stellate neurons of the neocortex. *J. Neurophysiol.*, 54:782–805, 1985.

[4] M.N. Shadlen and W.T. Newsome. Noise, neural codes and cortical organization. *Curr. Opin. Neurobiol.*, 4:569–579, 1994.

[5] M.N. Shadlen and W.T. Newsome. Reply: Is there a signal in the noise? *Curr. Opin. Neurobiol.*, 5:248–250, 1995.

[6] W.R. Softky and C. Koch. The highly irregular firing of cortical cells is inconsistent with temporal integration of random EPSPs. *Journal of Neuroscience*, 13(1):334–350, 1993.

[7] W.R. Softky. Simple codes versus efficient codes. *Curr. Opin. Neurobiol.*, 5:239–247, 1995.

[8] T.W. Troyer and K.D. Miller. A simple model of cortical excitatory cells linking NMDA-mediated currents, ISI variability, spike repolarization and slow AHPs. *Society for Neuroscience Abstracts*, 1995.

[9] T.W. Troyer and K.D. Miller. Physiological gain leads to high ISI variability in a simple model of a cortical regular spiking cell. *Neural Computation*, Vol. 9 No. 4, 1997. In Press.

33

ON THE GENERATION OF RANDOM DENDRITIC SHAPES

Jaap van Pelt,[1*] Alexander E. Dityatev,[2] and Andreas Schierwagen[3]

[1]Netherlands Institute for Brain Research
Meibergdreef 33, 1105 AZ Amsterdam, The Netherlands
j.van.pelt@nih.knaw.nl
[2]Molecular Neurobiology Center
University of Hamburg
Martinistrasse 52, D-20246 Hamburg, Germany
dityatev@plexus.uke.uni-hamburg.de.
[3]Department of Informatics
University of Leipzig
Augustusplatz 10/11, D-04109 Leipzig, Germany
schierwagen@informatik.uni-leipzig.dbp.de

1. INTRODUCTION

Dendritic branching patterns are complex and show a large degree of variation in their shapes. This variation can be found in typical shape parameters, such as the number, length and connectivity pattern (topological structure) of the constituent segments. Dendritic branching patterns emerge during neuronal development from the behavior of growth cones which determine the processes of branching and lengthening of segments (e.g., Bray, 1992). This dynamic behavior of growth cones is the result of cellular responses to the local environments (e.g., Kater et al., 1994; Letourneau et al., 1994). Many mechanisms are involved in growth cone behavior, making it plausible to hypothesize that dendritic arborizations emerge from stochastic behavior of growth cones. A crucial test for this hypothesis is to show that the characteristic variations in dendritic morphologies can be reproduced by a process of random branching. This study concentrates on the variation in the number of segments and their connectivity patterns in the trees. Metrical properties will be ignored and dendrites are reduced to their skeleton of segments and bifurcation points.

* correspondence author

2. MODELING DENDRITIC GROWTH

It is assumed that the growth of a dendrite proceeds by a sequence of branch events. At each branch event a new (terminal) segment is attached to one of the existing segments in the tree, thereby creating also a new bifurcation point. Randomness is assumed with respect to 1) the occurrence of a branch event in time, and 2) the selection of the branching segment from the present ones. Thus, the question focusses on whether a simple scheme of probabilities underlying these random choices can be found so as to accurately describe the variation in observed dendritic trees. Previous studies have already shown that the topological variation in dendritic trees from a variety of neuronal cell types and species can accurately be described by assuming randomness in the selection of the branching segment (Van Pelt et al., 1992; Dityatev et al., 1995). General agreement was found when branching was restricted to terminal segments while the probability for a terminal segment to be selected could depend slightly on its centrifugal order position in the tree (i.e., the number of bifurcation points on the path from the root to the terminal segment). Therefore, branching will in the following be restricted to terminal segments only.

2.1. Variation in the Number of Segments, *BE*-Growth Model

Variation in the number of segments per dendritic tree emerges when trees experience a variable number of branch events during outgrowth. In the so-called *BE*-growth model it will be assumed that branch events occur at random points in time. To this end, the developmental time period T will be divided into a number of N time-bins with not necessarily equal durations. In each time-bin i a terminal segment in the tree may branch with a probability given by $p_i = B / Nn_i^E$, with the scale factor B denoting the expected number of branch events in the full period of an isolated segment and the parameter E denoting the dependence of the branch probability of a terminal segment on the total number of terminal segments n_i in the growing tree. The number of time-bins is taken sufficiently large so as to make the branch probabilities p_i much smaller than one, and making the probability for more than one branch event per time bin in the tree negligibly small.

Examples of growth sequences are given in Figure 1 for parameter values $B=3$ and $E=0$. The sequences show the random occurrences of branch events in time as well as the variation in the number of terminal segments in the final trees. This variation is most

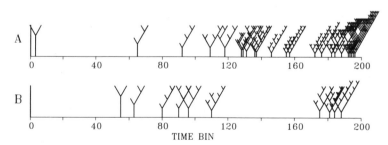

Figure 1. Growth of a branching pattern versus a time-bin scale. A tree is plotted at the time-bin in which a branch event actually has occurred. Trees are plotted in a standardized way, such that at each bifurcation point the largest subtree in the pair is pointing to the right. The full period is divided into $N=200$ time-bins. At each time-bin i, any of the terminal segments in the growing tree may branch with a probability $p_i = B / N n_i^E$, with n_i denoting the number of terminal segments in the tree. The two examples are calculated for the parameter values $B=3$ and $E=0$, making the branch probability per time-bin per terminal segment constant and equal to $p=0.015$.

clearly shown in Figure 2 in which the distribution functions are displayed of the terminal segment number in trees obtained for different values of the growth parameters B and E. Additionally, the sequences show that the number of branch events in the full period can be much larger than the value of $B=3$, because of the proliferation in the number of terminal segments during growth.

For $E=0$, all the distributions are monotonously decreasing with increasing number of terminal segments and have long tails for even small values of the parameter B. For $E>0$, the distributions become unimodal, and increasingly narrower with a disappearence of the long tails for increasing values of E.

2.2. Variation in Topological Structures, QS-Growth Model

As shown in previous studies (Van Pelt et al., 1992, Dityatev et al., 1995), the random selection of segments for branching is a sufficient condition to describe accurately the observed topological variation found in sets of dendritic trees with an equal number of terminal segments, i.e., having experienced an equal number of branch events. The question when these events have occurred in time, has not been discussed in these studies. In the so-called QS-model, the selection probability of a segment is taken to depend on the type of the segment (intermediate or terminal) and on its centrifugal order. The selection probability p_{term} for branching of a terminal segment at centrifugal order γ is defined by $p_{term} = C\,2^{-S\gamma}$, with parameter S modulating the dependence on centrifugal order, and C being a normalization constant to make the sum of the selection probabilities of all the segments in the tree equal to one. For $S=0$, all terminal segments have the same probability for being selected for branching. For $S=1$, the probability of a terminal segment decreases by a factor of two for each following centrifugal order. The selection probability of an intermediate segment p_{int} for branching relates to that of a terminal segment of the same order via $p_{int} = p_{term}\,Q\,/\,(1-Q)$. The parameter Q, having values between 0 and 1, roughly indicates the total branch probability for all intermediate segments in a tree, with for $Q=0$ no branching of intermediate segments, and for $Q=1$ branching of intermediate segments only. With the two parameters Q and S, a range of growth modes can be described, including the well known random terminal and random segmental branching modes. An accurate

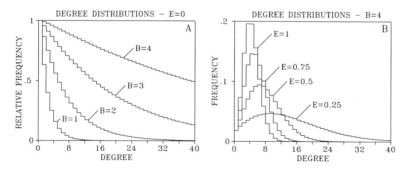

Figure 2. Distributions of the number of terminal segments per dendritic tree (*degree*) obtained by growth modes in which (A) each terminal segment has a constant branch probability per time-bin p, given by $p=B\,/\,N$, with $N=500$ and $B = 1, 2, 3$ or 4, respectively, and (B) the branch probability p_i per terminal segment per time-bin i depends on the total number of terminal segments n_i with $p_i=B\,/\,Nn_i^E$, $N=500$, $B=4$ and $E = 0.25, 0.5, 0.75$ and 1.0, respectively. The distributions for $E=0$ are monotonously decreasing, but become for $E > 0$ unimodal and narrower for increasing values of E.

description of the topological variability in dendrites from several neuron types and species could be obtained by assuming branching to occur at terminal segments only (i.e., $Q=0$), with possibly a slight dependence of the selection probability on centrifugal order (small or zero value for S) (Van Pelt et al., 1992, Dityatev et al., 1995). For $Q=0$, the QS-model reduces to the S-model.

2.3. Variation in Both the Number of Segments and Topological Structure, *BES*-Growth Model

In the *BE*-model all terminal segments have equal probability for branching and the topological variation, produced by the *BE*-model is similar to that produced by the random terminal growth mode $(Q=0, S=0)$. An accurate account of the topological variability can now be given in a combined *BES*-model by making the branch probability of a terminal segment p_i at time-bin i also dependent on the centrifugal order of the segment, like in the S-model, such that $p_i = C \, 2^{-S\gamma} \, B \, / \, N \, n_i^E$, with γ denoting the centrifugal order of the terminal segment and $C = n \, / \Sigma_{j=1}^{n} 2_j^{-S\gamma}$ being a normalization constant, with a summation over all n terminal segments. The normalization ensures that the summed branch probability per time-bin of all the terminal segments in the tree is independent of the value of S.

3. RESULTS

Several groups of observed dendrites have been analysed for their variation in terminal segment number. For each group, the model parameters B and E were optimized such that the mean and standard deviation of the model generated distribution optimally corresponded to the observed values. The shape of the model generated distribution has subsequently been tested statistically against the observed one by means of the Chi-square test. In none of the analysed groups did the shape of the optimized model distribution differ significantly from the observed distribution. Figure 3 shows the results for rat visual cortex pyramidal and multipolar nonpyramidal cells, and human dentate granule cells, based on the optimized parameter values for (B,E) of $(3.25,0.66)$, $(1.49, 0.42)$, and $(2.27,0.39)$, respectively. The figure demonstrates (i) the large variation in the number of

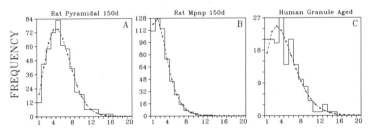

OBSERVED AND EXPECTED DISTRIBUTIONS

NUMBER OF TERMINAL SEGMENTS

Figure 3. Frequency distributions of the number of terminal segments per dendritic tree for (A) pyramidal cell basal dendrites and (B) multipolar nonpyramidal dendrites from 150 days old rat visual cortex (Uylings et al., 1990), and (C) dendrites of aged human fascia dentata granule cells (De Ruiter and Uylings, 1987). Each panel displays the observed frequency distribution, plotted as a continuous line histogram, and the expected distribution for the optimized growth model, plotted as a dashed curve.

terminal segments per dendritic tree, (ii) the differences in shapes of the terminal segment number distribution between different cell types, and (iii) the accuracy with which the growth model can reproduce the shape of the observed distributions.

4. DISCUSSION

A model is obtained for the growth of dendritic branching patterns that accurately describes the variation in the number of segments and their connectivity pattern, as found in observed dendritic trees. The assumptions in the *BE*-model result in a good description of the variation in the number of terminal segments per dendritic tree. The nonzero values for the optimized parameter E in the three examples in Figure 3 indicate that during growth of these cells, the branch probability per terminal segment per time-bin has declined with the increasing number of terminal segments. Constant branch probabilities are obtained for $E=0$, but result in terminal segment number distributions with unrealistically long tails (Figure 2A). Including a distinction of terminal segments for their centrifugal order position in the tree (*BES*-model) additionally gives a good description of the variation in topological tree types. These results provide a strong support for the random branching hypothesis of growth cones during neurite outgrowth.

ACKNOWLEDGMENTS

This work was supported by NATO grant CRG 930426.

5. REFERENCES

Bray, D. 1992. Cytoskeletal basis of nerve axon growth. In: P.C. Letourneau and S.B. Kater (eds), The nerve growth cone. Raven Press, New York, pp. 7-17.

De Ruiter, J.P. and H.B.M. Uylings. 1987. Morphometric and dendritic analysis of fascia dentata granule cells in human aging and senile dementia. Brain Res., 402, 217–229.

Dityatev, A.E., Chmykhova, N.M., Studer, L., Karamyan, O.A., Kozhanov, V.M., and H.P. Clamann. 1995. Comparison of the topology and growth rules of motoneuronal dendrites. J. Comp. Neurol., 363, 505–516

Kater, S.B., Davenport, R.W., and P.B. Guthrie. 1994. Filopodia as detectors of environmental cues: signal integration through changes in growth cone calcium levels. In: J. van Pelt, M.A. Corner, H.B.M. Uylings and F.H. Lopes da Silva (eds), The Self-Organizing Brain: From Growth Cones to Functional Networks, Progress in Brain Research, Vol 102. Elsevier, Amsterdam, pp. 49–60.

Letourneau, P.C., Snow, D.M., and T.M. Gomez. 1994. Growth cone motility: substratum-bound molecules, cytoplasmic $[Ca^{2+}]$ and Ca^{2+}-regulated proteins. In: J. van Pelt, M.A. Corner, H.B.M. Uylings and F.H. Lopes da Silva (eds), The Self-Organizing Brain: From Growth Cones to Functional Networks, Progress in Brain Research, Vol 102. Elsevier, Amsterdam, pp. 35–48.

Uylings, H.B.M., Van Eden, C.G., Parnavelas, J.G., and A. Kalsbeek. 1990. The prenatal and postnatal development of rat cerebral cortex. In: B. Kolb and R.C. Tees (eds). The cerebral cortex of the rat. MIT Press, Cambridge MA, pp. 35–76.

Van Pelt, J., and R.W.H. Verwer. 1986. Topological properties of binary trees grown with order-dependent branching probabilities. Bull. Math. Biol., 48, 197–211.

Van Pelt, J., Uylings, H.B.M., Verwer, R.W.H., Pentney, R.J., and M.J. Woldenberg. 1992. Tree asymmetry - a sensitive and practical measure for binary topological trees. Bull. Math. Biol., 54, 759–784.

34

IRREGULAR FIRING IN CORTICAL CIRCUITS WITH INHIBITION/EXCITATION BALANCE

C. van Vreeswijk and H. Sompolinsky

Racah Institute of Physics and Center for Neural Computation
Hebrew University
Jerusalem, Israel

ABSTRACT

Cortical neurons in the intact brain show irregular spiking patterns. The origin of this irregularity is unknown. We study theoretically the hypothesis that this irregularity is due to approximate balance between the inhibitory and excitatory inputs into the cortical neurons. We model the cortical circuits by sparsely connected networks of inhibitory and excitatory neurons with relatively strong synapses. We show that a state with balanced excitatory and inhibitory inputs into the cells emerges without fine tuning of the parameters. The balanced state is characterized by strongly chaotic dynamics, even with constant external input. Despite the highly non-linear dynamics of single neurons, such networks respond linearly to external inputs and track changing stimuli on a time scale much smaller than the integration time constant of the neuron.

Recent theoretical and experimental studies have focused on the source of irregular firing in the brain [1]. In vitro experiments show regular firing patterns of cortical neurons when they are stimulated by constant current [2] implying that the irregular spiking in vivo is due to fluctuating synaptic currents. The origin of these fluctuations is not yet known. Usually it is assumed that the individual synaptic inputs into a cell are stochastic. However, it is not clear what is the origin of this stochasticity. A second question is why the fluctuations in the individual synaptic inputs are not averaged out in the summation over the large number of synapses that project to a cell.

One possibility is that the fluctuating inputs are synchronized and therefore strongly affect the postsynaptic cell's potential. Model networks with a high degree of connectivity can display synchronized irregular firing [3]. However, such models yield states with strongly synchronized bursts. It is questionable whether cortical networks display this kind of behavior.

A second possibility is that the fluctuations of the individual synaptic inputs remain effective because of cancellation between the mean excitatory and inhibitory currents [4]. Here we address the questions: Can networks with deterministic dynamics generate a highly irregular state with balanced inhibitory and excitatory currents and low levels of synchrony?

How much fine tuning of the synaptic parameters in required to ensure this balance? What are the characteristics of this state and what are its functional implications?

To address these questions we consider a network of N_E excitatory and N_I inhibitory neurons [5]. The network also receives input from N_0 excitatory neurons outside of the network. We will use the subscripts E, I and 0 to denote the excitatory, inhibitory and external population, respectively. The connection are random but fixed in time. The probability that a neuron of the k-th population projects to a neuron of of population l is K/N_l, where K is the average number of synapses of each population that project to a cell. The synaptic strength of these projections is $J_{kl}/\sqrt{K}$. Here $k = E, I$ and $l = 0, E, I$. The parameters J_{k0} and J_{kE} are positive, while J_{kI} are negative.

The state of the neurons is described by a binary variable that denotes whether the neuron is active or quiescent. A neuron is active if its post-synaptic potential exceeds its threshold when the neuron is updated. Inhomogeneities between the neurons are modeled by choosing the thresholds of the neurons in population k from a uniform distribution between θ_k and $\theta_k + D$. The neurons are updated sequentially. A cell fires at its maximum rate if it is active every time it is updated. Excitatory neurons are updated once per time unit, inhibitory units every τ time units.

An important assumption of our model is that the total excitatory and inhibitory currents as well as the total current from outside of the network into a cell are large compared to the neuronal threshold. We model this by choosing thresholds θ_k that are of order 1 and assuming that the strength of individual synapses is of order $1/\sqrt{K}$ as stated above. Thus if $K = 1000$, roughly 30 excitatory inputs are needed to bring the neurons above threshold. We analyzed the properties of the network assuming a high degree of sparsity $K \ll N_k$.

To show that the stochastic nature of the firing is a property of the network, we assume a *constant* external input. We denote the input to the excitatory and inhibitory neurons as $J_{E0}m_0\sqrt{K}$ and $J_{I0}m_0\sqrt{K}$, respectively. Here m_0 is the mean rate of the input neurons.

We now outline the analysis of the stationary state of the network. If the external input is constant, $m_0(t) = m_0$, the network will settle into a stationary state in which the average rates of excitatory and inhibitory populations are constant, $m_k(t) = m_k$. Here m_k is the probability that a cell of population k updates to the active state.

In such a state a cell receive, on average, input from Km_k active cells in population k. Thus the average input, u_k, of cells in population k is given by

$$u_k = \sqrt{K} \left(\sum_{l=0,E,I} J_{kl}m_l \right).$$ (1)

Since the connectivity is sparse we can assume that the activity of the cells that project to a given cell are only weakly correlated. This means that the standard deviation of the input σ_k satisfies

$$\sigma_k^2 = \sum_{l=E,I} (J_{kl})^2 m_k.$$ (2)

Equations (1) and (2) imply that in general both the total excitatory and the total inhibitory synaptic input are of the order $\sqrt{K}$, while the fluctuations in these inputs are of order 1. To have stationary states in which the cells do not fire at their maximum rate or are completely quiescent, the net input has to be of the same order as the threshold. Thus, for large K, the

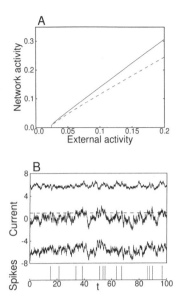

Figure 1. (A) Network rates as function of the input rates The excitatory rates (solid line) and the inhibitory rate (dashed line) are shown as function of the external rates. Above a minimum input that exceeds the threshold the network rates increase nearly linearly with in external rates. (B) Total excitatory input (upper trace), total inhibitory input (lower trace) and net input (middle trace) for a neuron in a large network with $K = 1000$. The inputs are shown in units of the threshold input (dashed line). In all figures we used $K = 1000$, $J_{E0} = J_{EE} = J_{IE} = 1$, $J_{I0} = 0.7$, $J_{EI} = -2.0$, $J_{II} = -1.7$, $\theta_E = 1.0$, $\theta_I = 0.7$, and $D = 0.3$.

total excitatory and inhibitory inputs have to nearly cancel each other. According to Eq. (1) this means that the network rates satisfy (to leading order in $1/\sqrt{K}$)

$$m_k = -\sum_{l=E,I} \mathbf{J}_{kl}^{-1} J_{l0} m_0. \tag{3}$$

Here $\mathbf{J}^{-1}$ is the inverse of the 2×2 matrix of connections J_{kl} between the network populations. The requirement that the network rates are positive leads to the constraint on the synaptic strengths $\sum_l \mathbf{J}_{kl}^{-1} J_{l0} < 0$.

The requirement that the total excitatory and inhibitory inputs cancel, leads according to Eq. (3) to a linear dependence of the network rates on the rate of the external input, even though the individual units do not have linear response characteristics. This linear dependence strictly holds only in the large K limit. However, Fig. 1A shows that this is also approximately true in networks with finite K. This figure shows the equilibrium rates as function of the input rate in a network with $K = 1000$. Figure 1B shows total excitatory, inhibitory and net input for a single neuron in the network. As explained, both the excitatory and inhibitory inputs are large compared to the threshold. However the average net input is below threshold. The fluctuations determine the precise timing of the threshold crossings, yielding a very irregular pattern of activity (lower panel) even though the external input does not fluctuate. This disorder is a signature of deterministic chaos.

In the balanced state different cells of the same population can have very different time averaged rates. Figure 2A shows the analytically determined rate distribution of the excitatory population. The rates have a broad unimodal distribution which is skewed and, for small m_k, extends up to $\sqrt{m_k}$. The distribution becomes more skewed as the average network rate

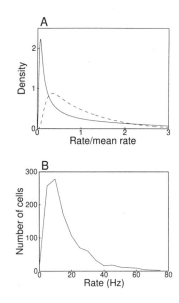

Figure 2. (A) Rate distribution for the excitatory population for different average rates. The external rates were adjusted so that the population rates are $m_E = 0.01$ (solid line) or $m_E = 0.03$ (dashed line). (B) Distribution of firing rates in the right prefrontal cortex of a monkey attending to a complex stimulus (light + sound) and executing a simple reaching movement. The rates were averaged over the duration of events that showed a significant increase in activity.

decreases. There are two sources for this distribution of the rates. The first source is due to the randomness of the connectivity. This causes the number of projections from each population that cells receive to be different. The second source is the inhomogeneity in the thresholds. Since the fluctuations in the input become small when the rate is low, only those neurons that have the lowest thresholds have a relatively high probability that their input will exceed the threshold. These cells will fire at a rate that is much above the average rate, most other cells will fire at a rate that is significantly below average. In Fig. 2B we show experimental data for the distribution of rates in the right prefrontal cortex of a monkey that is attending to a stimulus that consists of a sound and a light and executes a simple reaching movement [6]. The figure shows a broad unimodal distribution of the firing rates with most cells firing at a rate below the average rate and a small fraction of the neurons firing at much higher rate, in agreement with the theory.

We now address the potential the functional advantage of balanced networks. An important advantage of a balanced network is that it reacts on a short time scale to changes in the external input. This response is much faster than that for networks in which the excitatory and inhibitory inputs are of the same order as the neuronal threshold. This is shown in Fig. 3, which displays the response of the network to temporal changes in the external rate m_0. For comparison we also show the relatively slow response to this changing input in a model with threshold linear units, with synaptic strengths of order $1/K$. In both models the units have the same time constants and stationary rates.

Our network's ability to rapidly track changes in external inputs results from a combination of a large synaptic gain (of order $\sqrt{K}$) and the sequential updating. Because of the large synaptic gain a small change in the external rates results in a large change in the synaptic input, so that at a given time the cells that happen to be ready to update make a large change in

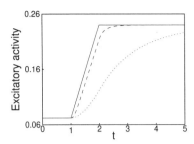

Figure 3. Comparison of the speed with which a balanced network and a network without balancing track changes the external rate. Between $t = 1.0$ and $t = 2.0$ the external rate is increased uniformly, after that it is kept constant again. The solid line shows the activity of a hypothetical network that tracks changes in the external rates infinitely fast. A balanced network (dashed line) tracks the external rates much faster than a network without balancing (dotted line).

their activity. Successive recruitment of groups of neurons leads to a rapid adjustment of the population rate to changes in input.

ACKNOWLEDGMENTS

We thank D. J. Amit, D. Hansel, and T. Sejnowski for extensive discussions. We are grateful to M. Abeles, H. Bergman, and E. Vaadia, for permission to present their data.

REFERENCES

[1] B. D. Burns and A. C. Webb, *Proc. R. Soc. Lond. B*194, 211 (1976);
R. J. Douglas and K. A. C. Martin, *Trends Neurosci.* 14, 286 (1991); W. R. Softky and C. Koch, *J. Neurosc.* 13, 334 (1993).
[2] G. R. Holt, W. R. Softky, C. Koch and R. J. Douglas, *J. Neurophysiol.* 75, 1806 (1996); Z.J. Mainen and T. Sejnowski. *Science* 268, 1503 (1995).
[3] D. Hansel and H. Sompolinsky, *Phys. Rev. Lett.* 68, 718 (1992); *J. Comp. Neurosci.* 3, 5 (1996); P. C. Bush and R. J. Douglas, *Neural Comp.* 3, 19 (1991).
[4] M. N. Shadlen and W. T. Newsome, *Curr. Opin. Neurobiol.* 4, 569 (1994); *Curr. Opin. Neurobiol.* 5, 248 (1995); W. R. Softky, *Curr. Opin. Neurobiol.* 5, 239 (1995); A. Bell, Z. Mainen, M. Tsodyks and T. Sejnowski, *Soc. Neurosci. Abs.* 20, 1527 (1994). D. J. Amit and N. Brunel, *Cerebral Cortex* (in press). M. Tsodyks and T. Sejnowski, *Network* 6, 111 (1995).
[5] C. van Vreeswijk and H. Sompolinsky, *Science* 274, 1724 (1996).
[6] Unpublished data from M. Abeles, H. Bergman, and E. Vaadia.

35

INTENSITY CODING IN AN OLFACTORY SENSORY NEURON

Influence of Neuron Structure and Auxiliary Cells

Arthur Vermeulen[1,2] and Jean-Pierre Rospars[1*]

[1]Laboratoire de Biométrie
Institut National de la Recherche Agronomique
78026 Versailles Cedex, France
[2]Laboratoire de Traitement d'Images et de Reconnaisance des Formes
Institut National Polytechnique
38031 Grenoble Cedex, France
(rospars@versailles.inra.fr)

ABSTRACT

Coding properties at steady-state of the receptor potential were analyzed in a biophysical model of an olfactory neuron with and without taking into account the auxiliary cells that surround it. It was found that the neuron model without auxiliary cells has generally a higher sensitivity and a narrower dynamic range. The dynamic range is wide when the input resistance of passive dendrite, soma and axon is small and the sensory dendrite is unbranched, whereas the sensitivity is high in the opposite conditions. Both coding properties are large for a long enough dendrite.

1. INTRODUCTION

In the present work we study how the receptor potential codes for odour intensity i.e. for the concentration of the odorant molecules. Odorant molecules activate receptor proteins borne by the membrane of sensory dendrites that ultimately trigger the opening of odorant-dependent ion channels. This opening corresponds to a conductance change that gives rise in turn to a membrane depolarization, called receptor potential. The receptor potential spreads passively to the other parts of the neuron - passive dendrite, soma and axon

* Corresponding author.

- and triggers the generation of action potentials at the axon initial segment. We have recently proposed a model[4,7] of the olfactory sensory neuron which describes these main
conversion steps. In the present investigation we are only concerned with the conversion
of the odorant-dependent conductance into the receptor potential. We continue our studies
of the influence of the neuron structure (i.e. morphology and electrical characteristics) on
its coding properties[7,8], and we start considering the influence of the auxiliary cells (also
called accessory cells), that surround the sensory neuron in insect sensilla.

2. MODELS STUDIED

2.1 Neuron Structure

The model of neuron (Fig. 1a) presents a sensory part and a non-sensory one. The
sensory part is a tree of N cylindrical dendrites, each of them with an electrotonic length L
(expressed in space constants $\lambda=(r_m/(r_i+r_e))^{1/2}$, see Fig. 1). The non-sensory part corresponds to passive dendrite, soma, axon and (when included) the auxiliary cells. Whatever
its shape, the non-sensory part can be described by its input resistance r_{in} (see section 3).
This structure is close to that actually observed in insects and vertebrates.

2.2 Auxiliary Cells

In "classical" models[6] the neuron is isolated from its auxiliary cells and the membrane
resting potential for elements of membrane (battery E_r) is the same all along the neuron. In
this case the receptor potential is generated by a battery E_d which corresponds to the equilibrium potential of the permeating ions. This battery is located on the sensory dendrites in series
with the odorant-dependent conductance g (see Fig. 1b). In reality, the olfactory sensory neurons are located in an epithelium. Because of differences in the ionic compositions of the
separated media, the electrical properties of the cell membranes at the external and internal
sides of the epithelium can be highly asymmetric. This is especially found in insect sensilla[3,5].
For this reason, Thurm[5] proposed a general sensory neuron model which includes two transepithelial current paths passing through respectively the sensory cell and the auxiliary cells.
Kaissling and Thorson[1,2,3] adapted this model to the moth sex-pheromone receptor cell. They

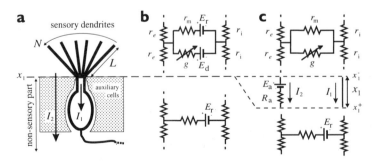

Figure 1. Model neuron (a) with N sensory dendrites and non-sensory part. (b,c) Electrical circuits of elements of
membrane for the models without (b) and with (c) auxiliary cells (dotted regions). Circuit (b) is a classical cable.
In (c), two transepithelial current paths are present, I_1 through the sensory cell and I_2 through the auxiliary cells
(simulated by battery E_a in series with resistance R_a). In the moth sex-pheromone receptor cell, no batteries are
present in the sensory dendrites. The conductance g is directly related to the odorant concentration.

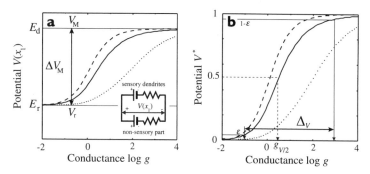

Figure 2. Example of membrane potential $V(x_1)$ (a) and normalized receptor potential V^* (b) for different values of the input resistance $r_{in}=0.1$ (dotted line), 1 (solid line) and 5 (dashed line), without taking into account the auxiliary cells. Inset in (a) shows the neuron equivalent circuit with sensory dendrites and non-sensory part replaced by their Thévenin equivalent circuits. Parameters: $N=L=1$.

modelled the auxiliary cells by a battery E_a in series with a resistance R_a and they suggested that no batteries are present in the sensory dendrites. Hence $E_d=0$ and the battery E_r is only present in the non-sensory part (see Fig. 1c).

3. RECEPTOR POTENTIAL

The membrane potential V as a function of the odorant-dependent conductance g is given by the solution of the cable equation[6]. The complexity of this solutions lead us to assume (i) steady-state conditions (e.g. the odour concentration is kept constant) and (ii) a uniform distribution of odorant molecules, receptors and ionic channels on the sensory dendrites[4]. Therefore the conductance g is the same all along these dendrites. The use of this spatially extended input distinguishes our work from previous ones using point stimulations[6]. The cable equation is solved in an original way by replacing successively the sensory dendrites and the non-sensory part by a Thévenin equivalent circuit which consists of a battery in series with a resistance (see inset in Fig. 2a). We obtain finally the membrane potential $V(x_1)$ at the junction of the two parts. It is sufficient to know this potential because the coding properties studied are the same at this point and at all points of the non-sensory part (including the axon initial segment).

Whatever the neuron structure, the potential V as a function of $\log g$ is always a sigmoid curve which varies between a minimum value V_r (obtained for $g=0$) and a maximum value V_M (obtained for $g \to \infty$). The receptor potential $V-V_r$ and its asymptotic value $\Delta V_M = V_M - V_r$ are illustrated in Fig. 2a at point x_1. The normalized receptor potential $V^* = (V - V_r)/\Delta V_M$ as a function of $\log g$ has the same shape (Fig. 2b) as $V(x_1)$ but no longer depends on the values of the batteries because it varies between 0 and 1. V^* at x_1 and at all points of the non-sensory part is given by

$$V^* = \frac{g\tanh\left(\sqrt{1+g}L\right)}{(1+g)\tanh\left(\sqrt{1+g}L\right)+\sqrt{1+g}\,/\,Nr_{in}},$$

(1)

for the model without auxiliary cells, and

$$V^* = \frac{\sqrt{1+g}\tanh\left(\sqrt{1+g}L\right)-\tanh(L)}{\sqrt{1+g}\tanh\left(\sqrt{1+g}L\right)+1/Nr_{\mathrm{in}}},$$

$$(2)$$

for the model with auxiliary cells, where r_{in} denotes the input resistance of the non-sensory part (expressed in units of $(r_i + r_e)\lambda$ and g is expressed in units of r_{m}^{-1} (see Fig. 1).

4. CODING PROPERTIES OF THE RECEPTOR POTENTIAL

4.1. Definition

The sigmoid curve of the receptor potential V as a function of $\log g$ is characterized by three coding parameters (see also Fig. 2) - its asymptotic value ΔV_{M}, its sensitivity (measured as the conductance $g_{V/2}$ at half maximal response, i.e. g for $V^*=0.5$) and its dynamic range Δ_V in log units (measured by the range of conductances between threshold $V^*=\varepsilon$ and saturation $V^*=1-\varepsilon$ with $\varepsilon=0.05$ for example). In the present work, we focus our attention on the last two parameters which are completely described by the normalized receptor potential V^* as given by Eqs. 1 and 2. Their dependence upon the neuron structure and the auxiliary cells is shown in Fig. 3.

4.2. Influence of Neuron Structure

Both sensitivity and dynamic range increase with L up to one space constant, whereas only sensitivity increases with r_{in}. When $r_{\mathrm{in}} \gg 1$ (e.g. in the case of a small soma), the model neuron has a high sensitivity but a small dynamic range. A high dynamic range may be obtained by decreasing r_{in} (i.e. increasing the size of the soma). The influence of

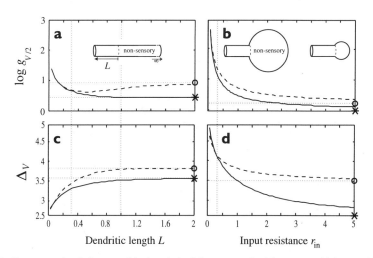

Figure 3. Coding properties. Influence of the length L of the sensory dendrites on sensitivity $g_{V/2}$ (a) and dynamic range Δ_V (c). Influence of the input resistance r_{in} of the non-sensory part (which depends for example on its radius, see inset in b) on sensitivity $g_{V/2}$ (b) and dynamic range Δ_V (d). Model without (solid line, star) and with (dotted line, circle) auxiliary cells. Star and circle correspond respectively to $L\to\infty$ and $r_{\mathrm{in}}\to\infty$. The case $r_{\mathrm{in}}\to\infty$ is mathematically equivalent to the commonly used point model neuron for which $V^*=g/(1+g)$. Parameters: $N=L=r_{\mathrm{in}}=1$.

the number of sensory dendrites N is not shown because it corresponds to a multiplication of r_{in} by N (see Eqs. 1 and 2) which means that increasing the number of sensory dendrites is equivalent to increasing the input resistance of the non-sensory part.

4.3. Influence of Auxiliary Cells

The coding properties are independent of the presence of the auxiliary cells for short sensory dendrites ($L<0.3$) or a small input resistance ($r_{in}<0.3$, e.g. a large soma with a single sensory dendrite). It becomes important to take the auxiliary cells into account when L and r_{in} increase. In this case, the neuron model with auxiliary cells has the widest dynamic range but the lowest sensitivity. The observed difference in dynamic range can be as large as one log unit (40%) for $L \geq 1$ and $r_{in}>5$ (e.g. a small soma). The differences in sensitivity are much smaller and generally less than half a log unit.

5. OPTIMAL NEURON STRUCTURE

The optimal neuron structure depends on the feature to optimize. If a high sensitivity is needed, i.e. if the neuron must detect low odour concentrations, the optimal structure corresponds to a long enough sensory dendrite ($L>1$) and a large r_{in} (when $N=1$, $r_{in}>5$ without auxiliary cells and $r_{in}>1$ with these cells). It is advantageous to increase the number of sensory dendrites ($N>1$) for small r_{in}, in particular for the model without auxiliary cells.

If a wide dynamic range is expected, i.e. if the neuron must discriminate over a wide range of stimuli, a single ($N=1$), long enough ($L>1$) sensory dendrite and a small input resistance ($r_{in}<<1$) is needed. Selecting auxiliary cells or not is a minor choice because r_{in} is small. However, better performances are obtained with auxiliary cells for $r_{in}>0.3$. This tuning of parameters seems to correspond to the moth sex-pheromone receptor neuron for which $N=1$, $L \approx 1$ and $0.4>r_{in}>0.8^{2.3}$ (calculated with a specific membrane resistance of $4 k\Omega cm^2$).

REFERENCES

1. Kaissling, K.-E. (1971) In Handbook of Sensory Physiology, Vol. IV, Chemical Senses, part I, Olfaction, L.M. Beidler (ed.) (Springer-Verlag, Berlin) pp. 351–431.
2. Kaissling, K.-E. (1987) R.H. Wright Lectures on Insect Olfaction. K. Colbow (ed.) (Simon Fraser University, Burnaby).
3. Kaissling, K.-E. and Thorson, J. (1980) Insect olfactory sensilla: structural, chemical and electrical aspects of the functional organization. In Receptors for Neurotransmitters, Hormones and Pheromones in Insects. D.B. Satelle, L.M. Hall and J.G. Hildebrand (eds.) (Elsevier/North Holland Biomedical Press, Amsterdam) pp. 261–282.
4. Rospars, J.-P., Lánský, P., Tuckwell, H.C. and Vermeulen, A. (1996) Coding of odor intensity in a steady-state deterministic model of an olfactory receptor neuron. J. Comput. Neurosci. 3, 51–72.
5. Thurm, U. and Küppers, J. (1980) Epithelial physiology of insect sensilla. In Insect Biology in the Future. M. Locke and D. Smith (eds.) (Academic Press, New York) pp. 735–764.
6. Tuckwell, H.C. 1988. Introduction to Theoretical Neurobiology, Volume 1: Linear Cable Theory and Dendritic structure. (Cambridge University Press, New York).
7. Vermeulen, A., Rospars, J.-P, Lánský, P. and Tuckwell, H.C. (1996) Coding of stimulus intensity in an olfactory receptor neuron: role of neuron spatial extension and dendritic backpropagation of action potentials. Bul. Math. Biol. 58, 493–512.
8. Vermeulen, A., Lánský, P., Tuckwell, H.C. and Rospars, J.-P. (1997) Coding of odour intensity in a sensory neuron. BioSystems 40, 203–210.

36

NEURAL NETWORK ANALYSIS OF SINGLE UNIT RECORDING FROM HUMAN BASAL GANGLIA

Robert Worth,[1] Samir Sayegh,[2] and Kishan Ranasingh [3]

[1]Department of Neurological Surgery
Indiana University Medical Center
Indianapolis, Indiana
[2]Department of Physics
[3]Department of Biology
Purdue University
Ft. Wayne, Indiana

INTRODUCTION

There has been a renewed interest in, and application of, the technique of posteroventral pallidotomy (PVP) for the treatment of certain patients with Parkinson's Disease (PD). Pallidotomy was initially developed by Leksell in Sweden in the middle fifties with good but variable results. With recent advances in imaging techniques, stereotactic surgery in general has become much more frequent. At the same time, the physiological recordings of Vitek, De Long, et al. in the thalamus and basal ganglia suggested these techniques could be extended to the surgical theater to produce refinement in the targeting procedures for PVP.

Maximal effectiveness requires that the lesion be made in the most posterior inferior part of the globus pallidus internus (Gpi). At the same time, injury to the optic tract or internal capsule which lie immediately inferior and posteromedial, respectively, to the lesion must be avoided. Magnetic resonance (MR) scanning can be used to compute a preliminary target in GPi. There is, however, distortion in MR scans which can lead to errors of up to 4 mm in targeting and which at present is irreducible by physical means. Even assuming technology is eventually available to deal with these scan-induced inaccuracies, individual variation in the normal microanatomy as well as pathophysiologic alterations induced by the PD will make targeting strictly on the basis of structural information uncertain for the foreseeable future. Furthermore, the problem of "brain shift" induces an additional source of error. This problem results from the fact that the anatomical targeting computed on the basis of brain position in preoperative imaging studies may not accurately reflect changes in the coordinates of intracranial contents which occur during surgery for example as a result of the evacuation of cerebrospinal fluid. Thus, until real-time intraoperative imaging is practical, some method on "on-line" correction is essential.

It is particularly fortunate, then, that De Long, et al. have demonstrated that the firing patterns of cells in the GPi are characteristic and distinguishable from those of globus pallidus externus GPe, optic tract or internal capsule. It is thus possible to record the firing of individual cells via microelectrodes inserted along the stereotactic trajectory at the time of PVP. At present, however, assessment of these signals is hampered by two factors. First, the operating room is an extremely noisy environment from the electrical standpoint and the signals are of very low amplitude, requiring a great deal of amplification. This leads to a great deal of difficulty separating legitimate signal from background noise. Secondly, there is no quantitative description of the differences in the signals in question. Thus, successful discrimination of the different anatomical regions on the basis of their "electronic signatures" requires many hours of experience on the part of the observer and even so is fraught with error.

The current study was undertaken in order to determine if neural network computerized processing of signal characteristics could provide objective data to support discrimination of the anatomical location of the recording site.

2. METHODS

During the performance of PVP, single unit action potentials from GPe and GPi were recorded via stereotactically implanted platinum/iridium microelectrodes obtained from Frederick Hare, Inc., Brunswick, ME. The analogue data are then digitized by a Vetter 4000A A-D converter indexed and stored in a Compaq Prolinea 590 computer. The digitized data was then transferred via writable CD-ROM and analyzed offline.

The analysis consisted in extracting multiple segments of the signal corresponding to each position of the electrode. Low pass filtering and discrete Fourier transform are then performed and the spectral density is binned in a 100-element vector that constitutes the input to an error back propagation neural network. The network is endowed with 30 hidden nodes with a hyperbolic tangent nonlinear function and 3 output nodes. The 3 output nodes correspond to a 3 way classification that includes Gpe, Gpi and so called "pauser" cells. The present analysis will concentrate on the discrimination between Gpe and Gpi and no further references to pauser cells will be made. It is sufficient therefore to think in terms of one output unit which turns on for Gpi (+1) and off for Gpe (-1).

The network is trained on multiple segments of data from one individual patient/procedure and two independent locations of the electrode given by fixed stereotactic coordinates. These two different locations are presumed to correspond to Gpe and Gpi either by imaging, by human aural discrimination or both. Again either or both may be considered to be the "teacher." Part of the difficulty of the task is the absence a definite standard. It is to be emphasized however that for each location several consecutive segments of data are analyzed. A certain degree of internal consistency can thus be achieved.

3. RESULTS

The network was repeatedly presented with data from three different patients. Internal consistency as explained above was almost always present. Indeed, as long as the recording electrode was in the same position, the network classification would remain constant. Typically 16 samples were used for training. For testing purposes, 8 samples each (16, total) of time series data were presented, identified respectively as GPe and GPi by a clinician experienced in the analysis of the intraoperative data. In a sample run, the system would correctly identify all eight members of the set of GPi neurons and seven of the eight GPe samples. The eighth GPe sample was identified as being different from GPi but the confidence level was too low for conclusive categorization as a GPe cell. The results were consistent for the individual patients and only slightly inferior for the 3 patients taken together. Exact numerical tabulation is not pursued at this time as one may need to first address the issue of the teacher.

4. DISCUSSION AND CONCLUSION

Assessment of single unit potentials using a neural network algorithm running on a PC can provide objective evidence to support the discrimination of a recording location in GPe from that in GPi. The initial results indicate that despite the fact that the Gpe and Gpi signal may be difficult to discriminate due to similar "fundamental frequencies" (not too far from 60 Hz), a combined signal processing and neural network approach may be a successful one. The implementation of a reliable real time system would be of enormous practical usefulness as it would increase the success rate of stereotactic pallidotomy which today depends to a certain degree on the patient's cooperation. The ability to routinely record from the basal ganglia during a procedure that is regaining great popularity may have fundamental implication for our understanding of the neurobiology of movement and movement disorders. One of the fundamental difficulties is that of the absence of a reliable teacher. This can be somewhat circumvented by the introduction of "voting" mechanism and that of "undecided" classification category. Present and future work concentrates on the 3 goals of standardization, development of a reliable real time system and acquisition of data for the development of a computational theory of basal ganglia in the human brain.

SECTION III

NETWORK

37

SELF-ORGANIZING CIRCUITS OF MODEL NEURONS

L. F. Abbott and O. Jensen

Volen Center for Complex Systems
Brandeis University
Waltham, MA
E-mail: abbott@volen.brandeis.edu
E-mail: jensen@volen.brandeis.edu

ABSTRACT

Activity can affect the growth patterns of neurons as well as modifying their membrane and synaptic conductances. Most models of activity-dependent plasticity have concentrated on one of these aspects, most commonly synaptic plasticity. We construct and study a model in which all three forms of plasticity are present. Starting from random initial conditions, an array of these model cells can develop spontaneously into a highly coupled circuit displaying a complex pattern of activity. Although all the model cells are described by identical equations, they differentiate within these circuits and individually show a variety of intrinsic characteristics.

INTRODUCTION

Activity plays an important role in developing and maintaining the functional characteristics of neurons. Activity-dependent modification of intrinsic membrane conductances[2,11], synaptic conductances[3,4] and growth[6,7] have all been reported. Interestingly, intracellular calcium is implicated as an important regulatory element in all of these cases. Separate models of these three forms of plasticity have been developed and studied[1,8,10,5,12] but the interplay between them has received less attention. Here we study a model of developing neural circuits that involves all three forms of plasticity. Our goal is to model the self-assembly of neural networks, that is, the spontaneous transformation of uncoupled arrays of inactive neurons into circuits of active, interacting cells.

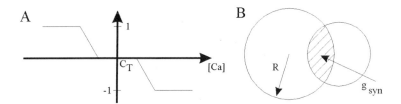

Figure 1. (A) The function $\sigma([Ca^{2+}])$ determining the regulation of the conductances and the size of the neuritic field. (B) The dendritic field of a cell is modeled as a circle. The synaptic conductance between the two cells is proportional to the overlap of the fields.

METHODS

The model is constructed by combining and expanding two lines of research. Previously, models have been constructed which have dynamically varying maximal membrane conductances controlled and regulated by the intracellular calcium concentration[1,8,10]. It has been shown that such model cells individually can assemble the conductances needed to achieve a particular target pattern of activity such as periodic bursting.

The model cell we use is based on the Morris-Lecar model[9] which has two active membrane currents, a calcium and a potassium current. In the original model, the maximal conductances for these two currents, g_{Ca} and g_K, are fixed constants. However, when modeling self-assembling neurons in a network it is essential that individual neurons take advantage of the entire range of parameter values if the complete circuit is to assemble properly. Hence, in our version, they are dynamic variables governed by

$$\tau_g \frac{dg_{Ca}}{dt} = \sigma([Ca^{2+}]) \text{ and } \tau_g \frac{dg_K}{dt} = -\sigma([Ca^{2+}]) \tag{1}$$

The time constant τ_g is long compared to the neuronal dynamics in the model, so that the adjustment of maximal conductances is a slow process. The function $\sigma([Ca^{2+}])$ is plotted in Fig. 1A. When $[Ca^{2+}]$ is high, indicating a large amount of neuronal activity, g_K will increase and g_{Ca} will decrease. This will decrease the activity until $[Ca^{2+}]$ enters the window range around C_T where $\sigma([Ca^{2+}])$ is zero. If the activity, and thus $[Ca^{2+}]$, is low, g_K will decrease and g_{Ca} will increase until equilibrium is again maintained. As a result, these model cells automatically adjust their conductances to maintain $[Ca^{2+}]$ within the window around C_T.

Van Ooyen and van Pelt[12] have constructed a simple model of neuronal growth and synaptogenesis, assuming these processes are also controlled by intracellular calcium. In their model, the dendritic fields of neurons are represented by circles with dynamically varying radii. By growing and shrinking, the model neurons can modify both the nature and strength of their connectivity. The synaptic conductance between two cells is proportional to the overlap of their dendritic fields (Fig. 1B). The rate of change of the dendritic radius is controlled by the intracellular calcium concentration according to the formula

$$\tau_R \frac{dR}{dt} = \sigma([Ca^{2+}]) \tag{2}$$

where τ_R is a long time constant and the function $\sigma([Ca^{2+}])$ as plotted in Fig. 1A (this is a slight modification of the function used in [12]). The important features are: 1) if $[Ca^{2+}]$ is low, $\sigma([Ca^{2+}])$ is positive and the dendritic field (R) grows. 2) if $[Ca^{2+}]$ is high, $\sigma([Ca^{2+}])$ is negative and the dendritic field (R) shrinks. 3) If $[Ca^{2+}]$ is within a window around an

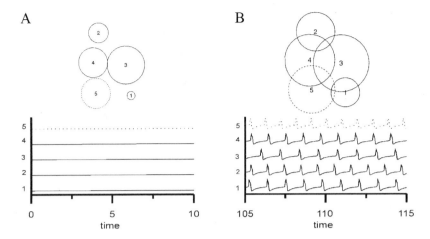

Figure 2. (A) The circuit is initialized with a set of cells having different locations and radii, but the same intrinsic conductances. (B) After a period of growth and adjustment of intrinsic conductances, the circuit equilibrates and all the cells are oscillating. The dotted line indicate an inhibitory cell, all others are excitatory.

equilibrium calcium concentration C_T, the radius remains constant. The time constants are chosen so that $\tau_R > \tau_g$. Hence growth is slower than conductance modification, both of which are slower than the membrane potential dynamics of the model neuron.

RESULTS

To study self-assembling circuits we begin by placing a number of excitatory and inhibitory model cells at random positions on a surface, giving them random small dendritic radii but the same conductances (Fig. 2A). As seen, the cells were silent and uncoupled at the beginning of the run. They then grew, developed connections, modified their conductances and went through cycles of shrinking and growing until each neuron had found a conductance and circuit configuration which placed its value of $[Ca^{2+}]$ within the window area of Fig. 1A. At this point (Fig. 2B) no more modification occurs unless the circuit is perturbed. It is surprising that the model consistently (though not infallibly) reaches an equilibrium configuration even in circuits with many more neurons.

Fig. 3A shows that the equilibrium conductances for the cells are all different; the neurons have differentiated. This variability is not due to the finite size of the equilibrium window in Fig. 1. Rather, individual neurons have different intrinsic properties because they have different synaptic connections and because their developmental histories differed. Note that cell #4 is the most excitable since it has the highest Ca^{2+} conductances and the lowest K^+ conductance. In Fig. 3B, the synaptic connections in the equilibrated network Fig. 2B, have been turned off to reveal the intrinsic properties of the cells. The network without synaptic conductances consists of 1 endogenous oscillator and 4 silent cells with cell #4 acting as a pacemaker for the oscillating network. In repeated runs, we have found a wide variety of similar circuits that show complex output patterns and spontaneously differentiated neurons using anywhere from 5 to 100 cells.

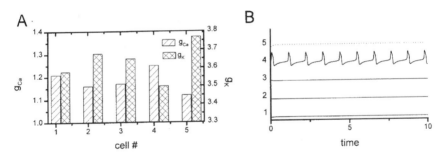

Figure 3. (A) After the network has equilibrated the cells have differentiated, expressing different intrinsic conductances. The histogram shows the maximal conductance of K^+ and Ca^{2+} currents for each cell in Fig. 2B. (B) The differentiated conductances give the cells different properties in the absence of input from other cells. When the cells in the equilibrated network are isolated, cell 4 is an endogenous oscillator and the rest of the cells are silent.

SUMMARY

The model we have presented is simple but nevertheless able to produce nontrivial circuits of differentiated neurons. This shows the power of combining mechanisms of activity/calcium-dependent regulation of membrane conductances, synaptic connections and patterns of growth.

REFERENCES

[1] Abbott, L.F. and LeMasson, G. 1993 Analysis of neuron models with dynamically regulated conductances. Neural Comp. Neural Comp. 5:823-842.

[2] Alkon, D.L. 1984 Calcium-mediated reduction of ionic currents: A biophysical memory trace. Science 226:1037-1045.

[3] Artola, A. and Singer, W. 1993 Long-term depression of excitatory synaptic transmission and its relationship to long-term potentiation. Trends Neurosci. 16:480-487.

[4] Baudry, M. and Davis, J.L., eds. 1991 Long-Term Potentiation. MIT Press, Cambridge MA.

[5] Byrne, J.H. and Berry W.O. 1989 Neural Models of Plasticity. Academic Press, San Diego.

[6] Fields, R.D., Neale, E.A. and Nelson, P.G. 1990 Effects of patterned electrical activity on neurite outgrowth from mouse neurons. J. Neurosci. 10:2950-2964.

[7] Kater, S.B. and Mills, L.R. 1991 Regulation of growth cone behavior by calcium. J. Neurosci. 11:891-899.

[8] LeMasson, G., Marder, E. and Abbott, L.F. 1993 Activity-dependent regulation of conductances in model neurons. Science 259:1915-1917.

[9] Morris, C. and Lecar, H. 1981 Voltage oscillations in the barnacle giant muscle fiber. Biophys. J. 35:193-213.

[10] Siegel, M., Marder, E. and Abbott, L.F. (1994) Activity-Dependent Current Distributions in Model Neurons. Proc. Natl. Acad. Sci. USA 91:11308-11312.

[11] Turrigiano, G., Abbott, L.F., Marder, E. 1994 Activity-dependent changes in the intrinsic electrical properties of cultured neurons. Science 264:974-977.

[12] van Ooyen, A. and van Pelt, J. 1994 Activity-dependent outgrowth of neurons and overshoot phenomena in developing neural networks. J. Theor. Biol. 167:27-44.

[13] Research supported by the Sloan Center for Theoretical Neurobiology at Brandeis University, NIMH-46742 and the W.M. Keck Foundation.

38

SPONTANEOUS REPLAY OF TEMPORALLY COMPRESSED SEQUENCES BY A HIPPOCAMPAL NETWORK MODEL

D. A. August and W. B. Levy

Department of Neurological Surgery
Box 420
University of Virginia Health Sciences Center
Charlottesville, VA
E-mail: august@virginia.edu, wbl@virginia.edu

ABSTRACT

Recent experimental evidence suggests that the hippocampus replays a temporally compressed version of recently-learned spatial sequence information during slow-wave sleep. This phase of sleep is characterized by intermittent episodes of high-frequency firing known as sharp waves. Here we partially characterize a simplified neural network model of hippocampal area CA3, based on integrate-and-fire cells, which is capable of recalling temporally compressed sequence information during brief periods of high activity.

1. INTRODUCTION

The hippocampus contains numerous "place cells", which fire whenever the animal is in a certain location in space [15]. Thus, spatial navigation can be thought of as a sequence learning problem, in which the animal learns a sequence of place cell firing. For some time, it has been known that place cells active during waking behavior are more likely to fire during slow-wave sleep (SWS) [16]. Recently, this finding has been extended to sequences of place cell firing. It has also been shown that the sequence replay during SWS is compressed in time [21, 18].

Accompanying slow-wave sleep are electophysiological events in the hippocampus known as sharp waves (SPWs) [5, 19]. The most obvious feature of the SPW is a generalized increase electrical field potential, which occurs intermittently and with a variable duration and frequency [3].

Computational Neuroscience
edited by Bower, Plenum Press, New York, 1997

Here, we show that a simplified model of hippocampal area CA3 can recall a temporally compressed sequence during a SPW-like event, and contrast these episodes of recall to the behavior of a non-learning network.

2. METHODS

Details of our methods appear elsewhere [1]. Briefly, the network consists of an input layer, analogous to the entorhinal cortex and dentate gyrus (EC/DG), connected one-to-one to a CA3-like layer. The CA3 layer has 1000 cells, each of which projects excitatory connections randomly and uniformly to 10% of the other units, reflecting the sparse recurrent connectivity observed in the hippocampus [7]. Each cell is modeled as an integrate-and-fire unit, with a membrane time-constant of 20 ms.

Activity is controlled by a feedback and a feedforward inhibitory interneuron. The output of the feedforward interneuron is proportional to the average EC/DG activity, and the output of the feedback interneuron is proportional to the average CA3-layer activity. We model inhibition as a shunting (divisive) effect, so that the synaptic current is the ratio of excitation to excitation plus inhibition.

Excitatory recurrent synapses are modified by an unsupervised Hebbian rule based on physiological observations of long-term potentiation [8]. For a synapse to modify, the postsynaptic cell must fire, but the strength and sign of modification is controlled by a 150 ms running average of presynaptic activity [9].

The input to the model is a simple sequence of EC/DG cell activity, in which the first pattern activates EC/DG units 1-10, the next activates units 2-11, and so on. Each pattern lasts for 20 ms, but because of the spatial overlap between patterns, each EC/DG unit remains active for 200 ms. We simulate a sequence with 100 patterns, lasting for a total of 2000 ms. The sequence is circular, with the (nominal) first and final patterns overlapping.

The simulation has two phases: training (learning), and testing (recall). During training, the input sequence arrives from the EC/DG layer, and recurrent connections can modify. Training continues until the average synaptic weight has stabilized (about 10 cycles of the input sequence). During testing, synaptic modification is switched off, inhibition is lowered by about 10-fold, and EC/DG units are activated randomly at an average rate of 1 Hz. The situation during testing is similar to SWS, during which the network may not re-experience a previously learned pattern exactly, but some amount of random activity will be present. Thus, we also refer to testing as modeled slow-wave sleep (mSWS).

3. RESULTS

Figure 1a and b show a network during the final cycle of learning, and during an equally long period of mSWS. The network exhibits two distinct episodes of spontaneous recall, one starting at roughly 100 ms and the other at 1400 ms. Each of the two recall episodes includes several temporally compressed repetitions of the learned sequence. Periods of recall also coincide with a substantial increase in the overall average network activity, as shown in Figure 1c. The frequency and/or duration of spontaneous recall events be changed by adjusting various network parameters. For example, lowering the amount of feedback inhibition produces longer episodes of recall, until finally, the sequence is recalled continually throughout mSWS. Raising

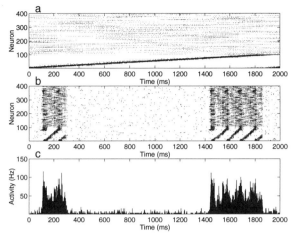

Figure 1. *(a) Cell firing at the end of training.* This is a rastergram of network activity during the tenth trial of learning, by which time the average synaptic weight has stabilized. Time is along the abscissa, and neuron number is along the ordinate. Thus, each row represents the activity of one cell over time, and each dot represents one spike. This figure illustrates two distinct patterns of firing. The first 100 cells are driven directly by the input sequence from the EC/DG layer, as shown by the dark diagonal band of firing. The remaining 900 cells (of which only 300 are shown for clarity) are driven by recurrent connections from within CA3 only. These units fire over discrete episodes, spanning different portions of the sequence. We call these cells "local context units", and hypothesize that they are roughly analogous to hippocampal place cells. *(b) Cell firing during mSWS.* Slow-wave sleep is modeled by disabling learning, lowering inhibition by approximately tenfold, and randomly activating EC/DG units at an average rate of 1 Hz. Notice the emergence of two distinct episodes of recall, each containing multiple repetitions of a temporally compressed version of the original sequence. *(c) Average network activity.* Shown here is the time course of the average network activity over all cells during the mSWS episode of (b). Although the network usually remains at low levels of activity, brief periods of high-frequency firing, reminiscent of sharp waves occur, and these are the spontaneous recall events.

inhibition produces more episodes of recall, but decreases the duration of each episode. Raising the rate of EC/DG firing has a similar effect, by virtue of the increase in feedforward inhibition.

Is learning required for the network to produce SPW-like events? We investigate this by comparing a "naive" network that has not learned a sequence to an experienced network that has been exposed to 10 cycles of the training sequence. Consistent with previous simulations [13], the naive network mostly exhibits stable non-periodic or limit cycle behavior, as shown in Figure 2a and b. However, by a change of parameters, the naive network can be made to exhibit intermittent bursts of high activity intermixed with periods of quiescence, as shown in Figure 2c. Although the time course of this pattern of firing is similar to that of spontaneous recall, its frequency profile differs considerably. As shown in Figure 2d, the power spectra of the naive networks contains multiple peaks on a relatively flat baseline while the power spectrum of activity during spontaneous recall decreases steadily with frequency.

Because the behavior of the naive and experienced networks differ, the ability to produce SPW-like episodes of high activity must be partly due training. We therefore examine how the final distribution of synaptic weights influences this behavior by deleting different sets of synaptic weights. Removing the synaptic weights *above* a certain value makes episodes of spontaneous recall less likely although it does not diminish their quality. Deleting the weights *below* a certain value decreases both the duration and the quality of these recall events (data not shown). However, recall is still maintained even with deletion of 50% of the weaker synapses.

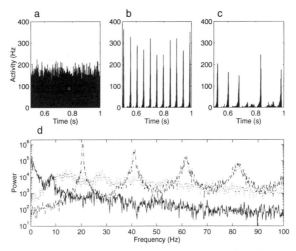

Figure 2. *(a) Stable aperiodic firing.* Shown here is the average activitity of a naive network over 500 ms, during which EC/DG units fire randomly at an average rate of 1 Hz. In this case, the network exhibits stable, aperiodic firing. *(b) Oscillatory firing.* Changing one parameter (in this case, changing the time-constant of inhibition from 2 to 7 ms) switches the network into oscillatory behavior. *(c) SPW-like activity.* When the inhibitory time-constant is further increased to 10 ms, the firing pattern becomes more complex, with brief, intermittent surges of activity, reminiscent of sharp waves. *(d) Power spectra.* Although the time course of average activity shown in (c) above bears some qualitative similarity to the time course of average activity during mSWS shown in Figure 1c, the power spectra are different. Whereas the power spectra of the naive networks in (b, dash-dot line) and (c, dotted line) are relatively flat with some peaks at their characteristic frequencies, the power spectrum of the experienced network during mSWS decays steadily with frequency (solid line).

4. DISCUSSION

Because of the autoassociative nature of CA3 [11, 17], random activity can trigger the recall of stored pattern(s) of activity. The amount of activity required may be slight. When CA3 is disinhibited with picrotoxin, for example, even a *single* cell can initiate a population burst [12]. Experimental and modeling efforts have suggested that decreased levels of acetylcholine enhance synaptic transmission, thereby facilitating autoassociative recall [6]. These findings are consistent with our result that spontaneous recall occurs in the setting of random input layer activity combined with lower inhibition.

Which synapses are involved in recall? We studied this issue by deleting different subsets of synaptic weights before mSWS. The larger weights serve to trigger spontaneous recall, while the smaller weights control the quality and the duration of each episode of spontaneous recall. Thus, recall requires more than just an interconnected group of highly potentiated synapses, as might be expected. Instead, recall makes use of all manner of synapses, both strong and weak. This is consistent with the recent suggestion that synaptic modification exerts a complex, time- and activity-dependent effect, rather than a fixed gain adjustment [10].

The power spectrum of average neural activity during spontaneous recall declines exponentially at low frequencies, in contrast to the relatively flat spectra of the naive network. A similar decay (the so-called "1/f" spectrum) has also been found in a Hopfield-type network model of neural activity during sleep [14], and in a randomly-driven network model of the cortex [20].

Memory consolidation is the process by which certain types of memory are transferred from intermediate-term storage in the hippocampus to long-term storage in the cortex [22]. The initial phase of consolidation has been described colloquially as the hippocampus "teaching" the cortex. The data of McNaughton and colleagues [21] suggest that this teaching happens during slow-wave sleep, when sequence information is replayed. Because recall presumably occurs during SPWs, the cortex will be "taught" under conditions of high activity. This high activity is ideal for inducing long-term potentiation [4], a well-known cellular model of learning [2].

In a simplistic way, our model reproduces many of the features of the behavior of area CA3 during memory consolidation. Under the influence of spontaneous input activity, the network can repeatedly recall a temporally compressed version of a learned sequence during transient periods of increased overall activity. Such compression will facilitate associations between temporally distant events.

ACKNOWLEDGMENTS

This work was supported by NIH MH10702 to DAA, by NIH MH48161, MH00622, EPRI RP8030-08, and Pittsburgh Supercomputing Center Grant BNS950001 to WBL, and by the Department of Neurosurgery, Dr. John A. Jane, Chairman.

REFERENCES

[1] D A August and W B Levy. Temporal sequence compression by a hippocampal network model. In *INNS World Congress on Neural Networks*, pages 1299–1303, Mahwah, NJ, 1996. Lawrence Erlbaum.

[2] T V P Bliss and T Lomo. Long-lasting potentiation of synaptic transmission in the dentate area of the anaesthetized rabbit following stimulation of the perforant path. *J. Physiol.*, 232:331–356, 1973.

[3] G Buzsaki. Hippocampal sharp waves: Their origin and significance. *Brain Res.*, 398:242–252, 1986.

[4] G Buzsaki. Two-stage model of memory trace formation: A role for 'noisy' brain states. *Neuroscience*, 31(3):551–570, 1989.

[5] G Buzsaki, Z Horvath, R Urioste, J Hetke, and K Wise. High-frequency network oscillation in the hippocampus. *Science*, 256:1025–1027, 1992.

[6] M E Hasselmo, E Schnell, and E Barkai. Dynamics of learning and recall at excitatory recurrent synapses and cholinergic modulation in rat hippocampal region CA3. *J. Neurosci.*, 15(7):5249–5262, 1995.

[7] N Ishizuka, J Weber, and D G Amaral. Organization of intrahippocampal projections originating from CA3 pyramidal cells in the rat. *J. Comp. Neurol.*, 295:580–623, 1990.

[8] W B Levy and O Steward. Synapses as associative memory elements in the hippocampal formation. *Brain Res.*, 175:233–245, 1979.

[9] W B Levy and O Steward. Temporal contiguity requirements for long-term associative potentiation/depression in the hippocampus. *Neuroscience*, 8(4):791–797, 1983.

[10] H Markram and M V Tsodyks. Redistribution of synaptic efficacy between neocortical pyramidal cells. *Nature*, 382:807–810, 1996.

[11] D Marr. Simple memory: a theory for archicortex. *Phil. Trans. Royal. Soc. Lond.*, 262:23–81, 1971.

[12] R Miles and R K S Wong. Single neurones can initiate synchronized population discharge in the hippocampus. *Nature*, 306:371–373, 1983.

[13] A A Minai and W B Levy. The dynamics of sparse random networks. *Biol. Cybern.*, 70:177–187, 1993.

[14] M Nakao, K Watanabe, T Takahashi, Y Mizutani, and M Yamamoto. Structural properties of network attractor associated with neuronal dynamics transition. In *Proceedings of the International Joint Conference of Neural Networks*, volume 3, pages 529–534. Inst. of Electrical and Electronic Engineers, 1992.

[15] J O'Keefe and L Nadel. *The Hippocampus as a Cognitive Map*. Oxford: Clarendon Press, London, 1978.

[16] C Pavlides and J Winson. Influences of hippocampal place cell firing in the awake state on the activity of these cells during subsequent sleep episodes. *J. Neurosci.*, 9(8):2907–2918, 1989.

[17] E T Rolls. Functions of neuronal networks in the hippocampus and cerebral cortex in memory. In R M J Cotterill, editor, *Models of Brain Function*, pages 15–33. Cambridge Univ. Press, 1989.

[18] W E Skaggs and B L McNaughton. Replay of neuronal firing sequences in rat hippocampus during sleep following spatial experience. *Science*, 271:1870–1873, 1996.

[19] S S Suzuki and G K Smith. Spontaneous EEG spikes in the normal hippocampus. I. Behavioral correlates, laminar profiles and bilateral synchrony. *Electroenceph. Clin. Neurophys.*, 67(4):348–359, 1987.

[20] M Usher, M Stemmler, C Koch, and Z Olami. Network amplification of local fluctuations causes high spike rate variability, fractal firing patterns and oscillatory local field potentials. *Neural Computation*, 6(5):795–836, 1994.

[21] M A Wilson and B L McNaughton. Reactivation of hippocampal ensemble memories during sleep. *Science*, 265:676–679, 1994.

[22] S Zola-Morgan and L R Squire. The primate hippocampal formation: Evidence for a time-limited role in memory storage. *Science*, 250:288–290, 1990.

39

A CORTICAL NETWORK MODEL OF COGNITIVE ATTENTIONAL STREAMS, RHYTHMIC EXPECTATION, AND AUDITORY STREAM SEGREGATION

Bill Baird

Department of Mathematics
University of California at Berkeley
Berkeley, CA
E-mail: baird@math.berkeley.edu

ABSTRACT

We have developed a neural network architecture that implements a theory of attention, learning, and trans-cortical communication based on adaptive synchronization of 5-15 Hz and 30-80 Hz oscillations between cortical areas. Here we present a specific higher order cortical model of attentional networks, rhythmic expectancy, and the interaction of higher-order and primary cortical levels of processing. It accounts for the "mismatch negativity" of the auditory ERP and the results of psychological experiments of Jones showing that auditory stream segregation depends on the rhythmic structure of inputs. The timing mechanisms of the model allow us to explain how relative timing information such as the relative order of events between streams is lost when streams are formed. The model suggests how the theories of auditory perception and attention of Jones and Bregman may be reconciled.

1. INTRODUCTION

We have shown how oscillatory associative memories may be coupled to recognize and generate sequential behavior, and how a set of novel mechanisms utilizing these complex dynamics can be configured to solve attentional and perceptual processing problems. Space here permits only a verbal overview of this work. For background and full treatment with mathematics and references see [1, 2]. Using dynamical systems theory, an architecture is constructed from recurrently interconnected oscillatory associative memory modules that model higher order sensory and motor areas of cortex. The modules learn connection weights between themselves which cause the system to evolve under a 5-20 Hz clocked sensory-motor processing cycle by a sequence of transitions of synchronized 30-80 Hz oscillatory attractors within the modules.

The architecture employs selective"attentional" control of the synchronization of the 30-80 Hz "gamma band" oscillations between modules to direct the flow of computation to recognize and generate sequences. The 30-80 Hz attractor amplitude patterns code the information content of a cortical area, whereas phase and frequency are used to "software" the network, since only the synchronized areas communicate by exchanging amplitude information. The system works like a broadcast network where the unavoidable crosstalk to all areas from previous learned connections is overcome by frequency coding to allow the moment to moment operation of attentional communication only between selected task-relevant areas.

The behavior of the time traces in different modules of the architecture models the temporary appearance and switching of the synchronization of 5-20 and 30-80 Hz oscillations between cortical areas that is observed during sensorimotor tasks in monkeys and humans. The architecture models the 5-20 Hz evoked potentials seen in the EEG as the control signals which determine the sensory-motor processing cycle. The 5-20 Hz clocks which drive these control signals in the architecture model thalamic pacemakers which are thought to control the excitability of neocortical tissue through similar nonspecific biasing currents that cause the cognitive and sensory evoked potentials of the EEG. The 5-20 Hz cycles are thought to "quantize time" and form the basis of derived somato-motor rhythms with periods up to seconds that entrain to each other in motor coordination and to external rhythms in speech perception[3].

The architecture illustrates the notion that synchronization of gamma band activity not only"binds" the features of inputs in primary sensory cortex into "objects", but further binds the activity of an attended object to oscillatory activity in associational and higher-order sensory and motor cortical areas to create an evolving attentional network of intercommunicating cortical areas that directs behavior. The binding of sequences of attractor transitions between modules of the architecture by synchronization of their activity models the physiological mechanism for the formation of perceptual and cognitive "streams" investigated by Bregman[4], Jones[3], and others. In audition, according to Bregman's work, successive events of a sound source are bound together into a distinct sequence or "stream" and segregated from other sequences so that one pays attention to only one sound source at a time (the cocktail party problem).

MEG tomographic observations show large scale rostral to caudal motor-sensory sweeps of coherent thalamo-cortical 40Hz activity across the entire brain, the phase of which is reset by sensory input in waking, but not in dream states. This suggests an inner higher order "attentional stream" is constantly cycling between motor (rostral) and sensory (caudal) areas in the absence of input. It may be interrupted by input "pop out" from primary areas or it may reach down as a "searchlight" to synchronize with particular ensembles of primary activity to be attended.

2. MISMATCH NEGATIVITY AND JONES THEORY OF DYNAMIC ATTENTION

Jones[3] has developed a psychological theory of attention, perception, and motor timing based on the hypothesis that these processes are organized by neural rhythms in the range of 10 to .5 Hz — the range within which subjects perceive periodic events as a rhythm. These rhythms provide a multiscale representation of time and selectively synchronize with the prominent periodicities of an input to provide a temporal expectation mechanism for attention to target particular points in time.

In this view, just as two cortical areas must synchronize to communicate, so must two nervous systems. Work using frame by frame film analysis of human verbal interaction [2], shows evidence of "interactional synchrony" of gesture and body movement changes and EEG of both speaker and listener with the onsets of phonemes in speech at the level of a 10 Hz "microrhythm" – the base clock rate of our models. Normal infants synchronize their spontaneous body flailings at this 10 Hz level to the mothers voice accents, while autistic and schizophrenic children fail to show interactional synchrony. Autistics are unable to tap in time to a metronome.

This implies a fundamental organization of neural processing in the brain around entrainment to levels of periodicity in the environment. For human communication in speech, poetry, music, dance, and sports, this is plausible. For most other vertebrates as well, an all-important part of the environment consists of the nervous systems of other animals with periodic locomotion systems. The survival of predator and prey alike depends on anticipating and countering each other's rhythmic locomotion patterns.

Neural expectation rhythms that support Jones' theory have been found in the auditory EEG. In experiments where the arrival time of a target stimulus is regular enough to be learned by an experimental subject, it has been shown that the 10 Hz activity *in advance of the stimulus* becomes phase locked to that expected arrival time. This fits our model of rhythmic expectation where the 10 Hz rhythm is a fast base clock that is shifted in phase and frequency to produce a match in timing between the stimulus arrival and the output of longer period cycles derived from this base clock.

Jones notes the ubiquitous evidence for neural rhythms and their entrainment to rhythmic motor and perceptual events, but provides no detailed physiological theory. We seek here to provide this theory. Our approach supplies a rhythmic expectancy and short term memory feature presently absent from the other approaches to stream formation by 40 Hz binding (Wang, and Brown and Cooke[5]). These have problems of biological plausibility with their representation of time and sequential grouping. A stimulus does not seem to be represented by continued activation or reactivation of the original 40 Hz stimulus response in primary auditory cortex, as suggested in these models. The gamma band response to a single auditory input onset lasts 100 - 150 ms, whereas the data of van Noorden[4] shows stream segregation is possible at stimulus repetition rates of up to one second. There is no 40 Hz activity available in primary cortex from the previous stimulus for the present input activity to synchronize with for sequential binding.

The "mismatch negativity" (MNN)[6] of the auditory evoked potential appears to be an important physiological indicator of the action of a neural expectancy system like that proposed by Jones. It has been localized to areas within primary auditory cortex by MEG studies[6] and it appears as an increased negativity of the ERP in the region of the N200 peak whenever a psychologically discriminable deviation of a repetitive auditory stimulus occurs. Mismatch is caused by deviations in onset or offset time, rise time, frequency, loudness, timbre, phonetic structure, or spatial location of a tone in the sequence.

A deviation in the direction of increase or decrease of frequency produces an expectancy mismatch. There is even a small mismatch signal to the expected event after a deviant event which indicates that some expectancy adjustment has already occurred. The mismatch is abolished by blockers of the action of NMDA channels[6] which are important for the synaptic changes underlying the kind of Hebbian learning which is used in the model. MEG studies

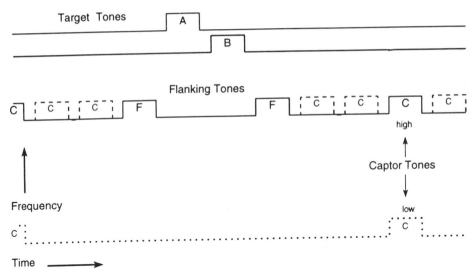

Figure 1. Stimuli of the Jones-Bregman experiment in the solid lines show the case where formation of two streams occurs. The rhythm of the captor stream (C) is slow (1:3) compared with the rate of the target tones (A,B). The rhythm of the captors can be made identical to the target rate by the addition of the captor tones shown as dashed lines. Then there is no performance improvement when the captor tone frequency is changed from low (dotted lines) to high frequency, and performance is worse than the no distractor control condition. Conclusion: no separate target stream is formed without the rhythmic distinction

show further that MNN occurs at the location in auditory cortex where the expected tone is represented instead of where the deviant tone is represented.

MNN is not a direct function of echoic memory because it takes several repetitions for the expectancy to begin to develop, and it decays in 2 - 4 seconds. It appears only for repetition periods greater that 50-100 msec and less than 2-4 seconds. Thus the time scale of its operation is in the appropriate range for Jones' expectancy system. Stream formation also takes several cycles of stimulus repetition to build up over 2-4 seconds and decays away within 2-4 seconds in the absence of stimulation. Those auditory stimulus features which cause streaming are also features which cause mismatch. This supports the hypothesis in the model that both phenomena are functionally interactive.

Finally, MNN can occur independent of attention – while a subject is reading or doing a visual discrimination task. This implies that the auditory system at least must have its own timing system that can generate timing and expectancies independent of other behavior. We can talk or do internal verbal thinking while doing other tasks. A further component of this negativity appears in prefrontal cortex and is thought by Nataanen to initiate attentional switching toward the deviant event causing perceptual "pop out"[6].

2.1. Jones - Bregman Experiment

Jones replicated and altered a classic streaming experiment of Bregman and Rudnicky[4], and found that their result depended on a specific choice of the rhythm of presentation. The experiment required human subjects to determine of the order of presentation of a pair of high target tones AB or BA of slightly different frequencies. Also presented before and after the target tones were a series of identical much lower frequency tones called the capture tones CCC and two identical tones of intermediate frequency before and after the target tones called the flanking tones F - CCCFABFCCC (see figure 1). Bregman and Rudnicky found that target order determination performance was best when the capture tones were near to the flanking tones in frequency, and deteriorated as the captor tones were moved away. Their explanation was that the flanking tones were captured by the background capture tone stream when close in frequency, leaving the target tones to stand out by themselves in the attended stream. When the captor tones were absent or far away in frequency, the flanking tones were included in the attended stream and obscured the target tones.

Jones noted that the flanking tones and the capture stream were presented at a stimulus onset rate of one per 240 ms and the targets appeared at 80 ms intervals. In her experiments, when the captor and flanking tones were given a rhythm in common with the targets, no effect of the distance of captor and flanking tones appeared. This suggested that rhythmic distinction of targets and distractors was necessary in addition to the frequency distinction to allow selective attention to segregate out the target stream. Because performance in the single rhythm case was worse than that for the control condition without captors, it appeared that no stream segregation of targets and captors and flanking tones was occurring until the rhythmic difference was added. *From this evidence we make the assumption in the model that the distance of a stimulus in time from a rhythmic expectancy acts like the distance between stimuli in pitch, loudness, timbre, or spatial location as factor for the formation of separate streams.*

2.2. Rhythmic Expectation in the Model

To implement Jones's theory in the model and account for her data, subsets of the oscillatory modules are dedicated to form a rhythmic temporal coordinate frame or time base by dividing down a thalamic 10 Hz base clock rate in steps from 10 to .5 Hz. Each derived clock is created by an associative memory module that has been specialized to act stereotypically as a counter or shift register by repeatedly cycling through all its attractors at the rate of one for each time step of its clock. Its overall cycle time is therefore determined by the number of attractors, and each cycle is guaranteed to be identical, as required for clock function, because it has strong attractors that correct the perturbing effect of noise. Only one step of the cycle can send output back to primary cortex - the one with the largest weight from receiving the most match to incoming stimuli. Each clock derived in this manner from a thalamic base clock will therefore phase reset itself to get the best match to incoming rhythms.

The match can be further refined by frequency and phase adjustment of the base clock itself, as is shown in the work of Large, and McCauley[1], who have configured such oscillators to track the varying tempo of a piano improvisation, and modelled psychological data on tempo discrimination. Three such counters are sufficient to model the rhythms in Jones' experiment as shown in the architecture of figure 2. The three counters divide the 12.5 Hz clock down to 6.25 and 4.16 Hz. The first contains one attractor at the base clock rate which has adapted to entrain to the 80 msec period of target stimulation (12.5 Hz). The second cycles at $12.5/2 = 6.25$ Hz, alternating between two attractors, and the third steps through three attractors, to cycle at $12.5/3 = 4.16$ Hz, which is the slow rhythm of the captor tones.

The modules of the time base send their internal 30-80 Hz activity to primary auditory cortex in 100msec bursts at these different rhythmic rates through fast adapting connections (which would use NMDA channels in the brain) that continually attempt to match incoming stimulus patterns using an incremental Hebbian learning rule. The weights decay to zero over 2-4 sec to simulate the data on the rise and fall of the mismatch negativity. These weights effectively compute a low frequency discrete Fourier transform over a sliding window of several seconds, and the basic periodic structure of rhythmic patterns is quickly matched. This serves to establish a quantized temporal grid of expectations against which expressive timing deviations in speech and music can be experienced.

Following Jones[3], we hypothesize that this happens automatically as a constant adaptation to environmental rhythms, as suggested by the mismatch negativity. Retained in these weights of the timebase is a special kind of short term memory of the activity which includes temporal information since the timebase will partially regenerate the previous activity in primary cortex at the expected recurrence time. This top-down input causes enhanced sensitivity in target units by increasing their gain. Those patterns which meet these established rhythmic expectancy signals in time are thereby boosted in amplitude and pulled into synchrony with the 30-80 Hz attentional searchlight stream to become part of the attentional network sending input to higher areas. In accordance with Jones' theory, voluntary top-down attention can probe input at different hierarchical levels of periodicity by selectively synchronizing a particular cortical column in the time base set to the 40 Hz frequency of the inner attention stream. Then the searchlight into primary cortex is synchronizing and reading in activity occurring at the peaks of that particular time base rhythm.

2.3. Cochlear and Primary Cortex Model

At present, we have modeled only the minimal aspects of primary auditory cortex sufficient to qualitatively simulate the Jones-Bregman experiment, but the principles at work allow expansion to larger scale models with more stimulus features. We simulate four sites in auditory cortex corresponding to the four frequencies of stimuli used in the experiment, as shown in figure 2. There are two close high frequency target tones, one high flanking frequency location (which includes the captor tones in one variant of the experiment shown in figure 1), and the low frequency location of the captor stream.

These cortical locations are modeled as oscillators with the same equations used for associative memory modules[1, 2], with full linear cross coupling weights. This lateral connectivity is sufficient to promote synchrony among simultaneously activated oscillators, but insufficient to activate them strongly in the absence of external input. This makes full synchrony of activated units the default condition in the model cortex, as in Brown's model[5]. Bregman[4] sees this as an assumption by the auditory system that all input is due to the same environmental source in the absence of evidence for segregation. Thus the background activation is coherent, and can be read into higher order cortical levels which synchronize with it.

Brown and Cooke[5] model the cochlear and brainstem nuclear output as a set of overlapping bandpass ("gammatone") filters consistent with auditory nerve responses and psychophysical "critical bands". A tone can excite several filter outputs at once. We approximate this effect of the gammatone filters as a lateral fan out of input activations with weights that spread the activation in the same way as the overlapping gammatone filters do.

Experiments show that frequencies within the 30-80 Hz gamma band vary within individuals on different trials of a task, and that neurotransmitters can quickly alter frequencies of neural clocks. Following the evidence that the oscillation frequency of binding in vision goes up with the speed of motion of an object, we assume that unattended activity in auditory cortex synchronizes at a default background frequency of 35 Hz, while the higher order attentional stream is at a higher frequency of 40 Hz. Just as fast motion in vision can cause stimulus driven capture of attention, we hypothesize that expectancy mismatch in audition causes the deviant activity to be boosted above the default background frequency to facilitate synchronization

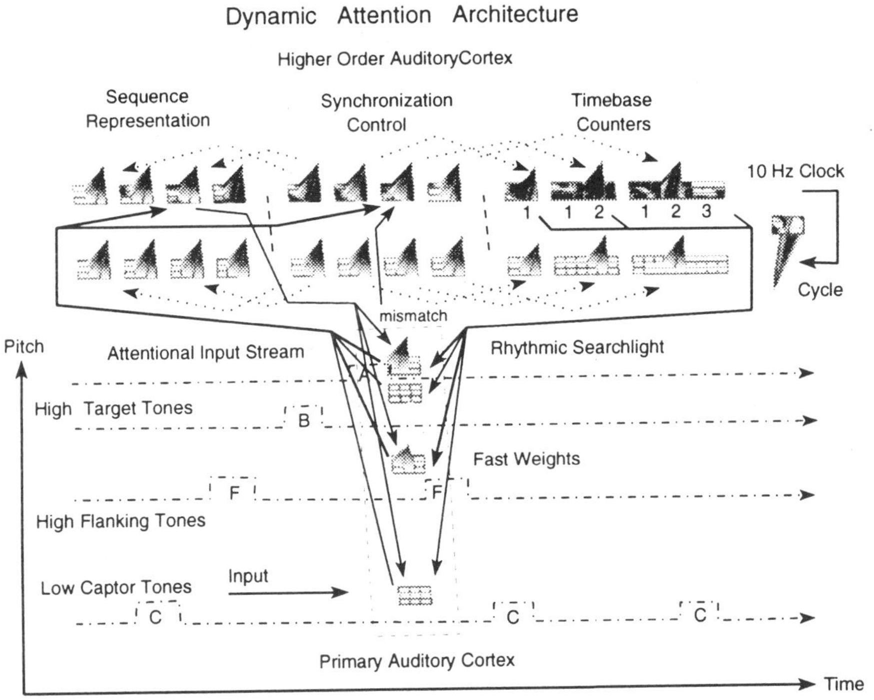

Dynamic Attention Architecture

Higher Order AuditoryCortex

Figure 2. Horizontally arrayed units at the top model higher order auditory and motor cortical columns which are sequentially clocked by the (thalamic) base clock on the right to alternate attractor transitions between upper hidden (motor) and lower context (sensory) layers to act as an Elman net[1, 2]. Three cortical regions are shown – sequence representation memory, attentional synchronization control, and a rhythmic timebase of three counters. The hidden and context layers consist of binary "units" composed of two oscillatory attractors. Activity levels oscillate up and down through the plane of the paper. Dotted lines show frequency shifting outputs from the synchronization control (attention) modules. The lower vertical set of units is a sample of primary auditory cortex frequency channels at the values used in the Jones-Bregman experiment. The dashed lines show the rhythmic pattern of the target, flanking, and captor tones moving in time from left to right to impact on auditory cortex. The case shown is where the flanking tones are in the same stream as the targets because the captor stream is at the lower sound frequency channel. At the particular point in time shown here, the first flanking tone has just finished, and the first target tone has arrived. Both channels are therefore active, and synchronized with the attentional stream into the higher order sequence recognizer.

with the attentional stream at 40 Hz. This models the mechanism of involuntary stimulus driven attentional "pop out". Multiple streams of primary cortex activity synchronized at different eigenfrequencies can be selectively attended by uniformly sweeping the eigenfrequencies of all primary ensembles through the passband of the 40 Hz higher order attentional stream to "tune in" each in turn as a radio receiver does.

Following, but modifying Brown and Cooke[5], the core of our primary cortex stream forming model is a fast learning rule that reduces the lateral coupling and spreads the resonant cortical gamma band frequencies (eigenfrequencies) between sound frequency channels that do not exhibit the same amplitude of activity at the same time. This coupling and eigenfrequency difference recovers between onsets. In the absence of lateral synchronizing connections or coherent top down driving, synchrony between streams is rapidly lost because of their distant resonant frequencies. Activity not satisfying the Gestalt principle of "common fate"[4] is thus decorrelated.

with the attentional stream at 40 Hz. This models the mechanism of involuntary stimulus driven attentional "pop out". Multiple streams of primary cortex activity synchronized at different eigenfrequencies can be selectively attended by uniformly sweeping the eigenfrequencies of all primary ensembles through the passband of the 40 Hz higher order attentional stream to "tune in" each in turn as a radio receiver does.

Following, but modifying Brown and Cooke[5], the core of our primary cortex stream forming model is a fast learning rule that reduces the lateral coupling and spreads the resonant cortical gamma band frequencies (eigenfrequencies) between sound frequency channels that do not exhibit the same amplitude of activity at the same time. This coupling and eigenfrequency difference recovers between onsets. In the absence of lateral synchronizing connections or coherent top down driving, synchrony between streams is rapidly lost because of their distant resonant frequencies. Activity not satisfying the Gestalt principle of "common fate"[4] is thus decorrelated.

Sequential Grouping by Coupling and Resonant Frequency Labels The trade off of the effect of temporal and sound frequency proximity on stream segregation follows because close stimulus frequencies excite each other's channel filters. Each produces a similar output in the other, and their activities are not decorrelated by coupling reduction and resonant frequency shifts. On the other hand, to the extent that they are distant enough in sound frequency, each tone onset weakens the weights and shifts the eigenfrequencies of the other channels that are not simultaneously active. This effect is greater, the faster the presentation rate, because the weight recovery rate is overcome. This recovery rate can then be adjusted to yield stream segregation at the rates reported by van Noorden[4] for given sound frequency separations.

In the absence of rhythmic structure in the input, the temporary weights and resonant frequency "labels" serve as a short term "stream memory" to bridge time (up to 4 seconds) so that the next nearby input is "captured" or "sequentially bound" into the same ensemble of synchronized activity. This pattern of synchrony has been made into a temporary attractor by the temporary weight and eigenfrequency changes from the previous stimulation. This explains the single tone capture experiments where a series of identical tones captures later nearby tones. For two points in time to be sequentially grouped by this mechanism, there is no need for activity to continue between onsets as in Browns model[5], or to be held in multiple spatial locations as Wang does[5].

Furthermore, the decorrelation rule, when added to the mechanism of timing expectancies, explains the loss of relative timing (order) between streams, since the lateral connections that normally broadcast actual and expected onsets across auditory cortex, are cut between two streams by the decorrelating weight reduction. Expected and actual onset events in different streams can no longer be directly (locally) compared. Experimental evidence for the broadcast of expectancies comes from the fast generalization to other frequencies of a learned expectancy for the onset time of a tone of a particular frequency (Schreiner lab - personal communication). By this same argument, MNN itself is affected by stream formation and should no longer occur at sound frequency locations that fall in separate streams when timing deviations occur from expectations in the other stream location. This is a testable prediction of the model providing a possible *physiological* indicator of stream formation.

When rhythmic structure is present, the expectancy system becomes engaged, and this becomes an additional feature dimension along which stimuli can be segregated. Distance from expected *timing* as well as sound quality is now an added factor causing stream formation by decoupling and eigenfrequency shift. Feedback of expected input can also partially "fill in" missing input for a cycle or two so that the expectancy protects the binding of features of a stimulus and stabilizes a perceptual stream across seconds of time.

2.4. Dynamic Attention Architecture

Figure 2 shows the architecture used to simulate the Jones-Bregman experiment. Our mechanistic explanation of Jones result is that the early standard target tones arriving at the 80 msec rate first prime the dynamic attention system by setting the 80 msec clock to oscillate at 40 Hz, then the slow captor tones at the 240 msec rate establish a background stream at 35 Hz with a rhythmic expectancy that is later violated by the appearance of the fast target tones. These are thereby driven into a separate stream by decorrelation and brought into the foreground frequency by the mismatch pop out mechanism. This allows the attentional stream into the Elman sequence recognition units, also primed to look for that rhythm, to synchronize and read in activity due to the target tones for order determination.

In this simulation, the connections to the first two Elman associative memory units are hand wired to the A and B primary cortex oscillators to act as a latching, order determining switch. If synchronized to the memory unit at the attentional stream frequency, the A target tone oscillator will drive the first memory unit into the 1 attractor which then inhibits the second unit from being driven to 1 by the B target tone. The second unit has similar wiring from the B tone oscillator, so that the particular higher order memory unit which is left in the 1 state after a trial indicates to the rest of the brain which tone came first. The flanking and high captor tone oscillator is connected equally to both memory units, so that a random attractor transition occurs before the targets arrive, when it is interfering at the 40 Hz attentional frequency, and poor order determination results. If the flanking tone oscillator is in a separate stream along with the captor tones at the background eigenfrequency of 35 Hz, it is outside the receiving passband of the memory units and cannot cause a spurious attractor transition. Thus, in the absence of a rhythmic distinction for the target tones, their frequency difference alone is insufficient to drive the formation of a separate stream, and the targets cannot be reliably discriminated.

Integration of Bregman and Jones Theories An important contribution of a mechanistic description is that it can sometimes reconcile and generalize functional descriptions that seem otherwise to be at odds. This architecture demonstrates mechanisms that integrate the theories of Jones and Bregman about auditory perception, since it models the primitive preattentive levels and the rhythmic and schema driven attentional levels. Stream formation is a preattentive process that works well on non-rhythmic inputs as Bregman asserts, but an equally primary and preattentive rhythmic expectancy process is also at work as Jones asserts and the mismatch negativity indicates. This becomes a factor in stream formation when rhythmic structure is present in stimuli as demonstrated by Jones.

In the model, both the rhythmic expectancy and stream forming processes are carried out by fast temporary weights that bridge time and space to organize input into perceptual sources. This is a form of short term memory, distinct from the long term learning that Bregman sees as required at the schema level of processing. The short term memory required for stream formation can itself be viewed as an "expectation" that the next stimulus within that sound frequency range should be part of the same stream. Thus expectancy by itself is not exclusively a schema level process, as Bregman might assert. In Jones' view, as captured in the model, the orientation of internal processing cycles to external rhythms is as basic as spatial orientation, and is not restricted to music perception as Bregman has suggested[4].

Top down attentional processes in the model may restructure or make use of this preattentive structuring both in the streaming and rhythmic expectation domains. In particular, they may be schema driven from the higher order sequence representation units where long and intermediate term memory weight changes may occur, as both Jones and Bregman would suggest is true in the brain. An important class of schemas are rhythmic patterns, as asserted by Jones. Detailed and complex rhythmic schemas may be learned by the Elman sequence learning part of the architecture[1, 2], and these allow attention to be directed according to the temporal knowledge of input classes, as well as spectral knowledge.

ACKNOWLEDGMENTS

Supported by ONR grant N00014-95-1-0744. It is a pleasure to acknowledge the invaluable assistance of Morris Hirsch and Walter Freeman.

REFERENCES

[1] B. Baird, T. Troyer, and F. Eeckman, Attentional network streams of synchronized 40 Hz activity in a cortical architecture of coupled oscillating associative memories. In: S. Levine, V. Brown, and T. Shirley, editors, *Oscillations in Neural Systems*. Laurence Erlbaum, NJ (1997, in press).

[2] B. Baird, T. Troyer, and F. H. Eeckman, Attention as selective synchronization of oscillating cortical sensory and motor associative memories. In: F. H. Eeckman, editor, *Neural Systems Analysis and Modeling*, pages 167–175, Kluwer, Norwell, MA, 1994.

[3] M. R. Jones and M. Boltz, Dynamic attending and responses to time. *Psychological Review*, 96:459–491, 1989.

[4] A. S. Bregman, *Auditory Scene Analysis*, Oxford Univ. Press, Oxford, 1992.

[5] G. Brown and M. Cooke, A neural oscillator model of auditory stream segregation. In: *IJCAI Workshop on Computational Auditory Scene Analysis*, 1996, in press.

[6] R. Naatanen, editor, *Attention and Brain Function*, Laurence Erlbaum, NJ, 1992.

40

CALCULATING CONDITIONS FOR THE EMERGENCE OF STRUCTURE IN SELF-ORGANIZING MAPS

Hans-Ulrich Bauer,[*] Maximilian Riesenhuber, and Theo Geisel

Institut für Theoretische Physik
SFB Nichtlineare Dynamik
Universität Frankfurt
Robert-Mayer-Str. 8–10, 60054 Frankfurt/Main, Germany
bauer@chaos.uni-frankfurt.de
geisel@chaos.uni-frankfurt.de
max@ai.mit.edu

INTRODUCTION

The self-organization of sensotopic maps in the brain has been modeled with various approaches, using models with linear [1, 2, 3, 4] and nonlinear [5, 6, 7] lateral interaction functions. In the realm of visual maps, all models are able to generate ocular dominance or orientation structure [8]. Models differ, however, with regard to more subtle effects, like the impact of input correlation on ocular dominance stripe width, the self-organization of oriented receptive fields from non-oriented stimuli or correlation functions, or the preferred angle of intersection between ocular dominance and orientation column systems. For a correct assessment of the behavior of particular models with regard to these or other phenomena, it is dangerous to rely on simulations only. Rather, analytic results on conditions for the pattern formation in map models are desirable. We present here a method for calculating such conditions for a map model with strong lateral nonlinearity, the high-dimensional version of Kohonen's Self-Organizing Map (SOM). Using this method we then analyze two relevant models, a SOM-model for the development of orientation maps and a SOM-model for the development of ocular dominance maps.

METHODS

A Self-Organizing Map (SOM) consists of neurons characterized by a position $\mathbf{r}$ in the map lattice plus a receptive field $\mathbf{w}_r$. A stimulus $\mathbf{v}$ is mapped onto that neuron $\mathbf{s}$ whose

---

[*] Present address: Center for Biol. & Comput. Learning and Dept. of Brain & Cognitive Sciences, Massachusetts Institute of Technology, Cambridge, MA 02142, U S A.

receptive field $\mathbf{w}_s$ matches $\mathbf{v}$ best. This amounts to a winner-take-all rule, i.e. a strong lateral nonlinearity which can be regarded as a consequence of lateral inhibition [9]. The map results as a stationary state of a self-organization process, which successively changes all receptive fields $\mathbf{w}_r$,

$$\Delta \mathbf{w}_r = \varepsilon h_{rs} (\mathbf{v} - \mathbf{w}_r), \tag{1}$$

following the presentation of stimuli $\mathbf{v}$. Here, ε controls the size of learning steps, h_{rs} denotes a neighborhood function, centered around the winning neuron $\mathbf{s}$ and usually chosen to be of Gaussian shape,

$$h_{rs} = e^{-\frac{\|r-s\|^2}{2\sigma^2}}. \tag{2}$$

h_{rs} enforces neighboring neurons to align their receptive fields. In this way the property of topography is imposed on the SOM.

In the most general formulation of a SOM, stimuli and receptive fields are activity resp. synaptic weight distributions over a layer of input channels. In this so-called "high-dimensional" variant, which allows for a simultaneous self-organization of map structure and the structure of individual receptive fields and which is analyzed in the following, the best-matching neuron $\mathbf{s}$ for a particular stimulus is determined by

$$\mathbf{s} = \arg\max_r (\mathbf{w}_r \cdot \mathbf{v}). \tag{3}$$

The relation of this high-dimensional SOM-variant to the less general "feature map" variant, which is based on (linear) features of stimuli and receptive fields, is discussed in [10, 11].

Exploiting the analogy to vector quantizers, and utilizing the concept of stochastic approximation, it has been argued [9] that the SOM approximately minimizes the distortion function

$$E_w = \sum_r \sum_{r'} \sum_{v' \in \Omega r'} (\mathbf{v}' - \mathbf{w}_r)^2 e^{\left(-\frac{r-r'^2}{2\sigma^2}\right)} \tag{4}$$

Ω_r denotes the set of stimuli $\mathbf{v}$ which are mapped onto node $\mathbf{r}$. By Eq.(3) Ω_r depends on the $\mathbf{w}_r$. Even though the SOM learning dynamics does not proceed along the gradient of this function (or any other energy function), the deviations become small in the limit of an ordered map with large values for σ [12]. On the other hand, in the limit of $\sigma \to 0$, the SOM approaches the LBG-algorithm for vector quantization, which performs gradient descent on E_w ($\sigma = 0$). Therefore, a sensible strategy to determine the final state of a SOM could be based on a comparison of E_w for various possible map states. An evaluation of E_w, however, involves the receptive fields $\mathbf{w}_r$; quantities for which an ansatz is in general difficult to make. We therefore replace E_w by the related distortion function

$$E_v = \sum_r \sum_{r'} \sum_{v' \in \Omega r'} \sum_{v' \in \Omega r} (\mathbf{v}' - \mathbf{v})^2 e^{\left(-\frac{r-r'^2}{2\sigma^2}\right)} \tag{5}$$

Under quite general assumptions, the minima of E_v coincide with those of E_w in the limit of $\sigma \to 0$, and the deviations are small otherwise. A great advantage of this replacement lies in the fact that the evaluation of Eq. (5) requires only an ansatz for the $\Omega_r(\mathbf{w}_r)$, not for the $\mathbf{w}_r$ themselves. The following examples show that making such an ansatz is relatively easy, whereby E_v can be explicitly calculated for different states of the map, and the state with the lowest distortion can be determined.

RESULTS

We now apply our method of analysis to two SOM-models for the development of topographic maps in the visual cortex [5, 15], which were previously accessible only numerically. In these models a sensory input space with one, resp. two, layers of $N \times N$ retinal channels is mapped onto a $N \times N$-neuron output layer, the cortical area.

SOM-Model for the Development of Orientation Maps

The first model is concerned with the development of orientation maps. Using ellipsoidal Gaussian activity distributions as stimuli (minor axis σ_1, major axis $\sigma_2 > \sigma_1$), simulations of this model led to maps with oriented receptive fields for substantially elongated stimuli [5]. Using rather circular stimuli ($\sigma_2 \approx \sigma_1$) non-oriented receptive fields were also observed [13]. In order to analyze this transition, we restrict the number of stimulus orientations to two (horizontal and vertical), and denote by $\Delta_{ij\|}$ the squared difference between two vertical stimuli, separated by distances i horizontally and j vertically, and summed over the input layer $\Delta_{ij\text{-}}$, $\Delta_{ij\text{—}}$ $\Delta_{ij|}$ defined analogously). Then the distortions $E_{v,\text{non-ori}}$ and $E_{v,\text{ori}}$ for the two states of the map amount to

$$E_{v,\text{non-ori}} = N^2 \sum_{i=-\infty}^{\infty} \sum_{j=-\infty}^{\infty} e^{-\frac{i^2+j^2}{2\sigma^2}} \left(\Delta_{ij\|} + \Delta_{ij|\text{-}} + \Delta_{ij\text{-}|} + \Delta_{ij\text{--}} \right),$$
(6)

$$E_{v,\text{ori}} = N^2 \sum_{i=-\infty}^{\infty} \sum_{j=-\infty}^{\infty} e^{-\frac{i^2+j^2}{2\sigma^2}} \left(\Delta_{2ij\|} + \Delta_{2ij\text{--}} + \tfrac{1}{2}\Delta_{2i+1j\|} + \tfrac{1}{2}\Delta_{2i-1j\|} + \tfrac{1}{2}\Delta_{2i+1j\text{--}} + \tfrac{1}{2}\Delta_{2i-1j\text{--}} \right).$$
(7)

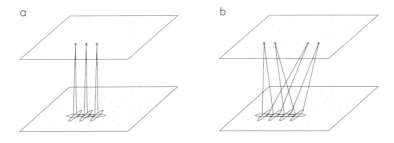

Figure 1. Illustration of possible tesselations of a reduced stimulus space in a SOM-model for the development of orientation maps. Stimuli are indicated as ellipses in the lower square (retinal space), map neurons as crosses in the upper square (orientation map). **a:** two stimuli with different orientations but located at the same position in retinal space are mapped to the same neuron (corresponding to a non-oriented receptive field). **b:** two stimuli with the same orientation but centered at different locations are combined (corresponding to oriented receptive fields). In the latter case we assume that neighboring neurons receive input from stimuli of the same orientation.

Equating (6) and (7) now yields the transition point, which in the limit of $\sigma_{1,2} \gg \sigma \gg 1$ is (after some calculations)

$$\sigma_{2,\text{crit}} \approx \sigma_1 + \sqrt{3}\sigma. \tag{8}$$

The condition (8) for the break of symmetry from non-oriented to oriented receptive fields is very well corroborated by numerical simulations using the reduced stimulus set (Fig. 2), as well as by additional simulations with the full stimulus set (all orientations and positions). This additive relation between σ_1 and $\sigma_{2,\text{crit}}$ deviates from the multiplicative relation found for a corresponding model in feature map approximation [14].

SOM-Model for the Development of Ocular Dominance Maps

Second, we analyze a SOM-model for the development of ocular dominance maps in the visual cortex [15]. Its input space consists of two retinal layers for the two eyes. Here, we assume stimuli to be an activity distribution over 3×3 input channels in one retina, complemented by a distribution of identical shape, but attenuated by a factor of c, in the other retina. The center position of the group of 3×3 channels, as well as the retina with the larger overall activation, are chosen at random for each stimulus. The degree c of correlation between retinae is the control parameter of the model. For $c \approx 1$, the weight vectors $\mathbf{w}_r$ develop symmetrically in both retinae. For each neuron $\mathbf{r}$, Ω_r contains two stimuli of opposite ocularity, but centered at the same retinal position (solution type **a**). With decreasing c, a symmetry breaking transition takes place, and neurons develop a preference for one or the other retina (ocular dominance). The Ω_r now contain two stimuli of the same ocularity, but centered at different (neighboring) positions. Depending on the clustering pattern of same-ocularity neurons in the map, different types of solutions can be distinguished. Here we consider the following arrangements: a chequerboard-type alternation (type **b**), or bands of length N in one direction, and widths 1 (type **c**), 2 (type **d**), or $N/2$ (type **e**) in the orthogonal direction. Type **e** corresponds to a degenerate solution, which can be found in models [17, 18], but not in the visual cortex.

Evaluating the distortions E_v for the different combinations of squared differences between stimuli, multiplied with the neighborhood factors

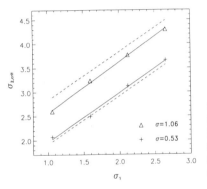

Figure 2. Critical value $\sigma_{2,\text{crit}}$ of the longer half-axis of elliptic stimuli for the occurrence of an orientation map, as a function of the shorter half-axis σ_1, for two exemplary values of the map neighborhood function width σ. Symbols indicate the results of simulations of SOMs, the solid line is a fit to these four points, resp. The dashed lines show the corresponding analytic results (8).

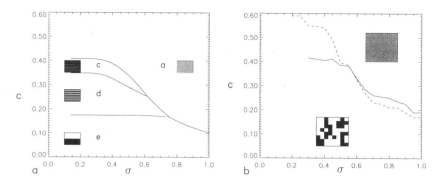

Figure 3. Phase diagrams for an SOM-model for the development of ocular dominance maps. **a:** analytical solution. Note, that at constant σ, the width of bands increases as c decreases. **b:** numerical solution. Solid line: multiplicative normalization, dashed line: subtractive normalization.

$$e^{-\frac{(\Delta i)^2+(\Delta j)^2}{2\sigma^2}},$$

we obtain the phase diagram depicted in Fig. 3a.

For large values of c, the type **a** solution prevails (no ocular dominance). At intermediate values for c, solutions **c** and **d** result, with increasing band width for decreasing c. For very small values of c, the degenerate type **e** solution with only two ocular dominance bands is attained. So the phase diagram of this ocular dominance model exhibits a transition scenario to broader ocular dominance bands with decreasing correlation parameter c, a result which coincides with the findings of a recent neuroanatomical experiment involving strabismic cats (corresponding to smaller values of c) and normal-sighted cats (corresponding to larger values of c) [16].

Simulations of this model corroborated the analytic results very well (Fig. 3b). The transition from non-ocular to ocular receptive fields, takes place very closely to the analytically predicted critical value of c. The transition takes place at about the same critical c, regardless of the type of normaization employed in the learning dynamics of the SOM (multiplicative or subtractive).

SUMMARY AND CONCLUSIONS

We expect that our method will also be helpful in future investigations of high-dimensional SOM-models for more complicated mapping problems, in the brain and also in technical domains of application, which as yet were handicapped by the need for very costly numerical simulations. The method was useful already in the investigation of the relation between oriented receptive fields and non-oriented stimuli in a SOM driven by On-center and Off-center inputs (see accompanying paper).

REFERENCES

1. C. von der Malsburg, Kybernetik **14**, 85 (1973); Biol. Cyb. **32**, 49 (1979).
2. K. D. Miller, J. B. Keller, M. P. Stryker, Science **245**, 605 (1989).

3. K. D. Miller, J. Neurosc. **14**, 409 (1994).
4. M. Miyashita, S. Tanaka, NeuroRep. **3**, 69 (1992).
5. K. Obermayer, H. Ritter, K. Schulten, Proc. Nat. Acad. Sci. USA **87**, 8345 (1990).
6. R. Durbin, G. Mitchison, Nature **343**, 644 (1990).
7. F. Wolf, H.-U. Bauer, T. Geisel, Biol. Cyb. **70**, 525 (1994).
8. E. Erwin, K. Obermayer, K. Schulten, Neur. Comp. **7**, 425 (1995).
9. T. Kohonen, *Self-Organizing Maps*, Springer, Berlin (1995).
10. M. Riesenhuber, H.-U. Bauer, T. Geisel, Biol. Cyb., in print (1996).
11. K. D. Miller, submitted to Neur. Comp. (1995).
12. E. Erwin, K. Obermayer, K. Schulten, Biol. Cyb. **67**, 47 (1992).
13. K. Obermayer, *Adaptive Neuronale Netze und ihre Anwendung als Modelle der Entwicklung kortikaler Karten*, infix Verlag, Sankt Augustin (1993).
14. K. Obermayer, G. G. Blasdel, K. Schulten, Phys. Rev. A **45**, 7568 (1992).
15. G. J. Goodhill, Biol. Cyb. **69**, 109 (1993).
16. S. Löwel, J. Neurosci. **14**, 7451 (1994).
17. C. von der Malsburg, *Neural and Electronic Networks*, Eds. Zornetzer et al., Academic Press, 421 (1994).
18. G. J. Goodhill, D. J. Willshaw, Network **1**, 41 (1990); P. Dayan, Neur. Comp. **5**, 392 (1993).

41

DYNAMICS OF SYNAPTIC PLASTICITY: A COMPARISON BETWEEN MODELS AND EXPERIMENTAL RESULTS IN VISUAL CORTEX

Brian Blais, Harel Shouval, and Leon N Cooper

The Department of Physics
The Institute for Brain and Neural Systems
Box 1843, Brown University
Providence, RI
E-mail: bblais@cns.brown.edu
E-mail: hzs@cns.brown.edu
E-mail: leon_cooper@brown.edu

ABSTRACT

Receptive fields in the visual cortex can be altered by changing the visual environment, as has been shown many times in deprivation experiments. In this paper we simulate this set of experiments using two different models of cortical plasticity, BCM and PCA. The visual environment used is composed of natural images for open eye and of noise for closed eyes. We measure the response of the neurons to oriented stimuli, and use the time course information of the neuronal response to provide a preliminary quantitative comparison between the cortical models and experiment.

1. INTRODUCTION

Two important characteristics of the visual response of most neurons in cat striate cortex are that they are binocular and show a strong preference for contours of a particular orientation [7]. Although some orientation selectivity exists in striate cortex prior to visual experience, maturation to adult levels of specificity and responsiveness requires normal contour vision during the first 2 months of life (for review see Fregnac and Imbert, 1984).

Ocular dominance is a measure of how effectively the neuron can be driven through the left and right eyes, respectively. Up to the level of the LGN, visual information originating from the two eyes is segregated in separate pathways, striate cortex is thus the first site where

Computational Neuroscience
edited by Bower, Plenum Press, New York, 1997

individual cells receive afferent projections from both left and right eyes, and in normal kitten striate cortex most of the visually responsive cells are binocular.

One of the most dramatic examples of cortical plasticity is the alteration of ocular dominance in kitten striate cortex in monocular deprivation, a procedure in which, one eye is deprived of patterned stimuli (by either suturing the eyelid closed or using an eye patch). In such an imbalanced visual environment, cells in kitten's striate cortex change from mostly binocular to almost exclusively monocular: in less than 24 hours most cells lose their responsive to stimulation through the deprived eye and can only be driven through the eye that remains open [10, 13].

The change induced by monocular deprivation is reversible. In the rearing condition called reverse suture there is an initial period of monocular deprivation: after the cortical neurons have become monocular, the deprived eye is opened and the other eye closed. In this situation the cortical neurons lose responsiveness to the newly closed eye, and become responsive to the newly opened eye [2]. Acute studies indicate that, as the ocular dominance of cortical responsiveness shifts from one eye to the other, there is rarely a period when cells can be strongly and equally activated by both eyes [10].

In this paper we model these experiments, looking specifically at the time course of neuronal response during visual deprivation. In this way we wish to directly compare two different learning rules with each other, and also with experiment. The visual environment to which open eyes are exposed is assumed to be composed of preprocessed natural images, whereas closed eyes receive noise as their inputs. The architecture assumed in these models is of single learning neurons, as a first step to understanding the behavior of networks.

We compare the two learning rules BCM [1, 8] and PCA [11]. Both of these rules have been shown to develop orientation selectivity in a realistic visual environment [9, 12]. Previous simulations of the deprivation experiments have been performed using simplified inputs for the BCM theory[3]. In this paper we use realistic inputs.

2. METHODS

The visual environment used here is described in Law and Cooper, 1994. The exact time course of these simulations is dependent on the parameters chosen; we have therefore examined these over a large parameter regime. At each set of parameters we performed all of the deprivation simulations, and compared their timing ratios, in order to test the robustness of our predictions. Examples of the simulations for PCA and BCM are shown in Figure 1.

Table 2 shows the specific form of the BCM learning rule, as well as the range of parameters, which was used. The precise time course of the neuron activity is dependent on the choice of the functional form of both the modification function, ϕ, and the sigmoid. Variations of these are currently being investigated.

We measure the response $Y(t)$ of the neurons using oriented stimuli. Of particular interest is characteristic half-rise (half-fall) time for the growth (decay) of neuronal response, referred to as either $t_{1/2}$ or simply $\mathcal{T}$. The half-time measurement provides a direct comparison between the BCM and PCA models. It also provides a way to quantitatively translate between simulation cycles and real time, yielding a quantitative comparison with experiment.

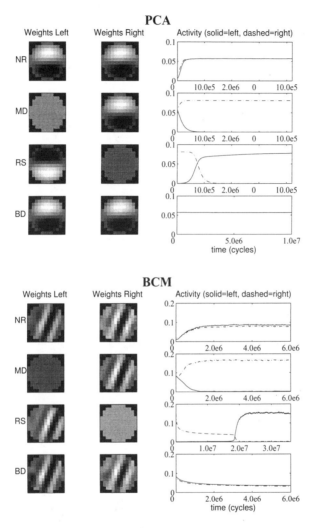

Figure 1. Example PCA (top) and BCM (bottom) Simulations. Left: Final weight configuration. Right: Maximum response to oriented stimuli, as a function of time. Simulations from top to bottom are as follows. Normal Rearing (NR): both eyes presented with patterned input. Monocular Deprivation (MD): following NR, one eye is presented with noisy input and the other with patterned input. Reverse Suture: following MD, the eye given noisy input is now given patterned input, and the other eye is given noisy input. Binocular Deprivation (BD): following NR, both eyes are given noisy input. **It is important to note that for BCM if Binocular Deprivation is run longer, selectivity will eventually be lost. This is not true for PCA**

Table 1. Summary of Experimental Results

Experiment	Reference	Half-Time $\mathcal{T}$
Monocular Deprivation	• OD changes were observed as **early as 6 h**[5, 10], **complete loss** of response to closed as early as **12 h**[10]	$\mathcal{T}_{\text{fall}}^{\text{MD}} \approx$ 6-12 h
Binocular Deprivation	• cortical response reduced **within 3 d** [4]	$\mathcal{T}_{\text{fall}}^{\text{BD}} <$ 3 d
Reverse Suture	• the time course for the reduction of response to the newly deprived eye was **similar to monocular deprivation**[10]	$\mathcal{T}_{\text{fall}}^{\text{RS}} \approx \mathcal{T}_{\text{fall}}^{\text{MD}}$
	• **At least 24 h** of reverse suture is required before the responses to the deprived eye reappears[10]	$\mathcal{T}_{\text{rise}}^{\text{RS}} \approx$ 1-4 d

3. RESULTS

3.1. Summary of Experimental Results

Table 1 summarizes the experimental results to which we can compare the two models. Since the exact results depend on when in the critical period the experiments were done, we look at *ratios* of the experimentally determined development times, so we can make a reasonable comparison with the theory. The values of $\mathcal{T}$, either half-rise times, $\mathcal{T}_{\text{rise}}$, or half-fall times, $\mathcal{T}_{\text{fall}}$, from Table 1 give us the following ratios that we can then use in our comparison: $(\mathcal{T}_{\text{fall}}^{\text{RS}} / \mathcal{T}_{\text{fall}}^{\text{MD}} \approx 1)$, $(2 < \mathcal{T}_{\text{rise}}^{\text{RS}} / \mathcal{T}_{\text{fall}}^{\text{MD}} < 16)$, and $(0.33 < \mathcal{T}_{\text{rise}}^{\text{RS}} / \mathcal{T}_{\text{fall}}^{\text{BD}} < 16)$.

3.2. PCA Results

We can use the full time-domain solution[14] of the PCA equations in order to explore the deprivation simulations analytically. Under the assumptions that natural scenes are dominated by the first eigenvector of the covariance matrix, $\mathbf{v}_1$ with eigenvalue λ_1, and that closed eyes are presented with noise with variance σ^2, we obtain the following equations for the time development of the weights

$$\mathbf{m}^{\text{MD}}(t) = \left(e^{\lambda_1 t} \mathbf{v}_1 \quad e^{\sigma^2 t} \mathbf{v}_1 \right) / (e^{2\lambda_1 t} + e^{2\sigma^2 t})^{1/2} \tag{1}$$

$$\mathbf{m}^{\text{BD}}(t) = (\mathbf{v}_1 \quad \mathbf{v}_1) / \sqrt{2} \tag{2}$$

$$\mathbf{m}^{\text{RS}}(t) = \left(e^{\sigma^2 t} \mathbf{v}_1 \quad e^{\lambda_1 t} \varepsilon \mathbf{v}_1 \right) / (e^{2\sigma^2 t} + \varepsilon e^{2\lambda_1 t})^{1/2} \tag{3}$$

Comparing the MD and RS solutions (Equations 1 and 3) one sees that the times for the decay and recovery of neuronal activity must be identical for each regardless of noise levels and input statistics. Therefore the PCA model predicts correctly the time for the neuronal activity to fall in these deprivation situations, but fails to properly predict the time for the recovery of activity in RS compared to MD.

Equation 2 implies that a neuron following Oja's rule, experiencing binocular deprivation following normal rearing, performs a random walk about the normal reared state. Thus PCA is again inconsistent with the experiment results.

3.3. BCM Results

In order to determine the dependence of the response half-times on model and input parameters we need to perform simulations over a range of those parameters and measure the

Table 2. Parameters for BCM

Learning Rule	$\dot{\mathbf{m}} = \eta c(c - \theta)\mathbf{d}$
	$\dot{\theta} = \frac{1}{\tau}(c^2 - \theta)$
Activation Rule	$c = \sigma(\mathbf{m} \cdot \mathbf{d})$
cortical sigmoid	$\sigma(-\infty) = -1$
	$\sigma(+\infty) = 50$
Initial threshold	$\theta_o = 0.73$
Input mean	$\langle \mathbf{d} \rangle = -3.3 \cdot 10^{-5}$
Input variance	$\mathrm{var}(\mathbf{d}) = 1.0$
RF Diameter	13 pixels
Retinal DOG ratio	3:1
Learning rate	$\eta = 5 \cdot 10^{-7}, ..., 5 \cdot 10^{-5}$
Memory constant	$\tau = 10, ..., 3510$
Noise Levels	uniform noise=
	[-.25:.25],...,[-2.5,2.5]

time $\mathcal{T}$ for each of these. The parameter regime is initially chosen to give *stable* simulations, and then explored more finely to determine regions with consistent time ratios. Stability is lost if the learning rate is too large, or if the memory constant, τ, is too large causing BCM sliding threshold, θ, to move too slowly.

The two most important parameters are the noise level to the closed eye and the memory constant, τ. We see from Figure 2 that the noise level has a dramatic effect on the time course of the neuronal response, and is different for each deprivation simulation. This allows us to possibly locate a range of noise levels for which the relative times are consistent with experiment. In the parameter regime tested we found surprisingly little effect from the memory constant on the time course. We are currently exploring possible reasons for this apparent difference between the natural scene environment and previous work done with abstract inputs[3], where the memory constant played a more important role in the timing.

4. SUMMARY AND CONCLUSIONS

It has been demonstrated that the PCA rule fails to account for the time course of the deprivation experiments. Most dramatic is the result that binocular deprivation to a PCA neuron gives only a random walk response. BCM on the other hand can account for the time course, within a particular parameter regime. This parameter regime can be summarized by the range of 3 parameters: memory constant $\tau = 500, ..., 3000$, learning rate $\eta = 4.5 \cdot 10^{-6}, ..., 6 \cdot 10^{-6}$, and noise level $\sigma = 0.88, ..., 1.4$ (or [-1.1:1.1],...,[-1.45:1.45]). Note that the valid parameter regime is *not* defined by a sharp boundary, but that near the edge the system is less robust.

One of the features of this approach is the direct calculation of the model parameters in terms of experimental quantities. Using the range of half-time values exhibited across the valid parameter regime for, say, monocular deprivation, and attributing that range to the known 6-12 hour range obtained from experiment, we can find a range for the possible values of τ in minutes. Doing this we find that τ ranges between 1-15 minutes. We are investigating how this time range might be affected by variations such as changes in the slope of the modification function, ϕ, at threshold and the form of the cortical sigmoid. A final determination of the allowed values of τ may give us clues to the cellular and molecular processes involved.

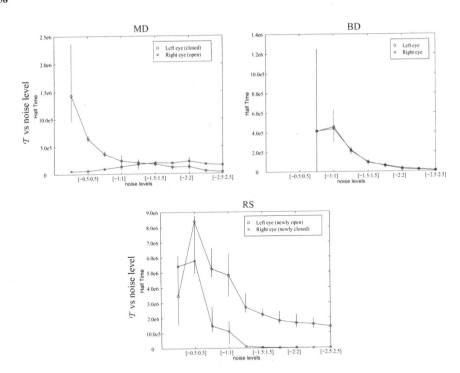

Figure 2. BCM Results: The effect of the noise level on the response half-times of deprivation simulations, Monocular Deprivation (MD), Binocular Deprivation (BD), and Reverse Suture (RS).

REFERENCES

[1] E. L. Bienenstock, L. N. Cooper, and P. W. Munro. Theory for the development of neuron selectivity: orientation specificity and binocular interaction in visual cortex. *Journal Neuroscience*, 2:32–48, 1982.

[2] C. Blakemore and R. R. van Sluyters. Reversal of the physiological effects of monocular deprivation in kittens: further evidence for sensitive period. *J. Physiol. Lond.*, 248:663–716, 1974.

[3] E. E. Clothiaux, L. N Cooper, and M. F. Bear. Synaptic plasticity in visual cortex: Comparison of theory with experiment. *Journal of Neurophysiology*, 66:1785–1804, 1991.

[4] R. Freeman, R. Mallach, and S. Hartley. Responsivity of normal kitten striate cortex deteriorates after brief binocular deprivation. *Journal of Neurophysiology*, 45(6):1074–1084, 1981.

[5] R. Freeman and C. Olson. Brief periods of monocular deprivation in kittens: Effects of delay prior to physiological study. *Journal of Neurophysiology*, 47(2):139–150, 1982.

[6] Y. Frégnac and M. Imbert. Development of neuronal selectivity in the primary visual cortex of the cat. *Physiol. Rev.*, 64:325–434, 1984.

[7] D. H. Hubel and T. N. Wiesel. Receptive fields, binocular interaction and functional architecture in the cat's visual cortex. *J. Physiol*, 160:106–154, 1962.

[8] N. Intrator and L. N. Cooper. Objective function formulation of the BCM theory of visual cortical plasticity: Statistical connections, stability conditions. *Neural Networks*, 5:3–17, 1992.

[9] C. Law and L. Cooper. Formation of receptive fields according to the BCM theory in realistic visual environemnts. *Proceedings National Academy of Sciences*, 91:7797–7801, 1994.

[10] L. Mioche and W. Singer. Chronic recording from single sites of kitten striate cortex during experience-dependent modification of synaptic receptive-field properties. *J. Neurophysiol.*, 62:185–197, 1989.

[11] E. Oja. A simplified neuron model as a principal component analyzer. *Journal of Mathematical Biology*, 15:267–273, 1982.

[12] H. Shouval and Y. Liu. How does retinal preprocessing Affect the Receptive field of a stabilized hebbian neuron. In J. M. Bower, editor, *The Neurobiology of Computation: The Proceedings of the Third Annual Computation and Neural Systems confrence*, pages 129–134. Kluwer, 1994.

[13] T. N. Wiesel and D. H. Hubel. Single-cell responses in striate cortex of kittens deprived of vision in one eye. *Journal of Neurophysiology*, 26:1003–1017, 1963.

[14] J. L. Wyatt and I. M. Elfadel. Time-domain solutions of Oja's equations. *Neural Computation*, 7(5):915–922, 1995.

42

PHYSICAL BASIS OF THE EPILEPSY OCCURRENCE

Vladimir E. Bondarenko

Institute of Biochemical Physics
Russian Academy of Sciences
Kosygin Street, 4, Moscow, Russia
E-mail: bond@sai.msu.su

ABSTRACT

A model of the epilepsy occurrence and spreading on the basis of the simple artificial neural network is presented. It is shown that the epilepsy-like phenomena can occur in the neural networks at the increasing of neuronal excitability or under the action of an external force. Dynamics of the quantitative EEG characteristics (correlation dimension, amplitude, largest Lyapunov exponent) is similar to one at the epilepsy occurrence. The results of epilepsy modelling are compared with the experimental investigations. Phenomenon of epilepsy is considered as a self-organization process which appears at the increasing of the energy transfer through the neural network.

1. INTRODUCTION

Particular attention to the chaos in the human brain and artificial neural networks is explained by the hope that the understanding of the human brain functions will be achieved just in this direction [1, 2, 3, 4]. At present, different chaotic solutions are obtained in neural network modelling [5, 6, 7, 8, 9, 10], but comparison of their quantitative characteristics with the human or animal EEGs can not be performed. Therefore, it is important to compare the dynamics of chaotic neural network with the real brain behavior. As an example the epilepsy is chosen for this study.

Phenomenon of epilepsy represents itself as synchronous large-amplitude non-linear oscillations which can occupy whole brain [11]. Usually, epilepsy begins from a small pathological brain region with enhanced synchronous activity and spreads then to another areas. This phenomenon is accompanied by infrequent periods of loss of consciousness and sometimes by convulsions. Experimental investigations of this disease show that during epileptic seizure strong decreasing of the EEG complexity is observed. So, the epileptic seizure decreases the correlation dimension of EEG from 6–8 in awake state to 2 when epileptic fit takes place [1, 2]. A lot of experiments on the artificial provocation of the epilepsy show that in this case the increasing of the K^+ ions density plays essential role which leads to the higher brain excitability [12, 13]. Epilepsy can be provoked by the action of the external electric field on the brain also [13, 14].

This paper presents the results of study of the neural network model which produces chaos similar to the human EEG [15]. It is shown that the increasing of the neural network asymmetry (increasing of the neural excitability) leads to the phenomenon which is similar to the epilepsy. In this case, in the model, we observe synchronous large-amplitude non-linear oscillations of all neuron outputs and strong decreasing of the correlation dimension. We consider also the action of the external sinusoidal force on the neural network which simulates both the source of pathological rhythmic activity inside the brain and the external action of the electric field. Two cases are studied: (1) simultaneous influence on all neurons; (2) the influence on one neuron in the neural network. It is shown that the action of the external force also leads to the decreasing of the degree of chaos, even though the only neuron is affected by the external action. Obtained values of the correlation dimension in the neural network model are close to ones which are observed in the human epilepsy study [1, 2].

2. NEURAL NETWORK MODEL AND METHOD OF ANALYSIS

Asymmetric analog neural network model with the time delay and under the external sinusoidal force is considered. It is described by the set of differential equations:

$$\dot{u}_i(t) = -u_i(t) - \sum_{j=1}^{M} a_{ij} f(u_j(t - \tau_j)) + e \cdot \sin \omega_e t, \qquad i, j = 1, 2, ..., M, \tag{1}$$

where $u_i(t)$ is the input signal of the ith neuron, M is the number of neurons, a_{ij} are the coupling coefficients between the neurons, τ_j is the time delay of the jth neuron output, $f(x) = c \cdot \tanh(x)$, e and ω_e are the amplitude and frequency of the external force, respectively. In this letter the case is studied when the value of τ_j is constant for all neurons ($\tau_j = \tau$). The coupling coefficients are produced by random numerical generator in the range from -2.0 to $+2.0$. In order to change the asymmetry of the neural network model and to increase the fraction of excitatory coupling coefficients the interval of a_{ij} variation is changed so that the averaged value

$$\bar{a} = \frac{1}{M^2} \sum_{i,j} a_{ij} \tag{2}$$

is varied from -1.0 to $+1.0$. The coefficient c is used in order to vary coupling coefficients between the neurons simultaneously.

Method of solution of Eqs. (1) is described in details in [15].

For the evaluation of the correlation dimension v which characterizes a number of excited degrees of freedom and complexity of the EEG the Grassberger–Procaccia algorithm is used with $N = 8192$ points. The largest Lyapunov exponent is defined as

$$\lambda = \lim_{t \to \infty} \lim_{D(0) \to 0} t^{-1} \ln[D(t)/D(0)], \tag{3}$$

where $D(t)$ and $D(0)$ are the distances between perturbed and unperturbed trajectories at the current and at the initial moments, respectively.

Along with the correlation dimension and largest Lyapunov exponent, Shannon entropy $S_{Sh} = -\sum_{k=1}^{K} p_k \ln p_k$ is used to characterize the degree of chaos of the neuron outputs. Here p_k is the probability to find the amplitude u_i in the kth subinterval with length Δ_{Sh} (interval of the amplitude variation is divided into $K = 128$ subintervals). To avoid uncertainty in the definition of S_{Sh}, renormalized Shannon entropy S_r is also used. S_r does not depend on the averaged effective energy of neuron outputs, because for the calculation of S_r we renormalize interval length Δ_{Sh}: $\Delta_r = 4\sqrt{E}\Delta_{Sh}$, where

$$E = \frac{1}{N} \sum_{i=1}^{N} \left(\frac{u_i - \bar{u}}{A} \right)^2. \tag{4}$$

Here $\bar{u}$ is the constant level of u, $A = 4096$ is the normalization constant.

3. SIMULATION RESULTS

Numerical solution of Eqs. (1) at $e = 0.0$ shows that presented neural network model produces constant, periodic or chaotic outputs, depending on the network parameters. Calculations of the correlation dimension and largest Lyapunov exponent show that their values approach the ones obtained in experimental investigations of the human and animal EEGs. Time delay τ plays an important role in controlling chaos. In presented neural network model the more time delay τ is the more chaotic output we obtain. If we make time normalization so that the main frequency of numerical solution to be equal to the frequency of the human α-rhythm (approximately 10 Hz), we obtain that the value of the largest Lyapunov exponent varies from small negative value of -0.02 s^{-1} to 4.8 s^{-1}. The experimental values of λ which are calculated from the human EEGs are in the range from 0.028 s^{-1} to 2.9 s^{-1} [1, 3]. Thus, solutions of Eqs. (1) represent the chaotic time series which have the same non-linear characteristics as the human EEGs.

Fig. 1 shows dependence of the neural network output characteristics without an external action (correlation dimension ν, largest Lyapunov exponent λ, Shannon entropy S_{Sh}, renormalized Shannon entropy S_r and averaged amplitude $\bar{u}_{sq}$) on the degree of neuronal excitability which is characterized by $\bar{a}$. Here

$$\bar{u}_{sq} = \frac{1}{AN} \sum_{j=1}^{N} \left(\sum_{i=1}^{M} (u_{ij} - \bar{u}_i)^2 \right)^{1/2}. \tag{5}$$

It is seen from Fig. 1 that there are three different intervals of $\bar{a}$ within which the neural network have different types of dynamics. If $\bar{a} = -0.5$ we obtain non-zero constant solution. Increasing $\bar{a}$, the solution becomes unstable, and we have a growth of the correlation dimension ν, largest Lyapunov exponent λ, Shannon entropy S_{Sh} and averaged amplitude $\bar{u}_{sq}$. When $\bar{a}$ reaches value -0.2 and up to $\bar{a} = +0.4$ we obtain the chaotic solutions with the correlation dimension 4–8 which are equal to ones obtained from the human and animal EEG analysis [1, 2, 3, 4]. The values of λ, S_{Sh}, S_r and $\bar{u}_{sq}$ do not change essentially on the interval $-0.2 < \bar{a} < +0.4$ too. When $\bar{a}$ exceeds $+0.4$ we observe decreasing of the correlation dimension to 1.0–2.2 and increasing of the averaged amplitude by factor 3–5 which are the features of the epileptic

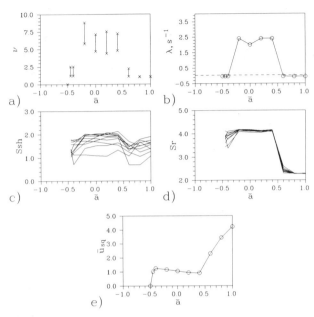

Figure 1. Correlation dimension v (a), largest Lyapunov exponent $\lambda(s^{-1})$ (b), Shannon entropy S_{Sh} (c), renormalized Shannon entropy S_r (d) and averaged amplitude $\bar{u}_{sq}$ (e) vs $\bar{a}$: M=10, c=3.0, τ=10.0.

seizures [1, 2, 14]. This also leads to the decreasing of λ and S_r that points out to the self-organization processes in neural network model. Note, that decrease of S_r is more clearly shows decrease of neural network output complexity than S_{Sh}.

Similar behavior is observed also in many non-linear physical systems when growth of energy transfer through the system leads to the replacement of the chaotic transport by the wave one. Such phase transition is the result of the instability development which yields new steady state, in general, with another features.

Numerical simulation of the external sinusoidal force action on the neural network shows that the neurons outputs can be controlled in both cases: the all-neurons action and the single-neuron action. As it is seen from Fig. 2, the all-neurons controlling of chaos is more effective than the single-neuron action. For the suppression of chaos in the case of the all-neurons action it is necessary the external force amplitude $e = 10.0$ that is an order of the output amplitude values (the largest Lyapunov exponent at the chaos suppression approaches zero). In the case of single-neuron action, suppression of chaos is achieved when $e = 20.0$. In both cases the correlation dimension decreases and achieves the values 1.1–2.5 that is the feature of the human epileptic seizures. The further increasing of the external force amplitude leads to the increasing of the outputs amplitudes also. Comparing these results with the experimental observations of the epilepsy arising, we can consider the all-neurons action as the case of large-area pathology. The single-neuron action of the external force simulates the spreading of synchronous pathological activity from the small area.

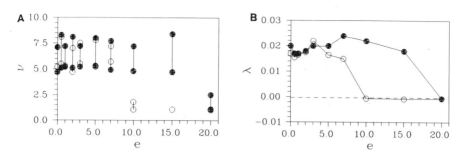

Figure 2. Correlation dimension v (a) and largest Lyapunov exponent λ (b) as functions of external force amplitude e: M=10, c=3.0, τ=10.0, ω_e=0.8; $\circ$ – the all-neurons action, $\bullet$ – the single-neuron action.

4. CONCLUSIONS

Thus, in this paper the mechanism of the epilepsy initiation and spreading is considered on the basis of the neural network model with the high-dimensional chaotic outputs which is similar to the human EEG. It is shown that the epilepsy can be considered as the self-organization process which arises at the increasing of the neuronal excitability (it is also observed in the experiments). Proposed model produces the epilepsy-like outputs under the action of the external sinusoidal force too. Calculated values of the correlation dimension $v = 1.1$–2.5 are close to the range of the observed ones $v = 2.0$–4.0 and the growth of the calculated output amplitudes has the same order as the experimental values.

Author thanks O. S. Bartunov and S. I. Blinnikov for support of this research, and E. V. Efremova for help in computer modelling.

REFERENCES

[1] A. Babloyantz and A. Destexhe, *Proc. Natl. Acad. Sci. USA* **83**, 3513 (1986).
[2] A. Destexhe, J. A. Sepulchre, and A. Babloyantz, *Phys. Lett. A* **132**, 101 (1988).
[3] A. C. K. Soong and C. I. J. M. Stuart, *Biol. Cybern.* **62**, 55 (1989).
[4] J. Röschke and J. Aldenhoff, *Biol. Cybern.* **64**, 307 (1991).
[5] W. J. Freeman, *Biol. Cybern.* **56**, 139 (1987).
[6] K. Aihara, T. Takabe, and M. Toyoda, *Phys. Lett. A* **144**, 333 (1990).
[7] M. Nakagawa and M. Okabe, *J. Phys. Soc. Japan* **61**, 1121 (1992).
[8] K. Nakamura and M. Nakagawa, *J. Phys. Soc. Japan* **62**, 2942 (1993).
[9] A. Destexhe, *Phys. Rev. E* **50**, 1594 (1994).
[10] A. Babloyantz and C. Lourenco, *Proc. Natl. Acad. Sci. USA* **91**, 9027 (1994).
[11] P. A. Schwartzkroin, *Epilepsy: models, mechanisms, and concepts* (Cambridge Univ. Press, Cambridge, 1993).
[12] S. J. Korn, J. L. Giacchino, N. L. Chamberlin, and R. Dingledine, *J. Neurophysiol.* **57**, 325 (1987).
[13] S. F. Traynelis and R. Dingledine, *J. Neurophysiol.* **59**, 259 (1988).
[14] M. Steriade and F. Amzica, *J. Neurophysiol.* **72**, 2051 (1994).
[15] V.E. Bondarenko, *Phys. Lett. A* **196**, 195 (1994).

43

RETRIEVAL PROPERTIES OF ATTRACTOR NEURAL NETWORKS INCORPORATING BIOLOGICAL FEATURES — A SELF-CONSISTENT SIGNAL-TO-NOISE ANALYSIS

Anthony N. Burkitt

Research School of Information Sciences and Engineering
Australian National University
Canberra, Australia

ABSTRACT

A network of integrate-and-fire excitatory neurons is investigated using the recently proposed self-consistent signal-to-noise analysis, which provides a new method for analyzing the behaviour of networks of neurons that have asymmetric synaptic matrices. The neural dynamics is described in terms of two continuous variables, namely the firing rate and the afferent current of each (excitatory) neuron. The afferent current is described by a differential equation that includes a decay term, the weighted inputs from other excitatory neurons, and a term that models the inhibitory interneurons. The effective inhibition chosen here depends upon both the level of activity of the excitatory neurons and the stored patterns, and it serves to control the activity of the excitatory neurons through a feedback process. The afferent current induces a spike rate described by the integrate-and-fire gain function with noise, thus providing a closed set of dynamical current-rate equations. Retrieval of a memory is characterized by the set of neurons associated with the retrieved pattern firing at a substantially higher rate than the remaining quiescent neurons for some macroscopic time. The quality of retrieval of the memory is characterized by the similarity of the evoked firing rates to the stored pattern.

1. INTRODUCTION

The properties of attractor neural networks have been extensively investigated using replica symmetric mean-field theory [1, 2, 3], which has enabled the capacity, quality of retrieval, and robustness to noise to be analyzed. This analysis has also been applied to analogue networks [4, 5, 6]. The mean-field approach, however, is limited in its applicability to networks for which a Lyapanov function exists. This limitation does not hold for the self-consistent signal-to-noise analysis of analogue networks by Shiino and Fukai [7], which provides a method for studying the properties of analogue neural networks that is also applicable to networks with asymmetric synaptic matrices for which there is no Lyapanov function. In this paper the self-consistent signal-to-noise analysis is extended to the study of analogue networks of excitatory neurons whose connections are given by a Hebbian synaptic matrix and where the effect of the inhibitory interneurons is modeled by an effective inhibition that depends upon the level of activity of the excitatory neurons. The network thus consists of *specific* neurons (i.e., the individual neurons are either excitatory or inhibitory), in accord with Dale's law, in contrast to standard Hopfield-type models, which do not possess such specificity. The inhibitory neurons, whose effects are incorporated by an effective inhibition term, serve to control the activity of the excitatory neurons through a feedback process.

2. THE MODEL

The neural dynamics is described in terms of two continuous variables: the firing rate v_i and the afferent current I_i of each (excitatory) neuron:

$$\mathcal{T}\frac{dI_i(t)}{dt} = -I_i(t) + \sum_{j(j\neq i)}^{N} \mathcal{T}J_{ij}v_j(t) - B\left(\{\eta_j^\mu\}, \{v_j(t)\}\right) \tag{1}$$

where $\mathcal{T}$ is the time decay constant, J_{ij} is the synaptic matrix, N is the number of neurons, and $\{\eta_j^\mu\}$ are the stored patterns. The function B represents the effective inhibition of the inhibitory interneurons. Amit and Tsodyks [8, 9] (hereafter referred to as ATI and ATII respectively) investigated a network of noisy integrate-and-fire neurons and showed that the description in terms of individual spikes can be converted into an effective description in terms of current dynamics driven by spike rates. The afferent current induces a spike rate described by a neural gain function, $v_i = \phi(I_i)$ thus providing a closed set of dynamical equations. The integrate-and-fire neural gain function [8] is used in this paper.

The synaptic matrix is chosen as:

$$J_{ij} = \frac{1}{\mathcal{T}Nf}\sum_{\mu}^{p}\eta_i^\mu\eta_j^\mu \tag{2}$$

where the η_i^μ, $\mu = 1,..,p$ are the random N-bit memorized patterns on the N excitatory neurons, of which Nf were ones (foreground) and $N(1-f)$ were zeros (background), and where f is the sparsity of the patterns. Information in this model is stored only on the synapses between excitatory neurons, in accord with the common view in neuroanatomy [10]. The effect of the inhibitory neurons upon the excitatory neurons is given by the function B.

Amit and Tsodyks [9] used mean-field techniques to investigate a particular model where the effective inhibition is

$$B_{Hopfield}\left(\{\eta_j^\mu\}, \{v_j\}\right) = \sum_{\mu}\left\{\eta_i^\mu\frac{1}{N}\sum_{j}^{N}v_j + f\frac{1}{N}\sum_{j}^{N}\frac{\eta_j^\mu}{f}v_j - f\frac{1}{N}\sum_{j}^{N}v_j\right\} \tag{3}$$

In this model the dynamics is described by a Lyapanov function. They found that such a network can lead to stationary (reverberating) activity states with low spike rates and that the various approximations involved in this description are quite realistic in typical cortical conditions.

The particular type of inhibition investigated here is that given by:

$$B\left(\{\eta_j^\mu\},\{v_j\}\right) = \sum_\mu \eta_i^\mu \frac{1}{N} \sum_j^N v_j \tag{4}$$

Such a form of effective inhibition could possibly arise as a result of LTP involving inhibitory interneurons, however a study of the mechanisms by which it could be produced lies outside the scope of the present investigations. This particular choice of effective inhibition is both simple and effective in facilitating the network retrieval of the stored patterns, as will be shown. The purpose of choosing this particular form of effective inhibition is to investigate, using the self-consistent signal-to-noise analysis, a simple form of the effective inhibition that can not be investigated by mean-field techniques. It is also of interest to consider the case when the inhibition contains random noise, which may be done using this approach.

The fixed-points of the dynamical equation for the neurons are given by:

$$I_i = \sum_j^N \mathcal{T} J_{ij} v_j - B\left(\{\eta_j^\mu\},\{v_j\}\right) \tag{5}$$

The associative storage of the patterns and their recall is then a result of the cooperative dynamics of the network, in which the activity of the excitatory neurons is controlled by the inhibitory neurons through a feedback process.

3. SELF-CONSISTENT SIGNAL-TO-NOISE ANALYSIS

The basis of this analysis is the systematic splitting of the I_i into signal and noise parts [7] and involves a renormalization of the signal part of the local field. Using the form of effective inhibition, equation(4), the local current I_i may be expressed in terms of the overlaps as:

$$I_i = \sum_\mu \eta_i^\mu \left(m_\mu^+ - m_0\right) - \alpha\phi(I_i) = (1-f)\sum_\mu \eta_i^\mu m_\mu - \alpha\phi(I_i) \tag{6}$$

which will be solved to give:

$$v_i = \phi(I_i) = \hat{\phi}\left((1-f)\sum_\mu \eta_i^\mu m_\mu\right) \tag{7}$$

where the usual notation [9] is used for the overlaps m_μ and m_0. The details of the analysis for this particular model may be found in Burkitt [11]. This technique provides a method for studying the properties of analogue neural networks that is also applicable to networks with asymmetric synaptic matrices, for which there is no Lyapanov function (the resulting equations are identical to those of the mean-field analysis for the case of symmetrical synaptic connections).

4. SOLUTION OF THE FIXED-POINT EQUATIONS

The equations for the fixed-point take a particularly simple form when the number of patterns p is kept fixed, as N increases (i.e., $\alpha = p/N \to 0$ as $N \to \infty$) since in this limit v_η becomes independent of z:

$$v_\eta = \phi\left((1-f)\sum_\mu \eta_\mu m_\mu\right) \tag{8}$$

In the case of the retrieval of a single pattern (i.e., with $m^\mu = m\delta^{\mu,1}$):

$$m = \phi((1-f)m) - \phi(0) \tag{9}$$

This expression simply represents the difference between the (uniform) rates of the neurons in the foreground (the first term) and in the background (the $\phi(0)$ term). The argument of the first term on the right-hand side is the afferent current in the foreground, which is uniform across the foreground neurons. The neurons in the background receive no afferent current and have a firing rate determined solely by the level of afferent noise.

In the case where $p \to \infty$ (i.e., $\alpha > 0$), the equations may be solved numerically [11]. The mean rate m_+ for neurons in the foreground of the retrieved pattern remains at a relatively constant value (approx. 0.19 for this particular set of neural parameters, corresponding to a firing rate of 95 spikes/sec) up to a loading of approx. 0.10, and the mean rate m_- of the background neurons remains extremely low. For larger values of α the rate m_+ falls and m_- increases, with the two rates coalescing as in ATII. The results indicate that there is a clearly defined critical capacity at which there is a sharp transition between a retrieval phase and a non-retrieval phase. The method also enables the robustness of the network to both noise and synaptic dilution to be investigated [11]. It is straightforward within the framework of the self-consistent signal-to-noise analysis to consider the deviation of the synapses from their Hebbian values. This was done by assuming that the deviations are random, and that they may be therefore included as a new source of noise whose effect is to increase the width of the Gaussian distribution of the noise. This enables the study of clipped synapses and the dilution of synapses in a straightforward way.

5. NUMERICAL SIMULATIONS

The results are compared with numerical simulations of the network in order to check the accuracy of the self-consistent signal-to-noise analysis, as well as checking the finite size effects and investigating the stability of the retrieved memories.

Since the equations (1) being simulated are completely deterministic differential equations in which the neuronal noise is incorporated into the gain function, the simulations are really determined by the initial conditions and the synaptic couplings. The initial conditions are determined by the stimulus, which consisted of incoming external currents of intensity 0.2 presented during the first ten steps of the integration. The simulations were carried out on a network of 1000 neurons with a varying number of stored patterns, from 10 to 800, $\alpha = 0.01 - 0.80$, and with a sparsity of $f = 0.05$. The neural parameters were those used in ATII, i.e., $\tau = 4.0$, $\theta = 0.25$, $v = 0.2$, and $\sigma = 0.0125$, and likewise the same integration step was used (0.2 of the time constant of the neuron).

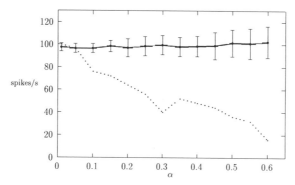

Figure 1. Results from the numerical simulation of the network with $f = 0.05$ and neural parameters as given in the text. The error bars show the standard deviation in the rates at each α. The dashed line shows the percentage of stimuli which resulted in the successful retrieval of the corresponding single stored pattern (a firing rate of 100 spikes/sec corresponds to a an overlap of $m = 0.2$).

One of the stored patterns was chosen as the stimulus, and an external current was imposed upon the foreground neurons of this pattern. As in ATII, at each storage level simulations were carried out with five sets of patterns, and for each set it was run five times, each with a stimulus based on a different stored pattern. A number of additional simulations were also carried out to investigate the error correction, the effect of $1/p$ corrections, and the stability of the network to variation of the stimulus strength and duration, as well as other parameters.

The most striking feature of the simulations was that the network reached a stable state in every simulation carried out, with no chaotic or oscillatory trajectories being observed. Moreover, when the network retrieved a single pattern (i.e., a memory), the convergence rate was very rapid. The results of the simulations agree with the self-consistent signal-to-noise analysis results for small loadings (i.e., for a relatively small number of patterns stored in the network, as compared to the number of neurons in the network). The simulations, however, do not show the discontinuous jump in the retrieval behaviour that the self-consistent signal-to-noise analysis results show as the loading is increased. Rather the network continues to retrieve not only the stimulus pattern but also other patterns – the overlap with the stimulus pattern is high, but also the overlap with one (or more) other patterns tends to increase. The network retrieval was fairly insensitive to the amplitude and duration of the afferent stimulus–current, both of which were varied over a range of values without appreciable effect upon the results. The retrieval was also rather insensitive to errors in the stimulus and simulations with varying levels of noise in the afferent current were carried out.

6. DISCUSSION

The self-consistent signal-to-noise analysis provides a new method for analyzing the behaviour of networks of neurons that have asymmetric synaptic matrices. The advantage of this analysis is that it enables the investigation of a much wider class of networks to which the replica method can not be applied, and in particular models that incorporate more biologically desirable features.

The approach adopted here of considering the effective interaction rather than modeling the dynamics of the inhibitory interneurons directly neglects the time course of the interneurons and assumes that the inhibitory interneurons operate very fast, so that their activity instantaneously reflects the activity of the excitatory neurons. Although this clearly involves an approximation to the neurophysiological situation, it may nevertheless partially capture the shorter transmission times for inhibitory signals that results from the more local nature of inhibitory neurons.

The fixed-point equations of the self-consistent signal-to-noise analysis provides an accurate analysis of the network retrieval properties at low loadings, and it can guide our understanding of the network properties at higher loadings. This enables more realistic networks to be investigated, not only using numerical simulations but also maintaining some analytic control of the network properties. The advantage of the self-consistent signal-to-noise analysis is that it enables the study of a much wider range of networks to which the mean-field replica method can not be applied. Moreover, it provides a very different framework to the mean-field replica method, and one that is more readily comprehensible in terms of the origin and interpretation of the resulting variables.

REFERENCES

[1] J.J. Hopfield, "Neural networks and physical systems with emergent collective computational abilities", *Proc. Natl. Sci.* **79** (1982) 2554–2558.

[2] D.J. Amit, H. Gutfreund, and H. Sompolinsky, "Statistical mechanics of neural networks near saturation", *Ann. Phys., NY* **173** (1987) 30–67.

[3] D.J. Amit, *Modeling Brain Function: The world of attractor neural networks*, (Cambridge University Press, 1989).

[4] R. Kühn, "Statistical mechanics for networks of analog neurons", In I. Garrido, *Statistical mechanics of neural networks*, (Springer, Berlin, 1990).

[5] R. Kühn, S. Bös, and J.L. van Hemmen, "Statistical mechanics for networks of graded-response neurons", *Phys. Rev.* **A43** (1991) 2084–2087.

[6] M. Shiino and T. Fukai, "Replica-symmetric theory of the nonlinear analogue neural networks", *J. Phys.* **A23** (1990) L1009–L1017.

[7] M. Shiino and T. Fukai, "Self-consistent signal-to-noise analysis and its application to analogue neural networks with asymmetric connections", *J. Phys.* **A25** (1992) L375–L381.

[8] D.J. Amit and M.V. Tsodyks, "Quantitative study of attractor neural network retrieving at low spike rates: I.", *Network* **2** (1991) 259–273.

[9] D.J. Amit and M.V. Tsodyks, "Quantitative study of attractor neural network retrieving at low spike rates: II.", *Network* **2** (1991) 275–294.

[10] E.R. Kandel and J.R. Schwartz, *Principles of Neural Science*, (Elsevier, Amsterdam, 1985).

[11] A.N. Burkitt, "Retrieval properties of attractor neural networks that obey Dale's law using a self-consistent signal-to-noise analysis", *Network* **7** (1996) 517–531.

44

MODELING DELAY-PERIOD ACTIVITY IN THE PREFRONTAL CORTEX DURING WORKING MEMORY TASKS

Marcelo Camperi and Xiao-Jing Wang

Physics Department
University of San Francisco
San Francisco, CA
Volen Center for Complex Systems
Brandeis University
Waltham, MA

1. INTRODUCTION

It is well known that during delayed-response task experiments, cells in the prefrontal cortex (PFC) of awake monkeys exhibit sustained activity during the delay period between the presentation of a transient stimulus and the arrival of a "go" signal [1, 2, 3, 4]. Presumably this persistent PFC network activity carries a *short-term memory* of the stimulus during the delay period (usually a few seconds). This information is then used by the animal to perform a behavioral response at the go signal.

In this work, we investigate by means of computational modeling a possible mechanism underlying the working memory processes in an oculomotor delayed-response experiment [3]. In the basic experimental paradigm, a trained monkey is forced to keep its gaze on a light spot at the center of a monitor. A cue in the form of a short (0.5 s) spot of light, is presented randomly in one of 8 equidistant positions over a circle centered at the fixation spot. This fixation signal is kept during the whole delay period (3-5 s) that follows the suppression of the cue signal. After this delay, the fixation spot is turned off. This constitutes the go signal for the monkey, who has then to saccade towards the original position of the cue. Since anticipatory responses and other possible causes of the sustained activity (e.g. motor-related activity) are discarded, correct performance is only possible based on an actual recall of the cue position [3].

Some PFC neurons where found to display *memory fields*, namely a given neuron shows strong persistent activity during the delay period only when the cue is close to a particular location in the visual field (say θ, corresponding to a preferred position on the cue circle). A *tuning curve* of a PFC cell is then defined as the mean firing rate during the delay period as a function of the cue position. Such curve shows a maximum at the preferred position θ, and a minimum (which is normally lower than the spontaneous firing rate of the cell) at the cue position 180° away from θ [3, 4].

In this paper we explore the possibility that the persistent delay-period activity can be maintained in the PFC by an interplay between network properties (local recurrent excitation and inhibition, [4]) and an intrinsic cellular property (bistability, [5, 6]). We show that a network model endowed with such properties is capable of displaying simultaneously a uniform rest state, and a continuum of activity profiles ("line attractors" [7, 8]). These network activity patterns can encode the cue position in a graded way. The transient cue and the subsequent go signal act as the switches between the rest state and an active state. Our model simulations produced tuning curves similar to experimental measurements from the PFC cells [3]. We also propose a "distractor experiment" which can be used to test the present network model.

2. THE MODEL

We use a simple model that captures the relevant architectural details of a local cortical circuit. The model is adapted from a previous one of the visual cortex [7]. The model consists of a network of N neurons that respond selectively to a transient cue. The neurons are parameterized by the the angle θ which represents the memory field or preferred-cue-location for the given neuron. This parameter ranges from $-\pi$ to π. Thus, if the transient cue is located at θ_0, the neuron with memory field centered at $\theta = \theta_0$ will be the one firing maximally during the delay period (we will refer to a neuron with memory field θ as a neuron *at* θ). In the large network limit, we can assume that the neurons cover uniformly all the angles. In this limit, the activity profile of the network can be represented by a continuous function $r(\theta, t)$, the average firing rate of the neuron with memory field in the neighborhood of θ, at time t. The interaction between a neuron at θ and one at θ' is given by a function of the form $W(\theta - \theta')$, thus depending on how far apart their respective memory fields are. The firing rate $r(\theta, t)$ obeys the following mean-field equation:

$$\tau_0 \frac{d}{dt} r(\theta, t) = f[r(\theta, t)] + g[I(\theta, t)] \tag{1}$$

where f is a given function of the firing rate, g is a gain function, and τ_0 is a characteristic time. In our scheme, the particular form of the functions f and g will define the model (g is chosen to be simply a linear gain function with a threshold). Moreover, I is the total input on each neuron, given by

$$I(\theta, t) = I^{ext}(\theta, t) + \int_{-\pi}^{\pi} \frac{d\theta'}{2\pi} W(\theta - \theta') r(\theta', t) \tag{2}$$

The term $W(\theta - \theta') r(\theta', t)$ in the integral provides the essential synaptic recurrent network interaction. The function W includes a global inhibition and a structured excitation, and is given by

$$W(\theta) = -W_I + W_E \left(\frac{1 + \cos(\theta)}{2} \right) \tag{3}$$

Here, the constants W_I and W_E measure the strengths of the inhibitory and excitatory interactions respectively. Note that the modulation function, $(1 + \cos\theta)/2$, was chosen for

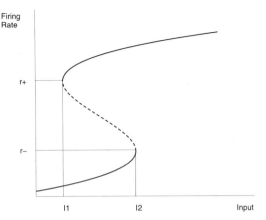

Figure 1. Single-cell steady-state curve.

its simplicity. Other reasonable functions with the same global characteristics would work as well. Furthermore, the external input $I^{ext}(\theta, t)$ includes the transient cue and a possible constant input. It is given by

$$I^{ext}(\theta, t) = I^0 + \Theta(t - t_0)\Theta(t_0 + \Delta t_0 - t)\left(\frac{1 + \cos(\theta - \theta_0)}{2}\right) \qquad (4)$$

where $\Theta(t)$, the Heaviside or step function, is used to indicate the fact that the cue stimulus starts at $t = t_0$, lasting for a time Δt_0. The modulation factor, similar to that used in (3), represents the cue being centered at $\theta = \theta_0$ (again the actual form is irrelevant, as long as it is strong enough and modulated with a peak at the cue location).

Finally, the function $f(r)$ is chosen to be

$$f[r] = -r + ar^2 - br^3, \qquad (5)$$

where a and b are parameters. This nonlinear form endows a single cell with a bistable input-output relationship. Indeed, the equation for a single cell with a constant input I is

$$\tau_0 \frac{d}{dt}r = -r + ar^2 - br^3 + I, \qquad (6)$$

The steady-state of this system is obtained by setting $dr/dt = 0$. Figure 1 shows the steady-state curve for this model for the set of parameter values chosen to fit the experimental data. We see that bistability is present at the single cell level: for a given value of the input I, such that $I_1 < I < I_2$, the cell can be in either one of two possible steady states (the middle state, represented by the dashed line, is unstable). Moreover, switching between the two steady states can be induced by a transient input.

3. SIMULATION RESULTS

The model – equation (1) – was simulated numerically. After the steady state for the given uniform input has been reached, we simulate the cue signal using a modulated transient

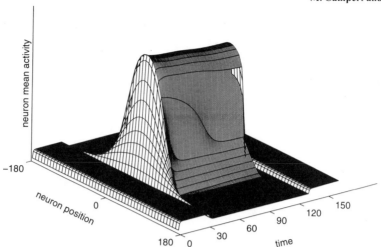

Figure 2. Typical time evolution for the model. Neuron positions are in degrees, while time is measured in units of τ_0, the characteristic time. The cue signal, centered at $\theta = 0^o$, is presented at $t \approx 30\tau_0$, and lasts for about $10\tau_0$. At $t \approx 125\tau_0$ the go signal is presented. Notice how the network acquires an activity profile that carries a memory of the cue location and how this shape persists through the delay-period.

input as described above. The go signal is simulated with a constant and uniform inhibition to all the neurons.

We observe that the network acquires a shape or distribution of activities that is independent of the details of the input, and remains so for as long as desired, after the cue input has disappeared. The memory of the position of the cue stimulus is preserved in the form of a network activity profile, as depicted in figures 2 and 3, which show our main results. This result is partly due to the intrinsic bistability of the neurons in the network: some of the neurons have been able to jump to the upper level of activity by transient cue inputs. However, they are sustained in the upper state only by the recurrent excitation in the network, without which the total inputs to those cells would have been below I_1 (cf. figure 1), hence neurons would not be able to stay active after the cue is turned off. Furthermore, the feedback inhibition is crucial in generating a nonuniform activity profile (hence a tuning curve for individual cells).

To further test the model, we simulate an experiment that would provide a tuning curve for a given cell, following [3]. We take a neuron with memory field at a given θ and register its delay-period activity for 8 different cues located at intervals of 45°. We notice that there is a small difference in the results depending on whether or not our cell has a memory field θ that *coincides* with one of the tested cue locations. Figure 4 shows two such tuning curves. We observe a remarkable similarity with the experimental tuning curves in [3]. In addition, we see that our tuning curves allow gaussian or near-gaussian fits, as was noted in [3]. Obviously, a large set of cues (not just 8) would produce a tuning curve that looks just like the network activity profile shown in figure 3. Therefore, it appears that an experimental tuning curve with more than 8 cue locations would shed more light on the nature of the process being studied.

4. A DISTRACTOR SIMULATION

It is interesting to ask how our model would behave when confronted with a distraction. That is, if a second cue is presented at a location different from the first one during the

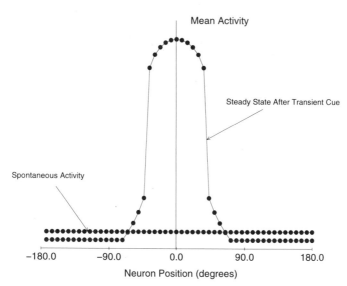

Figure 3. Two snapshots of the network activity profile, corresponding to the rest state before the cue stimulus, and persistent activity at a time between the disappearance of the cue and the go signal, respectively. Note that cells with preferred-cue-positions far away from the actual cue stimulus position have a suppressed level of activity.

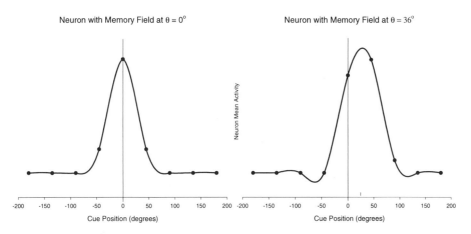

Figure 4. Two tuning curves, for a cell whose preferred-cue-position coincides with one of the used stimuli (left panel), or is different from all used stimuli (right panel). The cues were positioned at equal intervals of 45°, starting at -180°.

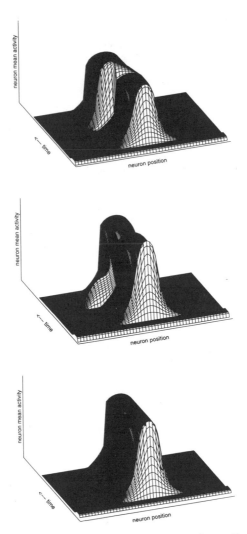

Figure 5. Effects of a distraction during the delay period, in the form of a cue identical in size and duration to the original one, but in different positions. The different rows represent distractors located at 45°, 90°, and 135° respectively, with respect to the location of the original cue, represented by the peak of activity.

delay period, would the network preserve the first stimulus, or would it be perturbed, thus "remembering" a wrong cue position? To our knowledge, this question has not been addressed experimentally, thus our simulated experiment will constitute a concrete prediction of the model.

    The results of our simulated experiment can be seen in figure 5. We see that the results show no appreciable shift for distractor cues that are far apart from the original cue, and just a very small shift for very close distraction cues (apart from a transient activity during the distraction cue). This was true even for distractor cues of considerably larger amplitudes than the original one. Too strong a distractor, however, could shift the peak of the activity profile (hence the memorized cue location). The reason for this behavior is clear if we look at figure 3. We see that there is a gap in the activity profile, created by intrinsic bistability, which *protects* the memory of the cue against distractors: in order to perturb this memory a large enough input is needed to overcome the gap. Furthermore, the robustness against distractors is reinforced by feedback inhibition sustained by persistent network activity, which renders the inactive cells even less excitable by the external inputs.

5. DISCUSSION

We have shown that a network of bistable neurons endowed with recurrent cortical excitatory interaction and feedback inhibition exhibits persistent activity after a transient input, with a network activity profile which can be related to tuning curves for individual cells. Thus, such a mechanism may play an important role in short-term memory processes in the prefrontal cortex.

Is intrinsic cellular bistability necessary to explain persistent activity of PFC neurons? Although previous modeling works and the present one postulate such a distinct neuronal property, no direct evidence on this issue has been reported. Activity patterns that coexist with the rest state can arise as a network dynamical property, even when bistability is not present at the single-cell level (see for example [9, 10]). In a different version of our network model, we found that, indeed, persistent network activity can still be reproduced without intrinsic bistability. However, the tuning curve shape and the response to a distractor by the network, were found to be different from the results reported here. Details on this and more studies on both models will be published elsewhere.

ACKNOWLEDGMENTS

This work was supported by the Alfred P. Sloan Foundation.

REFERENCES

[1] Fuster, J. and Alexander, G. Neuron activity related to short-term memory. *Science* 173, 652-654 (1971).
[2] Kubota, K. and Niki, H. Prefrontal cortical unit activity and delayed alternation performance in monkeys. *J. Neurophysiol.* 34, 337-347 (1971).
[3] Funahashi, S. *et al.* Mnemonic coding of visual space in the monkey's dorsolateral prefrontal cortex. *J. Neurophysiol.* 61, 331-349 (1989).
[4] Goldman-Rakic, P. Cellular basis of working memory. *Neuron* 14, 477-485 (1995).
[5] Marder, E. Plateaus in time. em Current Biology 1, 326-327 (1991).
[6] Booth, V. and Rinzel, J. A minimal, compartmental model for a dendritic origin of bistability of motoneuron firing patterns. *J. Comp. Neurosc.* 2, 1-14 (1995).
[7] Ben-Yishai, R. *et al.* Theory of orientation tuning in visual cortex. *Proc. Natl. Acad. Sci. USA*, 92, 3844-3848 (1995).
[8] Seung, S. How the brain keeps the eyes still. *Proc. Natl. Acad. Sci. USA*, in press.
[9] Amari, Shun-ichi. Dynamics of pattern formation in lateral-inhibition type neural fields. *Biological Cybernetics* 27, 77-87 (1977).
[10] Wilson, H., and Cowan, J. A mathematical theory of the functional dynamics of cortical and thalamic nervous tissue. *Kybernetik* 13, 55-80 (1973).

45

WHAT DO DYNAMIC RECEPTIVE FIELD PROPERTIES REVEAL ABOUT COMPUTATION IN RECURRENT THALAMOCORTICAL CIRCUITS?

John K. Chapin[1] and Miguel A. L. Nicolelis[2]

[1]Department of Neurobiology and Anatomy
Allegheny University of the Health Sciences
Philadelphia, Pennsylvania 19129
[2]Department of Neurobiology
Duke University
Durham, North Carolina 27710

I. ABSTRACT

Recurrently connected artificial neural networks can classify input patterns through iterative feedback reverberations. We have utilized simultaneous multi-single unit recordings to investigate sensory processing in the rat somatosensory thalamocortical system, which is recurrently interconnected. Punctate stimulation of facial whiskers produced a complex, spatiotemporally distributed series of response peaks across neuronal ensembles in both the ventroposterior thalamus (VPM) and somatosensory (SI) cortex. Correlation matrices relating the timing of these peaks in all histograms constructed using VPM vs. SI neurons revealed an orderly series of post-stimulus correlation "peaks and valleys", indicating that the dynamic properties of receptive fields in both VPM and SI are caused by reverberatory interactions between them. Canonical correlation analyses suggest that these interactions may reflect computational recognition of patterns of stimulus movement across the whiskers.

II. INTRODUCTION

What is the functional role of the large recurrently connected systems in the mammalian brain, in particular the thalamo-cortico-thalamic system? Though serial single unit recordings have provided a wealth of information on the relationship between individual neurons and external parameters, they can only indirectly reveal the neural information which is contained in the covariant interactions between the interconnected neurons. As a heuristic experiment on how to address this problem we have recently tested the efficacy of various neurophysiological techniques for analyzing how information is processed in simple artificial neural networks[7]. For example, autoassociative networks[8], can "classify" random input patterns by tranforming them into one of a few memorized output patterns. Since such nets tranform input patterns into output patterns, the correlation of any one node (i.e. "neuron") in the network with any one input channel is virtually irrelevant. In fact, when we performed traditional "single unit" analyses on such a network, they revealed typical center-surround-type "receptive fields" for each neuron, but this information was useless for explaining the true function of the network, which was pattern transformation. On the other hand, analysis of the same network using simultaneous recordings and multivariate statistical analysis techniques yielded a more accurate description of the information processing within the network. In particular, canonical correlation analysis (CCA) was found effective in defining how a set of arbitrary input patterns were transformed into different target output patterns. This involved making a n x m correlation matrix between the n inputs and the m outputs of the network, averaged over a large number of trials with different random input vectors. CCA was then used to derive a small set of underlying roots (i.e. factors) which explained most of variance in the data. The first 3 roots contained weightings of both inputs and outputs which correctly classified nearly all of the trials. Moreover, these weightings revealed information about the weights of the recurrent "synapses" in the network. This technique was also used to map temporal interactions through successive iterative reverberations of this network. Overall, these observations gave confidence that essentially linear multivariate statistical techniques such as CCA could be used to elucidate the functional transformations carried out by a nonlinear neural network.

To implement such a multivariate analytical approach for analysis of thalamocortical networks in awake animals, we simultaneously recorded up to 48 single neurons in the tri-geminal somatosensory thalamus (VPM) and SI cortex of awake rats during controlled punctate deflection of the facial whiskers. As part of a general strategy to use multivariate statistical techniques to analyze the ensemble properties of these neurons[6], CCA was used to analyze the transformation of information as it ascends from the VPM thalamus to the SI cortex.

III. METHODS

Detailed methods for our chronic multineuron recordings have been previously published[9–12]. Briefly, simultaneous recordings of the extracellular activity of up to 48 single neurons in Long-Evans rats were obtained through chronically implanted microwire electrodes (50 micron, teflon-coated stainless steel; NB Labs, Denison, TX). Arrays of 16 electrodes (0.2 mm spacing) were implanted in the whisker representation of the primary somatosensory (SI) cortex[3] and bundles of 8 electrodes were implanted in the ventral posteromedial (VPM) thalamus. During surgical implantation the final position of each electrode was verified by recording unit activity and mapping its receptive field (RF).[4] About 1 week after surgery the rats were placed in the experimental apparatus and signals from all of the microwires were simultaneously recorded using a 64 channel amplifier-filter-discriminator system, obtained from Spectrum Scientific (Dallas, TX). This employed digital signal processors (DSPs) to perform time-voltage discrimination (using 3 windows) of digitized (40 KHz) waveforms obtained from each microwire. The discrimination was validated on-line by plotting each sample waveform as a dot within an x-y space defined by components 1 and 2 of a principal components analysis (PCA) of typical spike waveforms.

The spatiotemporal patterning of whisker-sensory responses across the population of recorded neurons was quantitated by stimulating up to 20 different whiskers with a computer controlled vibromechanical stimulus probe. Stimuli consisted of single step deflections (3o, initially upward, 100ms duration) at a rate of 1 Hz. Post-stimulus time histograms from 300–600 deflections of each different whisker were used as the data base for this study. Statistical analyses of the correlated activity of these neurons were then carried out using software from CSS-Statistica (Tulsa, OK).

IV. RESULTS

The following illustrations utilize data recorded from a rat in which neuronal ensembles were simultaneously recorded in the VPM thalamus (8 neurons) and SI cortex (16 neurons) during 300–500 repeated deflections of each of 8 single whiskers on the face. The first aim was to investigate the spatial interactions between the VPM and SI ensembles: an 8x16 correlation matrix was constructed relating the post-stimulus responses (measured in the 5–15ms post-stimulus epoch) of the 8 VPM neurons to those of the 16 SI neurons, across all 8 whiskers. CCA was then used to define underlying canonical roots which could explain most of the variance in this correlation matrix. For each root, CCA defined two sets of weightings, one for the VPM neurons, and one for the SI neurons. These weightings were then used to calculate population vectors (PVs) from the original data. (A PV is defined here as the dot product of the weight matrix times the standardized average discharge of the neuronal ensemble during a particular time frame.) CCA attempts to weight the VPM and SI ensembles such that their PVs are as similar as possible. Here, the PVs were calculated to generate post-stimulus histograms for roots 1–3, for both the VPM and the SI. The VPM and SI PVs of the higher numbered roots (which explained the most variance) were quite similar. Moreover, the post-stimulus histograms defined by these PVs indicated distinct response patterns for each stimulated whisker. These roots, therefore, provided a weighting scheme for reliably predicting the identity of the stimulated whisker, based on the activity of either the VPM or the SI neuronal ensembles. Furthermore, they provided a weighting scheme for predicting the activity of either of these ensembles, based on the activity of the other. Thus, they defined the information transform between these two processing levels.

In order to understand this transform, however, an additional complication needed to be addressed: as we have previously reported, the whisker-sensory responses of neurons in both the VPM and SI of rats exhibit complex combinations of short and long latency responses (Nicolelis and Chapin, 1995). In particular, neurons which respond at short latencies (5–7ms) to stimulation of the caudal facial whiskers tend to also respond at longer latencies (15–30ms) to stimulation of the rostral whiskers. This caudal-to-rostral shifting of the receptive field may correspond to the rostral-to-caudal movement of whisker stimulation which occurs during active whisker protraction, which occurs when rats use their whiskers to explore the tactile environment. Our hypothesis about this receptive field shifting is it may be a manifestation of long-loop reverberatory interactions between the VPM thalamus and SI cortex. Thus, an important question to address here is whether this shifting is an important component of the overall temporal pattern of interaction between the VPM and SI, as characterized by CCA.

The smoothed 3D surface in figure 1 depicts a correlation matrix which illustrates the temporal pattern of correlation between the VPM and SI. Both axes of this matrix cover time from 1–30ms post-stimulus. While VPM activity is shown in the horizontal axis the corresponding SI activity is shown in the vertical axis. The amplitude of each point on the surface, as depicted in grayscale, is the correlation coefficient between the activity of all VPM neurons vs. all SI neurons, measured during stimulation of 4 different whiskers, over the post-stimulus time from 1–30ms (3ms bins). The overall temporal correlation pattern exhibited by this surface is consistent with a fairly regular damped reverberatory interaction between the VPM

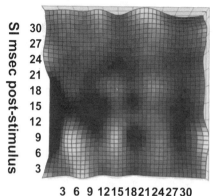

Figure 1. Correlation matrix relating 10 post-stimulus bins of histograms from VPM vs. SI. This smoothed 3D surface plot shows the pattern of correlations between each of the 10 post-stimulus bins (each covering 3ms) in PSTHs of 8 VPM neurons and 16 SI neurons which were simultaneously recorded in an awake rat. See text for details.

and the SI. Since this regular pattern could not be seen in the post-stimulus responses of any one neuron in this system, it is clearly an emergent function of the overall population. Each cycle of this reverberation appears to last about 9ms, with the thalamocortical phase lasting about 3ms and the corticothalamic phase about 6ms. These time delays are consistent with the known average conduction velocities between these structures. The corticothalamic phase presumably also includes cortico-cortical conduction.[5]

CCA was used to define the roots underlying the structure of this correlation matrix. Figure 2 shows the weightings of roots 1–3 for both the VPM (solid line) and SI (dotted line). In each, the weights are shown on the vertical axis, and the post-stimulus time on the horizontal axis. Root 1 appears to be concerned mainly with transmission of the initial sensory volley from the VPM to the SI: the largest VPM peak is at 6ms and the largest SI peak is at 9ms. This root appears to explain the peak in the lower left of the surface in figure 2, which is at 6ms on the VPM axis, and 9ms on the SI axis. In contrast, roots 2 and 3 exhibit high weightings for various longer latency responses.

Next, these roots were used to produce PVs specifying a magnitude function for each neuron. (Each PV = the dot product of each root's weight matrix and the neuron's 10 bins of post-stimulus response to each whisker). The cluster plot in figure 3 shows the PVs for roots 1 and 2 of the first 8 SI cortical neurons to stimulation of 4 different whiskers. Each of the 8 neurons are represented in four different locations on this plot by different shapes representing the four whiskers, according to the legend at right. This plot shows a clear separation, along the Root 2 axis, between two rostral whiskers (E5 and C6) and two more caudal whiskers (B4 and A3). Further separation is achieved by different roots: root 1 clearly separates A3 from the rest, and root 3 (not shown) separates E5 and C6. This CCA procedure, therefore, demonstrated that the complex temporal patterning of each neuron's post-stimulus response contains significant information on the identity of the stimulated whisker. In particular, the fact that root 2, which weights mainly longer latency responses, was found to separate rostral vs. caudal whiskers is consistent with our hypothesis that these longer latency responses reflect reverberant activity in the thalamocortical system which carries information about the timing of rostral whisker stimulation.

V. DISCUSSION

The results of this study provide a demonstration of the usefulness of simultaneously recording multi-neuron ensembles as opposed to the serial recording of single neurons.

The image axis labels:
SI msec post-stimulus — 30 27 24 21 18 15 12 9 6 3
VPM msec post-stimulus — 3 6 9 12 15 18 21 24 27 30

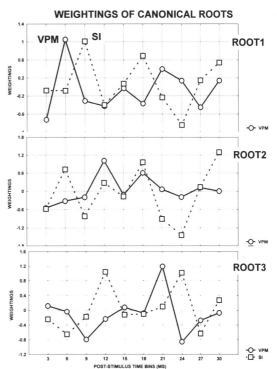

Figure 2. Canonical roots 1–3. To summarize the structure of the correlation matrix in fig. 1, CCA was used define a series of canonical roots which best capture the correlation between the different time bins. The weightings in Root 1 show that the greatest source of correlation relate the short latency responses in VPM (4–6ms) and SI (7–9ms). Root 2 shows a relationship between medium latency responses (12 & 21ms) in the VPM and long latency inhibitory (24ms) and excitatory (30ms) responses in the SI. Root 3 shows a relationship between medium and long latency responses (12 and 24ms) in the SI and long latency responses (21ms) in the VPM.

Furthermore, they show how multivariate statistical methods can be used to characterize the information contained in these neuronal populations. In particular, canonical correlation analysis (CCA) was shown to be an effective and useful technique for statistically defining the functional relations between reciprocally connected layers of neurons during periods of dynamic interaction.

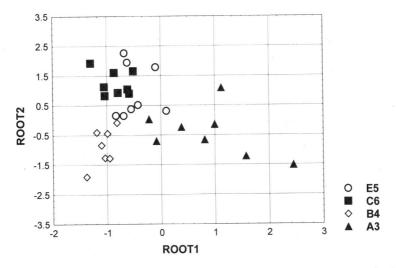

Figure 3. Responses of 8 single cortical neurons to 4 stimulated whiskers clustered in a space defined by canonical roots 1 and 2. See text for details.

That such dynamic interactions occur in the thalamic and cortical levels of the rat somatosensory whisker system was originally inferred from the fact that post-stimulus histograms of neurons in VPM neurons exhibited multiple response peaks, and that these responses define a pattern of rostral-to-caudal receptive field shifting[11]. The current results of constructing population correlation matrices between neuronal ensembles in the VPM and SI show that similar temporal complexity of sensory response is also present in the SI cortex, and that the VPM and SI responses are coordinated into a reverberatory thalamo-cortico-thalamic interaction.

CCA was used here to show that this overall reverberation is maintained by different neuronal groups at different times, and under different stimulus circumstances. The internal structure of the reverberatory interaction revealed by the CCA was again consistent with our hypothesis that the longer latency responses carry significant information from stimulation of the rostral whiskers to the caudal whiskers. This phenomenon may be relevant to the fact that rats commonly use active whisker protraction for tactile exploration of the environment[1,2]. Stationary objects encountered by such forward movement would first be felt by the rostral whiskers, and 10–20ms later, by the caudal whiskers. It is possible that these temporal interactions may play a role in the detection or perception that such tactile objects are stationary, even though the whisker movement causes activation of a defined temporal sequence of whisker sensory receptors.

ACKNOWLEDGMENTS

Supported by grants NS23722 and ONR N00014–95–1- 0246 to JKC.

VI. REFERENCES

1. Carvell GE, Simons DJ, (1990) Biometric analyses of vibrissal tactile discrimination in the rat. Journal of Neuroscience 10(8):2638–48
2. Carvell GE, Simons DJ, Lichtenstein SH, Bryant P (1991) Electromyographic activity of mystacial pad musculature during whisking behavior in the rat. Somatosensory & Motor Research 8(2):159–64
3. Chapin, J.K., and Lin, C-.S. (1984) Mapping the body representation in the SI cortex of anesthetized and awake rats. J. Comp. Neurol. 229:199–213
4. Chapin, J.K. (1986) Laminar differences in sizes, shapes, and response profiles of cutaneous receptive fields in the rat SI cortex. Exp. Brain Res. 262:
5. Chapin, J.K., Guise, J,L., Sadeq, M., and Woodward, D.J. (1987) Cortico-cortical connections within the primary somatosensory (SI) cortex of rat. J. Comp. Neurol. 263:326–346.
6. Chapin, J., M. Nicolelis, C.-H. Yu, and S. Sollot (1989) Characterization of ensemble properties of simultaneously recorded neurons in somatosensory (SI) cortex. Abstr. Soc. for Neurosci. Ann Mtg.
7. Chapin, J.K. and Nicolelis, M.A.L. (1995) Reverberatory interactions between VPM thalamus and SI cortex during processing of somatosensory stimuli in awake rats. Abst. in Soc. Neurosci. 51:8.
8. Hopfield, J. and Tank, D. (1986) Science. 233:625.
9. Nicolelis, M.A.L., Lin, C.-S., Woodward, D.J. and Chapin, J.K. (1993) Distributed processing of somatic information by networks of thalamic cells induces time-dependent shifts of their receptive fields. Proc. Natl. Acad. Sci. 90:2212–2216.
10. Nicolelis, M.A.L., Lin, C.-S., Woodward, D.J. and Chapin, J.K. (1993) Peripheral block of ascending cutaneous information induces immediate spatio-temporal changes in thalamic networks. Nature 361:533–536.
11. Nicolelis, M.A.L. and J.K. Chapin (1994) The spatiotemporal structure of somatosensory responses of many-neuron ensembles in the rat ventral posterior medial nucleus of the thalamus. J. Neurosci. 14(6):3511–3532.
12. Nicolelis, M.A.L., Baccala, L.A., Lin, R.C.S. and Chapin, J.K. (1995) Synchronous neuronal ensemble activity at multiple levels of the rat somatosensory system anticipates onset and frequency of tactile exploratory movements. Science, 268:1353–1358.

46

CORTICAL CIRCUITS FOR CONTROL OF VOLUNTARY ARM MOVEMENTS

Paul Cisek,[1,2] Daniel Bullock,[1] and Stephen Grossberg[1]

[1]Department of Cognitive & Neural Systems
Boston University
677 Beacon Street, Boston, Massachusetts 02215
[2]Département de Physiologie
Université de Montréal
2960 Chemin de la tour, C.P. 6128 Succursale Centreville, Montréal, Québec,
    H3C 3J7 Canada

ABSTRACT

A model of voluntary movement and motor proprioception is developed to address data from neurophysiology and psychophysics. Model elements correspond to cell types reported in cortical areas 4 and 5, basal ganglia, and spinal cord. The functional scope of the model includes voluntary movement, posture maintenance, exertion of forces against obstacles, passive relaxation, and load compensation. Movements are generated by composing complex commands in area 4, which effect gradual shifts of the limb's equilibrium posture via descending pathways. Movement direction is computed in area 5 using a representation of limb position based upon corollary discharges and spindle feedback. Predictions are made regarding cell responses in novel experimental paradigms.

1. INTRODUCTION

A model of pre- and post-central cortical areas is presented in an attempt to unify diverse data on the neural control of movement (see 1, 2 and 3 for detailed treatment). This model focuses upon single-joint voluntary reaching movements and posture maintenance, and also addresses issues of load compensation, passive relaxation, and exertion of forces against obstacles. The model system has the capacity to execute self-terminating voluntary reaches to targets while compensating for effects of static and inertial loads. If an obstacle prevents the target from being reached, the system will exert a force against it, while maintaining an accurate representation of limb position. When in a relaxed state, the system passively responds to externally imposed movement, releasing tension while keeping spindles loaded to maintain proprioception accuracy.

Computational Neuroscience
edited by Bower, Plenum Press, New York, 1997

The model's elements correspond to observed cell types in areas 4 and 5 and model simulations reproduce qualitatively the response profile, latency of activation, and kinematic and kinetic sensitivities of these cells. The model helps to clarify why many cells do not simply correlate with a single movement variable such as position or force, and why distinct regions of cortex contain cells with similar – but not identical – kinematic and kinetic sensitivities and response profiles. Because the model proposes distinct functional roles for physiologically-identified cell types, it generates predictions regarding novel experimental manipulations that should generate distinctive responses from these cell types.

2. MODEL DESCRIPTION

Figure 1 summarizes the model.

Model area 4 cells project a continuously computed descending command to spinal motor centers. This command consists of three signals: 1) a Desired Velocity Vector (DVV) projecting to dynamic gamma motoneurons; 2) an Outflow Position Vector (OPV) projecting to static gamma motoneurons; and 3) a signal to alpha motoneurons which assembles position and force components necessary to execute the desired kinematics (called the Outflow Force+Position Vector, or OFPV). Together, these signals specify a trajectory of limb configuration which is stabilized in part by the spring property of the muscles and by spinal mechanisms such as the stretch reflex. An efference copy of the OPV signal projects to anterior area 5, where it is combined with feedback "position error" signals from muscle spindles (routed through area 2) to compute a representation of the actual limb position (called a Perceived Position Vector, or PPV). This in turn is subtracted from a representation of target position (supplied by the parietal "where" pathway) to compute a Difference Vector (DV) in posterior area 5. The DV is gated by a GO signal from motor thalamus to generate a kinematic command in the form of the Desired Velocity Vector (DVV) in area 4. Finally, the DVV is integrated over time by the OPV cells, which specify the time-sequence of desired positions. As the limb moves toward the target, the DV is reduced and movement terminates when the DV has returned to baseline.

Several additional model elements and pathways extend the basic feedback-control system described above. The first of these is a reciprocal pathway from the area 5 PPV to

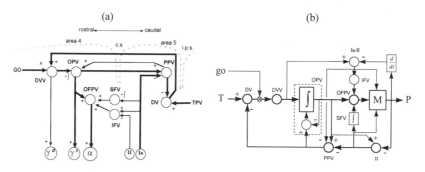

Figure 1. Model circuit. (a) Notation facilitating interpretation in terms of brain regions. Thick lines highlight the position feedback control aspect of the model. DVV – Desired Velocity Vector; OPV – Outflow Position Vector; OFPV – Outflow Force+Position Vector; SFV – Static Force Vector; IFV – Inertial Force Vector; PPV –Perceived Position Vector; DV – Difference Vector; TPV – Target Position Vector; γ^d – dynamic gamma motoneuron; γ^s – static gamma motoneuron; α – alpha motoneuron; Ia – primary spindle afferent; II – secondary spindle afferent; c.s. – central sulcus; i.p.s. – intraparietal sulcus. (b) The model expressed in control-theory notation. M – neuromuscular motor plant.

the area 4 OPV. When the limb is relaxed (i.e. when no GO signal is given), this pathway lets the limb comply with external forces by shifting the OPV command toward the actual limb position. This releases tension on the limb and keeps muscle spindles loaded and sensitive to stretch. In addition, this pathway contributes significant stability to the system during movement.

Two additional circuits enable the controller to compensate for static and inertial loads. The first operates by integrating muscle spindle "position error" signals and adding the result to the descending OFPV alpha motoneuron command. This creates a negative feedback loop which modifies muscle force until the trajectory perturbing effects of static loads have been counteracted. The second compensatory circuit extracts "velocity error" signals from the spindles and adds these to the OFPV. This generates phasic launching and braking pulses which help counteract the effects of inertia during movement.

The model has been described in detail elsewhere (see 1, 2 and 3). The present report summarizes the model and discusses how it can be used to interpret several physiological phenomena.

3. SIMULATIONS

Kalaska, Cohen, Hyde, & Prud'homme (4) have classified the response profiles of area 4 cells into four classes: "phasic reaction-time (RT)", "phasic movement-time (MT)", "tonic", and "phasic-tonic" (see Figure 2). Similar classifications have been proposed by others (5,6). Phasic cells tend to appear in more superficial layers of cortex while tonic and phasic-tonic cells in deeper layers and are presumably often pyramidal tract neurons. The tonic and especially the phasic-tonic cells have the highest degree of load-sensitivity. In the context of the model, we interpret these activities as neural correlates of the IFV, DVV, OPV, and OFPV, respectively. Figure 2 illustrates the qualitative similarity between the activity of these model elements and the neural activities during a simple target-reaching task.

Similar classes of responses are observed in area 5. Kalaska, Cohen, Prud'homme, & Hyde (7) describe three common activity types as: "phasic", "tonic", and "reversal"

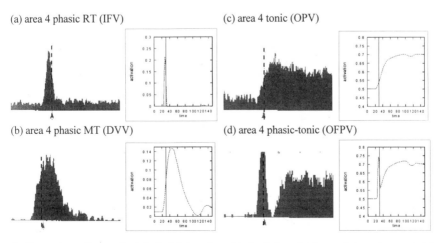

(a) area 4 phasic RT (IFV)   (c) area 4 tonic (OPV)

(b) area 4 phasic MT (DVV)   (d) area 4 phasic-tonic (OFPV)

Figure 2. Comparison of neurophysiological data and model simulations for four cell types in area 4: (a) phasic RT; (b) phasic MT; (c) tonic; (d) phasic-tonic. Histograms (reprinted with permission from 4) are aligned around the onset of movement which is indicated in both the data and the simulations by a vertical dashed line.

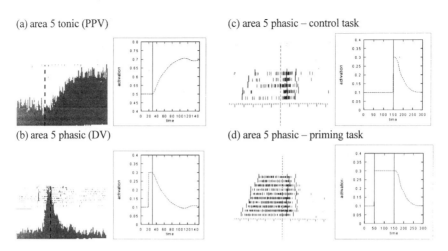

(a) area 5 tonic (PPV)   (c) area 5 phasic – control task

(b) area 5 phasic (DV)   (d) area 5 phasic – priming task

Figure 3. Comparison of neurophysiological data and model simulations for area 5 cells. (a-b) Cell activity during a simple reaching task. Histograms (reprinted with permission from 4) are aligned around the onset of movement which is indicated in both the data and simulations by a vertical dashed line. (c-d) Activity of an area 5 phasic cell during a control task (c) and a priming task (d) where the target presentation preceded the signal to begin the movement. Rasters (reprinted with permission from 8) are aligned around the time of the "go" stimulus, indicated by the vertical dashed line. In the simulation of priming, the first vertical line indicates the onset of the target stimulus and the second the onset of the "go" stimulus.

cells. The activities of phasic cells vary with direction while the activities of tonic cells correlate better with position, and neither class of cells exhibits load-sensitivity. The tonic cells are mostly "late" (i.e. the onset of their activity changes follows the onset of movement), while about half of the phasic cells are "early" (onset precedes the onset of movement). In the context of the model, we interpret the early phasic cells as neural correlates of the DV and the tonic cells as correlates of the PPV (we make no proposal here regarding the functional role of "reversal cells"). The proposal that area 5 phasic cells correspond to the DV is supported by studies showing that the activity of these cells can be "primed" by the presentation of the target before the signal to move is given (8,9). Figure 3 shows the qualitative comparison.

The proposed correspondence between model elements and cell activities in areas 4 and 5 is also supported by data concerning the onset of activity changes. During a voluntary movement, cells in posterior area 5 activate first (10). We suggest this early activity is that of the primable phasic cells which correspond to the DV. This is followed by activity changes in area 4, then by muscular action and movement, as in the model. Next, cells in area 2 show activities during movement. In the context of the model this may be interpreted as feedback signals from muscle receptors which are used to compute the PPV. This is consistent with the Fromm & Evarts (11) description of area 2 cells as signaling misalignment between desired and actual position, i.e. "position error". Finally, activity changes are observed in anterior area 5. We suggest these are changes in the "late" tonic activities which correspond in the model to updating the PPV.

4. PREDICTIONS

The model presented above consists of a set of hypotheses expressed through a computational formalism, namely a network whose interactions are governed by a system of

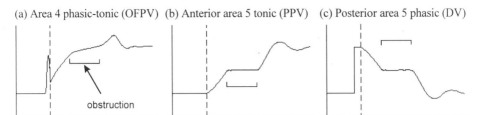

(a) Area 4 phasic-tonic (OFPV) (b) Anterior area 5 tonic (PPV) (c) Posterior area 5 phasic (DV)

Figure 4. Predicted response profiles of three cortical cells during an obstructed-reaching paradigm. The horizontal bracket indicates the period during which force is exerted against the obstacle, and the vertical dotted line indicates the onset of movement.Because area 4 phasic-tonic cells are proposed to assemble the descending alpha motoneuron command, it is expected that their activities will grow in proportion to the force being exerted against the obstacle. In contrast, if anterior area 5 tonic cells subserve the role of a Perceived Position Vector, their activities should not change while the limb is stationary. Likewise, if posterior area 5 phasic cells act as a Difference Vector then their activities should also stop changing during the period of obstruction, and decay to baseline only as the arm continues toward the target.

differential equations. Each of the hypotheses generates testable predictions, including predictions of the responses of specific cortical cell types during various novel experimental paradigms. One such paradigm can be described as follows: A monkey performs reaching movements to visually-specified targets using a manipulandum. During some movement trials, the manipulandum is constrained from reaching the target by an obstacle. The monkey is to exert a given level of force against the obstacle for a period of time, after which the obstacle yields and the target can be reached. The model predicts that cells in areas 4 and 5 will respond differently during the period that force is being exerted against the obstacle. Figure 4 illustrates the expected response profiles from area 4 phasic-tonic cells, anterior area 5 tonic cells, and posterior area 5 phasic cells.

5. CONCLUSIONS

Fetz (12) raised doubts as to whether movement parameters are recognizably coded in the activity of single neurons. Among the reasons that led to his skepticism are the observations that the function of movement control seems to be distributed throughout many places in cortex, including pre-motor, motor, and somatosensory cortices, with similar response properties in each. Furthermore, cells seem to change their coding properties depending on the conditions of the task.

The present model allows one to respond to these concerns without abandoning the idea that cell activity can be interpreted from a functional perspective. First, the fact that function is distributed among different cortical regions does not imply that these regions are in any way redundant. The present model rationalizes tonic and phasic representations in pre-central and post-central regions, and why they should be so heavily interconnected. Second, in the context of the model, it is not surprising that a cell which shows a strong correlation to muscle activity under conditions of external load should show a better correlation to position when the load is removed. This result need not suggest that cells change their coding properties with different movement conditions (12), but instead that a suitable control circuit can superimpose load compensation activity upon a positional code – as is the case with OFPV cells – without loss of function. Rather than treat the net command descriptively in terms of concepts like "virtual trajectories" (13), we have focused on the command as a dynamically assembled composite, and have accordingly specified how separate circuit elements can compensate for distinct sources of deviations from desired trajectories.

In conclusion, the model allows one to interpret neurophysiological results from a unified functional perspective, and to begin to make sense of a diverse set of data that appear puzzling when viewed piecemeal. Furthermore, because the model proposes specific functional roles for specific cell types in identified brain regions, it generates testable predictions for a number of novel experimental paradigms.

REFERENCES

1. D. Bullock, P. Cisek, S. Grossberg, Center for Adaptive Systems Technical Report, Boston University CAS/CNS TR-95–019, (1996).
2. P. Cisek, S. Grossberg, D. Bullock, Center for Adaptive Systems Technical Report, Boston University CAS/CNS TR-96–035, (1996).
3. P. Cisek, Ph.D. Dissertation. Boston University (1996).
4. J. F. Kalaska, D. A. D. Cohen, M. L. Hyde, M. J. Prud'homme, *J. Neurosci.* **9**, 2080 (1989).
5. P. D. Cheney, E. E. Fetz, *J. Neurophysiol.* **44**, 773 (1980).
6. C. Fromm, S. P. Wise, E. V. Evarts, *Exp. Brain Res.* **54**, 177 (1984).
7. J. F. Kalaska, D. A. D. Cohen, M. J. Prud'homme, M. L. Hyde, *Exp. Brain Res.* **80**, 351 (1990).
8. D. J. Crammond, J. F. Kalaska, *Exp. Brain Res.* **76**, 458 (1989).
9. J. F. Kalaska, D. J. Crammond, *Cerebral Cortex* **5**, 410 (1995).
10. P. Burbaud, C. Doegle, C. Gross, B. Bioulac, *J. Neurophysiol.* **66**, 429 (1991).
11. C. Fromm, E. V. Evarts, *Brain Research* **238**, 186 (1982).
12. E. E. Fetz, *Behav. Brain Sci.* **15**, 679 (1992).
13. E. Bizzi, N. Hogan, F. A. Mussa-Ivaldi, S. F. Giszter, *Behav. Brain Sci.* **15**, 603 (1992).

47

IDENTIFYING OSCILLATORY AND STOCHASTIC NEURONAL BEHAVIOR WITH HIGH TEMPORAL PRECISION IN MACAQUE MONKEY VISUAL CORTEX

U. Ernst, A. Kreiter,[1] K. Pawelzik, and T. Geisel

Institut f. Theor. Physik
Robert-Mayer-Straße 8, D-60054 Frankfurt
E-mail: {udo,klaus,geisel}@chaos.uni-frankfurt.de
[1] MPI für Hirnforschung
Deutschordensstraße 46, 60528 Frankfurt
E-mail: kreiter@mpih.mpg.d400.de

ABSTRACT

We characterize the network dynamics underlying multi-unit activities from area MT of macaque monkey in terms of a hidden markov model (HMM) and compare our results to those previously obtained from cat visual cortex. We find that the collective dynamics acts like a stochastic oscillator with a broad range of frequencies. Oscillatory and stochastic periods in the spike trains are detected with high temporal precision, and synchronous network events are identified and localized.

1. INTRODUCTION

Synchronization phenomena in the cortex are believed to underlie the perception of Gestalt properties. They provide a plausible mechanism for contour integration and figure-ground segregation which have been addressed as the binding and superposition problem [7]. While experiments in the cat provide some evidence for this notion [6], the relevance of synchronization in other species is not clear. Even the detection of synchronous states, which may be occluded by an incomplete or stochastic observation, is not trivial.

In this contribution we analyze multi-unit spike train data recorded simultaneously with two electrodes from area MT in the extrastriate visual cortex of macaque monkeys engaged in a

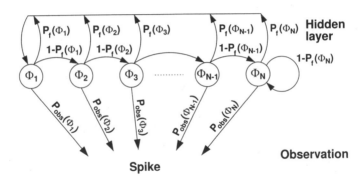

Figure 1. Hidden Markov model. The state of the network is described by a renewal process. In each time step (in our analysis $1ms$), the network returns with probability $P_f(\Phi_i)$ to the synchronous state Φ_1, otherwise it reaches the next state $P_f(\Phi_{i+1})$. In each state Φ_i, a spike can be detected with probability $P_{obs}(\Phi_i)$, thus the underlying dynamics is hidden to the observer.

fixation task. The neurons at both recording sites had partially overlapping receptive fields and were activated with moving bar stimuli. We refer to [1] for the details of these experiments.

The purpose of this work is to analyze the observed time series in terms of an underlying network dynamics. Due to experimental constraints, the observation of the network dynamics is limited in at least three different ways. First, we only can observe a small fraction of all cells engaged in the perception of the stimulus. Second, the recorded activity is likely to stem from a pool of cells rather than from only one cell (multi-unit-activity, MUA). Third, the spiking of one specific cell is a probabilistic process depending on the state of the network. Thus the real dynamics of the neuronal pool is somewhat hidden behind the position of the observer seeing only a part of it. This motivated us to the development of a hidden Markov method for the analysis of stochastic network oscillations depicted in Fig. 1. For the details see [2, 4, 5].

While the HMM was able to identify oscillatory and stochastic multistable behaviour in data from cat visual cortex, we show in this paper, that synchronization without oscillations is detected with high temporal precision in area MT of macaque monkey. Additionally, we quantify the degree of synchronization between two observation sites, and classify the dynamics in comparison with both renewal and Poissonian processes with burst-like spiking behaviour.

2. HIDDEN-MARKOV MODEL

The collective network dynamics is modeled as a renewal process with a sequence of states Φ_i, $i = 1...N$ (Fig. 1). The first state Φ_1 is called 'synchronous state'. In each but the last state the Markov process can return to the first state with probability $P_f(\Phi_i)$ or proceed to the next state with probability $1 - P_f(\Phi_i)$. In the last state Φ_N the system can return to Φ_1 with probability $P_f(\Phi_N)$ or rest in Φ_N with probability $1 - P_f(\Phi_N)$. The dynamics of this 'hidden layer' cannot be observed directly, but instead we assume that spikes occur stochastically with probability $P_{obs}(\Phi_i)$ depending on the actual state Φ_i. For constant $P_f(\Phi_i)$ and for episodes when the system remains in the last state Φ_N, the pulse sequence observed is equivalent to a Poissonian process. The latter justifies the term 'stochastic' state for Φ_N. In the following paragraphs, we refer to P_f and P_{obs} as transition and observation probabilities, respectively.

The only information we get from the experiment about this special kind of HMM is the spike sequence 'triggered' by the observation probabilities P_{obs}. Using the well-known iterative Baum-Welch-algorithm [3, 8], it is possible to estimate the unknown parameters P_f and P_{obs} from this sequence. Once P_f and P_{obs} were calculated, the following procedures are applicable.

- Estimation of the autocorrelation function out of P_f and P_{obs}, which, as a check for consistency, must reproduce the spike-spike autocorrelation.

- Estimation of the probability for being in state Φ_i at time t $P(\Phi(t) = \Phi_i)$ for all i and t. If $i = 0$, $P(\Phi(t) = \Phi_0)$ can help to detect the exact time of synchronization in the underlying network dynamics.

- Estimation of the state sequence with highest probability, answering the question 'Given the observed time series, which is the most likely state Φ_i the system was in at time t?'

- Assuming that two different sites obey the **same** hidden dynamics (full synchronization), but have different observation probabilities, we can calculate their hypothetical cross-correlogram. Comparing it to the real cross-correlogram of the two electrodes, the degree of synchronization can be quantified easily.

3. RESULTS

We used 14 recordings from 2 electrodes each consisting of 10 trials over $5000ms$. Additionally, we generated five sets of test data with the same mean firing rate, which we used to categorize the HMMs for the real data. These five sets were generated by a) a Poissonian process, b) a renewal process (integrate-and-fire neuron), c) a Poissonian process with burstlike firing activity, d) a renewal process with burstlike firing activity (Fig. 2c), and e) the original data with the interspike-intervals being scrambled in their order of appearance (Fig. 2d). The bursting activity was modeled in terms of spiking probabilities in intervals of $6ms$ after the occurrence of a synchronous event. These probabilities have been calculated directly from the original data.

The comparison between the autocorrelation functions (ACF) estimated by the model and the ACFs from the original data (Fig. 3) show that the algorithm converges to a well-defined solution, and that the model is a good representation of the measured data.

In Fig. 2a and b, we present two examples for the HMMs estimated from the real data. Both show a high firing probability shortly after the synchronous state and also a peak in the transition probabilities at the same position. These high observation and transition probabilities account for the bursting behaviour of the spike sequence, leading to a high probability to fire twice or more and to a high probability to return to the first state after a short period of time, respectively. The relatively flat tails with several small peaks suggest a broad range of frequencies. Additionally, we found some models having sharp or even wide peaks in the transition probabilities due to their preferred transition frequencies (data not shown).

Fig. 4 shows a part of two time series recorded simultaneously from two electrodes. For each of them, the state sequence with the highest likelihood and the probability for being in the synchronous state have been plotted. In particular, the two systems synchronize in the time interval $[2200ms, 2400ms]$, being in the first state Φ_1 at approximately the same time. The dynamics of synchronization can be understood by the particular form of the transition

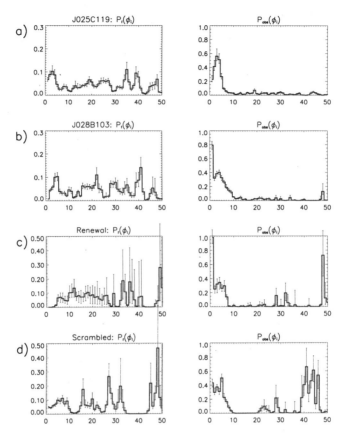

Figure 2. Hidden-Markov models for natural and synthetic data determined by the Baum-Welch algorithm. a-b) These models for two different recordings with electrode # 2 show an increased probability to return into the first state over their first 5-10 bins. Both show a broad range of typical frequencies. c) For comparison, we show a HMM estimated for a bursting renewal process and d) a HMM for synthetic data sampled from a random sequence of the original interspike-intervals of b). While the observation probabilities in b)-d) are not significantly different, c) and d) lack the increased probability to return to the first state in short intervals.

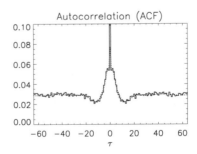

Figure 3. Autocorrelation functions (ACF). The solid line depicts the ACF from the spike train data, while the dotted line shows the ACF extrapolated from the HMM. The similarity of the curves indicates the successful fit of the original data by the HMM.

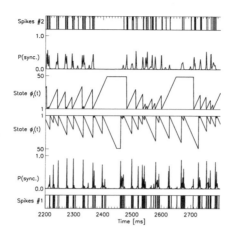

Figure 4. State sequence $\Phi_i(t)$ and probabilities $P(sync.) = P(\Phi_i(t) = \Phi_1)$ for being in the synchronous state Φ_1. The upper part and lower part of the figure refer to the first and second electrode, respectively. The spike train data from trial 5 is depicted at the upper and lower boundary, whereas the state sequence $\Phi_i(t)$ is plotted in the middle of the figure to allow visual detection of synchronous events. The synchronization probabilities $P(sync.)$ complete this picture.

probabilities in the HMM (Fig. 2b). Differing from a bursting Poissonian or renewal process (Fig. 2c and d), the dynamics being in the first states we find a high probability to return to the synchronized state. This allows two neuronal pools which are slightly out of tune to resynchronize with high temporal precision, i.e. if pool # 1 reaches the synchronous state some milliseconds in advance of pool # 2 (compare with Fig. 4 at $t = 2280ms$ or $t = 2340ms$).

To quantify the degree of synchronization, we have to compare the cross-correlogram for two electrodes $C(\tau)$ with the cross-correlation $C_{HMM}(\tau)$ which would have been measured if the two recording sites were fully synchronized, i.e. would have the same transition probabilities. This hypothetical cross-correlation can be calculated from the individual observation probabilities for each site via

$$C_{HMM}(\tau) = \sum_i \sum_j P_{obs}(\Phi_i)(\mathbf{M}^\tau)_{ij} P_{obs}(\Phi_j)\rho_s(\Phi_j), \tag{1}$$

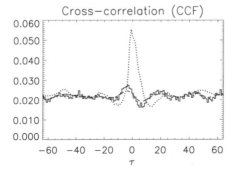

Figure 5. Cross-correlation functions (CCF). The solid line depicts the CCF from the spike train data, while the dotted line shows the CCF to be expected if the two dynamical processes at both recording sites were perfectly synchronous. The ratio of the modulation differences of both functions yield a synchronization level of 32%. The dashed line shows the CCF of $P(\Phi_1)$.

where M_{ij} denotes the transition matrix obtained from the individual transition probabilities averaged over the recording sites, and ρ_s the stationary state density of the Markov process, respectively. A quick estimate for the degree of synchronization is the ratio of the maximal amplitude modulations of C and C_{HMM} (Fig. 5). For the 14 recordings, we got the following results.

Recording #	1	2	3	4	5	6	7	8	9	10	11	12	13	14
Sync. [%]	65	47	52	64	68	60	48	46	21	61	28	21	16	14

From the level of synchronization, we can clearly identify the two stimulus paradigms. Recordings #1 to #7 (58% average sync.) have been obtained by stimulating with a single bar, and recordings #8 to #14 (30% average sync.) have been made while presenting a 'conflicting stimulus' consisting of two orthogonally oriented moving bars [1].

4. DISCUSSION

In conclusion, our application of the HMM-method for the identification of hidden synchronization dynamics [2, 4, 5] to spike trains from macaque monkeys characterizes the bursting behaviour as well as the broad spectra of frequencies of the observed neurons which is in contrast to the relatively regular firing patterns in cat visual cortex [6]. We also find that once synchronization takes place, the next synchronous event is hardly predictable. The comparison to synthetic data for several types of stochastic processes shows that a neuronal pool is able to re-synchronize its activity with the network dynamics. Furthermore, synchronization can be detected with a much higher temporal precision than if we would examine the spike train data directly. The degree of synchronization can be estimated easily from the HMMs. Finally our estimation of state sequences allows us to identify episodes of the spike trains where the underlying dynamics is stochastic, and where the dynamics is oscillatory. This opens the possibility to select relevant data for further processing.

REFERENCES

[1] A. Kreiter and W. Singer, Oscillatory neuronal responses in the visual cortex of the awake macaque monkey, Europ. J. Neurosci., **4**, 369 (1992).

[2] J. Deppisch, K. Pawelzik, and T. Geisel, Uncovering the synchronization dynamics from correlated neuronal activity quantifies assembly formation, Biol. Cyb. **71**, 387 (1994).

[3] L.E. Baum and T. Petrie, Statistical inference for probabilistic functions of finite state Markov chains, Ann. Math. Stat. **37**, 1554 (1966).

[4] J. Deppisch, H.-U. Bauer, T.B. Schillen, P. König, K. Pawelzik, and T. Geisel, Alternating oscillatory and stochastic states in a network of spiking neurons, Network **4**, 243 (1993).

[5] K. Pawelzik, H.-U. Bauer, J. Deppisch, and T. Geisel, How oscillatory neuronal responses reflect bistability and switching of the hidden assembly dynamics, In: J.E. Moody, S.J. Hanson, and R.P. Lippmann (eds), Proceedings of the NIPS'93, **5**, Morgan Kaufmann, San Mateo.

[6] C.M. Gray, P. König, A.K. Engel, and W. Singer, Oscillatory responses in cat visual cortex exhibit inter-columnar synchronization which reflects global stimulus properties, Nature **338**, 334 (1989).

[7] C.v.d. Malsburg, The correlation theory of brain function, Internal report 81-2, MPI for Biophysical Chemistry, Göttingen.

[8] L.R. Rabiner, A tutorial on hidden-Markov models and selected applications in speech recognition, Proc. IEEE **77**, 257 (1989).

48

A JOINT CROSS INTERVAL DIFFERENCE ANALYSIS FOR DETECTING COUPLING TRENDS BETWEEN NEURONS

Michelle A. Fitzurka[1] and David C. Tam[2]

[1]Department of Physics
Catholic University of America
Washington, DC 20064
fitzurka@stars.gsfc.nasa.gov
[2]Department of Biological Sciences and Center for Network Neuroscience
University of North Texas
Denton, Texas 76203
dtam@unt.edu

ABSTRACT

We present a new spike train analysis technique based on *interval difference* statistics. The *Joint Cross Interval Difference (JCID)* scatter plot is introduced to examine the dependency relationship between adjacent *CIDs* in a spike train allowing for inferences to be made about local cross train trends. Monotonically increasing, decreasing, constant, or alternatively varying *CIDs* will appear in separate quadrants in this *JCID* scatter plot. The method was applied to simulated spike trains in order to display the capabilities of this new technique. Repeated cross train trends between pairs of neurons can seen as clusters and bands of points in specific quadrants of this new scatter plot.

1. INTRODUCTION

Spike train analysis methods explore both the temporal relationships within single spike trains, and the temporal and spatial relationships across multiple spike trains. A spike train is the temporal record of the sequential firing of action potentials by a neuron. Each spike in the train represents the occurrence in time of an action potential.

Both *interval*[4,7,8] and *interval difference*[1,3] statistics have been defined for single spike trains. An *interspike interval (ISI)* denotes the time difference between consecutive spikes while an *interspike interval difference (ISID)* is defined as the difference between adjacent *ISIs* in a spike train. The relationship between adjacent *ISIDs* has been characterized in the *Joint Interspike Interval Difference (JISID)* scatter plot[3].

So far, however, only *interval*[5,6,9,10] but not yet *interval difference* statistics have been described between *two* spike trains. A *cross interval (CI)* measures the time difference between the occurrence of a spike in the reference train and the last or next spike in the compared train. In this paper, the *cross interval difference (CID)* is defined and the relationship between adjacent *CIDs* is characterized by these *JCID* plots.

2. METHODS

If spike trains $r(t)$ and $c(t')$ are taken to be the reference and compared spike trains respectively (see Fig. 1), they can be defined as:

$$r(t) = \sum_{n=1}^{n=N} \delta(t - t_n),$$

(1)

with N indicating the total number of spikes in the reference spike train, and t_n denoting the time of occurrence of the n-th spike, and $\delta(t)$ denoting the delta function, and as:

$$c(t') = \sum_{m=1}^{m=M} \delta(t' - t'_m),$$

(2)

with M indicating the total number of spikes in the compared spike train, t'_m denoting the time of occurrence of the m-th spike.

The magnitudes of the *pre-cross-interval (pre-CI)* $\tau'_{n,-1}]$ and the *post-cross-interval (post-CI)* $\tau prim_{n,+1}$ relative to the n-th reference spike are defined as follows:

$$\left| \tau'_{n,-1} \right| = \left| t'_m - t_n \right| = t_n - t'_m$$

(3)

and

$$\left| \tau'_{n,+1} \right| = \left| t'_{m+1} - t_n \right| = t'_{m+1} - t_n$$

(4)

which satisfies the condition:

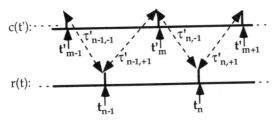

Figure 1. Segments of the reference, $r(t)$, and compared, $c(t')$, spike trains showing the *pre-cross-intervals (pre-CIs)* $\tau'_{n,-1}$ and $\tau'_{n-1,-1}$ and *post-cross-intervals (post-CIs)* $\tau'_{n,+1}$ and $\tau'_{n-1,+1}$ with respect to the n-th and $n-1$)-th reference spikes respectively. The *cross-interval-difference (CID)*, $\Delta\tau'_{n,+1}$ for the n-th reference spike and the *cross-interval-difference (CID)* $\Delta\tau'_{n-1,+1}$ for the $(n-1)$-th reference spike are defined within the text.

$$t'_m < t_n < t'_{m+1}$$

while that of the *pre-CI*, $\tau'_{n-1,-1}$ and the *post-CI*, $\tau'_{n-1,+1}$ relative to the $(n-1)$-th reference spike are:

$$\left|\tau'_{n-1,-1}\right| = \left|t'_{m-1} - t_{n-1}\right| = t_{n-1} - t'_{m-1} \tag{5}$$

and

$$\left|\tau'_{n-1,+1}\right| = \left|t'_m - t_{n-1}\right| = t'_m - t_{n-1}. \tag{6}$$

From these definitions then, the *cross interval differences (CIDs)*, can be defined. The *CID*, $\Delta\tau'_{n,+1}$ for the *n*-th reference spike is:

$$\Delta\tau'_{n,+1} = \left|\tau'_{n,+1}\right| - \left|\tau'_{n,-1}\right| = \left|t'_{m+1} - t_n\right| - \left|t'_m - t_n\right|$$
$$= \left(t'_{m+1} - t_n\right) - \left(t_n - t'_m\right) = t'_{m+1} - 2t_n + t'_m \tag{7}$$

and the *CID*, $\Delta\tau'_{n-1,+1}$ for the $(n-1)$-th reference spike is:

$$\Delta\tau'_{n-1,+1} = \left|\tau'_{n-1,+1}\right| - \left|\tau'_{n-1,-1}\right|. \tag{8}$$

Reference Fig. 1 to see these relationships.

A plot of adjacent *CIDs* can then be made by plotting the *CIDs* defined at the $(n-1)$-th and *n*-th reference spikes respectively, ($\Delta\tau'_{n-1,+1}$ vs. $\Delta\tau'_{n,+1}$), in the *Joint Cross Interval Difference (JCID)* plot. This is analogous to the *JISID* plot of adjacent *ISIDs* in a single spike train[3]. With this construction, consecutive cross train trends can be examined. Monotonically increasing or decreasing cross train trends over two *CIDs* will appear as points in the first and third quadrants respectively. Decreasing/increasing or increasing/decreasing cross train trends will appear as points in the second and fourth quadrants, respectively. Constant cross train trends over two *CIDs* will appear as points at the origin.

3. RESULTS

3.1. Burst Firing Example

To simulate burst firing, we create a synthetic spike train (spike train A) in which the *ISIs* mimic a burst of approximately equally spaced action potentials (25 msec *intraburst period*, 2.5 msec Gaussian variance) followed by a pause of no activity (125 msec *interburst period*, 12.5 msec Gaussian variance) followed by another similar burst. A dependent spike train (spike train B) is generated based on this bursting neuron in which a spike in spike train A would induce a spike in spike train B with a probability of 0.9 at a 3 msec latency with 0.3 msec variance

Figure 2 shows the *JCID* scatter plot for the coupling relationship between spike trains A and B, using spike train A and spike train B as the reference spike train in Figs. 2a and 2b, respectively. In both plots, the central cluster nearest the origin represents

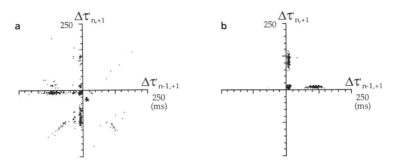

Figure 2. (a) The *JCID* scatter plot for this burst firing example with spike train A taken as the reference spike train with respect to spike train B. (b) The *JCID* scatter plot for this burst firing example with spike train B taken as the reference spike train with respect to spike train A.

the situation in the *intraburst* region where consecutive differences may be expected to be approximately zero.

The two prominent wings near the $-x$-axis and $-y$-axis in Fig. 2a, and near the $+x$-axis and $+y$-axis in Fig. 2b represent the transitions from the *interburst* to *intraburst* regions and from the *intraburst* to the *interburst* regions, respectively. The excess points in Fig. 2a correspond to drop out points due to the fact that there is only a 90% coupling probability between the two spike trains.

3.2. Sawtooth Firing Example

To simulate alternating increasing and decreasing intervals, we simulated a repeating sawtooth firing pattern (spike train A) with the following 8 *ISIs*: (25, 50, 87.5, 137.5, 200, 137.5, 87.5, 50) msec. Each of these intervals had a Gaussian variance of 2.5 msec.

A dependent spike train (spike train B) was generated based on this sawtooth firing pattern by each spike in spike train A inducing a spike in spike train B with a probability of 0.9 at a 3 msec latency with 0.3 msec variance.

Figure 3 shows the *JCID* scatter plot for the coupling relationship between spike trains A and B, using spike train A and spike train B as the reference spike train in Figs. 3a and 3b, respectively. In both plots, the 8 prominent clusters of points represent the transitions, either increasing or decreasing, from one interval to the next in the sawtooth series. Extra clusters in Fig. 3a reflect the 90% coupling probability between the two spike trains.

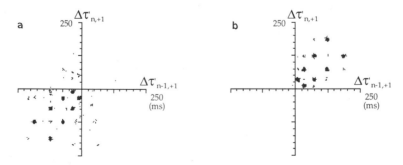

Figure 3. (a) The *JCID* scatter plot for this sawtooth firing example with spike train A taken as the reference spike train with respect to spike train B. (b) The *JCID* scatter plot for this sawtooth firing example with spike train B taken as the reference spike train with respect to spike train A.

DISCUSSION

We have applied this new *JCID* method to examine the dependency relationship between adjacent *CIDs* to show local cross train trends. By noting the distribution of points in the two examples illustrated above, these trends were revealed to be either monotonically increasing, decreasing or varying.

For instance, in Figs. 2a and 3a, the clusters lie mainly in the $(-x,-y)$ quadrant (quadrant 3), while in Figs. 2b and 3b, the clusters lie mainly in the $(+x,+y)$ quadrant (quadrant 1). This indicates repeated cross interval trends for 2 consecutive reference spikes (consecutively negative *CIDs* in the former, consecutively positive *CIDs* in the latter). This is expected for these simulated cases (bursting and sawtooth firing) since the secondary spike train B was generated from the primary spike train A with a given latency.

With this new technique, it is now possible to characterize serial changes in the coupling relationship between neurons. Additionally, whether this relationship is reciprocal can also be established (i.e., by comparing Fig. 2a with 2b or Fig. 3a with 3b). These types of changes in the coupling relationship that span two consecutive spikes are not seen at the level of conventional cross-correlation analysis. Therefore, the evolution of the coupling relationship, or coupling dynamics, can be represented graphically with this analysis.

ACKNOWLEDGMENTS

This research was supported in part by ONR grant number N00014–94–1–0686 to DCT.

REFERENCES

1. Fitzurka, M. A. and Tam, D. C. (1996a) First order interspike interval difference phase plane analysis of neuronal spike train data. In: *Computational Neuroscience.* (J. M. Bower, ed.) Academic Press Inc.: San Diego, CA, pp. 429–434.
2. Fitzurka, M. A. and Tam, D. C. (1996b) Second order interspike interval difference phase plane analysis of neuronal spike train data. In: *Computational Neuroscience.* (J. M. Bower, ed.) Academic Press Inc.: San Diego, CA, pp. 435–440.
3. Fitzurka, M. A. and Tam, D. C. (1995) A New Spike Train Analysis Technique for Detecting Trends in the Firing Patterns of Neurons. In: *The Neurobiology of Computation* (J. M. Bower, ed.) Kluwer Academic Publishers, Boston, MA, pp. 73–78.
4. Perkel, D. H., Gerstein, G. L. and Moore, G. P. (1967a) Neuronal spike trains and stochastic point process. I. The single spike train. *Biophysical Journal.* 7: 391–418.
5. Perkel, D. H., Gerstein, G. L. and Moore, G. P. (1967b) Neuronal spike trains and stochastic point process. II. Simultaneous spike trains. *Biophysical Journal*, 7: 419–440.
6. Perkel, D. H., Gerstein, G. L., Smith, M. S. and Tatton, W. G. (1975) Nerve-impulse patterns: A quantitative display technique for three neurons. *Brain Research*, 100: 271–296.
7. Rodieck, R. W., Kiang, N. Y.-S. and Gerstein, G. L. (1962) Some quantitative methods for the study of spontaneous activity of single neurons. *Biophysical Journal.*, 2: 351–368.
8. Smith, C. E. (1992) A heuristic approach to stochastic models of single neurons. In: *Single Neuron Computation* (T. McKenna, J. Davis and S. Zornetzer, eds.) Academic Press, San Diego, CA. pp. 561–588.
9. Tam, D. C. (1993) A multi-neuronal vectorial phase-space analysis for detecting dynamical interactions in firing patterns of biological neural networks. In: *Computational Neural Systems.* (F. H. Eeckman and J. M. Bower, eds.) Kluwer Academic Publishers, Norwell, MA. pp. 49–53.
10. Tam, D. C., Ebner, T. J., and Knox, C. K. (1988) Cross-interval histogram and cross-interspike interval histogram correlation analysis of simultaneously recorded multiple spike train data. *Journal of Neuroscience Methods*, 23: 23–33.

49

DO "LATERAL CONNECTIONS" IN THE CORTEX CARRY OUT TOPOLOGICAL INFORMATION?

F. Frisone, V. Sanguineti, and P. Morasso

DIST — Department of Informatics, Systems and Telecommunication
University of Genova
Via Opera Pia 13, 16145 Genova, Italy
F: +39 10 3532154
E-mail: friso@dist.dist.unige.it

ABSTRACT

The observed massive presence of non-local lateral connections in the cerebral cortex is not compatible with the implicit assumption of *flatness* of most models, including models of associative areas. We suggest a novel hypothesis about the functional role of lateral connections in such areas: they may reflect a topological representations of the *task* space. In particular, we show how the topologic information, supported by long-range connections in associative areas, can represent *spatial* or *metric* knowledge. The power of the mechanism is demonstrated by describing an activation dynamics and showing the formation of bands of ocular dominance.

1. INTRODUCTION

The main goal of this study is to investigate biologically realistic computational models of cortical maps, accounting for some recent experimental results and theoretical achievements in the fields of cognitive science (connectionist models, properties of distributed representations), of neuroanatomy/neurophysiology (geometry and role of lateral connections among cortical columns) and motor control (nature and emergent properties of visuomotor transformations).

Topologically, the human cerebral cortex is a highly folded sheet and it has a modular organization, characterized by a large number of columns communicating by means of *lateral* connections, parallel to the cortical surface [6], prompting the (wrong) view of a 2-D *lattice*.

However, the observed patterns of lateral connectivity show the existence of a large number of lateral connections that are not local and it is also known that such connections are very significant in representing complex high-dimensional input spaces; in fact, they are present in cortical areas sensitive to multimodal stimuli [12]. We suggest a novel hypothesis on the functional role of lateral connections, accounting for their observed complexity, which is motivated by the observation that higher-order lattices are plausible mechanisms for mapping n-dimensional spaces into a 2D substrate [1]: *non-local lateral connections in the cortex may represent higher dimensional stimuli so that maps that appear to be non-topographic are topological representations of the stimulus space.*

2. THE MODEL

In agreement with the assumption of the modularity of the cortex, we will model it as a 2-D grid of n processing elements PE, (corresponding to cortical columns) operating in parallel on a common afferent signal $\vec{x} \in \Re^n$:

$$Map \overset{\triangle}{=} (\mathcal{P}, C, \mathcal{W}) \tag{1}$$

where $\mathcal{P} = \{PE_i : i = 1, \ldots, n\}$ is a set of PE; $C = \{c_{ij} : i, j = 1, \ldots, n\}$ is the symmetric matrix of *lateral* connection weights; $\mathcal{W} = \{w_{ki} : k = 1, \ldots, m j = 1, \ldots, n\}$ is the asymmetric matrix of *external* connection weights. This reflects the fact that each PE must have an excitatory external input, coming from thalamo-cortical connections, and an excitatory lateral input, coming from cortico-cortical connections, which drive the dynamics of the PE together with a component of self-inhibition, ultimately determining its activity level V_i^x. In particular, we have used the following model equation:

$$\frac{dV_i^x}{dt} + V_i^x = f(h_i^{lat} + h_i^{ext}) \tag{2}$$

where $f()$ is a non-linearity applied to the total input for generality, h_i^{lat} and h_i^{ext} terms are intended to express the massive lateral (cortico-cortical) and external (thalamo-cortical) excitatory connections, respectively:

$$h_i^{lat} = \sum_j c_{ij} \cdot V_j \qquad h_i^{ext} = \sum_k w_{ik} \cdot V_k \tag{3}$$

As regards the external input, determined by the patterns of thalamo-cortical connections, we assume that it can be approximated by a rather broad receptive field $G_i(\vec{x})$ which is maximized by a preferred input $\vec{x} = \vec{\pi}_i^x$. (Typically, G_i is a broad Gaussian.) Response selectivity can be sharpened by a mechanism of local competition, without use of lateral inhibition, such as the mechanism of competitive distribution of activation. (It is assumed that any PE has a finite amount of excitation to distribute at any moment to the set of PEs which are the targets of its axon collaterals and such distribution is competitive in the sense that, the more active is a PE the larger is its excitation share.)

Learning is implemented by means of a Hebbian rule applied to both the lateral and external connections. In the latter case, in particular, this is equivalent to the *smooth tuning* of the receptive field centers:

$$\triangle \vec{\pi}_i^x = \varepsilon \cdot (\vec{x} - \vec{\pi}_i) \cdot V_i^x \tag{4}$$

A basic computational need is that the distribution of PEs must reflect the topology of the represented domains: *similar* input values correspond to adjacent (i.e. directly connected) PEs on the map, and vice versa adjacent PEs correspond to *similar* input values. The formation of the lateral connections is driven by two interactive adaptive processes by means of a

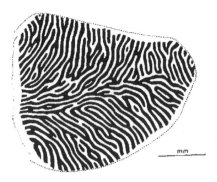

Figure 1. Ocular dominance bands from tangential sections of a monkey's right occipital lobe, visualized by the reduced silver method of Hubel et al., (1974). Alternate bands have been inked in (from Hubel & Freeman, (1977)).

Hebbian learning law. It makes sense to assume that the connections initially have a random topology within a predetermined range, and later on a pruning process takes place according to a self-organize Hebbian process to form global order: this is in agreement with a general underlying ecological pressure on brain formation [2, 7, 8] towards: minimization of volume and maximization of connectivity. In such a model the statistics of the number of lateral connections over the whole map is strongly indicative of the inherent dimensionality of the stimuli [3]. In particular the number of connections established is a way to represent into a 2-D lattice a higher dimensional space while maintaining its topology.

3. OCULAR DOMINANCE BANDS

A simple application of the method described in the previous section is related to the formation of ocular dominance bands in the primary visual cortex (see Figure 1). In striate cortex, inputs from the two eyes converge onto single cells, accomplishing a sort of reduction in redundancy as two separate views of the world are collapsed into one. A remarkable feature of the mapping is the continuity across the cortical surface which preserves neighborhood relationships at the expense of retinotopic position. We simulated the process of formation of such bands, in the context of the cortical model outlined in the previous section, by assuming that the afferent signals to the cortical sheet are characterized by the 3-D feature vector $\vec{x}$ where the first two components correspond to retinal coordinates and the third one is a binary variable which identifies the dominating eye (+1 for the left eye and -1 for the right eye). Therefore, we may see the process as the compression of a particular 3-D manifold onto a 2-D substrate. At the end of training, the first two coordinates of the prototype vectors $\vec{\pi}^x$ identify the retinotopic coordinates of the receptive field centers and the third one identifies which eye the given PE is sensitive to. Irregular banded patterns are found drawing the contour of the curves of level for a definite value of threshold of the activation functions V_i^x centered on the prototypes $\vec{\pi}^x$. Figure 2 show the result for the right eye sensitive cells.

Figure 2. Irregular banded pattern for the right eye sensitive cells.

4. CONCLUSIONS

A Hebbian learning paradigm based on distributed spatial representation has been outlined which provides an unifying framework for dealing with open issues in modelling cortical maps such as the topologic representation of multidimensional stimuli [10]. The topologic aspect is essential for representing knowledge of *spatial/metric* type, like the concepts of *continuity* and *straightness* [11].This necessity is particularly important in the case of coordinate transformations which are necessary for planning complex movements [4, 9].

REFERENCES

[1] V. Braitenberg. *Vehicles - Experiments in Synthetic Psychology*. MIT Press, Cambridge, MA, 1984.

[2] C. Cherniak. Component placement optimization in the brain. *Journal of Neuroscience*, 14:2418–2427, 1994.

[3] F. Frisone, F. Firenze, P. Morasso, and L. Ricciardiello. Application of Topology-Representing Networks to the Estimation of the Intrinsic Dimensionality of Data. In *ICANN95- Int. Conf. on Artificial Neural Networks*, volume 1, pages 323–327, Paris, October 1995.

[4] A.P. Georgopoulos, M. Taira, and A. Lukashin. Cognitive neurophysiology of the motor cortex. *Science*, 260:47–51, 1993.

[5] D.H. Hubel and D.C. Freeman. Short communications: Projection into the visual field of ocular dominance columns in macaque monkey. *Brain Res.*, 122:336–343, 1977.

[6] E. I. Knudsen, S. du Lac, and S.D. Esterly. Computational maps in the brain. *Annual Review of Neuroscience*, 10:41–65, 1987.

[7] G. J. Mitchison. Neuronal branching patterns and the economy of cortical wiring. *Proc. of the Royal Society*, B:151–158, 1991.

[8] G. J. Mitchison. Axonal trees and cortical architecture. *Trends in Neurosciences*, 15:122–126, 1992.

[9] P. Morasso and V. Sanguineti. Self-organizing body-schema for motor planning. *Journal of Motor Behavior*, 26:131–148, 1995.

[10] V. Sanguineti, P. Morasso, and F. Frisone. Cortical maps of sensorimotor spaces. In P. Morasso and V. Sanguineti, editors, *Self-organization, Computational Maps, and Motor Control*, pages 1–36. Elsevier Science Publishers, Amsterdam, 1997. in press.

[11] D. M. Wolpert, Z. Ghahramani, and M. I. Jordan. Are arm trajectories planned in kinematic or dynamic coordinates? an adaptation study. *Experimental Brain Research*, 1995. in press.

[12] E. Zohary. Population coding of visual stimuli by cortical neurons tuned to more than one dimension. *Biological Cybernetics*, 66:265–272, 1992.

50

INFORMATION BASED LIMITS ON SYNAPTIC GROWTH IN HEBBIAN MODELS

Allan Gottschalk

University of Pennsylvania
Philadelphia, Pennsylvania 19104

1. INTRODUCTION

Hebb's seminal concept that synaptic strength varies as a function of correlated pre- and post-synaptic activity has impacted pharmacologic, physiologic, and computational models of the nervous system [Hebb, 1949]. Specifically, he proposed that correlated pre- and post-synaptic activity leads to an increase in synaptic strength. When computational models incorporate Hebbian synapses, some limit must be placed on synaptic growth if well-defined solutions are to be produced, and it may be necessary to incorporate some means by which synaptic strength can also decrease. Strategies to limit synaptic growth have included some type of limit on individual or total synaptic size. Another approach incorporates a computationally [Bienenstock, Cooper, and Munro, 1982] motivated "sliding threshold" for synaptic modification where the sliding threshold is stimulus dependant.

In particular, Bienenstock, Cooper and Munro (BCM) in a computational model of ocular dominance column formation found it necessary to introduce a stimulus dependant "sliding threshold" to account computationally for the full range of results from monocular deprivation experiments. Their "sliding threshold" is a function of average post-synaptic neural activity. Here, an information theory oriented analysis of a simple visual system model reveals how such dynamic "sliding thresholds" must emerge in the process of balancing synaptic resource use with the information flux they generate.

2. MODEL/ANALYSIS

Consider the simple visual system model of Fig. 1. Members of the stimulus ensemble are detected by an array of sensors (ganglion cell receptive fields (RFs)) whose outputs $\mathbf{g}$ are transformed by Φ and corrupted by noise $\mathbf{n}$ to produce $\mathbf{h}$. If the signal and noise are Gaussian and independent, and Φ is linear, an expression for the mutual information $I(\mathbf{g};\mathbf{h})$ between $\mathbf{g}$ and $\mathbf{h}$ can readily be obtained [Gottschalk, 1996]. When $I(\mathbf{g};\mathbf{h})$ is optimized with respect to Φ, it is necessary to either modify the cost function $I(\mathbf{g};\mathbf{h})$ or to adjoin constraints to it in order to obtain well-defined solutions. Examples of the types of

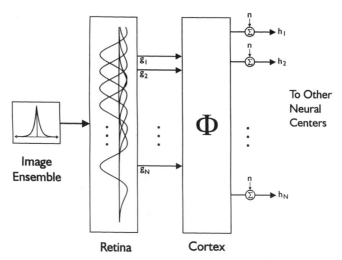

Figure 1. Simple visual system model.

terms which can be incorporated include total synaptic use $\mathrm{tr}[\Phi\Phi^T]$ or total neural output power $\mathrm{tr}[\Phi C\Phi^T]$, where $\mathrm{tr}[\bullet]$ denotes the matrix trace, and C is the correlation matrix of sensor outputs ($C = E[\mathbf{gg}^T]$, where $E[\bullet]$ is the expectation operator). Synaptic resources or neural output power can be incorporated respectively by generating equations of the form

$$I(\mathbf{g};\mathbf{h}) - \alpha\,\mathrm{tr}[\Phi\Phi^T] \tag{1}$$

$$I(\mathbf{g};\mathbf{h}) - \beta\,\mathrm{tr}[\Phi C\Phi^T] \tag{2}$$

which are optimized with respect to Φ. These equations represent modifications of the cost function and include the meta-parameters α and β, which specify the allowed trade-off between information and synaptic size or neural output power. Equations formed by adjoining constraints on synaptic size and neural output to the cost function *via* Lagrange multipliers are similar to Eqs. 1 and 2, with the exception that the Lagrange multipliers appear in place of the meta-parameters α and β. Combinations which include up to all four conditions are also possible. Costs/constraints on individual neurons and synapses can also be included, and, particularly when a fixed meta-parameter is included, the optimal solutions differ little from the problems described above. Analytic solutions of these equations for simple systems can be obtained algebraically. However, the solutions for more complex problems must be obtained using techniques for numerical optimization [Gottschalk, 1996].

3. RESULTS

Results pertaining to the specifics of the RF shape and efficiency produced by the optimization have already been presented [Gottschalk, 1994 and 1996]. For the one-dimensional case when only the cost of neural output power is considered, the elements of Φ at the optimum can be chosen randomly as long as the neural output due to the stimulus P is chosen so that

$$P = 1/(2\beta) - \sigma^2, \quad \beta < 1/(2\sigma^2) \tag{3}$$

where σ^2 is the power of the neural noise. Note that this solution is independent of the structure of the correlation matrix, and that if β is not sufficiently small, it is more efficient to set the elements of Φ to zero. When the tradeoff between synaptic size and information flux is considered, the elements of Φ at the optimum are proportional to the eigenvector of the correlation matrix C with the largest eigenvalue, and this vector should be adjusted so that the neural output power due to the stimulus is given by

$$P = \lambda/(2\alpha) - \sigma^2, \ \lambda > 2\alpha\sigma^2 \tag{4}$$

where λ is the largest eigenvalue of the correlation matrix C. Thus, considerations of synaptic size lead to a more specific solution which depends critically upon the structure of the correlation matrix. However, as before, if synaptic use is too costly relative to the information flux, the most efficient solution is to set the elements of Φ to zero.

Consider how synaptic strength is optimally adjusted to balance information flux with the synaptic cost. Fig. 2 shows the *change* in synaptic strength as a function of the stimulus power for a one-dimensional system. This change is relative to the synaptic size obtained for a stimulus with power P_0. At all points along the curve, the sign of the pre- and post-synaptic correlation remains unchanged. Thus, the change in synaptic strength depends on prior activity and whether the new stimulus conditions warrant an increase or decrease in synaptic strength in order to maintain the balance between information flux and neural resource use. The horizontal portion of the curve at low stimulus power levels represents the setting of the synapse to zero because the information obtained from that synapse no longer justifies the expenditure of synaptic resources. When the cost of neural output power is considered instead of synaptic size, the result is considerably simpler since Eq. 3 indicates that only two levels of synaptic strength are possible, and any changes in synaptic size depend upon which of these values is associated with stimulus power P_0.

A more elaborate sliding threshold can be obtained when limitations on *both* synaptic size and neural output power are included. The optimal solution for Φ in the one-dimensional case is still proportional to the eigenvector of C with the largest eigenvalue. But, now this vector is adjusted so that the neural output power due to the stimulus is given by

$$P = \tfrac{1}{2}\lambda/(\beta\lambda+\alpha) - \sigma^2, \ \lambda > 2\alpha\sigma^2/(1-2\beta\sigma^2) \tag{5}$$

where λ, again, is the largest eigenvalue of the correlation matrix C. Again, Φ is nonzero only if synaptic size or neural output power is not too costly. The corresponding alterations in synaptic strength are similar to those depicted in Fig. 2, except that synaptic strength does not continue to monotonically increase with increases in stimulus power, reflecting the additional tradeoff between information flux and neural output power specified by the meta-parameter β.

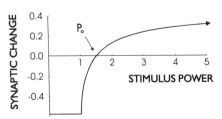

Figure 2. Change in synaptic strength as a function of stimulus power level when the trade-off between information flux and synaptic size is optimized. Changes are with respect to the synaptic strength obtained for a stimulus of power level P_0.

Of particular interest and importance is the fact that including limits to both synaptic size and neural output permits the derivation of a function which can be seen to be *exactly* the hyperbolic ratio equation ($y=V_{MAX} x^2/(x^2 + x_{1/2}^2)$) with a response exponent of n=2 [Gottschalk et al., 1993]. To see this, consider total neural output power P_{TOT} which is the sum of the neural output power due to the stimulus P, and the neural noise σ^2. P_{TOT} then becomes

$$P_{TOT} = 1/(2\beta) \, \lambda/(\lambda + \alpha/\beta) \qquad (6)$$

where $1/(2\beta)$ V_{MAX}, α/β ~ $x_{1/2}$, and the response exponent of two emerges because λ represents a component of the stimulus power and is, therefore, proportional to the square of the stimulus.

Solutions for higher dimensional cases corresponding to multiple output neurons and/or those where explicit constraints on neural resources are adjoined to the cost function with Lagrange multipliers can also be obtained. For the higher dimensional cases the solution depends little upon the structure of the correlation matrix if only neural output power is considered. When synaptic size is considered by itself or along with neural output power, the optimal Φ is chosen from the space spanned by the eigenvectors associated with the largest eigenvalues. The optimal solution is obtained by adjusting the magnitude of the corresponding basis vectors using equations like those given in Eqs. 3–5. When explicit constraints are present, one observes reallocation of these fixed pools of resources among the different neurons as the properties of the stimulus ensemble are varied.

4. DISCUSSION

One of the apparent paradoxes associated with the use of Hebbian synapses to explain physiological phenomena is the fact that in examples like those cited above, synaptic strength appears to decrease despite the presence of correlated pre- and post-synaptic activity. This has led to model refinements which consider not only correlations in pre- and post-synaptic activity, but also the prevailing level of neural activity. Thus, if correlations are present, but are not as pronounced as before, synaptic size will decrease. Eventually, equilibrium is achieved and the current neural activity level becomes the new reference point. As shown here, such a "sliding threshold" can be derived from considerations of efficient information processing, and has a ready interpretation. The nervous system will continue to allocate resources in the form of increasing synaptic strength as long as the corresponding increase in information flux is sufficiently large. This sufficiency is controlled by the corresponding meta-parameter, and/or specified by the corresponding Lagrange multiplier. Thus, the presence of pre- and post-synaptic correlation only indicates the direction to go if information flux is to be increased, it does not indicate whether doing so would be worthwhile. In certain cases, the information flow through the synapse does not justify the expenditure of the resources to maintain it, and such synapses are optimally set to zero.

This information oriented approach to Hebbian synaptic modification helps to indicate why synapses initially formed during development may be strengthened, decreased in strength, or set to zero. However, these same principles may also underlie many more dynamic aspects of cortical function as synaptic strength is dynamically adjusted in response to alterations in the properties of the stimulus. One of the best know examples of this is the response of the visual system to alterations in stimulus contrast, a process where the experimental data is often fit by the hyperbolic ratio equation [Albrecht and Hamilton, 1982]. This suggests that this S-shaped nonlinearity serves to adjust output so that, on average, the use of synaptic resources is balanced by the quantity of information available at the output of the neuron. More persistent alterations in stimulus activity should induce a reallocation of the available neural resources which would be reflected in adjustments to the parameters governing the shape of the hyperbolic ratio equation. An example of this type of adjustment may be the lateral shifts in the contrast response function induced by persistent alterations in the level of stimulus contrast [Ohzawa, Sclar, and Freeman, 1985].

ACKNOWLEDGMENTS

Supported in part by EY10915.

REFERENCES

Albrecht, D.G., and Hamilton D.B. Striate cortex of monkey and cat: Contrast response function. J. Neurophysiol. 48:217–237, 1982.

Bienenstock, E.L., Cooper, L.N., and Monroe P.W. Theory for the development of neuron selectivity: orientation specificity and binocular interaction in visual cortex. J. Neurosci. 2:32–48, 1982.

Gottschalk, A., McLean, J., Palmer, L.A. Contrast response in the context of a cortical model which maximally preserves information transfer to higher visual centers. Invest. Opthal. Vis. Sci. 34:909, 1993.

Gottschalk, A. Limits on image representation in early vision. In: Computation in Neurons and Neural Systems, Eeckman, F.H. (ed.), Boston: Kluwer Academic Publishers, 1994.

Gottschalk, A. Maximizing mutual information in a simple model of early vision. Submitted for publication, 1996.

Hebb, D.O. The Organization of Behavior. New York, John Wiley, 1949.

Ohzawa, I., Sclar, G., and Freeman, R.D. Contrast gain control in the cat's visual cortex. J. Neurophysiol. 54:651–667, 1985.

51

AN ASSOCIATIVE MEMORY MODEL WITH PROBABILISTIC SYNAPTIC TRANSMISSION

Bruce Graham and David Willshaw

Centre for Cognitive Science
University of Edinburgh
2 Buccleuch Place
Edinburgh, UK

ABSTRACT

The associative net model of heteroassociative memory with binary-valued synapses has been extended to include recent experimental data that indicates that in the hippocampus one form of synaptic modification is a change in the probability of synaptic transmission [2]. Pattern pairs are stored in the net by a version of the Hebbian learning rule that changes the probability of transmission at synapses where the presynaptic and postsynaptic units are simultaneously active from a low, *base* value to a high, *modified* value. Numerical calculations of the expected recall response have been used to assess the performance for different values of the *base* and *modified* probabilities. If there is a cost incurred with generating the difference between these probabilities, then the optimal difference is around 0.4. Performance can be greatly enhanced by using multiple cue presentations during recall.

1. INTRODUCTION

Information is stored in neural net models of associative memory by alteration of the strengths of the synaptic connections between units according to a local learning rule. These synaptic strengths are used to weight the input signals during pattern recall. Analysis of this type of model is relevant to neurobiological research, where long-term changes in synaptic strength have been suggested as the basis for learning and memory in the brain [1]. The exact nature of these changes is not completely understood and there may well be different synaptic learning rules in different parts of the brain. Recent data suggest that in the hippocampus one form of synaptic modification is a change in the probability of synaptic transmission [2]. We show here how the associative net can be used with this form of synaptic modification to give a stochastic net.

Computational Neuroscience
edited by Bower, Plenum Press, New York, 1997

The associative net model of heteroassociative memory [3] consists of a layer of input units connected to a layer of output units by feedforward connections. In the standard, deterministic net pairs of patterns with binary valued components are stored using a *clipped Hebbian* learning rule whereby the strength of the synaptic connection between an input unit and an output unit that are both active for the same pattern pair is modified from 0 to 1. During recall, a previously stored input pattern is placed on the input units as a cue. An output unit will receive one unit of activation from each active input unit to which it is connected by a modified synapse. A threshold-setting strategy (such as *winners-take-all*) is used to set the activity of the output units on the basis of their summed input activations.

The associative net is unique amongst associative neural network models in having binary-valued synapses. These synapses can be interpreted in terms of probabilities of synaptic transmission. In the standard model, a weight of 0 corresponds to 0% success in transmission, whereas a weight of 1 corresponds to 100% success. In the stochastic net the *clipped Hebbian* learning rule still operates but now synaptic modification involves changing the probability of transmission from a *base* value to a *modified* value. For example, a *base* synapse could have a probability (P_b) of 0.2 of transmitting a signal from an input to an output unit, while the corresponding probability for a *modified* synapse (P_m) could be 0.8. During recall, the summed inputs to an output unit will be affected by the probabilistic transmission through the synapses. Here we investigate the effects of different values of P_b and P_m on the recall properties of this stochastic associative net.

2.   METHODS

We have carried out numerical calculations of pattern recall from a net with $N_A = 8000$ input units and $N_B = 1024$ output units. Pattern pairs consisting of input patterns each containing $M_A = 240$ active input units and output patterns each containing $M_B = 30$ active output units were stored in the net. Recall of an output pattern was achieved by using it's associated input pattern as a cue and a *winners-take-all (WTA)* thresholding strategy. This involves each output unit measuring the weighted sum of its inputs *(dendritic sum)*. A unit's *dendritic sum* is different each time a cue is presented due to the probabilistic transmission. The same threshold is applied to all output units. It is lowered until M_B output units have a *dendritic sum* that is greater than or equal to the threshold. These units are made active and constitute the recalled pattern.

Recall performance is measured by pattern overlap and net capacity. Overlap is the correlation between the recalled pattern and the original stored pattern (1=perfect recall). Capacity is the maximum number of pattern pairs that can be stored so that all output patterns can be recalled reliably (at most one unit in error). With deterministic transmission ($P_b = 0$ and $P_m = 1$), this net has a capacity of 3576 pattern pairs.

The main point of interest is the effect of different values of P_b and P_m on recall performance. In particular, the same difference in probabilities can be achieved for different values of P_b and P_m. Three schemes for setting the probability difference were investigated: *Low*: P_b varied, P_m fixed at 0.9; *High*: P_b fixed at 0.1, P_m varied; *Both*: both P_b and P_m varied, equidistant from 0.5.

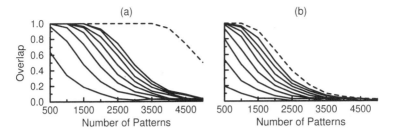

Figure 1. Recall overlap as a function of the number of stored pattern pairs for (a) a fully connected and (b) a partially connected net. The dashed line in each graph is the deterministic net. The solid lines have $P_m = 0.9$, with P_b from 0.1 to 0.8 in increments of 0.1, from right to left.

3. RESULTS

3.1. Recall overlap

When this stochastic associative net is fully connected, the only source of noise during recall, apart from the other stored patterns, is the probabilistic transmission at the synapses. Even for a large difference in probabilities (e.g. $P_b = 0.1$, $P_m = 0.9$), recall performance is considerably impaired compared to the deterministic net. Even so, many patterns can be stored and recalled without error. Examples of recall overlap when $P_m = 0.9$ and P_b is varied are shown in Figure 1.

Partial connectivity introduces another noise component during recall. The results shown here are when each output unit receives connections from a random selection of 60% of the input units. In this situation the performance of the deterministic net is degraded and is not significantly better than the stochastic net for high probability differences. This is shown by Figure 1(b).

3.2. Capacity

Figure 2 compares the capacity obtained at different probability differences, when the differences are set using the three schemes given above. For the fully connected net, consistently better performance is achieved when P_m is high (the *Low* scheme) (Fig. 2(a)). An optimal probability difference can be obtained by assuming there is a cost incurred that is proportional to the difference. The optimum can be found by considering the relative capacity, which is the net capacity divided by the probability difference. For the fully connected net using the *Low* scheme, the optimal difference is around 0.4 (Fig. 2(b)).

The capacity and relative capacity of the partially connected net are shown in Figure 2(c)&(d). Now the *High* scheme (low P_b) consistently provides a higher capacity than the other methods. The optimal probability difference is also now higher than for the fully connected net, being around 0.5 for the *High* scheme (Fig. 2(d)).

3.3. Multiple cueing

Performance can be improved by presenting the input cue many times so that each output unit can calculate an average *dendritic sum* before the WTA threshold is applied. Figure 3(a)

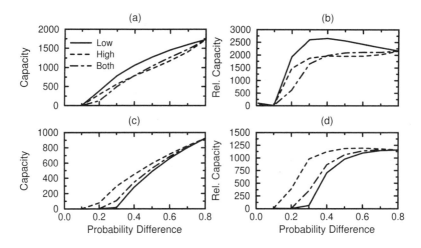

Figure 2. Capacity and relative capacity as functions of probability difference. (a),(b) Fully connected. (c),(d) Partially connected.

shows the effect of multiple cueing on pattern overlap. Near optimum performance is obtained after 100 cue presentations.

The effect of multiple cueing on net capacity is shown in Figure 3(b). Capacity is highly dependent on the probability difference for only one cue presentation. The capacity for a probability difference of 0.8 (1729) is 4.5 times the capacity for a difference of 0.2 (387) after one cue presentation. However, after 1000 cue presentations the capacity at 0.8 (3204) is only 1.1 times the capacity at 0.2 (2964). Even a difference of 0.1 provides a large capacity if cues are presented 1000 times. If there is a cost associated with each cue presentation, then five presentations is optimal for a difference of 0.1. A single presentation is optimal for greater differences.

3.4. Input Coding Rate

The effect of different input coding rates on net capacity is shown by Figure 4. Altering the input coding rate, M_A, alters the optimal probability difference. Higher values of M_A are less sensitive to the difference and so have a lower optimum. Lower values are more sensitive due to the smaller sample size these input patterns provide to output units during recall.

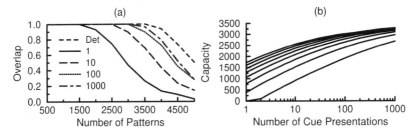

Figure 3. Multiple cueing with the fully connected net. (a) Recall overlap when the input cue is presented 1, 10, 100 or 1000 times ($P_b = 0.1$ and $P_m = 0.9$). (b) Capacity as a function of the number of cue presentations for the *Low* probability difference scheme ($P_m = 0.9$; P_b ranges from 0.1 to 0.8 going from top to bottom).

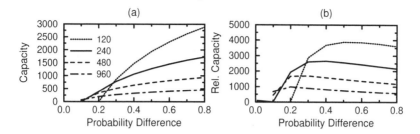

Figure 4. Capacity and relative capacity for one cue presentation to a fully connected net with input coding rates, M_A, of 120, 240, 480 and 960.

4. SUMMARY

An associative net with probabilistic transmission functions well as a memory device. The optimum difference between the *base* and *modified* probabilities depends on net connectivity and pattern coding rate, but is generally small (≈ 0.4). Coincidentally, this corresponds to the magnitude of change seen experimentally [2]. Multiple cue presentation allows high performance with very small probability differences.

ACKNOWLEDGMENTS

We acknowledge the MRC for financial support under program grant PG 9119632

REFERENCES

[1] Bliss, T.V.P. and Collingridge, G.L., *Nature*, 361:31-39, 1993.
[2] Stevens, C.F. and Wang, Y., *Nature*, 31:704-707, 1994.
[3] Willshaw, D.J., Buneman, O.P. and Longuet-Higgins, H.C., *Nature*, 222:960-962, 1969.

52

GUSSTO: A NEURAL NETWORK MODEL OF GUSTATORY PROCESSING IN THE RAT NTS AND PBN

Frank W. Grasso[1,2*] and Patricia M. Di Lorenzo[2]

[1]Boston University Marine Program
MBL
Woods Hole, Massachusetts 02543
fgrasso@hoh.mbl.edu
[2]Department of Psychology
P.O. Box 6000, SUNY Binghamton
Binghamton, New York 13902
diloren@bingvmb.cc.binghamton.edu

0. ABSTRACT

The GUSSTO model is presented to explain the earliest stages in rat gustation. The model incorporates a diverse population of receptor cell types which supply input to two serial recurrent networks of model neurons representing the Nucleus and Tractus Solitarius (NTS) and the Para-Brachial Nucleus of the pons (PbN) in the brain stem. The data compression imposed by this architecture (many receptors to few principal neurons) uses the dynamics of recurrent connections in the NTS and PbN to produce a hierarchical and distributed representation of taste quality and intensity. In GUSSTO the NTS and PbN layers possess distinct representations that emphasize different aspects of gustatory quality. The model dynamics suggest subsequent processing levels may receive the information made explicit in both the NTS and PbN representations from the PbN output alone. Single model neurons in each layer show best stimuli and the population a range of tuning broadness across model neurons comparable to their biological counterparts. The GUSSTO model provides a distributed neural mechanism that can account for these and other salient single cell properties in the NTS and PbN while the population of each layer encodes properties quality and intensity of taste stimuli.

* To whom correspondence should be addressed (At the MBL).

1. INTRODUCTION

In all mammalian sensory systems the building blocks of perception are organized from raw stimulus energy by neural machinery somewhere between the receptors and higher brain centers. The Gustatory Unit Stimulus Specific Topographical Organizer (GUSSTO) model suggests that in taste the mechanism by which this may be accomplished is a distributed dynamic data compression from a diverse set of chemo-receptors. The construction of the model and the results contrast with the usual independent channel (non-recurrent) or labeled line theories of gustatory brainstem mechanisms (Frank and Pfaffmann 1969, Scott and Giza 1990, Nagai et. al. 1992). In these independent-channel models the diversity of response profiles at each level of neural processing represents noise that must be suppressed for signal clarity whereas in GUSSTO the diversity of response profiles represents a central feature of the neural code of stimulus quality and quantity.

GUSSTO models the rat gustatory brainstem and peripheral receptors. The model focuses on the roles of two gustatory brainstem nuclei; the gustatory portion of Nucleus of the Solitary Tract (NTS) and the Parabrachial Nucleus of the pons (PbN). The NTS receives the primary receptor input from the anterior two thirds of the tongue via the corda tympanii nerve. The two nuclei are anatomically connected in series there is no evidence of feedback from the PbN to the NTS (Travers 1993, Travers et. al. 1994). From the PbN taste information proceeds via the thalamus to the cortex which makes the NTS and PbN a likely sites for the construction of taste primitives. We have neurophysiological (Frank and Pfaffmann 1969, Doetsch and Erickson 1970, Di Lorenzo and Hecht 1993), and neuroantomical (Travers 1993, Travers et. al. 1994) data to constrain models of both these structures. The diversity of receptor types and recurrent network structure in GUSSTO permits the emergence of dynamic model neurons with complex response profiles which can be directly and quantitatively compared to response properties reported in neurophysiological studies of brainstem gustatory neurons.

The challenge we set for ourselves with GUSSTO was two fold. The first was to determine the extent to which a distributed model of the rat gustatory system could account for general issues of quality and intensity coding within the constraints of networks of model neurons scaled to the size of the real NTS and PbN and an input which statistically captures the character of the NTS's corda tympanii input. The second was to determine if known response properties of real single neurons in the NTS and PbN could be mirrored by GUSSTO network neurons in a network that was capable of coding taste quality and intensity.

2. METHODS

2.1 Model Description

Figure 1 shows a schematic of the network architecture. The model consists of a sheet of 4032 "receptor cells" which are each sensitive to a unique combination of levels of the four primary tastants (Salt, Sweet, Bitter and Acid). The proportions of these sensitivities across the input in the model match the known composition of the receptor fibers from the corda tympanii nerve: the input to the model NTS and PbN layers captures the broad statistical composition of the receptor sheet as a whole using values of receptor sensitivity taken from the rat gustatory neuorphysiology literature. The input itself represents

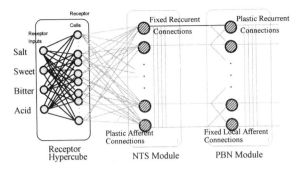

GUSSTO Neural Network Schematic

Figure 1. GUSSTO network architecture. Inputs to the net are four floating point numbers representing the four primary tastants. These are represented in the activity of a population of receptor cells each of which gives a unique response to the partiular input combination. The recptor cells connect to all NTS module element which compete in weight space to represent the input space through recurrent connections. The tatse-space mapping in the NTS layer is topographically projected onto the PbN layer with fan out. Reccurent inhibitory connections in the PbN layer produce a different representation of the taste space. (See text for details).

(molar) levels of the four tastants (as four floating point numbers) present on "the tongue" at any given time step. (The model could have been generalized to a greater number of primary tastants but we choose the four "classic" tastants in order to obtain results that would be directly comparable to the neurophysiological literature.). The NTS and PbN layers are represented as Kohonen-like recurrent neural nets (Kohonen, 1983) of excitatory projection neurons and local inhibitory interneurons (225 model neurons of each kind in each layer). The local connectivity of these model neurons is scaled to match that of the rat NTS and PbN (radii of 1/5 the layer). Each neuron is modeled as an integrate and fire element with a threshold. Outputs of the principal neurons in each layer (as are the model receptor cells) are floating point numbers representing a frequency code. Detailed biophysical models of the receptor cells (which have been described in the extensive literature) have not been incorporated to keep the model simple. The responses of the network elements evolve in time incorporating delayed signals from previous inputs carried by the recurrent connections. The weights of the connections between the receptor sheet and the NTS and the intrinsic connections of the PbN are initially random but plastic. They self organize to stable maps in response to inputs according to a Hebbian rule synaptic modification rule (see the next section).

2.2 Model Training

Each NTS element receives input all receptor elements. These connections are plastic and modifiable by a Hebbian rule. The lateral connections in the NTS are of fixed strength. In the PbN layer the afferent (excitatory) inputs from the NTS are of fixed strength and the inhibitory lateral connections within the layer are plastic and modified by and anti-Hebbian rule. During training the responses to a given input were allowed to evolve in the net for 32 time steps, after which the respective Hebbian or anit-Hebbian rule was applied. All weights on a given element were then normalized to unity before a response to a new input was computed. The training set consisted of all combinations of the primary tastants in log-step proportions to allow the networks to form representations of all the primary tastants and all their mixtures in different proportions. This was continued until the responses of the network to the four primary tastants at 0.0001, 0.001 and 0.01 M concentrations did not change between three consecutive presentations of the training set. This training was intended as a method to arrive at a mature connectivity and not as a model for the development of the rat gustatory system.

3. RESULTS

Such "mature" nets were tested for their ability to code the quality and quantity of the stimulus present at the input. Across element patterns of activation were compared for pure tastants and mixtures at concentrations varying over 4 log units. The similarity of these response patterns were compared using various standard measures of similarity (traditional cluster analysis and multi dimensional scaling). Dendrograms of the similarity showed that mature NTS and PbN nets were competent to code both stimulus intensity and quality for the primary tastants in a way that was comparable to taste cell neurophysiology and rat psychophysics. The clusters formed segragated by quality first with intensity coded within a quality. Sweet was most different from the other qualities and salt and bitter, while distinct, were more similar to one another than to acid.

This discrimination was most pronounced in the NTS layer while in the PbN layer mixtures of these primaries were brought into sharper contrast. That this capacity to code in both layers was due to the network dynamics is demonstrated by the observation that switching off the lateral inhibitory connections (reducing the net to a pure feed-forward net) abolished the capacity to code primary tastant quality and intensity in both layers. Further, the capacity to code quality and intensity occurred in arbitrary subsets of network neurons (in either layer) which decayed with the size of the set.

The response profiles of individual NTS and PbN model neurons showed dynamics and quality/intensity profiles that were qualitatively similar to those from neurophysiological recordings of the NTS and PbN. Most model neurons had a "best" stimulus drawn from the primary tastants and model neurons with mixtures as their best stimuli were rare (<2%). The breath of tuning (Smith and Travers 1979) in the entire model NTS population was unimodal with mean indicating narrow tuning comparable to that of measured rat NTS populations.

We looked at the correlation between the NTS and PbN across element response patterns during the evolution of a response. We noted that early in the response the NTS and PbN arrays were correlated (point-to-point; on average r=0.3) and that this correlation dropped monotonically later in the response. This correlation is not surprising given the topographical mapping of the NTS to PbN connection in the GUSSTO model. The loss of this correlation later in the response however has functional implications for taste coding which will be taken up in the discussion.

4. DISCUSSION

The power of GUSSTO to explain single cell response profiles and global taste coding points out recurrent neural networks as a class of models which might lead to a deeper understanding the neural basis of early taste processing. Its formulation, which allows direct comparison of model cell responses to neurophsyiological data, suggests that much productive work can still be done within this framework. By refining the model specification to become more 'biological' in detail, neurphysiologically verifiable experiments which inform our understanding the global coding of taste are suggested and the validity of the model itself can be directly tested; if the ability to code taste quality and intensity is lost by the inclusion of a biological constraint the model is wrong.

In the neurophysiology of taste coding the model supports a new perspective for the role of the single neuron. In contrast to independent channel models GUSSTO explains the breadth of response profiles observed in real NTS and PbN neurons as an integral part of the coding process, not as noise. As with all distributed models each element contributes to the network response and the building blocks of perception become a continuum of across neuron patterns of activation in each layer.

The model is able to reliably code both stimulus quality and intensity over 4 log units in spite of an anatomically accurate 140 fold convergence between the receptor and NTS layers. It does this by forming maps of the entire taste-space (as represented in the receptor layer) in the NTS and PBN which capture the variability in the training set to the limit determined by the (fixed) number and resolution of the elements. This limit is much smaller than the size of the input space which raises the question of why there should be so much convergence from the corda tympanii nerve to the NTS. The explanation suggested by our model is that by keeping the diverse information available in the periphery the rat can learn to "tune into" novel tastes at the level of the NTS and PbN that come to posses biological significance. The animal could for example learn to discriminate finer gradations of "salty mixtures" while loosing sensitivity in one or more of the other primary tastants by adjusting (but not erasing) its NTS and PbN taste mappings. This explanation of peripheral receptor diversity seems more appealing than a noisy design suggested by labeled line theories.

Given the serial nature of the NTS-PbN connections and the absence of feedback it has been traditional to view the NTS as a "relay" rather than an active processing station. Our simulation results which show the correlation of the NTS and PbN activity early in the response followed by a phase of near-zero correlation coupled with the relative specialization of the NTS and PbN for 'pure tastants' and 'mixtures' respectively suggests an alternative view of the role of the NTS. Given that the axons of the real PbN principal neurons are the path for the information contained in the NTS response to reach the thalamus and higher centers it is tempting to suggest a temporal segregation of the NTS and PbN codes. That is, early in the response the PbN axons might carry, as in the GUSSTO model, the NTS primary tastant across fiber code and then later, after the PbN local circuitry has had its effect, the PbN mixture specific across fiber code. At present the early correlation is low and we do not want to overemphasize this speculation as it pertains to the gustatory system. As a temporal coding mechanism however it is interesting in itself and we plan future simulation studies which will investigate the possibility of increasing this early correlation while maintaining the PbN layer's ability to produce a distinct code which represents different features of the same input information at different times.

REFERENCES

Di Lorenzo, P. and G. Hecht (1993) Perceptual Consequences of Electrical Stimulation in the Gustatory System. Behavioral Neurosci. 107:130–138.

Doetsch, G. and R. Erickson (1970) Synaptic Processing of Taste-Quality Information in the Nucleus Tractus Solitarius of the Rat. J. Neurophys. 33:490–497.

Frank, M. and C. Pfaffmann (1969), Taste Nerve Fibers: A Random Distribution of Sensitivites to Four Tastes. Science 164:1183–1185.

Kohonen, T. (1983) Self Organized Formation of Topologically Correct Feature Maps. Biol. Cybern. 43:50–69.

Nagai, T., Y. Taakashi, H. Katayama, M Adachi and K. Aihara (1992) A novel method to analyze response patterns of taste neurons by artificial neural networks. NeuroReport 3:745–748.

Scott, T. and B. Giza (1990) Coding Channels in the Rat Taste System. Science 249:1585–1587.

Smith D. and J. Travers (1979) A metric for the breath of tuning of gustatory neurons. Chem. Senses Flavor 4:215–229.

Travers, S. (1993) Orosensory Processing in Neural Systems of the Nucleus of the Solitary Tract. In S. Simon and S. Roper (Eds.) Mechanisms of Taste Transduction, CRC Press, Raton, pp. 339–394.

Travers S, D. Becker, C. Halsell, M. Harrer, and J. Travers (1994) Functional Organization of the Orally-Responsive NST. In: Proceedings of the 11th International Symposium on Taste and Olfaction, K Kurihara, N Suzuki, H Ogawa (Eds). Springer-Verlag, Tokyo, pp. 396–401.

53

SINGLE CELL AND POPULATION ACTIVITY IN A STATISTICAL MODEL OF THE HIPPOCAMPAL CA3 REGION*

T. Gröbler and G. Barna

Department of Biophysics
KFKI Research Institute for Particle and Nuclear Physics
Hungarian Academy of Sciences
P.O. Box 49, H–1525
Budapest, Hungary
E-mail: grobler@sunserv.kfki.hu

ABSTRACT

A statistical model is given to describe the electrical activity patterns of large neural populations of the hippocampal CA3 region. The population model incorporates basic electrophysiological properties of hippocampal pyramidal and inhibitory neurons. Population activities as well as underlying single cell voltages are simulated during normal and epileptiform activities in the CA3 region of the hippocampus. It is demonstrated that our model can reproduce electrophysiological phenomena characteristic to both single cell and population activities. Specifically, the intrinsic burst response of individual pyramidal cells to injected currents, fully synchronized population bursts, sustained multiple population bursts, synchronized synaptic potentials, and low amplitude population oscillation were obtained.

1. INTRODUCTION

Modeling the function of large neural populations is becoming extremely difficult as more and more anatomical, biochemical, biophysical, and electrophyisiological data are available about the behavior of individual neurons and their synaptic connections. Current mathematical models of neural networks either lack the biological details of these mechanisms or are unable to treat more than a few hundreds of neurons [8].

*This work was supported by the OTKA grant F014020, by the exchange program between the Hungarian Academy of Sciences and the Consiglio Nazionale delle Ricerche (Italy), and by the Fogarty International Research Collaboration Award, HHS Grant No. 1 R03 TW00485-01.

Both experiments and theoretical studies suggest the existence of a universal synchronization mechanism in the hippocampal CA3 region. Synaptic inhibition regulates the spread of firing of pyramidal neurons. Collective network properties have been studied successfully by Traub and Miles [8, 9]: they found a few different epileptic phenomena such as synchronized bursts, synchronized multiple bursts and seizure-like events. At increasing levels of inhibition, different phenomena have been obtained, ranging even to normal, non-epileptic activity such as synchronized synaptic potentials (SSP).

In our study, a statistical model, based on the kinetic population model by Ventriglia [11, 12, 13, 14, 15], is given to describe the above phenomena of the hippocampal CA3 region. The approach proposed here is a statistical model which considers the probability distribution of the possible states of *any number* of neurons instead of treating every individual neuron separately. Thus the model uses the notion of neural continuum instead of a system of separate neurons to describe the interactions within and between neural populations. Interactions take place between infinitesimal parts of the neural continuum rather than between individual neurons. The spatial localization of neurons is preserved and their electrophysiological properties can be arbitrarily detailed. The model is described in [2].

The skeleton population model has been made *scalable* in order to describe both global behavior of large populations and local phenomena. In other words, the continuous model is discretized in such a way that the first and second moments of the p.d.f's are preserved even at different discretization units. A detailed mathematical discussion of the discretization method is presented elsewhere [4].

The size, connection patterns, and other anatomical details of the model are taken from available data on rat hippocampus. Electrophysiological properties of the cells should also be incorporated into the model. It is also required that single cell activities should be observable during the simulations. Thus the model has been extended to describe characteristic ionic currents and bursting properties of hippocampal pyramidal cells. Moreover, individual cell activities can also be traced by running a single cell model in the environment of the surrounding population.

Computer simulations were run to model rhythmic activity in the CA3 region of the hippocampal slice. Preliminary results were presented in [3] and [2]. Here it is demonstrated that our model can reproduce the electrophysiological phenomena characteristic to both population and single cell activity.

2. BASIC MODEL PROPERTIES

The model describes neural population activity in terms of two variables: the probability density functions (p.d.f's) of (i) the neurons and (ii) spikes travelling between the neurons. The state space consists of the two-dimensional space coordinate (for both neurons and spikes), a subthreshold membrane potential coordinate (for neurons), and an intracellular calcium-concentration coordinate (for pyramidal neurons only). Stochastic effects are taken into account by additive noise terms, expressing diffusion of neurons over the membrane potential and diffusion of spikes in space.

The neural continuum has the following properties: The different neural populations (pyramidal and inhibitory cells) have their own neuronal fields which can interact through the emission and absorption of spikes (action potentials). The neurons are fixed in space while spikes can freely travel among them. The change of subthreshold membrane potential of neurons is determined by the post-synaptic potentials caused by absorbed spikes, and by different ionic currents. While inhibitory neurons only have leakage channels, pyramidal cells also contain Ca-, and Ca-dependent K-channels, responsible for burst generation and afterhyperpolarization. Since firing is governed by a separate mechanism, Na-, and delayed rectifier K-channels are not included. Instead a dynamic threshold mechanism, related to Na-conductance, determines the proportion of neurons that will fire. Firing is followed by a refractory period after which a calcium concentration dependent return potential is restored and intracellular calcium concentration is increased. Stochastic aspects of spike spreading and of membrane potential changes are accounted for by a diffusion process. The qualitative and quantitative details of the above properties are taken from anatomical [1, 7, 6, 5] and electrophysiological [10, 8, and references therein] data on the CA3 region of rat hippocampus. The model equations are presented elsewhere [3, 2].

3. SIMULATIONS

The response of CA3 pyramidal cells to injected currents, namely *intrinsic burst* discharges, were reproduced by the single cell model [3, 2]. Some characteristic features of the physiological responses reproduced there were (i) summation of spike afterdepolarization due to the increase of intracellular Ca-concentration; (ii) Ca-spike; (iii) intrinsic burst followed by a long (1000 msec) after-hyperpolarization (AHP); (iv) the ability to prevent full burst generation by properly timed hyperpolarizing input.

The total suppression of inhibition leads to synchronized population activity in consequence of the excitatory interactions between pyramidal cells, as comparative experiments and simulations with networks also demonstrate [8, 9]. In our population model, *fully synchronized bursts* without inhibition, and their propagation in space, were also demonstrated [3, 2].

The role of inhibitory and excitatory interactions is studied in the simulations presented here. The full model contains three cell types: pyramidal cells and two types of inhibitory cells, one producing fast IPSP's, and the other producing slow IPSP's. A range of epileptiform and non-epileptic rhythms have been obtained. For classification of these behaviors, the measure of synchronization is defined as the percentage of simultaneously (within 3 ms) firing pyramidal cells. The underlying single cell activities can be studied by collecting the values of synaptic inputs from the simulation of the population model and then running the single cell model using this data as 'average synaptic input' of the subpopulation containing the given cell. Thus the model also offers the possibility to follow the activity of an 'average cell' at any point of the continuum. In the figures, the percentage of firing cells and the voltage trace of the average cell is shown for the different phenomena.

When excitatory interactions between pyramidal cells are sufficiently strong, minimal inhibition leads to *synchronized multiple bursts* (Fig. 1).If excitatory synapses are weaker, the afterbursts disappear and synchronized bursts can be observed (see [3, 2]).

Increasing the strength of fast inhibitory synapses, the synchronization of pyramidal cell bursts is gradually abolished. Partially synchronized bursts may correspond to physiological sharp waves *in vivo* [9]. If the degree of synchronization is low, individual cells very rarely produce a burst and *synchronized synaptic potentials (SSP)* can be observed (Fig. 2).

Further strengthening of inhibitory connections abolishes pyramidal cell firing see Fig. 4/C2 in [9]. Adding, however, a strong bias current to pyramidal cells, low amplitude population oscillation emerges (Fig. 3). The frequency of this oscillation is about a magnitude higher than that of SSP. Individual pyramidal cells rarely and irregularly fire single action potentials, no bursts can be found.

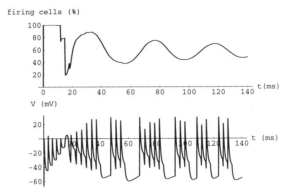

Figure 1. Synchronized multiple burst in the population model (cf. [8, Fig. 6.9A]).

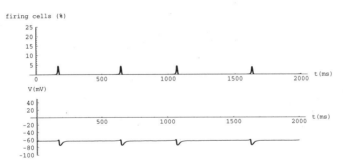

Figure 2. Synchronized synaptic potentials (cf. [9, Fig. 4/C1]).

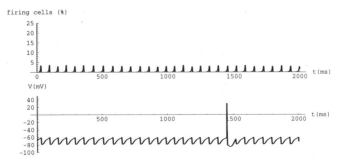

Figure 3. Low amplitude population oscillation (cf. [9, Fig. 4/A2]).

4. DISCUSSION

We can conclude that the statistical model presented here is capable of reproducing a range of epileptiform and non-epileptic population rhythms of the CA3 region of the hippocampus. The synaptic strengths of excitatory and inhibitory synapses have been varied. The degree of pyramidal cell synchronization has been studied while it was also possible to monitor single cell activity underlying population behavior. Phenomena that were reproduced included the intrinsic burst response of individual pyramidal cells to injected currents, fully synchronized population bursts, sustained multiple population bursts, synchronized synaptic potentials, and low amplitude population oscillation.

Since the model allows for any extension of both the channel properties of single cells and the connection patterns within and between different populations, more complex phenomena such as population activity during physiological theta and sharp waves or a learning mechanism can also be incorporated.

ACKNOWLEDGMENTS

The authors thank Péter Érdi, Francesco Ventriglia, János Tóth, and Attila Gulyás for many fruitful discussions.

REFERENCES

[1] D. G. Amaral, N. Ishizuka, and B. Claiborne. Neurons, numbers, and the hippocampal network. In Storm-Mathsien, J. Zimmer, and O. P. Ottersen, editors, *Progress in Brain Research*, volume 83, pages 1–11. Elsevier Sci. Publ., 1990.

[2] P. Érdi. Rhythmogenesis in single cells and population models: Olfactory bulb and hippocampus. *BioSystems*, 1997.

[3] T. Gröbler and G. Barna. A statistical model of the CA3 region of the hippocampus. In R. Trappl, editor, *Cybernetics and Systems '96*, pages 503–507. Austrian Society for Cybernetic Studies, Vienna, 1996.

[4] T. Gröbler, G. Barna, and P. Érdi. Modelling rhythmic activity in the CA3 slice: II. Statistical population model. *Biological Cybernetics* (submitted).

[5] A. I. Gulyás, personal communication.

[6] A. I. Gulyás, R. Miles, N. Hájos, and T. F. Freund. Precision and variablity in postsynaptic target selection of inhibitory cells in the hippocampal CA3 region. *Eur. J. Neurosci.*, 5:1729–1751, 1993.

[7] A. Sík, N. Tamamaki, and T. F. Freund. Complete axon arborization of a single CA3 pyramidal cell in the rat hippocampus, and its relationship with postsynaptic parvalbumin-containing interneurons. *Eur. J. Neurosci.*, 5:1719–1728, 1993.

[8] R. D. Traub and R. Miles. *Neuronal Networks of the Hippocampus*. Cambridge University Press, New York, 1991.

[9] R. D. Traub and R. Miles. Modeling hippocampal circuitry using data from whole cell patch clamp and dual intracellular recordings in vitro. *Seminars in The Neurosciences*, 4:27–36, 1992.

[10] R. D. Traub, R. K. S. Wong, R. Miles, and H. Michelson. A model of a CA3 hippocampal pyramidal neuron incorporating voltage-clamp data on intrinsic conductances. *J. Neurophysiol.*, 66:635–650, 1991.

[11] F. Ventriglia. Kinetic approach to neural systems I. *Bull. Math. Biol.*, 36:534–544, 1974.

[12] F. Ventriglia. Kinetic theory of hot neural systems. *Cybernetics and Systems*, 18:147–155, 1987.

[13] F. Ventriglia. Computational simulation of activity of cortical-like neural systems. *Bull. Math. Biol.*, 50:143–185, 1988.

[14] F. Ventriglia. Towards a kinetic theory of cortical-like neural fields. In F. Ventriglia, editor, *Neural Modeling and Neural Networks*, pages 217–249. Pergamon Press, 1994.

[15] F. Ventriglia and P. Érdi. Statistical approach to the dynamics of cerebral cortex: learning aspects. In J. Rose, editor, *Cybernetics and Systems: The Way Ahead*, volume 1, pages 443–447. Thales, Lytham St. Annes, 1987.

54

A MODEL OF MONOCULAR CELL DEVELOPMENT BY COMPETITION FOR TROPHIC FACTOR

Anthony E. Harris,[1] G. Bard Ermentrout,[2] and Steven L. Small[1]

[1]Intelligent Systems Program
Center for the Neural Basis of Cognition
Department of Neurology
University of Pittsburgh
Pittsburgh, PA
[2]Center for the Neural Basis of Cognition
Department of Mathematics and Statistics
University of Pittsburgh
Pittsburgh, PA

ABSTRACT

Recent experimental evidence has shown that application of certain neurotrophic factors (NTs) to the developing primary visual cortex prevents the development of ocular dominance (OD) columns. One interpretation of this result is that afferents from the lateral geniculate nucleus (LGN) compete for postsynaptic trophic factor in an activity dependent manner. Application of excess trophic factor eliminates this competition, thereby preventing monocular cell development. We present a model of monocular cell development, incorporating Hebb-like synaptic modification and activity-driven competition for NT, which accounts for the following results: 1) monocular cells form normally when available NT is below a critical amount, 2) monocular cells form in the presence of positive inter-eye correlations, while being entirely self-normalizing in that no normalization of synaptic strengths is necessary to enforce the competition, and 3) monocular cells are prevented in a local neighborhood in which excess NT has been added. The model integrates several disparate neurobiological findings into a cohesive framework, and makes predictions concerning the quantitative dependence of monocular cell development on trophic factor availability.

Computational Neuroscience
edited by Bower, Plenum Press, New York, 1997

1. INTRODUCTION

Several lines of experimental data ([7], [8]) and theoretical modeling [3] suggest that ocular dominance (OD) column development in primary visual cortex depends on activity-dependent competition among axons from the lateral geniculate nucleus (LGN). Recent experimental evidence has shown that the infusion of excess neurotrophins (NTs) prevents OD column development [2]. It has been suggested that LGN axons compete for NT, a competition eliminated by exogenous NT application.

We present a model of monocular cell development incorporating Hebb-like synaptic modification and competition among afferents for NT. The model has three essential parts: First, synaptic strengths increase in part due to Hebb-like long-term potentiation (LTP), and decrease due to heterosynaptic long-term depression (LTD). Second, positive feedback exists between the rate of synaptic strength increase and the rate of trophic factor uptake. Finally, afferents compete for a limited amount of trophic factor. The model accounts for the prevention of monocular cell development with excess NT. In addition, it accounts for monocular cell development with positive inter-eye correlations without any weight normalization procedures. We first present the equations governing synaptic strength development. Next we present phase-plane diagrams showing spontaneous symmetry-breaking of the afferent weights when NT is low and prevention of segregation when NT is high. We give a bifurcation diagram demonstrating monocular cell development with a limited NT supply, but binocular cell development above a critical amount. Finally we discuss explicit predictions of the model.

2. MODEL

The model studies a single, cortical neuron receiving inputs from each eye. The neurons in the model are standard linear units:

$$ev = \sum_j w_j^r a_j^r + w_j^l a_j^l \tag{1}$$

where v is post-synaptic voltage, $w_j^{r,l}$, $a_j^{r,l}$ are respectively, synaptic strength from and activity of the jth input from the right or left eye. In all that follows, $j = 1$. We formulate the equations in terms of mass-action kinetics for the input from the right eye (with identical equations for the left eye but with $r's$ and $l's$ interchanged). There is a fixed pool of trophic factor distributed over all inputs, and each connection has a fixed amount of material from which to add connection strength (which we set to 1 for all synapses):

$$N^f + (n^r + n^l) = N \tag{2}$$

$$w^r + f^r = 1 \tag{3}$$

where the total amount of trophic factor available to be distributed over inputs is N, the amount currently taken up by the synapse is $n^{r,l}$, and N^f is the free NT left (in eqn. 2). In eqn. 3, w^r is the current amount of connection strength, and f^r is the free store of synaptic raw material at that synapse.

The free synaptic material is converted to connection strength by the simple kinetic scheme:

$$f^r = (1-w^r) \quad \overset{K^+(n^r,v,a^r)}{\underset{K^-(v)}{\rightleftharpoons}} \quad w^r \tag{4}$$

The rate constants depend on the potential, the inputs, and the current amount of synaptic NT. The NT obeys a simple kinetic equation as well:

$$N^f = N - (n^r + n^l) \quad \overset{w^r}{\underset{\beta_2}{\rightleftharpoons}} \quad n^r \tag{5}$$

so that the stronger the weight, the faster the uptake rate of free NT. Note that the rate constants $K^\pm(\cdot)$ and w_r are either functions or dynamic variables. The forward rate constant for increase of connection strength is a simple Hebbian rule, modulated by the current amount of NT:

$$K^+(n^r,v,a^r) = n^r v a^r \tag{6}$$

This dependence of LTP on trophic factor is consistent with recent experimental work [4]. Using the definition of v from equation (1), and averaging over the input ensemble (see [5]), we obtain:

$$K^+(n^r,v,a^r) = n^r(C^{rr}w^r + C^{rl}w^l) \tag{7}$$

where $C^{rr} = \langle a^r a^r \rangle$ is the intraeye correlation, and $C^{rl} = \langle a^r a^l \rangle$ is the intereye correlation. Note that all correlations are positive. The backward rate constant depends only on the post-synaptic potential v, thus being an implementation of heterosynaptic LTD [1]:

$$K^-(v) = \beta_1 v = \beta_1(w^r a^r + w^l a^l) = \beta_1(w^r + w^l) \tag{8}$$

after averaging, with all constants absorbed into β_1. Substituting the above into the equations describing mass-action kinetics in (4) and (5), we obtain:

$$\dot{w}^r = n^r(C^{rr}w^r + C^{rl}w^l)(1-w^r) - \beta_1(w^r + w^l)w^r \tag{9}$$

$$\dot{n}^r = (N - (n^r + n^l))w^r - \beta_2 n^r \tag{10}$$

(and identical equations for the left eye with $r's$ and $l's$ interchanged).

3. RESULTS

We first show two phase-plane diagrams of the weight space trajectories. In Figure 1, the available NT is set low; in Figure 2, NT is above a critical amount. In Figure 1, two representative trajectories are shown. Beginning from random initial conditions of weights and trophic factor, the trajectories converge to the asymmetric state where one weight value (i.e., from one eye) is high and the other is low, corresponding to a monocular cell. Note also that, in contrast to a monocular cell developing under a Hebbian rule with subtractive normalization

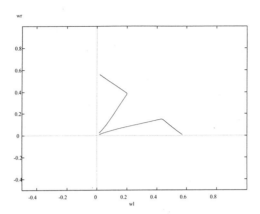

Figure 1. Trajectories from 2 random initial conditions with a low amount of NT ($N = 1.0$). Other parameters are: $C^{rr} = C^{ll} = 0.9$, $C^{lr} = C^{rl} = 0.2$, $\beta_1 = 0.6$, $\beta_2 = 0.05$. Weights from each eye grow at a similar rate at first, then one eye comes to dominate over the other, resulting in a monocular cell.

(necessary to generate a monocular cell with positive inter-eye correlations [6]), the weights are not pinned to their minimal or maximal values. Instead, the final weights depend on parameter values, in particular, the correlations between the two eyes. Thus, our model makes the more plausible prediction that final weight values depend on the correlational structure of the inputs, i.e., that synaptic weights vary continuously with changes in the environment. In Figure 2, when there is a high amount of NT present, trajectories from random initial conditions converge to a symmetric steady state, (or a binocular cell), in qualitative agreement with the results in [2].

Next, we present the bifurcation diagram of one weight as a function of the fixed supply of NT (Figure 3). For low values of NT, two stable states are present, one high and the other low. As NT is increased, a new branch appears, corresponding to a midrange weight value of a symmetric steady-state. Two asymmetric unstable branches separate this symmetric steady-state from two asymmetric, stable branches. For realistic initial conditions (weights small and almost equal), the trajectories will never end in the basins of attraction of these asymmetric branches. This is consistent with the computational experiments suggesting that for large NT

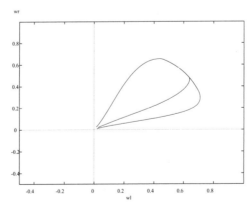

Figure 2. Trajectories from 3 random initial conditions with an excessive amount of NT ($N = 3.0$). Remaining parameters are same as before. Weights from each eye converge to the same value, resulting in a binocular cell.

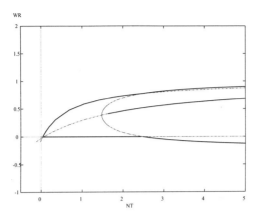

Figure 3. Bifurcation diagram of w^r as function of N. Heavy lines show stable steady-states (ss); dashed lines are unstable. Note stable, symmetric ss for high amounts of NT, which becomes unstable below a critical point and asymmetric ss becomes stable. Unstable, asymmetric ss appear above the critical point, which separate asymmetric from symmetric stable ss. These ensure that only symmetric ss is reached for plausible initial conditions.

values, a binocular cell develops, while for small NT values, a monocular cell develops.

4. CONCLUSION

The present model suggests that interactions between activity-dependent weight modification and competition for trophic factor generate monocular cells. The essential features are: 1) a positive feedback interaction between rate of connection strength increase and rate of neurotrophin uptake, and 2) stabilization of this feedback by competition for a fixed amount of trophic factor. Addition of excess trophic factor eliminates the competition, thereby preventing monocular cell development.

Our model accounts for several pieces of experimental data, while making testable predictions. We have accounted for the finding that excess NT administration prevents OD column formation. In addition, monocularity develops with positive inter-eye correlations, by mechanisms supported by experimental data: namely, Hebb-like learning rules, competition for a limited supply of trophic factor [2], and enhancement of connection strength increase by trophic factor uptake [4]. Our model predicts that trophic factor uptake rate is enhanced by increased synaptic weight, i.e. that afferents with the largest connection strength "win" the competition for trophic factor.

REFERENCES

[1] T. H. Brown, A.H. Ganong, E.W. Kairiss, and C.L. Keenan. Hebbian synapses: Biophysical mechanisms and algorithms. *Annual Review of Neuroscience*, 13:475–511, 1990.

[2] R. J. Cabelli, A. Hohn, and C.J. Shatz. Inhibition of ocular dominance column formation by infusion of nt-4/5 or bdnf. *Science*, 267:1662–1666, 1995.

[3] E. Erwin, K. Obermayer, and K. Schulten. Models of orientation and ocular dominance columns in the visual cortex: A critical comparison. *Neural Comp*, 7:425–468, 1995.

[4] M. Korte, P. Carroll, E. Wolf, G. Brem., H. Thoenen, and T. Bonhoeffer. Hippocampal long-term potentiation is impaired in mice lacking brain-derived neurotrophic factor. *Proceedings of the National Academy of Sciences, USA*, 92:8856–8860, 1995.

[5] K. D. Miller, J.B. Keller, and M.P. Stryker. Ocular dominance column development: Analysis and simulation. *Science*, 245:605–615, 1989.

[6] K. D. Miller and D.J.C. MacKay. The role of constraints in hebbian learning. *Neural Comp*, 6:100–126, 1994.

[7] C. J. Shatz, S. Lindstrom, and T.N. Wiesel. The distribution of afferents representing the right and left eyes in the cat's visual cortex. *Brain Res*, 131:103–116, 1977.

[8] S. M. Sherman, R.W. Guillery, J.H. Kaas, and K.J. Sanderson. Behavioral, electrophysiological, and morphological studies of binocular competition in the development of geniculocortical pathways of cats. *J Comp Neurol*, 158:1–18, 1974.

55

SIMULATED RETINAL CENTER/SURROUND ARTIFICIAL NEUROPROCESSING USING ANALOG VLSI

Todd A. Hinck and Allyn E. Hubbard

Department of Electrical and Computer Systems Engineering
College of Engineering
Boston University
Boston, Massachusetts 02215

ABSTRACT

We designed an array of CMOS circuits to realize part of a model of a biological retina that acts as a preprocessing stage in a Boundary Contour System/Feature Contour System[1]. The basis for the circuitry is a shunting neuron equation. The "membrane" voltage of this silicon neuron hyperpolarizes as a function of the surround excitation and depolarizes as a function of the center excitation. The electronics approximates a Difference of Gaussians operator for edge enhancement. That quantity is scaled by an approximation to a Sum of Gaussians operator. Thus, the output reports on contrast that is normalized by the image's absolute intensity in a local region.

1. INTRODUCTION

Synthetic Aperture Radar (SAR) is capable of producing image data with potentially high spatial resolution, but the images present difficulty for interpretation by human observers and automatic recognition systems. One major problem is image speckle, which is generated by coherent processing of radar signals. Another problem is that the signal magnitude can have four orders of dynamic range. Ultimately, these problems might be solved by mimicking the biological retina, whose structure has evolved to handle a large dynamic range.

The Boundary Contour System/Feature Contour System[1] (BCS/FCS), a neural network-based image processing system, makes objects more obvious to human observers than can be seen in the original SAR image if it is simply converted from its digital form and presented on a display screen. The BCS/FCS system requires considerable computation power, and one goal of the present work is to achieve speedup via special electronics.

Computational Neuroscience
edited by Bower, Plenum Press, New York, 1997

Alternatively, these electronics could be part of a screen driver system that would high-light certain features of the image.

The first stage of the BCS/FCS is primarily responsible for edge enhancement and normalization by the local ambient image intensity. The enhancement and normalization operations are realized using a distance-dependent shunting neuron equation. Each pixel of the unprocessed SAR image is presented to an artificial neuron in a feed-forward neural network which computes the equilibrium value of a differential equation defining the ac-tivities of the ON and OFF preprocessing cells. This paper will focus only on the ON cell case. Thus, ON cell Activation is defined as:

$$\frac{d}{dt}x_{ij}^g = \underbrace{-D(x_{ij}^g - E)}_{Tonic} + \sum_{(p,q)}[\underbrace{(U - x_{ij}^g)I_{pq}C_{ijpq}^g}_{Center} - \underbrace{(L + x_{ij}^g)I_{pq}S_{jpq}^g}_{Surround}] \tag{1}$$

Each cell's activation in the neural network at the position (i,j) and at spatial scale g is denoted by x_{ij}^g in (1). The term I_{pq} is the input at the (p,q) position of the neural network. U is the polarization constant that bounds the upper limit of the cell's activity while L is the hyperpolarization constant that bounds the lower limit of the cell's activity. The tonic level is determined by the difference between the cell's activity and E, the baseline activ-ity level, multiplied by the activation decay constant D. The convolution kernels C_{ijpq}^g and S_{ijpq}^g are Gaussian functions:

$$C_{ijpq}^g, S_{ijpq}^g = \frac{1}{2\pi\sigma_{c(s)g}^2}\exp\left\{-\frac{(i-p)^2 - (j-p)^2}{2\sigma_{c(s)g}^2}\right\} \tag{2), (3}$$

Solving equation (1) for the equilibrium solution yields:

$$X_{ij}^g = \frac{DE + \sum_{(p,q)}(UC_{ijpq}^g - LS_{ijpq}^g)I_{pq}}{D + \sum_{(p,q)}(C_{ijpq}^g + S_{ijpq}^g)I_{pq}} \tag{4}$$

Equation (4) contains a difference of Gaussians term in the numerator and a sum of Gaus-sians term in the denominator.

2. METHODOLOGY

The circuit[2] in Figure 1 is equivalent to the following differential equation:

$$C_m\frac{d}{dt}V = (V^+ - V)g^+ + (V^- - V)g^- + (V^p - V)g^p \tag{5}$$

The terms g^+, g^-, and g^p are conductances. They can vary as a function of voltage or time, producing a number of effects, for example a neural impulse. A particular parallel circuit branch draws increased current (shunting) when its conductance increases. An alternative way to view the effect is that an increased conductance in a particular branch tends to pull the node voltage towards the value of the battery in that branch.

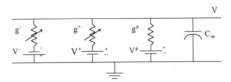

Figure 1. Circuit representation of cell membrane potential.

There are strong similarities between equation (1) and the Hodgkin and Huxley cell membrane equation. We set the g+ conductance to be proportional to the input at the (ith,jth) position. We set the g- conductance proportional to the output of the Gaussian-filter that operates on image data in the neighborhood (pth,qth). Thus, each cell's conductances hyperpolarize or depolarize the membrane potential as a function of the center field and the surround field. Assuming the equilibrium case, the capacitive element in Figure 1 is eliminated. Replacing the Hodgkin-Huxley notation with the BCS/FCS notation yields the circuit pictured in Figure 2A. Essentially, the cell voltage is controlled via conductances that are proportional to the convolution of the image data with the appropriate kernels.

3. CIRCUITS AND SYSTEMS

Instead of the circuit shown in Figure 2A, we used the circuit shown in Figure 2B. That circuit is smaller and with modified conductance functions can approximate the functionality of the original circuit. The latter circuit will have a new resting level and different sensitivities to illumination.

3.1. Realization of Variable Conductances

A transconductance operational amplifier (TAMP) multiplies a differential voltage by a gain term (G_m) and outputs a current (I_{out}). A TAMP operated in a linear range is governed by the following equation:

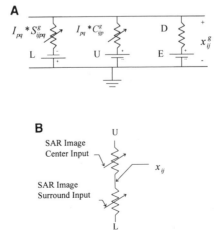

Figure 2. (A) BCS/FCS ON Cell equivalent circuit. (B) Circuit without tonic branch.

$$I_{out} = G_m(V^+ - V^-)$$ (6)

Note we use V and I to represent voltage and current respectively. By connecting the I_{out} terminal to the V^- terminal the TAMP can provide a good approximation to a variable conductance. Our TAMP is based on a CMOS Four-Quadrant Multiplier with bias feedback[3] that multiplies a voltage of either sign by another voltage of either sign and outputs a current proportional to their product.

3.2. Realization of Convolution Function

To implement the surround filter with an exponential convolution function a resistive grid was used. The grid was comprised of series resistors with parallel resistors to ground[4]. The configuration is that of a lossy cable. Since the spatial spread of the center filter was very small, its series resistors were omitted.

3.3. Systems Tested

To implement mathematical models using electronic circuits requires approximations. Therefore, we simulated three different systems of artificial center-surround processing, the first of which was defined by equations 1–3, whereas the second system used an exponential rather than a Gaussian convolution function, and the third system was our circuit implementation. Both the original and second system were simulated using MATLAB while the third system was simulated in SPICE 3 after extraction from a layout that describes our silicon chip.

Each system tested was a one-dimensional array comprised of 52 artificial neurons (AN). The simulation work tested these three system's output responses using a spatial impulse and a special pattern containing several orders of magnitude of image intensity information. The equilibrium outputs are the DC voltage response for the circuit and the solution to equation (4) for systems I and II. Table 1 gives the parameters used in the systems simulations.

4. RESULTS

4.1. Spatial Impulse Response

By applying a single input at a centralized AN, we determined the system's spatial impulse response. The response should resemble a "Mexican Hat" pattern that has been associated with the center-surround processing[5]. This pattern is typical of the DOG operation

Table 1.

Name	Description	System I & II value	System III value	Eqn.
σ_{cg}	Center blurring space constant	0.3	n/a	2,3
σ_{xg}	Surround blurring space constant	1.2	n/a	2,3
U	Depolarization constant	1.0	2.0	1,4
L	Hyperpolarization constant	1.0	2.0	1,4
D	Activation decay	2000	*see note	1,4
E	ON baseline activity level	0.5	*see note	1,4

*NOTE—There was no separate tonic branch in the circuit.

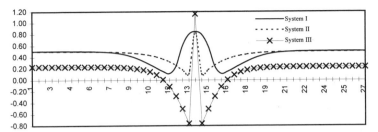

Figure 3. Spatial impulse responses.

because the surround Gaussian filter response is subtracted from the center Gaussian filter response. Typically, the center Gaussian filter magnifies the input and has a small spatial spread, whereas the surround Gaussian filter has a large spatial spread, but a smaller scale factor.

Figure 3 shows the output response of all three systems to a spatial impulse delivered to the middle of the test array. The system with the Gaussian filter (System I) produces a true "Mexican Hat". It has a high peak center surrounded by two dipping sides lobes. By comparison, the system using an exponential filter (System II) resembles a "Mexican Hat" whose response is sharper than that of the Gaussian filter. We attribute this sharpness to the form of the convolution function. Away from its center point a Gaussian function first decays slowly, but then it's magnitude goes quickly to zero. The exponential function decays at the same rate at every point relative to the center of the distribution. We also used Figure 3 to check the baseline (tonic) activity levels of the systems. In Figure 3, the outputs of system I and II start at roughly 0.5. This value corresponds to the parameter E, which was set to 0.5. However, the circuit has no tonic branch. Instead, we designed the circuit to have an inherent E parameter that is roughly 0.22V.

4.2. Four Orders of Magnitude Input Pattern

Typically, SAR produces images with four orders of magnitude of variation in intensity levels. Therefore useful systems must provide both compression of a large dynamic range input signal and also contrast enhancement of local image structure. A particular sample image acquired from the Navy had intensities that varied from a low value of 49 to a maximum of 25,254. Comparable input levels combined with quantifiable edges were produced using the input staircase function shown in Figure 4. We set the first twelve AN at 10,000 and then stepped up or down every six AN after that. To obtain the input to the circuit, the input pattern was linearly scaled into a current such that the voltage values on the resistive grid were between 0.0 V and 5.0 V.

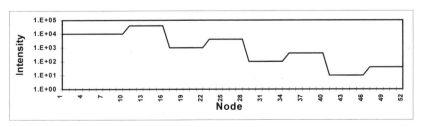

Figure 4. Input pattern.

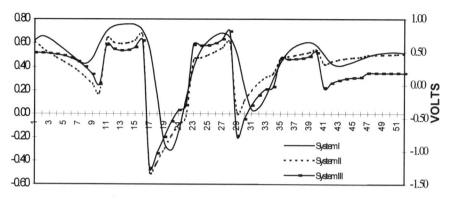

Figure 5. Outputs from spatial arrays comprising System I, II, and III.

Figure 5 depicts the equilibrium output responses of all three systems. System II and system III responses are qualitatively similar. However, they have differing resting responses to low signal levels as well as differing sensitivities to small changes around a low ambient level. System I and II responses are qualitatively dissimilar, in that the edge responses of the former are much more smooth. However, the resting levels and sensitivities to low-level contrast changes are similar.

5. DISCUSSION

Our circuit nominally produces the difference of exponentials scaled by a sum of exponentials. The former operation produces edge enhancement, while the latter produces a local normalization of the input image intensity. Thus, four orders of magnitude of input intensity are condensed it into a range between roughly ± one volt, while still retaining edge-enhancement functionality.

The exponential convolution function is not generally used in digital image processing algorithms. However, it is simpler to implement than the Gaussian function in electronics[6]. Differences between systems tested are fundamentally due to the intrinsic difference between the Gaussian and exponential filters. However, the two systems with an exponential convolution kernel (Figure 5) also show differences between them. For example, the circuit (system III) demonstrates a clear edge response at node 47. This peak is a result of a smaller equivalent value of D set by the circuit's electronics. The other systems use a large D parameter that makes the smaller intensities insignificant in the numerator and denominator of equation 4. In addition, the tonic levels also differ. Both effects occur because the circuit of Figure 2B cannot duplicate both the tonic level and the sensitivity to input changes produced by the circuit of Figure 2A. However, the performance of the circuit is arguably superior, because edges are more strongly emphasized.

ACKNOWLEDGMENT

This work was supported by ONR MURI N00014–95–1–04–09, A MURI Center for Automated Vision and Sensing Systems.

REFERENCES

1. Grossberg, E. Mingolla and J. Williamson, 1995, Synthetic Aperture Radar Processing by a Multiple Scale Neural System for Boundary and Surface Representation, *Neural Networks*, 8(7):1005–1028.
2. A.L. Hodgkin and A.F. Huxley, 1952, A Quantitative Description of Membrane Current and its Application to Conduction and Excitation in Nerve, *J. Physiol.*, 117:500–44.
3. Shen-Iuan Liu, Yuh-Shyan Hwang, 1994,CMOS Four-Quadrant Multiplier Using Bias Feedback Techniques, *J. Solid-State Circuits*, 29(6):750–752.
4. C.A. Mead, 1989, *Analog VLSI and Neural Systems*, Addison-Wesley, Reading MA.
5. D. Marr, 1982, *Vision*, W.H. Freeman and Company, San Francisco CA.
6. K. Boahen and A. Andreou, 1992, A Contrast Sensitive Retina with Reciprocal Synapses in *Advances in Neural Information Processing*, Volume 4 (J. Moody, ed.), Morgan Kauffman, San Mateo CA.

56

INTERNEURON PLASTICITY IN ASSOCIATIVE NETWORKS

Hajime Hirase[*] and Michael Recce

Department of Anatomy and Developmental Biology
University College London
London WC1E 6BT, United Kingdom
{hhirase,recce}@anat.ucl.ac.uk

ABSTRACT

Associative network models with binary synapses are widely studied as a biologically plausible memory mechanism. These models often include a single interneuron, used to set a global threshold for a network of sparsely interconnected principal cells, and the storage capacity improves with the use of a multi-step recall process (Gardner-Medwin, 1976). We demonstrate that the inclusion of non-saturating modifiable Hebbian synaptic weights in the projection from the interneuron to the principal cells drastically improves the performance of the network. These synaptic weights reduce the influence of the principal cells that are active in a disproportionate number of memory events.

1. INTRODUCTION

The associative memory, proposed by Willshaw and co-workers consists of binary-valued neurons, with binary-valued, clipped synapses (Willshaw *et al.*, 1969). Memory events are stored by extrinsically activating principal cells, the number of active cells in any one event is constant, and the set of active cells is selected randomly (i.e. maximising the storage capacity).

Recall is initiated by extrinsically activating a partial set of active cells (*seed pattern*), that provide positive feedback to other cells in the event (*correct cells*). A global threshold determines the set of active cells, including additional correct cells and fewer other cells (*spurious cells*). Gardner-Medwin (1976) demonstrated that multi-step recall (threshold sequence) increases the storage capacity. In this case, the extrinsic input is maintained throughout recall. Gibson and Robinson (1992) developed a theory that more

---

[*] Current address: Center for Molecular and Behavioral Neuroscience, 197 University Ave., Rutgers University, Newark, NJ 07102

accurately predicts the number of correct (c) and spurious (s) cells, but only for a transiently applied input. Applying this theory we demonstrated that the highest storage capacity occurs when the global threshold is a linear function of the network activation (Hirase and Recce, 1996). This threshold can be modelled by a single interneuron, that sums the activation of the principal cells and projects to each principal cell.

The capacity limit results from overlap between stored events. Gardner-Medwin (1989) distinguished between *common cells* that participate in a disproportionate number of events and *distinct cells* that are in fewer events and provide more effective inputs to correct cells. A natural strategy is to promote distinct cells and inhibit common cells until later steps in the recall process. We describe here how the capacity of the network can be increased by Hebbian modification of the projection from the interneuron to each principal cell.

2. MODEL AND ANALYSIS

The network consists of N principal cells and one interneuron. The principal cells make R recurrent connections to other randomly selected cells. The network learns M events, each with W randomly selected, principal cells. The recurrent synapses are initially zero, and are modified by a clipped Hebbian rule (see figure 1).

The interneuron has a fixed weight (value of one) synaptic connection from every principal cell, hence its activity is proportional to the number of active principal cells. The interneuron also projects to each principal cell and this synaptic weight increases each time that the principal cell is part of an event, and thereby co-active with the interneuron. Synaptic weights do not change during recall.

For a principal cell, the modified interneuron input is analogous to a cell-specific threshold function, that varies with the number of events the cell experienced during learning (*unit usage = m*). We first find this cell-specific threshold function and then define a method for changing the synaptic weights of the interneuron projection. By extending prior work (Buckingham and Willshaw, 1993) to fit the present model, the probability (ρ) that a recurrent projection to a principal cell becomes effective is:

$$\rho(m) = 1 - \left(\frac{W}{N}\right)^m$$

(1)

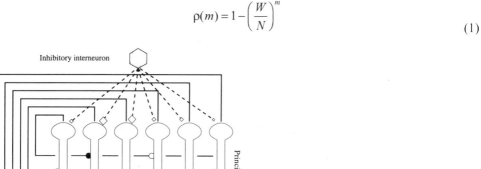

○ ineffective synapse   —— axon collateral
● effective synapse   ◇ inhibitory synapse

Example Events
0 1 1 0 0 0
0 0 1 0 1 0
1 1 0 0 0 0

Figure 1. A schematic diagram of the associative memory network ($N=6$, $R=2$, $M=3$, $W=2$). Each of the principal cells sends its output via the collaterals (thick line). Some of the synapses are effective (black circle) as a result of learning the example events. The interneuron provides an inhibitory input to each principal cell. Inputs to the inhibitory cell are omitted in this figure.

The probability that a principal cell participates in one event is W/N, so the probability P_e that a principal cell experiences m events out of M events is binomially distributed with W/N, i.e. $P_e = B(W/N,M,m)$. Based on this distribution, the probability that a spurious cell with m unit usage receives exactly r effective inputs given w active correct cells is:

$$P(D_{spur} = r|m) = B\left(\rho(m)\frac{R}{N}, w, r\right)$$

(2)

We choose the threshold $T(m)$ to keep spurious firing probability small (i.e. T is the minimum value that satisfies the following condition:

$$1 - \sum_{o=0}^{T(m)} P(D_{spur} = i|m) < 0.0001$$

(3)

In Figure 2A the calculated threshold levels are plotted, for a range of different activations, as a function of the unit usage for an example network ($N=6000$; $R=3000$, $W=150$). Since the threshold (T) function is linearly dependent on unit usage (Figure 2A) it is approximated to be $\alpha(w)m+\beta(w)$, where the slope α and y-intercept β are both functions of the network activation w. These slopes and intercepts are plotted for a range of activation levels in Figure 2B, and both have a linear dependence on activation ($\alpha(w) =Aw + B$; $\beta(w) = Cw+D$). By substitution:

$$T_i = (Am + C)w + Bm + D$$

(4)

The output of a principal cell is the step function (θ) of the recurrent inputs minus the threshold. Therefore D is the independent offset threshold of the cell, B depends only on the prior unit usage of the cell (not the current input), and $(Am+C)w$ is input from the interneuron. In this interpretation, C is the initial value of each synaptic projection from the interneuron to a principal cell and with each stored event involving that principal cell

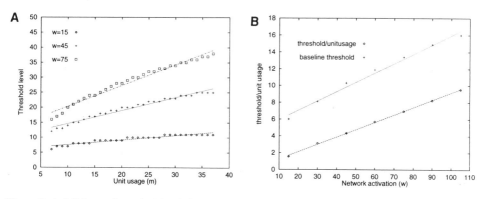

Figure 2. A: Minimum theoretical threshold levels to inhibit spurious firing are plotted against the unit usage of the cell. The level threshold changes according to the current activity of the network (w). w consists of correct cells only. The plots indicate that the three levels of network activation ($w=15,45$, and 75) can be approximated to a linear function. The network configuration is $N=6000$, $R=3000$, $W=150$. B: The slopes and intercepts of the lines in figure 2A are plotted against the activation level of the network (w). A least square linear fit to the data is included in the plots.

the synaptic weight increases by A. In the example plotted in the figure, the values of the constants are $A = 8.8 \cdot 10^{-3}$, $B = 3.8 \cdot 10^{-2}$, $C=0.11$, and $D=4.4$.

In order to estimate the relative contributions of each term in the threshold function, we compute the expectation value of m ($\langle m \rangle = WM/N$). Near the capacity limit M = 1400, $\langle m \rangle = 35$, $w=W=150$, $A\langle m \rangle w = 46.2$, $Cw=16.5$, $B\langle m \rangle = 1.3$ and $D = 4.4$. For all values of activation (after application of the 10% seed pattern) and for all values of $\langle m \rangle$ the $B\langle m \rangle$ term is smallest. Furthermore, B does not reflect inputs from interneuron plasticity, so in the following simulations B is set to zero.

3. SIMULATION RESULTS

To test the effect of interneuron plasticity we evaluated the storage capacity of an example network ($N=6000$, $R=3000$ $W=150$), seeded with 10% correct cells). The interneuron projection starts with a synaptic weight of $C=0.11$ to each principal cell and increments it by $A = 3.8 \cdot 10^{-2}$ each time that the principal cell is active in a stored event.

The average recall quality of 100 patterns was evaluated with increasing numbers of stored events. Following Gardner-Medwin (1989) the recall quality Q was defined to be $Q = \{I_0 - I_c\}/I_0$, where $I_0 = NH(W/N)$ is the total amount of information required to specify an event, $I_c = wH(s/w) + (N-w)H((w-c)/(N-w))$ is the information required to correct the event, and $H(x)$ is the Shannon entropy.

As shown in Figure 3B, without modifying the interneuron synapses performance of this network deteriorates after 950 patterns have been stored. Figure 3A shows the average performance of this network as a function of the parameter D after 1400 patterns have been stored. The value of D has been varied since part of the calculated offset B was set to zero. Figure 3B compares the best performance achieved without interneuron learning to the capacity with the best value of D from figure 3A.

4. DISCUSSION

We have described how Hebbian plasticity, in the projection from a single interneuron to a sparse network of principal cells, can reduce the effect of overlap on the

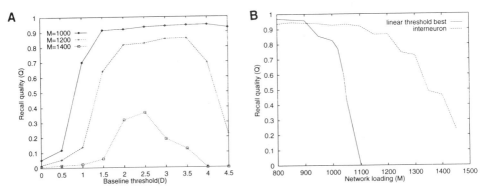

Figure 3. Simulation results for the example network ($N=6000$, $R=3000$, $W=150$). A: Averaged recall quality is plotted against the value of baseline threshold (D in the text). The network configuration is the same as in figure 2. B: Performance comparison between global thresholding and interneuron learning. Average recall quality is plotted against loading of the network (number of learned event.) The network with interneuron learning has 50% more capacity.

performance of an auto-associative memory. In the present analysis we found a threshold for individual cells that depend on the unit usage of the cell. An implication of this threshold is that the best values of the constants (A, C, and D) will not depend on M. Therefore the constants depend only on the topology of the network and on W, parameters that may be set genetically.

The initial findings described here are confined to a single network topology, but it is expected that these results can be supported by theory. In prior work (Hirase and Recce, 1996) we have observed that spurious cells are active during the initial stages of the optimal threshold sequence, suggesting that the threshold criteria used in this paper may be overly restrictive. Also the networks were only tested on recall with a fractional, but correct seed pattern and the work needs to be extended to include corrupted seed inputs.

Other approaches have sought to increase the capacity by introducing real-valued recurrent synapses (Hopfield, 1982). However, the learning rule for these synapses is non-Hebbian. Furthermore, many levels of synaptic weight in the interneuron projection add less complexity than modification to the recurrent projections.

There is initial evidence for plasticity in projections to interneurons (Ouardouz and Lacaille, 1995), but the present work suggest that further experimental determination of the conditions underlying interneuron synaptic change, and the number of distinct levels of interneuron synaptic weights, may lead to a better understanding to the neuronal basis of auto-associative memory.

REFERENCES

Buckingham, J. T. and Willshaw, D. J. (1993) On setting unit thresholds in an incompletely connected associative net, *Network: Computation in Neural Systems* **4** 441–459

Gardner-Medwin, A. R. (1976) The recall of events through the learning of associations between their parts, *Proc. R. Soc. Lond.* **B 194** 375–402

Gardner-Medwin, A. R. (1989) Doubly modifiable synapses: a model of short and long term auto-associative memory, *Proc. R. Soc. Lond.* **B 238** 137–154

Gibson, W. G. and Robinson, J. (1992) Statistical analysis of the dynamics of a sparse associative memory. *Neural Networks* **5** 645–661

Hirase, H. and Recce, M. (1996) A search for the optimal thresholding sequence in an associative memory, *Network: Computation in Neural Systems* **7** 741–756

Hopfield, J. J. (1982) Neural networks and physical systems with emergent collective computational abilities, *Proc. Nat. Acad. Sci.* **79** 2554–2558

Ouardouz, M. and Lacaille, J-C. (1995) Mechanism of selective long-term potentiation of excitatory synapses in stratum oriens/alveus interneurons of rat hippocampal slices, *J. Neurophysiol.* **73** 810–819

Willshaw, D. J., Buneman, O. P., and Longuet-Higgins, H. C. (1969) Non-holographic associative memory. *Nature*, **222** 960–962

57

PARAMETERS OF LTP INDUCTION MODULATE NETWORK CATEGORIZATION BEHAVIOR

Karl Kilborn, Don Kubota, Gary Lynch, and Richard Granger

Center for the Neurobiology of Learning and Memory and
Department of Information and Computer Science
University of California
Irvine, CA
E-mail: kkilborn@ics.uci.edu
E-mail: dkubota@ics.uci.edu
E-mail: glynch@ics.uci.edu
E-mail: granger@ics.uci.edu

ABSTRACT

Long-term potentiation (LTP) can only be induced at a synapse up to a fixed maximum extent; the magnitude of this *ceiling* has been found to be modulated by both endogenous and exogenous factors. The effect of the LTP ceiling on network-level categorization performance in a simulated layer of cortical neurons is investigated. The simulation is presented with random-dot stimuli and the similarity of its outputs is used to determine how it classifies its inputs. It was found that a higher ceiling leads to wider category formation, thus suggesting additional computer simulations and human experiments to test the connection between this LTP parameter and categorization performance.

1. INTRODUCTION

Much research has investigated the connection between long-term potentiation (LTP) and declarative memory, but most work has concentrated on the recognition and association aspects of memory. LTP may also, however, play a role in category formation.[1]

Recent findings in hippocampal slice preparations reveal that the maximum degree of elicited LTP as well as the per-stimulus increment can be regulated by both endogenous and exogenous factors[2,3] and the same pharmacological agents that increase measured LTP in slice preparation increase the learning rate in rats.[4]

Computer simulations presented by Hasselmo and Barkai show that cholinergic modulation of LTP can prevent trained patterns from interfering with one another.[3] Mathematical analysis presented by Kilborn further predict that serotonergic modulation of LTP might consequently affect the formation of categories by changing the maximum amount of LTP that can be elicited.[5]

In this paper we test the latter prediction (about LTP ceiling) using the random dot experimental paradigm first introduced by Posner[6] and still used to test categorization performance in humans (see, for example, ref. 7). Although the random dot test may putatively appear a visual task, it can also be cast as a general pattern classification problem where the patterns are encoded in a distributed representation. Random dot tests provide easily-manipulable artificial stimuli that can also provide information about category breadth under various conditions.[8]

2. MODELS OF CATEGORIZATION

Smith and Medin describe three fundamental models of category formation: classical, probabilistic, and exemplar-based.[9] Connectionist models suggest a fourth, distributed memory, theory of category formation.[10] In such a model, categories exhibit elements of both the probabilistic and exemplar-based theories. A representation of a prototype will emerge from repeated presentations of exemplars, but outliers within a particular category may form their own representation distinct from the prototype.

A model presented by Granger *et al*[11] and Ambros-Ingerson *et al*[1] is a biological simulation of olfactory bulb and layer II olfactory cortex that similarly exhibits elements of both the probabilistic and exemplar-based theories. Early responses to a stimuli correspond to a broad categorization and later responses correspond to finer distinctions and ultimately individual exemplars. In this paper we concentrate on the early responses and how they can be altered by changing the maximum degree of potentiation (ceiling-naive ratio) of LTP.

3. CORTICAL MODEL

To test the effect of ceiling size on category formation, we devise a simplified version of the model of the olfactory cortex presented by Ambros-Ingerson *et al*.[1] Rather than looking at hierarchical clustering suggested by olfactory cortex's feedback to bulb, we consider only initial responses due to the bulb's feedforward activation of cortical cells, which are divided into "patches" by feedback interneurons.

Inputs are presented to the olfactory bulb where their firing rate is transformed into a proportion of mitral/tufted cells firing in each glomerulus; the mitral/tufted cells then make sparse, random contact with cortical cells in layer II. Meanwhile, inhibitory interneurons ensure that once a cortical cell within a patch fires, all cells whose level of activation is not sufficiently close to the firing cell will be inhibited by the interneuron.[12] Figure 1A illustrates the network.

As each stimulus is presented to the network, activated synapses of cells that fire increase their degree of LTP by a fixed increment up to a limit (ceiling) past which they make no further change.

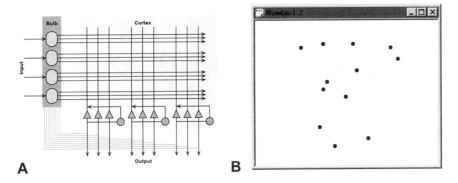

Figure 1. (A) Schematic of olfactory bulb and cortex model (B) Sample input

4. METHODS

Prototypes are generated by randomly placing twelve dots onto a 50×50 grid. This grid is then translated into a 2500-dimensional input vector where each element corresponds to a position on the grid. Elements corresponding to a dot position have value 1 and elements corresponding to the dot's neighbors have a value of $\exp(-r/2)$, where r is the Euclidean distance from the dot position to the grid position. Figure 1B shows an example input to the network.

In order to ensure that the the vector representation of the inputs exhibits the same similarity properties as the original dot patterns, a distance metric was developed to quantify the similarity of any two patterns, whether they came from the same prototype or not. We chose to measure the minimum sum of the distances between corresponding points in the two patterns across all possible correspondences.

Given a random dot pattern P, we want to determine how close it is to pattern Q. If P consists of n dots, whose positions are $\vec{p}_1, \vec{p}_2, \ldots, \vec{p}_n$ where each $\vec{p} = (p^x, p^y)$, the two-dimensional location of the dot, then its distance to Q is $d(P,Q) = \sum_{i=1}^{n} \sqrt{(p_i^x - q_{j_i}^x)^2 + (p_i^y - q_{j_i}^y)^2}$ where the j_i constitute one permutation of 1 through n. Then, $D(P,Q)$ is the minimum value of $d(P,Q)$ for all $n!$ possible permutations of the j_i.

There is a strong monotonic association between the Euclidean distance between the generated input vectors and the metric D distance between two random-dot patterns. Figure 2 plots the relation in a typical set of patterns with varying distortion from the prototype ($\rho = 0.94$).

We construct a model where the olfactory bulb consists of 2500 glomeruli which innervate four mitral/tufted cells each, for a total of 10,000 afferents to the cortex, which consists of 512 cells organized into 16 patches of 32 cells each. Then, we present the network with a experimental paradigm similar to that of Homa and Vosburgh,[8] namely a training set of eighteen patterns consisting of two low-, two medium-, and two high-distortion exemplars generated from each of three prototype patterns and a test set of 69 patterns consisting of the three original prototypes, the eighteen training patterns, another eighteen new patterns generated in a similar way as the training patterns, and 30 patterns that were not generated from the prototypes. Training patterns are presented to the naive network under four LTP ceiling-to-naive ratios within a biologically-plausible range: 4:3 (33% potentiation), 3:2 (50%), 2:1 (100%), and 3:1

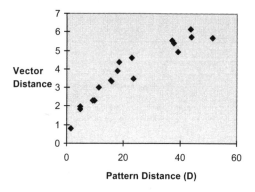

Figure 2. Dot pattern distance (D) versus Euclidean distance of encoded vectors

(200%). Five different simulations under each condition are conducted using different random seeds to distribute the synapses across the cells. Results represent the average of these runs.

Category membership is determined by comparing patterns of cortical cell firings. Output patterns of test inputs are compared to output patterns of training inputs using a nearest-neighbor algorithm. The output of a test input is said to identify category A if, among all the output patterns of the training inputs, an output of A was both closest and within a specified threshold θ. If an output of a test pattern is not within θ of any output of the training patterns, then the network is said to fail to recognize the category of the test input. Although the choice of θ is arbitrary, relative comparisons among the different ceiling-to-naive ratios hold. For this experiment, $\theta = 4.9$.

5. RESULTS

Among novel stimuli, the model shows best recognition performance for the prototype, and declining performance for progressively larger distortions (see Figure 3). Further, as predicted by Kilborn,[5] a larger LTP ceiling size yields better recognition of the high distortion patterns.

It can also be seen that a lower ceiling leads to a failure to identify some exemplars as belonging to any of the trained categories, typically the higher distortions. This "pattern exclusion" disappears as the ceiling size increases. Conversely, a higher ceiling tends to identify some random, untrained patterns as belonging to one of the trained categories. This "random inclusion" is minimal at lower ceilings.

Thus we observe a tradeoff in ceiling size. While too low of a ceiling will induce a narrow category that will fail to include many members near the boundary, too high of a ceiling will induce a broad category that will include too much.

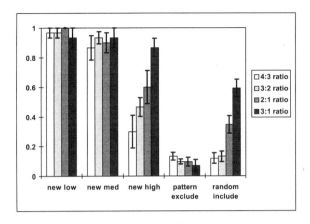

Figure 3. Categorization performance at different LTP ceilings

6. SUMMARY

This experiment suggests another: whether effects of manipulating LTP parameters can be seen at the human categorization level. As the model's stimuli are the same as a human's, quantitative as well as qualitative comparisons can be made between the cortical model of LTP and experiments performed on humans. Although there is much to be learned about how input categorization at the network level translates into behavior, an initial hypothesis would be that pharmacological agents which interfered with LTP elicitation during learning would show a greater degradation of performance on recognizing high-distortion patterns than low-distortion patterns than would controls, while at the same time perform less well on recognizing new distortions than ones seen during training than controls.

Also, recent work that compares recognition and classification performance between anterograde amnesic and normal subjects suggests that there may be two separate systems that subserve recognition and classification memory.[7] In other words, category classification does not appear to depend on declarative recall of individual training exemplars.

An alternative hypothesis is that the same system does subserve both forms of memory, but damage to the system can eliminate one form while leaving the other relatively intact. In the model of olfactory cortex forwarded by Ambros-Ingerson et al,[1] for instance, damage to the feedback mechanisms would, within the same layer of cells, permit early (general category) identification to remain largely intact while causing subsequent (finer category or even individual instance) identification to fail. This model of category learning and the random-dot paradigm can help investigate this hypothesis.

REFERENCES

[1] Ambros-Ingerson, J., Granger, R., and Lynch, G. (1990). Simulation of paleocortex performs hierarchical clustering. *Science*, **247**:1344-1348.
[2] Arai, A. and Lynch, G. (1992). Factors regulating the magnitude of long-term potentiation induced by theta pattern stimulation. *Brain Research*, **598**:173-184.
[3] Hasselmo, M. E. and Barkai, E. (1995). Cholinergic modulation of activity-dependent synaptic plasticity in the piriform cortex and associative memory function in a network biophysical simulation. *Journal of Neuroscience*, **15**:6592-6604.

[4] Staubli, U., Rogers, G., and Lynch, G. (1994). Facilitation of glutamate receptors enhances memory. *Proceedings of the National Academy of Sciences*, **91**:777-781.

[5] Kilborn, K., Lynch, G., and Granger, R. (1996). Effects of LTP learning rules on response selectivity of simulated cortical neurons. *Journal of Cognitive Neuroscience*, **8**:328-343.

[6] Posner, M. I., Goldsmith, R., and Welton, Jr., K. E. (1967). Perceived distance and the classification of distorted patterns. *Journal of Experimental Psychology*, **73**:28-38.

[7] Knowlton, B. J. and Squire, L. R. (1993). The learning of categories: parallel brain systems for item memory and category knowledge. *Science*, **262**:1747-1749.

[8] Homa, D. and Vosburgh, R. (1976). Category breadth and the abstraction of prototypical information. *Journal of Experimental Psychology: Human Learning and Memory*, **2**:322-330.

[9] Smith, E. E. and Medin, D. L. (1981). *Categories and Concepts*. Harvard.

[10] Knapp, A. G. and Anderson, J. A. (1984). Theory of categorization based on distributed memory storage. *Journal of Experimental Psychology*, **10**:616-637.

[11] Granger, R., Ambros-Ingerson, J., and Lynch, G. (1989). Derivation of encoding characteristics of layer II cerebral cortex. *Journal of Cognitive Neuroscience*, **1**:61-87.

[12] Coultrip, R., Granger, R., and Lynch, G. (1992). A cortical model of winner-take-all competition via lateral inhibition. *Neural Networks*, **5**:47-54.

58

MEMORY AND EPILEPTOGENESIS IN COMPLEX BIOLOGICAL AND SIMULATED SYSTEMS

J. C. Klopp,[1,2] P. Johnston,[3] E. Halgren,[4,5] K. Marinkovic,[4,5] and V. I. Nenov[1,4]

[1]Brain Monitoring and Modeling Laboratory
 Div. Neurosurgery, UCLA
[3]University of Northumbria, Newcastle, England
[2]Interdepartmental Neuroscience Ph.D. Program, UCLA
[4]Brain Research Institute, UCLA
[5]INSERM U97, Paris, France

1. ABSTRACT

Oscillations of neural activity may bind widespread cortical areas into a neural representation that encodes disparate aspects of an event. In order to test this theory we have turned to data collected from complex partial epilepsy (**CPE**) patients with chronically implanted depth electrodes. Data from regions critical to word and face information processing was analyzed using spectral coherence measurements. Similar analyses of intracranial EEG (**iEEG**) during seizure episodes display HippoCampal Formation (**HCF**)—NeoCortical (**NC**) spectral coherence patterns that are characteristic of specific seizure stages (Klopp et al. 1996). We are now building a computational memory model to examine whether spatio-temporal patterns of human iEEG spectral coherence emerge in a computer simulation of HCF cellular distribution, membrane physiology and synaptic connectivity. Once the model is reasonably scaled it will be used as a tool to explore neural parameters that are critical to memory formation and epileptogenesis.

2. INTRODUCTION

Neuroscientists have long pursued an understanding of the biological basis of memory. Many experiments have been designed to tease apart the steps by which memories are encoded, consolidated and retrieved. Theoretical models attempt to integrate experimental findings. Computational simulations claim to represent fundamental memory mechanisms in a highly abstracted form. Yet despite this effort the enigma that enshrouds the basic mechanisms of memory remains largely intact.

Computational Neuroscience
edited by Bower, Plenum Press, New York, 1997

It may seem that the diseases classified as epilepsy are far removed from memory mechanisms. However, this is not the case. A connection between memory consolidation and the phenomenon of kindling was first made by Goddard (Goddard et al. 1975). Kindling is the result of daily sub-convulsive electrical shocks delivered to specific regions of a rat's brain, commonly the amygdala and HCF. Once kindled an animal exhibits a long lasting susceptibility to seizure activity. This was interpreted as a crude form of learning, or perhaps more accurately an undesirable side effect of normal memory mechanisms. A subtle form of kindling may occur in humans through HCF-NC reciprocal connections. In a sense, mechanisms that establish memories under normal conditions could be usurped by hyper-excitable tissue into the production of epileptogenesis.

Neural models provide an integrated schema proceeding from molecular sub-synaptic events through physiology and anatomy to behavior. However, most models have not been made computational and many computational models have not been subjected to rigorous neural constraints. We are working toward combining human memory and epilepsy data into a computational neural model. iEEG from human patients provide a test bed against which the model's performance can be evaluated. With a biologically realistic computational memory model we may begin to understand under what conditions normal memory mechanisms contribute to epileptogenesis.

2.1. Basics of Memory

One can begin to understand a process by looking at examples of its dysfunction. Different theorists have viewed amnesia as arising due to either a loss of the memory trace (Rolls 1989), or the loss of active encoding / retrieval processes (Warrington et al. 1982). However, the dynamics of the amnesic syndrome observed in patients with medial temporal lobe (**MTL**) lesions suggest that the memory deficit can not easily be accounted for in terms of a loss of encoding or retrieval processes. Retrograde amnesia can not be due to a defective encoding procedure as the memories which are lost were encoded prior to the onset of amnesia. Conversely, as remote memories are spared, defective retrieval processes can not account for retrograde amnesia unless a retrieval process is postulated that is specific to recent memory. Evidence from transient global amnesia shows that the HCF is necessary at encoding and retrieval (Halgren 1994a). Taken together this suggests that new memories are initially stored within the HCF and over time a transfer to the NC for long term storage occurs.

A number of recent neural models have identified the HCF as a temporary storage site for the declarative memory trace (Treves et al. 1994; McLelland et al. 1995). Generally, these models consider the HCF to constitute an associative network in bi-directional communication with the NC. Temporary traces are set up in the HCF that assist the NC in reconstructing recent memories, and / or in consolidating neocortical traces of long term memory. A number of these models have specifically identified the auto-associative network of pyramidal cells within the CA3 field as a prime candidate for the substrate of rapid short-term declarative memory formation and storage (Read et al. 1994).

2.2. Basics of Epilepsy

The most prevalent form of epileptic syndrome is complex partial epilepsy. Complex Partial Seizures (**CPS**), also known as psychomotor seizures, account for 40% of all epilepsy cases (McNamara 1994) and are characterized by complicated illusory phenomena and semi-purposeful complicated motor acts. CPS often start in, or spread through, limbic structures of the MTL. Ammon's horn sclerosis, a degeneration of principal cells in the HCF, is commonly associated with seizure activity. The CPS reflects hyper-synchronous activation of cortico-limbic neuronal networks. Most commonly, these networks include neurons in both the association NC and the HCF including the hippocampus proper, dentate gyrus and parahippocampal gyrus.

2.3. The Connection between Memory and Epilepsy

Several lines of evidence suggest that CPE and declarative memory are closely related phenomena. Both often involve the HCF, a neural structure that has the lowest threshold to seizure activity of any cortical site. It is generally agreed that lesions within the MTL, particularly the HCF, result in the purist and most severe cases of amnesia (Zola-Morgan et al. 1986). Whilst similar memory deficits may also occur as a result of lesioning in other brain areas, for instance damage to the diencephalon, mecencephalon or medial telencephalon also result in a severe memory impairment (Squire 1987), these deficits are less specific than MTL amnesia.

Basic memory formation is hypothesized to involve some form of synaptic long term potentiation (LTP). Experimental evidence in animal models indicates that LTP is also induced in the HCF by seizure discharges (Stewart 1994). Seizure activity in human subjects can be evoked by local electrical stimulation. Although these patients may not subjectively realize they are producing seizure activity, a profound inability to learn new material occurs during the after discharge (Halgren et al. 1991). This suggests that memory and seizure pathways overlap as they can not simultaneously occur.

Reorganization of synaptic pathways is generally accepted as the basis for the development of epilepsy in an otherwise normal brain (Houser 1992). Evidence of reorganization resulting from seizure activity is seen in prolonged GABA responses from dentate granule cells (Williamson et al. 1995), variations in cell density (Masukawa et al. 1995) and single mossy fiber axonal arborization patters (Isokawa et al. 1993).

3. METHODS

Medically intractable CPE patients may undergo iEEG monitoring for localization of seizure onset. Once implanted these patients participate in a variety of memory and recognition tasks.

3.1. Methods for Depth EEG Data Acquisition

64 channels of iEEG (0.1–50 Hz, 166 samples/sec at 12 bits) were recorded using a common reference from each of 4 subjects during word (n=280) and face (n=240) Delayed Recognition (**DR**) tasks (see Halgren et al. 1994b for details of the DR task). Artifact rejection removed trials during which eye-blinks and or epileptogenic spikes occurred. Face-selective event related potentials were present in the fusiform gyrus leads of all subjects.

3.2. Methods for iEEG Data Analysis

Human iEEG data was transferred to a Sun workstation and analyzed using S-Plus software (MathSoft, StatSci Division, Seattle, WA). Exploratory data analyses using spectral coherence measurements were performed. Coherence is defined as a function of the power spectral outputs for the two channels at any given frequency f:

$$C_{xy}(f) = |S_{xy}(f)|^2 / S_x(f)*S_y(f), \tag{1}$$

where x and y represent two simultaneously acquired iEEG time series. Coherence *may* be interpreted as a measure of the functional relationship between brain areas (Rappelsberger 1994).

In this work, spectral coherence measurements for multiple frequency bands (1–3 Hz, 4–7 Hz, 8–12 Hz, 15–25 Hz & 35–45 Hz) were estimated from moving windows of 120 msec duration. Each window was offset by 10% (12 msec). P values (uncorrected for multiple tests) were derived using a T statistic to compare pre-stimulus reference windows to subsequent windows. Bootstrap statistical methods (Effron et al 1991) were also employed to account for non-normal distributions common in spectral data. Due to the greater computational speed of the traditional analytic statistics and the apparent lack of difference in results between the two statistical methods we primarily used the T statistic.

3.3. Methods for Computational Modeling

A preliminary computational model was constructed using Genesis (**GE**neral **NE**ural **SI**mulation **S**ystem) software (Bower et al. 1995). The model currently consists of compartmental representations of pyramidal and interneuron basket cells that contain active and passive conductances and Hebbian-type channels capable of changing synaptic efficacy. The cellular distribution and connectivity in the model is a scaled down version of the hippocampal CA3 region. Further layers representing the major cytoarchitectural divisions of the human hippocampal formation and adjacent cortices will be added with future development.

It would be premature to make claims based on the performance of the model at this stage as it simulates only a few hundred cells. A proper cell ratio is crucial to the functionality of the network because it allows for an appropriate sparsness of synaptic connectivity to be built into the model. Development of a larger scaled version of the model is currently underway and will implement the parallel version of Genesis. Parallel Genesis was developed by Nigel Goddard at the Pittsburgh Supercomputing Center and allows Genesis simulations to be run on parallel processing super computers.

4. RESULTS OF IEEG ANALYSES

In the four patients analyzed transient fluctuation of coherence between lingual and fusiform neural activity correlated with a stage of the DR task that overlaps with common evoked response potentials. Furthermore, the response differed depending on the type of task. Word stimuli evoked a mild increase (p<0.005, uncorrected) in coherence. Face stimuli in all subjects evoked a statistically-significant (p<0.005, uncorrected) decrease in coherence. Coherence fluctuation for word and face stimuli was maximal at approximately 200–330 msec after stimulus onset. No significant change in 35–45 Hz power was found in the fusiform / lingual derivation during the same epoch, nor were consistent coherence changes observed between either lingual or fusiform gyrus and the posterior hippocampus during this period.

Multiple pairings of next neighbor coherence estimates along the longitudinal length of a single depth probe within the hippocampus show interesting evoked, transient alterations of coherence. In one patient a highly significant increase in coherence occurred for windows centered on 120–250 msec post-stimulus presentation for faces exclusively. In another example, next neighbor contact pairs in the posterior hippocampus showed a long lasting coherence inversion starting at windows centered at 80 msec for faces and 180 msec for words. A probe implanted in the nearby fusiform gyrus showed no change for the same task.

Patients displayed novel-repeated differences over a wide number of depth electrode contacts that generally were similar between frequency bands. However, this variation was specific to individual patients and commonalties between patients are speculative at this stage of data analysis.

5. CONCLUSION

The pattern of coherence fluctuation over time and between neighboring and distant iEEG sites provides an enticing new measure of the cooperation or coordination between distinct neural populations. These results suggest that a wide-spread wide-band decrease in spectral power occurs during event encoding, and a focal phasic decrease occurs in high-frequency coherence localized to fusiform / lingual gyri. Furthermore, the coherence decrease occurs at the moment that the fusiform gyrus makes a specific contribution to face-encoding (Halgren et al. 1994b). These observations do not seem consistent with some current theories regarding the role of 35–45 Hz activity in binding experiential elements. However, one subject did yield a transient increase in gamma band activity and coherence that better fits to expectations of the binding hypothesis.

Our Genesis model will be refined and extended to simulate realistic neuronal activity of the human medial temporal lobe. Analyses of the model's performance will be evaluated using spectral measures similar to those used in the human iEEG. This will allow us to make a critical judgment of the model's performance in regard to how well it mimics its biological template. As a platform for simulation, Genesis holds the potential for portability to massively parallel processing machines. This is a critical factor for addressing the computational problems of a large scale simulation.

6. REFERENCES

Bower J & Beeman D: The Book of Genesis. Springer-Verlag New York, Inc. 1995.

Effron B & Tibshirani R: Statistical data analysis in the computer age. Science 253:390–395, 1991.

Goddard GV & Douglas RM: Does the engram of kindling model the engram of normal long term memory? The Canadian Journal of Neurological Sciences, pp385–394, 1975.

Halgren E, et al: Memory dysfunction in epilepsy patients as a derangement of normal physiology. Advance in Neurology 55:385–410, 1991.

Halgren, E: Physiological integration of the declarative memory system. *The Memory System of the Brain*, ed. J. Delacour, (World Scientific, New York), pp. 69–155, 1994a.

Halgren E, et al: Spatio-temporal stages in face and word processing: 1&2. J Physiology (Paris) 88:51–80,1994b.

Houser CR: Morphological changes in the dentate gyrus in human temporal lobe epilepsy. Epilepsy Res Suppl 7:223, 1992.

Isokawa M, et al: Single fiber axonal systems of human dentate granule cells studied in hippocampal slices from patients with temporal lobe epilepsy. Journal of Neuroscience, 13(4):1511–22, 1993.

Klopp J, Nenov V, Simon S, Jones C: One-dimensional strobomaps and coherence maps characterize human depth EEG during seizure activity. Complex Systems Conference Proceedings, pp 34–44, June 1996.

Masukawa LM, et al: Longitudinal variation in cell density and mossy fiber reorganization in the dentate gyrus from temporal lobe epileptic patients. Brain Research 678(1–2):65–75, 1995.

McClelland J, McNaughton B & O'Reilly R: Why there are complementary learning systems in the hippocampus and neocortex. Psychological Review 102(3):419–57, 1995.

Mcnamara J: Cellular and molecular basis of epilepsy, The Journal of Neuroscience 14(6):3413–3425, 1994.

Rappelsberger P, et al: Calculation of event-related coherence: A new method to study short lasting coupling between brain areas. Brain Topography 7(2):121–127, 1994.

Read W, Nenov V, Halgren E: Role of inhibition in memory retrieval by hippocampal area CA3. Neuroscience and Biobehavioral Reviews 18(1):55–68, 1994.

Rolls E: Functions of neural networks in the hippocampus and neocortex in memory. *Neural Models of Plasticity: Experimental and Theoretical Approaches*, eds. J.H. Byrne & W.O. Berry (Academic Press, New York) pp240–65, 1989.

Squire LR: Memory and Brain. Oxford University Press, New York, 1987.

Stewart C, Jeffery K, Reid I: LTP-like synaptic efficacy changes following electroconvulsive stimulation. Neuroreport 5(9):1041–4, 1994.

Treves A & Rolls E: A computational analysis of the role of the hippocampus in memory. Hippocampus, Vol. 4, pp. 374–92, 1994.

Warrington EK & Wieskrantz L: Amnesia: a disconnection syndrome. Neuropsychologica 17:233–48, 1982.

Williamson A, Telfeian AE, Spencer DD: Prolonged GABA responses in dentate granule cells in slices isolated from patients with temporal lobe sclerosis. Journal of Neurophysiology 74(1):378–87, 1995.

Zola-Morgan S. & Squire LR: Memory impairment in monkeys following lesions limited to the hippocampus. Behavioral Neuroscience 100:155–60, 1986.

59

MODULATING THE CALCIUM DEPENDENT POTASSIUM CONDUCTANCE IN A MODEL OF THE LAMPREY CPG

Anders Lansner,[1] Jeanette Hellgren Kotaleski,[2] Maria Ullström,[2] and Sten Grillner[2]

[1]Department of Numerical Analysis and Computing Science
Kungl. Tekniska Högskolan
ala@nada.kth.se
[2]Department of Neuroscience
Karolinska Institutet
Stockholm, Sweden

1. INTRODUCTION

The lamprey is a primitive water-living vertebrate that moves by means of undulatory swimming. It is of particular interest as an experimental model for the neural generation of locomotion [Grillner et al., 1995]. A major advantage of this system is that the motor pattern underlying swimming can be elicited in an isolated piece of spinal cord. Being one of the best characterized vertebrate neuronal systems, the lamprey spinal CPG has been the subject of a number of modelling and simulation studies.

1.1. The Local CPG Network

The local, "segmental", rhythm generating network can be viewed as two reciprocally inhibiting halves, each one comprised of a set of premotor interneurons, i.e. excitatory interneurons (EIN), contralaterally and caudally projecting inhibitory interneurons (CCIN) and lateral inhibitory interneurons (LIN). Recent studies have indicated that the LIN is not of primary importance for burst generation [Fagerstedt et al., 1995]. In the model presented here we have only included EIN and CCIN neurons.

The rhythm generating network is activated by glutamate projections from the brainstem. Already the first modelling attempts demonstrated that moderate to high frequency bursting can readily be produced by such a network [Grillner et al. 1988]. Effects of 5-HT seen during fictive swimming could be reproduced. A following series of simulations using biophysically detailed compartmentalized Hodgkin-Huxley type model neurons replicated many aspects of the rhythm generation seen experimentally [Ekeberg et al., 1991; Wallén et al., 1992; Brodin et al., 1991; Hellgren et al., 1992; Tråvén et al., 1993].

1.2. The Distributed CPG Network

The entire distributed CPG forms a column-like structure which extends along most of the spinal cord. An EIN projects in its local neighborhood about two segments both rostrally and caudally, whereas a CCIN produces contralateral inhibition extending about the same distance rostrally, but caudally some twenty segments. This CPG produces a wave of motor activity that travels along the body, propelling the animal through the water. The length of this wave is about the same as the body length over a substantial range of swimming speeds.

A "continuous" network model of the above type, with the basic connectivity patterns of the local CPG extended longitudinally and lacking discrete segment boundaries, was first used in a neuro-mechanical model of lamprey swimming behavior [Ekeberg, 1993]. A similar network has also been simulated using biophysically detailed model neurons [Wadden et al., 1997]. This model produced an appropriate bursting frequency range and a flexible and stable phase lag. However, the phase lag showed a tendency to increase with frequency.

1.3. 5-HT Mediated AHP Modulation

We report here the effects, on local rhythm generation as well as on intersegmental coordination, of incorporating 5-HT (serotonin) mediated modulation of the slow AHP into the previous CPG model [Lansner et al., 1996]. The modulatory influence, via a calcium dependent potassium channel, of 5-HT on locomotion is well established from invertebrates to primitive vertebrates and mammals, but its precise mechanisms of action are still unclear [Harris-Warrick and Cohen, 1985; Wallén et al., 1989; Christenson et al., 1991]. In the lamprey, descending fibers from reticulospinal 5-HT releasing neurons and a dense plexus of 5-HT releasing midline neurons are likely to be involved in producing such effects.

2. DESCRIPTION OF THE NETWORK MODEL

The simulation model used in this study is comprised of simplified spiking and adapting model units that represent at the same time the EIN and the CCIN neurons in the real network. The two different types of premotor interneurons are known to differ slightly in their firing patterns. Yet, our simulation results demonstrate that this simplification serves well as a first approximation. The units have fast dynamics incorporating a passive membrane time constant and a slow adaptation corresponding to the calcium dependent slow AHP. This adaptation is modelled with a simple first order dynamics including a parameter for rate of build-up and a time constant for decay.

A local half-segment of the CPG is comprised of one EIN-CCIN model unit with self-excitation. The 5-HT neurons are not explicit in the model; just their proposed modulatory action on the adaptation properties of the pattern generating premotor unit. This modulation is simulated by changing the rate of build-up of the slow adaptation. A full local CPG consists of two half-segments with reciprocal inhibition between them (fig. 1A).

The distributed CPG was simulated with eighty units on each side. EIN connections extended two segments rostrally and caudally and the lack of excitatory inputs at the rostral and caudal ends was compensated for by making the remaining connections stronger. The reciprocal CCIN connections extended two segments rostrally and twenty segments caudally and were not "end-compensated". Except for the end-compensation, all

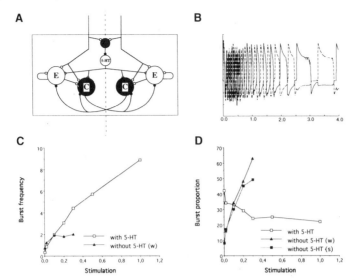

Figure 1. (A) Local CPG network with excitatory interneuron (E), contralaterally inhibitory interneuron (C), 5-HT neurons and brainstem drive. (B) Sample of burst generation at high to low levels of stimulation. Population EIN-CCIN activity on y-axis, with solid and dashed lines for right and left sides respectively, and arbitrary time on the x-axis. (C) Burst frequency and (D) burst proportion as a function of tonic stimulation. Curves in (C) and (D) are with and without 5-HT modulation and with weak (w) and strong (s) reciprocal inhibition.

connections of the same type were equally strong irrespective of extent. This conforms with experimental data on strong functional intersegmental coupling [Mellen et al., 1995].

3. SIMULATION RESULTS

The simulated network was examined with respect to burst generation properties as well as intersegmental co-ordination.

3.1. Local CPG Burst Generation

The network was activated by means of a constant input to the EIN-CCIN units. At the same time, the rate of build-up of adaptation was increased linearly reflecting a decreased 5-HT release. This produces a more pronounced AHP which builds up quicker to terminate the burst. A stable bursting with an adequate frequency range of 1:10 or more (fig. 1B-D) resulted. Without AHP modulation, burst proportion (the ratio of burst duration to cycle duration) increased markedly (fig. 2D) and bursting was produced only at lower levels of stimulation (fig. 2C).

The burst generating mechanism differs from those considered previously in that accumulated slow AHP is the dominant burst terminating factor. In fact, a small self-exciting EIN population is itself capable of producing adequate bursting over the entire frequency range.

3.2. The Distributed CPG and Intersegmental Co-ordination

We introduced the modulatory effects on the AHP in the distributed CPG model. This resulted in stable bursting over a ten-fold frequency range. With a small added extra drive to the most rostral part of the simulated cord, a constant phase lag of around 1 % resulted over this frequency range (fig. 2A-C). An isolated spatially extended EIN population could do most of the job with the reciprocal inhibition merely serving to produce alternation between the two sides. Weak inhibition gave alternation without other effects on bursting. Strong inhibition prevented overlap of bursts on opposite sides.

The simulated network can operate also without caudally projecting reciprocally inhibiting connections. However, when introduced, these have a stabilizing effect on the system, especially at higher frequencies, and give a bias for about 1 % phase lag. It is possible to control the phase lag and even to reverse it by extra excitation and inhibition to the tail and head parts respectively of the simulated CPG (fig. 2D). The simulated network reacts fast, within one or two cycles, and with only a brief transient activity to sudden changes in brainstem drive (fig. 2E). This network model can be used to simulate experiments like local blocking of activity or reciprocal inhibition (fig. 2F).

Recently, these results have been replicated in a network of graded output units used in a previous model [Ullström et al., 1997]. We have further initiated a study of the 5-HT modulation in a network comprised of detailed model neurons. Preliminary results indicate that the effects are qualitatively the same as in the simplified model used here.

4. DISCUSSION

The model of the lamprey spinal swimming CPG presented here adds a new kind of spinal burst generating mechanism in which 5-HT mediated modulation of the slow AHP plays a key role. Simulation of this pattern generator demonstrates a good bursting frequency range as well as an adequate control of burst proportion. The network conforms with biology in that it lacks segmental boundaries and has long-range longitudinal connectivity.

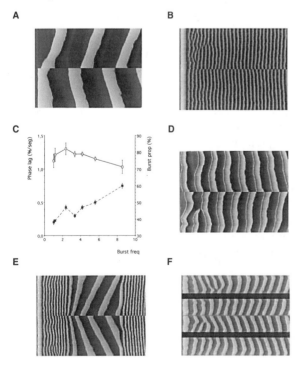

Figure 2. Activity, frequency range and burst proportion. In A, B, D-F, activity is shown as light and inactivity/inhibition as dark. Time is along the horizontal dimension. Activity in the rostral part is shown centrally with the right and left sides unfolded upwards and downwards respectively. (A) Slow swimming. (B) Fast swimming. (C) Phase lag (upper curve) and burst proportion (lower curve) vs. bursting frequency. (D) Backward swimming. (E) Changing brainstem drive high – low – high. (F) Simulated blocking of activity in eleven segments located halfway between head and tail.

Our model predicts a reduction of 5-HT release with bursting frequency, and thus with shorter burst duration. This is in accordance with the effects on firing patterns seen with bath application of 5-HT. Details of how 5-HT levels change with activation of the spinal locomotor network are, however, not yet known. Further, brainstem control of 5-HT releasing neurons is suggested here. Another attractive possibility is local control, e.g. by CCIN inhibition of local 5-HT neurons. This possibility may be investigated by further modeling and experimental analysis. It should also be noted that several other modulators modify the AHP amplitude and could be of relevance here.

The CPG model presented is capable of generating intersegmental co-ordination with a phase lag independent of bursting frequency over an adequate bursting frequency range. It should be noted, however, that the constant phase lag resulted only with end-compensated excitatory connectivity and a slight constant head bias. This may be a reasonable description of the intact cord, but hardly for a cut cord piece maintained under *in vitro* conditions. The issue of end effects in cut cord pieces need further examination and the proposed role of AHP modulation remains hypothetical until further experimental results on the role of 5-HT in the spinal locomotor network is available.

5. CONCLUSIONS

The modified network model presented here matches well experimental data on burst generation and intersegmental co-ordination in the lamprey. It takes biological detail better into account compared to e.g. previous coupled oscillator models [Cohen et al., 1992] in that a discrete segmented structure is replaced by a non-segmented, "continuous", network structure with strong interactions over long distances.

6. REFERENCES

Grillner S., Deliagina T., Ekeberg Ö., El Manira A., Hill R. H., Lansner A., Orlovsky G. N., and Wallén P. (1995). "Neural networks that co-ordinate locomotion and body orientation in lamprey." *Trends Neurosci.*, 18(6), 270–279.

Fagerstedt P., Wallén P., and Grillner S. (1995). "Activity of interneurons during fictive swimming in the lamprey." *4th IBRO Word Congress of Neuroscience*, Kyoto, Japan, 346.

Ekeberg Ö., Wallén P., Lansner A., Tråvén H., Brodin L., and Grillner S. (1991). "A computer based model for realistic simulations of neural networks. I: The single neuron and synaptic interaction." *Biol. Cybern.*, 65, 81–90.

Wallén P., Ekeberg Ö., Lansner A., Brodin L., Tråvén H., and Sten G. (1992). "A Computer-Based Model for Realistic Simulations of Neural Networks. II: The Segmental Network Generating Locomotor Rhythmicity in the Lamprey." *J. Neurophysiol.*, 68, 1939–1950.

Brodin L., Tråvén H., Lansner A., Wallen P., Ekeberg Ö., and Grillner S. (1991). "Computer simulation of N-Methyl-D-Aspartate receptor induced membrane properties in a neuron model." *J. Neurophysiol.*, 66(2), 473–484.

Hellgren J., Grillner S., and Lansner A. (1992). "Computer Simulation of the Segmental Neural Network Generating Locomotion in Lamprey by using Populations of Network Interneurons." *Biol. Cybern.*, 68, 1–13.

Tråvén H., Brodin L., Lansner A., Ekeberg Ö., Wallen P., and Grillner S. (1993). "Computer Simulations of NMDA and non-NMDA Receptor-Mediated Synaptic Drive: Sensory and Supraspinal Modulation of Neurons and Small Networks." *J. Neurophysiol.*, 70, 695–709.

Ekeberg Ö. (1993). "A combined neuronal and mechanical model of fish swimming." *Biol. Cybern.*, 69, 363–374.

Wadden T., Hellgren Kotaleski J., Lansner A., and Grillner S. (1997). "Intersegmental coordination in the lamprey: simulations using a network model without segmental boundaries." *Biol. Cybernetics.* In press.

Lansner A., Ekeberg Ö., and Grillner S. (1996). "Realistic modeling of burst generation and swimming in lamprey." Neurons, Networks, and Motor Behavior, P. Stein, D. Stuart, S. Grillner, and A. Selverston, eds., Tucson, Arizona. To appear.

Harris-Warrick R. M., and Cohen A. H. (1985). "Serotonin modulates the central pattern generator for locomotion in the isolated lamprey spinal cord." *J. Exp. Biol.*, 116, 27–46.

Wallén P., Buchanan J., Grillner S., Christenson J., and Hökfelt T. (1989). "The effects of 5-hydroxytryptamine on the afterhyperpolarisation, spike frequency regulation and oscillatory membrane properties in lamprey spinal cord neurons." *J. Neurophysiol.*, 61, 759–768.

Christenson J., Wallén P., Brodin L., and Grillner S. (1991). "5-HT Systems in a Lower Vertebrate Model: Ultrastructure, Distribution, and Synaptic and Cellular Mechanisms." *In* Volume Transmission in the Brain: Novel Mechanisms for Neural Transmission, K. Fuxe and L. F. Agnati, eds., Raven Press, New York, 159–170.

Mellen N., Kiemel T., and Cohen A. H. (1995). "Correlational analysis of fictive swimming in the lamprey reveals strong functional intersegmental coupling." *J. Neurophysiol.*, 73(3), 1020–1030.

Ullström M., Lansner A., Hellgren Kotaleski J., and Grillner S. (1997). "Significance of modulated adaptation for rhythm generation and intersegmental co-ordination in lamprey." In preparation.

Cohen A. H., Ermentrout G. B., Kiemel T., Kopell N., Sigvardt K., and Williams T. L. (1992). "Modelling of Intersegmental Coordination in the Lamprey Central Pattern Generator for Locomotion." *Trends Neurosci.*, 15, 434–438.

60

NONLINEAR NETWORK MODELS OF THE OCULOMOTOR INTEGRATOR

D. D. Lee, B. Y. Reis, H. S. Seung, and D. W. Tank

Bell Laboratories
Lucent Technologies
Murray Hill, NJ
E-mail: ddlee@physics.bell-labs.com
E-mail: reis@physics.bell-labs.com
E-mail: seung@physics.bell-labs.com
E-mail: dwt@physics.bell-labs.com

ABSTRACT

The neural integrator of the oculomotor system is modeled as a network of spiking, conductance-based model neurons. The static function of the integrator, holding the eyes still when the head is fixed in space, is the focus of the modeling. The synaptic weight matrix, which is of outer product form, is tuned by minimizing the mean squared drift velocity of the eyes over a range of eye positions, leading to an approximate line attractor dynamics. The conductance-based model is reduced to a rate-based one to simplify the tuning procedure.

1. INTRODUCTION

In the time intervals between saccades, the eyes are held still by a neural network integrator located in the hindbrain [1, 2]. With the head fixed in space, constant neural activity in the integrator is necessary for gaze stabilization. The pattern of activity is different for each eye position, even though the visual and vestibular inputs to the integrator remain the same. The origin of the integrator's ability to maintain multiple persistent patterns of activity is not obvious, since its constituent neurons do not appear to possess long intrinsic persistence times. A network mechanism is thus needed to explain the observed long time constants associated with gaze stabilization in the dark.

Computational Neuroscience
edited by Bower, Plenum Press, New York, 1997

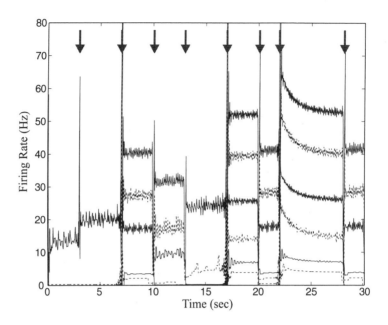

Figure 1. Spiking conductance-based model of the integrator with 40 neurons. The firing rates of six of the neurons are shown being driven by pulses of input from saccadic command neurons at the times indicated by the arrows.

2. CONDUCTANCE-BASED MODEL

Previous models of the integrator have relied on positive feedback in networks of completely linear neurons [3, 4, 5, 6]. In contrast, the present work uses conductance-based model neurons as described by Hansel and Sompolinsky [7]. Their nonlinear current-discharge relationships qualitatively match intracellular measurements of real integrator neurons [8, 9]. Synapses were modeled as current sources, with a time dependence described by the relation $e^{-t/\tau_1} - e^{-t/\tau_2}$, where $\tau_1 = 150$ ms and $\tau_2 = 20$ ms. The synaptic currents were assumed to sum linearly and the synaptic strengths were tuned by an optimization procedure that is described in the next section.

Figure 1 depicts the response of the model network to short pulses of input from saccadic command neurons at the times indicated by the arrows. Shown are the instantaneous firing rates (1/ISI) of the neurons. Eye position is not shown in the figure, but was modeled as a weighted linear sum of the rates of the neurons. Thus the firing rates and eye position all increase and decrease in steps, and are roughly constant in time between the quick saccadic movements that take place at the arrows. The reason for the "sag" in response to the second to last saccadic input will be explained in a later section.

The model captures two essential aspects of integrator operation that are observed experimentally. The first is that the firing rate of each individual neuron is directly related to eye position, except when the neuron is below threshold. The second is that the eye position signal is intrinsically generated by the network itself, not read out from some other area in the brain. This is clear from the fact that the inputs to the model network only briefly pulse during saccades and otherwise remain constant over time, yet the network maintains persistent neural activity between saccades that is proportional to the different eye positions.

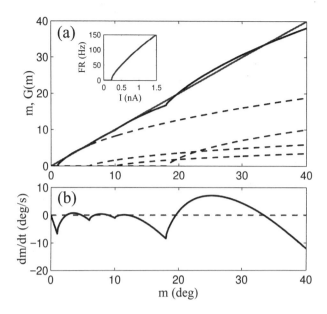

Figure 2. a) Approximation of a line attractor using 4 nonlinear neurons. The inset shows the rate-current relationship of an individual neuron. b) The resulting drift velocity of the eye as a function of eye position.

3. RATE-BASED MODEL

The task of tuning the synaptic weights of the conductance-based model was simplified by reducing it to a rate-based model. To carry out this reduction, the steady-state current-discharge relationship of an individual conductance-based model neuron was measured from numerical simulations and is shown in the inset to Fig. 2(a). Thus the biophysical complexities of the conductance-based model were reduced to a simple functional relationship between the instantaneous firing rate v_i of a neuron and the sum total u_i of its synaptic input currents

$$v_i = g(u_i) \, . \tag{1}$$

The dynamics of the synaptic currents u_i, in turn, depend linearly on the firing rates v_j of the presynaptic neurons

$$\tau_s \frac{du_i}{dt} + u_i = \sum_{j=1}^{N} W_{ij} v_j + h_i \, . \tag{2}$$

The synaptic weight W_{ij} describes the influence of neuron j firing on another neuron i. In addition, the feedforward inputs coming from outside the integrator onto neuron i are lumped together in the h_i term. The time scale τ_s of all the synapses was taken to equal $\tau_1 = 150$ ms in the conductance-based model. This reduced rate model is an excellent approximation to the conductance-based model when the firing rates change slowly.

The synaptic weight matrix of the model was taken to be the outer product form $W_{ij} = \xi_i \eta_j$. Such a form emerges naturally when a Hebb-like learning rule is used to train the network, as will be discussed in a future publication. This type of weight matrix causes all dynamical trajectories to be attracted to a line in state space of the form

$$u_i = m\xi_i + h_i \, . \tag{3}$$

Relaxation onto this line occurs on the short time scale τ_s. Once the network is on the line, its dynamics is described by the scalar variable m which serves as the internal representation of eye position. The dynamical equation for m can be found by substituting Eq. (3) into Eqs. (1) and (2) to yield

$$\tau_s \frac{dm}{dt} = \sum_{i=1}^{N} \eta_i g(m\xi_i + h_i) - m$$

$$\equiv G(m) - m .$$ (4)

The tuning procedure adjusts the synaptic weights such that the right hand side of Eq. (4) is very small, i.e. $G(m) \approx m$. Then the decay rate $dm/dt \approx 0$ and the variable m is approximately constant in time.

Eye position can be read out linearly from the network,

$$E = \sum_i \eta_i v_i .$$ (5)

If $G(m) \approx m$ in Eq. (4), then $E \approx m$ when the network is on the line of Eq. (3). Then the actual eye position E is equal to the internal representation m in the integrator. Furthermore,

$$v_i \approx g(E\xi_i + h_i)$$ (6)

when the network is on the line, so that the relationship between v_i and E is approximately linear above the firing threshold with a slope proportional to ξ_i. In other words, the position sensitivity of neuron i is directly proportional to the parameter ξ_i.

4. OPTIMAL APPROXIMATION

The tuning procedure is to minimize a cost function equivalent to the mean square drift velocity averaged over a range of m,

$$\left\langle \left| \sum_{i=1}^{N} \eta_i g(m\xi_i + h_i) - m \right|^2 \right\rangle_m .$$ (7)

The ξ_i and h_i parameters, and the range of m are fixed so that the ranges of firing rates, currents, and eye positions seen in the model correspond roughly to what is seen experimentally. The minimization is done with respect to the η_i parameters only.

If $g(u)$ were linear, then η_i could be tuned so that $G(m) = m$ in Eqs. (4) and (7) exactly. Such linearity has been the basis of previous network models of the integrator [3, 4, 5, 6]. This implies that the drift velocity dm/dt is zero at all positions m and that every point on the line is a fixed point of the dynamics. Thus Eq. (3) describes an attractive line of fixed points, otherwise known as a *line attractor* [2].

Fig. 2 illustrates what happens to the line attractor when $g(u)$ is nonlinear. The upper part of the figure shows the optimal approximation $G(m) \approx m$ achievable with four nonlinear neurons. It is clear that superpositions of scaled and translated versions of the current-discharge relationship $g(\xi_i m + h_i)$ produce a nonlinear curve that cannot exactly match the linear function m over the whole continuous range of m. Instead, the optimal η_i generates a nonzero drift velocity dm/dt from Eq. 4 that is shown in Fig. 2(b). In this case, the integrator network

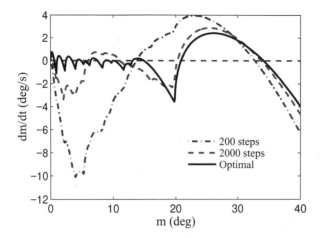

Figure 3. Training a network of 40 neurons using gradient descent. The drift velocity as a function of eye position is shown after 200 and 2000 iterations along with the optimal solution.

contains a finite number of fixed points corresponding to the zero crossings in Fig. 2(b) and therefore is only an approximate line attractor.

The form of the optimal approximation can vary, depending on biological assumptions which enter through the choice of the parameters ξ_i and h_i and the constraints on the η_i optimization. Currently we are investigating a wide variety of model networks resulting from different biological assumptions. Fig. 2 shows just one example, which was chosen to illustrate a point about the interplay between recruitment and sublinearity. The ξ_i were chosen randomly within a positive range, giving all neurons the same on-direction. The inputs h_i were randomly chosen such that the thresholds of the neurons were distributed over only the lower half range of m for which the cost function was minimized. Furthermore, the η_i were constrained to be nonnegative resulting in a nonnegative least squares optimization that can be done by standard methods. This constraint along with the positivity of the ξ_i gives rise to a network with only excitatory connections.

Figure 2 illustrates the importance of nonlinear recruitment in compensating for the sublinear behavior of neurons above threshold. This recruitment can also be seen in Fig. 1 as firing rates increase. In the region where there are no thresholds, the network is unable to compensate by recruiting additional neurons and the drift velocities become significantly worse. This results in the large drift at high firing rates which is seen in the simulation of the conductance-based model in Fig. 1.

5.   LEARNED APPROXIMATION

The least squares optimization of Eq. (7) was calculated using a nonbiological algorithm. As will be shown elsewhere, a biologically plausible synaptic learning rule can perform the same optimization in an online fashion. However, such learning rules generally lead to solutions that are only approximately optimal. A solution qualitatively similar to these suboptimal solutions is obtained by performing gradient descent on the cost function in Eq. (7) and stopping short of the true minimum. Figure 3 shows the result of 200 and 2000 projected gradient descent steps on a network of 40 neurons. Notice that the drift velocity tends to be greater and the number of fixed points is less than in the optimal solution. Although the quality of the solution improves as more gradient descent steps are taken, the rate of improvement slows over time because the cost function in Eq. (7) is ill-conditioned.

6. DISCUSSION

Our model suggests that a biological approximation to line attractor dynamics has two sources of error. One source is the finite error of the best approximation by nonlinear neurons. The other is the suboptimality of learned approximations. Both types of error lead to a drift velocity that depends on eye position. In our current research, we are trying to understand these two sources of error and their relative contribution in real networks.

Measurements of eye position in the dark generally reveal a position-dependent drift velocity [10], which can be compared with model results like those shown in Fig. 2(b). Our model is able to relate the physiological properties of neurons and synapses in the integrator with a behavioral quantity, the drift velocity of the eyes. Making experimentally testable predictions, however, is complicated because the drift velocity can vary depending on the biological assumptions that are made.

We are also exploring the effects of dilute connectivity on our model, along with the role of inhibition and excitation. The present model is able to qualitatively integrate its inputs in the sense that velocity-coded inputs cause changes in position. But achieving quantitatively correct linear integration of input signals from nonlinear neurons requires further tuning of the system [11, 12]. Furthermore, modeling integrator function in conditions when rates vary quickly, as during saccades and high velocity VOR, will require model neurons that exhibit spike-frequency adaptation [13].

ACKNOWLEDGMENTS

We are grateful to B. Shraiman for his assistance in the early stages of this investigation. We would also like to thank E. Aksay, R. Baker, B. Mensh, and H. Sompolinsky for helpful discussions.

REFERENCES

[1] D. A. Robinson. Integrating with neurons. *Annu. Rev. Neurosci.* 12, 33 (1989).

[2] H. S. Seung. How the brain keeps the eyes still. *Proc. Natl. Acad. Sci. USA* 93, 13339 (1996).

[3] S. C. Cannon, D. A. Robinson, and S. Shamma. A proposed neural network for the integrator of the oculomotor system. *Biol. Cybern.* 49, 127 (1983).

[4] S. C. Cannon and D. A. Robinson. An improved neural-network model for the neural integrator of the oculomotor system: more realistic neuron behavior. *Biol. Cybern.* 53, 93 (1985).

[5] D. B. Arnold and D. A. Robinson. A learning network model of the neural integrator of the oculomotor system. *Biol. Cybern.* 64, 447 (1991).

[6] D. B. Arnold and D. A. Robinson. A neural network model of the vestibulo-ocular reflex using a local synaptic learning rule. *Phil. Trans. R. Soc. Lond.* B337, 327 (1992).

[7] D. Hansel and H. Sompolinsky. Chaos and synchrony in a model of a hypercolumn in visual cortex. *J. Comput. Neurosci.* 3, 7 (1996).

[8] M. Serafin, C. de Waele, A. Khateb, P. P. Vidal, and M. Mühlethaler. Medial vestibular nucleus in the guinea-pig. I. Intrinsic membrane properties in brainstem slices. *Exp. Brain Res.* 84, 417 (1991).

[9] S. du Lac and S. G. Lisberger. Membrane and firing properties of avian medial vestibular nucleus neurons in vitro. *J. Comp. Physiol.* A176, 641 (1995).

[10] K. Hess, H. Reisine, and M. Dursteler. Normal eye drift and saccadic drift correction in darkness. *Neuroophthalmol.* 5, 247 (1985).

[11] F. Miles and S. Lisberger. Plasticity in the vestibulo-ocular reflex: a new hypothesis. *Annu. Rev. Neurosci.* 4, 273 (1981).

[12] S. du Lac, J. Raymond, T. Sejnowski, and S. Lisberger. Learning and memory in the vestibulo-ocular reflex. *Annu. Rev. Neurosci.* 18, 409 (1995).

[13] R. Quadroni and T. Knöpfel. Compartmental models of type A and type B guinea pig medial vestibular neurons. *J. Neurophysiol.* 72, 1911 (1994).

61

A SIMPLIFIED HIPPOCAMPAL MODEL THAT LEARNS AND USES THREE KINDS OF CONTEXT

William B. Levy

Department of Neurological Surgery
University of Virginia Health Sciences Center
Charlottesville, Virginia 22908
wbl@virginia.edu

1. ABSTRACT

Context plays a critical role in cognition. Previously Hirsh ('74), Kesner & Hardy (83), and Gray ('82) proposed that the hippocampus could learn context. Presented here is a simple hippocampal model that learns and uses three types of context: a context coming from the past, a context coming from the present, and a context concerned with the future. In all three of these situations, context is encoded by neurons that fire in a way analogous to hippocampal place cells. When these firing patterns do not appear, the network seems incapable of solving context dependent problems.

In psychology, the importance of context arises before the turn of the century (Boring, '50), in particular Titchner advocated a critical role for context in perception. Context is important to the networks that learn language in the schemes of Pollack ('90), Elman ('90), Jordan ('86), and Mozer ('92). Although not part of the usual terminology, context is at the heart of frames and schemas used by other cognitive psychologists. More to the point here, context learning seems to be part of hippocampal function (Hirsh, '74, Kesner & Hardy, '83, Gray, '82). Context learning is compatible with the Cohen and Eichenbaum theory of flexible memory (Cohen, '84; Eichenbaum et al., '92). Hirsh places the use of context at the center of proper encoding and recall of long-term memory. Context specifies the location of long-term memory storage. In this view, context is equivalent to episodic memory. Moreover, episodic memory associates disparate objects and events from single experiences; unfortunately, it is a lack of episodic memory that so hampers patients like H. M. and R. B. And, finally, even though not an explicit part of O'Keefe and Nadel's ('78) cognitive mapping theory (but see Nadel and Willner, '80), a coding that is analogous to hippocampal place cells (we call them context cells) — a coding that can be used to get from point A to point B — are context-based codes when viewed within the function of our model of the hippocampus.

To appreciate the role of context in memory, picture this one situation. You go to the hippocampal conference at Grand Cayman and for the first time you meet John Smith, a scientist from Seattle. A year or two later you visit the NIH and you see a vaguely familiar face; it's John Smith but you cannot remember his name. (As always striving for politeness as well as wishing to avoid embarrassment, you struggle to come up with a name to match the face.) If you can only remember the place, the circumstances, the episode where you met him, then you will have a chance of remembering the name. You well up a vague association of the conference room where you met and at the same time comes the hotel, the beach, and then ... "John, what a surprise seeing you here! How are you?"

In other words, the storage of unique events is intimately associated with the surrounding circumstances (context). Of course, the idea of context-dependent memory is a couple of thousand years old as exemplified by the Roman's method of loci for memorizing long speeches. One sequence (the speech) is learned by associating it with another sequence of patterns (the sequence of statues you pass as you walk through a well-known museum) by using each successive statue and its locus as the context for successive words and phrases in the speech.

Less grandiose forms of context are useful in many other types of cognitive processes. Thus, many cortical regions would need to produce context codes. But, context-based codes do seem particularly important for hippocampal functions including setting up appropriately retrievable stores of memories.

2. THE SIMPLIFIED MODEL

The problem of creating a code that appropriately reflects context can be seen as quite challenging when we consider the arbitrariness of patterns that might be associated - people with places or rhetoric on taxation to build some triremes with statues of Greek gods. Thus, it is pleasantly surprising that such a difficult problem can be solved (or at least begun to be solved) so simply. The extremely simplified, CA3-inspired network model which we have been studying (Levy, '89; Levy et al., '95; Levy & Wu, '96; Wu et al., '96) spontaneously and adaptively produces codes for context.

A generally accepted, gross hippocampal computational architecture (e.g. Levy, '89; Eichenbaum & Buckingham, '91; O'Reilly & McClelland, '94; Hasselmo & Schnell, '94) has three parts: 1) an input layer; 2) a recoder inspired by the hippocampal CA3 region, and 3) a decoder of the recoded signals inspired by the hippocampal CA1 region and its output targets. A CA3-like structure alone is all that is needed to create context codes, and to make this point, we do not include an explicit CA1 in our recent models (see Fig. 1a).

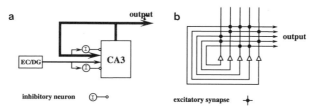

Figure 1. The Model. a. In the model the input layer is a combination of the entorhinal cortex and dentate gyrus. Accompanying this feedforward excitation is a proportional feedforward inhibition. The strong excitation of the network results from the recurrent connections, which is also accompanied by a feedback inhibition. The output of the network is the state of the excitatory CA3 cells themselves, and this is decoded by a simple cosine comparison. b. The recurrent excitatory synapses are sparse and randomly placed.

Figure 2. Pictorial representation of the enviroment. There are two sequences of 12 input patterns to learn. Each sequence contains three orthogonal segments. The two sequences share a subsequence of three patterns.

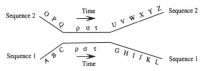

The specification of the CA3-like portion of the network has four essential aspects: (1) sparse recurrent excitatory connectivity (Fig. 1b) that produces more overall excitation than the external input; (2) a neuronal delay of at least one time step in converting an input to an output (i-o); (3) an associative modification rule that spans at least the i-o time step; and (4) some generic feedback inhibition that narrowly, but imperfectly compared to competitive networks, bounds total activity.

We have studied three abstract prediction problems to show that the network can learn and use three types of context: context coming from the past, context coming from the present, and context coming from the future. Context past is needed to solve sequence disambiguation (Minai et al., '94; Levy et al., '95; Wu et al., '96) (Fig. 2 and Fig. 3a). Context present is needed to solve the configural learning problem of transverse patterning (Levy et al., '96; Wu et al., '97). Context future is needed to solve goal finding problems (Levy et al., '95: Wu & Levy '96; Levy & Wu, '97) (Fig. 2 and Fig. 3b). In goal finding, a desired characteristic of the goal, but not its entire code, is part of the input. Note in Figure 3b how the network, when given a small fraction of the goal code, is able to overcome its natural tendencies (as shown by the sequence disambiguation result of Fig. 3a) and produce the appropriate path to the goal.

As opposed to simple sequence completion problems, these three context dependent problems require the network to construct cell firing patterns that we call local context firings. E.g., compare Fig. 1a and 1b of Wu et al. ('97) which shows two networks — the Fig.

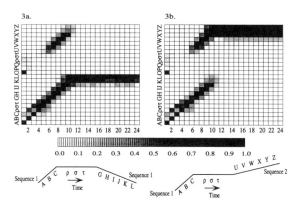

Figure 3. Sequence prediction after learning the two partially overlapping sequences illustrated in Fig. 2. These similarity matrices use the cosine function. They compare the network states generated by each full pattern sequence after learning (ordinate) with the network states generated over time during testing (abscissa). Decode the output during testing by finding the darkest square in each column; the letter on the ordinate is the decoded answer. 3a. Sequence completion in the disambiguation problem is made difficult by the shared subsequence [ρστ]. Here we show the similarity values (used for decoding) when the network is transiently given pattern A. Appropriately enough for the learning and for the starting point, the states go to pattern L, a pattern essentially orthogonal to the representation of the other learned goal pattern, Pattern Z. 3b. When the same network is given the same transient input but two neurons of goal Z are also turned on, the network produces a sequence of representations leading to this partially specified goal. To create this path, the network must produce a novel sequence that appropriately combines its knowledge of the two separately learned sequences.

1b network learns the configural learning problem of transverse patterning, while the other does not. The repetitive cell firings of the recurrently driven neurons of Fig. 1b of Wu et al., ('97) are examples of local context neurons, the hypothesized analog of hippocampal place cells.

Finally, using as its inputs the compressed spontaneous replay of learned sequences coming out of the hippocampus (August & Levy, '96, '97), sequences of context codes could be learned by the cerebral cortex. These sequences — representing longer time spans — would be associated together by the cerebral cortex followed, recursively, by further compression in CA3 and association in cerebral cortex.

In summary, the CA3 computational model learns and uses three different kinds of context. The model reveals particular cell firing patterns that are, by hypothesis, the code for context. Further, we conjecture that this model is the appropriate neural basis of hippocampal theories that learn context and teach context to the cerebral cortex.

3. ACKNOWLEDGMENTS

This work was supported by NIH MH48161 and MH00622, by EPRI RP8030-08, and Pittsburgh Supercomputing Center Grant BNS950001 to WBL, and by the Department of Neurosurgery, Dr. John A. Jane, Chairman.

4. REFERENCES

August. D. A. & Levy, W. B (1996) Temporal sequence compression by a hippocampal network model. INNS World Congress on Neural Networks, 1299–1304.

August, D. A. & Levy, W. B (1997) Spontaneous replay of temporally compressed sequences by a hippocampal network model. CNS*96, this proceedings.

Boring, E.G. (1950) A history of experimental psychology. 2nd edition. New York:Appleton Century Crofts.

Cohen, N.J. (1984) Preserved learning capacity in amnesia: Evidence for multiple memory systems. In: Neuropsychology of Memory (L.R. Squire & N. Butters, Eds.), New York:Guilford Press, pp. 83–103.

Eichenbaum, H. & Buckingham, J. (1991) Studies on hippocampal processing: Experiment, theory, and model. In: Neurocomputation and Learning: Foundations of Adaptive Networks. (M. Gabriel & J. Moore, Eds.), Cambridge: MIT Press, pp. 171–231.

Eichenbaum, H., Otto, T. & Cohen, N.J. (1992) The hippocampus - what does it do? Behav. Neural Biol., 57:2–36.

Elman, J.L. (1990) Finding structure in time. Cog. Sci., 14:179–211.

Gray, J.A. (1982) The neuropsychology of anxiety: an enquiry into the functions of the septo-hippocampal system. Oxford University Press:New York.

Hasselmo, M.E. & Schnell, E. (1994) Laminar selectivity of the cholinergic suppression of synaptic transmission in rat hippocampal region CA1: computational modeling and brain slice physiology. J. Neurosci., 14:3898–3914.

Hirsh, R. (1974) The hippocampus and contextual retrieval of information from memory. Behav. Biol., 12:421–444.

Jordan, M.I. (1986) Attractor dynamics and parallelism in a connectionist sequential machine. In: Proceedings of the Eighth Conference of the Cognitive Science Society. Hillsdale, NJ: Lawrence Erlbaum Assoc. Inc., pp 531–546.

Kesner, R.P. & Hardy, J.D. (1983) Long-term memory for contextual attributes: Dissociation of amygdala and hippocampus. Behav. Brain Res. 8, 139–149.

Levy, W.B (1989) A computational approach to hippocampal function. In: Computational Models of Learning in Simple Neural Systems. (R.D. Hawkins & G.H. Bower, Eds.), New York: Academic Press, pp. 243–305.

Levy, W. B & Wu, X. B. (1996) The relationship of local context codes to sequence length memory capacity. Network 7, 371–384.

Levy, W. B & Wu, X. B. (1997) Predicting novel paths to goals by a simple, biologically inspired neural network. CNS*96, this proceedings.

Levy, W. B, Wu, X. B. & Baxter R. A. (1995) Unification of hippocampal function via computational/encoding considerations. In: Proceedings of the Third Workshop: Neural Networks: from Biology to High Energy Physics. International J. of Neural Sys. 6 (Supp.), 71–80.

Levy. W.B, Wu, X. B. & Tyrcha, J.M. (1996) Solving the transverse patterning problem by learning context present: A special role for input codes. INNS World Congress on Neural Networks, 1305–1309.

Minai, A. A., Barrows, G. L., & Levy, W. B (1994) Disambiguation of pattern sequences with recurrent networks. INNS World Congress on Neural Networks, IV-176–181.

Mozer, M.C. (1992) Induction of multiscale temporal structure. In: Advances in Neural Information Processing Systems, 4. (J.E. Moody, S.J. Hanson & R.P. Lippmann, Eds.), San Mateo, CA:Morgan Kauffman, pp. 275–282.

Nadel, L. & Willner, J. (1980) Context and conditioning: A place for space. Physiol. Psychol. 8, 218–228

O'Keefe, J & Nadel, L. (1978) The Hippocampus as a Cognitive Map. Oxford: Oxford Univ. Press.

O'Reilly, R.C. & McClelland, J.L. (1994) Hippocampal conjunctive encoding, storage, and recall: avoiding a tradeoff. Hippocampus 4, 661–682.

Pollack, J.B. (1990) Recursive distributed representations. Artificial Intelligence 46:77–105.

Wu, X. B., Baxter, R. A. & Levy, W. B (1996) Context codes and the effect of noisy learning on a simplified hippocampal CA3 model. Biol. Cybern. 74, 159–165.

Wu, X. B. & Levy, W. B (1996) Goal finding in a simple, biologically inspired neural network. INNS World Congress on Neural Networks, 1279–1282.

Wu, X. B., Tyrcha, J. M. & Levy, W. B (1997) A special role for input codes in solving the transverse patterning problem. CNS*96, this proceedings.

62

SHORT TERM MEMORY FUNCTION IN A MODEL OF THE OLFACTORY SYSTEM

Christiane Linster[*] and Michael E. Hasselmo

Department of Psychology
Harvard University
33, Kirkland Street, Cambridge, Massachusetts 02138
linster@berg.harvard.edu; hasselmo@berg.harvard.edu

ABSTRACT

In a computational model of the olfactory bulb and piriform cortex, we show how the self-organized representation of olfactory stimuli, in the form of activated subsets of pyramidal cells, spreads from rostral to caudal olfactory cortex with increasing stimulus familiarity. In this framework, we propose a mechanism for the performance of olfactory delayed match-to-sample function, based on short-term synaptic changes during stimulus presentation.

1. INTRODUCTION

The olfactory bulb and the piriform (olfactory) cortex constitute the primary olfactory pathways in vertebrates. Most experimental data suggest that the olfactory bulb (OB) plays a crucial role in feature extraction, noise reduction and contrast enhancement (for reviews see Holley, 1991; Scott et al., 1993), whereas more cognitive functions such as the associative storage of olfactory information would be located in the olfactory cortex (Haberly and Bower, 1989). Computational models of the OB and the insect antennal lobe have demonstrated the capability of this circuitry to perform feature extraction in a highly unstable molecular input space (White et al. 1990; Linster and Masson 1996; Masson and Linster 1996; Linster and Gervais 1996; Linster and Hasselmo 1996; Li and Hopfield 1989). Computational models of piriform cortex on the other hand have demonstrated the capability of this structure to act as an associative memory (Haberly and Bower 1989; Hasselmo 1993). We have recently developed a combined model of the olfactory bulb and

---

[*] To whom correspondence should be addressed

Computational Neuroscience
edited by Bower, Plenum Press, New York, 1997

the piriform cortex, allowing us to study the respective roles of the two structures in olfactory processing (Linster et al. 1995). Here, we show how associative learning during successive sniff cycles leads to a spread of the olfactory information from rostral to caudal piriform cortex pyramidal cells. We propose a mechanism for delayed match-to-sample in an olfactory short-term memory task based on short-term potentiation of association fibers in the piriform cortex.

2. METHODS

We utilize a model combining elements of the previously developed models of olfactory bulb and piriform cortex (Linster and Gervais, 1996; Linster and Hasselmo, 1996; Hasselmo, 1993; Barkai et al., 1994). In this model, connections from the olfactory bulb to the piriform cortex are made by way of mitral cell synapses onto pyramidal cells and feedforward local interneurons (Figure 1). Mitral cell axons project onto the piriform cortex with a connection probability and strength that decreases from rostral to caudal piriform cortex (Haberly 1985; Wilson and Bower 1992). As a consequence, rostral pyramidal cells and interneurons can be directly driven by mitral cell input, whereas more caudal pyramidal cells are not driven by mitral cell input alone. In the model, before learning, the intrinsic connections between pyramidal cells are weak and uniformly distributed and pyramidal cells need strong input from the OB in order to increase their spiking probability. In both the olfactory bulb and the piriform cortex model, parameters are chosen so as to reproduce available electrophysiological data showing field potentials induced by electrical stimulation, single unit activity in the OB and PC, and EEG recordings from both structures (Linster and Gervais 1996; Linster et al. 1995; Barkai et al. 1994). Feedback connections from pyramidal cells onto granule cells have been included in the model; these are essential for the synchronization of the two structures.

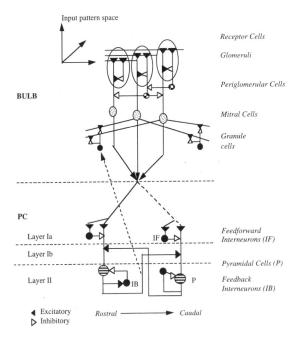

Figure 1. Schematic representation of the model. Olfactory receptor cells synapse with mitral cell primary dendrites and periglomerular cells (PG) in the olfactory bulb glomeruli, where PG cells and mitral cells interact. Mitral cell secondary dendrites interact with other mitral cells via inhibitory interneurons, the granule cells. Mitral cell axons connect to piriform cortex pyramidal cells and feedforward interneurons with a decreasing probability from rostral to caudal piriform cortex regions. Two types of interneurons interact with pyramidal cells via GABA-A and GABA-B type synapses. Pyramidal cells interact with other pyramidal cells through modifiable excitatory connections.

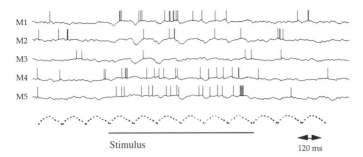

Figure 2. Across fiber pattern of mitral cell activities. The membrane potentials and action potentials of 5 mitral cells (M1-M5) in response to an olfactory stimulus are shown. Dotted lines show time course of low noise input mimicking respiration, over which olfactory stimuli (solid line) are added.

3. RESULTS

3.1. Learning during Successive Sniff Cycles

In a first step, we have investigated the learning of olfactory stimuli conveyed to the pyramidal cells from the mitral cells after preprocessing in the olfactory bulb. In response to an olfactory stimulus, an across fiber pattern of mitral cell activities is conveyed to the piriform cortex (Figure 2).

A learning rule which takes into account pre- and postsynaptic spikes in pyramidal cells strengthens association fibers between pyramidal cells during presentation of an input stimulus during successive sniff cycles (Barkai et al., 1994; Hasselmo and Barkai, 1995). The connection strengths are normalized in such a way that the total presynaptic strength of each pyramidal cell stays constant. In addition, the total sum of connection strength between all pyramidal cells is limited to an upper value in order to ensure stability of the system. Due to the higher connection probability of mitral cells towards more rostral pyramidal cells, intrinsic connections between these cells get strengthened during the first few sniff cycles. When rostral cells are driven to high spiking frequencies by olfactory input, they start to activate more caudal cells and the feedforward connections between these cells are strengthened (Figure 3).

Eventually, if the learning process lasts a few sniff cycles, intrinsic connections from rostral to caudal and between caudal cells are strengthened. In other words, the representation of olfactory stimuli in the piriform cortex model spreads from rostral to caudal regions of the piriform cortex during subsequent sniff cycles (Figure 4 and Figure 5A,B).

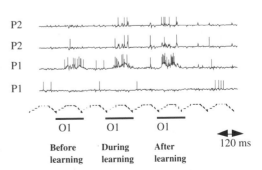

Figure 3. Learning of olfactory stimuli in the combined model of olfactory bulb and cortex. The neural representation of an olfactory stimulus spreads from rostral (P1) to caudal (P2) pyramidal cells during learning. Before learning, only P1 cells respond with increased spiking frequencies to input (O1). After learning, intrinsic connections between pyramidal cells are strengthened and P2 cell start responding.

Novel stimuli lead to patterns of spiking pyramidal cells in the rostral area, in response to input from the OB. After several sniff cycles, more caudal pyramidal cells start firing in response to the stimulus, and additional patterns of activity appear in the caudal area. As a consequence, more caudal pyramidal cells will fire more strongly in response to familiar stimuli than in response to novel stimuli.

3.2 Rapid Learning of Previously Stored Patterns in a Delayed-Match-to-Sample Task Can Result in Short Term Memory of the Sample Stimulus without Sustained Neural Activity

As shown above, the representation of stimuli (the distribution of active pyramidal cells) spreads from rostral to caudal when learning occurs during several sniff cycles. After learning during several sniff cycles, the synaptic connections in the rostral region are mostly saturated, and any additional learning will result in strengthening of synaptic connections in the caudal region. In a delayed match-to-sample task, one of several highly familiar stimuli (O1) is presented for a short duration (one cycle) (Figure 5). After a delay, a second stimulus, also highly familiar, is presented, which is either the same stimulus (O1-match) or a different stimulus (O2-non-match). During the first presentation, intrinsic sy-

A

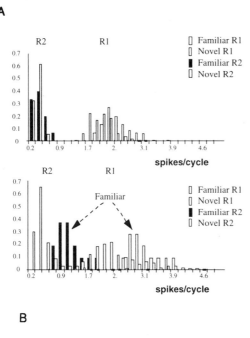

B

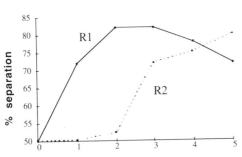

Figure 4. The neural representation of olfactory stimuli spreads from rostral to caudal pyramidal cells during subsequent sniff cycles. A: Histograms of spikes/sniff cycle in rostral (R1) and caudal (R2) pyramidal cells in response to presentation of familiar and novel stimuli after learning occurs during a single sniff cycle (top graph), and after learning occurs during four cycles (lower graph). The results are based on 100 different network realizations. Each simulation run had different, randomly chosen stimuli each activating 10% of the mitral cells in the model. B: The ratio of correct choices (% separation) between familiar and novel stimuli based on numbers of spikes in rostral (R1) and caudal (R2) pyramidal cells increases with learning during multiple cycles. The choice is made as a function of the total number of spikes elicited in rostral and caudal pyramidal cells.

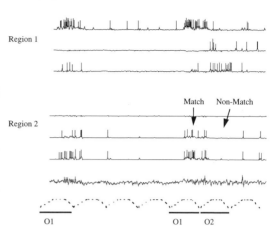

Figure 5. Delayed match-to-sample based on spike rates of pyramidal cells. After exposure to a highly familiar pattern (O1) during a single sniff cycle, a subsequent presentation of O1 results in increased spiking activity in rostral (Region 1) and in caudal (Region 2) pyramidal cells (match) while the presentation of a different, also highly familiar stimulus, (O2) does not lead to increased spiking in caudal pyramidal cells (non-match). Dotted lines show time course of low noise mimicking respiration, to which olfactory stimuli (solid lines) are added.

naptic connections between caudal pyramidal cells are strengthened. This short presentation of one of the stimuli results in short term synaptic changes, strongly activating caudal pyramidal cells. A comparison between the pyramidal cell responses in the caudal region to the sample (match) or to one of the other stimuli (non-match) yields over 85% percent correct choices after learning of the sample stimulus when number of spikes are counted.

If we assume passive synaptic decay with a long time constant (1 min.) for non reinforced synapses, we can obtain deterioration of the short term memory as seen in behavioral experiments (Ravel et al. 1994) using an olfactory delayed match-to-sample task (Figure 6).

4. DISCUSSION

We here propose a mechanism for olfactory short term memory based on short-term synaptic changes in the piriform cortex. In our model, the representation of olfactory stimuli (i.e. the distribution of activated pyramidal cells) spreads from rostral to caudal regions of the piriform cortex with increasing stimulus familiarity. This result is based mainly on the observation that the afferent fiber inputs from the olfactory bulb mitral cells to the pyramidal cells decrease towards caudal regions of the piriform cortex (see Haberly, 1985 for review). Interestingly, recent observations using optical recording methods have shown that the distribution of the signal evoked by olfactory stimulation in the piriform cortex differs between novel and previously reinforced stimuli (Litaudon et al. 1995).

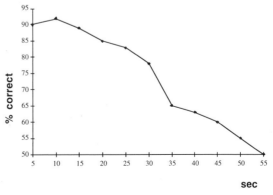

Figure 6. Time course of delayed match-to-sample Synaptic weights which are not reinforced decay with a long (1 min.) time constant. This leads to a decay in time of the correct choice (after presentation during a single sniff) between the matched or non-matched stimulus when this choice is based on the total number of spikes elicited in the caudal pyramidal cells.

Based on the fact that most olfactory short-term memory paradigms use highly familiar stimuli (Ravel et al., 1994; Schoenbaum and Eichenbaum, 1995), we here show how a mechanism based on short-term synaptic changes, using highly familiar stimuli, can solve a short term memory task. Such a short-term potentiation of association fibers, but not afferent fibers has been observed experimentally in brain slices of olfactory cortex (Kanter and Haberly 1990). In contrast to other models of short term memory (Zipser, 1991; Linster and Masson, 1996) our model does not necessitate sustained neural activity during the delay. Electrophysiological studies have investigated the neural responses in the piriform cortex during similar olfactory memory tasks (Schoenbaum and Eichenbaum, 1995) and have found a wide diversity of responses patterns during different phases of the behavioral task. Although it has been shown that several cortical regions can retain specific information about stimuli through persistent pyramidal cell activity (Fuster and Jervey 1982) in delayed match-to-sample tasks, this type of mechanism is not the only possible means for maintaining trial-specific information.

REFERENCES

Barkai, E. Horwitz, G., Bergman, R.E. and Hasselmo, M.E. (1993) Long-term potentiation and associative memory function in a biophysical simulation of piriform cortex. *Soc. Neurosci. Abstr.* 19: 376.3.

Fuster, J.M. and Jervey, J.P. (1981) Inferotemporal neurons distinguish and retain behaviorally relevant features of visual stimuli. *Science*, 212: 952–955.

Haberly, L.B. (1985) Neuronal circuitry in olfactory cortex: anatomy and functional implications. *Chem. Sens.* 10: 219–238.

Haberly L.B. and Bower J.M. (1989) Olfactory cortex: model circuit for study of associative memory. *Trends in Neuroscience*, 12: 258–264.

Hasselmo, M.E. (1993). Acetycholine and learning in a cortical associative memory *Neural Comp.* 5: 32–44.

Hasselmo, M.E. and Barkai, E. (1995) Cholinergic modulation of activity-dependent synaptic plasticity in rat piriform cortex. *J. Neurosci.* 15: 6592–6604.

Holley, A. (1991) Neural coding of olfactory information. In Smell and Taste in Health and Disease, edited by T.V. Getchell et al. New York, Raven Press: 329–343.

Kanter, E.D., Haberly, L.B. (1990) NMDA-dependent induction of long-term potentiation in afferent and association fiber systems of piriform cortex in vitro. Brain Res. 525:175–179.

Li, Z. and Hopfield, J.J. (1989) Modeling the olfactory bulb and its neural oscillatory processings. *Biol. Cybern.*, 61:379–392.

Linster C., Hasselmo M.E. and Gervais R. (1995) *Soc. of Neuroscience Abst.*

Linster, C. and Hasselmo, M. (1996) Modulation of inhibition in a model of olfactory bulb reduces overlap in the neural representation of olfactory stimuli, *Beh. Brain. Res.* (accepted for publication).

Linster, C. and Masson, C. (1996) A neural model of olfactory sensory memory in the honeybee's antennal lobe. *Neural Computation*, 8 (1) : 94–114.

Linster, C. and Gervais, R. (1996) Investigation of the role of interneurons and their modulation by centrifugal fibers in a neural model of the olfactory bulb. *Journal of Computational Neuroscience*, 3: 225–246..

Masson, C. and Linster, C. (1996) Towards a cognitive understanding of odor discrimination: combining experimental and theoretical approaches, *Behavioral Processes* 35: 63–82.

Ravel, N., Elaagouby, A. and Gervais, R. (1994) Scopolamine injection into the olfactory bulb impairs short-term olfactory memory. *Behav. Neurosci.*, 108: 317–324.

Schoenbaum, G. and Eichenbaum, H. (1995) Information coding in the rodent prefrontal cortex. I. Single-neuron activity in orbitofrontal cortex compared with that in pyriform cortex. *J. Neurophysiol.* 74: 733–750.

Scott, J.W., Wellis, D.P., Rigott, M.J. and Buonoviso, N. (1993) Functional organization of the main olfactory bulb. *Micros. Res. Tech.*, 24:142–156.

White, J., Hamilton, K.A., Neff, S.R. and Kauer, J.S. (1992) Emergent properties of odor information coding in a representational model of the salamander olfactory bulb. *J. Neurosci.*, 5: 1772–1780.

Wilson, M.A. and Bower, J.M. (1992) Cortical oscillations and temporal interactions in a computer simulation of piriform cortex. *J. Neurophysiology*, 67: 193–198.

Zipser, D. (1991) Recurrent network model of the neural mechanism of short-term active memory. *Neural Comp.* 3: 179–193.

63

MODELING THE MCN1-ACTIVATED GASTRIC MILL RHYTHM

The Interaction between Fast and Slow Oscillators

Yair Manor, Farzan Nadim, Michael P. Nusbaum, and Eve Marder

Brandeis University
Waltham, Massachusetts

INTRODUCTION

The gastric mill rhythm in the stomatogastric ganglion (STG) of the crab *Cancer borealis* is a network-generated oscillation with a period of 7 to 15 s. This rhythm can be elicited by tonic stimulation of modulatory commisural neuron 1 (MCN1)[1]. At the heart of the circuit (Figure 1A) is a half-center oscillator, consisting of the lateral gastric neuron (LG) and interneuron 1 (Int1).

When MCN1 is silent, LG shows subthreshold activity and Int1 fires action potential bursts that are time-locked with another STG rhythm, the faster pyloric rhythm[2] (Figure 1B). Action potentials in MCN1 elicit EPSPs in LG, Int1 and the dorsal gastric neuron (DG). The MCN1 excitation of these neurons results in a rhythm in which LG bursts in

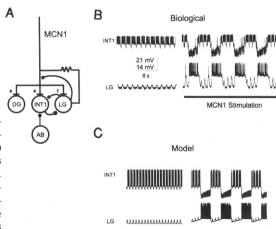

Figure 1. A, Schematic diagram of the gastric mill circuit. B, Stimulation of MCN1 activates the gastric mill rhythm (period 7–15 s) in the crab. When MCN1 is not active, LG is quiescent whereas Int1 fires in pyloric time. When MCN1 is stimulated, LG and Int1 oscillate as a half-center oscillator. C, the (canonical) model was tuned to reproduce the known physiological behavior. (A & B adapted from Coleman et al.[3]).

antiphase with Int1 and DG. DG receives no synaptic input from either LG or Int1, nor does it make synaptic connections back to the circuit (Figure 1A). LG presynaptically inhibits the STG terminals of MCN1 and is also electrically coupled to MCN1[1].

On the basis of the anatomical and physiological data, Coleman et al.[3] suggested the following verbal model for the MCN1-activated gastric mill rhythm. When MCN1 is activated, it excites Int1 rapidly, and produces a slower excitation of LG. Because of the fast excitation, Int1 bursts and through inhibition prevents LG from bursting at the same time. LG continuously receives excitation from MCN1, however, and so it slowly depolarizes. Eventually, LG escapes from Int1 inhibition and starts to fire a burst. The LG burst inhibits Int1, which stops firing and hyperpolarizes. At the same time, LG presynaptically inhibits the terminals of MCN1, thereby shutting off the chemical excitation to itself, Int1 and DG. LG remains bursting, partially because it continues to receive electrical excitation from MCN1. The electrical coupling by itself is not sufficient to sustain the LG firing, and as the effect of the chemical excitation from MCN1 wanes, the LG burst terminates and the cycle repeats.

We built a biophysical model of this network to explore the sensitivity of the rhythm to the various components of the circuit, and thereby to determine the important factors in the generation and modulation of this gastric mill rhythm. This model led to a surprising result that the strength of the pyloric rhythm-timed inhibition to Int1 was critical for setting the gastric mill period.

METHODS

We constructed a computational model based on the verbal model suggested by Coleman et al.[3] MCN1 was modeled with two compartments, one representing the axonal terminals and the second representing the rest of the neuron. We used Hodgkin-Huxley type neurons that were capable of producing action potentials. These neurons were connected using decaying chemical synapses. We also added a rhythmic inhibition to Int1 (absent in the verbal model of Coleman et al.[3]). All simulations were carried out with the NEURON[4] package.

RESULTS

The parameters of the model were tuned such that known physiological behaviors were reproduced. For example, without MCN1 stimulation, Int1 fired in pyloric time and LG showed subthreshold activity (compare Figure 1B and 1C). When MCN1 was stimulated, LG and Int1 oscillated in alternation, with a period of 7 s (Figure 1C). The strength of the electrical coupling between MCN1 and LG was tuned such that it could not sustain firing in LG without additional chemical excitation (not shown). We refer to the set of parameters of Figure 1C as the *canonical* model.

To examine the sensitivity of the gastric mill cycle period to parameters, we varied each parameter of the canonical model. The results are summarized in Figure 2. Each group in the histogram represents a different parameter. The horizontal line shows the gastric mill cycle period value for the canonical model. Within each group, the parameter was varied by −20% to +20% of the canonical model. Model oscillations were somewhat sensitive to the strength of the reciprocally inhibitory synapses between LG and Int1, and relatively insensitive to the electrical coupling conductance, the MCN1 synaptic strengths and time constants. By far, we found that the gastric mill cycle period was most sensitive

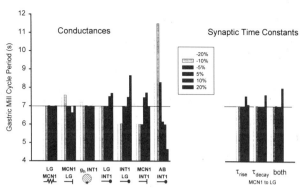

Figure 2. Sensitivity of the model gastric mill period to variation in maximal conductances and synaptic time constants. Each parameter (group) was varied incrementally up to '''20% of its canonical value. Horizontal line represents the gastric mill cycle period of the canonical model.

to the synaptic conductance of the pyloric inhibition to Int1. Specifically, as the synaptic conductance decreased, the gastric mill cycle period became longer, up to a point where the gastric mill oscillations were disrupted and replaced by tonic activity in Int1 and a quiescent state of LG (not shown).

These results suggested that the pyloric inhibition has a crucial role in the generation of the MCN1-activated gastric mill rhythm. Following this prediction, we examined the variation of the gastric mill cycle period as a function of the pyloric period. Figure 3A shows the traces of the model oscillations for pyloric periods of 900 ms and 1300 ms. The gastric mill period was longer for the longer pyloric period. To assess the effect of pyloric period on the gastric mill period, a 100 s simulation was run at a fixed pyloric period, and the gastric mill periods were measured. This procedure was repeated for different pyloric periods, from 0.5 s to 3 s with increments of 50 ms (Figure 3B). The solid lines represent integer multiples of the pyloric period. All gastric mill periods fell along these lines, showing that the gastric mill rhythm oscillated with a period that was an integer multiple

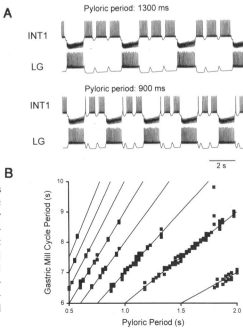

Figure 3. The period of the model gastric mill rhythm is an integer multiple of, and varies linearly with, the pyloric period. A, Traces of the model gastric mill oscillations for pyloric periods of 900 ms and 1300 ms. B, Gastric mill period is plotted versus pyloric period. Solid lines represent integer multiples of the pyloric period. When the model pyloric period was varied from 0.5 to 2 s, the gastric mill period increased in a *piecewise* linear fashion (i.e., at transition points the gastric mill period switched from n×pyloric periods to (n−1)×pyloric periods) and was restricted to 6 to 10 s.

of the pyloric period (P_{pyl}). There were transition regions where the gastric mill period switched from $n \times P_{pyl}$ to $(n-1) \times P_{pyl}$. In these regions, the model would produce oscillations at both periods. Thus, the gastric mill period increased in a *piecewise* linear fashion with the pyloric period. The integer relationship between the model gastric mill and pyloric periods could be explained by the strong likelihood that LG would make the transition from its inhibited phase to the burst phase when Int1 was inhibited by the pyloric circuit (thereby releasing LG from inhibition). This can also be seen in Figure 3A, where the transition from Int1 burst to LG burst occurred always at the same pyloric phase (as did the transition from LG burst to Int1 burst).

SUMMARY

In contrast with intuitions derived from physiological data alone, we suggest that inhibition from a fast oscillator onto one of the elements of the half-center is essential to produce the rhythmic alternation between LG and Int1. The gastric mill rhythm may be dramatically affected by the synaptic strength of the pyloric inhibition to Int1 (as seen in Figure 2), and to a lesser extent by alteration of the pyloric period (see Figure 3B). This provides a mechanism in which the gastric mill and the pyloric rhythms can be separately modulated when necessary, but still interact with each other. The predictions of this model are now beeing tested in the physiological preparation.

ACKNOWLEDGMENTS

This research was supported by NS17813, MH46742, and the Sloan Center for Theoretical Neurobiology at Brandeis University.

REFERENCES

1. M.J. Coleman and M.P. Nusbaum, 1994, Functional consequences of compartmentalization of synaptic input, *J. Neurosci.* **14**:6544–6552.
2. J.M. Weimann, P. Meyrand and E. Marder, 1991, Neurons that form multiple pattern generators: identification and multiple activity patterns of gastric/pyloric neurons in the crab stomatogastric system, *J. Neurophysiol.* **65**:111–122.
3. M.J. Coleman, P. Meyrand and M.P. Nusbaum, 1995, A switch between two modes of synaptic transmission mediated by presynaptic inhibition, *Nature* **378**:502–505.
4. M. Hines, 1993, NEURON- a program for simulation of nerve equations, in: *Neural Systems: Analysis and Modeling*, (F. Eeckman ed.), pp. 127–136, Kluwer Acad. Pub, Boston.

64

MODELING DYNAMIC RECEPTIVE FIELD CHANGES IN PRIMARY VISUAL CORTEX USING INHIBITORY LEARNING

Jonathan A. Marshall and George J. Kalarickal

Department of Computer Science
CB 3175, Sitterson Hall, University of North Carolina
Chapel Hill, North Carolina 27599–3175
marshall@cs.unc.edu, kalarick@cs.unc.edu

ABSTRACT

The position, size, and shape of the visual receptive field (RF) of some primary visual cortical neurons change dynamically, in response to artificial scotoma conditioning in cats[7] and to retinal lesions in cats and monkeys.[3] The "EXIN" learning rules[6] are used to model dynamic RF changes. The EXIN model is compared with an adaptation model[11] and the LISSOM model.[9,10] To emphasize the role of the lateral inhibitory learning rules, the EXIN and the LISSOM simulations were done with *only* lateral inhibitory learning. During scotoma conditioning, the EXIN model without feedforward learning produces centrifugal expansion of RFs initially inside the scotoma region, accompanied by increased responsiveness, without changes in spontaneous activation. The EXIN model without feedforward learning is more consistent with the neurophysiological data than are the adaptation model and the LISSOM model. The comparison between the EXIN and the LISSOM models suggests experiments to determine the role of feedforward excitatory and lateral inhibitory learning in producing dynamic RF changes during scotoma conditioning.

1. INTRODUCTION

1.1. Dynamic Receptive Fields

In experiments using artificial scotoma conditioning[7] and retinal lesions,[3] primary visual cortical neurons corresponding to a particular region of visual space were deprived of visual stimulation, while primary visual cortical neurons corresponding to a surrounding region received visual stimulation. In response to these manipulations, a variety of dynamic changes occurred in the position, size, and shape of the receptive field (RF) of some

Computational Neuroscience
edited by Bower, Plenum Press, New York, 1997

primary visual cortical neurons. For example, after 15 minutes of artificial scotoma conditioning, the RF size of some primary visual cortical neurons corresponding to the inside of the scotoma expanded by a factor of five in area; subsequently, after 15 minutes of normal stimulation, the RFs returned to their original size.[7]

Dynamic receptive fields are of interest for several reasons. They reveal some of the ways in which visual systems may adaptively overcome damage from lesions or scotomas. In addition, they reveal some of the functional organization of visual cortex. Dynamic visual receptive fields might also be related to the dynamic response properties found in other cortical areas, such as the tactile RF expansion/contraction found in somatosensory cortex in response to intracortical microstimulation.[8]

1.2. Three Models of Dynamic RFs

A neural network model that exhibits similar dynamic RF changes was formulated and tested. The EXIN model[6] is compared with the "LISSOM" model[9,10] and an "adaptation" model.[11] The EXIN model[6] uses an instar feedforward excitatory learning rule[5] and an outstar lateral inhibitory learning rule[6] (see Section 2.1).

The LISSOM model[9,10] uses synaptic modifications to account for changes in RFs. The LISSOM model uses feedforward excitatory, lateral excitatory, and lateral inhibitory learning rules with weight normalization. All of the LISSOM learning rules are instar rules[5] (see Section 2.2).

In the adaptation model,[11] the RF changes occur as a result of adaptive modifications in the sensitivity of *single* neurons, rather than as a result of modifications in the synaptic weights between *pairs* of neurons (see Section 2.3).

The three models were simulated, and their performance was compared with the experimental neurophysiological data.[3,7] To emphasize the role of lateral inhibitory learning, the EXIN model was simulated without feedforward learning, and the LISSOM model was simulated without feedforward and lateral excitatory learning.

2. METHODS

A patch of primary visual cortical neurons, arranged in a 30×30 grid of spatial positions, was simulated. The position of each neuron's RF corresponded to the neuron's position in the grid. Adjacent RFs initially had more than 50% spatial overlap.

In *scotoma/normal* conditioning, random stimulation in the whole visual field except in an "artificial scotoma" region was followed by whole-field stimulation.

Three sets of learning/adaptation rules, described below, were used. The critical difference between the three sets of rules is how they model the changes in lateral interactions within the cortex.

2.1. EXIN Network Simulation of Dynamic RFs

In the EXIN model[6] of dynamic RFs, cortical interactions are mediated by lateral inhibitory pathways. The weights Z_{ij}^{-} of lateral inhibitory pathways from neuron i to neuron j vary according to an asymmetric anti-Hebbian outstar learning rule:

$$(d/dt)\, Z_{ij}^{-} = \delta\, g(x_i)\, (-Z_{ij}^{-} + q(x_j)), \qquad (1)$$

where $\delta > 0$ is a small learning rate constant and g and q are half-rectified nondecreasing functions. (The EXIN model can also accommodate lateral excitatory pathways and has been used in models of other visual cortical functions with lateral excitation; however, the lateral excitatory pathways are not strictly necessary in the EXIN model of dynamic RFs.)

In an outstar learning rule,[5] *pre*synaptic activity "enables" the learning at a synapse; when the learning is enabled, the weight tends to become proportional to the postsynaptic activity. In an instar learning rule, *post*synaptic activity enables the learning; when the learning is enabled, the weight tends to become proportional to the presynaptic activity.

2.2. Comparison with LISSOM Model

In the LISSOM model[9,10] intracortical interactions are mediated by both lateral excitatory and lateral inhibitory pathways. The weights of both lateral excitatory and lateral inhibitory pathways change according to an instar learning rule:

$$Z_{ij}(t+1) = (Z_{ij}(t) + \xi\, x_i\, x_j) / (\textstyle\sum_k (Z_{kj}(t) + \xi\, x_k\, x_j)^2)^{1/2} \tag{2}$$

where Z_{ij} is the pathway weight and x_i is the learning rate for each type of pathway. The LISSOM rules also incorporate a spatial distance function, so that the net effect of learning between two nearby neurons is Hebbian, while the net effect of learning between two distant neurons is anti-Hebbian.

Sirosh et al.[10] modeled RF changes during scotoma conditioning using the LISSOM model. During scotoma conditioning, neurons with RFs straddling the scotoma boundary are active, and the LISSOM feedforward instar learning causes the feedforward connection weights from the scotoma region to these active neurons to weaken. During RF measurement after scotoma conditioning, these neurons are insensitive to the previously occluded regions and hence exert less inhibition on neurons that were inactive during conditioning. Thus, the neurons that were inactive during the conditioning show increased responsiveness and RF expansion. Changes in the feedforward connection weights are sufficient to produce RF expansion during scotoma conditioning.

To compare the consequences of an instar and an outstar inhibitory learning rule, the EXIN and the LISSOM models are simulated with *only* the lateral inhibitory learning.

2.3. Comparison with Adaptation Model

In the adaptation model, intracortical interactions are mediated by fixed (nonadapting) lateral inhibitory pathways. A neuron's ability to fire decreases/increases after a period of activity/inactivity, without any synaptic changes. Neuronal adaptation is modeled by the value T_i, for neuron i, which varies according to

$$(d/dt)T_i = -\eta\, T_i + (\tau - T_i)\, x_i \tag{3}$$

where η and τ are positive constants and T_i is bounded within $[0, \tau]$. Neuronal activation is inversely proportional to the adaptation parameter. Thus, periods of neuronal activation/inactivation cause the adaptation parameter to increase/decrease, thereby decreasing/increasing the neuron's responsiveness.

3. RESULTS

3.1. RF Expansion during Scotoma Conditioning

During scotoma conditioning, the EXIN inhibitory learning rule decreases the weights of inhibitory pathways from neurons with RFs outside the scotoma to those with RFs inside the scotoma. This decrease in strength of inhibitory pathways results in a centrifugal expansion of RFs that were originally inside the scotoma (Figures 1(a) and 2(a)).

The LISSOM inhibitory learning rule decreases the strength of lateral inhibitory pathways from neurons with RFs within the scotoma region to those with RFs outside the scotoma region, during scotoma conditioning. This becomes manifested as centripetal expansion of the RF of neurons whose RF was initially outside the scotoma region (Figures 1(b) and 2(b)).

Scotoma conditioning, in the adaptation model, causes neurons whose RF is in the scotoma region to increase their responsiveness and those whose RF is outside the scotoma region to decrease their responsiveness. Thus, the neurons whose RF is outside the scotoma region exert less inhibition on those whose RF is inside the scotoma region. This results in expansion of the RF of neurons whose RF is initially within the scotoma region (Figures 1(c) and 2(c)).

In the absence of any visual stimulation, all neurons in the adaptation model tend to become equally responsive. Thus, the RF expansion following scotoma conditioning is di-

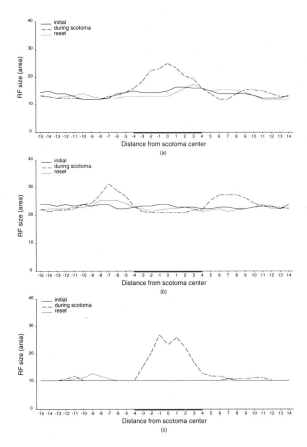

Figure 1. RF size as a function of position. The RF area before scotoma conditioning (solid line), after scotoma conditioning (dashed line), and after re-conditioning with normal stimuli (dotted line) in the EXIN network (a), the LISSOM network (b), and the adaptation network (c), are shown as a function of the position of Layer 2 neurons relative to the scotoma center (15, 15). The scotoma is a square of size 9 × 9. In the graph, the scotoma edges are at 4 and –4. The RF area shown is for a one-dimensional cross-section through Layer 2: neurons (15, 0)–(15, 29). The RF area of a Layer 2 neuron is the number of locations at which the test stimulus evokes a response in the Layer 2 neuron. The thick line on the abscissa represents the scotoma region.

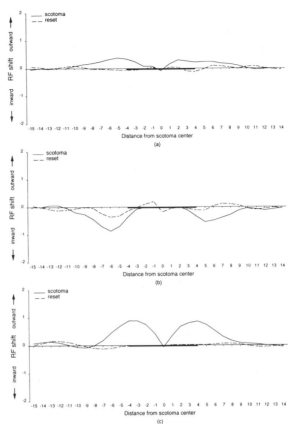

Figure 2. RF shift as function of position. Shift in RF center after scotoma conditioning (solid line) and after re-conditioning with normal stimuli (dashed line) with respect to the initial RF centers is shown as a function of distance of Layer 2 neurons from scotoma center for the EXIN network (a), the LISSOM network (b), and the adaptation network (c) as in Figure 1. The RF shift shown is for a one-dimensional cross-section through Layer 2: neurons (15, 0)–(15, 29). Positive and negative shifts represent a shift away from and toward the center of the scotoma, respectively. The RF center of a Layer 2 neuron is the center of moment of the neuron's responsiveness to input at different positions within its RF.

minished by a period of no visual stimulation (Figure 3). However, Pettet & Gilbert[7] reported that RF expansion in cat primary visual cortex persisted in the absence of visual stimulation.

3.2. Comparison of Models with Neurophysiological Data

During scotoma/normal conditioning, a computational simulation of the EXIN model produced the following effects, corresponding closely to the experimental neurophysiological results reported by Pettet & Gilbert[7] and Darian-Smith & Gilbert:[3] (1) centrifugal expansion of RFs that were initially *inside* the scotoma region; (2) the greatest expansion for RFs closest to the scotoma boundary and inside it; (3) increased response from the area of the initial RF[4] without changes in spontaneous activity; (4) RF contraction to original size during normal stimulation; and (5) negligible RF contraction in the absence of stimulation.

The adaptation model[11] produced several effects similar to those produced by the EXIN model (compare Figure 1(a) with 1(c) and Figure 2(a) with 2(c)). However, the adaptation model produced the following effects inconsistent with the EXIN model and neurophysiological data: (1) changes in spontaneous activity because of adaptation; and (2) RF changes in the absence of stimulation.

In response to scotoma/normal conditioning, the LISSOM model[9,10] with only lateral inhibitory learning produced the following effects inconsistent with neurophysiological

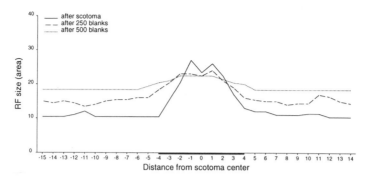

Figure 3. Blank screen causes RF changes in the adaptation model. The RF area after scotoma conditioning (solid line), after 250 steps of conditioning with blank stimuli (dashed line), and after 500 steps of conditioning with blank stimuli (dotted line) in the adaptation network, are shown as a function of the position of Layer 2 neurons relative to the scotoma center (15, 15). See Figure 1 for conventions.

data: (1) centripetal expansion of RFs that were initially *outside* the scotoma region; and (2) the greatest expansion for RFs closest to the scotoma boundary and outside it.

4. SUMMARY AND CONCLUSIONS

In scotoma/normal conditioning, the three models reproduce the neurophysiological experimental data to varying degrees; the EXIN model is closest to experimental data, followed by the adaptation model and then the LISSOM model.

The adaptation model differs from the EXIN model in its responses to blank stimuli and change in spontaneous neural activation after artificial scotoma conditioning. The LISSOM model with only lateral inhibitory learning and the EXIN model with only lateral inhibitory learning differ in the locus and direction of the effects of artificial scotoma conditioning.

Simulating the EXIN model and the LISSOM model with only lateral inhibitory learning shows that the EXIN *outstar* inhibitory learning rule is *sufficient* and the LISSOM *instar* inhibitory learning rule is *insufficient* to model RF expansion during scotoma conditioning. Thus, in the LISSOM model, other learning, such as feedforward learning, is necessary and should be faster than the inhibitory learning to overcome the opposite effects of the inhibitory learning.

The role of feedforward excitatory learning rule in producing fast (on the order of minutes or hours) RF changes in adult animals may be very limited. Restricted retinal lesion in cats produced RF changes in area 17 neurons within hours *only if* the non-lesioned eye was *closed*.[2] This result is contrary to the prediction of a model with fast *instar* feedforward excitatory learning rule, because active neurons would weaken their connections from the lesioned region, regardless of whether the other eye is open or closed, to produce RF shifts. Furthermore, fast *instar* feedforward excitatory learning rule predicts that the active neurons would weaken their connections from the closed eye.

The validity of the EXIN and LISSOM lateral inhibitory rules can be tested by intracellularly inactivating a test neuron during whole field stimulation with bars. In this situation, the feedforward weights would remain close to their equilibrium values in both models. The EXIN outstar inhibitory rule predicts that after conditioning the test neuron would show increased responsiveness and an RF expansion, because the inhibitory weights from the active neurons to the test neuron decrease. On the other, hand the LISSOM inhibitory rule would cause the lateral inhibition from the test neuron to the active neurons to decrease, resulting in an decrease in the responsiveness and RF size of the test neuron. A similar conditioning experiment in *Aplysia* showed increased responsiveness of the test neuron.[1]

These modeling results reveal some aspects of the functional organization and the plausible rules underlying cortical dynamic RFs and point to new experiments to validate the proposed rules.

ACKNOWLEDGMENTS

Supported by the Office of Naval Research (ONR N00014-93-1-0208) and the Whitaker Foundation.

REFERENCES

1. Carew, T.J., Hawkins, R.D., Abrams, T.W., & Kandel, E.R. (1984) A test of Hebb's postulate at identified synapses which mediate classical conditioning in *Aplysia*. *The Journal of Neuroscience*, **4**, 1217–1224.
2. Chino, Y.M., Kaas, J.H., Smith, E.L. III, Langston, A.L., & Cheng, H. (1992) Rapid reorganization of cortical maps in adult cats following restricted deafferentation in retina. *Vision Research*, **32(5)**, 789–796.
3. Darian-Smith, C. & Gilbert, C.D. (1995) Topographic reorganization in the striate cortex of the adult cat and monkey is cortically mediated. *The Journal of Neuroscience*, **15(3)**, 1631–1647.
4. DeAngelis, G.C., Ohzawa, I., Anzai, A., & Freeman, R.D. (1994) Does an artificial scotoma induce plasticity of receptive field structure for neurons in the visual cortex? *Investigative Ophthalmology & Visual Science*, **35(4)**, 1468.
5. Grossberg, S. (1972) Neural expectation: Cerebellar and retinal analogs of cells fired by learnable or unlearned pattern classes. *Kybernetik*, **10**, 49–57.
6. Marshall, J.A. (1995) Adaptive perceptual pattern recognition by self-organizing neural networks: Context, uncertainty, multiplicity, and scale. *Neural Networks*, **8**, 335–362.
7. Pettet, M.W. & Gilbert, C.D. (1992) Dynamic changes in receptive-field size in cat primary visual cortex. *Proceedings of the National Academy of Science of the USA*, **89**, 8366–8370.
8. Recanzone, G.H., Merzenich, M.M., & Dinse, H.R. (1992) Expansion of the cortical representation of a specific skin field in primary somatosensory cortex by intracortical microstimulation. *Cerebral Cortex*, **2**, 181–196.
9. Sirosh, J. & Miikkulainen, R. (1994) Cooperative self-organization of afferent and lateral connections in cortical maps. *Biological Cybernetics*, **71**, 66–78.
10. Sirosh, J., Miikkulainen, R., & Bednar, J.A. (1996) Self-organization of orientation maps, lateral connections, and dynamic receptive fields in the primary visual cortex. In J. Sirosh, R. Miikkulainen, and Y. Choe, *Lateral Interactions in the Cortex: Structure and Function*. UTCS Neural Networks Research Group, Austin, TX, 1996. Electronic book, ISBN 0–9647060–0–8, http://www.cs.utexas.edu/users/nn/web-pubs/htmlbook96.
11. Xing, J. & Gerstein, G.L. (1994) Simulation of dynamic receptive fields in primary visual cortex. *Vision Research*, **34**, 1901–1911.

65

NONLINEAR DYNAMICS IN A COMPOUND CENTRAL PATTERN GENERATOR

Masakazu Matsugu[1] and Chi-Sang Poon[2]

[1] Imaging Research Center
Canon Inc., Ohta-ku
Tokyo, 146, Japan
[2] Harvard-MIT Division of Health Sciences and Technology
Rm 20A-126
Massachusetts Institute of Technology
Cambridge, MA

ABSTRACT

We studied the nonlinear dynamical behavior of several compound central pattern generators in the form of a half-center network oscillator coupled with an endogenous pacemaker, as exemplified by the respiratory motor generator in the mammalian neonate. Using pacemaker inputs with varying amplitudes, frequencies and phases, we demonstrated several pathologic oscillatory patterns including recurrent apnea (intermittent cessation of oscillation), quasi-periodic fluctuations and chaos. The apneic pattern can be attributed to decreased excitation (e.g., decreased overall chemoreceptor activity), unbalanced excitation (e.g., unbalanced tonic inputs to inspiratory and expiratory related neurons), or disparity between the intrinsic oscillatory frequency and pacemaker frequency. Results may have important implications in the pathogenesis of abnormal respiratory pattern associated with sudden infant death syndrome.

1. INTRODUCTION

In animal nervous systems, central pattern generators (CPGs) are composed of specialized groups of neurons that produce rhythmic activities (Koppel, 1988; Selverston and Mazzoni, 1989). In general there are two basic mechanisms of rhythmogenesis: the spontaneous bursting activity of pacemaker cells or oscillations produced by the collective interaction within a network of non-pacemaker neurons (Friesen and Stent, 1977; Delcomyn, 1980). Many CPGs are built upon either of these mechanisms, but there are exceptions where both may co-exist.

For example, respiratory rhythm in mammals is produced by a CPG composed of an interconnecting neural network in the brain stem. A basic mechanism of rhythmogenesis in the adult respiratory CPG is reciprocal inhibition between inspiratory and expiratory related neurons, with general excitation from chemoreceptor afferents and an adaptation effect in some neurons (Richter et al., 1986). These neural processes undergo progressive maturation during infancy, where respiratory rhythm may also be driven partly by endogenous pacemaker neurons with conditional bursting properties (Arata et al., 1990; Smith et al., 1991). During the neonatal developmental period, therefore, respiratory rhythm may be influenced by the nonlinear interaction between the network CPG and the pacemaker(s) if they are strongly coupled together and are of comparable potency.

To examine the effects of such nonlinear interaction in compound CPGs and, in particular the neonatal respiratory pattern generator, we focused on a simple model of the neonatal respiratory CPG in the form of a neural network oscillator coupled with a pacemaker. We investigated how the behavior of such compound neural oscillators differs from those of isolated pacemakers or network oscillators.

2. THE NETWORK MODEL

The oscillator model (Figure 1(a)) used in this study is composed of two mutually inhibiting types of neuron - the inspiratory (I) and expiratory (E) neurons - with adaptive properties in the I neuron. A virtual interneuron F, which provides negative recurrent feedback for the I neuron, accounts for the adaptation properties of the I neuron. This model may be considered as a minimal model of the mammalian respiratory rhythm generator (Duffin, 1991). The equation describing the activity of any neuron in the model is:

$$T_i \frac{dx_i}{dt} + x_i = R_i + \sum_j C_{ij} g(x_j - H_j) + S_i \tag{1}$$

where x_i is the current activity of the ith neuron; R_i is its resting activity (-10 for the I and E neurons, 0 for neuron F); H_i is its threshold (0 for all neurons); C_{ij} is the strength of the connection from the jth to the i th neuron; and $g(z)$ is a nonlinear function (e.g., $g(z) = max(0,z)$) used to model the activation threshold for the model neurons. The network is driven by both constant excitation input and periodic pacemaker inputs of varying amplitudes, frequencies and phases. For simplicity, the inputs to oscillator neurons I and E from driving source D are assumed to be of the form:

$$S_p = A_p \sin(2\pi f_p t + \phi_p) + DC_p \tag{2}$$

where index p denotes either I or E neuron.

To explore the general characteristics of compound oscillators we investigated other types of network oscillator under similar driving conditions, including a half-center oscillator (Calabrese, 1995) composed of two Morris-Lecar type neurons (Wang & Rinzel, 1992; Skinner et al., 1994) and a three-phase model of the respiratory pattern generator (Gottschalk et al., 1994) composed of five distinct respiratory neurons(Figure 1 (b)). A key common feature among the three models is the existence of at least two reciprocal inhibitory neurons which contribute network oscillation. In our forced Gottschalk model, only early inspiratory and expiratory neurons, both reciprocally inhibiting each other, are fed by sinusoidal inputs as in the Duffin model. More detailed explanation on simulation condition and results will be found elsewhere (Matsugu, Duffin, and Poon, 1997).

(a)   **(b)**

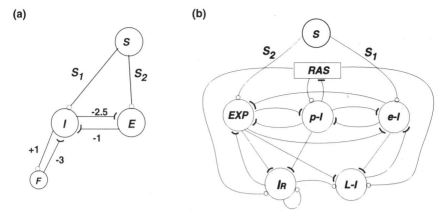

Figure 1. Compound CPG models. (a) Forced Duffin model. I: inspiratory neuron, E: expiratory neuron, S: oscillatory source; (b) Forced Gottschalk model. RAS: reticular activating system. Other labels denote following neurons. EXP: expiratory, p-I: post-inspiratory, e-I: early-inspiratory, L-I: late-inspiratory, I_R: ramp-inspiratory neurons. Connection strengths and other details should be referred to Gottschalk et al. (1994).

3.   RESULTS

3.1.   Spontaneous oscillations under steady (DC) inputs

One of the first and more elementary features of the CPG model was that spontaneous rhythmicity was contingent upon adequate excitation from an external drive (such as the central and peripheral chemoreceptors inputs or metabolic inputs). Sustained respiratory rhythm was produced only when both the I and E neurons were suitably excited with the relative magnitudes of excitation lying within a given range (Figure 2, solid line); network oscillation is constrained to a limited range of DC ratio, DC_E/DC_I. An interesting implication of this finding is that cessation of oscillation may result from the lack of general excitation as well as preferential excitation of the I or E neurons.

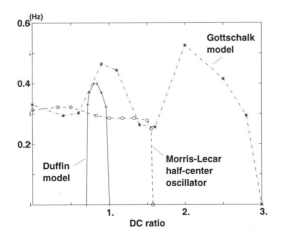

Figure 2. Spontaneous frequencies under varying DC ratios defined by the ratio of input DC levels to reciprocally inhibiting neurons (i.e., I and E neurons in the forced Duffin model, *early-I* and *Expiratory* neurons in the forced Gottschalk model, and two of mutually inhibiting neurons in the forced half-center oscillator).

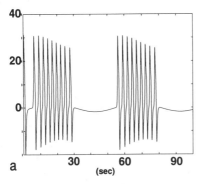 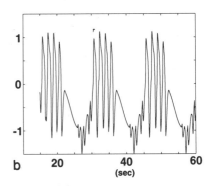

Figure 3. Intermittent cessation or suppression (apnea) of oscillation. (a) Forced Duffin model. DC levels to *I* and *E* neurons are 50 and 49.9, respectively, Input frequency is 0.02 Hz (spontaneous frequency: 0.4 Hz) and amplitude is 0.5. (b). Forced Gottschalk model. DC levels to *early-I* and *Expiratory* neurons are 1.0 and 1.1, respectively. Input frequency is 0.067 Hz (spontaneous frequency: 0.44 Hz).

3.2. Recurrent cessation of oscillation under low-frequency periodic inputs

We obtained recurrent apneic patterns (Figure 3 (a)) when the network was fed by periodic inputs with extremely low frequencies (e.g., one tenth or less of the intrinsic oscillation frequency) and intermediate amplitudes (as compared with the magnitude of the DC component). In this case, the oscillation pattern exhibited an alternating period of oscillation with spontaneous frequency followed by a period of quiescence (apnea); the frequency of this alternating oscillatory phase and non-oscillatory phase was exactly the frequency of the periodic input. The apneic pattern occurred only when the input amplitude was in the intermediate range - an exceedingly strong pacemaker input may totally entrain the network whereas a weak input may not produce any appreciable effect on the spontaneous oscillation of the network CPG. It turned out that unbalanced DC excitations of the *I* and *E* neurons (i.e., the DC ratio is marginally inside or outside the allowed range for network oscillation) could render the CPG prone to recurrent apnea when subjected to oscillatory inputs, whereas balanced tonic inputs may preclude such pathologic state.

A possible implication of this finding is that decreased or unbalanced chemoafferent excitation of the neonatal respiratory CPG may lead to life-threatening respiratory instability during the developmental period when the respiratory rhythm is partly driven by pacemaker inputs. Indeed, decreased chemoreceptor activity has been found to be associated with babies who died of sudden infant death syndrome (Kinney et al., 1995).

3.3. Entrainment, quasi-periodicity and chaos

Finally, with strong (supracritical) pacemaker input and/or weak spontaneous network oscillation, the network CPG is found to be entrained to the pacemaker. However, for subcritical pacemaker inputs the resulting rhythm exhibits various forms of distortion including quasiperiodicity and chaos. Weakly chaotic behavior (Figure 5 (a)) was observed only when the network CPG was driven by multiple periodic sources (e.g., independent oscillatory signals of different frequencies), and even so the bifurcation to chaos was highly non-robust and was readily abolished (Figure 5 (b)) by changes in the amplitude or frequency of the periodic drives. Specifically, the entrained oscillations resulting from anti-phase (i.e., inputs to *I* and *E* neurons are 180 deg. out of phase) inputs or single oscillatory inputs were both stronger (Matsugu,

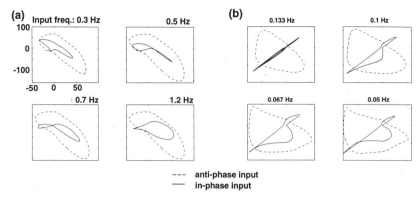

Figure 4. Trajectories in the phase plane of *I* and *E* neurons activities under varying input frequencies and phase relationship (e.g., in-phase or anti-phase) for (a) forced Duffin model and (b) forced Morris-Lecar half-center oscillator. In both models, much greater waveform distortion is observed for trajectory under in-phase inputs, while the amplitude and waveform stability is greater for anti-phase inputs.

Duffin, and Poon, 1997) and more stable (Figure 4) than those resulting from in-phase inputs (i.e., no phase lag between the two inputs) under similar driving input conditions.

This behavior was evidenced from the fact that the critical amplitude for entrainment is lower for anti-phase inputs than for in-phase inputs (Matsugu, Duffin, and Poon, 1997) and can be intuitively deduced by considering that the agonist-antagonist pattern of anti-phase inputs is intrinsically more compatible with network oscillation in reciprocally inhibiting neurons.

3.4. Comparison to other models

The above behaviors were consistently reproduced in other oscillatory neural networks under similar driving inputs to reciprocally inhibitory neurons. In particular, intermittent suppression or cessation of oscillation was also provoked in the forced half-center oscillator (not presented due to space limitation) model and the forced Gottschalk model (Figure 3 (b)) as well under oscillatory inputs of extremely low frequency as compared with the spontaneous oscillation frequency.

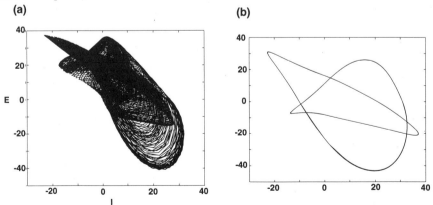

Figure 5. Chaotic and quasi-periodic oscillations. Weakly chaotic oscillation (Lyapunov exponent: + 0.06) was induced only when *I* and *E* neurons received oscillatory inputs (amplitude: 21) of different frequencies (e.g., 3 Hz to *I* and 4.5 Hz to *E* respectively). (b) The chaotic behavior is abolished when the sinusoidal input amplitude is increased to 50.

The spontaneous frequencies under varying DC excitation ratios in these models (Figure 2; dash dot and dashed lines) revealed DC ratio cutoff only in the high-end. The result implies the robustness of these models against the variability of tonic excitation and is in contrast to the Duffin model which show both low and high-ends (Figure 2; solid line). The result also supports the existence of apneic behaviors under oscillatory inputs of low frequency, which occur only when the instantaneous DC ratio exceeds the high-end cutoff, thus breaking down the condition for network oscillation. The resumption of network oscillation occurs when the instantaneous DC ratio falls below the high-end cutoff. The greater stability of these models may be partly due to the adaptive property exhibited by many types of neurons, whereas only one neuron (i.e., I neuron) has this effect in the Duffin model.

Greater stability and greater amplitudes of induced waveform under varying frequency of anti-phase inputs were also observed in the forced Gottschalk model (not presented) and forced half-center oscillator composed of Morris-Lecar type neurons (Figure 4 (b)) as well. The results suggest that anti-phase excitation is a more favorable (and hence, probably more likely) arrangement than in-phase excitation in a compound CPG.

4. CONCLUSION

Our simulation results revealed several neurophysiologic and neuroanatomic conditions that may be crucial for the maintenance of stable spontaneous or entrained oscillations in a network CPG driven by constant and/or periodic inputs.

Our general finding is that the neural network oscillator interacts more harmoniously with anti-phase inputs having relative antagonistic effects on the reciprocally inhibiting neurons than otherwise. In addition, we demonstrated several pathologic states in such compound oscillators, including apnea (e.g., intermittent cessation or suppression of oscillation) and weakly chaotic behaviors.

The simulation results are in good agreement with a wide variety of experimental findings concerning the normal behavior of the respiratory CPG, and may shed important light on its abnormal behavior in certain pathologic states.

ACKNOWLEDGMENTS

This investigation was supported in part by Office of Naval Research grant N00014-95-1-0414; National Science Foundation grant BCS-9216419; and National Institutes of Health grant HL52925.

REFERENCES

[1] Arata, A., Onimaru, H., Homma, I. (1990) Respiration-related neurons in the ventral medulla of newborn rats in vitro. Brain Res. Bull. 24: 599-604.
[2] Calabrese, R.L. (1995) Half-center oscillators underlying rhythmic movements. In: Arbib, M.A. ed. The Handbook of Brain Theory and Neural Networks. MIT Press, Cambridge, MA. pp. 444-447.
[3] Delcomyn, F. (1980) Neural basis of rhythmic behavior in animals. Science 210:492-498.
[4] Duffin, J. (1991) A model of respiratory rhythm generation. Neuroreport 2: 623-626.
[5] Friesen, W.O. and Stent, G.S. (1977) Generation of a locomotory rhythm by a neural network with recurrent cyclic inhibition. Biol. Cybern. 28:27-40.

[6] Gottschalk, A., Ogilvie, M.D., Richter, D.W., and Pack, A. (1994) Computational aspects of the respiratory pattern generator. Neural Computation 6:56-68.

[7] Kinney, H.C, Filiano, J.J., Sleeper, L.A., Mandell, F., Valdes-Dapena, M., and White, W.F. (1995) Decreased muscarinic receptor binding in the arcuate nucleus in sudden infant death syndrome. Science, 269: 1446-1450.

[8] Koppel, N. (1988) Toward a theory of modelling central pattern generators. In: Cohen, A.H., Rossignol, S., and Grillner, S. eds. Neural Control of Rhythmic Movements in Vertebrates. John Willey & Sons, New York. pp. 369-413.

[9] Matsugu, M., J. Duffin, C.-S. Poon (1997) submitted to J. of Comput. Neurosci.

[10] Richter, D.W., Ballantyne, D., and Remmers, J.E. (1986) How is the respiratory rhythm generated ? A model. News Physiol. Sci. 1:109-112.

[11] Selverston, A. and Mazzoni, P. (1989) Flexibility of Computational Units in Invertebrate CPGs. In: Durbin, R., Miall, C., and Mitchison, G. eds. The Computing Neuron. Addison-Wesley Pub. Ltd., Reading, MA. pp.205-228.

[12] Skinner, F.K., Kopell, N., and Marder, E. (1994) Mechanisms for oscillation and frequency control in reciprocally inhibitory model neural networks. J. of Computational Neurosci. 1:69-88.

[13] Smith, J.C., Ellenberger, K., Ballanyi, D., Richter, D.W., and Feldman, J.L. (1991) Pre-Bötzinger complex: a brain stem region that may generate respiratory rhythm in mammals. Science 254:726-729.

[14] Wang, X.J. and Rinzel, J. (1992) Alternating and Synchronous Rhythms in Reciprocally Inhibitory Model Neurons. Neural Computation 4: 84-97.

66

CONTROL OF CA3 PLACE FIELDS BY THE DENTATE GYRUS: A NEURAL NETWORK MODEL

Ali A. Minai

Complex Adaptive Systems Laboratory
Department of Electrical and Computer Engineering and Computer Science
University of Cincinnati
Cincinnati, OH

ABSTRACT

A very interesting aspect of hippocampal anatomy is the presence of two pathways projecting from the entorhinal cortex (EC) to the CA3 region — one directly via the perforant path (PP), and the other through the dentate gyrus (DG) using the mossy fibers of the granule cells. This implies that the place fields of the CA3 arise from the joint influence of EC and DG. We hypothesize that the DG plays a modulatory role in this scheme, serving to enhance discrimination during the learning of new place codes. Drawing in part on some receny experimental findings, we model a mechanism whereby DG neurons accomplish pattern separation by modulating the balance of dendritic and somatic inhibition in granule cells. Our results are consistent with a variety of observations in the literature, including the following: 1) DG lesions do not abolish CA3 place fields but disrupt spatial memory; 2) Even similar environments produce different place fields in CA3 but not in the EC. We show that DG modulation allows the model hippocampus to control spatial discrimination, and produces realistic place fields.

1. INTRODUCTION

Marr (1969) was the first to propose that the dentate gyrus might function as a pattern separator for afferent input from the entorhinal cortex, projecting it to CA3 for storage via the mossy fiber (MF) path. Following the discovery of the direct perforant path (PP) projection from EC to CA3 (Yeckel and Berger, 1990), it was suggested that this might represent a cuing input for pattern retrieval in CA3 (Rolls, 1989). Analysis of this two-path system for the storage of random patterns demonstrated its ability to handle the conflicting requirements of pattern separation (for storage) and pattern recognition (for retrieval) (Treves and Rolls, 1992; O'Reilly and McClelland, 1994). The possible role of stochastic quantal secretion for pattern separation in the DG was studied by Gibson et al. (1991), also under random pattern and connectivity assumptions. The present paper addresses the issue of DG-based pattern separation explicitly in the context of place-field type patterns, and using mechanisms suggested by recent experimental results.

The two-path projection from EC to CA3 implies that CA3 place fields are formed by a conjunction of PP and MF effects. Most of the sensory information comes via the PP input, and there is evidence that this projection alone is sometimes sufficient to maintain place-specific activity in CA3 (McNaughton et al., 1989; Knierim and McNaughton, 1995). However, the MF pathway appears to be essential for adequate spatial performance. One clue to its function comes from the report (Quirk et al., 1992) that EC place fields of the same cell in two similar environments are similar, but those of a CA3 cell are totally different. This clearly suggests context-specific recoding of EC information by the MF pathway.

The primary hypothesis studied in this paper is that the DG enhances discrimination in EC place codes by changing the balance of dendritic and somatic inhibition on the granule cells. This is based on indirect evidence (Moser, 1996) that, during exploratory behavior, there is an increase in dendritic inhibition and a decrease in somatic inhibition in the rat DG. The DG is known to have extensive and complex inhibitory subsystems (Buckmaster and Schwartzkroin, 1994), and is an ideal site for the modulation of EC information en route to CA3, functioning much like a hidden layer in a feed-forward neural network.

2. METHODS

We hypothesize that the dentate gyrus operates in two modes: 1) During the *quiescent* mode — identified with retrieval—, the response of DG granule cells to stimulation from EC is very low, so that, as a whole, the signal from DG to CA3 is a low amplitude, non-specific noise-like one. Thus, CA3 neurons respond primarily to PP stimuli acting as cues for retrieval without MF influence. 2) The *alert* mode is used for the formation of new place representations (or recall of old ones from intact cues). During this mode, the DG granule cells 1) increase dendritic inhibition; and 2) decrease somatic inhibition. Thus, only those granule cells which get several simultaneously active inputs will fire, but will fire strongly. This will strongly bias *a few CA3 cells at a time*, causing the diffuse PP-based place-fields in CA3 to become localized via a competitive inhibitory mechanism. The assumption is that the DG can be switched between the two regimes based on recognition or other motivational contexts. The existence of a strong subcortical input to the DG and CA3 makes this at least plausible. There is already evidence that cholinergic influences from the septal region modulate synaptic excitability in the cortex to help in memory storage (Hasselmo et al., 1992). Such a mechanism could operate during the alert mode to increase the influence of the DG bias and preclude interference from the mnemonic system (Hasselmo, 1993).

2.1. Entorhinal Cortex (EC)

The model for EC place fields is purely phenomenological, based on the known charac-teristics of these fields (broad spatial coverage, noisiness, dependence on sensory cues, etc.) Thus, the response of EC cells is modeled with broadly tuned, noisy place fields centered at random coordinates in the environment. Each EC cell, i, is assigned a center, $\bar{c} = (c_i^x, c_i^y)$. The output of cell i at time t is defined by the following stochastic prescription:

$$z_i(t) =$$
$$\exp(-a_i(x(t) - c_i^x + q_x)^2 - b_i(y(t) - c_i^y + q_y)^2 +$$
$$d_i\sqrt{a_i}(x(t) - c_i^x)\sqrt{b_i}(y(t) - c_i^y)) \tag{1}$$

where $x(t)$ and $y(t)$ are the animal's coordinates at time t, a_i and b_i are parameters which control the size and shape of the elliptical place field, d_i is an orientation parameter with values between 1 and -1, and q_x, q_y are 0-mean uniform random variables with variance σ_q, which essentially controls the accuracy with which an EC cell discerns the animal's location. Usually, σ_q will be small.

2.2. Dentate Gyrus (DG)

The activation to DG granule cell i is given by:

$$y_i(t) = \sum_{j \in EC} w_{ij} f(z_j(t) - \phi\theta_i) \tag{2}$$

where w_{ij} is the synaptic weight from EC cell j to cell i, $z_j(t)$ is the output of j at time t, θ_i is a synaptic threshold representing dendritic inhibition, and ϕ is an indicator variable for the alert mode. The output is then calculated as

$$z_i(t) = f(\tanh((G_{DG} + \phi g_{DG})y_i(t)))) \tag{3}$$

where G_{DG} is a small fixed gain parameter and g_{DG} a much larger gain increment representing the decrease in somatic inhibition during the alert mode. $f()$ is a rectifier function: $f(x) = x$ if $x > 0$ and 0 otherwise.

2.3. CA3

Only the extrinsic afferents to CA3 are included in this model, since it focuses on pattern formation. The activation of CA3 pyramidal cell, i, is modeled as:

$$y_i(t) = \sum_{j \in EC} w_{ij} z_j(t) + \sum_{j \in DG} w_{ij} z_j(t) + \sum_{j \in CA3} w_{ij} z_j(t-1) \tag{4}$$

where w_{ij} are synaptic weights and $z_j(t)$ presynaptic cell outputs. The output is then calculated as

$$z_i(t) = f(\tanh(G_{CA3}(y_i(t) - \alpha(t)\theta_i))) \tag{5}$$

where G_{CA3} is a fixed gain parameter and θ_i is the firing threshold. The $\alpha(t)$ parameter, which models normalizing inhibition, is defined as:

$$\alpha(t) = \frac{1}{N_{CA3}} \sum_{i \in CA3} y_i(t) \tag{6}$$

Simulations were carried out in $L \times L$ square environments. All place fields were systematically mapped out at all L^2 locations. Parameters were set to reasonable values without any explicit optimization.

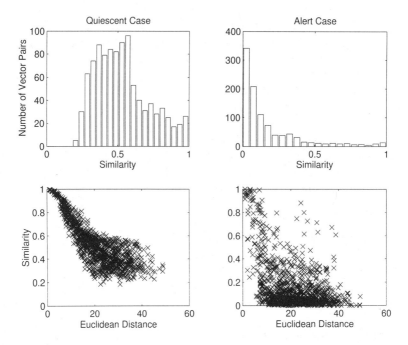

Figure 1. The top two graphs show the distribution of similarity values between 1000 pairs of CA3 place code vectors in the quiescent and alert modes. The lower two graphs plot these similarities as a function of euclidean distance between the locations in each pair. A thresholding effect is clear in the bottom right graph.

3. RESULTS

The results demonstrate that modulating the balance of inhibition can significantly enhance place discrimination both within and across similar environments. Figure 1 shows the effect of mode switching on the similarity of place codes in a simulated 40×40 grid environment with random EC fields and a $200/1000/200$ network. It plots the similarity between CA3 place codes for 1000 randomly chosen pairs of locations (out of a possible $2,558,400$ pairs). Similarity between two place code vectors is measured by the cosine of the angle between them:

$$s(A,B) = \frac{A.B}{|A||B|} \tag{7}$$

where A and B are place code vectors. This precludes a globally attenuated version of code A as being considered different from A.

In the quiescent mode, with the dentate gyrus effectively out of the loop, similarity values are roughly a linear function of distance between locations. In the alert mode, however, there is a clear thresholding effect, indicating that the environment is divided up into well differentiated local neighborhoods. The overall distribution of similarity also changes from a broad one centered around 0.5 to one skewed strongly towards low similarity — again indicating an increase in spatial discrimination on average.

Figure 2 shows how the hypothesized mechanism can enhance discrimination of similar environments. Similar environments are produced by first generating a random 40×40 environment and then perturbing its EC place field centers to produce another. The similarity of

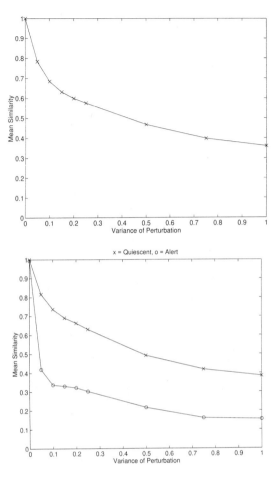

Figure 2. The graph on the left shows average similarity between EC representations of two environments as a function of their (simulated) sensory similarity. The graph on the right shows the corresponding CA3 similarities between environment pairs in the quiescent and alert modes. Note how alertness increases discrimination between very similar environments.

the two environments is quantified by the variance of the applied perturbation. Similarity values are then calculated for the corresponding locations in the environments, and averaged to give the overall similarity of the CA3 place representation. It can be seen that, in the quiescent mode, discrimination in CA3 is very similar to that in the EC as a function of environment similarity. In the alert mode, however, discrimination between similar environments is enhanced greatly.

4. SUMMARY

This paper has presented a simple neural network model of how changing the degree of dendritic and somatic thresholds in the dentate gyrus can lead to the enhancement of spatial discrimination in the hippocampus. Clearly, however, this simplistic mechanism needs to be augmented in complex ways to account for the experimental data on place code stability and plasticity in response to cue manipulation, etc. In particular, the issues of context-dependent discrimination and path integration need to be addressed. Interesting ideas on these are now beginning to emerge in the literature (Touretzky and Redish, 1996; McNaughton et al., 1996).

REFERENCES

[1] Buckmaster, P.S., and Schwartzkroin, P.A. (1994) Hippocampal mossy cell function: A speculative view. *Hippocampus* **4**: 393-402.

[2] Gibson, W.G., Robinson, J., and Bennett, M.R. (1991) Probabilistic secretion of quanta in the central nervous system: Granule cell synaptic control of pattern separation and activity regulation. *Phil Trans. R.. Soc. Lond. B* **332**: 199-220.

[3] Hasselmo, M.E. (1993). Acetylcholine and learning in a cortical associative memory. *Neural Computation* **5**: 32-44.

[4] Knierim, J.J., and McNaughton, B.L. (1995) Differential effects of dentate gyrus lesions on pyramidal cell firing in 1- and 2-dimensional spatial tasks. *Soc. Neurosci. Abstr.* **21**: 940.

[5] Marr, D. (1969) Simple memory: A theory for archicortex. *Phil. Trans. R. Soc. Lond. B)* **262**: 23-81.

[6] McNaughton, B.L., Barnes, C.A., Meltzer, J., and Sutherland, R.J. (1989). Hippocampal granule cells are necessary for normal spatial learning but not for spatially-selective pyramidal cell discharge. *Exp. Brain Res.* **76**: 485-496.

[7] McNaughton, B.L., Barnes, C.A., Gerrard, J.L., Gothard, K., Jung, M.W., Knierim, J.J., Kudrimoti, H., Qin, Y., Skaggs, W.E., Suster, M., and Weaver, K.L. (1996) Deciphering the hippocampal polyglot: The hippocampus as a path integration system. *J. Exper. Biol.* **199**: 173-185.

[8] O'Reilly, R.C., and McClelland, J.L. (1994). Hippocampal conjunctive encoding, storage and recall: Avoiding a tradeoff. *Tech. Rep. PDP.CNS.94.4*, PDP and CNS Group: Pittsburgh, PA.

[9] Quirk, G.J., Muller, R.U., Kubie, J.L., and Ranck, J.B., Jr. (1992). The positional firing properties of medial entorhinal neurons: Description and comparison with hippocampal place cells. *J. Neurosci.* **12**: 1945-1963.

[10] Rolls, E. (1989). The representation and storage of information in neuronal networks in the primate cerebral cortex and hippocampus. In: **The Computing Neuron**, R. Durbin, C. Miall, and G. Mitchison (eds.) 125-159, Addison-Wesley.

[11] Thompson, L.T., and Best, P.J. (1989). Place cells and silent cells in the hippocampus of freely-behaving rats. *J. Neurosci.* **9**: 2382-2390.

[12] Touretzky, D.S., and Redish, A.D. (1996) Theory of rodent navigation based on interacting representations of space. *Hippocampus* **6**: 247-270.

[13] Treves, A., and Rolls, E.T. (1992). Computational constraints suggest the need for two distinct input systems to the hippocampal CA3 network. *Hippocampus* **2**: 189-200.

[14] Yeckel, M.F., and Berger, T.W. (1990). Feedforward excitation of the hippocampus afferents from the entorhinal cortex: Redefinition of the role of the trisynaptic pathway. *Proc. Nat. Acad. Sci.* **87**: 5832-5836.

67

A SIMPLE MODEL FOR CORTICAL ORIENTATION SELECTIVITY

Trevor Mundel,[1] Alexander Dimitrov,[2] and Jack D. Cowan[2]

[1] Department of Neurology
University of Chicago Hospitals
Chicago, IL
[2] Department of Mathematics
University of Chicago
Chicago, IL
E-mail: mundel@math.uchicago.edu
E-mail: a-dimitrov@uchicago.edu
E-mail: cowan@math.uchicago.edu

ABSTRACT

A simple mathematical model for the large-scale circuitry of primary visual cortex is introduced. It is shown that a basic cortical architecture of recurrent local excitation and lateral inhibition can account quantitatively for such properties as orientation tuning. Non-local coupling between similar orientation patches, when added to the model, can satisfactorily reproduce such effects as non-local iso-orientation suppression, and non-local cross-orientation enhancement. Following this an account is given of perceptual phenomena such as the direct and indirect tilt illusions.

1. INTRODUCTION

The edge detection mechanism in the primate visual cortex (V1) involves at least two fairly well characterized circuits. There is a local circuit operating at sub-hypercolumn dimensions comprising strong orientation specific recurrent excitation and weakly orientation specific inhibition. The other circuit operates between hypercolumns, connecting cells with similar orientation preferences separated by several millimeters of cortical tissue. This circuit provides local processes with information about the global nature of stimuli and has been invoked to explain a wide variety of context dependent visual processing. A good example of this is the tilt illusion (TI), where surround stimulation causes a misperception of the angle of tilt of a grating.

Computational Neuroscience
edited by Bower, Plenum Press, New York, 1997

The interaction between such local and long-range circuits has also been investigated. Typically these experiments involve the separate stimulation of a cells receptive field (the *classical* receptive field or "center") and the immediate region outside the receptive field (the *non-classical* receptive field or "surround"). In the first part of this work we present a simple model of cortical center–surround interaction. Despite the simplicity of the model we are able to quantitatively reproduce many experimental findings. We then apply the model to the TI. We are able to reproduce the principle features of both the direct and indirect TI with the model.

2. METHODS

2.1. Principles of Cortical Operation

Recent work with voltage-sensitive dyes[1] augments the early work of Hubel and Wiesel[2] which indicated that clusters of cortical neurons corresponding to cortical columns have similar orientation preferences. Thus the the appropriate units for an analysis of orientation selectivity are the localized clusters of neurons preferring the same orientation. We view the cortex as a lattice of hypercolumns, in which each hypercolumn comprises a continuum of iso-orientation patches distinguished by their preferred orientation ϕ. The population model we adopt throughout this work is a simplified form of the Wilson–Cowan equations.

2.2. Local Model

Our local model is a ring ($\phi = -90^o to + 90^o$) of coupled iso-orientation patches and inhibitors with the following characteristics

- Weakly tuned orientation biased inputs to V1. These may arise either from slight orientation biases of lateral geniculate nucleus (LGN) neurons or from converging thalamocortical afferents

- Sharply tuned (space constant $\pm 7.5^o$) recurrent excitation between iso-orientation populations

- Broadly tuned inhibition to all iso-orientation populations with a cut-off of inhibition interactions at between 45° and 60° separation

The principle constraint is that of a critical balance between excitatory and inhibitory currents as described in recent theoretical studies.[4,5] We implement this critical balance by explicitly tuning the strength of connection weights between excitatory and inhibitory populations so that the system state is subcritical to a bifurcation point with respect to the relative strength of excitation/inhibition.

2.3. Horizontal Connections

Experimental work in the tree shrew[6] and preliminary work in the macaque (Blasdel, personal communication) indicate that visuotopic connection (connections between edge detectors along an axis parallel to the detectors preferred orientation) is the predominant pattern

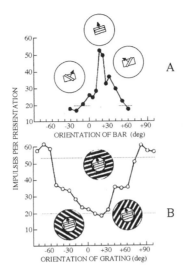

Figure 1. Non-local effects on orientation tuning - experimental data. Response to constant center stimulation at 15° and surround stimulation at angles $[-90^o, 90^o]$ (open circles), Local tuning curve (filled circles). Redrawn from Blakemore and Tobin[7]

of long-range connectivity. This connectivity pattern allows for the following reduction in dimension of the problem for certain experimental conditions.

Consider the following experiment. A particular hypercolumn designated the "center" is stimulated with a grating at orientation ϕ resulting in a response from the ϕ-edge detector. The region outside the receptive area of this hypercolumn (in the "surround") is also stimulated with a grating at some uniform orientation ϕ' resulting in responses from ϕ'-edge detectors at each hypercolumn in the surround. In order to study the interactions between center and surround, then to first order approximation only the center hypercolumn and interaction with the surround along the ϕ visuotopic axis (defined by the center) and the ϕ' visuotopic axis (once again defined by the center) need be considered. In fact, except when $\phi = \phi'$ the effect of the center on the surround will be negligible in view of the modulatory nature of the horizontal connections detailed above. Thus we can reduce the problem (a priori three dimensional — one angle and two space dimensions) to two dimensions (one angle and one space dimension) with respect to a fixed center. This reduction is the key to providing a simple analysis of complex neurophysiological and psychophysical data.

3. RESULTS

3.1. Center–Surround Interactions

A typical example of the surround suppressive effect is shown in figure 1. Further examples of surround suppression can be found in the paper of Sillito et al.[8]. Figure 2 depicts simulations in which long-range connections to local inhibitory populations are strong compared to connections to local excitatory populations.

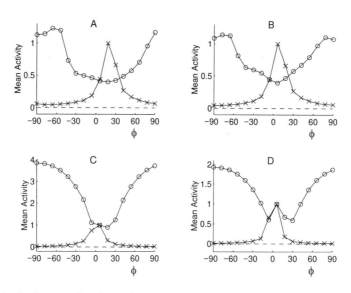

Figure 2. Non-local effects on orientation tuning . (o-o-o) = Center response to preferred orientation at different surround orientations, (x-x-x) = Center orientation tuning without surround stimulation, (—) = Center response to surround stimulation alone. A - response of population with 20° orientation preference. B, C and D - response of populations with 5° orientation preference.

3.2.   The Tilt Illusion

The tilt illusion (TI) is one of the basic orientation-based visual illusion. A TI occurs when viewing a test line against an inducing grating of uniformly oriented lines with an angle of θ between the orientation of the test line and the inducing grating. Two components of the TI have been described[9], the *direct* TI where the test line appears to be repelled by the grating–the orientation differential appears increased, and the *indirect* TI where the test line appears attracted to the orientation of the grating–the orientation differential appears decreased. Figure 3 depicts a typical plot of magnitude of the tilt effect versus the angle differential between inducing grating and test line reproduced from Wenderoth and Beh[10].

The TI thus provides compelling evidence that local detection of edges is dependent on information from more distant points in the visual field. It is generally believed, that the direct TI is due to lateral inhibition between cortical neurones.[9,11] It has been postulated that the

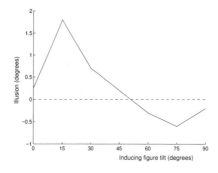

Figure 3. Direct (positive) and indirect (negative) tilt effects

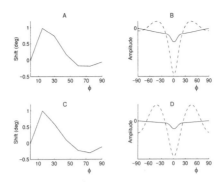

Figure 4. Model simulations of the tilt effect (A and C). B and D show the corresponding kernels mediating long-range interactions. Solid lines indicate the absolute kernel and dashed lines indicate the effective kernel

indirect TI occurs at a higher level of visual processing. We show here that both the direct and indirect TI are a consequence of the lateral and local connections in our model.

In figure 4 we give examples of the TI obtained from the model system. The effective kernels for long-range interactions are obtained by filtering the absolute kernels with the local filter which has a band-pass characteristic. It is this effective kernel which determines the tilt effect in keeping with our simulations and analysis which show that orientation preference is determined at the small amplitude linear stage of system development.

4. SUMMARY

We have shown that a very simple center–surround organization, operating in the orientation domain can successfully account for a wide range of neurophysiological and psychophysical phenomena, all involving the effects of visual context on the responses of assemblies of spiking neurons. We expect to be able to show that such an organization can be seen in many parts of the cortex, and that it plays an important role in many forms of information processing in the brain.

REFERENCES

[1] Blasdel, G.G., *Orientation selectivity, preference, and continuity in monkey striate cortex*, J. Neurosci. **12** No 8, 3139–3161 (1992).

[2] Hubel, D.H. and Wiesel, T.N., *Receptive fields, binocular interaction and functional architecture in the cat's visual cortex*. J. Physiol. Lond. **160**, 106–154, (1962).

[3] Victor, J.D., Purpura, K., Katz, E. and Mao, B., *Population encoding of spatial frequency, orientation and color in macaque V1*, J. Neurophysiol., **72** No 5, (1994).

[4] Tsodyks, M.V. and Sejnowski, T.,*Rapid state switching in balanced cortical network models*, Network, **6** No 2, 111–124, (1995).

[5] Vreeswijk, C. and Sompolinsky, H., *Chaos in neuronal networks with balanced excitatory and inhibitory activity*, Science **274**, 1724–1726, (1996).

[6] Fitzpatrick, D.,*The Functional Organization of Local Circuits in Visual Cortex: Insights from the Study of Tree Shrew Striate Cortex*, Cerebral Cortex **6**, 329–341, (1996).

[7] Blakemore, C. and Tobin, E.A.,*Lateral Inhibition Between Orientation Detectors in the Cat's Visual Cortex*, Exp. Brain Res., **15**, 439–440, (1972).

[8] Sillito, A.M., Grieve, K.L., Jones, H.E., Cudeiro, J. and Davis, J.,*Visual cortical mechanisms detecting focal orientation discontinuities*, Nature, **378**, 492–496, (1995).

[9]   Wenderoth, P. and Johnstone, S.,*The different mechanisms of the direct and indirect tilt illusions*, Vision Res., **28** No 2, 301–312, (1988).

[10] Wenderoth, P. and Beh, H.,*Component analysis of orientation illusions*, Perception, **6** 57–75, (1977).

[11] Carpenter, R.H.S. and Blakemore, C.,*Interactions between orientations in human vision*, Expl. Brain. Res., **18**, 287–303, (1973).

68

DYNAMICAL BEHAVIOR OF NETWORKS OF ACTIVE NEURONS

Sean D. Murphy[1,3] and Edward W. Kairiss[1,2]

[1]Neuroengineering and Neuroscience Center
[2]Department of Psychology
Yale University, New Haven, Connecticut 06520
[3]Department of Bioengineering
University of Pennsylvania
Philadelphia, Pennsylvania 19104

1. INTRODUCTION

Common features of activity are seen in biological neural networks in different species and different neural structures, such as partially synchronized oscillations. More detailed distinct categories (or *modes*) of biological neural network activity may be identifiable by combining traditional techniques in analyzing dynamical systems with additional techniques that are particular to parameterizing biological neural networks. Once the basic dynamical modes present in biological networks are identified, physiological data on network and multi-network scales may be more easily classified and understood in relation to behavior. Some dynamical modes might be better suited to implementing spatial filters, such as might be expected in cortical area V1 of the visual system. Other dynamical modes might be suited to replaying or identifying spatio-temporal signals, such as in the auditory cortex. Other modes might be useful for storing and retrieving associative mnemonic representations, such as those that might be found in the hippocampus.

Beyond the identification of different modes of dynamical activity in different parts of the nervous system, there is the problem of understanding how different areas of the nervous system collectively establish a global mode of dynamical activity. To some degree, the modes in different parts of the nervous system are constrained and controlled by the need for communication between and among many brain areas at once. Therefore, it should not be surprising to find dynamical modes of activity present that exist primarily for interareal communication and that are not useful for the local computation taking place per se (Wilson and Bower 1991).

Very little is known about the detailed spatio-temporal features of activity of biological neural networks. Although single-unit record and multi-unit recording studies have yielded some information about the large-scale patterns of activity in the mammalian brain (e.g. Wilson and McNaughton, 1994), even the best studies have only been able to record from extremely small percentages of cells. While computational modeling of biological neural networks cannot replace data from biological experiments, one of the things it can do is provide hypotheses and principles about detailed activity in biological neural networks that can greatly enhance the usefulness of limited recording techniques. For example, computational models of biological neural networks could demonstrate classes, or "motifs" of emergent activity that might not be deducible from biological data alone, but might be detectable with available data. In turn, once the biological data is seen to have certain characteristics that are similar to that found in models, it may be seen to have additional characteristics that are not found in the model. By embellishing the model to reproduce those additional characteristics, its potential for further useful predictions increases.

The primary goals of this study are (1) to identify different modes of dynamical activity exhibited by a model network, and (2) show how these different modes are dependent on the basic network features. Some of the specific questions addressed include:

- How much variability is there in the spatio-temporal patterns of activity?
- How does the variability in spatio-temporal patterns depend on different parameters of intrinsic connectivity, afferent connectivity, stimulus pattern, and connection weights?
- What modes and/or patterns of activity exhibit the highest CV values for interspike intervals? Which exhibit the lowest?
- Are there examples of synchronous and partially synchronous oscillations, and what combination of parameters tends to favor synchronicity?
- Under what combinations of parameters do the afferents obviously influence the spatial pattern of network activity, if any? Under what parameters do they not?

2. METHODS

The architecture examined in this study is an abstract computational model of a network of units whose characteristics are common to most physiological neural networks. The particular features selected for study represent some of the important properties that distinguish biological neural networks from artificial ones.

2.1. Single Neuron Representation

Two different cell types were used in the simulation, excitatory and inhibitory. The excitatory cell model (a single RC compartment) incorporated Hodgkin-Huxley Na and K channels adapted from Traub et.al. (1991). The combinations of values for R_m, C_m, and conductance levels for the active channels were determined by two constraints: an average cell firing rate of 10Hz, and the requirement that the interspike intervals of a cell would have a coefficient of variation of approximately 0.7. This value is intermediate between the levels above 1.0 that are considered important to cortical information (e.g. Softky and Koch, 1992), and a level of close to 0.0 which would be a purely regular response to synaptic events. This CV level was achieved through an iterative process that involved a balancing of the active Na and K conductances (cf. Bell et.al. 1995), as well as manipulation of cell Rm and Cm and excitatory and inhibitory synaptic conductance magnitude.

Because sparse connectivity is a prominent feature of the cortex (Abeles 1991) and because it was anticipated in advance that sparse connectivity might play an important role in governing network dynamics, the goal of maximum of 5% inter-connectivity was established for excitatory-to-excitatory cells. With the constraints of cell number (see below) and sparse connectivity in place, each excitatory cell received 180 excitatory connections and 45 inhibitory connections. All synaptic activations were generated by dual-exponential rise-and-decay functions with an open time-constant of 1ms and close constant of 2ms, with the exception of the inhibitory-to-excitatory synapses, which have an open time-constant of 1ms and a close time-constant of 4ms. These values correspond with data from slice preparations (Komatsu et.al. 1988) and identical values have been used in similar models of the cortex (Wilson and Bower 1991, 1992), where it was found that these rates provided the basis for accurate reproduction of population EEG recordings. Afferents to the cell were kept at 10Hz for excitatory connections and 20Hz for inhibitory connections. These firing rates were arbitrary, but based on a compromise between the lower rates seen in the hippocampus (Wilson and McNaughton 1994) and the higher rates seen in the cortex (e.g., Bair et.al. 1994).

Typical values of the time constant for excitatory cells generated by the above procedure ranged around 80 ms, although in practice, the Rm and Cm values were randomized in the network simulation by +/− 80% of these values. In the cortex, it is generally considered to be the case that inhibitory cells have a higher firing rate (Scharfman, 1992; Wilson and McNaughton 1994; Wilson and McNaughton 1993), and a smaller cell body than excitatory cells. The inhibitory parameters were chosen without as much emphasis on CV, the main constraint being a firing rate of approximately 20Hz when stimulated with 180 excitatory synapses each driven at an average rate of 10Hz. The time constant of the inhibitory cells was 8 ms.

2.2. Network Architecture

For reasons of simplicity and practicality, a total network size of approximately 5000 cells was desired, with 3600 being excitatory and 900 being inhibitory. This ratio of 3:1 excitatory to inhibitory cells is in rough agreement with the anatomy of the mammalian cortex (e.g. Abeles 1991; Braitenberg and Schuz 1991). A 60x60 network of excitatory cells and a network of 30x30 inhibitory cells were superimposed across a 1000 micron virtual square in a 2-dimensional toroidal geometry. One hundred afferent axons were connected to the network, with each axon diverging to 1440 synaptic connections Each excitatory cell received a total of 40 afferent connections, and the average firing rate per afferent was approximately 20Hz. Each excitatory cell sent and received 180 connections to other excitatory cells. Each excitatory cell sent 45 connections to the inhibitory cell population. Each inhibitory cell sent 180 connections to terminals of excitatory cells. Axonal conduction velocities ranged from 1 to 10 m/sec. The resulting ratio of 45/180 or 1:4 for afferent-to-intrinsic excitatory connections for the excitatory cells in the network corresponds loosely to the 15% estimate of Braitenberg and Schuz (1991). Overall, counting afferent and intrinsic excitatory and inhibitory connections, excitatory cells received 265 synaptic connections. Inhibitory cells each received in the collective phase of a population of neurons all firing at approximately the same rate 180 excitatory connections from excitatory cells.

2.3 Experimental Design

Our goal was to sample and examine a restricted region of parameter space. We identified four basic parameters for analysis:

A. Intrinsic connectivity pattern. We constructed 8 types, ranging from small uniform "disks", to heterogeneous patterns for the excitatory and inhibitory subnetworks.

B. Afferent connectivity pattern, which could have global or local distributions.

C. Afferent stimulus pattern, including uniform random, non-uniform random, and burst-like.

D. Connectivity strengths. There were 27 types, representing combinations of weak, medium, and strong connectivity for each of the excitatory-excitatory, excitatory-inhibitory, and inhibitory-excitatory connection types.

This combination of parameters gave 1296 unique networks, each of which was simulated for 500 milliseconds of virtual time. During the simulation the total number of afferent synaptic events stimulating the network was approximately 1.4×10^6. Simulation results were analyzed with respect to spike event statistics (interspike interval histograms, firing rate analyses, power spectra) and spatial distributions of activity over the excitatory and inhibitory networks.

3. RESULTS

Four basic dynamical modes were apparent from an overview of the results: network-level oscillations, correlated spatial patterns in both networks, clustering in the excitatory network without any spatial structure in the inhibitory network, and run-away activity in both networks. There is variability within each basic mode, and two of the four basic modes that have large memberships can be further subdivided into multiple subtypes.

3.1. Oscillatory Modes

Approximately 45% of all simulations exhibited oscillatory activity, which usually fell in the 30 to 50 Hz range. The most important determining factor for oscillation was afferent connectivity; random connectivity usually resulted in oscillatory behavior, whereas local connectivity tended to generate stable patterns of clustered activity.

3.2. Correlated Cluster Modes

Approximately 50% of all simulations displayed some form of spatial clustering, which could take many forms. Some involved excitatory clusters of varying size, surrounded by inhibitory clusters. These were usually generated by the presence of local afferent connectivity and random stimulus patterns, and the size of the clusters was determined by the intra-network synaptic strengths. Examples of some of the observed spatial structures are shown in Figure 1.

3.3. Other Modes

Some simulations showed clustering in one (usually excitatory) network, with diffuse activity in the other. This mode emerged with strong intrinsic excitatory connections, weak global inhibition, and burst-like input patterns. Afferent connectivity patterns had little influence. Runaway activity patterns were also observed with strong excitation and

trails (as wave cycles)   tiny clusters   small clusters

medium clusters   large clusters   mix of cluster sizes

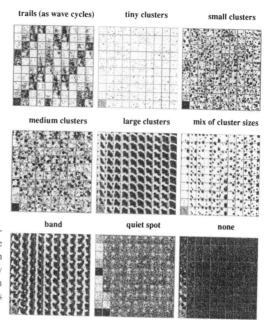

band   quiet spot   none

Figure 1. Spatial Clustering Modes in the Excitatory Network. Each of the nine frames contains the results of a different 500 ms simulation. Each frame consists of 100 "snapshots" of the excitatory network, taken 5 ms apart, arranged top to bottom (first) and left to right. Each dot therein represents the firing of a cell in the 60x60 excitatory network.

weak (surround) inhibition. Finally, a small percentage (<2%) of simulations exhibited a transition from one mode type to another during the 500 ms of simulation time. For example, large clusters could lead to diffuse inhibition and excitatory band patterns, or oscillatory activity could give way to either clustering or runaway activity.

4. DISCUSSION

Each of the 4 parameter groups studied plays a role in influencing the overall patterns of activity in this class of network. At a selected region in this pattern space, however, only one or two parameters may be dominant, in the sense that small changes to a parameter will have a significant effect on network dynamics. The major roles of the key parameters are:

A. Afferent connectivity: this plays a major role in determining whether the resulting dynamics are oscillatory, or whether stable clusters of activity emerge.

B. Intrinsic connectivity: in networks with oscillatory dynamics, this governs whether wave-cycle oscillations or homogenous oscillations emerge. In networks with clustered activity, it determines characteristics of the clusters such as the sharpness of the cluster boundaries.

C. Stimulus pattern: influences synchrony of oscillation modes and shape and contrast of clusters

D. Connection weights: influence the frequency of oscillations and size of clusters

One of our original motivations in this work was the hypothesis that groups of neurons might demonstrate reverberating activity. Our analyses to date have not attempted to detect this in the existing simulations. However, note that there are many examples of self-sustained localized clustering of excitatory activity that are clearly not driven by afferent stimulation. These are particularly apparent in simulations with local afferent connectivity, random input patterns, strong excitatory-to-excitatory connections, and intrinsic connectivity patterns that promote local positive feedback with surround inhibition.

Our network had cells with a broad range of time constants, and thus the observed range of network oscillations (30 to 50 Hz) is less likely to be a function of individual cell time constants, and probably represents an emergent property of this class of network. Similarly, the results were largely the same over a broad range of axonal conduction velocities, suggesting a relative insensitivity of network dynamics to this parameter.

In summary, networks of this class demonstrate a remarkable heterogeneity of dynamical modes, many of which are surprisingly sensitive to afferent input. There is a wide range of different spatio-temporal features that are present in the activity of the network that are not directly related to either the cell properties or the network properties, and can therefore be considered emergent properties. Emergent features analogous to those seen here may play a critical role in organizing activity in biological networks, particularly for the generation and long-term storage of representations. Future studies will address the possibility that use-dependent synaptic plasticity might embed spatiotemporal patterns in the network, and that these patterns could form the basis of a memory storage system.

ACKNOWLEDGMENTS

This work was supported by the Yale Neuroengineering and Neuroscience Center.

REFERENCES

Abeles, M. (1991) Corticonics: Neural circuits of the cerebral cortex. Cambridge University Press, Cambridge.

Bair, W., Koch, C., Newsome, W., and Britten, K. (1994) Power spectrum analysis of bursting cells in area MT in the behaving monkey. Journal of Neuroscience. 14:2870–2892.

Bell AJ, Mainen ZF, Tsodyks M, Sejnowski TJ (1995) 'Balancing' of conductances may explain irregular cortical spiking. Technical Report no. INC-9502, Institute for Neural Computation, UCSD, San Diego CA.

Braitenberg, V. and Schuz, A.(1991) Anatomy of the Cortex. Springer-Verlag, Berlin.

Komatsu, Yl, Nakajima, S., Toyama, K., and Fetz, E. (1988) Intracortical connectivity revealed by spike-triggered averaging in slice-preparations of cat visual cortex. Brain Research 442:359–362.

Scharfman, H. (1992) Differentiation of rat dentate neurons by morphology and electrophysiology in hippocampal slices: granule cells, spiny hilar cells and aspiny 'fast-spiking' cells. Epilepsy Research - Supplement. 7:93–109.

Softky WR, Koch C (1992) Cortical Cells Should Fire Regularly, But Do Not. Neural Computation, 4:643–646.

Traub RD, Wong RKS, Miles R, Michelson H (1991) A Model of a CA3 Hippocampal Pyramidal Neuron Incorporating Voltage-Clamp Data on Intrinsic Conductances. Journal of Neurophysiology, 66:635–650.

Wilson, M.A., and Bower, J. (1991) A computer simulation of oscillatory behavior in primary visual cortex. Neural Comp 3, 498–509.

Wilson, M.A., and Bower, J. (1992) Cortical Oscillations and Temporal Interactions in a Computer Simulation of Pyriform Cortex. J. Neurophysiology 67:981–995.

Wilson, M.A., and McNaughton, B.L. (1993) Dynamics of the Hippocampal Ensemble for Space. Science 261:1055–1058.

Wilson, M.A. and McNaughton, B.L. (1994) Reactivation of Hippocampal Ensemble Memories During Sleep. Science. 265:676–679.

69

FUNCTIONAL SIGNIFICANCE OF SYNAPTIC DEPRESSION BETWEEN CORTICAL NEURONS

S. B. Nelson, J. A. Varela, Kamal Sen, and L. F. Abbott

Volen Center
Brandeis University
Waltham MA
E-mail: nelson@binah.brandeis.edu
E-mail: varela@binah.brandeis.edu
E-mail: ksen@binah.brandeis.edu
E-mail: abbott@volen.brandeis.edu

ABSTRACT

Intracortical synapses exhibit several forms of short-term plasticity that cause synaptic efficacy at any given time to depend on the previous history of presynaptic activity. We have measured synaptic transmission between layer 4 and layer 2/3 in slices of rat visual cortex and used the data to construct an accurate mathematical description of intracortical short-term synaptic plasticity. These data show rapid synaptic facilitation and three forms of synaptic depression differing in their rates of onset and recovery. The dominant effect seen is overall synaptic depression that causes steady-state synaptic efficacy to decrease as a function of presynaptic firing rate. At high rates, the steady-state efficacy is inversely proportional to firing rate which implies that cortical synapses do not convey information about the magnitude of sustained high firing rates. However, this same dependence means that, for transient signals, synapses convey information about fractional rather than absolute changes in presynaptic firing rates. We explore the functional significance of this result including its implications for spike-rate adaptation and mechanisms that produce directional selectivity in visually responsive neurons.

1. INTRODUCTION

Although synaptic efficacies in neural network models are often characterized by fixed weights, it has long been known that synapses do not transmit with constant strength but rather with an efficacy that depends on the history of presynaptic activity. This effect has been studied at cortical synapses predominantly by comparing responses to pairs of stimulation pulses[1-4]. To explore the functional consequences of the activity dependence introduced by various forms of short-term plasticity, it is important to be able to predict synaptic efficacy for an arbitrary sequence of presynaptic action potentials. For this reason, we measured postsynaptic responses in layer 2/3 of rat visual cortex slices to random presynaptic spike trains evoked by extracellular simulation in layer 4. We then constructed a mathematical model that accurately fit these data and used it to study the functional implications of the synaptic depression that these synapses display prominently[5,6] (related work has been done in Refs. 7-9). Unlike inhibition or intrinsic adaptation, synaptic depression does not reduce the overall responsiveness of a neuron to its synaptic inputs but rather decreases the gain at specific highly active synapses. This allows the neuron to be more responsive to more slowly firing afferents. Our measurements and model make this qualitative statement more precise, indicating that the cortical gain control provided by synaptic depression is precisely tuned to convert absolute presynaptic rate changes into postsynaptic conductance transients that are proportional to relative or fraction rate changes. In addition, the temporal characteristics of synapses exhibiting depression may serve other computational roles, for example, providing a mechanism for generating the directional selectivity seen in neurons of the primary visual cortex.

2. METHODS

We have measured and characterized synaptic transmission along a major excitatory pathway (layer 4 to layer 2/3) in slices of rat primary visual cortex. These slices (400μm thickness) were prepared and data were taken using standard methods as described in[5,6]. Trains of electrical stimuli (10-150 μA, biphasic, 80 μs) were applied via a monopolar stimulating electrode placed in layer 4. Stimulus pulses were delivered in either a random sequence drawn from a Poisson distribution, or at regular intervals. Postsynaptic responses were measured in layer 2/3 immediately above the stimulation site in two ways; either as field potentials or as postsynaptic currents measured intracellularly. Both methods yielded similar results when fit as described below.

We have constructed a mathematical description that can accurately predict the amplitude of the responses to individual stimulation pulses occurring in an arbitrary sequence. The amplitude A of the response is written as the product of four factors

$$A = A_0 F D_1 D_2 D_3 \, . \tag{1}$$

A_0 is a constant that determines the overall magnitude of the response. F accounts for a synaptic facilitation. Initially, $F = 1$, but after each presynaptic spike, F is augmented to $F + f - 1$ where $f > 1$ is a factor controlling the onset of facilitation. Between presynaptic spikes, F relaxed back to one, obeying the differential equation

$$\tau_F \frac{dF}{dt} = 1 - F \, . \tag{2}$$

The depression factors, D_i for $i = 1, 2, 3$, obey similar equations but with different time constants

$$\tau_i \frac{dD_i}{dt} = 1 - D_i \, . \tag{3}$$

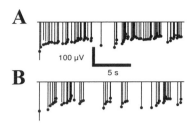

Figure 1. Data, fits and predictions for field potentials recorded in layer 2/3 evoked by stimulation of layer 4. The sticks show the peak amplitude of the responses to individual stimulation pulses in a random Poisson train. Dots are the amplitudes predicted by the model using the parameters $f = 1.63$, $\tau_f = 167$ ms, $d_1 = 0.59$, $\tau_1 = 28$ ms, $d_2 = 0.57$, $\tau_2 = 475$ ms, $d_3 = .984$, $\tau_3 = 5.22$ s. A) Measured and fitted response to a 4 Hz Poisson train that was used to determine the best fitting parameters. B) Measured responses to a 2 Hz train and predictions of the model using the parameters determined by the fit in A.

Following a spike, each D_i is decremented by a multiplicative factor $D_i \rightarrow d_i D_i$ with $0 \leq d_i < 1$. Facilitation is described additively as in references[10,11] while depression is described multiplicatively to prevent A from becoming either negative or infinite[6]. We find the best values of the parameters A_0, f, τ_f, d_i and τ_i (for $i = 1, 2, 3$) by exhaustive search and error minimization. A typical data sequence and fit, along with characteristic fitting parameters, is given in Fig. 1.

The dominant form of short-term plasticity seen in these data is depression that sets in with a d-factor in the range .65 to .86 and a time constant τ between 300 ms and 550 ms. For the discussion and examples shown here, we have simplified the mathematical description by including only this single form of depression. However, the results do not change dramatically if we include all four types of short-term plasticity. All the simulations discussed use a single compartment integrate-and-fire neuron with a resting potential of -70 mV and a membrane time constant of 30 ms. Synapses were represented as conductance changes with a reversal potential of 0 mV that decay exponentially to zero with a time constant of 2 ms.

3. RESULTS

Figure 1 is a typical example that illustrates how well the mathematical description fits the data. In this figure, the mathematical model accounts for 95% of the variance seen in the data and the average discrepancy between the fit and the data is 6%. Having such a mathematical description available means that we can study the functional implications of the synaptic depression that is the dominant effect seen in the data. We have done this in Ref. 5. Here we will review some of the results of Ref. 5 and then present some additional work.

As discussed in the Methods section, it is possible, for many purposes, to replace the four forms of short-term synaptic plasticity we see with a single, overall depression. In this case, the single form of depression is described by a multiplicative onset factor, d and a recovery time constant τ. We have found that[5]:

1) The steady-state amplitude $A(r)$ in response to a periodic sequence of presynaptic spikes at rate r is

$$A(r) = \frac{1 - \exp(-1/r\tau)}{1 - d\exp(-1/r\tau)} \tag{4}$$

which varies approximately as $1/r$ for rates greater than 10-20 Hz. The total synaptic input arising from a set of afferents firing steadily at rate r is roughly proportional to this amplitude times r. Because $A(r) \propto 1/r$ at large rates, the product $A(r)r$ is approximately independent of r so these synapses convey little information about the magnitude of steady, high-frequency presynaptic firing rates. For example, the postsynaptic response to a set of afferents firing steadily at 50 Hz is not very different from that evoked when those afferents fire at a sustained rate of 100 Hz.

2) The transient response that occurs when a set of afferents suddenly switches from a firing rate r to a rate $r + \Delta r$ is proportional to $A(r)\Delta r$ which is, in turn, proportional to $\Delta r/r$ for rates above 10-20 Hz. As a result, these synapses perform an interesting transformation on presynaptic firing-rate changes. Absolute changes in presynaptic firing rates are converted into relative or fraction changes, introducing a scale invariance into neuronal responses. A postsynaptic neuron will respond similarly to equal percentage changes of presynaptic firing rates independent of what those rates are. For example, a change of presynaptic firing rate from 25 to 50 Hz produces virtually the same response as a change from 50 to 100 Hz.

3) Most network models use the approximation that neurons respond to a synaptically weighted sum of the firing rates of their afferents. Neurons described by such models are insensitive to input firing-rate changes that do not affect this sum. When synaptic depression is included in the model, neurons become sensitive to changes in the pattern of afferent firing even if these do not affect the steady-state synaptically weighted firing-rate sum. This result can be interpreted in either of two ways: the increased sensitivity to transients may be a significant source of the variability seen in cortical neurons[12] or it may represent a form of neuronal coding[13] that can be accessed only through the effects of synaptic depression.

4) Even though synaptic depression reduces the ability of a synapse to transmit information about high sustained firing rates, it does not eliminate the possibility of generating tuned steady-state neuronal responses. Depression may broaden these responses but this broadening is offset by an increased sensitivity to transients.

Fig. 2 shows an interesting consequence of synaptic depression when considered in conjunction with the spike-rate adaptation displayed by many cortical neurons[14]. Without synaptic depression, the postsynaptic depolarization produced by presynaptic input from a model neuron that does not adapt is vastly greater than that produced by an adapting presynaptic cell. However, with depression included, the difference is quite small. Thus it is possible that the degree of synaptic depression and the amount of spike-rate adaptation seen in cortex are matched so that presynaptic neurons can produce a given degree of transient postsynaptic depolarization using the minimum number of spikes.

Many neurons in primary visual cortex are sensitive to visual images moving in a particular direction. Discussion and models of directionally selective neurons[15-19] have relied on coupling a temporal phase shift to a spatial phase shift between two classes of inputs to the directionally-selective neuron. Various mechanisms for generating the temporal phase shift have been proposed, including delays in cortical feedback[18,19] and lagged thalamic responses[15]. Synaptic depression offers another alternative. At frequencies between about 0.5 and 10 Hz, the synaptic depression we have discussed generates a roughly $90°$ phase shift in the postsynaptic response to a sinusoidal oscillation of presynaptic firing rates. When coupled to a $90°$ spatial phase shift this can produce a model of a directionally-selective neuron as shown in Fig. 3.

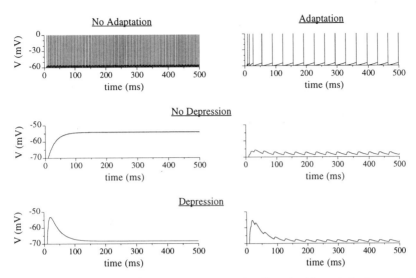

Figure 2. A comparison of postsynaptic responses to presynaptic spike trains (shown in the top row) without (left column) or with (right column) spike-rate adaptation. The middle and bottom rows show the resulting postsynaptic depolarization without and with synaptic depression respectively.

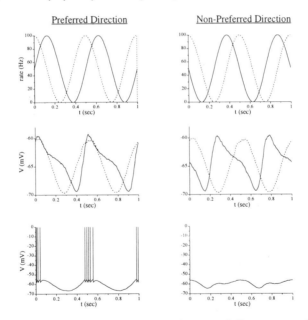

Figure 3. A model of directional selectivity. The firing rates of two sets of afferents to a model neuron are shown in the top row. These firing rates correspond to inputs evoked by a sinusoidal grating moving in either the preferred direction for the cell (left column) or in the opposite direction (right column). In the model, these two sets of inputs activate two different types of synapses. The solid curves corresponds to presynaptic rates at synapses that exhibit depression while the rates shown by the dashed curves are for synapses without any depression. The middle row shows the postsynaptic depolarizations produced by these two sets of afferents. The depolarization shown by the dashed curve corresponding to nondepressing synapses is roughly proportional to the presynaptic firing rate. The solid curve, corresponding to depressing synapses is distorted and shifted by 90° relative to the presynaptic firing rate due to the temporal effects of synaptic depression. The two contributions to the depolarization of the postsynaptic cell add in phase for motion in the preferred direction and out of phase for motion in the opposite direction and, as a result, the model neuron fires action potentials only in response to motion in the preferred direction (bottom row).

4. SUMMARY

Because synaptic depression and other forms of short-term synaptic plasticity introduce a spike train history dependence into synaptic transmission, they have rather profound implications for neuronal coding and information processing. Synaptic depression implies that transient responses are of particular significance and that these may encode information about scale-invariant fractional changes rather than absolute changes of firing rate. With synaptic depression, transient neuronal responses may have access to richer, and/or noisier neural encoding schemes with heightened sensitivity. Synaptic depression and spike-rate adaptation may be matched for optimal information transmission of these transient signals. Finally, synaptic depression is a candidate for the mechanism generating temporal phase shifts that allow neurons in primary visual cortex to be directionally selective.

REFERENCES

[1] R. Deisz and D. Prince, 1989, Frequency-dependent depression of inhibition in Guinea-pig neocortic in vitro by GABAb receptor feedback on GABA release, *J. Physiol.* **412**:513-541.

[2] S.B. Nelson and D. Smetters, 1993, Short-term plasticity of minimal synaptic currents in visual cortical neurons, *Soc. Neurosci. Abst.* **19**:629.

[3] A.M. Thomson and J. Deuchars, 1994, Temproal and spatial proerties of local circuits in neurocortex, *Trends Neurosci.* **17**:119-126.

[4] K.J. Stratford, K. Tarczy-Hornoch, K.A.C. Martin, N.J. Bannister, J.J.B. Jack, 1996, Excitatory synaptic inputs to spiny stellate cells in cat visual cortex, *Nature* **382**:258-261.

[5] L.F. Abbott,J.A. Varela, K. Sen and S.B. Nelson, 1997, Synaptic Depression and Cortical Gain Control. *Science* (in press).

[6] J.A. Varela, K. Sen, J.A Gibson, J. Fost, L.F. Abbott and S.B. Nelson, 1996) A quantitative description of short-term plasticity at excitatory synapses in visual cortex, (submitted).

[7] H. Markram, M. Tsodyks, 1996, Redistribution of synaptic efficacy between neocortical pyramidal neurons, *Nature* **382**:807-810.

[8] M.V. Tsodyks, H. Markram, 1996, Plasticity of neocortical synapses enables transitions between rate and temporal coding, in: *Lecture Notes in Computer Science* (C. von der Malsburg, W. von Seelen, J.C. Vorbruggen and B. Sendhoff ed), p. 445-450, Springer, Berlin.

[9] M.V. Tsodyks, H. Markram, 1997, Neurotransmitter release probability determines the nature of the neural code between neocortical pyramidal neurons, *Proc. Natl. Acad. Sci.* (in press).

[10] Magleby KL, Zengel JE (1975) A quantitative description of stimulation-induced changes in transmitter release at the frog neuromuscular junction. *J. Gen. Physiol.* **80**:613-638.

[11] K. Sen, J.C. Jorge-Rivera, E. Marder and L.F. Abbott, 1996, Decoding Synapses, *J. Neurosci.* **16**:6307-6318.

[12] W.R. Softky, C. Koch, 1992, Cortical cells should fire regularly, but do not, *Neural Comp.* **4**:643-646.

[13] H.B. Barlow, 1989, Unsupervized learning, Neural Comp. 1:295-311.

[14] The model of adaptation we use was constructed by X.J. Wang.

[15] A.B. Saul and A.L. Humphreys, 1992, Evidence of input from lagged cells in the lateral geniculate nucleus to simple cells in cortical area 17 of the cat, *J. Neurophysiol.* **68**:1190-1208.

[16] B. Jagadeesh, H.S. Wheat and D. Ferster, 1993, Linearity of summation of synaptic potentials underlying direction selectivity in simple cells of the cat visual cortex, *Science* **262**:1901-1904.

[17] L.L. Kontsevich, 1995, The nature of the inputs to cortical motion detectors, *Vision res.* **35**:2785-2793.

[18] H. Suarez, C. Koch and R. Douglas, 1995, Modeling direction selectivity of simple cells in stiate visual cortex within the framework of the canonical microcircuit, *J. Neurosci.* **15**:6700-6719.

[19] R. Maex and G.A. Orban, 1996, Model circuit of spiking neurons generating directional selectivity in simple cells, *J. Neurophysiol.* **75**:1515-1545.

[20] Research supported by the Sloan Center for Theoretical Neurobiology at Brandeis University, National Science Foundation grants NSF-IBN-9421388, NSF-DMS-9503261 and NSF-IBN-9511094, a Sloan Research Fellowship and the W.M. Keck Foundation.

70

HOW TRANSMISSION DELAYS AND NOISE MODIFY THE SIMPLE AND LARGE NEURAL NETWORKS DYNAMICS

Khashayar Pakdaman, Joël Pham, Eric Boussard, and Jean-François Vibert

B3E, ESI, INSERM U444, ISARS
Faculté de Médecine Saint-Antoine
Université Pierre et Marie Curie, Paris VI
27 rue Chaligny, 75571 Paris Cedex 12, France
pakdaman@b3e.jussieu.fr

In the nervous system, information transfer is partly mediated by action potentials travelling along axons connecting neurons. Conduction velocity ranges from 20 to 60 m/s, leading to non negligible transmission delays, from milliseconds to hundreds of milliseconds. These delays, referred to as inter-neural delays (INDs) reflect axonal propagation time, synaptic delay, etc., can greatly affect the behavior of living neural networks. The effect of IND on the dynamics of a single neuron receiving recurrent excitation after a controlled delay, and on those of large fully interconnected excitatory networks were described and analyzed in two previous papers (Pakdaman et al., 1996; Vibert et al., 1996). Noise, as INDs, is also omnipresent in the nervous system, and can greatly modify the behavior of neurons and neural networks (Segundo et al., 1994). In this paper, we are interested on how noise affects the IND effect on both the single neuron with recurrent excitation and the excitatory neural network.

1. SINGLE NEURON WITH RECURRENT EXCITATION

Diez-Martìnez and Segundo (1983) gave experimental evidences that the IND can affect the dynamics of a single neuron with a recurrent excitatory connection using the pacemaker neuron in the crayfish stretch receptor organ and having each spike trigger electronically a brief stretch. They showed that the time separating the spike from the stretch, called "delay", was strongly influential, and that small changes led to markedly different outcomes. As the delay was increased the discharge patterns went from pacemaker spike trains to multiplets separated by silent intervals to still longer burst and longer silent intervals. It was hypothesized that neuronal adaptation combined with the delay produced the observed behavior. This hypothesis was tested using models of increasing complexities (Pakdaman and Vibert, 1995; Pakdaman et al., 1996).

The simplest model was that of a neuron with a sigmoidal transfer function and a charging/discharging time constant, but no adaptation to repeated stimuli. In this case, the dynamics were not sensitive to the delay. Two models exhibited adaptation to repeated stimuli, one a leaky integrator, and the other a conductance-based. Neither of these models exhibited saturation when recurrent excitation was introduced. Their dynamics were in fact similar to those in the crayfish preparation, both exhibiting pacemaker firing for short delays, multiplets or burst for intermediate delays, and rapid regular firing for long delays. Simulations were therefore compatible with the hypothesis that neuronal adaptation and the IND greatly influence the behavior of a single neuron with recurrent excitation (Pakdaman et al., 1996). The leftmost part of Figure 1, shows an example of the behavior of a leaky integrator model of a pacemaker neuron (natural period: 32 ms) with a recurrent excitation whose IND increases from bottom to top. Neurons are leaky integrator models simulated using XNBC (Vibert et al, 1997).

To mimic the synaptic noise usually impinging neurons in a network, gaussian noise was added to the membrane potential. Figure 1 illustrates the various changes in the firing pattern of a neuron with a recurrent excitatory connection, brought about by the adjunction of noise of increasing intensity.

It can be seen that for short delays (Fig. 1-lower trace) and in the absence of noise, the neuron fires with a period close to the natural period. In presence of random perturbations, the pattern is that of a noisy pacemaker. The recurrent excitatory loop has little influence when the delay is short. Intermediate delays are characterized by periodic successions of multiplets during which the spikes are separated by an interval close to the delay. Two successive multiplets are separated by an interval longer than the natural period (left part of 7 ms delay trace in Fig. 1). The number of spikes during the multiplet increases with the delay. Even though the system displays a similar firing pattern in presence of moderate perturbations, the number of spikes during the multiplet becomes highly variable (middle part of the same trace in Fig. 1). Increasing noise intensity in-

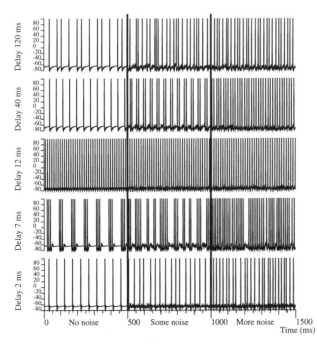

Figure 1. Temporal evolution of the membrane potential of five simulated neurons with a recurrent connection with delay increasing from bottom to top . The delay value is indicated on the right of the figure. The natural period of this pacemaker is 42 ms. From 0 to 500 ms, the system was noise free. A moderate noise was added at 500 ms (some noise), and its intensity was increased at 1000 ms (more noise).

creases the number of spikes during the multiplet and shortens the inter-multiplet intervals. In this regime, most interspike intervals are close to the delay, a few exceptions close to the natural period (right part of the same trace). In the 12 ms delay trace in Fig. 1 the neuron fires with a period equal to the delay. The firing pattern is extremely robust against perturbations impinging upon the membrane potential. This is because in this system, each firing is triggered by the arrival of the EPSP, at a phase when the soma membrane potential is not close enough to the threshold, for the noise to be influential. The influence of such rapidly increasing EPSPs have been described in excitatory networks without delay (Mirollo and Strogatz, 1990; Catsigeras and Budelli, 1992) where they play an important role in synchronizing the network.

For long delays close to, but still shorter than the natural period, the noise-free firing is composed of intervals equal to the delay, as in the previous case (40 ms trace in Fig. 1). However, this pattern is not robust against random perturbations. The adjunction of moderate noise, transforms the firing pattern to that of a regular succession of doublets. The mechanism underlying this change of behavior has not been completely elucidated yet but is related to the fact that the noisy pacemaker has a mean interspike interval which is shorter than the natural firing period of the noise-free pacemaker. Thus for appropriate range of parameters the delay may become longer than the mean interspike interval. Increasing the noise intensity destroys this regular pattern as can be seen on the right part of the 40 ms trace in Fig. 1, where the neuron displays a rapid irregular firing.

For delays longer than the natural period, the noise-free spike train is a periodic succession of multiplets composed of interspike intervals close to the natural period and close to the difference between the delay and the natural period, as shown in the left part of the upper trace in Fig. 1. This pattern is not robust against the adjunction of moderate noise (middle part). It is transformed into an irregular succession of long and short intervals. Increasing noise intensity shortens the intervals (left part).

Figure 2 represents the effect of noise on a unit with a recurrent excitation after an intermediate delay inducing multiplets. The top row (A) is the control situation, without noise, while in row B the noise amplitude (its standard deviation) is set to a medium value and in row C, to a high value. The left column represents the temporal evolution of the unit membrane potential. The center column represents phase maps of the unit (the threshold as a function of the membrane potential), and the left column, the temporal evolution of the interspike interval. Note that without noise both long and short intervals are almost constant. When noise is added, long intervals are the most affected, since their mean value decreases and their variability increases. This phenomenon is enhanced again when noise increases, since the longer intervals no longer constitute a secong interval population. The phase maps show that the whole graph is confined toward the upper values of threshold. When the neuron presents a multiplet pattern of discharge, the noise reduces the long interspike intervals while it does not affect the short ones.

2. EXCITATORY NEURAL NETWORKS

Several types of networks containing from 2 to 1000 neurons were simulated using XNBC. These networks were fully connected through excitatory synapses. We were interested by both the global network activity and the fact that the unit discharge pattern was regular (pacemaker) or not. Initial conditions were determinant for the network behavior: some networks were initially synchronized (oscillating), others not. In a previous work (Vibert et al., 1994) we showed that networks whose discharge was synchronous displayed unsteady os-

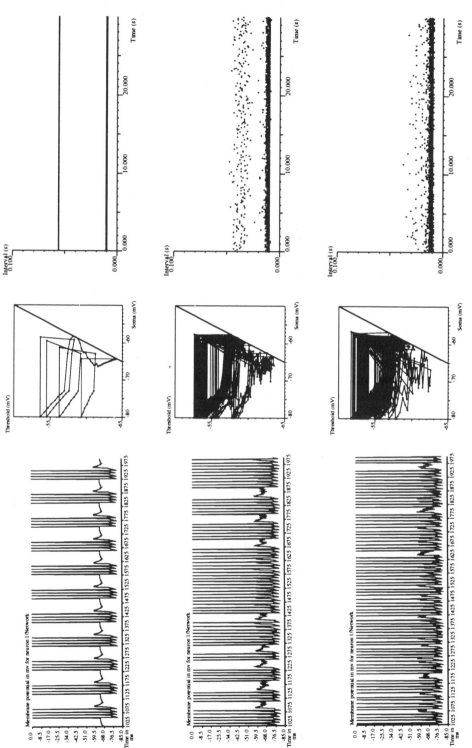

Figure 2. Effect of noise on the interval interspike distribution of a neuron with recurrent inhibition with intermediate delay (8 ms). From top to bottom: no noise, some nose and more noise. From left to right: membrane potential versus time, Threshold versus membrane potential (phase map), interspike interval versus time.

cillations when the interneural transmission delay was short, and became steady when delay increased. A sharp increase of delay switched the network behavior from the non synchronized state to the synchronized state. The presence of pacemaker units did not change the result. When pacemaker units were present, they induced a natural tonic activity at the very beginning of the simulation, and then the synchronization of pacemaker units was rapid. Without pacemaker, background noise was needed to maintain a tonic activity. In the absence of noise, delay modification leads to abrupt switch from a given pattern of synchronization to another one (doublets, triplets, multiplets, ...) (Vibert et al., 1994).

Networks of units were simulated, assuming that all delays were set at a given value, and changed during the simulation. Figure 3 shows the effect of both noise and delay on such a network. At the simulation start, all the neurons are at rest. This initial setting provides a sufficient condition for the network to display an overall oscillatory activity. The bottom row of Fig. 3 shows that in this particular configuration the network is periodically active, with bursts of activity of about 50 % of units simultaneously active. When the delay is increased, the discharge pattern changes, and all units are synchronized during short repetitive periods. When the delay is decreased back to the initial value, the period of the network discharges lengthens, but does not return to the initial pattern in this particular case (sometimes it does as shown in Vibert et al, 1994). For short delays, EPSPs arrive early in the refractory period, and thus are not capable of generating a new spike. An increase in the delay causes major changes in the neuron's behavior. The EPSPs arrive later in the refractory period and can generate spikes, as long as the threshold fatigue is low.

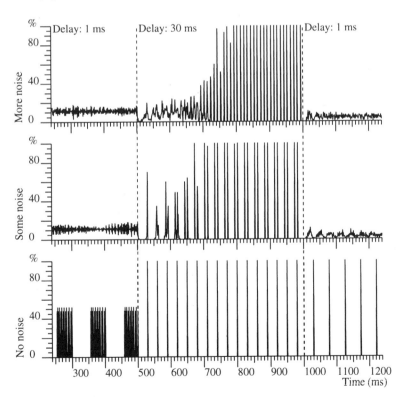

Figure 3. Effect of delay and noise on a neural network of 100 units. Noise increases from bottom to top. IND is increased at 500 ms and decresead back at 1000 ms. Abscissae: time in ms. Ordinates: percentage of of active units in the network.

The firing pattern of the neurons changes. Due to threshold fatigue, the neuron fires in multiplets. In a synchronized network and in the absence of a background noise, the delay is an effective bifurcation control parameter. Its increase leads to periodic behaviors with increasing lengths, up to a point where the delay is long compared to the refractory period. At this point the activity switches to a robust period-one cycle. This phenomenon is fully reversible (for more details, see Vibert et al., 1994). In this case, the behavior of the network is similar to that of a single neuron with recurrent excitation.

When a Gaussian background noise is added to the network, with short delay, the noise induces irregularities in the pacemaker period, desynchronizing units and leading to a tonic and apparently random activity. When delay is increased, the network displays phasic activity, with a discharge pattern dependent on both the noise amplitude and the delay value. A fully connected excitatory network of neurons can transform a Gaussian noise input signal into strong and robust phasic synchronized discharges, the number of multiplets and the time intervals separating them depend on the delay in the network. Fig. 3-middle row shows a noise amplitude leading to multiplets, after a transient phase during which the networks becomes immediately phasic, but with discharge amplitudes gradually increasing. With more noise (Fig. 3-top row), the transient phase increases, the network stays tonic a longer time, and then becomes phasic with higher frequency than with low noise. Properties described for the single neuron with recurrent excitation are also observed at the network level, unraveled by the sharp increase of delay. In both noise conditions, the network returns back to a low but tonic activity with a transient phase before returning back to the initial level of tonic activity (not visible on Fig. 3). Noise greatly modifies the network behavior, and can induce multiplets in an otherwise regularly firing network.

3. CONCLUSION

Excitatory networks were found in several places in the central nervous sustem. Respiratory rhythmogenesis could closely depend on such networks (Fortin and Champagnat, 1993). Both intrinsic neuron properties and network related parameters can be efficiently tuned to avoid saturation, even in fully connected networks. This property seems important for the respiratory function since it allows to transform sporadic random activity into a continuous tonic activity necessary to feed up the respiratory network whose role is to shape the respiratory rhythm. Our results suggest that in a homogeneous excitatory network subject to noisy inputs, the delay may be considered as an important parameter affecting the coherence of the activity throughout the network. It should be noted that the delay is a physiologically relevant parameter that can be dynamically tuned to control the network activity. In fact, in addition to the delay due to the action potential propagation and chemical synaptic transmission, delay might be introduced by the rise time of the synaptic currents (Brown, 1988) or the intrinsic properties of the postsynaptic membrane. Among the intrinsic properties of membranes, the transient outward current named I_A which is particularly important in brainstem regions may introduce delays lasting more than several hundred milliseconds between a depolarizing event and the first consequent action potential (Connor and Stevens, 1971; Storm, 1988). These results obtained using simulations were confirmed by a theoretical study showing that IND strongly influences the network activity (Pakdaman et al., 1995a,b). Noise was recognized since a long time as a mean to displace a dynamical system from local attractors, and thus a factor linearizing theses systems (Vibert et al, 1981, Segundo et al., 1995). We showed here that associated with delay, noise can also, unexpectedly, regularize the network discharge pattern.

4. ACKNOWLEDGMENT

This work was supported in part by a DRET grant (contract 94/2526A/DRET).

5. REFERENCES

Brown D.A.: M-current: an update. Trends in Neurosciences, 11: 294–299 (1988)

Connor J.A., Stevens C.F.: Inward and delayed outward membrane currents in isolated neural somata under voltage clamp. Journal of Physiology, 213: 1–19 (1971)

Diez-Martinez O., Segundo, J.P. Behavior of a single neuron in a recurrent excitatory loop. Biol. Cybern, 47:33–41 (1983) .

Fortin G., Champagnat J. Spontaneous synaptic activities in rat nucleus tractus solitarius neurons in vitro. Bran Res. 630: 125–135 (1993)

Pakdaman K., Grotta-Ragazzo C., Malta C.P., Vibert J-F. Effect of delay on the boundary of the basin of attraction in a system of two neurons. Publicações Universidade de São Paulo IFUSP/P-1169: 1–25 (1995a)

Pakdaman K., Grotta-Ragazzo C., Malta C.P., Vibert J-F. Delay-induced transient oscillations in a two-neuron network. Publicações Universidade de São Paulo Technical report IFUSP/P-1181: 1–23 (1995b)

Pakdaman K., Vibert J-F. Modeling excitatory networks. J. SICE. 34: 788–793 (1995c)

Pakdaman K., Vibert J-F, Boussard E., Azmy N. Single neuron with recurrent excitation: Effect of the transmission delay. Neural Networks, in press (1995d)

Segundo J.P., Vibert J-F, Pakdaman K., Stiber M., Diez Martínez, O. Noise and the neurosciences: a long history with a recent revival (and some theory). In: Pribram K. (Eds). "Origins: Brain & Self Organization". Lawrence Erlbaum Associates Pub.; 299–331 (1994)

Storm J.F.: Temporal integration by a slowly inactivating K^+ current in hippocampal neurons, Nature, 336: 379–381 (1988)

Vibert J-F, J.P., Segundo. Slowly-adapting stretch-receptor organs: periodic stimulation with and without perturbations. Biol. Cybernetics, 33: 81–95 (1979)

Vibert J-F, Pakdaman K., Azmy N. Inter-neural delay modification synchronizes biologically plausible neural networks. Neural Networks, 7: 589–607, (1994).

Vibert J-F., Pham J., Pakdaman K, Azmy N. XNBC: A simulation tool for neurobiologists. In: The neurobiology of Computation. J. Bower Ed. Kluwer Academic Pub., Boston (USA); 346–352 (1995)

Vibert J-F, Pakdaman K., Boussard E., Av-Ron E. XNBC: a simulation tool. Application to the study of neural coding using hybrid networks. BioSystems, 40: 211–218 (1997)

71

A SIMPLE MODEL FOR DEVELOPMENT AND FUNCTION OF LONG-RANGE CONNECTIONS IN NEOCORTEX

K. Pawelzik and T. Sejnowski[1]

Institut f. Theor. Physik
Robert-Mayer-Straße 8, D-60054 Frankfurt
E-mail: klaus@chaos.uni-frankfurt.de
[1]The Salk Institute for Biological Studies
CNL, San Diego, CA
E-mail: terry@salk.edu

ABSTRACT

We present a model for the development of long-range horizontal connections in cortex. Fixed short-range interactions with local excitatory feedback and an inhibitory surround induce localized attractors ('activity blobs') that depend sensitively on small variations of the input. This provides the basic mechanism for the putative role of the long-range horizontal connections: to switch between different attractors and to organize rapidly the grouping of distant neurons. We show that Hebbian adaptation of sparse long-range horizontal connections under these conditions is sufficient to break the initial symmetry of homogeneous connectivity. This can induce a strong dependency of the local responses on the input from outside the classical receptive fields. This grouping mechanism may explain a wide range of physiological and psychophysical observations.

1. INTRODUCTION

Recent experiments in primary visual cortex have shown that long-range horizontal connections preferentially connect neurons of similar orientation [1]. These connections appear as elongated arrangements of patches which are often aligned parallel to the same orientation in visual space [2, 3]. New physiological observations may be a consequence of these axon collaterals. The effect of stimulation outside the classical receptive field were originally reported to be suppressive [4], but more recent experiments in cat and monkey primary visual cortex [5, 6] have revealed that responses may depend on the features of the

Computational Neuroscience
edited by Bower, Plenum Press, New York, 1997

surround stimulus and may even increase the response to stimuli within the classical receptive field. Finally, many psychophysical experiments have been reported, such as 'pop out' and facilitation [7] that may be supported by the long-range horizontal connections.

In the present study we have examined to what extent a simple nonlinear model for the visual cortex can explain these phenomena. In particular we were interested in the interaction of dynamical activity and the self-organization of long-range connections through a Hebbian learning rule. We found that simple assumptions were sufficient to account for receptive field properties observed in cortex including non-classical receptive fields. Furthermore, we found that these connections can contribute substantially to orientation preference maps. The model also offers a surprisingly simple explanation for several novel aspects of cortical reorganization [8] and may provide a mechanism for the persistence of orientation preference maps in reverse occlusion experiments where the thalamo-cortical connections do not contribute [9].

2. MODEL

The model consists of a two-dimensional layer of continuous mean-field neurons at locations x whose activities m represent rates of firing of cortical neurons

$$\dot{m}(x,t) = -m(x,t) + g(h(x,t)), \tag{1}$$

where g denotes a simple threshold linear function, i.e. $g(h) = 0$ if $h < \theta$ and $g = \alpha(h - \theta)$ otherwise. The input

$$h(x) = \sum_{x'} w(x,x')m(x') + h^{ext}(x) \tag{2}$$

consists of lateral and external contributions. The lateral weights w are modeled to be composed of two parts: fixed short range connections w^s and adaptive long range connections w^l:

$$w(x,x') = w^s(x,x') + w^l(x,x'). \tag{3}$$

The short-range interaction has a mexican hat shape

$$w^s(x,x') = A \exp{-(x-x')^2/2\sigma_e^2} - B \exp{-(x-x')^2/2\sigma_i^2} \tag{4}$$

which induces localized blobs of activity. These attractors served as the basis of several cortical models (see e.g. [11, 12, 13]) and have recently been rediscovered [14, 17] as an explanation of orientation tuning. Further, we assume that the strengths of sparse long-range connections change according to a Hebbian rule on a time scale which is assumed to be long compared to the dynamics, i.e. $\Delta w^l(x,x') \propto m(x)m(x')$ with $\sum_{x'} w(x,x') = const$.

Finally we assume that the input fields are large and strongly overlapping. This means we present stimuli $s(y)$ to an input surface at locations y which then are projected to the neuronal sheet: $h^{ext}(x) = \int f(x-y)s(y)dy$ where $f(x) = \exp(-x^2/2\sigma_r^2)$.

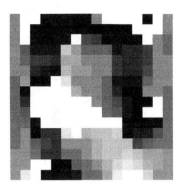

Figure 1. Orientation preference map emerging from horizontal interactions in a layer of 16×16 neurons with large isotropic input fields f. The parameters were: $\sigma_r = 7$, $\sigma_e = .7$, $\sigma_i = 2.5$, $\theta = .08$, $\alpha = \theta$. 10000 Stimuli were bars of four orientations presented at random locations (periodic boundary conditions).

3. RESULTS

Our simulations show that these simple assumptions produce spontaneous symmetry breaking of the long-range weights. In particular, we find that development using elongated stimuli induces patterns of patchy connections which form elongated groups, as observed in visual cortex [2, 3].

The sparse connections allow particular arrangements of localized attractors to be selected by visual input. For stimuli with low orientation contrast (short bars), these sparse connections provide a weak bias for which small perturbations are effective for breaking the translational symmetry of the localized attractors. This implies that the horizontal connections not only contribute to orientation preference but may even induce orientation selectivity, as shown in Fig. 1 where the input fields f are isotropic.

Finally, for some neurons we find that surround suppression of responses to strong stimuli within the receptive field [4] and surround facilitation for weak central input [6] may simultaneously emerge in our model from the interactions of localized attractors without any further assumptions (compare e.g. [16, 18]).

4. DISCUSSION

Our model assumes that the rate dynamics of a local cortical circuit may sensitively depend on perturbations which are transmitted by long range horizontal connections. Under these conditions a simple Hebb-rule for the development of long range connections leads to elongated patterns of clustered connections. We showed that these connection patterns are sufficient for generating orientationally tuned responses because they can amplify and sharpen orientations of stimuli which are retinotopically projected onto the cortical surface. Given that the orientation information is not available from anisotropies of afferent connections (which is the case during reverse-occlusion experiments) this mechanism may still provide orientation tuning which in turn can serve to break the symmetries for the development of new afferents (see however [19]). In this way our model provides a possible explanation for the recently observed precise restoration of orientation preference maps during reverse occlusion experiments [9, 10]

In this paper we considered the effect of excitatory long rage connections on abstract neurons which locally where both, excitatory and inhibitory, depending on the 'mexican hat' type of interaction. More realistic models which include populations of excitatory and inhibitory cells show a more complex dynamics. In particular recent analysis revealed that the net effect of the long range connections on the excitatory population of a local circuit may change from enhancing to suppressing depending on the total strength of input to the local circuit [20]. Preliminary results show that the development of long rang connections in this situation is even more robust than in the present model since this differential effect serves as gain control.

ACKNOWLEDGMENTS

Fruitful discussions with Misha Tsodyks and Juan Lin are gratefully acknowledged.

REFERENCES

[1] (1993) R.Malach, Y.Amir, M.Harel, and A. Grinvald, Relationship between intrinsic connections and functional architecture revealed by optical imaging and in vivo targeted biocytin injections in primate striate cortex. PNAS **90**, pp10469-10473, 1993.

[2] (1995) D. Fitzpatrick, The organization of local circuits in tree shrew striate cortex, Cold Spring Harbour Symposium Proceedings.

[3] (1995) K. E. Schmidt, S.Löwel, R. Goebel, W. Singer, Perceptional grouping criterion of colinearity is reflected by anisotropic long range tangential connections in cortex, submitted.

[4] (1992) J.J. Knierim and D.C. van Essen, Neuronal responses to static texture patterns in area VI of the alert macaque monkey, Jour. Neurophys. 67, pp 961-980.

[5] (1995) A.M. Sillito, K.L. Grieve, H.E.Jones, J.Cudeiro, J.Davis, Visual cortical mechanisms detecting focal discontinuities, Nature **378**, 492-496.

[6] (1995) M. Weliky, K. Kandler, D. Fitzpatrick, L.C. Katz, Visual cortical excitation and inhibition: common relationship to orientation columns, preprint.

[7] (1995) U.Polat, D. Sagi, The architecture of perceptual spatial interactions, Vision Res. 34, pp 73-78.

[8] (1995) A.Das, C.D. Gilbert, Long-range horizontal connections and their role in cortical reorganization revealed bu optical recording of cat primary visual cortex, Nature 375, pp780-784.

[9] (1994) D.-S. Kim, T. Bonhoeffer, Reverse occlusion leads to a precise restoration of orientation preference maps in visual cortex, Nature 370, pp370-372.

[10] (1996) I. Gödecke, T. Bonhoeffer, Development of identical orientation maps for two eyes without common visual experience, Nature 379, pp251-254.

[11] H.R.Wilson, J.D.Cowan, A mathematical theory of the functional dynamics of cortical and thalamic nervous tissue, Kybernetik **15**, 55-80 (1973).

[12] D.J.Willshaw, C. von der Malsburg, How patterned neural connections can be set up by self-organization, Proc. R. Soc. London B, 194:431-445, 1976.

[13] T. Kohonen, Self-organized formation of topologically correct feature maps, Biol. Cybern. 43, 59-69, 1982.

[14] (1995) R. Ben-Yishai, R. Lev Bar-Or, and H. Sompolinsky, Theory of orientation tuning in visual cortex, PNAS 92, pp 3844-3848.

[15] (1995) M. Tsodyks and T. Sejnowski, Associate memory and hippocampal place cells, Internat. Jour. Neur. Sys., in press.

[16] (1995) M. Stemmler,M. Usher, and E. Niebur, Lateral interactions in primary visual cortex: A model bridging physiology and psychophysics. Science 269: 1877-1880.

[17] (1995) D. Sommers, S.B. Nelson, and M. Sur, An emergent model of orientation selectivity in cat visual cortical simple cells, Jour. Neurosci. 15, 5448-5465.

[18] (1995) L.J.Toth, D.C.Sommers, S.Chenchal Rao, E.V.Todorov, D-S. Kim, S.B.Nelson, A.G.Siapas, M.Sur, Dyamic regulation of activity in visual cortex by long-range horizontal connections, preprint.

[19] F.Wolf, H.-U. Bauer, K.Pawelzik and T.Geisel, Organization of the visual cortex, Nature 382, pp306-307.

[20] K.Pawelzik, U. Ernst, T.Geisel, Orientation contrast sensitivity from long-range interactions in visual cortex, NIPS96, in the press.

72

SHORT-TERM CORTICAL REORGANIZATION IN A LATERAL INHIBITORY NEURAL NETWORK

Rasmus Petersen and John Taylor

Centre for Neural Networks
King's College London
London WC2R 2LS
UK

1. INTRODUCTION

Cortical receptive fields (RFs) are dynamic, both on long and short time scales; reviewed by [10]. Long-term reorganization is associated with amputation [11], behavioral training [9] and has been the subject of several modelling studies [7, 12, 13] based on competitive Hebbian learning. Short-term reorganization, on the other hand, has received less theoretical attention.

Within minutes of digit amputation, RFs previously restricted to the amputated skin surface expand and move onto nearby skin [2]. This phenomenon has sometimes been called *unmasking*, since phenomenologically it seems to involve unmasking of normally subthreshold inputs. Similar changes have been recorded after retinal denervation [4, 5, 6, 8] and also after simulated retinal denervation [14].

Unmasking would seem to involve some change, either in cortical circuitry or in the cortical input. Perhaps the simplest hypothesis is that cortex is locally disinhibited [3]. The aim of the current research is to investigate how the type of neural circuitry assumed in competitive Hebbian models of long-term self-organization responds to such inhomogeneous "perturbations".

2. MODEL

Competitive Hebbian models rely intimately upon lateral inhibition – *i.e.*, recurrent interactions are net excitatory at short distances, net inhibitory ones at longer distances. Provided that the recurrent currents are strong relative to the afferent ones, such nets naturally tend to form "bubbles" of activity, centerd around that cortical point receiving (locally) maximal afferent input. We study Amari's model [1]:

$$\tau\frac{\partial u(x,y_c,t)}{\partial t} =$$

$$-u(x,y_c,t) + \int dx'\, w(x-x')\, f[u(x',y_c,t)] + S(x,y_c,t) - h(x) \quad (1)$$

$u(x,y_c,t)$ is the membrane potential at cortical point x given a stimulus centerd at sensory point y_c at time t; w is the lateral weight, assumed to be a smooth function that depends only upon cortical distance; $f[u]$ is a binary output function that maps positive u to 1 and negative u to 0; $S(x,y_c,t)$ is the (smooth) net afferent input to cortical point x given a stimulus centerd at sensory point y_c at time t; τ the membrane time constant; finally, $h(x) > 0$ is a (smooth) tonic input, which models (tonic) disinhibition. We assume that cortex is topographically organized, isotropic and homogeneous: this implies that $S(x,y_c) \equiv S(x-y_c)$ and that $S(z)$ is a symmetric, unimodal function with a maximum at $z = 0$.

Our aim is to solve this equation for the steady-state RF properties (RF size and RF position). In general, this is difficult. However, we can progress by considering disinhibition as a *perturbation*: $h(x) = h - \varepsilon\tilde{h}(x)$. This seems a reasonable approach, since unmasking merely distorts network topography – it does not qualitatively change how the network responds to stimulation. Our work stands on the shoulders of Takeuchi-Amari theory [15].

3. ANALYSIS

Let $\bar{u}(x,y_c)$ be the equilibrium membrane potential due to a stimulus centerd at the sensory point y_c. Then we can define the RF at cortical point x to be the set of sensory points $\{y_c | \bar{u}(x,y_c) > 0)\}$. Due to the topography assumption, we can study RF properties by considering the two *RF border functions* $r_i(x)$, defined

$$\bar{u}(x,r_i(x)) = 0 \qquad i = 1,2$$

Assuming that stimulation evokes a single bubble of activity, equation (1) therefore implies that

$$\bar{u}(x,r_i(x)) = W(\ell_i(x)) + S(x-r_i(x)) - h + \varepsilon\tilde{h}(x) = 0 \qquad (2)$$

where $\ell_1(x) = x - r_2^{-1}(r_1(x))$ is the PF size when the network is stimulated at $r_1(x)$ and $\ell_2(x) = r_1^{-1}(r_2(x)) - x$ is that for $r_2(x)$. We consider an inhomogeneous perturbation $\varepsilon\tilde{h}(x)$ and seek solutions for the RF borders r_i and bubble sizes ℓ_i, as truncated power series $r_i^{hom} + \varepsilon\tilde{r}_i$ and $\ell_i^{hom} + \varepsilon\tilde{\ell}_i$ respectively. From [15], the ε^0 terms are known: $\ell_i^{hom}(x) = l_0$; $r_i^{hom} = x \mp \frac{1}{2}r_0$; $r_0 = l_0$. Substituting these expressions in equation (2) shows that

$$W\left(l_0 + \varepsilon\tilde{\ell}_i(x)\right) + S\left(\pm\frac{1}{2}r_0 - \varepsilon\tilde{r}_i(x)\right) - h + \varepsilon\tilde{h}(x) = 0$$

Expanding to second order accuracy in ε:

$$W'\tilde{\ell}_i(x) \mp S'\tilde{r}_i(x) + \varepsilon\tilde{h}(x) = 0 \qquad (3)$$

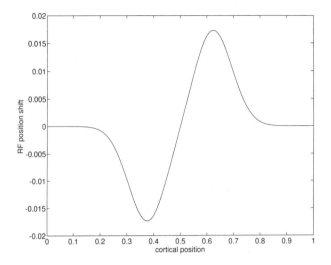

Figure 1. The change in RF position $\tilde{m}(x)$ due to gaussian shaped disinhibition.

where $W^1 \equiv w(l_0)$ is the first derivative of W, evaluated at l_0 and S^1 is the first derivative of S, evaluated at $\frac{1}{2}r_0$. Expanding $\tilde{\ell}_i$ shows that

$$\tilde{\ell}_1(x) = E^{-1}\tilde{r}_2(x) - \tilde{r}_1(x)$$
$$\tilde{\ell}_2(x) = \tilde{r}_2(x) - E\tilde{r}_1(x)$$

Substituting these expressions in equation (3) removes the ℓ_i dependency: some algebraic manipulation then yields expressions for the first order RF border perturbations $\tilde{r}_i$. Thence the RF position solution $\widetilde{m^{-1}}(x) = \frac{1}{2}\tilde{r}_1(x) + \frac{1}{2}\tilde{r}_2(x)$ is

$$\tilde{m}(x) = \frac{-W^1}{2S^1(S^1 + 2W^1)}\left(E^{-1} - E\right)\tilde{h}(x) \tag{4}$$

where $Ef(x) \equiv f(x + l_0)$ and $E^{-1}f(x) \equiv f(x - l_0)$. Since $S^1 < 0$, the qualitative form of the solution depends mainly on the sign of W^1. Provided that lateral input is strong, compared to afferent input, it is necessary for bubble stability that $W^1 < 0$ [1]. m is plotted for such a case in figure 1: RFs for cortical points on the flanks of the disinhibition shift away from the sensory region corresponding to the center of the disinhibition; as in the data cited above.

The RF size solution $\tilde{r} = \tilde{r}_2(x) - \tilde{r}_1(x)$ is

$$\tilde{r}(x) = \frac{1}{S^1(S^1 + 2W^1)}\left\{W^1\left(E + E^{-1} - 2\right) - 2S^1\right\}\tilde{h}(x) \tag{5}$$

Assume strong lateral interactions, so that $W^1 < 0$. The (positive) factors proportional to $-2W^1$ and $-2S^1$) cause RF expansion for neurons at the center of the disinhibited region, as in the data cited above. However, there is also a negative factor, proportional to $W^1(E + E^{-1})$, that causes RF *contraction* for neurons on the flanks of the disinhibitory region. The strength of the contraction effect depends upon S^1/W^1 – roughly speaking, the ratio of afferent to lateral input. If $|S^1| \gg |W^1|$, the positive effect of S^1 is so dominant that the contraction effect is virtually lost; figure 2. Otherwise, the expansion-plus-contraction occurs; figure 3.

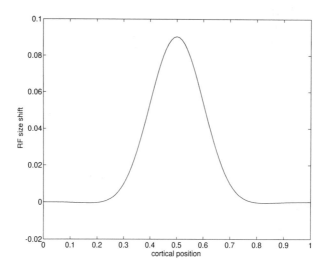

Figure 2. Change in RF size $\tilde{r}(x)$ due to gaussian shaped disinhibition: dominant S^1. Parameters such that $W^1 = -0.4, S^1 = -1.6$.

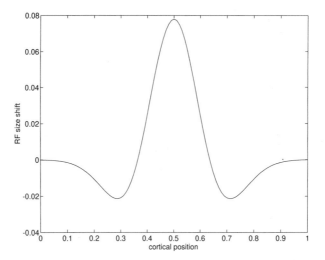

Figure 3. Change in RF size $\tilde{r}(x)$ due to gaussian shaped disinhibition: dominant W^1. Parameters such that $W^1 = -4.0, S^1 = -0.7$.

4. CONCLUSION

This paper has extended previous analytical work on homogeneous, lateral inhibitory networks by Amari and his colleagues to address perturbing inhomogeneities.

It is striking that even a mechanism as simple as tonic disinhibition can have rich effects on RF properties, due to recurrent feedback. In particular, our analysis reveals a surprising RF contraction effect. In principle, other mechanisms could be explored in a similar manner.

These results suggest that it might be interesting to measure RFs not only inside the artificial scotoma but also outside it, and around its borders.

REFERENCES

[1]   S. Amari. *Biol. Cybern.*, 27:77–87, 1977.

[2]   M.B. Calford and R. Tweedale. *Nature*, 332:446–448, 1988.

[3]   B. Chapman and L.S. Stone. *Neuron*, 16:9–12, 1996.

[4]   Y.M. Chino, J.H. Kaas, E.L. Smith III, A.L. Langston, and H. Cheng. *Vis. Rev.*, 32:789–796, 1992.

[5]   C. Darian-Smith and C.D. Gilbert. *J. Neurosci.*, 15:1631–1647, 1995.

[6]   C.D. Gilbert and T.N. Wiesel. *Nature*, 356:150–152, 1992.

[7]   K.A. Grajski and M.M. Merzenich. *Neural Computation*, 2:71–84, 1990.

[8]   S.J. Heinen and A.A. Skavenski. *Exp. Brain Res.*, 83:670–674, 1991.

[9]   W.M. Jenkins, M.M. Merzenich, M.T. Ochs, T. Allard, and E. Guic-Robles. *J. Neurophysiol.*, 63:82–104, 1990.

[10] J.H. Kaas. *Annu. Rev. Neurosci.*, 14:137–167, 1991.

[11] M.M. Merzenich, R.J. Nelson, M.P. Stryker, M.S. Cynader, A. Schoppmann, and J.M. Zook. *J. Comp. Neurol.*, 224:591–605, 1984.

[12] J.C. Pearson, L.H. Finkel, and G.M. Edelman. *J. Neurosci.*, 7:4209–4223, 1987.

[13] R.S. Petersen and J.G. Taylor. In D.S. Touretzky, M.C. Mozer, and M.E. Hasselmo, editors, *Neural Information Processing Systems 8*, pages 82–88. MIT, 1996.

[14] M.W. Pettet and C.D. Gilbert. *Proc. Natl. Acad. Sci. USA*, 89:8366–8370, 1992.

[15] A. Takeuchi and S. Amari. *Biol. Cybern.*, 35:63–72, 1979.

73

SIMULATION OF SPONTANEOUS ACTIVITY GENERATION IN AN EXCITATORY NETWORK INVOLVED IN THE CONTROL OF THE RESPIRATORY RHYTHM

Joël Pham, Khashayar Pakdaman, and Jean-Francois Vibert

B3E, ESI, INSERM U444, ISARS, Faculté de Médecine Saint-Antoine
Université Pierre and Marie Curie, Paris VI.
27 rue Chaligny, 75571 Paris Cedex 12, France
Tel: 33 (1) 44 73 84 36, Fax: 33 (1) 44 73 84 54
E-mail: pham@b3e.jussieu.fr

1. INTRODUCTION

Some neural circuits of the brainstem display a spontaneous activity even when they are isolated from their afferences (Bennett and St. John 1985, Ezure 1990). Such a spontaneous activity, related to the control of the respiratory rhythm, was observed by Fortin & Champagnat (1993) in slices of rat brainstem at the level of the Nucleus Tractus Solitarius (NTS). This spontaneous activity had a low frequency (10 to 15 synaptic events per second) and was generated by a network consisting of neurons connected only by excitatory connections. No pacemaker units were observed in this network. Fortin and Champagnat noted that Excitatory Post-Synaptic Potentials (EPSPs) were still occurring in the presence of tetrodotoxin (TTX), though action potentials were no longer generated in the cells, but with a lower frequency (1 to 1.5 EPSP per second). Fortin and Champagnat hypothesized that the spontaneous activity displayed by the brainstem slices was the result of the amplification by the excitatory connections of the low frequency background synaptic activity observed in the presence of TTX.

The aim of the present study is to investigate whether such a mechanism of spontaneous activity generation is possible or not. To this end, based on the experimental observations of Fortin and Champagnat, a model of excitatory network was developed. The neurons are described by a leaky-integrator model and are characterized by a long-lasting post-spike refractoriness. All the connections between the neurons in the network are excitatory. A background activity is generated by stimulating the neurons by a Gaussian noise. Simulations of this model indicate that the activity resulting from the mechanism proposed by Fortin and

Champagnat depends on the network connectivity. A spontaneous activity of low frequency as observed in the brainstem slices is possible only if the mean number of connections per neuron is low.

2.   THE MODEL

The behavior of a neuron i is described by a membrane potential V_i and a firing threshold θ. Their time course is presented in Figure 1A. The firing threshold (thick line) remains constant. When the membrane potential overtakes the threshold an action potential is produced. After a spike, the membrane potential is reset to an hyperpolarization $U(0)$ and V_i tends exponentially, with a specified time constant τ_m, to a resting potential U_r. A connection between two neurons i and j is defined by a binary variable W_{ji} taking on the value $+1$ if the connection from neuron j to neuron i exists and 0 otherwise. In a network made of N neurons each connection W_{ji} has a probability $p = K/(N-1)$ to exist and a probability $1-p$ not to exist. K is the mean number of connections that a neuron emits and receives. K characterizes the network connectivity and is referred to as the connectivity parameter. Self-connections are not allowed ($W_{ii} = 0$). At each iteration t the membrane potential V_i of neuron i is given by the following equations:

$$\begin{cases} V_i(t) = B_i(t) + U(t-T_i) + \sum_{j=1}^{N} W_{ji} \sum_{r=T_i}^{t} a_j(r-d) V_{psp}(t-r) \\ \\ U(t-T_i) = U_r + (U(0) - U_r) e^{-(t-T_i)/\tau_m} \end{cases} \tag{1}$$

T_i is the time at which the neuron fired its last spike. V_{psp} is the depolarization caused by an EPSP. d is the transmission delay. $a_j(t)$ takes on the value $+1$ if neuron j fired at time t and zero otherwise. $B_i(t)$ is the realization of a centered Gaussian random variable of standard deviation $\sigma = 2.5$mV. This background noise makes the neurons spontaneously discharge. The inter-spike interval histogram of an isolated neuron stimulated by the Gaussian noise is presented in Figure 1B. The mean rate of spike occurrence is 1.2 events per second. The curve indicates the theoretical distribution of the inter-spike intervals. Just after an action potential, the probability of spiking is very low because of the long-lasting post-spike hyperpolarization. This probability gradually increases as the membrane potential rises back toward the resting value. When the potential is close to this resting value, the probability of spiking due to noise is the same at each iteration and the probability of observing long inter-spike intervals decreases exponentially with the interval duration.

3.   RESULTS

Network simulations were performed using XNBC a workstation for the simulation of biologically plausible neural networks (Vibert *et al.*, 1994, 1995, 1997). Networks of size N ranging from 2 to 5000 were simulated. The behavior of a network was characterized by the dynamics of its activity level. The activity level is defined at a given time t by the fraction of the neurons which are spiking. When the size N was greater than 100 the network behavior was independent of N. Four different types of network dynamics could be distinguished, depending on the value of the connectivity parameter K.

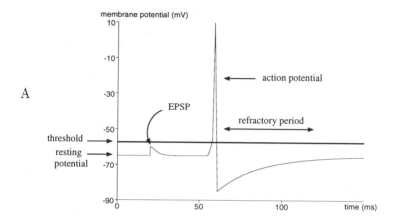

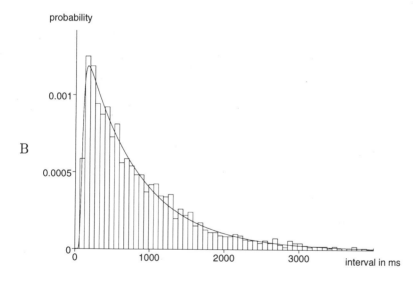

Figure 1. The neuron model. A: Time course of the membrane potential of a neuron (thin line) and of the firing threshold (thick line). At 55 ms a spike occurred. B: inter-spike interval histogram of an isolated neuron stimulated by a centered Gaussian noise of standard deviation 2.5 mV (number of spikes: 1173, mean interval: 830 ms).

3.1. Low values of K

Figure 2A represents the activity level time course of network consisting of $N = 200$ neurons over one second. The neurons are not connected ($K = 0$). Each neuron discharge independently of the others. The activity level value at a given time is not dependent of its previous values. When K varied from 0 to 2, networks behaved as if $K = 0$. The activity level fluctuated randomly around a low mean value (from 1.2 to 2.5 s^{-1}). There was no correlation between the activity level at time t and the activity level at time $t + rd$ with $r > 1$. For low values of the connectivity parameter K the activity was not spread within the network.

3.2. Intermediate values of K

When K increased from 2 to 7, the mean activity level of the networks gradually increased from 2.5 s^{-1} to 11.7 s^{-1}. There was a synchronization of the spikes of the neurons, increasing with K. Figure 2B represents the time course of the activity level of a network made of 200 neurons with $K = 4$. The dynamics of the activity are not simple fluctuations around a low value. Spiking tends to occur every delay d. There is a synchronization of the activity of the neurons. The activity level dynamics consist of a succession of activity bursts. A burst is ignited by a few spikes due to the background noise and occurring within a short time interval. The activity is then spread within the network, it is amplified by the excitatory connections until too many neurons are in a refractory state. Then, the activity progressively decreases toward zero and dies out. At the end of a burst an important fraction of the neurons are in a refractory state and their probability of spiking because of the noise is weak. This probability progressively increases as the membrane potential rises. Therefore, the inter-burst interval is determined by the duration of the refractory period and by the level (standard deviation) of the noise. The characteristics of the bursts are related to the connectivity parameter K. Figure 2C shows the time course of the activity level of a network with $K = 7$. Both the amplitude and the duration of the bursts increase with K. When K was less than 8, if the background noise was suppressed during the simulation, the activity disappeared.

3.3. Critical values of K

When K was close to 8 the behavior on the network depended of their connection matrix. Some networks exhibited only bursts of activity as for $K = 7$. Other networks, after a few iterations, maintained a sustained activity every delay: the activity level never went back to zero. It fluctuated between 0.3 and 1. Once ignited, this kind of activity was maintained even when the background noise was suppressed during the simulation. The time course of the activity level of such a network is presented in Figure 2D. Some networks were able to switch from one type of activity to another under the influence of the noise. Similar switching from one dynamics to the other have been described and analyzed by Van Ooyen and Van Pelt (1992).

3.4. High values of K

When K was greater than or equal to 9, the network always evolved toward a state of sustained activity. Figure 2E represents the activity level time course of a network with $K = 10$. The activity is sustained, the spikes occurring synchronously every delay. The sustained activity was more regular for high values of K: when K was greater than 12, the

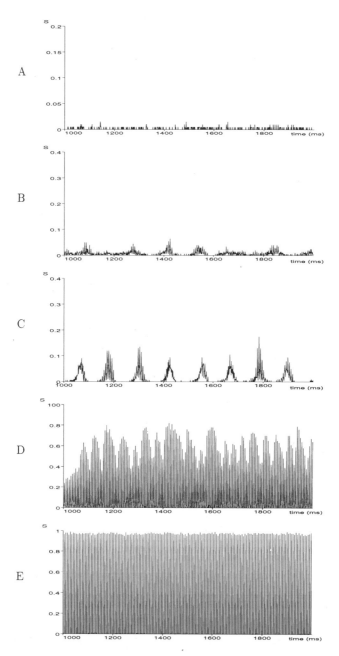

Figure 2. Activity level dynamics. Time courses of network activity level for different values of the connectivity parameter *K*. A: K = 0, B: K = 4, C: K = 7, D: K = 8, E: K = 10. The activity level (S) is the fraction of spiking neurons in the network.

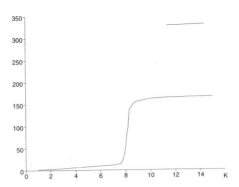

Figure 3. Mean spike rate. Mean spike rate of networks as a function of the connectivity parameter K. The mean spike rate is expressed in number of spikes per neuron and per second.

activity level became almost constant every delay. For high values of K, networks tended to behave as if they were fully connected.

3.5. Summary of the results

Figure 3 represents the mean of the spike rates (number of spikes per second) of neurons of networks with $N = 200$ plotted as a function of the connectivity parameter. The mean spike rate was computed for each value of K by averaging the spike rates, over five seconds of simulation, of all the neurons of thirty networks. Figure 3 shows that there is a sharp transition from a low mean spike rate to a high mean spike rate when K crosses a threshold value K_c close to 8. When K is less than K_c the activity is noise dependent: if the background noise is suppressed during the simulation the activity dies out. When K is greater than K_c the network is able to maintain a sustained activity every delay, or every half of the delay for K greater than 10, even when the background noise is suppressed. The sustained activity does not depend on the level of noise.

4. CONCLUSION

Simulations of randomly connected excitatory networks have shown that two different types of activity could result from the amplification of a low frequency background activity of the individual neurons: noise dependent or noise independent activity. Noise independent activities are not likely to occur in the brainstem slices studied by Fortin and Champagnat (1993). Indeed, they lead to high spike rates which were not observed experimentally. Noise dependent activities are more plausible for the NTS. With noise dependent activities, the background activity rate is amplified less than ten times by the excitatory connections. Such order of amplification was observed experimentally. Thus, our study shows that the mechanism of spontaneous activity generation proposed by Fortin and Champagnat is plausible, if the connectivity of the network is low.

REFERENCES

[1] Bennett F.M., St-John W.M. (1985) Function in ventilatory control of respiratory neurons at the pontomedullary junction. *Respir. Physiol.* **61**, 153-166.

[2] Ezure K. (1990) Synaptic connections between medullary respiratory neurons and considerations on the genesis of respiratory rhythm. *Prog. Neurobiol.* **35**, 429-450.

[3] Fortin G., Champagnat J. (1993) Spontaneous synaptic activities in rat nucleus tractus solitarius neurons in vitro. *Brain Res.* **630**, 125-135.

[4] Van Ooyen A. and Van Pelt J. (1992) The emergence of long-lasting transients of activity in simple neural networks. *Biol. Cyber.* **67**, 269-277.

[5] Vibert J.-F., Pakdaman K., Cloppet F. and Azmy N. (1994). NBC: a workstation for biological neural network simulation. In Skrzypek J (Eds) *Neural Network Simulation Environments*, 113-132. Kluwer Academic Publishers, Boston (USA).

[6] Vibert J.-F., Pham J., Pakdaman K. and Azmy N. (1995). XNBC: a simulation tool for neurobiologists. In: *The neurobiology of computation.* J. Bower Ed. Kluwer Academic Pub., Boston (USA).

[7] Vibert J.-F., Pakdaman K., Boussard E., Av-Ron E. (1997) XNBC: a simulation tool. Application to the study of neural coding using hybrid networks. *BioSystems* **40**, 211-218.

74

A CRITICAL ANALYSIS OF AN ATTRACTOR NETWORK MODEL OF SCHIZOPHRENIA

Alastair G. Reid

Centre for Cognitive Science
University of Edinburgh
Scotland, UK
E-mail: al@cns.ed.ac.uk

ABSTRACT

Hoffman and Dobscha (1989) (H+D hereafter) used an attractor (Hopfield) neural network to model the effects of over-pruning in the human cerebral cortex in an attempt to demonstrate an underlying brain mechanism for schizophrenia. One of their main findings is the emergence of autonomous regions of activity in the network unrelated to input—they call these 'parasitic foci'. I have looked at the analytical work on attractor networks and show that parasitic foci are overlaps in spin glass states which automatically exist in the network, and that pruning is equivalent to raising the 'temperature' in the stochastic state update equation. In addition I show that overloading rather than pruning the network will give the same qualitative results. This analysis is supported by various computer simulations. The conclusion is that while parasitic foci may form a very weak analogy with schizophrenic symptoms, their pathogenesis cannot be attributed to one particular process in the model. Even if the analogy holds the model has still not helped to elucidate the brain mechanism which underlies schizophrenia. I also discuss the inadequacy of the Hopfield network as a model of biological processes and schizophrenia.

1. INTRODUCTION

H+D hypothesized that pathological over-pruning in the frontal cortices gives rise to the positive symptoms of schizophrenia. To support this view they devised a neural network model which when over-pruned gave various output patterns that were claimed to represent schizophrenic symptomatology. H+D used a 100-unit stochastic Hopfield-type network organized as a 10x10 grid and pruned according to the following rule:

Prune connection from unit i to unit j if
$$w_{ij} < (p \times distance\ between\ unit\ i\ and\ unit\ j)$$

Computational Neuroscience
edited by Bower, Plenum Press, New York, 1997

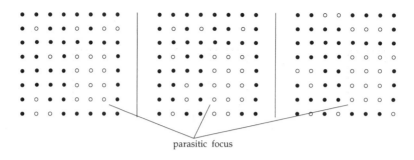

Figure 1.

p is the pruning coefficient and values ranged from 0.6 to 1.0. The network stored 9 memories. Inputs to the networks consisted of the stored memory patterns with either every fifth or every third bit flipped. This gave two sets of input patterns with respectively Hamming distances of 20 and 33 from the stored memories. These inputs were intended to represent two different levels of ambiguity for the memory model to cope with. Over-pruned networks showed the following three 'pathological' output states:

- **generalizations:** An amalgamation of memory fragments

- **loose associations:** Output with more than 10 bits different from input and not a generalization

- **parasitic foci:** Certain populations of neurons converging on the same non-memory output regardless of input. These are determined by comparing all the loose association outputs for one run of a network. An area of overlap which is at least a block of 12 units in size (either 3x4 or 2x6 units) occurring between all the output states examined is said to be a parasitic focus. The figure below represents three loose association end-states of a network for three different inputs. The white dots represent units with the same activation across all three output patterns, black dots are units with different activations.

It was claimed that these network activities correspond to psychotic phenomena such as thought disorder and auditory hallucination.

I have replicated H+D's original pruning experiments. I extended my simulation to include:

1. unpruned networks
2. pruned networks without self-weighted terms (i.e. $w_{ii} = 0$)
3. randomly pruned networks
4. unpruned overloaded networks
5. pruned and unpruned networks storing 5 patterns

Simulation results are the average of 10 different networks for each degree of pruning or overloading. Pruning coefficients ranged from 0.6 to 1.05; the number of patterns stored in overloaded networks ranged from 10 to 19.

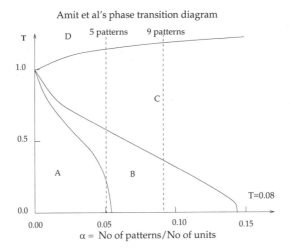

Figure 2. Amit et al.'s phase transition diagram.

2. RESULTS

Type of Network	No of parasitic foci
9-patterns pruned	24
9-patterns unpruned	31
Overloaded	13
5-patterns pruned	33
5-patterns unpruned	1

1. Unpruned 9-pattern networks produced parasitic foci to the same extent as pruned networks.

2. The removal of self-weighted terms produced fewer loose associations but did not affect the number of parasitic foci.

3. Randomly pruned networks functioned poorly as memories when the number of connections removed was the same as the number lost in the rule-based case (60-85% loss). They did not retrieve successfully any memories and produced no parasitic foci. When the number of connections lost randomly was lower (20-55% loss) the results obtained were similar to the rule-based case.

4. Overloaded networks showed very similar behavior to pruned networks but produced fewer parasitic foci.

2.1. Analysis

We can consider H+D's model in terms of the phase transition diagram for the stable states of a Hopfield network produced by Amit et al. (in Hertz, Krogh and Palmer (1991)) This diagram shows the stability in the $T - \alpha$ plane of the states of a stochastic attractor network. In region A the desired memories are the most stable states, although mixture states are also stable. In region B the desired memories are also stable but spin glass states are more stable.

In region C only the spin glass states are stable and in region D no states are stable.

In analytic terms H+D's *generalizations* are mixture states; *loose associations* are spin glass states; and *parasitic foci* are parts of spin glass state outputs from a particular network which overlap with each other to a significant degree.

In a stochastic attractor network S_i, the state of unit i, is set by:

$$Prob(S_i = 1) = \frac{1}{1 + e^{\left(-\frac{2}{T}h_i\right)}} \tag{1}$$

Where T is the 'pseudo-temperature' and h_i the net input to unit i. The diagram represents phase transitions in terms of the *normalized* value of T and α. Normalized in the sense that it is usual to include a $\frac{1}{N}$ term in the equation for h_i, and the stochastic update equation used by H+D used $\frac{1}{T}$ rather than the usual $\frac{2}{T}$ in the exponential term. If T_{norm} is the usual (normalized) form of T and T_{hoff} the form of T used in their simulations then the following equivalence holds:

$$T_{norm} = \frac{2}{N} T_{hoff} \tag{2}$$

Thus the value $T = 4$ in H+D's model is equivalent to a normalized value of $T = 0.08$. **Pruning the network is equivalent to weak dilution.** Under this situation the relative concentration of connections c is given by:

$$c = 1 - \left(\frac{\text{Number of connections removed}}{\text{Total number of connections}} \right) \tag{3}$$

the input h_i in the pruned case then becomes:

$$h_i = c\Sigma w_{ij}^{\text{unpruned}} S_j \tag{4}$$

$$h_i = c h_i^{\text{unpruned}} \tag{5}$$

This is equivalent to increasing T in the state update equation by $\frac{1}{c}$. The effect of increasing T, in terms of phase transitions, is to move the stable states of the system away from region A and into regions B and C where spin glass states predominate. This accounts for the gradual deterioration of a network's performance with pruning. Similarly, however, if a network is overloaded then clearly the size of α is increased and the effect is to move along the x-axis of the phase transition diagram. Again the stable states of the system move from region A to regions B and C

The performance of pruned and overloaded networks is shown in the graphs. Each point plotted is the average of 10 simulations with the same parameters. HD33hits represents correct retrieval of memories or generalizations with inputs of hamming distance 33 away from the stored memories. HD33loose represents retrieval of loose associations of hamming distance 33 away from the stored memories. Likewise for HD20hits and HD20loose

—— = HD33hits   -.-. = HD20hits   .... = HD33loose   - - - = HD20loose

The graphs verify the analytical results. We see a deterioration in memory performance and an increase in the number of spin glass states (or loose associations) with increased pruning or overloading. Both pruned 9-pattern networks and networks overloaded with more than 9 patterns never have stable states in region A of the phase transition diagram and so never function well as memories. 5-pattern unpruned networks functionas near perfect memories and 9-pattern unpruned networks function as moderate to poor memories, again as predicted by the phase transition diagram.

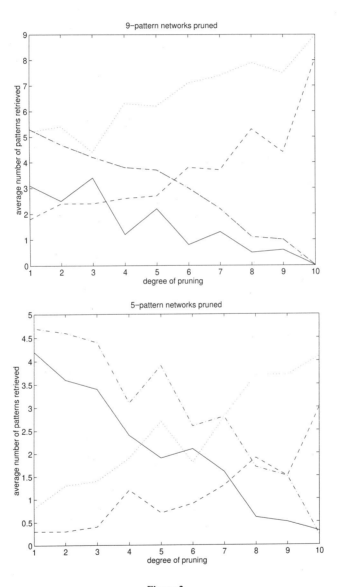

Figure 3.

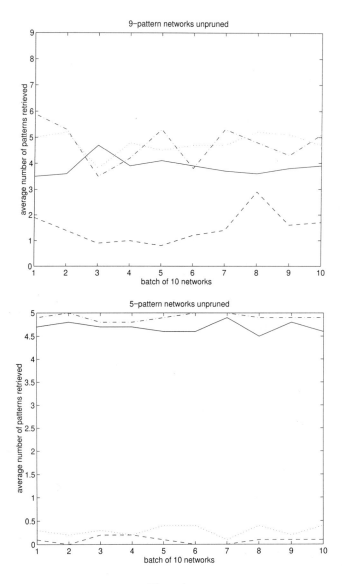

Figure 4.

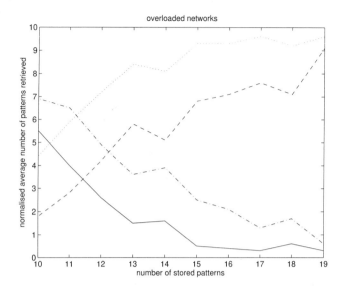

Figure 5.

3. SUMMARY

There are problems with H+D's methodology and with their model.

- Specifically it is unusual to include self-weighted terms in a Hopfield net as these increase the number of spurious patterns.

- H+D have used 9-pattern networks which have parasitic foci even when unpruned. While 5-pattern networks show genuine transitions to pathological states when pruned, these states can also be obtained by overloading the network rather than pruning it. Thus *pruning is not the only cause of parasitic foci*. This seriously undermines the use of this model to show a potential brain mechanism for schizophrenia.

- The motivation for the pruning rule is unclear. Pruning and synapse formation are both *developmental processes* but synapse formation is not included in the model. The pruning rule uses fixed weights which implies that pruning takes place at an instant in time, and after the cognitive processes which the model represents have been formed i.e. in the mature brain. Pruning is neither instantaneous nor a mature process in real brains.

- Allowing connection strengths to be symmetric (i.e. $w_{ij} = w_{ji}$) is also a problem in terms of the biological realism of the model.

- Another problem is that such a net is essentially a model of auto-associative long-term memory (LTM). If schizophrenic symptomatology arises through the breakdown of LTM function then we should see deficits in LTM occurring with these symptoms. This is certainly not the case.

- The main fault is that the model *lacks biological realism* and this arises entirely from using a Hopfield net to model biological processes.

While parasitic foci are real phenomena in a Hopfield net it seems unlikely that they actually occur in this way in the brain. However, the idea that alterations in the nature of the attractors that exist in real brain dynamics could describe the pathogenesis of schizophrenic symptoms remains very exciting. With a greater attention to anatomic and physiologic detail a model could be constructed to investigate this in more detail and give us further insight into the mechanisms of this illness.

REFERENCES

[1] Hertz, J; Krogh, A; Palmer, RG (1991): *Introduction to the theory of neural computation*. Pubs Addison Wesley.
[2] Hoffman, RE; Dobscha, SK (1989): Cortical Pruning and the Development of Schizophrenia: A Computer Model. *Schizophrenia Bulletin* Vol 15 No 3 pp477-489.

75

ON-CENTER AND OFF-CENTER CELL COMPETITION GENERATES ORIENTED RECEPTIVE FIELDS FROM NON-ORIENTED STIMULI IN KOHONEN'S SELF-ORGANIZING MAP

Maximilian Riesenhuber,[*] Hans-Ulrich Bauer, and Theo Geisel

Institut für Theoretische Physik
SFB Nichtlineare Dynamik
Universität Frankfurt
Robert-Mayer-Str. 8—10, 60054 Frankfurt/Main, Germany
max@ai.mit.edu, {bauer, geisel}@chaos.uni-frankfurt.de

INTRODUCTION

The self-organization of sensotopic maps, in particular of visual maps, continues to be an area of great interest in computational neuroscience. In order to distinguish between the different map formation models and between the specific self-organization mechanisms they assume, their behavior with regard to an as large number of physiological, anatomical or theoretical constraints as possible has to be investigated. An interesting case in point is the development of oriented receptive fields from stimulus distributions or stimuli with rotational symmetry, i.e., without orientation. This physiologically quite plausible symmetry breaking phenomenon has been observed in several models for the self-organization of receptive fields [1] orientation maps [2,3] which assumed a competition of On-center and Off-center cells,with rotational symmetry of the stimulus autocorrelation function. Whereas these models are characterized by a linear kernel operating on the afferent activity distribution and mediating the lateral interaction, a major competing model, Kohonen's Self-Organizing Map (SOM, [4]) employs a strongly non-linear lateral interaction function. This nonlinearity is presumably responsible for the successful reproduction of several properties of visual maps in respective SOM-models, like the widening of ocular dominance bands as a consequence of decreased stimulus correlation [5,6], or the pre-

[*] Present address: Center for Biol. & Comput. Learning and Dept. of Brain & Cognitive Sciences, Massachusetts Institute of Technology, Cambridge, MA 02142, U S A.

Computational Neuroscience
edited by Bower, Plenum Press, New York, 1997

ferred angle of intersection between iso-orientation and iso-ocularity bands [7, 8]. However, the investigation of SOM-models with regard to the formation of oriented receptive fields upon stimulation with non-oriented On-center and Off-center stimuli turned out to be difficult, not the least because these models are numerically expensive to simulate.

The recent development of an analytical method to solve (simplified) SOM models (see accompanying paper) now allowed us to calculate conditions for the above mentioned symmetry breaking phenomenon in a SOM. We then used these results as a guiding line for the choice of parameters for simulations. In the remainder of this summary we give a brief sketch of the model, of the analytical results and of first numerical results.

METHODS

Self-Organizing Maps (SOM)

A Self-Organizing Map (SOM) consists of neurons characterized by a position $\mathbf{r}$ in the map lattice plus a receptive field $\mathbf{w}_r$. A stimulus $\mathbf{v}$ is mapped onto that neuron $\mathbf{s}$ whose receptive field $\mathbf{w}_s$ matches $\mathbf{v}$ best,

$$\mathbf{s} = \arg\max_r \left(\mathbf{w}_r \cdot \mathbf{v}\right). \tag{1}$$

This amounts to a winner-take-all rule, i.e. a strong lateral nonlinearity which can be regarded as a consequence of lateral inhibition [4]. The map results as a stationary state of a self-organization process, which successively changes all receptive fields $\mathbf{w}_r$,

$$\Delta\mathbf{w}_r = \epsilon h_{rs} \left(\mathbf{v} - \mathbf{w}_r\right), \tag{2}$$

following the presentation of stimuli $\mathbf{v}$. Here, ϵ controls the size of learning steps, h_{rs} denotes a neighborhood function, centered around the winning neuron $\mathbf{s}$ and usually chosen to be of Gaussian shape,

$$h_{rs} = e^{-\frac{\|r-s\|^2}{2\sigma^2}}. \tag{3}$$

h_{rs} enforces neighboring neurons to align their receptive fields. In this way the property of topography is imposed on the SOM.

The Model

The set-up of our present model (Fig. 1) is very close to the set-up of the competing model by Miller [2] (the activation and adaptation rules are quite different, though!).

The activity distribution in the On-center and Off-center layers is assumed to result from spontaneous localized activity in retinal ganglion cells and is modeled as a peak of excitation in one of the layers (say, the On-center layer as in Fig. 1), complemented by an annulus-shaped activity distribution in the other layer. Peak and annulus result from the positive and negative parts of a "difference of two Gaussians" Mexican hat function (widths σ_1, σ_2, relative amplitude k of negative Gaussian). Position and sign (peak in the On-center or Off-center layer) of the stimuli are chosen at random. Note that the SOM-

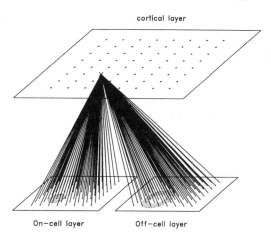

cortical layer

On−cell layer   Off−cell layer

Figure 1. A cartoon of the model. Two two-dimensional input layers, modeling On- and Off-center afferents, resp., feed to a two-dimensional cortical layer. The model is fully connected, ie., each cortical neuron receives input from all On-center and Off-center neurons. The gray areas indicate an exemplary stimulus, in this case with a central peak in the On-cell layer, and an annulus in the Off-cell layer.

model in the present high-dimensional formulation does not require the a priori extraction of certain stimulus and receptive field features, as does the low-dimensional Self-Organizing *Feature* Map (SOFM) variant.

RESULTS

Analysis of the Simplified Model

An analysis of high-dimensional SOM-models like the present one can be performed using a distortion measure [6],

$$E_v = \sum_r \sum_{r'} \sum_{v' \in \Omega r'} \sum_{v' \in \Omega r} (v' - v)^2 e^{\left(-\frac{\|r - r'\|^2}{2\sigma^2} \right)}$$

(4)

which evaluates different possible states of a map via the different ways the stimuli are distributed among the neurons in these states (Ω_r denotes the subset of stimuli which are mapped onto node **r**).

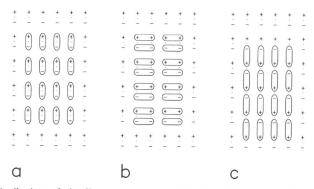

a   b   c

Figure 2. Possible distributions of stimuli among neurons. **a**: Stimuli of different signs, but at equal positions are combined in Ω_r (no orientation). **b**: Stimuli of same sign, but at neighboring positions are combined in Ω_r (oriented receptive fields, but with preference for On-cell layer or Off-cell layer). **c**: Stimuli of opposite signs, at neighboring positions, are combined in Ω_r (Oriented receptive fields, with On-cell and Off-cell layer symmetry).

Here, we simplify the analysis considerably by assuming that the stimulus center positions are restricted to a grid in the input layers which is of the same extension as the cortical layer grid. So there are twice as many stimuli as cortical neurons. For symmetry reasons, each cortical neuron is bestmatching for two stimuli. Depending on the relation between these two stimuli, three qualitatively different map solutions can occur (Fig. 2). Assuming furthermore that the stimuli only have an extension of 3×3 grid points, with the central position being the peak position, with amplitude $v_{peak} = 1$, and the surrounding 8 positions making up the annulus, with amplitude $v_{annulus}$, we can evaluate Eq. (4) for these three cases. The resulting analytical phase diagram for the globally stable state is depicted in Fig. 3a. Simulations of the model corroborated these results (see Fig. 3b).

Simulation Results for the Full Model

The phase diagrams for the simplified version of the model showed that the intended orientation structure occurs only at very small values of the neighborhood width σ. Consequently, we now chose such small values also for simulations of the full version of the

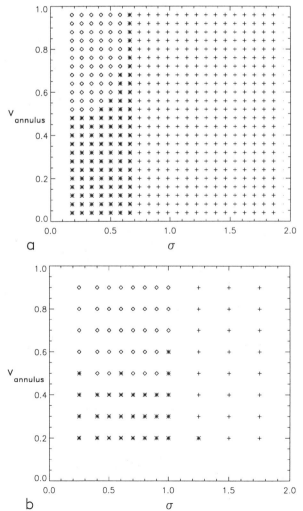

Figure 3. Phase diagrams for the different possible states of the maps. +: no orientation, ◊: orientation, On-cell Off-cell layer asymmetry (cf. Fig. 2b), *: orientation, On-cell Off-cell layer symmetry (cf. Fig. 2c). **a**: analytical phase diagram, **b**: numerical phase diagram, with 8×8-maps, 5×10^5 learning steps, $\epsilon = 0.2 \rightarrow$ 0.01.

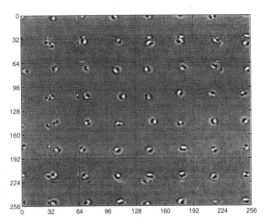

Figure 4. Final state of a 8×8-neuron SOM-map, stimulated with difference-of-Gaussians stimuli in 32×32 input layers, width of stimuli $\sigma_1 = 1.6$, $\sigma_2 = 2.4$, k =0.5, width of SOM neighborhood function $\sigma = 0.25$, 2×10^5 learning steps, $\epsilon = 0.1 \rightarrow 0.01$, periodic boundary conditions. For each neuron, the difference between On-center and Off center cell layer connection strengths is shown as a gray value image. The gray background means no connection strength, the black and white regions indicate the extension of the receptive fields.

model. In order to also be able to retain the rather small size of the stimuli, measured on the length scale of the cortical grid, we discretized the input layers with a finer grid (32×32) than the cortical grid (8×8 for our first simulations, described here). An exemplary resulting map is depicted in Fig. 4. The oriented structure of the receptive fields can clearly be seen. Neighboring receptive fields show roughly a continous change of preferred orientations and retinal position (periodic boundary conditions!), albeit with numerous distortions due to the small value of σ enforcing this continuity.

SUMMARY AND CONCLUSIONS

The results demonstrate that the self-organization of oriented receptive fields from a competition between non-oriented stimuli in On-center and Off-center cell layers can also be achieved in the highly nonlinear SOM-model for map development, an important, as yet missing, piece in the puzzle of map formation phenomena and map model accomplishments. The results further show that a fully connected ''high-dimensional'' SOM model can generate maps with receptive field structures which cannot be modeled using the popular feature map approximation: In the latter class of models the emergence of novel features in the receptive fields that are not present in the stimulus set, eg., the emergence of orientation from non-oriented stimuli, is impossible.

REFERENCES

1. R. Linsker, Proc. Nat. Acad. Sci. USA **83**, 8390 (1986).
2. K. Miller, J. Neurosc. **14**, 409 (1994).
3. M. Miyashita, S. Tanaka, NeuroRep. **3**, 69 (1992).
4. T. Kohonen, *Self-Organizing Maps*, Springer, Berlin (1995).
5. G. J. Goodhill, Biol. Cyb. **69**, 109 (1993).
6. H.-U. Bauer, M. Riesenhuber, T. Geisel, Phys. Rev. **E 54**, 2807–2810 (1996)
7. E. Erwin, K. Obermayer, K. Schulten, Neur. Comp. **7**, 425 (1995).
8. F. Hoffsümmer, F. Wolf, T. Geisel, S. Löwel, K. Schmidt, Proc. CNS 1995 Monterey, eds. J. Bower, Kluwer, Boston (1996), p. 197

76

SYNERGISM OF CELLULAR AND NETWORK MECHANISMS IN RESPIRATORY PATTERN GENERATION

Ilya A. Rybak,[1] Julian F. R. Paton,[2] and James S. Schwaber[1]

[1]DuPont Central Research
E. I. du Pont de Nemours & Co.
Experimental Station E-0328, Wilmington, Deleware 19880–0328
[2]Department of Physiology, School of Medical Sciences
University of Bristol
University Walk, Bristol BS8 1TD, United Kingdom

1. INTRODUCTION

Our ultimate goal is to study the mechanisms of cross-level integration and synergy of network and intrinsic neuronal properties, and the role of this integration in the behavior of biological neural networks and systems. In the present work, we have selected a relatively simple mammalian system, which produces the respiratory rhythm and pattern, as an example for investigation of the above issues.

The central respiratory pattern generator (CRPG) is located in a relatively small area of the venrolateral medulla. The respiratory cycle has been considered to comprise three phases: inspiratory, postinspiratory (stage I expiration) and late expiratory (stage II expiration)[1,2,16] which are seen in the phrenic neurogram (Fig. 1). The respiratory neurons are usually classified into types or groups, depending on their firing pattern and phase relative to the phrenic cycle, for example: early inspiratory (early-I); ramp-inspiratory (ramp-I); late inspiratory (late-I); post-inspiratory (post-I); stage II expiratory (E2); pre-inspiratory (pre-I)[1,2,6,10,11,16](see Fig. 1). The typical discharge patterns recorded from respiratory neurons include: (1) an adapting (or decrementing) firing pattern showing a decrease in spike frequency during the burst; (2) a ramp (or augmenting) firing pattern in which the frequency increases with time. The adapting bursts are typical for early-I and post-I neurons; the ramp firing patterns are observed in some inspiratory (e.g. ramp-I) and expiratory (e.g. E2) neurons (Fig. 1). The ramp firing patterns usually occur when neurons are released from inhibition and often follow a post-inhibitory rebound excitation. Interestingly, a period of silence characteristically follows the rebound spike(s) (Fig. 1 for ramp-I neuron).

Computational Neuroscience
edited by Bower, Plenum Press, New York, 1997

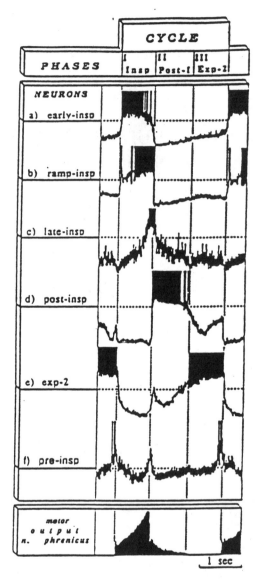

Figure 1. Normalized membrane potential trajectories and firing response patterns of different respiratory neurons[16].

   To date network models of respiratory rhythmogenesis have been based on relatively simple models of single neurons[6–12]. These models did not explore the role of intrinsic neuronal properties and did not allow a comparison of simulated versus experimentally recorded membrane trajectories and firing patterns of single respiratory neurons. For example, previous models failed to reproduce the specific physiological ramp firing patterns of respiratory neurons using pure network mechanisms (e.g. collateral self-excitation or mutual synaptic excitation between the neurons of the same group[6,10,11]). In order to create a more realistic and predictive models we have developed models of respiratory neurons in the Hodgkin-Huxley tradition and incorporated all available biophysical data of mammalian respiratory neurons. Using these neuron models as elements we have constructed a model of CRPG and analyzed the integrative role of intrinsic and network mechanisms in the genesis and control of the respiratory oscillations.

2. MODELS OF SINGLE RESPIRATORY NEURONS

Our respiratory neuron models are typical single compartment models of the Hodgkin-Huxley type. We used a large body of data about the existence of different voltage- and time-dependent ionic channels in respiratory neurons. Specifically, transient potassium (A-type, K_A), calcium-dependent potassium ($K_{AHP}(Ca)$), high-threshold (Ca_L) and transient low-threshold (Ca_T) calcium channels have been reported[3-5,13,14]. The formal descriptions of channel kinetics were drawn from the models of rat thalamic relay and cortical pyramidal neurons by Huguenard and McCormick[15]. The specific parameters of respiratory neurons: values of membrane area, input resistance membrane capacitance, etc. were taken or drawn from measurements in neurons in cardio-respiratory regions of the brainstem[5,13,14,17].

Two types of single neuron models (type I and type II) were structured that accurately reproduced the adapting and ramp firing activity patterns of respiratory neurons. Both types of neuron models contained: fast sodium (Na), delayed rectified potassium (K_{DR}), A-potassium (K_A) and calcium-dependent potassium ($K_{AHP}(Ca)$) channels but differed in type of calcium channels. The combination of $K_{AHP}(Ca)$ and Ca_L channels in neuron type 1 provided an adapting firing pattern of response to stepwise synaptic excitation. Consequently, neuron type 1 was used for simulating all adapting neurons (e.g. early-I and post-I).

Neuron type II was constructed by replacing the high-threshold Ca_L channels in neuron type I with the low-threshold Ca_T channels. We found that the combination of the low-threshold Ca_T channels with $K_{AHP}(Ca)$ channels in neuron type II allowed frequency augmenting (ramp) firing patterns following release from hyperpolarization under conditions of constant synaptic excitatory drive to the neuron. These types of ramp firing patterns closely reflected firing patterns of ramp-I and E2 neurons recorded intracellulary *in vivo* (Fig. 1). Neuron type II model was used in our respiratory network models for simulating ramp-I and E2 neurons which show similar ramp firing patterns during inspiration and stage II expiration respectively.

Thus, we found that neural models incorporating known membrane properties of respiratory neurons and intrinsic dynamics of different calcium channels can produce both the adapting and ramp firing patterns observed in respiratory neurons. We also hypothesized that the main difference between the respiratory neurons that adapt (e.g. early-I, and post-I) and the respiratory neurons that show ramp firing patterns (e.g. ramp-I and E2) may reflect the ratio between two types of calcium channels: Ca_L channels predominate in the former, whereas Ca_T channels prevail in the latter respiratory neuron types.

3. MODEL OF CRPG

Our model of the central respiratory pattern generator (CRPG) was developed employing experimental data and current hypotheses for respiratory rhythmogenesis. The model included a network of respiratory neuron types (early-I; ramp-I; late-I; post-I; E2; con-E2; pre-I) and simplified models of lung and pulmonary stretch receptors (PSR) which provided feedback to the respiratory network. (Fig. 2A). Model behavior (trajectories of membrane potentials of all neurons and integrated phrenic activity) is shown in Fig. 2B.

A

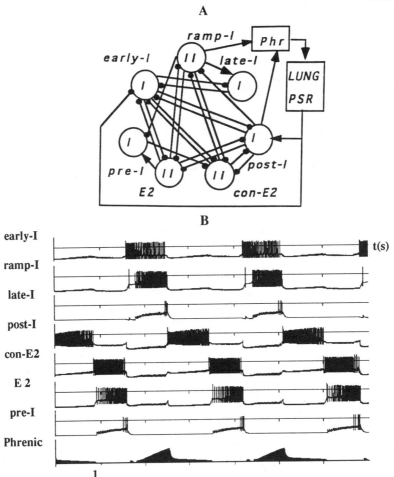

Figure 2. The model of CRPG. A. Schematics of the model. (The large circles represent neurons. The numbers inside the circles indicate the neuron types. The small filled circles are inhibitory synapses. The arrows are excitatory synapses. Tonic inputs to neurons are not shown). B. Model performance.

The mechanism used for the inspiratory off-switching (between inspiration and expiration) operates via the late-I neuron, which is considered to be an inspiratory off-switching neuron[18]. The late-I neuron depolarizes during inspiration due to both increasing excitation from ramp-I neuron and decreasing inhibition from adapting early-I neuron (Fig 2A,B). When membrane potential reaches firing threshold the late-I neuron fires and inhibits the early-I neuron; the latter, in turn, disinhibits the post-I neuron, which initiates expiration. The expiratory off-switch mechanism is based on the pre-I neuron which depolarizes slowly during the late expiratory phase (because of synaptic excitation from the ramp firing E2 neuron) and then generates a short discharge which terminates expiration.

Our CRPG model demonstrates a stable respiratory rhythm and physiologically plausible membrane trajectories and firing patterns of respiratory neurons (compare Fig. 1 with Fig. 2B). Disconnection of PSR feedback in the model produced an increase in the duration and amplitude of ramp-I neuron and phrenic discharges[18]; and decrease in the durations of post-I neuron bursts, post-inspiratory phase[16] and entire expiratory interval[18].

4. ROLE OF INTRINSIC NEURONAL PROPERTIES IN GENERATION AND CONTROL OF THE RESPIRATORY PATTERN

Phase switching mechanisms in our CRPG model are dependent upon both network and intrinsic neuronal properties. Intrinsic properties (e.g. specific dynamics of $[Ca^{2+}]_{in}$ and calcium and calcium dependent potassium channels) determine the specific firing patterns (e.g. adapting, ramp firing) of respiratory neurons and control the rates of spiking frequency changes. This, in turn, influence the durations of respiratory phases. It follows that intrinsic properties can control respiratory phase durations. We have focused on the roles of dynamics of both $[Ca^{2+}]_{in}$ concentration and calcium dependent potassium conductance $gK_{AHP}(Ca)$ in different respiratory neurons for controlling the respiratory phase durations and phase switching. Fig. 3 simultaneously shows the membrane potential trajectories of neurons in our CRPG model and dynamics of $[Ca^{2+}]_{in}$ and $gK_{AHP}(Ca)$ in early-I, ramp-I, post-I and E2 neurons (shown below the membrane potential trajectories of corresponding neurons).

The timing of the inspiratory off-switch and the duration of the inspiratory phase depend on both the adaptive characteristics of the early-I neuron (which inhibits the late-I neuron) and the ramp firing activity of the ramp-I neuron (which excites the late-I neuron). All these characteristics are defined by $[Ca^{2+}]_{in}$ and $gK_{AHP}(Ca)$ dynamics in these neurons. Thus, blocking of calcium influx or $[Ca^{2+}]_{in}$ accumulation in the early-I neuron eliminates its adaptive properties and ceases the respiratory oscillations in the inspiratory phase ("apneusis"; Fig. 4A). In contrast, blocking of calcium influx or accumulation in ramp-I neuron causes high frequency firing patterns instead of slow increase in spiking frequency. The duration of inspiration decreases significantly. The resultant breathing pattern does not contain complete inspiratory phases, but only short gasping-like discharges of inspiratory neurons (Fig. 4B).

The switching between postinspiration and late expiration in the model depends on adaptation of firing of the post-I neuron. The latter in turn is defined by $[Ca^{2+}]_{in}$ and

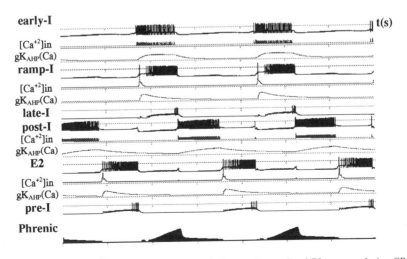

Figure 3. The trajectories of $[Ca^{2+}]_{in}$ and $gK_{AHP}(Ca)$ in early-I, ramp-I, post-I and E2 neurons during CRPG performance (are shown below the corresponding membrane potential trajectories)

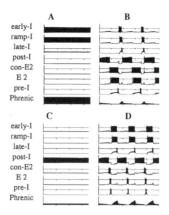

Figure 4. Performance of CRPG when calcium accumulation was blocked in different respiratory neurons: A. in the early-I neuron; B. in the ramp-I neuron; C. in the post-I neuron; D. in the E2 neuron.

$gK_{AHP}(Ca)$ dynamics in this neuron. Blocking of calcium influx or accumulation in the post-I neuron abolishes its adaptive properties and ceases the respiratory oscillations in the postinspiratory phase ("postinspiratory apnea"; Fig. 4C).

The expiratory off-switch in the model depends on ramp firing characteristics of the E2 neuron, which are defined by $[Ca^{2+}]_{in}$ and $gK_{AHP}(Ca)$ dynamics in this neuron. Blocking of calcium influx or accumulation in the E2 neuron causes high frequency firing patterns instead of a slow increase of spiking frequency. Consequently, the duration of the late expiratory phase decreases significantly. The resultant breathing pattern contains only two complete phases: inspiration and post-inspiration, and short bursts of the E2 neuron (Fig 4D).

Thus: (i) the duration of inspiration and the timing of the inspiratory off-switch are controlled by $[Ca^{2+}]_{in}$ and $gK_{AHP}(Ca)$ dynamics in the early-I and ramp-I neurons; (ii) the duration of the postinspiratory phase and timing of the switch between postinspiration and expiration are controlled by $[Ca^{2+}]_{in}$ and $gK_{AHP}(Ca)$ dynamics in the post-I neuron, whereas the duration of the late expiratory phase and timing of the expiratory off-switch are controlled by $[Ca^{2+}]_{in}$ and $gK_{AHP}(Ca)$ dynamics in the E2 neuron.

This study represents the first attempt to develop CRPG models based on Hodgkin-Huxley type models of single respiratory neurons. Our models clearly show that the intrinsic properties of respiratory neurons are integrated with the network properties of CRPG at the cellular level, to provide the specific firing patterns of respiratory neurons (e.g. ramp firing pattern), and at the network level, to provide switching between respiratory phases and to control the duration of different phases of the respiratory pattern.

REFERENCES

1. D. W. Richter and D. Ballantyne, D. A three phase theory about the basic respiratory pattern generator. In: M. Schlafke, H. Koepchen, and W. See (Eds.) Central Neurone Environment, Berlin: Springer, 1983, pp. 164–174.
2. D. Richter *et. al.*, How is the respiratory rhythm generated? A model. *News Physiol. Sci.* 1 (1986) 109–112.
3. J. Champagnat and D. W. Richter, The roles of K$^+$ conductance in expiratory pattern generation in anaesthetized cats. *J. Physiol. Lond.* 479 (1994) 127—138.
4. O. Pierrefiche *et al.*, Calcium-dependent conductances control neurones involved in termination of inspiration in cats. *Neurosci. Letters*, 184 (1995)101–104.
5. D. W. Richter *et al.*, Calcium currents and calcium-dependent potassium currents in mammalian medullary respiratory neurons. *J. Physiol.* 470 (1993) 23–33.

6. S. M. Botros and E. N. Bruce, Neural network implementation of the three-phase model of respiratory rhythm generation. *Biol. Cybern.* 63 (1990)143–153.

7. J. A. Duffin, A model of respiratory rhythm generation. *Neuroreport* 2 (1991) 623–626.

8. J. A. Duffin *et al.*, Breathing rhythm generation: focus on the rostral ventrolateral medulla. *News in Physiol. Sciences.* 10 (1995) 133–140.

9. S. Geman and M. Miller, Computer simulation of brainstem respiratory activity. *J. Appl. Physiol.* 41 (1976) 931–938.

10. A. Gottschalk *et al.* Computational aspects of the respiratory pattern generator. *Neural Comput.* 6 (1994) 56–68.

11. M. D. Ogilvie *et al.*, A network model of respiratory rhythmogenesis. *Am. J. Physiol.* 263 (1992) R962-R975.

12. J. E. Rubio, A new mathematical model of the respiratory center. *Bull. Math. Biophys.* 34 (1972) 486–481.

13. J. Champagnat *et al.*, Voltage-dependent currents in neurons of the nuclei of the solitary tract of rat brainstem slices. *Pflügers Arch.* 406 (1986) 372–379.

14. M. S. Dekin and P. A. Getting, In vitro characterization of neurons in the ventral part of the nucleus tractus solitarius. II. Ionic basis for repetitive firing patterns. *J. Neurophysiol.* 58 (1987) 215–229.

15. J. R. Huguenard and D. A. McCormick, Vclamp and Cclamp. A Computational Simulation of Single Thalamic Relay and Cortical Pyramidal Neurons. Neural Simulation Instruction Manual, 1991.

16. D. W. Richter, Rhythmogenesis of respiratory movements. In: D. Jordan (Ed.) Central Control of Autonomic Nervous System, Harwood Academic, 1996 (in press).

17. F. Kreuter *et al.*, Morphological and electrical description of medullary respiratory neurons of the cat. *Pflügers Arch.* 372 (1977) 7–16, 1977.

18. M. I. Cohen, Neurogenesis of respiratory rhythm in the mammal. *Physiol. Rev.* 59 (1979) 1105–1173.

77

REORGANIZATION OF OCULAR DOMINANCE COLUMNS

O. Scherf, K. Pawelzik and T. Geisel

Max-Planck-Institut für Strömungsforschung
Bunsenstraße 10, D-37073 Göttingen
SFB "Nichtlineare Dynamik"

ABSTRACT

Formation of ocular dominance columns in many models can be analyzed in terms of the spectra of eigenvalues of the linearized dynamics. Depending on the noise level and on boundary conditions, the corresponding modes however, may reorganize in favor of patterns with a lower spatial frequency. Here we argue, that this nonlinear dynamics generates patterns with a length scale which is bounded by the leftmost part of the positive spectrum of the linearization and we demonstrate that this value is indeed approached in very long simulations. Our analysis may provide a new explanation for the correlation dependency of ocular dominance patterns recently observed in cat visual cortex.

1. INTRODUCTION

Recent experiments [1] demonstrate that ocular dominance (OD) patterns in the cat systematically depend on the interocular correlation. In particular this evidence shows that increasing correlation between the input from both eyes leads to more narrow stripes [1]. These experiments provide important data which quantitatively underline the activity dependent selforganization of the visual cortex. The formation of patterns of feature representations has been modeled at different levels of abstraction. Examples of relatively detailed models can e.g. be found in [4, 3] while mathematically more abstract models based on Kohonen's selforganizing feature map [5] (SOFM) and the elastic net [2] (EN) are e.g. discussed in [8] and [6] respectively.

Recently we presented an abstract model for the self-organization of receptive fields which avoids the physiologically most implausible aspects of the SOFM and the EN ("winner takes it all" and "elastic force", respectively). This convolution model (CON) is cast into the mathematical form of convolutions [11] and strictly generalizes the SOFM. CON also approximates the pattern formation of the EN in the important limiting case of weak elasticity forces [11].

Computational Neuroscience
edited by Bower, Plenum Press, New York, 1997

In most models of OD, pattern formation is the result of mode amplification which is evident from a linear stability analysis [4, 8, 11]. As discussed in [9], in the case of shrinking receptive field sizes and low noise level, the Fourier mode which is growing fastest, governs the typical length scale of the emerging pattern ("mode selection"). In this paper we apply our model to OD development. We will argue that in the case of high noise level and constant under-critical receptive field width, the wave length of the resulting OD pattern is determined by a two step pattern formation process. At the beginning the linear dynamics dominates and the spectrum determines the wave initial length of the map. Then, during the nonlinear phase of the dynamics, reorganization to longer wave lengths may occur. We will show that the wave length of the stationary OD pattern is bounded by the leftmost part of the positive spectrum of the linearized dynamics. We also derive an upper bound for the wave number from an integrable version of the convolution model. After the reorganization, the correlation dependency of wave length is consistent with the experiment, just like in the mode selection scenario [9]. We will propose a critical experiment which should decide which of the two scenarios correctly describes the pattern formation process of the OD pattern in the primary visual cortex.

2. OUTLINE OF THE MODEL

We consider ocular dominance formation as a Hebbian adaptation of the weights $\vec{W}_{\mathbf{r}} = (W_{\mathbf{r}}^L(x,y), W_{\mathbf{r}}^R(x,y))$ from the left and right retinal surface (L and R respectively) onto a neuron at position $\mathbf{r}$ in a cortical sheet $\mathcal{N}$. The mean weight dynamics is given by

$$\frac{\partial}{\partial t}\vec{W}_{\mathbf{r}} \propto \int_X P(\vec{A})(\vec{A} - \vec{W}_{\mathbf{r}})\,e_{\mathbf{r}}(\vec{A})\,d\vec{A} \tag{1}$$

where P denotes the distribution of the stimuli $\vec{A} = (A^L(x,y), A^R(x,y))$, which we assume to exist. $e_{\mathbf{r}}$ denotes the excitation of neuron $\mathbf{r}$ and the term $-\langle \vec{W}_{\mathbf{r}} e_{\mathbf{r}} \rangle$ describes the weight decay which simply guarantees weight normalization. In order to analyze the above model we consider only the centers of gravity and the averaged inter-ocular differences of $\vec{W}_{\mathbf{r}}$ and $\vec{A}$, respectively. We obtain 3-dimensional feature vectors $\vec{v}$ and $\vec{w}$ in the so called "pizza box" [8]. In this representation the mean weight dynamics Eq.(1) reads

$$\frac{\partial}{\partial t}\vec{w}_{\mathbf{r}} \propto \int_M p(\vec{v})(\vec{v} - \vec{w}_{\mathbf{r}})\,e_{\mathbf{r}}(\vec{v})\,d\vec{v}, \tag{2}$$

where p is the projection of the full stimulus distribution P into the "pizza box" $\mathcal{M}$. Note that this reduction involves no approximation and therefore, a structure which appears in this representation should necessarily also emerges in the full problem. Inter-eye correlations are reflected by the variance of p along the ocularity coordinate of the "pizza box": $\langle v_z^2 \rangle \propto (1 - 2\kappa)^2$, with the correlation strength κ (for a more detailed discussion of the derivation of the feature space, representation, see [9]). The activation $e_{\mathbf{r}}$ in general might depend on more than the centers of gravity and the inter-ocular differences of the stimuli. Here, however, we use the physiologically plausible assumptions that i) there is a spread of activity in the neural area $\mathcal{N}$ caused by lateral excitatory connections of width $\sigma_{\mathcal{N}}$, ii) that the input to a neuron $\mathbf{r}$ not only comes from the overlap of its input field with the stimulus, but that neighboring neurons also contribute corresponding to their respective input fields, and iii) that that the activation in the cortical sheet is the result of competition for activity (see e.g. [12] and references therein). We combine these ingredients using a convolution

$$e_{\mathbf{r}}(\vec{v}) = \int_{\mathcal{N}} g(\mathbf{r}')h(\mathbf{r} - \mathbf{r}')d\mathbf{r}',$$

where

$$h(\mathbf{r} - \mathbf{r}') = \frac{1}{2\pi\sigma_{\mathcal{N}}^2} \exp\left(-\frac{(\mathbf{r} - \mathbf{r}')^2}{2\sigma_{\mathcal{N}}^2}\right), \quad g(\mathbf{r}) = \frac{\exp\left(-d_{\mathbf{r}}^2(\vec{v}, \vec{w})/2\sigma_{\mathcal{M}}^2\right)}{\int_{\mathcal{N}} \exp\left(-d_{\mathbf{r}'}^2(\vec{v}, \vec{w})/2\sigma_{\mathcal{M}}^2\right)d\mathbf{r}'}.$$

with

$$d_{\mathbf{r}}^2(\vec{v}, \vec{w}) = \int_{\mathcal{N}} h(\mathbf{r} - \mathbf{r}')(\vec{v} - \vec{w}_{\mathbf{r}'})^2 d\mathbf{r}'$$

Note, that the distance measure $d_{\mathbf{r}}(\vec{v}, \vec{w})$, which results from principles i) and ii), differs from the commonly used euclidian distance. As an important property we obtain an integrable model with an energy function, which becomes minimized during the adaption process:

$$E = -\sigma_{\mathcal{M}}^2 \int_{\mathcal{M}} p(\vec{v}) \ln\left(\int_{\mathcal{N}} \exp\left(-\frac{d_{\mathbf{r}}^2(\vec{v}, \vec{w})}{2\sigma_{\mathcal{M}}^2}\right) d\mathbf{r}\right) d\vec{v}. \tag{3}$$

3. THEORY OF THE REORGANIZATION PROCESS

The stability of the homogeneous solution $\vec{w}_0$ of Eq.(2), for which all neurons are binocular is analyzed in terms of the eigenvalues $\lambda(\mathbf{k})$ of the linearized and Fourier transformed dynamics: $\partial_t(\vec{w}_0 + \vec{\delta}) \approx \mathcal{L}_{\vec{w}_0}\vec{\delta} \Rightarrow \partial_t\hat{\delta}_z(\mathbf{k}) \approx \lambda_z(\mathbf{k})\hat{\delta}_z(\mathbf{k})$, where $\mathbf{k}$ denotes the wave-vector. For the integrable version of the convolution model we obtain [10]:

$$\lambda_z(\mathbf{k}) = -1 + \langle v_z^2 \rangle \exp\left(-\mathbf{k}^2\sigma_{\mathcal{N}}^2\right) \frac{(1 - \exp\left(-\mathbf{k}^2\sigma_{\mathcal{M}}^2\right))}{\sigma_{\mathcal{M}}^2}.$$

In the case of undercritical parameters $\sigma_{\mathcal{N}}$ and $\sigma_{\mathcal{M}}$, the initially structureless binocular solution is immediately unstable. The perturbation which corresponds to the largest positive $\lambda(\mathbf{k})$ grows exponentially fastest and, hence, in the beginning determines the wave number of the emerging pattern. But this pattern does not necessarily persist. Due to the noise, the wave number moves in steps towards lower frequencies and stops near the left boundary of the positive spectrum (see Fig 1). This instability behavior is called long-wave instability and is well known in hydrodynamics [7]. In order to understand this reorganization, we investigate the energy minima of the ansatz $\vec{w}_{\mathbf{r}} = (r_x, r_y, a\cos(kr))$ with respect to the amplitude a as a function of the wavenumber k. Fig. 2 shows, that this ansatz is an upper bound for the wave number of the stationary solution with the lowest energy. Instead of using the expansion of the energy function for small amplitudes, which leads to very complicated terms, we approximate in each

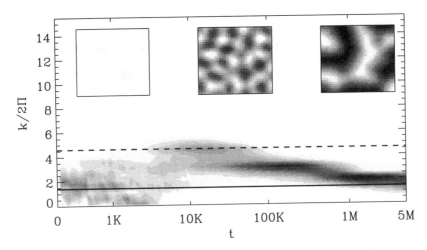

Figure 1. Evolution of ocular dominance patterns in a simulation of 32x32 neurons (open boundary conditions, $\sigma_{\mathcal{M}} = 0$, method from [5] with learning rate 0.1). The inputs were samples from a normal distribution with variance $\langle v_z^2 \rangle = \sigma_s^2 (1 - 2\kappa)^2 \pi/6$ at $\kappa = 0.0$. $\sigma_{\mathcal{N}}$ was constant at 1.5 (neural units) during the 5×10^6 iterations. The gray scaled power spectra of the OD patterns (inset) are normalized. First the modes with the greatest eigenvalue (broken line) were amplified. Then the pattern reorganizes and the frequency shifts via frequencies compatible to the boundary conditions towards the left edge of the positive spectrum (line). Note the logarithmic time-scale.

interval of half wave length the harmonic functions in the distance measure $d_{\mathbf{r}}(\vec{v}, \vec{w})$ *after* convolving with the neigbourhood function h:

$$
\cos(kr) \approx
\begin{cases}
1 - \left(\frac{r - n\lambda}{\lambda/4}\right)^2 & \forall r \in \left[-\frac{\lambda}{4} + n\lambda, \frac{\lambda}{4} + n\lambda\right], \quad n \in \mathbb{Z} \\
-1 + \left(\frac{r - (2n+1)\lambda/2}{\lambda/4}\right)^2 & \forall r \in \left[\frac{\lambda}{4} + n\lambda, \frac{3\lambda}{4} + n\lambda\right], \quad n \in \mathbb{Z}
\end{cases}
$$

Using this approximation, we get for small $\sigma_{\mathcal{R}}, \sigma_{\mathcal{N}}, k$ [10]:

$$
k \propto \frac{1}{\sqrt{\langle v_z^2 \rangle}}
$$

Hence, the reorganization scenario shows a similar correlation dependency of the stationary wave number as in the mode selection case [9].

4. CONCLUSION

We have shown that at fixed undercritical receptive field and neighborhood sizes during development, reorganization of the ocular dominance pattern may occur. The correlation dependency of the wave number of the stationary patterns is in agreement with the experimental findings [1]. Thus, the reorganization scenario provides a new explanation for the self-organization behavior in the primary visual cortex. In principle it is possible to decide whether the mode selection scenario, as in [9] or the reorganization scenario describes self-organization of the ocular dominance columns correctly. The first case postulates shrinking receptive fields and predicts constant periodicity once the pattern has formed. The second case which is analyzed in this contribution, presumes constant receptive field sizes, at first a correlation

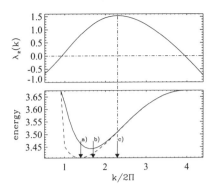 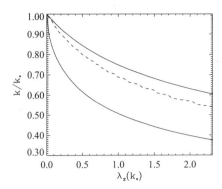

Figure 2. Left: Spectrum of eigenvalues (top) and energy of stationary solutions (broken line) compared to our harmonic ansatz (line). The optimal wave number a) is always bounded by the minimum of our ansatz b). c): position k_* of maximal eigenvalue $\lambda_z(k_*)$. Right: The wave number of the ansatz (upper line) and the left edge of the positive spectrum (lower line) are bounds for the wave number of the stationary solution (broken line).

independent lengthscale and then a growing wave length of the pattern during development. Experimentally observing the dynamics of ocular dominance pattern formation e.g. with optical recording will resolve this issue.

ACKNOWLEDGMENTS

We acknowledge fruitful discussions with S. Löwel and G. Goodhill and support from the DFG (Pa 569/1-1 and SFB 185).

REFERENCES

[1] (1994) S. Löwel and W. Singer, Oculardominance column development: strabismus changes the spacing of adjacent columns in cat visual cortex. J. Neurosci. **14**, 7451.

[2] (1987) R. Durbin and D. Willshaw, An analogue approach to the travelling salesman problem using an elastic net method, Nature **326**, 689-691.

[3] (1993) G. Goodhill, Topography and oculardominance: a model exploring positive correlations. Biol. Cyb. **69**, 109-118.

[4] (1983) A. F. Häussler and C. von der Malsburg, Development of retinotopic projections: an analytical treatment. J. Theoret. Neurobiol. **2**, 47-73.

[5] (1984) T. Kohonen, *Self-Organisation and Associative Memory*, Springer-Verlag, 1984.

[6] (1994) G. J. Goodhill and D. J. Willshaw, Elastic Net Model of Ocular Dominance: Overall Stripe Pattern and Monocular Deprivation, Neural Comp. **6**, 615-621.

[7] (1990) P. Manneville, *Dissipative Structures and Weak Turbulence*, Academic Press, 1990.

[8] (1992) K. Obermayer, G. G. Blasdel, and K. Schulten, Statistical-mechanical analysis of self-organization and pattern formation during the development of visual maps, Phys. Rew. A **45**, 7568-7589.

[9] (1996) K. Pawelzik, O. Scherf, F. Wolf, and T. Geisel, Correlation Dependence of Ocular Dominance Patterns Requires Receptive Field Shrinkage during Development, in *Computational Neuroscience*, Academic Press, 1996, 239-244.

[10] O. Scherf, K. Pawelzik, and T. Geisel, in preperation.

[11] (1994) O. Scherf, K. Pawelzik, F. Wolf, and T. Geisel, Unification of complementary feature map models, in *ICANN 94*, Springer-Verlag, 1994, 338-341.

[12] (1995) D. C. Somers, S. N. Nelsen, and M. Sur, An emergent model of orientation selectivity in cat visual cortical simple cells, J. Neurosci. **15**, 5448-5465.

78

EFFECT OF BINOCULAR CORTICAL MISALIGNMENT ON NETWORKS OF BCM AND OJA NEURONS

Harel Shouval, Nathan Intrator,* and Leon N Cooper

Departments of Physics and Neuroscience
The Institute for Brain and Neural Systems
Box 1843, Brown University
Providence, RI

ABSTRACT

A two-eye visual environment, composed of natural images, is used in training a network of interacting BCM and Oja (PCA) neurons. This work is an extends our previous single cell model [10] to networks of interacting neurons. We study the effect of misalignment, between the synaptic density functions connecting both eyes to each single neuron, on the formation of orientation selectivity and ocular dominance. We show that for the BCM rule a natural image environment with binocular cortical misalignment is sufficient for producing networks of orientation selective cells with varying ocular dominance. Oja neurons in contrast are always perfectly binocular.

1. INTRODUCTION

We have recently shown [10] a BCM neuron, trained in a binocular natural image environment can develop both orientation selectivity and varying degrees of ocular dominance. We have also shown in that study that PCA [6] neurons can not develop ocular dominance as a result of the invariance of the two-eye correlation function to a two-eye parity transformation. Furthermore, due to a slight bias in the natural image environment, the first principal component always prefers the horizontal direction.

In this paper we extend this study to networks of interconnected neurons. We concentrate here on the question of whether the network interactions can alter single cell receptive field properties. In particular we want to examine whether a network of interacting Oja neurons

---

*On leave, School of Mathematical Sciences, Tel-Aviv University

Computational Neuroscience
edited by Bower, Plenum Press, New York, 1997

can develop cells which are not perfectly binocular and are selective to orientations other than the horizontal. Furthermore we want to see if network interaction can alter receptive field structure and what is the effect of synaptic density misalignment on ocular dominance.

2. DETAILS OF THE MODEL

We have used the same visual environment described in our single cell study [10]. It is composed of a set of 24 natural images scanned at a 256×256 pixel resolution. We have avoided man-made objects, because they have many sharp edges, and straight lines, which make the formation of oriented receptive fields easier. The effect of the center surround retinal and LGN projections is modeled by convolving the images with a difference of Gaussians (DOG) filter, with a center radius of one pixel ($\sigma_1 = 1.0$) and a surround radius of three ($\sigma_2 = 3$). We have also used a symmetrized visual environment in order to avoid the slight horizontal bias found in the natural environment [11]. The symmetrized environment environment was created by rotating 12 of our images by 90, 180, and 270 degrees, thus reducing the horizontal bias of the natural environment.

In the lateral interaction network we have used, the activity of the neuron at point x ($c(x)$) which is also a function of the activity of it's neighbors and is given by

$$c(x) = \sigma \left(\sum_{x'} I(\mathbf{x} - \mathbf{x}') \sigma(\mathbf{m}(\mathbf{x}') \cdot \mathbf{d}) \right), \tag{1}$$

where $\mathbf{m}(\mathbf{x}')$ is the synaptic weight vector connecting the input $\mathbf{d}$ to the neuron at point $\mathbf{x}'$. The transfer function σ may be non symmetric around 0 to account for the fact that cortical neurons show a low spontaneous activity, and can thus fire at a much higher rate relative to the spontaneous rate, but can go only slightly below it. This asymmetry of σ does not alter the results obtained with Oja neurons but is essential for BCM neurons.

We have used a balanced DOG for the lateral interaction matrix (I) given by:

$$I(\mathbf{x}) = (1/2\pi\sigma_E^2)\exp(-\frac{1}{2}\left(\frac{x}{\sigma_E}\right)^2) - (1/2\pi\sigma_I^2)\exp(-\frac{1}{2}\left(\frac{x}{\sigma_I}\right)^2).$$

Where σ_E and σ_I are the length scales of the excitation and inhibition respectively.

The BCM theory [2] was introduced to account for the striking dependence of the sharpness of orientation selectivity on the visual environment. We have used a variation, due to Intrator and Cooper (1992), of the BCM synaptic modification rule given by:

$$\dot{m}_j(\mathbf{x}) = \eta\phi(c(\mathbf{x}), \tilde{m}(\mathbf{x}))d_j, \tag{2}$$

where the neuronal activity c is defined in equation 1, $\phi(c(\mathbf{x}), \tilde{m}(\mathbf{x})) = c(\mathbf{x})(c(\mathbf{x}) - \tilde{m}(\mathbf{x}))$, $m_j(\mathbf{x})$ are the synaptic weights, d_j the inputs, η is the learning rate and $\tilde{m}(\mathbf{x})$ is the modification threshold. The modification threshold, $\tilde{m}(\mathbf{x})$, is a nonlinear function of some time averaged measure of cell activity, given by

$$\tilde{m}(\mathbf{x}) = E[c^2(\mathbf{x})], \tag{3}$$

where E denotes the expectation over the visual environment.

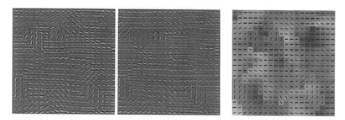

Figure 1. A Network of 20 ∗ 20 BCM neurons in a shifted non rotated environment. RF size is 16 shift is 12. A total linear shift of 75 %. The network interaction function I has $\sigma_1 = 1$ $\sigma_2 = 3$. On the left the feed forward receptive-fields of the left eye are presented and in the center the receptive-fields of the right eye. On the right a summary of receptive field properties is presented; the orientations of the bars represent the optimal orientation each neuron and the length represents the degree of each neuron's selectivity. The gray-scale of the background codes for Ocular Dominance; dark-dominated by left eye and light color by right eye. This network has varying degrees of ocular dominance and various optimal orientations.

The second learning rule that examined in this paper, has been proposed by Oja (1982), and has the form:

$$\dot{m}_i(\mathbf{x}) = \eta[d_i c(\mathbf{x}) - c^2(\mathbf{x}) m_i(\mathbf{x})] \qquad (4)$$

This learning rule has been shown to converge to the principal component of the data and is thus often termed the PCA learning rule. Many other Hebbian rules are also dominated by the first principal component of the data although they do not always exactly converge to it and may have a different normalization.

3. RESULTS

Figure 1 depicts results of a typical BCM network. The receptive fields displayed in such networks are very similar to those obtained for single cells [10]. They are orientation selective to all orientations and show various degrees of ocular dominance. The degree of ocular dominance depends on the overlap between the receptive fields of the two eyes, larger overlaps producing more binocular cells. The receptive fields vary continuously across the network forming patches of similar ocular-dominance and orientation selectivity, which resemble ocular-dominance and orientation columns in visual cortex.

In networks of Oja neurons the receptive fields obtained were also similar to those obtained by single Oja neurons. In a symmetrized environment the neurons can become selective to different orientations (see Figure 2 left) as in the single neuron case [11], whereas in the non-symmetrized environment (Figure 2 right) they are only horizontal [10]. Thus we find that the type of network interaction examined here is not sufficient for breaking the horizontal bias, instead, as can be seen on the bottom right of figure 2, the receptive fields become orthogonal to each other by shifting the phase rather than the angle. Figure: 3 depicts an example of a network of Oja neurons trained with a misaligned visual environment as used for the BCM neurons. This network is perfectly binocular and all cells are horizontal. In all cases examined, the cells were perfectly binocular. The network interaction has also failed to break this symmetry.

This result can be shown to follow from the same type of two-eye parity symmetry as in the single neuron case ([10], Appendix). The only difference between the single cell and

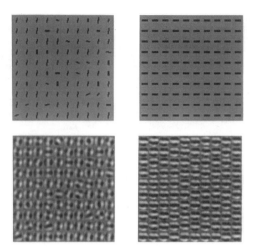

Figure 2. Small 10 * 10 Networks of interacting PCA neurons with an interaction term I as in the BCM network. Receptive fields size is 16 and the receptive fields from both eyes are completely overlapping. On the top the summary of receptive field properties (as for the BCM network above) are presented, below the feed-forward receptive fields are presented. On the left we present a network trained with the symmetrized input environment. Neurons are selective to various orientations. On the right results with a non symmetrized environment are presented, cells are horizontal despite the network interactions.

network result is that for a network all two-eye RF's have the same sign under parity, that is all R.F's are either symmetric or anti-symmetric to the two-eye parity transform.

4. DISCUSSION

We have extended our single cell results [10] to networks of connected neurons. The networks have been trained with a natural scene environment that was preprocessed with a DOG filter. The two eyes were exposed to small partially overlapping portions of these images. As in our single cell study we have shown that for BCM neurons, orientation selective cells with varying degrees of ocular dominance evolve. The degree of ocular dominance depends mostly on the overlap between the inputs. Furthermore, structures resembling ocular-dominance and orientation columns are formed in this network.

As far as we know this is the first network model, trained in a natural image environment, which develops concurrently both ocular dominance and orientation selectivity. The model

Figure 3. Receptive fields of PCA neurons in a shifted non rotated environment RF size is 16 shift is 12. A total linear shift of 75%. The network interaction function I has $\sigma_E = 1$ $\sigma_I = 3$. The receptive fields are perfectly binocular and horizontal.

presented recently by Olshausen and Field (96) develops orientation selective neurons from a natural image environment, however it does not attempt to model the binocular aspects of cortical cells or the organization of these receptive fields across the cortical sheet.

For PCA neurons, the type of network interactions examined here, fail to significantly alter the single cell receptive field structure. Neurons remain perfectly binocular and unless the environment is artificially symmetrized they are also horizontal.

Although PCA neurons, trained in a natural environment, fail to develop receptive fields that are similar to cortical receptive fields, some generalizations of the Oja rule [7] or other models that depend on higher order moments [1, 8] may indeed produce similar results. This is an open question until these models are tested with binocular natural image environments.

ACKNOWLEDGMENTS

The authors thank the members of Institute for Brain and Neural Systems for many fruitful conversations. This research was supported by the Charles A. Dana Foundation, the Office of Naval Research and the National Science Foundation.

REFERENCES

[1] Bell, A. J. and Sejnowski, T. J. (1995). An information-maximisation approach to blind separation and blind deconvolution. *Neural Computation*, 7(6):1129–1159.

[2] Bienenstock, E. L., Cooper, L. N., and Munro, P. W. (1982). Theory for the development of neuron selectivity: orientation specificity and binocular interaction in visual cortex. *Journal Neuroscience*, 2:32–48.

[3] Intrator, N. and Cooper, L. N. (1992). Objective function formulation of the BCM theory of visual cortical plasticity: Statistical connections, stability conditions. *Neural Networks*, 5:3–17.

[4] Linsker, R. (1986). From basic network principles to neural architecture: Emergence of orientation selective cells. *PNAS*, 83:7508–7512,8390–8394,8779–8783.

[5] Miller, K. D., Keller, J., and Stryker, M. P. (1989). Ocular dominance column development: Analysis and simulation. *Science*, 240:605–615.

[6] Oja, E. (1982). A simplified neuron model as a principal component analyzer. *Journal of Mathematical Biology*, 15:267–273.

[7] Oja, E. and Karhunen, J. (1995). Signal seperation by nonlinear Hebbian learning. In Attikiouzel, Y., Marks, R., Fogel, D., and Fukuda, T., editors, *Computational Intelligence - a Dynamics System Perspercitve*. IEEE, Press.

[8] Olshausen, B. A. and Field, D. J. (1996). Emergence of simple cell receptive field properties by learning a sparse code for natural images. *Nature*, 381:607–609.

[9] Sejnowski, T. J. (1977). Storing covariance with nonlinearly interacting neurons. *Journal of Mathematical Biology*, 4:303–321.

[10] Shouval, H., Intrator, N., and N Cooper, L. (1996). BCM network develops orientation selectivity and ocular dominance from natural scenes environment. *Vision Research*, page Submitted.

[11] Shouval, H. and Liu, Y. (1996). Principal component neurons in a realistic visual environment. *Network*, 7:3. In Press.

[12] von der Malsburg, C. (1973). Self-organization of orientation sensitive cells in striate cortex. *Kybernetik*, 14:85–100.

79

A MODEL FOR DURATION CODING IN THE INFERIOR COLLICULUS

S. Singh and D. C. Mountain

Boston University Hearing Research Center and
Department of Biomedical Engineering
Boston, Massachusetts

ABSTRACT

We have implemented a biophysically-plausible model of duration-sensitive cells in the inferior colliculus which process the output of an auditory nerve model using temporal interactions between excitation and inhibition. Duration tuning results from temporal coincidences between an offset response and a delayed onset response. Model cells tuned to a range of best durations were tested with signals having durations between 2 ms and 40 ms. The model accurately simulated the duration tuning seen in the physiological data of Casseday *et al.* (1994) and also successfully simulated the effects of inhibitory blockers.

I. INTRODUCTION

Sound duration is a biologically significant feature, especially for communication sounds. The auditory system of amphibians, avians and mammals respond to sounds differently depending upon their duration (Casseday *et al.*,1994). These responses are also dependent on the silent interval in a sound pattern. Sound duration is an important feature in the communication calls of the squirrel monkey and song recognition in song-birds. The encoding of duration has been found to be carried out by the central auditory system. In particular, researchers have pin-pointed particular neurons in different areas in the auditory system which are responsible for duration sensitivity. Neurons tuned for sound duration have been found in the frog's midbrain (Gooler and Feng ,1992) as well as in the bat.

Some Inferior Colliculus (IC) neurons respond specifically to sounds over a limited range of durations and different neurons in the IC exhibit a range of best durations where the best duration of a neuron is defined as that duration to which the neuron responded with the maximum number of spikes. Casseday *et al* (1994) found such duration sensitive cells in the IC but none at the lower levels of the auditory system. Their data led them to two conclusions. First, there is a population of IC neurons which tuned to signal duration similar to the way auditory neurons are tuned to specific ranges of frequency. Secondly,

Computational Neuroscience
edited by Bower, Plenum Press, New York, 1997

these duration sensitive cells first arise in the IC, and they occur due to interaction of excitation and inhibition. Also their results suggest that duration tuning is produced by coincidence between an offset response, due to rebound from inhibitory inputs, and a delayed excitatory input (Casseday et al., 1994).

Casseday et al. (1994) also tested the effect on duration tuning of blockers for the inhibitory neurotransmitters, GABA and glycine. Application of these blockers had a significant effect on the duration sensitivity of the cell. They found that application of biculculine, a GABA antagonist eliminated duration tuning. Strychnine, a blocker of glycine, almost completely eliminated duration tuning. This suggests that GABAergic and glycinergic inputs in the IC contribute significantly to duration tuning.

The goal of the present study was to investigate the postulated mechanisms of duration tuning in the IC through the use of a computational model. Since the details of the duration-tuning pathway are yet to be worked out in detail, the model is somewhat abstract, but includes an explicit model for the auditory nerve as well as postulated inhibitory and excitatory pathways to the IC. A preliminary report on this work has appeared previously (Singh and Mountain, 1996).

II. METHODS

The model consists of two major subsystems, the auditory-nerve (AN) model and the duration-coding (DC) network model. The AN model is a simplified version of the model developed by Mountain et al. (1991) and is used to represent adaptation in the IHC-auditory nerve synapse. The DC network consists of three components: delayed-onset detection, offset detection and coincidence detection. The model subsystems were implemented using the simulation package for dynamic systems called SIMULINK (The MathWorks, Inc.). Runge-Kutta fifth order integration was used to solve the set of model equations. In the model individual action potentials are not generated. Instead, the outputs of the model subsystems are deterministic and continuous and represent the average instantaneous firing rate (spikes per second) for a population of cells with similar physiological properties.

For computation efficiency, basilar membrane (BM) mechanics and inner hair cell (IHC) transduction were not simulated. Since the test stimuli used in the physiological experiments were pure-tone bursts at the best frequency of the cell under study, the inputs to the AN model were rectangular pulses, chosen to resemble the expected IHC receptor potential. To account for the nonlinearity observed in BM mechanics, the amplitude of the pulses were calculated by using a slope of 0.2 mV/dB for the relationship between IHC receptor potential and sound pressure level. The rise time of the IHC receptor potential was not simulated as it is short (~0.2 ms) compared to time constants in subsequent model stages.

The overall structure of the AN model is shown in Figure 1. It consists of a cascade of a 2-state Boltzmann model for the IHC Ca^{++} conductance with two stages of automatic gain control (AGC). The Ca^{++} activation (P_{Ca}) was computed by:

$$P_{Ca} = 1/(1 + \exp(-0.8(V_m - 4))) \tag{1}$$

where V_m is the IHC receptor potential in mV. The AGC circuits represent two stages of depletion processes such as neurotransmitter depletion and inactivation of postsynaptic receptors. See Mountain and Hubbard (1996) for a discussion of the relation between the AGC circuit to traditional neurotransmitter depletion models.

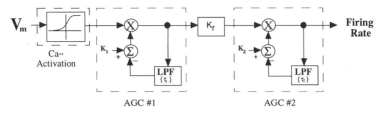

Figure 1. Block diagram of the auditory nerve model. Parameters for AGC #1: $K_1 = 1$, $\tau_1 = 22$ ms, gain = 5. Parameters for AGC #2: $K_2 = 2.23 \times 10^4$, $\tau_2 = 300$ ms, gain = 300. $K_r = 0.07$.

The model parameters were estimated by first fitting the model's response to the AN adaptation data obtained by Westerman (1984) and then adjusting the parameters to also reproduce the physiological results obtained by Robin *et al.* (1990) which emphasize the recovery of the onset response from adaptation. The parameters can be varied by about 10 percent and still reproduce both types of data.

The DC model receives its input from the AN model and its output is also deterministic and is expressed in terms of firing rate (spikes per second). The overall block diagram of the DC network is shown in Figure 2. We will refer to the implementation of the three key processes, offset detection, delayed-onset detection and coincidence detection, as "cells", however, the mapping of these processes on to actual biological cells is not straight forward. For example, a single process may involve more than one cell type since both excitation and inhibition are required for offset and delayed onset detection. It may also be that parts of more than one process may occur in a single cell, for example the duration sensitive cells in the IC may perform coincidence detection along with a portion of the offset detection process.

In our model, the output of the AN model is fed to two parallel pathways, one of which represents the delayed-onset cell while the other represents the offset detector. The delayed onset cell, as the name suggests, produces a delayed response to the stimulus onset. The offset detector responds at the cessation of stimulus. The outputs of these two cells serve as inputs to the coincidence detector cell. The output of the coincidence detector mimics the responses of the observed duration-sensitive cells in the IC. Duration tuning arises because coincidence will only occur if the duration of the sound stimulus is approximately equal to the delay in the onset branch. Since the experimental data are typically reported in terms of the number of spikes per stimulus, the equivalent response of the DC network is computed by integrating the output of the coincidence cell.

The delayed onset and the offset cells responses are both assumed to be the result of the combination of excitatory and inhibitory currents. The excitatory and inhibitory con-

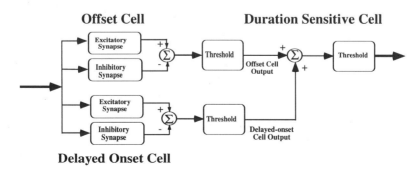

Figure 2. Block diagram of the duration-coding model.

ductances are represented as low-pass filters (LPFs), each with a different time constant. The time constants of both LPFs in the delayed onset cells are slower than the corresponding time constants in the offset cells. The LPFs were all fourth-order LPFs, but differed slightly in their transfer functions. The LPF for the excitatory pathway in the delayed-onset cell has the transfer function (H_{de}):

$$H_{de} = A_{de}/((\tau_{de1}+1)^3(\tau_{de2}+1))\tag{2}$$

and the transfer function for the inhibitory pathway (H_{di}) is:

$$H_{di} = A_{di}/(\tau_{di1}+1)^4\tag{3}$$

In the delayed onset cell, the time constant for the inhibitory filter is much faster than the time constant of the excitatory filter so initially the sum of the currents is below the threshold (T_d). When the onset response of the AN model declines, the inhibitory current drops faster than the excitatory current and the total current can cross threshold. In the offset cell, the time constants of the two filters are comparable and the resulting response is produced only after cessation of the stimulus.

The transfer function representing the excitatory conductance (H_{oe}) in the offset cell is:

$$H_{oe} = A_{oe}/((\tau_{oe1}+1)^2(\tau_{oe2}+1)^2)\tag{4}$$

and the transfer function for the inhibitory pathway (H_{oi}) is:

$$H_{oi} = A_{oi}/((\tau_{oi1}+1)^3(\tau_{oi2}+1))\tag{5}$$

The time constants of the LPFs representing the excitatory and the inhibitory conductances in the offset cell are small so that the interaction of these two currents produces a brief response at the offset of the stimulus. The gain associated with the inhibitory current is higher than the gain associated with the excitatory current to ensure that, within the duration of the tone burst, the amplitude of the inhibitory response is always greater than that of the excitatory response. The threshold (T_o) of the offset cell is set so that the offset response crosses threshold over the range of durations to which the duration-sensitive cell to be modeled shows a response.

The coincidence detector at the output of DC the network is modeled by a summer followed by a threshold. As the duration of the tone burst approaches the duration to which the duration sensitive cell is maximally sensitive, the responses from the offset cell and the delayed onset cell begin to coincide in time and the resulting sum crosses threshold. At its maximal sensitivity, the two responses coincide exactly producing the largest response. The DC network can easily be tuned to different best durations by changing the time constants of the postsynaptic excitatory and inhibitory currents.

To simulate the effects of the blockers of the inhibitory transmitters, we eliminated the inhibitory pathways in the duration coding model. This was achieved by reducing the gain of the appropriate transfer function to zero when the blockers are to be simulated. To simulate the effect of the blocker of glycine, the inhibitory pathway of the delayed onset cell was suppressed. This reduced the delayed onset cell to the excitatory pathway. The effect of blocking GABA was simulated by eliminating the inhibitory pathway in the offset cell.

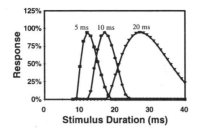

Figure 3. Duration tuning curves for model cells tuned to 5 ms, 10 ms, and 20 ms.

III. RESULTS

Figure 3 shows the simulated responses of three different duration coding cells. In this case, they are tuned to ''best durations'' of 5 ms, 10 ms and 20 ms for stimuli 40 dB above threshold. The amplitudes of the responses have been normalized in order to compare the shape of the tuning curves. The model is simulated for acoustic stimuli whose durations range from 1 ms to 40 ms in steps of 1 ms. Table 1 gives the values of the parameters used for each of the duration coding cells.

Figure 4 also illustrates the responses of the model with the same parameters but in the presence of inhibitory blockers. In the case of strychnine, a blocker of glycine, duration tuning almost completely disappears in both the experimental data and in the model simulation. This occurs because the delayed onset cell continues to respond throughout the duration of the stimulus and this response is always greater than the firing threshold,

Table 1. Model parameters for duration-coding cells.

Parameter	Best duration		
	5 ms	10 ms	20 ms
τ_{de1}	1.0	2.2	4.0
τ_{de2}	2.0	4.4	10.0
τ_{di1}	0.1	0.1	0.2
T_d	10	10	10
A_{de}	0.25	0.25	0.25
A_{di}	1.2	1.2	1.2
τ_{oe1}	0.25	0.25	0.5
τ_{oe2}	2.0	4.4	10.0
τ_{oi1}	0.05	0.05	0.1
τ_{oi2}	1.0	2.2	4.0
T_o	150	130	65
A_{oe}	1	1	1
A_{oi}	1.5	1.5	1.5

Simulations were also performed for pulses of different amplitudes. The response for pulse amplitudes over 40 dB were the same as those shown in Figure 3. This is because the AN model achieves its maximum firing rate for a sound level of about 40 dB. For input amplitudes lesser than 40 dB, the response preserved the tuning shape but decreased in magnitude. A comparison of the simulation data with the experimental results obtained by Casseday *et al.* (1994) are also shown in bottom curve of Figure 4. The model responses fit the physiological data well. The maximal response of this model cell occurs at 10 ms and the cell responds if the duration of the input pulse is between 4 and 17 ms.

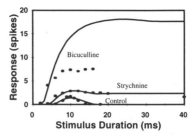

Figure 4. Comparison of model and experimental duration tuning curves without inhibitory blockers (lower curve), with strychnine (middle curve) and with bicuculline (upper curve). Model data are plotted as solid lines and experimental data are plotted as circles.

hence, the output does not drop to zero for longer duration pulses. However, for stimulus durations near the best duration, the offset response sums temporally with the delayed onset response to give an output response which is still maximal.

In the case of bicuculline, which blocks the GABAergic input, the duration tuning of the cell is completely eliminated. Without bicuculline, the inhibitory current in the model coincides to a large extent with the excitatory current with the result that the total current is normally much smaller in magnitude than either the inhibitory or excitatory currents. Hence, blocking the inhibitory current in the offset cell has a much larger effect than eliminating the inhibitory current in the delayed onset cell. When the GABAergic input is blocked, the output response of the duration-sensitive cell in the model is much larger and since the time constant of the excitatory conductance is smaller than the one in the delayed onset cell, threshold is crossed more rapidly resulting in a tuning curve which exhibits a response even for small pulse durations.

The model has a slightly wider tuning curve for the case with no blockers than observed in the experimental data shown, but accurately simulates the case when the inhibitory transmitter glycine is blocked by strychnine. When the other inhibitory transmitter, GABA, is blocked by bicuculline, the response of the model is much larger than the response observed in the experimental data. However, the shape of the tuning curve is similar and the result is a loss of duration tuning. The differences between model and experiment may be due to the use of the subtractive model for inhibition instead of shunting inhibition.

IV. DISCUSSION

The derivation of the Duration Coding Network is presented in this paper from a signal processing point of view, however, the model is also plausible from a biophysical point of view. The model is based the combination of an offset response, a delayed response to stimulus onset and coincidence detection in a fashion similar to that proposed by Casseday *et al.* (1994). The effects of the two inhibitory neurotransmitters, GABA and glycine are explicitly included and each neurotransmitter can be suppressed by removing the inhibitory conductances which are modeled as low pass filters.

The delayed onset cell in our model makes use of a temporal interaction between inhibitory and excitatory inputs to generate a response which represents the onset of the pulse delayed in time so that it coincides in time with the offset response. A wide range of first-spike latencies have been observed in the IC (Langner. and Schreiner, 1988) which supports this aspect of the model. In addition, the total synaptic current in the offset cell model has been found to closely resemble the currents observed using voltage clamping in duration-sensitive cells in the IC by Casseday *et al.* (1994), suggesting a plausible biophysical foundation for the offset cell model.

It remains to be seen, if and how the components of this model map onto actual biological processes and cells, but we have demonstrated that a network model of duration-sensitive cells can easily be developed in which each cell has a different best duration. In the mean time, this network already has significant potential as a building block in automatic recognition systems for processing of complex acoustic signals.

ACKNOWLEDGMENTS

Supported by the Office of Naval Research

REFERENCES

Casseday, J.H., Ehrlich, D. and Covey, E. (1994) Neural tuning for sound duration: role of inhibitory mechanisms in the inferior colliculus. Science, 264: 847–852.

Gooler, D.M., and Feng, A.S. (1992) Temporal coding in the frog auditory midbrain: the influence of duration and rise-fall tone on the processing of complex amplitude-modulated stimuli. J. Neurophys. 67:1–22.

Langner, G. and Schreiner, C.E. (1988) Periodicity coding in the inferior colliculus of the cat. I. Neuronal mechanisms. J. Neurophys. 60: 1799–1822.

Mountain, D.C. Wu, W. and Zagaeski, M. (1991) A calcium-current inactivation model of auditory nerve adaptation. ARO Abstracts, 14: 153.

Mountain, D.C. and Hubbard, A.E. (1996). Computational analysis of hair cell and auditory nerve processes. In: *Auditory Computation, Springer Handbook of Auditory Research*, H.L. Hawkins, T.A McMullen, R.R. Popper, and R.R Fay. eds, Springer Verlag, New York pp 121–156.

Robin, D.A., Abbas, P.J. and Hug, L.N. (1990). Neural responses to auditory temporal patterns. J. Acoust. Soc. Am. 87: 1673–1682.

Singh, S. and Mountain, D.C. (1996) A model for duration tuning in the inferior colliculus. ARO Abstracts 19: 92.

Westerman, L.A. (1985) Adaptation and recovery of auditory nerve responses. Special Report ISR-S-24. Institute of Sensory Research, Syracuse University, Syracuse, NY.

80

A LOCAL CIRCUIT INTEGRATION APPROACH TO UNDERSTANDING VISUAL CORTICAL RECEPTIVE FIELDS

David C. Somers, Emanuel V. Todorov, Athanassios G. Siapas, and Mriganka Sur

Department of Brain and Cognitive Science
MIT
Cambridge, MA
E-mail: somers@ai.mit.edu
E-mail: emo@ai.mit.edu
E-mail: thanos@ai.mit.edu
E-mail: msur@wccf.mit.edu

ABSTRACT

The traditional concept of the receptive field (e.g., [4, 6]) holds that each portion of the receptive field (RF), in response to a stimulus element, has unitary (excitatory or inhibitory) influence on neuronal response. Here, we argue: i) receptive field components naturally have dual or vector (both excitatory and inhibitory) influence; ii) neuronal integration is better understood in terms of local cortical circuitry than single neurons. Using a large-scale model of primary visual cortex, we demonstrate that the net effect of a given stimulus element within either the classical or extraclassical RF can switch between excitatory and inhibitory as global stimulus conditions change. We analyze and explain these effects by constructing self-contained modules (via a novel technique) which capture local circuit interactions. These modules illustrate a new vector-based RF analysis which unifies notions of classical and extraclassical RF, treating long-range intracortical inputs on equal footing with thalamocortical inputs.

1. INTRODUCTION

Neuronal receptive fields in primary visual cortex (V1) have not only "classical" regions, where visual stimuli elicit responses (presumably through thalamocortical axons), but also have "extraclassical" regions, where stimuli largely modulate responses evoked by other stimuli (presumably via long-range intracortical or inter-areal axons) [3, 16]. The traditional view of integration holds that each portion of a neuron's receptive field in response to a given stimulus element has <u>either</u> an excitatory <u>or</u> an inhibitory (i.e., a scalar) influence [4, 6, 15]. Although this approach has substantial explanatory power, it cannot account for phenomena in which the net effect of a stimulus element in a given portion of the receptive field appears to switch between excitatory and inhibitory as global stimulus conditions change [8, 10, 17, 19]. Two such phenomena, involving local and long-range integration respectively, are paradigmatic. First, increasing the luminance contrast of an oriented visual stimulus causes responses in primary visual cortex to initially increase, but subsequently saturate and even decrease ("super-saturate") [1, 10, 11] (see data of [10] in fig 1a). Second, adding a distal stimulus facilitates responses to a weak central stimulus, but suppresses responses to a strong stimulus [8, 9, 19, 17] (see data of [8, 17] in fig 1b).

Our goal is to develop an expanded notion of the visual cortical receptive field which can explain stimulus-dependent responses such as these. Three basic features of cortical anatomy, which are overlooked by the traditional receptive field view, are central to the expanded view: i) receptive field regions (via either thalamocortical or long-range intracortical axons) drive both excitatory and inhibitory cortical neurons [13, 25]; ii) different portions of the receptive field provide converging inputs to a shared population of cortical neurons [3, 16]; and iii) these neurons form dense, recurrent local connections [3, 7, 25]. Based on this anatomy, we propose that: i) each RF region in response to a given stimulus has both excitatory and inhibitory influences on neuronal responses which in general cannot be reduced to a scalar quantity but rather should be considered separately (i.e., RF input is a vector); ii) receptive field inputs are integrated by the local cortical circuitry; and iii) the net effect of a receptive field input depends both on the excitatory-inhibitory bias of the afferent inputs and on how other receptive field regions activate the local cortical circuitry. First we demonstrate this approach by capturing the paradoxical local and long-range phenomena within a large-scale visual cortical model, and later we present an analytic explanation. In contrast, prior computational investigations of local circuit influences either have captured anatomical details only in simulations with little formal analysis [21, 22] or have oversimplified local cortical excitatory and inhibitory interactions in order to obtain closed-form (scalar) analysis [5].

2. METHODS

Cortical circuitry under a 2.5mm by 5mm patch of primary visual cortex was represented by a model with 20,250 spiking cortical neurons and over 1.3 million cortical synapses. Neurons were organized into a 45 by 90 grid of "mini-columns" based on an orientation map obtained by optical recording of intrinsic signals of cat visual cortex (data from [23]). Each mini-column contains 4 excitatory and 1 inhibitory neurons modeled separately as "integrate-and-fire" neurons with realistic currents and experimentally-derived intracellular parameters [12] (see methods of [21] for equations and parameters). Intracortical connections provide short-range excitation (connection probabilities fall linearly from $\rho_{excit-excit} = 0.1$, $\rho_{excit-inhib} = 0.1$ at distance zero to $\rho = 0$ at $d = 150\mu m$), short-range inhibition (linear from $\rho_{inhib-excit} = 0.12$, $\rho_{inhib-inhib} = 0.06$ at $d = 0$ to $\rho = 0.5\rho_{peak}$ at $d = 500\mu m$; $\rho = 0$ elsewhere), and long-range excitation (linear with orientation difference ϕ between pre- and post-synaptic columns, from $\rho = 0.005$ at $\phi = 0°$ to $\rho = 0.001$ at $\phi = 90°$). Peak synaptic conductances, by source, onto excitatory cells are $g_{excit} = 7nS$, $g_{inhib} = 15nS$, $g_{lgn} = 3nS$, and $g_{long} = 1.2nS$ and

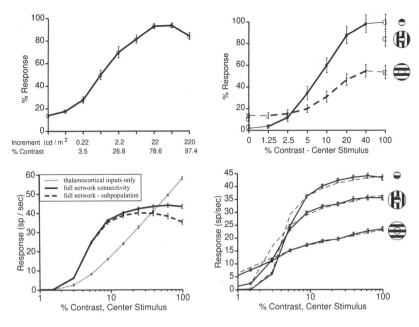

Figure 1. Experimental (a,b) and Simulation Results (c,d) for "super-saturating" contrast response functions (a,c) and surround facilitation/suppression of contrast responses (b,d). Solid and dashed lines in (d) are model and module responses, respectively.

onto inhibitory cells are $g_{excit} = 1.5nS$, $g_{inhib} = 1.5nS$, $g_{lgn} = 1.5nS$, and $g_{long} = 1.2nS$. Cortical magnification is 1 mm/deg, cortical RF diameters are roughly $0.75°$, and thalamocortical spikes are modeled as Poisson processes. Each thalamic neuron projects to cortical neurons over an area 0.6mm^2 and responds linearly with log stimulus contrast. Results are averaged over 20 networks constructed with these probability distributions.

Additional analysis was performed by constructing self-contained modules which capture local circuit properties of the large-scale models. Given a local neuronal population P whose mean firing rate $\mathbf{M} = \mathbf{F}(\mathbf{I}_d, \mathbf{I}_l)$ is a function $\mathbf{F}$ of the long-distance (intracortical and thalamocortical) inputs $\mathbf{I}_d$ and local (intracortical) inputs $\mathbf{I}_l$, we want to construct a closed system (module) whose response approximates $\mathbf{M}$ as a function of $\mathbf{I}_d$ only. All bold face quantities denote vectors with components corresponding to excitatory and inhibitory populations. Local inputs are defined as arriving from within a radius R, which is chosen to minimize approximation error. Module construction is only possible if $\mathbf{I}_l$ can be expressed as a function of $\mathbf{M}$ and $\mathbf{I}_d$. To that end we use a local homogeneity assumption $\mathbf{M} = \mathbf{I}_l$, i.e. neurons within R (not just P) have mean firing rates $\mathbf{M}$. Thus the module output $\mathbf{M}^*$ is the solution of $\mathbf{M} = \mathbf{F}(\mathbf{I}_d, \mathbf{M})$. This equation can be solved numerically if we model the response functions of integrate-and-fire neurons [18, 24]. Here, we compute $\mathbf{M}^*$ by simulating a module composed of excitatory and inhibitory neurons, in which neurons receive the same average number and strength of synapses as neurons in P receive from within the radius R. The homogeneity assumption is equivalent to isolating P and compensating for the "cut" connections from R by adding extra connections within P. Inhibition is treated as purely local (long-distance inhibition can be addressed by doubling the system dimensions). The radius R that minimizes approximation error is a balance between two conflicting constraints: homogeneity of local firing, which favors smaller R, and inclusion of cortical inhibition, which favors bigger R.

3.   RESULTS

Physiological responses to oriented grating stimuli of differing contrasts within the classical RF are captured by the model (see fig 1c). The responses shown here and below are for the excitatory subpopulation.   Responses saturate at contrast levels below which thalamic responses saturate [11], can decline for high contrasts (super-saturation) [1, 10, 11], and have firing rates well below maximal cellular firing rates [12].   Inhibitory neurons, on average, saturate at higher contrasts than do excitatory neurons (not shown). While preserving classical RF properties, our model also captures paradoxical extraclassical RF modulations [8, 9, 19, 17].   The modulatory influence of (fixed contrast) "surround" gratings on responses to optimal orientation "center" stimuli shifts from facilitatory to suppressive as center stimulus contrast increases (see fig 1d; see also [22]). These effects emerge from the local intracortical interactions (as will be shown below) and do not require synaptic plasticity or complex cellular properties.   Our model is the first to provide a unified account of these classical and extraclassical RF phenomena.

We understand the integration of classical and extraclassical RF influences by analyzing local circuitry as a unit. Neuronal responses in the model depend not only on thalamocortical and long-range intracortical inputs [3, 16], but also on recurrent local inputs. We simplify analysis by isolating nonlinear local interactions within a closed system (module) which receives only long-distance (thalamocortical and long-range intracortical) inputs and generates approximately the same mean responses as a local neuronal population embedded in the model. This task is non-trivial, because intracortical connections form a continuum. Simply isolating a small group of cells (together with the connections among them) will remove many local connections from across the group boundary, and thus lead to inaccurate responses.   The module we construct preserves the distribution of cellular properties and interactions within the local population, and compensates for the missing local connections by making extra connections within the isolated group (see methods).   This module will produce correct responses whenever mean firing rates are locally homogeneous. Note that the method can easily incorporate multiple distinct neuronal subpopulations (e.g. cell types and/or layers), and multiple sources of long-distance input (e.g. feedback projections). This technique differs from "mean-field" approximations (e.g., [20]) in that analysis is local and does not require oversimplification of cellular and network properties.

We construct a module consisting of two interacting homogeneous populations, excitatory and inhibitory neurons (see methods). Afferent inputs to the module excite both neuronal populations and thus must be treated as two-dimensional vectors; this contrasts with standard single neuron RF analyses in which inputs are scalars [4, 5, 6, 15]. Thalamocortical and long-range intracortical inputs are combined linearly (summed) for each subpopulation. Since these two input sources activate excitatory and inhibitory neurons in different proportions, the corresponding input vectors have different angles; vector magnitudes vary directly with stimulus strength. Module responses are a function of the summed input vectors, and mean firing rates of the module's excitatory neurons are completely characterized by the surface plotted in figure 2a. Increasing the contrast of the classical RF stimulus (in the absence of extraclassical stimulation) scales inputs to both cell populations, defining a straight line in the input plane (bottom plane of 2a). Presentation of a fixed surround stimulus activates long-range intracortical inputs; the effect of these inputs can be understood as a simple translation of the contrast input line via vector addition (surround stimulus effects mediated by feedback projections from area V2 can be treated similarly). Contrast response functions (CRFs) predicted by the module are obtained by projecting the resulting input line onto the surface.

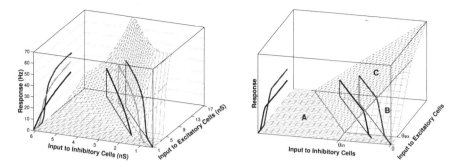

Figure 2. Responses of self-contained module (a) and a linear approximation of module (b). Axes represent total excitatory input to the two module populations, in units of average synaptic conductance. Total long-distance input converging on the center of the full model is plotted in the input plane for all stimulus conditions (medium - center only; light - orthogonal surround; dark - iso-orientation surround). Surround stimulation provides a vector input that translates the thalamic input line (which represents the set of vectors for all center contrasts). Module response curves are obtained by surface projection (also shown on backplane and in dashed lines of fig 1d).

These predicted CRFs are also shown as the dashed lines in figure 1d. Note that they closely approximate the CRFs generated by the model for all tested stimulus conditions (see fig 1c,d) as well as experimental CRFs (see fig 1a,b). Thus, the paradoxical classical and extraclassical RF integration phenomena are captured by local circuit interactions alone.

Local interactions are described by module response surface shapes. The surface shape shown in figure 2a is characteristic of a large class of recurrently connected excitatory-inhibitory circuits and can be thought of as providing generalized gain control. Note that integrate-and-fire neurons have approximately threshold-linear feedforward responses (fig 1c), and thus the module output (fig 2a) is a smoothed version of an underlying piecewise-linear surface. This underlying surface can be obtained from a simplified module, composed of interconnected threshold-linear neurons - a typical example is shown in fig 2b. Assume excitatory and inhibitory neurons have thresholds θ_{ex}, θ_{in}, and gains K_{ex}, K_{in}; total afferent inputs to the two populations are $I_{ex} = M_t T_{ex} + M_h H_{ex}$, $I_{in} = M_t T_{in} + M_h H_{in}$, where M_t, H_t are thalamic and long-range horizontal inputs, and $T_{ex}, T_{in}, H_{ex}, H_{in}$ are the corresponding synaptic efficacies. The synaptic weights among excitatory (e) and inhibitory (i) cells in the module are $W_{ee}, W_{ei}, W_{ie}, W_{ii}$. Then the mean firing rates in the module satisfy the following piecewise-linear system of equations: $M_{ex} = K_{ex}(I_{ex} + W_{ee}M_{ex} - W_{ie}M_{in} - \theta_{ex})$, $M_{in} = K_{in}(I_{in} + W_{ei}M_{ex} - W_{ii}M_{in} - \theta_{in})$. The response surface in fig 2b is $M_{ex}(I_{ex}, I_{in})$, as obtained from the above system. The surface has three planar regions, corresponding to (A) no excitatory firing, (B) recurrent self-excitation with no inhibition, and (C) balanced (competing) excitatory and inhibitory firing. Response saturation occurs when the contrast input line crosses region (B) and is parallel to the contours in region (C), i.e. $\theta_{in}/\theta_{ex} > T_{in}/T_{ex} = (W_{ii}+1/K_{in})/W_{ie}$ (shown with red curve). Super-saturation results from increasing the slope of the contrast input line, so that $T_{in}/T_{ex} > (W_{ii}+1/K_{in})/W_{ie}$. The surround facilitation/suppression effect (compare blue curve to red curve) is obtained when the translation vector resulting from surround stimulation has a bigger slope than the contrast input line, i.e. $H_{in}/H_{ex} > T_{in}/T_{ex}$. This corresponds to the physiological prediction that long-range intracortical inputs are less biased towards excitatory (vs. inhibitory) neurons than are thalamocortical inputs.

4. CONCLUSIONS

Modularity has long been proposed as a means of resolving the complexity of cortical function [14, 2]. Here we have constructed modules (corresponding to dense local cortical circuitry) which are quasi-autonomous: their response properties, as studied in isolation, are preserved in the larger system. Our modular analysis illustrates an expanded concept of the cortical receptive field: each portion of the RF has a dual excitatory-inhibitory influence whose net effect on a neuron depends on how other RF components activate the recurrent local cortical circuitry. This vector-based RF integration fully encompasses the traditional (scalar) view as a special case. Furthermore, this approach unifies notions of classical and extraclassical RFs by showing how long-range inputs can be considered on equal footing with thalamocortical inputs and how the effects of both can be analyzed together. Based on this analysis we predict that for different types of stimulation (involving, for example, luminance, orientation, or motion contrast), the influence of extraclassical stimulation shifts from facilitatory to suppressive as center RF drive increases. Since the properties of neurons and connections in visual cortex exploited here are common to other cortical areas, vector-based integration appears well-suited to other cortex as well.

REFERENCES

[1]   Bonds, A.B. *Visual Neurosci.* **6**, 239-255 (1991).
[2]   Douglas, R.J., Martin, K.A.C., Whitteridge, D. *Neural Comp* **1**, 480-488 (1989).
[3]   Gilbert, C.D. & Wiesel, T.N. *J. Neurosci.* **3**, 1116-1133 (1983).
[4]   Hartline, H.K. *Am. J. Physiol.* **130**, 700-711 (1940).
[5]   Heeger, D.J. *Visual Neurosci.* **70**, 181-197 (1992).
[6]   Hubel, D.H. & Wiesel, T.N. *J. Neurophysiol.* **148**, 574-591 (1959).
[7]   Kisvarday, Z.F., Martin, K.A.C., Freund, T.F., Magloczky, Z.F., Whitteridge, D., and Somogyi, D. *Exp. Brain Res.* **64**, 541-552 (1986).
[8]   Knierim, J.J. & Van Essen, D.C. *J. Neurophysiol.* **67**, 961-980 (1992).
[9]   Levitt, J.B. & Lund J.S. *Soc. Neurosci. Abstr.* **20**, 428 (1994).
[10] Li, C.Y. & Creutzfeldt, O.D. *Pflugers Arch.* **401**, 304-314 (1984).
[11] Maffei, L. & Fiorentini, A. *Vision Res.* **13**, 1255-1267 (1973).
[12] McCormick, D.A., Connors, B.W., Lighthall, J.W. & Prince, D.A. *J. Neurophysiol.* **54**, 782-806 (1985).
[13] McGuire, B.A., Gilbert, C.D., Rivlin, P.K. & Wiesel, T.N. *J. Comp. Neurol.* **305**, 370-392 (1991).
[14] Mountcastle, V.B. in *The Mindful Brain* (eds Edelman, G.M. & Mountcastle, V.B.) 7-50 (MIT Press, Cambridge, MA, 1978).
[15] Jones, J.P. & Palmer, L.A. *J Neurophysiol.* **58**, 1187-1211 (1987).
[16] Rockland, K.S. & Lund, J.S. *Science* **215**, 1532-1534 (1982).
[17] Sengpiel, F., Baddeley, R.J, Freeman, T.C.B., Harrad, R., & Blakemore, C. *Soc. Neurosci. Abstr.* **21**, 1649 (1995).
[18] Siapas, A.G., Todorov, E.V, & Somers, D.C. *Soc. Neurosci. Abstr.* **21**, 1651 (1995).
[19] Sillito, A.M., Grieve, K.L., Jones, H.E., Cudeiro, J., & Davis, J. *Nature*, **378**, 492 (1995).
[20] Ben-Yishai, R., Lev Bar-Or, R. & Sompolinsky, H. *Proc. Natl. Acad. Sci. U.S.A.* **92**, 3844-3848 (1995).
[21] Somers, D.C., Nelson, S.B. & Sur, M. *J. Neurosci.* **15**, 5448-5465 (1995).
[22] Stemmler, M., Usher, M. & Niebur, E *Science* **269**, 1877-1880, (1995).
[23] Toth, L.J., Rao, S.C., Kim, D.-S., Somers, D., and Sur, M. *Proc. Natl. Acad. Sci. USA*, **93**, 9869–9874 (1996).
[24] Tuckwell, H.G. *Stochastic Processes in the Neurosciences* (Soc. for Indust. & Appl. Math., Philadelphia, PA, 1989).
[25] White, E.L. *Cortical Circuits* 46-82 (Birkhauser, Boston, 1989).

81

SIMULATING A NETWORK OF CORTICAL AREAS USING ANATOMICAL CONNECTION DATA IN THE CAT

F. T. Sommer[1] and R. Kötter[2]

[1]Department of Neuroinformatics
University of Ulm
D-89069 Ulm, Germany
Friedrich.Sommer@uni-tuebingen.de
[2]Center for Anatomy and Brain Research
Heinrich Heine University Düsseldorf
PO Box 101007, D-40001 Düsseldorf, Germany
RK@hirn.uni-duesseldorf.de

ABSTRACT

To what extent does corticocortical connectivity alone provide an explanation of the macroscopical flow of activity recorded in the cerebral cortex of anesthetized cats?

Different connectivity structures between cortical areas have been incorporated into a simple cortex model with binary activity states. We compare connectivity models (nearest-neighbor and nearest-neighbor-or-next-door-but-one) with a connectivity based on a database collated from anatomical tracer studies. The agreement with functional neuronographic experiments produced by systematic connections is compared with random connectivity constrained by equal connection distributions. All systematic connectivity structures reproduce the experimentally observed activity patterns in the cerebral cortex of the cat significantly better than random connections (mean error percentage 20 % versus 40 %). However, real anatomical connectivity exhibits the most significant performance difference versus random connectivity. In addition, the simulation reproduces typical topographic features of electrophysiologically demonstrated activity spread in the cerebral cortex of the cat.

1. INTRODUCTION

Systematic application of tracer techniques in neuroanatomical research has revealed a wealth of association fibres connecting distinct areas in the cerebral cortex of mice, rats, cats, and macaques. The complexity of these association fibres calls for higher

Computational Neuroscience
edited by Bower, Plenum Press, New York, 1997

order analysis tools that can unravel general principles of organization and function of these networks.

Previous investigations concentrated on the structural organization of association fibre networks. Analysis of laminar patterns of fibre termination provided evidence that all major cortical sensory systems show hierarchical arrangements of cortical areas and that some systems can be divided further into several "streams of processing."[1,2] Optimization analysis of the topological organization of cortical areas indicates prominent connectivity between cortical motor and somatosensory systems, while frontal and allocortical areas form the most central network of interconnected processing units.[1,3]

Association fibres provide the most massive and direct route of information transfer between cortical areas. Therefore, it is likely that the structural organization of these fibre networks reflects functional interactions between cortical areas. In order to substantiate this notion, we propose a simulation model to correlate the structural organization of association fibre networks with functional maps of the cerebral cortex.

2. METHODS

2.1. The Connectivity Data

Recent publications present accumulated databases of corticocortical connections that were identified by anatomical tracer studies in different species.[2,4] A database of corticocortical connectivity in the cat contains 1139 connections between 65 areas collected from about 75 published studies.[4] The connection strengths in this database were graded from 0 to 3 according to Table 1.

The topography of the traced corticocortical connectivity was found to differ from simple local connectivity models such as the nearest-neighbor (connectivity model 1) or nearest-neighbor-or-next-door-but-one (connectivity model 2).[4]

2.2. The Electrophysiological Data

The aim of this paper is to relate the anatomical connectivity data in the cat with relevant functional data. Because of the limited amount of PET and fMRT data we turned to the older but extensive topographic data obtained by functional neuronography:[5,6] Saturated strychnine sulfate was applied to small patches of cortex inducing stable and reproducible patterns of cortical activity recorded with bipolar electrodes in a total of 18 anesthetized cats. The resulting activity was mapped topographically marking active and silent areas. The remainder of the areas had not been explored or showed variable activity.

Detailed studies of the flow of stimulation-evoked cortical activity have been published for interconnected areas in the suprasylvian gyrus of the cat.[7]

Table 1. Grading of connection strength in the corticocortical connectivity database

$s_{ij} = 0$	unreported connection
$s_{ij} = 1$	sparse or weak connection
$s_{ij} = 2$	intermediate strength or no strength information available
$s_{ij} = 3$	dense of strong connection

2.3. Basic Assumptions of the Model

In the following, it will be tested how close a network model of cortical areas with anatomical connectivity structures can reproduce the macroscopical flow of cortical activity measured in experiments. Cortical areas have been defined primarily on the basis of singular structural features such as their cellular and fiber architecture.[8] The structural homogeneity of an area implies a degree of functional homogeneity as reflected in the localization of sensory, motor, or speech areas. In the simulations we regard cortical areas as homogeneously activated.

By far the largest contingents of afferents to any cortical area originate from other cortical areas and from thalamic nuclei.[9] Corticocortical afferents are assumed to be excitatory since long axons originate only from pyramidal neurons. It is known that anesthesia impedes the transfer of activity from the periphery to the cerebral cortex predominantly at the level of the thalamus.[10] Therefore, cortical stimulation and recordings in anesthetized animals may give an impression of activity flow relying primarily on corticocortical connectivity. Based on these assumptions we wondered whether the activity spread caused by stimulation of small cortical areas in anesthetized cats would be replicated by selective activation of a network of homogeneous units that were linked according to known corticocortical connectivity.

2.4. The Network Model

Each model unit describes the activity of one cortical area as "active" or "silent" by a binary variable. Thus, cortical activity is represented by a binary pattern $x \in \{0,1\}^n$ with the number of areas n=65. A connection from area i to area j is denoted with s_{ij}. From traced corticocortical connectivity we determined interaction values between areas as $w_{ij} = (s_{ij})^\gamma$, where the exponent γ can be used to align the numerical difference between the degrees of connection strength. For the connectivity models interactions between areas satisfying the corresponding neighbor condition were set to a value of one in a binary connection matrix: $w_{ij} \in \{0,1\}$.

In each of the experiments described in[6] the initial pattern x(0) contained a single active component: the strychninized area. It is used in the following parallel iterative update prescription:

$$x(r+1)_j = H[\sum_{i=1}^n w_{ij} x(r)_i - \Theta] \ \forall \ j,$$

where Θ is the activation threshold of a unit and H[x] is the Heaviside function. For a chosen value α of active units the threshold was aligned so that the network reaches a steady state smoothly, i.e., $\Theta(r+1) = \max \{\Theta : |x(r)| < |x(r+1)| \leq \alpha\}$.

Following activation of individual units we observed a time-dependent spread of activity. The final activity states reached by this prescription were compared with the activity maps drawn in ref 6. Figure 1 gives an example of a spread of activation sequence obtained by a simulation.

Possible error types are "miss" errors if the simulation results in silent areas that have been reported to be active, and "add" errors if activity is found in areas reported to be silent. Areas with unknown or variable activity were disregarded in the error statstics. For corticocortical connectivity a fixed set of parameters γ and the intra-interactions $b := s_{ii}$ $\forall$ i were chosen as to produce good agreement with the physiological observations. In the connectivity models we used b=1. The parameter α was aligned for each initial pattern individually.

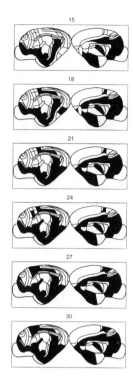

Figure 1. Snapshots from a simulated spread of activation sequence starting with stimulation in area 36. Left pictures show parietal views and right pictures show medial sections of the cat cortex. The small numbers above the pictures denote the time step.

3. RESULTS

We investigated the global match between the experimental results from[6] with our simulations using the three different connectivity structures. Non-binary interaction values have been generated with the parameters $\gamma = 2$ and b=10 from the corticocortical connectivity. However, variations of these parameters turned out to produce qualitatively similar results: the simulations where not very sensitive on the grading of connection strengths. The error percentages achieved by the three connectivity models is shown in Figure 2. Experiments with frontal, parietal, and occipital stimulation are better reproduced by traced corticocortical connections, whereas experiments with temporal stimulation are better fit by the connectivity models. However, with the same overall mean errors of roughly 20% the different connectivity structures appear to perform equally well.

In order to judge the simulation performance with a systematic connectivity structure, we need to know how much it outperforms models with random connections. We carried out additional simulation experiments using random connections with connection densities as in the systematic connectivity structures. Each curve in Figure 3 displays the average error in an ensemble of 20 random connection configurations. The most significant difference between random and systematic connections is obtained by anatomically identified connecticity depiced in Figure 4 where the performance of random connections is displayed with mean deviation: Obviously, the probability is very low that random connections can produce similar or better results than really existing connections.

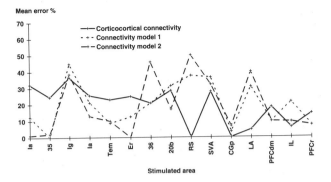

Figure 2. Performance of the different connectivity models. For corticocortical connectivity we used $\gamma = 2$ and $b = 10$. The mean values are 21 %, 20 % and 21 % (in the legend from top to down).

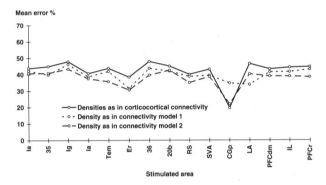

Figure 3. Performance of random connectivity with connection densities as in the connectivity models. The mean values are 41 % , 35 % and 37 % (in the legend from top to down).

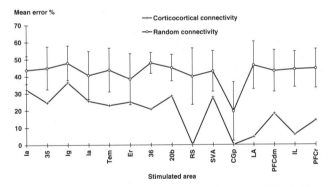

Figure 4.   Performance of corticocortical connectivity versus random connectivity (with standard deviations).

There are three basic types of problems causing discrepancies between simulations and experiments: a) the electrophysiological data are not appropriately transfered in the model, b) the available anatomical data are incomplete or inaccurate, c) the simplifying assumptions in the model design might not be correct.

The stimulated areas located near the Sylvian sulcus and the basal aspect of the brain (corresponding to experiments displayed as the six leftest points in the figures could not be identified unambiguously from the lateral and medial drawings of the cat brain shown in ref. 6. In particular, for corticocortical connectivity this problem of type a) could account for the comparatively high deviations in this range.

With respect to problems of type b) a careful look at the available studies of the cat cortex reveals a more detailed architecture than had been provided in the connectivity database, particularly, again for the ventral areas. Among the problems of type c) are the diversity among cortical areas (we used the same input/output functions), the laminar preferences of corticocortical connections (we did not differentiate between so-called "forward" and "backward" connections), and the limited selectivity of anesthetic agents (impact on cortical and thalamic firing patterns was ignored). At a first stage, we prefered to keep the model simple rather than to introduce a complexity that would have been hard to evaluate.

In addition to the global results displayed in the Figures 2–4 we found a striking correspondence of individual topographic features in the simulations using corticocortical connectivity with electrophysiological data. Among the well studied properties of the cerebral cortex of the cat is the massive projection from area 7 to area 5, which is thought to be the anatomical substrate of the short-latency posterior-to-anterior suprasylvian EPSPs described in.[7] Even though the connection strengths in the database of [4] do not reveal a strong asymmetry of the fiber connections between areas 7 and 5 the simulations consistently produced a spread of activity in the suprasylvian gyrus in the posterior-to-anterior direction which can be observed in Figure 1 beginning with the 18th time step.

A related feature is the selective spread of activitiy in the suprasylvian and posterior ectosylvian gyri (cf. stimulation in areas 36 and 20b). Again, the simulations with traced connectivity reproduced this topographic pattern and revealed a low threshold for the activation of areas EPp and 7 in the nomenclature of [4].

The search for a correlation between structure and function of the brain will be severely supported by recent development of functional imaging techniques, such as PET, fMRT, or MEG, allowing recordings from awake animals with highly enhanced spacial and temporal resolution. Unforunately, for the cat no such experiments have been carried out so far. Recordings of the activation induced by magnetic intragranular stimulation during light anesthesia of the animal with such imaging techniques would be very interesting. The presented method will be applied to anatomical and physiological data of the macaque monkey.

4. SUMMARY

1. The complexity of corticocortical connections requires computer-assisted analysis to reveal its influence on the cortical flow of activity. We suggest a straightforward approach to this problem by examining how the macroscopic activity flow observed in physiological experiments can be simulated using systematic connectivity structures between cortical areas.

2. All considered systematic connectivity structures reproduce neuronographic experiments with comparable overall mean errors of approximately 20%. They perform clearly better than simulations with random connections leading to errors of roughly 40%.

3. Traced corticocortical connectivity exhibits the most significant performance difference between systematic and random connectivity. The simulation performance does not critically depend on the precision of the grading of connection strengths from experimental tracer studies.

5. REFERENCES

1. Young, M. P., Scannell, J. W., Burns, G. A. P. C., Blakemore, C. Analysis of Connectivity: Neural Systems in the Cerebral Cortex. Rev. Neurosci., 5:227–249, 1994.
2. Felleman, D. J., Van Essen, D. C. Distributed hierarchical processing in the primate cerebral cortex. Cerebral Cortex, 1:1—47, 1991.
3. Young, M. P., Scannell, J. W., O'Neill, M. A., Hilgetag, C. C., Burns, G., Blakemore, C. Non-metric multidimensional scaling in the analysis of neuroanatomical connection data from the cat visual system. Phil. Trans. R. Soc., 348:281–308, 1995.
4. Scannell, J. W., Blakemore, C., Young M. P. Analysis of connectivity in the cat cerebral cortex. Journal of Neuroscience, 15(2):1463—1483, 1995.
5. McCulloch, W. C. The functional organization of the cerebral cortex. Physiol. Rev., 24:390–407, 1944.
6. MacLean, P. D., Pribham, K. H. Neuronographic analysis of medial and basal cerebral cortex I. cat. Journal of Neurophysiology, 16:312–323, 1953.
7. Amzica, F., Steriade, M. Disconnection of intracortical synaptic linkages disrupts synchronization of a slow oscillation. J. Neuroscience 6(15):4658–4677, 1995.
8. Jones, E. G. Brodmann's areas. In: Adelman, G., Encyclopedia of Neuroscience. Birkhaeuser, Basel, 1987, p. 180–181.
9. White, E. L. Cortical Circuits. Birkhaeuser, Basel, 1989.
10. Angel, A. The G. L. Brown lecture. Adventures in anaesthesia. Exp-Physiol. 1(76):1–38, 1991.

82

A COMPUTATIONAL AND EXPERIMENTAL STUDY OF REBOUND FIRING AND MODULATORY EFFECTS ON THE LAMPREY SPINAL NETWORK

Jesper Tegnér[1,2]*, Jeanette Hellgren-Kotaleski[1,2], Anders Lansner[2], Sten Grillner[1]

[1]Nobel institute for Neurophysiology
Department of Neuroscience
Karolinska Institute
S-171 77 Stockholm, Sweden
E-mail: {jespert,jeanette,ala}@nada.kth.se
[2]SANS — Studies of Artificial Neural Systems
Dept. of Numerical Analysis and Computing Science
Royal Institute of Technology, S-100 44 Stockholm, Sweden
E-mail: {Sten.Grillner}@neuro.ki.se

ABSTRACT

Computer simulation with simplified but fairly detailed model neurons can be powerful in evaluating to what extent known cellular and synaptic properties can account for the behavior seen at the system level and also allow analysis of the relative functional role of different components and mechanisms which might be difficult to address only experimentally. We have evaluated the role of low voltage-activated (LVA) calcium channels both on the single cell and network level in the lamprey locomotor circuitry. A previous computer simulation model has been extended with LVA calcium channels using the m^3h form, following a Hodgkin–Huxley paradigm. Experimental data from mainly lamprey neurons [9] was used to provide parameter values of the single cell model.

---

*To whom correspondence should be addressed,

Computational Neuroscience
edited by Bower, Plenum Press, New York, 1997

1. INTRODUCTION

To understand the cellular basis of motor behavior in mammals, the lamprey nervous system with comparatively few neurons has been used extensively as an experimental model [6]. The behavior underlying locomotor activity is now understood with respect to basic features of the neural network. The modulatory effects of GABAergic neurons on the activity of the spinal circuitry has been analyzed experimentally [9, 13]. The somatodendritic effects of $GABA_B$ receptors is due to both a low voltage activated (LVA) and a high voltage activated (HVA) calcium current on spinal interneurons [9] which reduces the tendency for rebound firing and decreases the amplitude of the slow afterhyperpolarization (sAHP) which is responsible for the adaptation in the system. On the network level $GABA_B$ receptor activation decrease the bursting frequency [13]. In this study, see also [11], we have (i) modeled the LVA calcium current, (ii) simulated whether the earlier cellular findings could be explained by the presence of an LVA calcium current and also (iii) experimentally compared the effects of a modulation of the sAHP and LVA calcium current on the network, and finally, (iv) tested to what extent the network effects of $GABA_B$ receptors are due to an LVA calcium conductance.

2. SINGLE CELL SIMULATIONS

We have extended the compartmentalized Hodgkin-Huxley cell model of intermediate complexity (5 compartments) used in earlier lamprey simulations (SWIM simulator) with an LVA calcium channel using the m^3h format in addition to Na^+, K^+, Ca^{2+} and $K_{(Ca)}$ channels. The excitatory (NMDA and AMPA/kainate channels) and inhibitory synaptic interactions synapses are placed on the dendritic compartments. The parameters in the original cell model was tuned to account for the firing and plateau properties in spinal neurons [4]. As tuning criteria for the LVA calcium channel we used data on thalamic relay cells and thalamic nuclear reticular neurons [2, 3, 8, 10] and data from spinal lamprey neurons [9] on size of the hyperpolarization induced by the negative pulse necessary for triggering a rebound spike as well as experiments with variable duration of the current pulse. After the model neuron was tuned in this manner using a subset of experimental data, it was tested whether the model could account for the remaining parts of the single cell data.

2.1. Rebound Spikes and Rebound Train

If a neuron is equipped with an LVA calcium conductance an inhibitory input can remove the inactivation and elicit a rebound depolarization eventually leading to one or several spikes (rebound train, see Figure 1A). Moreover, the closer the holding potential is to the threshold for eliciting an action potential, the likelihood to elicit a large rebound depolarization or spike increases [9]. Consequently, it is important to test whether the model neuron can account for these aspects.

The holding potential threshold for the occurrence of a rebound spike pulse amplitude and duration was around -55 mV in the cell illustrated in Figure 1A1–3. Note also that the spike occurs earlier when the cell is more depolarized (Fig 1A3 compared to Fig. 1A2). The simulation (Fig. 1B) matches the experiments in that the spike also occurs earlier when the model cell is more depolarized (compare the dashed and top solid line in Fig. 1B), since the difference between the holding potential and spike threshold has been reduced and the degree of activation of the LVA calcium channel is increased. For determining whether a single or several rebound

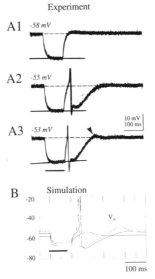

Figure 1. Role of holding potential for the rebound, comparison between experimental data and simulation **A** A lamprey neuron is held at three different holding potentials by continuous current injection in DCC mode. Note that the latency for the spike at -55 mV (A2) is longer as compared to that when the cell is held at -53 mV (A3). **B** The simulated neuron is held at -53 mV (dashed line), -55 mV (middle solid line) and -58 mV (solid line). The bar below the voltage traces indicates the duration of the negative current pulse.

spikes (rebound train, Fig. 2B) would occur for a given stimulus in the simulations, a key factor was the difference in potential between the reversal potential of the sAHP ($K_{(Ca)}$) and the membrane holding potential. Figure 2B2 shows a parameter plot of the holding potential and the reversal potential of the sAHP for the simulation model. A rebound train occurs in the upper part of the parameter space in which the model cell is more depolarized and/or the sAHP reversal potential is more negative, while in the lower part there is only a single rebound spike. The sAHP has to be large enough to remove the inactivation of the LVA current. Moreover, since the reversal potential for the sAHP was important for the occurrence of rebound firing the effect of increasing and decreasing the maximal conductance of the sAHP was also tested. A reduction or increase of the conductance of the sAHP by 20 % modifies the location of the boundary line (solid line, Fig. 2B2) separating the parameter plot into single spikes and rebound trains. This implies, that neurons are less likely to respond with a rebound train as the sAHP is reduced and this could be of importance for the regularity of the burst pattern [5]. The $GABA_B$ receptor induced reduction of the LVA current during a single negative current step and during a sinusoidal current stimulation was also simulated. The experimental effects of baclofen ($GABA_B$ receptor agonist) could be simulated if the conductance of the LVA calcium channels were reduced (not illustrated).

An experimental finding which appeared enigmatic was that neurons held at a more depolarized level showed a progressive depolarization during the negative current step. The simulations showed that essentially, this could be due to a larger degree of activation (m) leading to a larger $m^3 hg$ factor during the negative current step (compare the slopes Fig. 2A and Fig. 2B).

In conclusion, when the model cell is extended with an LVA calcium channel it accounts for a number of experimentally established properties, including rebound spike trains, the positive slope during the negative current steps at depolarized levels, the holding potential range of the rebound depolarization and the modulatory effects of $GABA_B$ receptor activation.

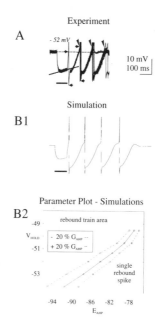

Figure 2. Factors responsible for the occurrence of a train of rebound spikes. **A** A 100 ms current pulse elicits a train of rebound spikes in this lamprey neuron. Note the positive slope on the voltage during the negative current pulse in the model neuron as indicated by a solid line. **B** Note the similar positive slope during the negative current pulse as compared to the experiment in A. **C** Parameter plot of the holding potential and reversal potential of the sAHP versus the occurrence of a train of rebound spikes or not. The line connecting the open circles (o) separates the parameter areas between rebound train and single spikes. The dashed lines indicate how the boundary between single spikes and rebound trains is translated if the maximal conductance of the sAHP is increased (+20 %) or decreased (-20 %).

3.   THE NETWORK — EXPERIMENTS AND SIMULATION

The single segment model of the spinal cord consists of a population of neurons (5 CCIN, 7 EIN) with parameters (cell sizes, synaptic conductances, delays) normally distributed with a standard deviation of 15% [7].

3.1.   Role of LVA channels during fictive locomotion

It is known from previous experimental studies that activation of $GABA_B$ receptors via the agonist baclofen reduces the burst frequency when the network is driven by NMDA or kainate. Experimentally, the $GABA_B$ receptor activation can (i) reduce the sAHP (ii) reduce the LVA calcium currents and (iii) through presynaptic inhibition reduce the amplitude of the EPSP and IPSP in the synaptic transmission. We experimentally compared the effects of reducing the sAHP by blocking the $K_{(Ca)}$ channels with apamin and LVA calcium channels (baclofen, $GABA_B$ receptor agonist). The effect of apamin (Fig. 3A2) was slight in the high NMDA frequency range compared to baclofen (Fig. 3A3) which is in accordance with earlier studies on the sAHP [5, 12].

This means that the $GABA_B$ receptor induced frequency is not essentially due to the sAHP mechanism but instead to either the reduction of the current through the LVA calcium channels and/or the presynaptic inhibition [1]. If LVA calcium channels in the simulations are added to all cells the bursting frequency is increased for a given NMDA level (Fig. 3B). The

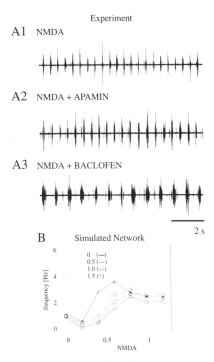

Figure 3. The effect of the LVA calcium currents during fictive locomotion in experiments and simulations. **A** 150 μM NMDA is used to induce rhythmic activity in the lamprey spinal cord. Extracellular recordings are made from a ventral root. Addition of 2.5 μM apamin (A2), which markedly reduces the slow afterhyperpolarization has a small effect on the burst frequency whereas 20 μM baclofen (A3) affects the burst pattern markedly. **B** Addition of an LVA calcium (a factor of 0–1.5) conductance increase the burst frequency in the simulated network driven by NMDA.

simulations with the NMDA driven network makes it likely that LVA calcium channels plays a major role in this respect.

4.   CONCLUSION

Even though the cell model has certain limitations (does not represent a full dendritic tree, lack of data on channel distribution), it provides insight to the earlier single cell experiments and also makes it likely that the LVA calcium channels are important for the frequency regulation in lamprey network.

ACKNOWLEDGMENTS

We are indebted to Dr. Örjan Ekeberg for developing the SWIM simulation software (see the Web site http://www.nada.kth.se/sans/) and for providing useful help in the initial phase of this project during which the code was modified. This work was supported by the Medical Research Council (proj. no. 3026), the Swedish Natural Science Research Council (proj. no. B-AA/BU03531), the Swedish National Board for Industrial and Technical

Development, NUTEK, (proj. no. 8425-5-03075) and the Swedish Society for Medical Research.

REFERENCES

[1] Simon Alford and Sten Grillner. The involvement of $GABA_B$ receptors and coupled G-proteins in spinal gabaergic presynaptic inhibition. *J. Neurosci.*, 12(11):3718–3728, 1991.

[2] Thierry Bal and David McCormick. Mechanisms of oscillatory activity in guinea-pig nucleus reticularis thalami *in vitro*: a mammalian pacemaker. *J. Physiol.*, 468(0):669–691, 1993.

[3] D.A. Coulter, J.R. Huguenard, and D.A. Prince. Calcium currents in rat thalamocortical relay neurons: kinetics properties of the transient low-threshold current. *J. Physiol.*, 414(0):587–604, 1989.

[4] Örjan Ekeberg, Peter Wallén, Anders Lansner, Hans Tråvén, Lennart Brodin, and Sten Grillner. A computer based model for realistic simulations of neural networks. I: The single neuron and synaptic interaction. *Biol. Cybern.*, 65(2):81–90, 1991.

[5] Abdel El-Manira, Jesper Tégner, and Sten Grillner. Calcium-dependent potassium channels play a critical role for burst termination in the locomotion network in lamprey. *J. Neurophysiol.* 72(4):1852–1861, 1994.

[6] Sten Grillner, Tanja Deliagina, Örjan Ekeberg, Abdel El Manira, Russell H. Hill, Anders Lansner, G.N Orlovsky, and Peter Wallén. Neural networks that co-ordinate locomotion and body orientation in lamprey. *Trends Neurosci.*, 18(6):270–279, 1995.

[7] Jeanette Hellgren, Sten Grillner, and Anders Lansner. Computer simulation of the segmental neural network generating locomotion in lamprey by using populations of network interneurons. *Biol. Cybern.*, 68:1–13, 1992.

[8] J.R. Huguenard and D.A. McCormick. Stimulation of the currents involved in rhythmic oscillations in thalamic relay neurons. *J. Neurophysiol.*, 68(4):1373–1383, 1992.

[9] Toshiya Matsushima, Jesper Tegnér, Russell Hill, and Sten Grillner. $GABA_B$ receptor activation causes a depression of Low- and High-Voltage-Activated Ca^{2+} currents, Postinhibitory Rebound, and Postspike After-hyperpolarization in lamprey neurons. *J. Neurophysiol.*, 70(6):2606–2619, 1993.

[10] D.A. McCormick and J.R. Huguenard. A model of the electrophysiological properties of thalamocortical relay neurons. *J. Neurophysiol.*, 68(4):1384–1400, 1992.

[11] Jesper Tegnér, Jeanette Hellgren-Kotaleski, Anders Lansner, and Sten Grillner. Low voltage activated calcium channels in the lamprey locomotor network-simulation and experiment. *Submitted*, 1996.

[12] Jesper Tegnér, Anders Lansner, and Sten Grillner. Modulation of burst frequency by calcium-dependent potassium channels in the lamprey locomotor system-dependence of the activity level. *Submitted*, 1996.

[13] Jesper Tegnér, Toshiya Matsushima, Abdel El Manira, and Sten Grillner. The spinal GABA system modulates burst frequency and intersegmental coordination in the lamprey: differential effects of $GABA_A$ and $GABA_B$ receptors. *J. Neurophysiol.*, 69(2):647–657, 1993.

83

MODELING VISUAL CORTICAL CONTRAST ADAPTATION EFFECTS

E. V. Todorov,[1] A. G. Siapas,[1] D. C. Somers,[1] and S. B. Nelson[2]

[1] Department of Brain and Cognitive Sciences
MIT
Cambridge, MA
[2] Department of Biology
Brandeis University
Waltham, MA
E-mail: emo@ai.mit.edu
E-mail: thanos@ai.mit.edu
E-mail: somers@ai.mit.edu
E-mail: nelson@binah.cc.brandeis.edu

ABSTRACT

We demonstrate model visual cortical circuits which exhibit robust contrast adaptation properties, consistent with physiological observations in V1. The adaptation mechanism we employ is activity-dependent synaptic depression at thalamocortical and local intra-cortical synapses. Model contrast response functions (CRF) shift so that cells remain maximally responsive to changes around the recent average stimulus contrast level. Hysteresis effects for both stimulus contrast and orientation are achieved; orientation hysteresis is weaker, and depends exclusively on intracortical adaptation. Following stimulation of the receptive field (RF) surround, RFs dynamically expand to "fill in" for the missing stimulation in the RF center; in our model this expansion results from adaptation of local inhibitory synapses, triggered by excitation from long range horizontal projections. All adaptation effects are achieved using the same synaptic depression mechanisms.

1. INTRODUCTION

Nearly all neurons in the primary visual cortex (V1) exhibit reduced responsiveness after exposure to high contrast stimuli [1, 3, 8, 11, 14]. This contrast adaptation appears to have functional utility as a cortical gain control mechanism: a neuron's contrast response function rapidly shifts so that the neuron remains maximally responsive to both contrast increments and decrements around the recent mean contrast level [14]. Adaptation effects occur rapidly and can be observed after even a single, brief (50 msec) stimulus presentation [3]. Recovery from adaptation occurs more slowly and hysteresis effects are readily observable in contrast response functions [3, 12]. Adaptation effects also depend on other stimulus properties, such as orientation. Given a fixed contrast level, hysteresis effects occur in the orientation domain; however, orientation hysteresis effects are smaller than contrast hysteresis effects for the same firing levels, indicating that different mechanisms may underlie the two effects [3].

The mechanisms underlying contrast adaptation, despite intense study, remain unclear. This phenomenon almost certainly arises within cortex since it exhibits inter-ocular transfer [8, 16] and since lateral geniculate (LGN) neurons lack contrast adaptation [3, 11, 14]. The hypothesis that "fatigue" of cortical neurons leads to adaptation [2] can be excluded since pharmacological induction of firing by glutamate iontophoresis does not induce adaptation in a cell and similarly blockade of the cell's response with GABA does not abolish adaptation effects [19]. Rather, adaptation may be a network effect since adaptation effects depend primarily on the mean stimulus contrast level over a recent time window. Several authors [3, 7, 14] have suggested that contrast adaptation effects may be accounted for by an inhibitory process, possibly involving divisive normalization. However, inhibitory mechanisms have not found experimental support. Blockade of GABA-mediated inhibition does not disrupt adaptation [3, 5] and the suggestion that divisive normalization is achieved via shunting inhibition is inconsistent with intracellular recordings in visual cortex [6]. Here, we explore a different hypothesis, that short-term changes in synaptic transmission properties at thalamocortical and intracortical synapses can account for cortical adaptation effects. Specifically, we explore synaptic changes that depend on pre-synaptic activity only.

Short-term adaptation effects following stimulation of the RF surround have also been observed [15, 4]. Prolonged exposure to an "artificial scotoma" stimulus (that covers an area outside a large central blank region) causes increased responsiveness in cells whose RFs lie well within the blank region. Testing with minimal bar stimulation also reveals a substantial expansion of the classical RF [15]. Reverse correlation analysis of white noise stimulation indicates that normally subthreshold RF regions become suprathreshold, but suggests that the effects are better characterized as a uniform gain increase than as a RF expansion [4]. It has been suggested that this phenomenon is due to changes in the input from long-range horizontal projections [15]. Since most of the expansion is localized to the scotoma region, we think that a more likely candidate mechanism is a change in the local circuit gain, triggered by inputs from horizontal projections. We demonstrate that synaptic depression mechanisms which achieve normal contrast adaptation can also produce local circuit gain changes consistent with the artificial scotoma results.

2. METHODS

Computer simulations were performed on model circuits composed of interacting excitatory and inhibitory cortical neurons which received feedforward excitation from LGN neurons. Peak conductances at thalamocortical excitatory, intracortical excitatory, and intracortical inhibitory synapses varied approximately inversely with the recent average firing rates of the pre-synaptic neurons. Cortical neurons were implemented as "improved integrate-and-fire" neurons [20] with membrane time constant and resistance values chosen to match values found experimentally for fast-spiking (inhibitory) and regular–spiking (excitatory) neurons [9]. Conductance changes (for a single spike event) were modeled as α-functions, and parameter values were similar to those used in [17]. In all cases the probability of a synaptic connection between two V1 neurons was 0.1. LGN responses increased linearly with log contrast (up to 50Hz at max contrast) and individual responses were described by Poisson processes. Ten LGN cells converged onto each V1 cell.

Synaptic adaptation was modeled as a reduction of post-synaptic conductance (and thus PSP size), dependent upon pre-synaptic firing (as observed in [13]). Each synapse had a synaptic efficacy, w, and an adaptation level, $AD(t)$, associated with it. If the pre-synaptic cell generated an action potential at time t, the peak of the conductance change in the post-synaptic cell was $w(1 - AD(t))$. The adaptation level increased after each spike according to $AD(t) = AD(t) + (AD_{max} - AD(t))AD_{inc}$ and passively decayed with time constant $\tau : AD(t+1) = AD(t)e^{-1/\tau}$. The constants AD_{max} and AD_{inc} corresponded to the maximum adaptation level $(AD_{max} < 1)$ and to the percent increase for each spike, respectively. This mechanism is consistent with two possible physiological phenomena: i) for each action potential the synapse releases a unit volume of some substance, which acts to increase the local concentration of that substance; AD_{max} is the maximum concentration level that can be achieved before diffusion or uptake mechanisms balance the unitary increase completely; ii) the adaptation level is increased by a constant amount for each vesicle release, but release probability decreases with adaptation. Although we modeled adaptation as decreasing PSP size, it is possible (and mathematically consistent with our model) that it actually corresponds to a decrease in release probability for synapses with multiple release sites. Note that for a fixed firing rate M of the pre-synaptic cell over a sufficiently long period of time, $AD(t)$ will asymptote to $A(M)$ satisfying the equation $(AD_{max} - A(M))AD_{inc}M = A(M)(1 - e^{-1/\tau})$, or $A(M) = AD_{max}M/(AD_{inc}M + 1 - e^{-1/\tau})$.

Due to the computational load of time-varying synapses, the size of the cortical circuits was restricted to include only the cortical dimensions needed to explore the data set at hand. As a result, three different circuits were employed. The first model (CRFs, Gain Control) was a homogeneous population of 200 excitatory and 50 inhibitory neurons, corresponding to a small population in V1 cells with similar RF position and orientation tuning. The second model (Hysteresis) consisted of 1000 excitatory and 250 inhibitory neurons, with overlapping RFs, and orientation tuning ranging from 0 to 180 degrees. In the hysteresis model, the total LGN input to each cortical neuron exhibited a modest orientation bias that was sharpened by intracortical connections (see [17]). Cortical neurons were organized into orientation columns and received substantial input from both cortical inhibitory and cortical excitatory neurons. Both sets of inputs came most densely from the same or nearby orientation columns with the inhibition from a somewhat broader set of orientations. The third model (artificial scotoma) represented visual space rather than orientation and long-range horizontal excitatory projections, that equally contacted excitatory and inhibitory cells, were added.

Stimuli were chosen to represent the average effects of moving, oriented sine-wave gratings. In order to reduce computational times, LGN neuronal firing rates were held at constant contrast-dependent level for the course of a simulation trial. It is assumed that moving gratings stimulate all LGN cells equally, resulting in a constant adaptation level for thalamocortical synapses when contrast is held fixed and only orientation is varied. Adaptation and recovery effects occurred at both thalamocortical and intracortical synapses. Stimuli were presented 10 times each and spike totals were used to generate contrast response functions. The CRFs for the excitatory population are presented. For contrast hysteresis studies, optimally oriented stimuli were presented first in order of increasing contrast and then in order of decreasing contrast, one presentation in each direction for 10 loops. Similarly, in the orientation hysteresis studies, high contrast stimuli were presented, rotating from non-preferred to preferred and then back.

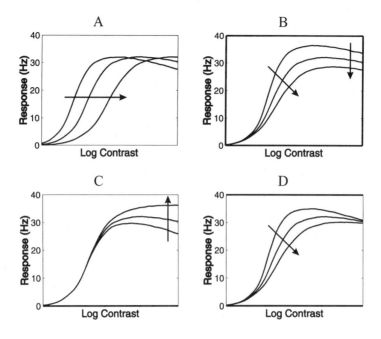

Figure 1. Synaptic depression shifts CRFs.

3. RESULTS

Before we can approach the phenomenon of contrast adaptation, we need a detailed model that produces plausible (i.e. saturating) contrast response functions without adaptation. (Note that the response of an isolated neuron to injected current saturates only at unphysiologically high firing rates). We have recently developed such a model [18] which we briefly describe here. LGN input is biased towards the excitatory (rather than inhibitory) cortical cells, and as a result they (on the average) respond at lower contrast levels. Thus, the steeply increasing portion of the response function is dominated by recurrent self-excitation. Saturation results from a balance between excitation and inhibition, once the inhibitory population enters the linear range of its response.

We now consider how the response of this recurrent system changes when different groups of synapses are modified. Figure 1 summarizes the effects of separately modifying thalamocortical (A), cortical excitatory (B), cortical inhibitory (C), and all cortical (D) synapses, by +/- 10%. In each panel, the arrow marks the direction of decreasing synaptic strength, and the curve in the middle is the same (non-adapted) CRF. Decreasing the efficacy of thalamocortical synapses (A) results in an almost pure rightward shift of the CRF. This pure rightward shift is also achieved in a model of Heeger [7] by increasing the size of a divisive inhibitory term. Our mechanism has the advantages that: i) it is independent of cortical inhibitory signaling and thus is compatible with inhibitory blockade studies; and ii) since LGN neurons themselves do not adapt to contrast, an invariant measure of stimulus contrast is always available at the site of adaptation. Decreasing the strength of intracortical excitation (B) has two effects: i) the slope of the steep part of the CRF decreases, since it is determined by the strength of recurrent self-excitation; ii) saturation occurs at a lower level, since it is the outcome of a balance between recurrent excitation and inhibition. Decreasing intracortical inhibition (C)

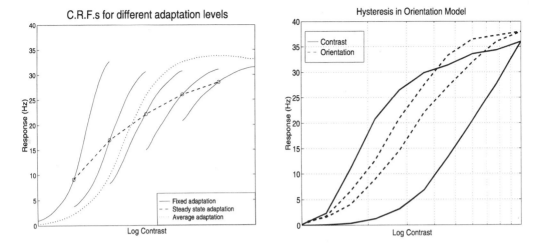

Figure 2. Contrast Gain Control and Hysteresis.

mostly affects the saturating part of the CRF - there is less saturation, and it occurs at a higher level (the steep part of the CRF may also be affected if extra inhibitory subpopulations are added). If we now modify simultaneously both intracortical excitatory and inhibitory synapses (D), the effects on the saturating part of the CRF from (B) and (C) essentially cancel each other, and the slope changes in the steep part of the CRF. Thus, adaptation at thalamocortical synapses shifts the CRF horizontally, while adaptation at intracortical synapses modifies the slope of the steep part of the CRF. For different synaptic efficacies in the recurrent circuit, depression at intracortical synapses can also result in upward and downward shifts of the CRF. Adaptation at intracortical excitatory synapses can also yield rightward CRF shifts, provided that spontaneous cortical excitation contributes to "resting" responses.

Combination of these adaptation mechanisms into a single model yields robust contrast gain control. We use Model 1, with adaptation parameters $(AD_{max}, AD_{inc}, \tau)$ the same for all synapses in the model, to simulate the experiment of [14]. For each of 5 contrast levels (uniformly spaced along the log contrast axis), we present the corresponding contrast until adaptation at all synapses asymptotes. The asymptotic cortical responses are shown by the dashed CRF in Figure 2a. After presenting each of the 5 contrasts, we "freeze" all synapses, and measure the responses to several contrast levels, centered around the adaptation contrast level (solid CRFs). In agreement with the reports in [14], for a range of contrast adaptation levels the CRF shifts so as to center the steepest part of the curve at or near that contrast level. This permits the circuit to be very sensitive to contrast increments or decrements around the recent average contrast level. Rightward (and leftward) shifts of the CRF due to thalamocortical synaptic changes are most prominent in this contrast gain control. Upward or downward shifts of the curves are also observed [1]. In our model these shifts result from adaptation at intracortical synapses. For higher contrasts, the adapted CRFs have lower slopes, which is a result of the adaptation of cortical excitatory synapses (see Fig 1b).

Stimulus presentation order has also been used to isolate adaptation mechanisms [3]. Presentation of optimal orientation stimuli in order from lowest contrast to highest contrast and back yields hysteresis effects, provided that a full cycle can be performed rapidly [3]. This experiment is simulated using Model 2, which includes orientation tuning. Our model 2 achieves this effect (see fig 2b) with higher responses exhibited for the march to higher

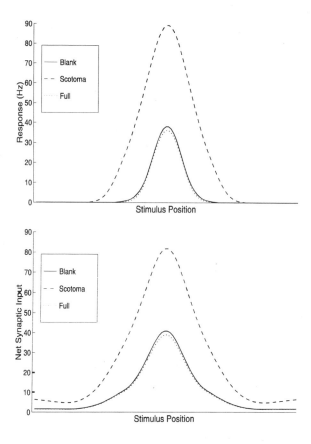

Figure 3. "Artificial Scotoma" Results.

contrasts. This hysteresis results because the level of average recent contrast, as encoded in synapses, is higher on the way down (and thus synaptic efficacy is lower) than on the way up. Contrast hysteresis effects reflect efficacy changes at all synapses. Orientation hysteresis effects have also been observed [3]. Presentation of a high contrast stimulus that shifts from non-preferred orientations to the optimal orientation and then back exhibits higher responses on the way to the optimal response than on the return (see fig 3). This effect is weaker than contrast hysteresis, even for identical levels of firing [3]. In our model, this difference in hysteresis effects reflects a difference in affected mechanisms. With orientation hysteresis, thalamocortical synapses adapt uniformly regardless of the stimulus orientation and thus all hysteresis effects result from selective adaptation of cortical synapses. Because our model cortical cells receive excitation most strongly from like tuned neurons and inhibition from a broader distribution of orientation, when the recent stimulus history is biased toward non-preferred orientations (on the way to preferred) a greater proportion of inhibitory than intracortical excitatory synapses is adapted and thus responses are higher. The converse is true on the way down.

Simulation results for the artificial scotoma stimulus are presented in figures 3a and 3b respectively - the receptive field measured by neuronal firing expanded, and the response in the center increased (3a), while the receptive field measured intra-cellularly (to reveal subthreshold responses) scaled up (3b). During presentation of the artificial scotoma stimulus, excitation from long-range horizontal projections weakly activated cells in the RF center. Since horizontal projections were not biased, and inhibitory cells are easier to activate, the resulting firing rates were 1 Hz for excitatory, and 5 Hz for inhibitory cells. (Results are consistent with our long-range model [18] in this volume). This small difference was greatly amplified by the synaptic depression mechanism, which is very sensitive to differences at low firing rates — local excitatory synapses were depressed by about 5%, and local inhibitory synapses by 25%. The difference in depression levels disrupted the balance between excitation and inhibition in the central region of the model, resulting in more net synaptic input (3b) which leads to wider classical receptive fields (3a) due to the thresholding of the neuronal firing mechanism. Note that adaptation to a full field stimulus does not have the same effect, since direct LGN stimulation activates both excitatory and inhibitory cells, which results in balanced depression at intracortical excitatory and inhibitory synapses.

3.1. Conclusions

In summary, we proposed a model of the local neuronal circuitry in V1, that relies on a bias of thalamocortical projections towards excitatory cortical neurons to achieve contrast saturation. Our model achieves many fundamental effects of contrast adaptation by utilizing pre-synaptic activity-dependent depression of synaptic efficacy. Such depression effects have been recently observed in V1[13], and have been implied in contrast adaptation[10]. This explanation has the advantage that it is "parsimonious" while also being consistent with data that indicate multiple mechanisms of contrast adaptation (e.g., contrast and orientation hysteresis), operating at multiple sites. Indeed stimulus adaptation effects are ubiquitous within sensory cortices; synaptic depression appears to be a well-suited candidate for addressing these broader phenomena.

REFERENCES

[1] D.G. Albrecht, S.B. Farrar, & D.B. Hamilton (1984) J. Physiol. 347: 713-739.
[2] C. Blakemore, R.H.S. Carpenter, & M. Georgeson (1970). Nature 228: 37-39.
[3] A.B. Bonds (1991) Vis. Neurosci. 6: 239-255.
[4] G.C. DeAngelis, A. Anzai, I. Ohzawa, & R.D. Freeman (1995). Proc. Natl. Acad. Sci (USA) 92: 9682-9686.
[5] E.J. DeBruyn & A.B. Bonds (1986). Brain Research. 383: 339-342.
[6] R.J. Douglas, K.A.C. Martin, & D. Whitteridge (1988). Nature 332: 642-644.
[7] D.J. Heeger (1992). Vis. Neurosci. 9: 181-197.
[8] L. Maffei, A. Fiorentini, & S. Bisti (1973). Science. 182: 1036-1038.
[9] D.A. McCormick, B.W. Connors, J.W. Lighthall, and D.A. Prince, D.A. (1985). J. Neurophysiol., 54: 782.
[10] J. McLean & L.A. Palmer. (1996) Invest. Opthalmol. & Vis. Sci. Suppl. 37(3): 2197.
[11] J.A. Movshon & P. Lennie (1979). Nature 278: 850-852.
[12] S.B. Nelson (1991). J. Neurosci. 11: 344-56.
[13] S.B. Nelson, J.A. Varela, K. Sen, & L.F. Abbott (1996). CNS96 Proceedings, Submitted.
[14] I. Ohzawa, G. Sclar, & R.D. Freeman (1985). J. Neurophysiol. 54: 651-667.
[15] M.W. Pettet & C.D. Gilbert (1992). Proc. Natl. Acad. Sci (USA) 89:8366-8370.
[16] G. Sclar, I. Ohzawa, & R.D. Freeman (1985). J. Neurophysiol. 54: 666-673.
[17] D.C. Somers, S.B. Nelson, & M. Sur (1995) J. Neurosci. 15: 5448-5465.
[18] D.C. Somers, E.V. Todorov, A.G. Siapas, & M. Sur (1996) CNS96 Proceedings, this volume.
[19] T.R. Vidyasagar (1990). Neuroscience 36: 175-179.
[20] Worgotter, F. & Koch, C. (1991). J. Neurosci. 11:1959.

84

AN ENGINEERING PRINCIPLE USED BY MOTHER NATURE: USE OF FEEDBACK FOR ROBUST COLUMNAR DEVELOPMENT

K. P. Unnikrishnan[1,2] and H. S. Nine[2]

[1]Computation and Neural Systems, 139-74
California Institute of Technology
Pasadena, CA
E-mail: `unni@hope.caltech.edu`
[2]Computer Science, 480-106-285
GM Research Labs
Warren, MI
E-mail: `hnine@nine.agn.net`

ABSTRACT

In spite of the inherent variability in parameters that govern development, the columnar structures in mammalian sensory systems are surprisingly robust. For example, the average width of ocular dominance columns in cats is about 400 μm, with very little variability from animal to animal. In engineering, the effect of appropriate feedback on stable, dynamical systems is to make them robust with respect to noise, including parameter variations. The question we ask (and answer) in this paper is, if during neural development, mother nature is cleverly using this engineering principle. Through computer simulations of a biologically plausible model we demonstrate that, during the development of ocular dominance columns, this is indeed the case. Transient neuron populations such as the subplate may play a major role in the initial formation of these feedback circuits. For cleverly using feedback, mother nature also gets a bonus: the synaptic computations in the circuits are completely local and hence independent of the time constants associated with the dendritic arbors of post-synaptic (layer 4) neurons that may still be growing!

1. INTRODUCTION

Columnar structures are ubiquitous features of mammalian sensory systems. These structures allow relevant features to be naturally represented along common dimensions. For

Table 1. Constancy (robustness) in ocular dominance column width in spite of the large variability in thalamic and cortical parameters. The labels within parenthesis are those used in the Models section. Data is for cats.

Parameter	Range (µm)	Source
Local correlation in LGN activity (R_C)	225 - 1500	[28]
Geniculo-cortical arbors X:	450 - 750	[33]
(R_G) Y:	800 - 1000	[33]
Excitatory intra-cortical interactions (R_E)	350 - 530	[34]
Inhibitory intra-cortical interactions (R_I)	570 - 750	[34]
Ocular dominance column width	450 ± 40	[35]
(ODC width)	400 ± 10	[36]

perceptual consistency, these structures in different members of the same species need to be very similar. If for example, the average width of columns in the visual pathway were very different in two cats, a (white) rabbit may look the size of a mouse to one and the size of a polar bear to the other. On the other hand, the parameters that govern the development can not be identical across all members of a species. Even in individual members, there is randomness in the processes and the parameters may vary during the course of development. So the question remains, in spite of inherent biological variability, how does mother nature achieve such robust (consistent) development?

In this paper we take up this question in relation to the development of ocular dominance columns (ODCs) in the mammalian visual system. Table 1 shows the variability in some of the parameters that influence the development and the constancy (robustness) in the average width of columns in spite of this.

One feature of the mammalian visual system responsible for this robustness may be the feedback pathways present during development. Feedback is another ubiquitous feature of mammalian sensory systems. In the adult visual system most of the cortical areas have reciprocal connections. The relay neurons in the Lateral Geniculate Nucleus (LGN) receive about 20 times as many synapses from cortical feedback projections as from retinal inputs [1]. In the primary visual cortex (V1), only about 5% of the the input to the layer 4 neurons come from the LGN [2]; the rest coming from feedback and lateral interactions. Most of these feedback pathways are in place and operational before the development of ocular dominance columns. The question then becomes: how do they contribute to robust ODC development?

2.   DEVELOPMENT OF APPROPRIATE CIRCUITS

Figure 1 gives the calendar of events for the development of LGN; layers 4, 6 and subplate* of the primary visual cortex.

The LGN neurons are born between embryonic days 20 and 30 (E20 - E30). By embryonic day 40 (E40), the geniculo-cortical axons grow into the subplate (SP) and wait there for about two weeks. By E55, these axons have invaded the cortical plate and by E65 (also post-natal day 0, P0), they have extensive branching in layer 4.

---

*Subplate is a transient neuron population that forms below the developing cortical plate, adjacent to layer 6. Its *structural* role in the development of connections between the thalamus and cortex has been the subject of many articles. (See [3] for a review.) In the conclusion section, we propose a *functional* role for subplate and similar transient neuron populations.

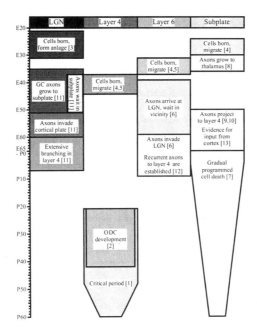

Figure 1. Developmental time-table of the cat visual system. *GC axons*, Geniculo-cortical axons; *Enn*, embryonic day *nn*; *Pnn*, post-natal day *nn*. Published sources for data: [1] Hubel and Wiesel (1970); [2] LeVay et al. (1978); [3] Shatz (1983); [4] Luskin and Shatz (1985); [5] Shatz and Luskin (1986); [6] McConnell and Shatz (1988); [7] Shatz et al., (1988); [8] McConnel et al., (1989); [9] Friauf et al. (1990); [10] Friauf and Shatz (1991); [11] Ghosh and Shatz (1992); [12] Larry Katz (private communication); [13] Friauf et al. (1990).

The layer 4 neurons have already migrated to their appropriate positions and matured by this time. Between P5 and P8, they send transient axonal projections to the subplate. By P10, they send axonal projections to layers 5 and 6 [4]. These projections mature by P15. The more numerous axonal projections are to layers 2+3. These start around P10 and mature by P30.

The layer 6 neurons migrate to their positions by E40, and by P10 their axons have invaded LGN. Around P0, they also form transient axonal projections to the subplate. The axonal projections to layer 4 start around P0 and the density of axonal arbors in layer 4 increase dramatically till around P20 [5, 6]. The apical dendrites of layer 6 pyramidal cells have extensive branching in layer 4, where they could get inputs from stellate cells and even LGN axons.

The subplate neurons are born and functional by E30, and by E40 a few (about 10%) of their axons grow to LGN. By E60, they have fairly extensive axonal projections to layer 4. Around this time, some of their dendritic arbors are in the nearby layer 6. After P10, subplate undergoes a gradual, programmed cell death. The end of critical period (P60) coincides with the elimination of subplate.

The neurons of layers 5 and 2+3 mature by birth. Hence by P20, when the ocular dominance columns begin to form, there are a couple of feedback loops that are functioning.

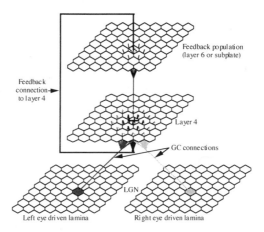

Figure 2. Schematic of the model. *GC*, Geniculo-cortical; Layer 4 to Layer 6 (or Subplate) connections may be via other layers (for example 2+3 and 5). The feedback layer (*FDBK*) receives spatially integrated layer 4 activity. *GC* connection strengths are the only parameters that change during the formation of ocular dominance columns are the GC connection strengths.

3.   COMPUTATIONAL MODEL

To simulate parts of the circuitry that exists at P20, and to investigate the role of feedback during the development of ocular dominance columns, we designed the network, shown schematically in Figure 2.   LGN is represented by two 96x96 arrays of neurons, arranged on a hexagonal grid. These neurons are linear and are driven by locally-correlated retinal activity in the interval [0,1].[*]   The extent of local correlations are controlled by the local correlation radius, R_C. Neuronal activities between the two LGN laminae are usually uncorrelated. Layer 4 is represented by another 96x96 hexagonal array of neurons. The LGN to layer 4 connectivity is controlled through the geniculo- cortical arbor radius R_G and the intra-cortical interactions in layer 4 are implemented using 'Mexican-hat' type functions. The excitatory radius R_E and the inhibitory radius R_I characterize these interactions. The activity $V_j(t)$ of the $j^t h$ layer 4 neuron at time t is given by the equation

$$V_j(t) = \sum_i G_{ij}(t) \cdot L_i(t) + \sum_k M_{kj} \cdot V_k(t-1) \tag{1}$$

where G_{ij} is the thalamo-cortical synaptic strength between the $i^t h$ LGN neuron and the $j^t h$ layer 4 neuron, M_{kj} represents the strength of interaction between the $j^t h$ and $k^t h$ layer 4 neurons, and L_i is the activity of the $i^t h$ LGN neuron.[†]

The feedback neuronal population, consisting of the subplate and layer 6, is represented by a third 96x96 hexagonal array. These neurons receive spatially integrated activity of layer

---

[*]A note on array size and shape: During simulations, we use periodic boundary conditions. For the range of parameters we have used in simulations, an array size of 96x96 is large enough to avoid wrap-around artifacts. In addition, this array size is large enough to contain several columns. Locally correlated retinal activity is explicitly generated. On a square array, using Euclidian distances, this is not possible, while on a hexagonal array it can be done (see [7] for details). A hexagonal array also simulates the close-packing of retinal ganglion cells.

[†]Sometimes, a non-linear version of eqn. (1), with RMS values of the first term $(G \cdot L)$, is used to speed up simulations; the final results are very similar. The Mexican hat interaction functions M are designed so that the ratio of net excitation to net inhibition is 2:1.

4 neurons and the extent of this spatial integration is determined by the feedback "dendritic" arbor radius R_{FD}. The activity of the lth feedback neuron at time t is given by the equation

$$F_l(t) = \sum_j Q_{jl} \cdot V_j(t) \qquad (2)$$

where the connections Q_{jl} have a fixed Gaussian profile. The feedback neurons project to the dendritic arbors of layer 4 neurons and their activity F is used for modifying the thalamo-cortical synaptic strengths. The extent of these feedback projections are controlled by the feedback "axonal" arbor radius R_{FA}. For computational simplicity, $R_F D$ and $R_F A$ are combined to a single parameter R_F. *

3.1. Algorithm for synaptic modification (Alopex)

The Alopex algorithm [13] is used to modify the thalamo-cortical synaptic strengths. The change in G_{ij} at time t is given by the equation

$$\Delta G_{ij}(t) = \lambda \cdot \Delta L_i(t) \cdot \sum_l \Delta F_{lj}(t), \qquad (3)$$

where

$$\Delta L_i(t) = L_i(t-1) - L_i(t-2)$$
$$\Delta F_{lj}(t) = F_{lj}(t-1) - F_{lj}(t-2)$$

In the equations above, ΔL_i is the change in activity of the ith LGN neuron and ΔF_{lj} is the change in activity of the lth feedback neuron that projects back onto the jth layer 4 neuron, and λ is a constant. The summation is over all the feedback neurons that project onto the jth layer 4 neuron.[†]

It is useful to contrast the Alopex algorithm with the commonly used version of the Hebb rule. Hebb uses correlations between pre- and post-synaptic activities for modification of synaptic strengths. Alopex uses correlations between *changes* in pre- and post-synaptic activities, where the 'post-synaptic activity' could be due to inputs from feedback neurons. Alopex is better suited for simulating structures with feedback than the Hebb rule.[‡] Computer simulations using detailed models of LGN and cortical neurons have shown the biological plausibility of Alopex [15, 16]. Due to space limitations it will not be presented here. Changes could also be reflected in the neuronal activities themselves. When layer 5 neurons or hippocampal pyramidal neurons are stimulated at distal dendritic sites, their responses appear as derivatives in the soma [17, 18]. In these circuits, even though the individual synapses may be computing Hebb-type correlations, the overall network will be computing Alopex-type correlations.

---

[*] In designing the above network, we have made some simplifications and assumptions. We assume that the feedback F is weak compared to L and V so that it is omitted from eqn. (1). The strong projection from layer 6 and a weak projection from subplate to the LGN are not modeled. Previous results on hierarchical networks [8, 9, 10, 11, 12] suggest that these feedbacks would help convergence rate of our model, without affecting the final outcome.

[†] In most of the simulations, the total synaptic strength on an individual layer 4 neuron is constrained to be $\leq$ a constant J.

[‡] If we use the Hebb rule instead of Alopex, the system shows very weak converge within a very narrow range of parameters. In most cases the synapses saturate without producing any useful columns. See [7, 14] for details.

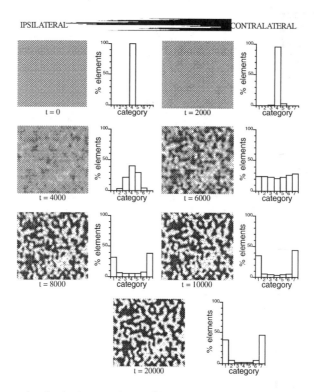

Figure 3. Development of ocular dominance columns of layer 4 neurons. We also show "Hubel and Wiesel" type plots on the side of each picture to show segregation (1 - completely dominated by one eye; 7 - completely dominated by the other eye). The average column width at 20,000 iterations is 410 μm.

4. RESULTS

Figure 3 shows the development of ocular dominance columns in the model network. The qualitative shape and the average column width[§] in the final, fully developed patterns match very well the columns seen in normal cats. We emphasize that when biologically realistic parameter values are used, the model renders columns of appropriate width.

In subsequent computer simulations, we varied the parameters over ranges found in nature (see Table 1). Table 2 shows a summary of our simulation results. In the first four rows, the parameters R_C, R_G, R_E, and R_I were varied over a very wide range. The average column width is clearly insensitive to variations in these parameters. It depends almost linearly on R_F.

Our model can also replicate the effects of monocular deprivation, reversed monocular deprivation and strabismus. The model also scales well: When different cortical distances (50 - 100 μ m) are assigned to individual pixels, columns of appropriate width form in the model. The average column width is also insensitive to the exact normalization algorithm used to keep the synaptic weights within bounds. These can not be described in detail here due to space limitations and will be taken up in a subsequent paper [14]. See also [7]. We leave the

---

[§]Since the columns are not oriented along any preferred direction, a 2D fourier transform of the final pattern does not give an accurate column width. We have used a method that calculates the width at each point in an orthogonal direction to the column boundary. Details of this method will be provided elsewhere [14].

Table 2. Summary of parametric study. The model parameter, indicated in the first column is varied over the indicated range, holding all other parameters at their base value. The average and deviation in the width of ODCs are given in subsequent columns

Parameter	Range, μm	Column width, μm							
		$R_C=0$	$R_C=200$	$R_C=400$	$R_C=600$	$R_C=800$	$R_C=1000$	$R_C=1200$	$R_C=1400$
R_C	0–1400	*	*	380	410	410	410	410	450
				±150	±210	±180	±180	±190	±220

		$R_G=0$	$R_G=200$	$R_G=400$	$R_G=600$	$R_G=800$	$R_G=1000$
R_G	0–1000	390	400	400	410	400	360
		±110	±120	±120	±180	±180	±90

		$R_E=250$	$R_E=350$	$R_E=450$	$R_E=550$	$R_E=650$
R_E	250–650	590	450	410	440	460
		±490	±310	±180	±150	±150

		$R_I=600$	$R_I=800$	$R_I=1000$	$R_I=1200$
R_I	600–1200	430	410	440	410
		±220	±180	±220	±180

		$R_F=100$	$R_F=200$	$R_F=300$	$R_F=400$
R_F	100–400	140	410	670	960
		±50	±150	±320	±600

reader with the following figure that illustrates the robustness with respect to initial starting conditions.

5.   DISCUSSION

Through computer simulations, we have previously investigated the role of feedback pathways in cognitive, attentional, and perceptual tasks [8, 9, 10, 11, 12]. Recently there has been quite a bit of theoretical and experimental activity in this direction [20, 21, 22, 23, 24, 25]. But to our knowledge, no one has seriously looked at the role of feedback during development. The modeling efforts reported in this paper grew out of our belief that feedback may be playing a critical role during the development of sensory systems. Simulations of the model described above suggest that this is indeed the case. In engineering systems, the effect of appropriate feedback on stable systems is to make them robust with respect to noise [26, 27]. The robustness of our model shows that in the developing visual system, feedback plays a very similar role.

In our opinion, many of the feed-forward models also incorrectly use the Hebb-rule. For modifying synaptic strengths, these models correlate pre-synaptic activity with some form of the post-soma potential (see [28] for example). For this to be biophysically plausible, the dendritic arbors of 'plastic' neurons (in this model, the layer 4 neurons) need to be isopotential over temporal windows comparable to the time-constants of NMDA receptors ($\sim 200\ ms$). The only detailed study that sheds light on this question shows that cells, especially those with long dendrites, are never isopotential [29]. In our model, since the post-synaptic

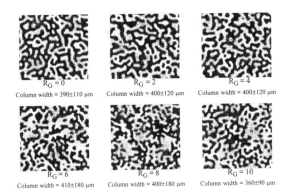

Figure 4. Robustness wrt the geniculo-cortical arbor radii. All other parameters held constant. The value of R_G and average column widths are given under each figure.

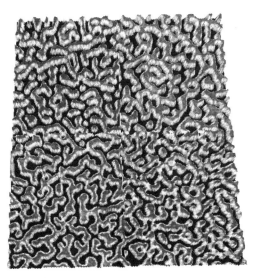

Figure 5. Robustness wrt initial starting conditions. A, B, C, and D show the columns (after 20,000 iterations) for four runs with different initial GC synaptic weights. Parameter values are the average ones from Table 2. Identical LGN input sequences were used in all four simulations.

depolarizations are provided by the feedback signals, the computations are completely local and independent of dendritic time-constants. This may be another critical role of feedback pathways. During development, the neurons are still growing and any mechanism that depends on the characteristics of these growing neurons is bound to be disastrous.

Our final comment is regarding the *functional* role of transient neuronal populations during development. This is still considered a mystery (see [30] for a discussion). Our simulation results suggest that the functional role of these neuronal populations may be to provide locally integrated neuronal activity from 'higher' centers to sites of synaptic plasticity so that (i) the computations for synaptic modification can be completely local, independent of dendritic time constants and (ii) the developmental process can be robust wrt noise and parameter variations.

Finally, there are numerous experimental predictions that arise from our model and we briefly mention a few. Our computer simulations predict that local inactivation of subplate or layer 6 will disrupt the feedback pathway and will result in the local disruption of ocular dominance columns. Recent results of Ghosh and Shatz essentially show this [31]. The feedback parameter R_F in our model has contributions from the (polysynaptic) layer 4 to layer 6 pathway, the recurrent (monosynaptic) layer 6 to layer 4 pathway and any similar circuitry that may transiently be completed via the subplate. Our simulations predict that (pharmacologically) alterations in any of the above (and only those of the above) would alter the width of ocular dominance columns. Our simulations also show that, due to feedback in the system, development of ocular dominance columns is *not* critically dependent on the correlation in LGN activity. Hence alterations at the retinal or LGN level that change this parameter will *not* alter the average width of ocular dominance columns.*

REFERENCES

[1] J. A. Robson, *J. Comp Neurol.*, 216, 89, 1983.
[2] A. Peters and B. R. Payne, *Cereb. Cortex*, 3, 69, 1993.
[3] K. L. Allendoerfer and C. J. Shatz, *Ann Rev Neurosci*, 17, 185, 1994.
[4] G. Meyer and R. Ferres-Torres, *J. Comp. Neur.*, 228, 226, 1984.
[5] E. M. Callaway and L. C. Katz, *J. Neurosci.*, 12, 570, 1992.
[6] E. M. Callawy, J. L. Lieber, and K.A. Reese *Preprint*, 1996.
[7] H.S. Nine, *PhD thesis, Univ. Michigan*, 1995.
[8] E. Harth and K. P. Unnikrishnan, *Intl. J. Psychophysiol.*, 3, 101, 1985.
[9] E. Harth, K.P. Unnikrishnan, and A.S. Pandya, *Science*, 237, 187, 1987.
[10] E. Harth, A. S. Pandya, and K. P. Unnikrishnan, *Conc. Neurosci.*, 1, 53, 1990.
[11] J. Janakiraman, and K.P. Unnikrishnan, *Proceedings of CNS93*, 215, 1993.
[12] P.S. Sastry, S. Singh, and K.P. Unnikrishnan, *Submitted*, 1996.
[13] K.P. Unnikrishnan, and K.P. Venugopal, *Neural Comp.*, 6, 469, 1994.
[14] K.P. Unnikrishnan, and H.S. Nine, *Submitted*, 1996.
[15] N.S. Sekar and K.P. Unnikrishnan, *Abstr. Lrn. Mem. Mtg, CSH Lab.*, 50, 1992.
[16] K.P. Unnikrishnan, and N.S. Sekar, *Soc. Neurosci. Abstr.* 19, 241, 1993.
[17] R. Yuste, M.J. Gutnick, D. Saar, K. Delayne, and D.W. Tank, *Neuron*, 13, 23, 1994.
[18] N. Spruston, Y. Schiller, G. Stuart, and B. Sakmann, *Science*, 268, 297, 1995.
[19] P.A. Anderson, J. Olavarria, and R.C. Vansluyters, *J. Neurosci.*, 8, 2183, 1988.
[20] D. Mumford, in: M.A. Arbib, ed., *The handbook of brain theory and neural networks* (MIT Press), 1995.
[21] C. Koch, *Neuroscience*, 23, 399, 1987.
[22] P. Sadja and L.H. Finkel, *J. Cog. Neurosci.* 7, 267, 1995.
[23] S. Grossberg, *Am. Scientist.*, 83, 438, 1995.

---

*Recent results from ferrets show that the waves of activity seen in the developing retina disappear before the formation of ocular dominance columns [32]. The small amount of correlation needed in our model can easily be generated by spontaneous activities of LGN neurons.

[24] A. M. Sillito, et al., *Nature*, 369, 479, 1994.

[25] J. Yan and N. Suga, *Science*, 273, 1100, 1996.

[26] K. S. Narendra and A. M. Annaswamy, *Stable Adaptive Systems* (Prentice Hall), 1989.

[27] B. C. Kuo, *Automatic control systems* (Prentice Hall), pp. 6-11, 1987.

[28] K.D. Miller, J.B. Keller, and M.P. Stryker, *Science*, 245, 605, 1989.

[29] T.H. Brown, et al., in *Single Neuron Computation*, T. McKenna, J. Davis, S. F. Zornetzer, eds., (Academic Press), 1992.

[30] C. J. Shatz et. al., *CSH Symp. Quan. Biol.*, 40, 269, 1990.

[31] A. Ghosh and C.J. Shatz, *Science*, 255, 1441, 1992.

[32] R.O.L. Wong, M. Meister, and C.J. Shatz, *Neuron*, 11, 923, 1993.

[33] A.L. Humphrey, et al., *J Comp Neur*, 233, 159, 1985.

[34] Y. Hata, et al., *J Neurophysiol*, 69, 40, 1993.

[35] S. Lowel, *J Neurosci*, 14, 7451, 1994.

[36] Y.C. Diao, et al., *Exp Brain Res*, 79, 271, 1990.

85

LOCALIZED NEURAL NETWORK CONTROL OF SPRING ACTUATED LEG

Tom Wadden* and Örjan Ekeberg

SANS–Studies of Artificial Neural Networks
NADA–Department of Numerical Analysis and Computing Science
Royal Institute of Technology, S-100 44
Stockholm, Sweden
E-mail: tomw@nada.kth.se, orjan@nada.kth.se

INTRODUCTION

For any animal to move about in an unstructured and dynamic environment it must have an adaptable locomotor control system. In legged vertebrates much of this control is handled by local circuits in the spinal cord [3]. The object of a neural controller is to lead the mechanical system through the proper motion pattern. This system must strive to be energy efficient, a task which is enabled through the use of springs in the form of muscle tendons [1]. Our simulated model incorporates local circuitry, series elasticity, sensory feedback and higher level control.

As a first step towards dynamically stable quadruped locomotion we have designed and simulated a single legged system, comprising of both neural and mechanical components. The combined neuro-mechanical system produces stable stepping over a range of frequencies. The leg was also able to cope with landing from different heights and under varying loads, an important property for the quadruped system presently under development.

NEURAL NETWORK

A non-spiking neuron model is used as previously reported in Ekeberg (1993). Each neuron represents a population of functionally similar neurons and its output is representative of the mean firing frequency of the population. Each neuron acts as a "leaky integrator" with a saturating transfer function.

On the basis of CPG theory and with experience from modeling such local circuitry in the lamprey, a primitive eel-like vertebrate, we have designed a modular one-legged neural

---

*Author of correspondence

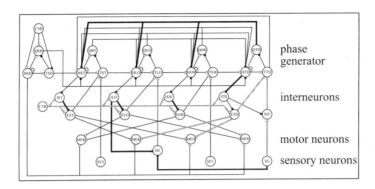

Figure 1. The complete neuronal network responsible for generating and adapting locomotor movements during the stepping cycle.

controller (figure 1). This network is composed of four locomotory phases, (where the phases correspond to stance, lift-off, swing and touchdown), a sensory module, a motor module and a separate module for stable standing. The four phase system is capable of operating independent of the other systems, through which it interacts via interneurons and a higher level control neuron.

Although local circuitry is capable of producing basic movement patterns, signals from the brain and periphery play a significant role in the timing and strengths of these movements. Speeding up, slowing down, and walking over small obstacles are just some of the movements we often coordinate without actually thinking about it. With this in mind we wanted to see what affects we could make on the local system by the addition of higher order control neurons and sensory feedback. The idea was to regulate the frequency of the local phase generator with a single control neuron CTR, which would exert its influence via a group of interneurons. The same interneurons relay phase information to the four motor neurons, i.e. hip and knee, flexors and extensors.

The sensory system consists of three sensory neurons. Two of which give feedback on hip position, i.e. one for hip extension, and one for hip flexion. The third sensory neuron is active when the leg is in contact with the ground. It signals when it's time to transfer activation from the lift-off/touchdown to the swing/stance phase.

Apart from phase transitions during the stepping cycle, a network controller should be able to switch between stepping and standstill. We have therefore added an extra module to our stepping generator which terminates stepping in favor of stable standing.

MUSCULO-SKELETAL SYSTEM

The system consists of a body, thigh and shank. As this study focuses on the single legs contribution to quadruped locomotion, we substitute the other three legs with a support. Single legged balance control was not within the scope of this study.

There are two antagonistic muscle pairs in the leg model. These are knee flexor and extensor, hip flexor and extensor. Each muscle is modeled as a spring and damper system, with muscle activation directly controlled by motor neuronal output. We assume a linear relationship between motoneuronal activity and the muscular spring constant [4, cf.]. It should

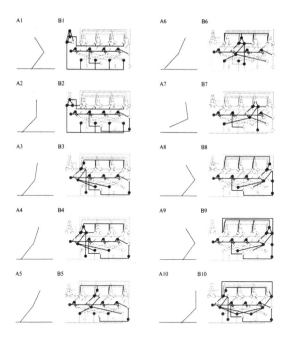

Figure 2. Step cycle progression through one phase starting with the standing phase. The position of the leg is shown in frames A1–10, with the corresponding neural activity seen in frames B1–10.

be noted that this arrangement makes it possible for the neural network not only to control the static torque, but also the stiffness of each joint.

RESULTS

Simulations were performed on three levels starting with the isolated nervous system, then the isolated mechanical system and finally the combined neuro-mechanical system.

We isolate the nervous system by decoupling it from the mechanical system and inhibiting sensory feedback. Stepwise increases in the input to this system from the control neuron resulted in successive increases in neural network frequency in the range 0.4 to 7.0 Hz.

To find the eigenfrequency of the mechanical system we removed the neural controller. The leg was suspended, so that it could not contact the ground, and then the hip was extended and the knee slightly flexed. While the hip was released to swing freely its' flexor and extensor muscles were given an equal amount of tonic excitatory drive. Stepwise increases in drive resulted in a linear increase in eigenfrequency from 1.8 to 8.8 Hz. Coupling the neural and mechanical systems resulted in a rhythmic and stable stepping pattern. Figure 2 shows a sequence of one complete phase of the step cycle. Frames A1–10 represent the leg movement while B1–10 show the corresponding neural patterns of activity. The leg starts off in standstill which is the most stable configuration. From standing the neural system proceeds into the stance phase (B3–4). Stepping in the stance phase continues until the four-phase generator switches phase. This phase time is shortened if the sensory neuron signals that the thigh nears the extreme rear position. Similar mechanisms result in a transfer from lift-off (B4–5) to swing (B7) to touchdown (B8–9), and then stance again (B10). The speed of this process is dependent upon the excitatory drive from the control neuron and gives a fourfold range of

A

B   C

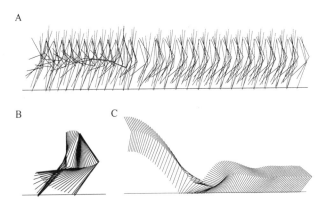

Figure 3. Stick figure representation of leg locomotion. **A**: A 10 second simulation of fast walking started from time 0 (trace every 0.05 seconds). Starting directly with a high control neuron input effects the initial stages of walking. **B**: Single step cycle for pattern **A** above and for figure 2. **C**: Leg landing with extra load (note: no motion in x direction, bias given to x position of consecutive traces in order to facilitate viewing).

stepping frequencies. Figure 3**a** shows fast walking over a ten second period and figure 3**b** a single step cycle pattern.

We next tested how the leg handles landing from different heights under varying loads. The leg landed successfully from heights greater than its own length and with twice its normal body weight (figure 3**c**). This is important in the quadruped as each leg must be able to provide support under different loads as well as during impact on landings.

CONCLUSION

We have presented a localized neural network for generating stepping movements in a single leg. The combined neuro-mechanical system works in synergy to produce a range of stepping frequencies. The spring actuated system also enables landing under varying conditions. This neural controller is modular so that it can be used in a multi-legged system and controlled from higher level networks.

REFERENCES

[1] ALEXANDER, R. M. (1988). *Elastic mechanisms in animal movement*. Cambridge: Cambridge University Press.
[2] EKEBERG, Ö. (1993). A combined neuronal and mechanical model of fish swimming. *Biol. Cybern.* **69**: 363–374.
[3] GRILLNER, S. (1996). Neural networks for vertebrate locomotion. *Scientific American* **274**: 64–69.
[4] TAX, A. A. M. AND DENIER VAN DER GON, J. J. (1991). A model for neural control of gradation of muscle force. *Biol. Cybern.* **65**: 227–234.

86

BURSTING AND OSCILLATIONS IN A BIOPHYSICAL MODEL OF HIPPOCAMPAL REGION CA3: IMPLICATIONS FOR ASSOCIATIVE MEMORY AND EPILEPTIFORM ACTIVITY

Gene V. Wallenstein and Michael E. Hasselmo

Department of Psychology and Program in Neuroscience
Harvard University
Cambridge, MA

ABSTRACT

A detailed biophysical model of hippocampal region CA3 was used to show how septal cholinergic and GABAergic modulation, through three distinct mechanisms, interacts with intrinsic and synaptic conductances to influence population behavior. A dissection of each mechanism demonstrates a continuum of population firing activity ranging from fully-synchronized behavior to a mixture of repetitive bursting and oscillations in reduced subsets of neurons, ideal for forming accurate associations during a learning and recall task. Rhythmic modulation (GABAergic) is shown to play a role in the formation of place fields and the model's capacity for learning sequence information. Such modulation is also shown to account for the phase precession of place units relative to the theta rhythm as an organism passes through the associated place field of the cell.

1. INTRODUCTION

There has been considerable theoretical investigation into the role of hippocampal region CA3 in associative memory [10, 13, 15]. In addition to a long history of behavioral observations relating hippocampal function to declarative memory (see 21, for review), the CA3 region, in particular, has two properties which make it especially attractive for the study of associative memory in general. First, it contains extensive recurrent excitatory synapses between pyramidal cells [1] and second these synapses are, in part, mediated by N-methyl-D-aspartate

(NMDA) receptor-gated channels [17] which have been implicated in long-term potentiation (LTP). The CA3 region has also been shown to be capable of exhibiting epileptiform-like activity where synchronous bursting in pyramidal cells occurs throughout a spatially extended portion of tissue [23]. Considering these observations, how is a stable balance achieved in the hippocampus between conditions which foster accurate associative memory and those that may lead to epileptiform events? Below, we show through a series of simulations, that transitions between these two markedly different functional states can occur with changes in a single parameter intrinsic to pyramidal cells.

2. METHODS

The model consisted of 500 pyramidal cells and 100 inhibitory interneurons. Each pyramidal cell was a reduced Traub model [24, 25], which consisted of a fast sodium current ($I_{Na(fast)}$), a delayed rectifier ($I_{K(DR)}$), a high-threshold calcium current (I_{Ca}), two calcium-dependent potassium currents ($I_{K(AHP)}$ and $I_{K(Ca)}$), a transient potassium current ($I_{K(A)}$) and a potassium leak current ($I_{K(leak)}$). Each of these currents were located at the soma and proximal dendrites, while the distal dendrites contained I_{Ca}, $I_{K(AHP)}$, $I_{K(Ca)}$, and $I_{K(leak)}$. Calcium buffering was performed in each compartment using a first-order diffusion process [25]. Each interneuron consisted of $I_{Na(fast)}$ and $I_{K(DR)}$ located at the soma, with the four remaining compartments (2 basal and 2 apical) being passive.

Recurrent excitatory synapses located at the apical dendrites of pyramidal cells included Hebbian modification of NMDA receptor-mediated conductances in conjunction with a voltage-dependent Mg^{2+} block of these channels based on previous modeling [27]. The network also included recurrent inhibitory synapses (GABA-A and GABA-B) at the soma and proximal dendrites of interneurons. Feedforward inhibition of pyramidal cells occurred via GABA-A receptors situated at the soma and proximal dendrites, while slower, GABA-B receptor-mediated inhibition was located at distal dendrites [6]. Feedforward excitation occurred through NMDA and AMPA receptors located at the soma of interneurons. Both cholinergic [9, 14] and GABAergic [8] projections from the medial septum were also included in the model.

3. RESULTS

3.1. Transitions Between Epileptic and Learning States

To test the manner in which ACh affects associative memory in region CA3 a simple pattern of afferent input was presented to the model. Figure 1 shows the membrane potential measured at the soma of a pyramidal cell in response to afferent input. The two ACh effects of interest, suppression of excitatory recurrent synapses and suppression of the adaptation (AHP) current, were manipulated during the learning period and restored to baseline during retrieval. With the maximal conductances underlying $I_{K(AHP)}$ reduced to 30 % of the original value, the model exhibited a rapid recruitment of pyramidal cells into globally-synchronous bursting at approximately 0.5 - 2 Hz. This behavior was found to result from changes in the bursting refractory period of pyramidal cells. With a reduction in the AHP, more pyramidal cells were available for EPSP-induced bursting until a critical number of cells became entrained and the

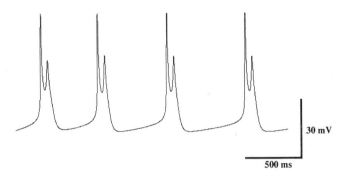

30 mV

500 ms

Figure 1. The membrane potential measured at the soma of a single pyramidal cell in response to afferent input. Each input pattern was delivered simultaneously to 50 different pyramidal cells in the network.

entire network quickly synchronized. Population activity of this sort resembles epileptiform-like behavior observed in hippocampal slice preparations [23].

This result demonstrated that an additional biophysical mechanism was needed to constrain the spread of excitatory bursting between pyramidal cells. A natural choice for such a mechanism was the known cholinergic suppression of recurrent excitatory synapses in stratum radiatum of region CA3 [10]. This effect was included in the model by decreasing the maximum conductance underlying the pyramidal cell synaptic current I_{AMPA} to 42 % of its normal value (1.5 nS), simulating a 38 % decrement in unitary EPSP height under 20 micromolar carbachol [10]. During learning, this resulted in the attenuation of both globally- and partially-synchronous pyramidal cell bursting in neurons not related to the input pattern, thus promoting accurate recall performance.

These results demonstrate the regulatory property of cholinergic modulation in CA3 as illustrated in the $g_{K(AHP)}$ - g_{AMPA} parameter space representation shown in Figure 2. By adjusting the maximum conductances underlying these two currents, a diverse variety of population behaviors emerged with substantially different functional implications associated with each.

3.2. Recalling Sequence Information and the Phase Precession Effect

Recent electrophysiological experiments have shown that the GABA-B receptor agonist baclofen selectively decreases the amplitude of excitatory and inhibitory post-synaptic potentials (PSPs) in hippocampal CA1 pyramidal cells induced by Schaffer collateral stimulation but has a much weaker effect on perforant path transmission [2, 4, 12; also see 11]. The suppression of excitatory and inhibitory synaptic transmission by GABA-B receptor activation is an interesting effect in the hippocampus because it is possible that such modulation occurs in a rhythmical fashion during theta activity. In the present model, total (across the entire network) GABA-B receptor-mediated conductance does, indeed, rise and fall with theta oscillations with highest values in the early to middle portion of each cycle that decay steadily until the beginning of the next cycle.

A completion task was employed to determine if rhythmical GABAergic modulation might affect the ability of the model to learn sequence information. After a learning period in which a series of patterns was presented to the network, a subset of a full sequence was shown to test if the model could complete the remaining sequence. During the learning period, it became apparent that many of the cells which fired independently of the input pattern for a

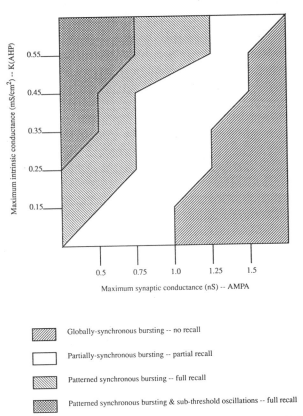

Figure 2. The $g_{K(AHP)}$ - g_{AMPA} parameter space: Qualitatively different modes of population firing activity were obtained depending on the values for the maximum conductances underlying $I_{K(AHP)}$ and I_{AMPA}.

brief period of time (ie. during a single pattern in the sequence) in the initial stages of learning began to fire continuously over a portion of the full sequence pattern after several exposures. Thus, with repeated exposure to the full sequence, these cells learned to respond to particular subsequences. This behavior is somewhat analogous to the development of place-specific firing of CA1 and CA3 neurons observed during spatial learning tasks [13, 16, 18].

Sequence recall was tested by placing the simulated rat at a random location on the track and determining if the remaining sequence (navigational path) was completed. Because GABA-B receptor-mediated effects modulated pyramidal cell firing with the theta rhythm, cells directly related to the sequence pattern as well as those exhibiting place cell behavior tended to preferentially fire toward the latter portion of each theta cycle. It has been observed recently in hippocampal recordings that as a rat enters a place field, activity in the associated cell typically begins at the same point in a theta cycle, and advances toward earlier phases as the rat exits the field [19, 20]. In our model, the initial cell firing was always dominated by afferent activity related to the present location because intrinsic fibers at recurrent collaterals were suppressed early in each theta cycle. Information about future locations became available later in each cycle once this suppression was attenuated. At the start of the next theta cycle, the present location of the rat was again updated by a dominant afferent activity pattern and the scenario repeated. This typically resulted in a place cell firing late in a theta cycle as the rat initially entered that cell's field. Several locations were then rapidly recalled late in a theta cycle with sustained place cell firing across a contiguous segment of locations. As the

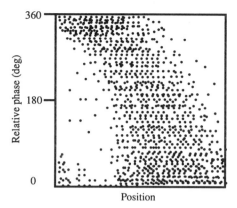

Figure 3. A representative firing pattern from a single place cell as the simulated rat passed through the place field of the cell. Note the systematic advance from firing late in a theta cycle when the rat first entered the field, to firing earlier in the cycle as the field is exited.

rat exited a place field, activity of the associated place cell was then usually shunted during the early portion of the next theta cycle due to GABA-B receptor-mediated suppression of EPSPs at recurrent collaterals. Thus, the phase of the theta cycle at which a place cell fired, systematically advanced as the rat passed through the cell's place field as shown in Figure 3.

4. DISCUSSION

The two predominant sources of neuromodulation from the medial septum to the hippocampus are cholinergic and GABAergic. Considering that the cholinergic effects on pyramidal cell activity are believed to develop slowly [7], and operate on a time scale greater than a theta cycle, it is unlikely that such modulation shapes population activity in a periodic manner. GABAergic modulation, on the other hand, may indeed influence hippocampal pyramidal cell activity on a time scale consistent with the theta rhythm. This is because in addition to the rhythmic septal influence, CA3 interneuron firing time-locked to theta oscillations [26], provides a local source of GABA and the time constants governing the synaptic kinetics for both GABA-A and GABA-B receptor activation fall within the approximate period of a typical theta cycle.

We have shown that dynamical transitions between different states of collective activity are governed primarily by the two cholinergically modulated mechanisms of controlling the AHP intrinsic to pyramidal cells and the degree of excitatory synaptic coupling among them. In addition to learning spatial patterns, the model was also found to be capable of learning sequence information when rhythmic GABAergic modulation was included in the simulation. This periodic change in the relative contributions of afferent and intrinsic information (via the suppression of EPSPs and IPSPs) to population dynamics was also found to be critical for the development of place fields and serves as a possible mechanism supporting the phase precession effect (Figure 3). Taken together, these effects suggest that selective GABA-B receptor antagonists such as phaclofen and CGP 36742 should attenuate the development of place fields in vivo and reduce an animal's capacity for sequence learning.

REFERENCES

[1]   Amaral, D.G. and Witter, M.P. (1989) The three-dimensional organization of the hippocampal formation: a review of anatomical data. Neurosci., 31:571-591.

[2]   Ault, B. and Nadler, J.V. (1982) Baclofen selectively inhibits transmission at synapses made by axons of CA3 pyramidal cells in the hippocampal slice. J. Pharmacol. Exp. Ther., 223:291-297.

[3]   Bernardo, LS and Prince, DA (1982) Cholinergic pharmacology of mammalian hippocampal pyramidal cells. Neurosci., 7:1703-1712.

[4]   Colbert, C.M. and Levy, W.B. (1992) Electrophysiological and pharmacological characterization of perforant path synapses in CA1: mediation by glutamate receptors. J. Neurophysiol., 63:1-8.

[5]   Cole, A.E. and Nicoll, R.A. (1984) Characterization of a slow cholinergic post-synaptic potential recorded in vitro from rat hippocampal pyramidal cells. J. Physiol., 352:173-188.

[6]   Doi, N., Carpenter, D.O. and Hori, N. (1990) Differential effects of baclofen and gamma-aminobutyric acid (GABA) on rat piriform cortex pyramidal neurons in vitro. Cell. Molec. Neurobiol., 10:559-564.

[7]   Dutar, P, Bassant, ME, Senut, MC and Lamour, Y, (1995) The septohippocampal pathway: Structure and function of a central cholinergic system. Physiol. Rev., 75:393-427.

[8]   Freund, T.F. and Antal, M. (1988) GABA-containing neurons in the septum control inhibitory interneurons in the hippocampus. Nature, 336:170-173.

[9]   Frotscher, M. and Leranth, C. (1985) Cholinergic innervation of the rat hippocampus as revealed by choline acetyltransferase immunocytochemistry: a combined light and electron microscopic study. J. Comp. Neurol., 239:237-246.

[10] Hasselmo, M.E., Schnell, E. and Barkai, E. (1995) Dynamics of learning and recall at excitatory recurrent synapses and cholinergic modulation in rat hippocampal region CA3. J. Neurosci., 15:5249-5262.

[11] Howe, J.R., Sutor, B. and Zieglgansberger, W. (1987) Baclofen reduces post-synaptic potentials of rat cortical neurons by an action other than its hyperpolarizing action. J. Physiol., 384:539-569.

[12] Kamiya, H. (1991) Some pharmacological differences between hippocampal excitatory and inhibitory synapses in transmitter release: an in vitro study. Synapse, 8:229-235.

[13] Levy, W.B. (1989) A computational approach to hippocampal function. In Computational Models of Learning in Simple Neural Systems, R.D. Hawkins and G.H. Bower, (Eds.), Academic Press, Orlando, FL., 243-305.

[14] Madison, D.V., Lancaster, B. and Nicoll, R.A. (1987) Voltage clamp analysis of cholinergic action in the hippocampus. J. Neurosci., 7:733-741.

[15] Marr, D. (1971) Simple memory: A theory for archicortex. Phil. Trans. Roy. Soc. B, 262:23-81.

[16] McNaughton, B.L., Barnes, C.A. and O'Keefe, J. (1983) The contributions of position, direction and velocity to single unit activity in the hippocampus of freely-moving rats. Exp. Brain Res., 52:41-49.

[17] Numann, R.E., Wadman, W.J. and Wong, R.K.S. (1987) Outward currents of single hippocampal cells obtained from the adult guinea-pig. J. Physiol., 393:331-353.

[18] O'Keefe, J. and Dostrovsky, J. (1971) The hippocampus as a spatial map: preliminary evidence from unit activity in freely moving rats. Brain Res., 34:171-175.

[19] O'Keefe, J. and Recce, M.L. (1993) Phase relationship between hippocampal place units and the EEG theta rhythm. Hippocampus, 3:317-330.

[20] Skaggs, W.E., McNaughton, B.L., Wilson, M.A. and Barnes, C.A. (1996) Theta phase precession in hippocampal neuronal populations and the compression of temporal sequences. Hippocampus, 6:149-172.

[21] Squire, L.R. (1992) Memory and the hippocampus - a synthesis from findings with rats, monkeys and humans. Psychol. Rev., 99:195-231.

[22] Stewart, M. and Fox, S.E. (1990) Do septal neurons pace the hippocampal theta rhythm?. Trends Neurosci., 13:163-168.

[23] Swartzwelder, H.S., Lewis, D.V., Anderson, W.W. and Wilson, W.A. (1987) Seizure-like events in brain slices: suppression by interictal activity. Brain Res., 410:362-366.

[24] Traub, R.D., Miles, R. and Wong, R.K.S. (1989) Model of the origin of rhythmic population oscillations in the hippocampal slice. Science, 243:1319-1325.

[25] Traub, R.D., Wong, R.K.S., Miles, R. and Michelson, H. (1991) A model of a CA3 hippocampal pyramidal neuron incorporating voltage-clamp data on intrinsic conductances. J. Neurophysiol., 66:635-650.

[26] Ylinen, A., Soltesz, I., Bragin, A., Penttonen, M., Sik, A. and Buzsaki, G. (1995) Intracellular correlates of hippocampal theta rhythm in identified pyramidal cells, granule cells, and basket cells. Hippocampus, 5:78-90.

[27] Zador, A., Koch, C. and Brown, T.H. (1990) Biophysical model of a Hebbian synapse. Proc. Natl. Acad. Sci., 87:6718-6722.

87

CORTICAL SYNCHRONIZATION MECHANISM FOR "POP-OUT" OF SALIENT IMAGE CONTOURS

Shih-Cheng Yen and Leif H. Finkel

Department of Bioengineering and Institute of Neurological Sciences
University of Pennsylvania
Philadelphia, Pennsylvania 19104
syen@jupiter.seas.upenn.edu
leif@jupiter.seas.upenn.edu

ABSTRACT

We present a model based on long-range intra-cortical connections which computes the salience of contours in a visual scene. The model accounts for a number of psychophysical and physiological results on contour salience, and provides a mechanism for several of the Gestalt laws of perceptual organization. In the model, cells lying on smooth contours facilitate each other, and strongly facilitated cells enter a "bursting" model. Horizontal connections allow bursting cells to synchronize, and perceptual salience is defined by the level of synchronized activity. In particular, we propose that the intrinsic properties of synchronization account for the increased salience of smooth, closed contours

1. INTRODUCTION

It has been suggested that cells in the supragranular layers of visual cortex with long-range horizontal connections might play a role in extracting salient features in a scene (Gilbert, 1992; Field *et al.*, 1993; Kovács and Julesz, 1993). These cells have been shown to be sensitive to stimuli outside the classical receptive field, allowing contextual information to influence the response of the cell (Nelson and Frost, 1985; Kapadia *et al.*, 1995). Similar cells in the supragranular layers of striate cortex have been observed to burst rapidly (Gray and McCormick, 1996) and could be involved in the temporal binding of contour elements. We present a model for computing the perceptual salience of contours that incorporates these two findings and is able to account for a number of physiological and psychophysical results (Polat and Sagi, 1993, 1994; Kapadia *et al.*, 1995; Field *et al.*, 1993; Kovács and Julesz, 1993, 1994; Kovács *et al.*, 1996).

2. MODEL ARCHITECTURE

Linear quadrature steerable filter pyramids (Freeman and Adelson, 1991) are used to model the response characteristics of cells in primary visual cortex. Steerable filters are computationally efficient as they allow the energy at any orientation and spatial frequency to be calculated from the responses of a set of basis filters. The fourth derivative of a Gaussian and its Hilbert transform were used as the filter kernels to approximate the shape of the receptive fields of simple cells.

Model cells are interconnected by long-range horizontal connections in a pattern similar to the co-circular connectivity pattern of Parent and Zucker (1989), as well as the "association field" proposed by Field *et al.* (1993). For a cell of orientation θ_A at location "A", there is a "preferred" orientation at location "B", ϕ_B, given by the tangent to the unique circle which passes through both "A" and "B", and whose tangent at "A" agrees with the local orientation, θ_A, at "A". The connection weights between the cell with orientation θ_A at "A" and the oriented cells at "B" peak at ϕ_B, and decrease with increasing angular difference between the two orientations. These excitatory connections are confined to two regions, one flaring out along the axis of orientation of the cell (co-axial), and another confined to a narrow zone extending orthogonally to the axis of orientation (transaxial). There is physiological and anatomical evidence consistent with the existence of both sets of connections (Rockland and Lund, 1983; Lund *et al.*, 1985; Nelson and Frost, 1985; Kapadia *et al.*, 1995; Fitzpatrick, 1996). The connection field is shown in Figure 1a. As observed physiologically, these excitatory connections only facilitate cells that receive local supra-threshold input. If the local orientation activity distribution at "B" peaks at ϕ_B, the cell with orientation θ_A at "A" will be strongly facilitated. As the local orientation at "B" deviates from ϕ_B, the degree of facilitation decreases. The "preferred" orientation at "B" can thus be thought of as providing "support" for the orientation, θ_A, at "A". Connection weights decrease for positions with increasing angular deviation from the orientation axis of the cell, as well as positions with increasing distance, in agreement with the physiological and psychophysical findings.

It is estimated that 20% of the horizontal connections target inhibitory cortical cells (McGuire *et al.*, 1991), and inhibition also occurs through di-synaptic pathways. Intra-cellular recordings have shown that stimulation of the horizontal connections produces short-latency EPSPs followed by longer-latency IPSPs (Weliky *et al.*, 1995). Since the mechanisms underlying long-distance inhibition are less well documented, we have chosen a simple descriptive mechanism. Each facilitated cell inhibits all other cells within the range of its long-distance connections such that only the elements that are strongly facilitated survive. This is consistent with neurophysiological evidence from Kapadia *et al.* (1995).

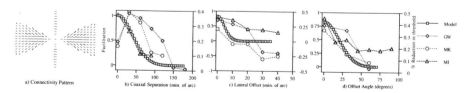

Figure 1. From left to right: Connectivity pattern of a horizontally oriented cell. Length of line indicates connection strength; The results are compared to the psychophysical data from 3 subjects (GW, MK, MI) reported in Kapadia *et al.* (1995).

In the model, cells that are strongly facilitated are assumed to enter a "bursting" mode, similar to the behavior of "chattering" cells described by Gray and McCormick (1996). We assume that bursting cells are able to synchronize with other bursting cells. Cells that enter the "bursting" mode are modeled as homogeneous coupled neural oscillators with a common fundamental frequency but different phases (Kopell and Ermentrout, 1986; Baldi and Meir, 1990). The phase of each oscillator is modulated by the phase of the other oscillators to which it is coupled. Oscillators are coupled only if the corresponding cells have strong, reciprocal, connections. A set of coupled oscillators together represent a contour. Since oscillators on different contours are not generally inter-connected, each contour in the scene can synchronize independently, and cells representing different contours to be discriminated and grouped together. It has been postulated that gamma frequency (40 Hz) oscillations often observed in the cortex could be responsible for generating perceptual binding across different cortical regions (Gray and Singer, 1995). Recent studies have questioned the functional significance and even the existence of these oscillations (Ghose and Freeman, 1992; Bair *et al.*, 1994). We use neural oscillators only as a simple means of computing synchronization and assume that synchronization may achieved through a different mechanism in cortex.

The salience of a contour can only be computed when all the oscillators on the contour are synchronized. Salience is represented by the sum of the activities of all the synchronized elements. The longer the chain of synchronized elements, the more perceptually salient it is. Synchronization occurs in parallel over the whole scene and the longest synchronized chain in the scene is identified as being the most salient, and the network selects it as its output. This allows us to compare the results of the model with a number of psychophysical findings. All simulations were conducted with the same parameter set.

3. RESULTS

3.1. Co-Axial Connection Pattern (Kapadia *et al.*, 1995)

The experiments of Kapadia *et al.* (1995) provide a test of the connection architecture. The degree of facilitation of post-synaptic cells was measured as a function of the position and orientation preference of a horizontally connected pre-synaptic cell. Figure 1 shows the effect of varying the co-axial distance, off-axis misalignment, and angular orientation of the pre-synaptic cell. The pre-synaptic cell is assumed to have constant activity, and so the changes in facilitation are due only to the differences in the connection weights between the post-synaptic cell and pre-synaptic cells at different positions and orientations. The results show good qualitative agreement with the experimental findings. The results are also in qualitative agreement with those of Polat and Sagi (1993). Our simulations do show an overestimation of the activation for closely spaced elements (Figure 1b) that would be compensated by local, short-range inhibition, which is not included in this simulation. Such inhibition would correspond to Polat and Sagi's (1993, 1994) observation of increased contrast detection thresholds at small separations. There is also some discrepancy between the model and the data at large off-axial misalignments. At large offsets the results of Kapadia *et al.* (1995) show that the influence of the surround becomes largely inhibitory. In the model, the facilitation at large lateral offsets becomes very weak and would be overwhelmed by the longer latency inhibition that is also not included in this particular simulation.

3.2. Contrast Sensitivity Modulation (Kovács and Julesz, 1993, 1994)

The trans-axial connections in our model may underlie a surprising psychophysical observation of Kovács and Julesz (1993, 1994). They measured changes in contrast sensitivity to a low contrast Gabor target placed at various locations inside and outside a circular and elliptical contour. The contour itself was formed from aligned Gabor patches. They found a sharp peak in contrast sensitivity at the center of the circle and at the two foci of the ellipse. In addition, contrast sensitivity was elevated at distances approximately 2λ on each side of the contour, while the sensitivity on the contour itself was greatly decreased as compared to the sensitivity to the target in the absence of the contour. Figure 2b,c shows the contrast sensitivity maps from Kovács and Julesz (1993). Figure 2a shows a simplified "silhouette" of the connectivity pattern for a horizontally oriented cell in our model. The gray level represents the connection weights; dark regions are facilitatory, and white regions are inhibitory. Since the psychophysical experiments were carried out using low-contrast probes oriented parallel to the closest element on the contour, only the trans-axial connections are likely to be stimulated. We have thus left out the co-axial connections for simplicity. The cell is surrounded by an inhibitory region at very close distances, corresponding to the intra-filter inhibition observed by Polat and Sagi (1993, 1994), and Kapadia *et al.* (1995). If a number of these silhouettes are placed along a circular or elliptical contour, their excitatory regions superpose. The resulting map of facilitatory regions resembles the experimental findings (Figure 2d,e). Note especially the peak in the center of the circle and the two peaks in the ellipse due to the trans-axial facilitatory connections. The trans-axial connections are usually strong enough to be facilitatory only out to about 2λ, but the superposition of the subthreshold facilitatory connections combine at the center of the circle and at the two foci of the ellipse to become much stronger. Since the range of facilitation observed in Polat and Sagi's (1993, 1994) experiments scales with size of the Gabor elements, this would also explain the similarities across scale in Kovács and Julesz's (1994) data.

3.3. Extraction of Salient Contours (Field *et al.*, 1993)

Using the same methods as Field *et al.* (1993), we tested the model's ability to extract contours embedded in noise (see Figure 3). We generated stimulus arrays of 256 oriented Gabor elements. Pairs of stimulus arrays were presented to the network, one array contained a contour composed of 12 Gabor elements, the other contained only randomly oriented elements. For each stimulus, the network determines the "salience" of all contours, and selects the contour with the highest salience. Of the two stimuli in each pair

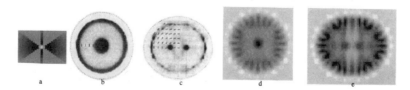

Figure 2. a) Simplified silhouette showing intra-filter inhibition and the facilitatory connections for a horizontally oriented cell. Dark regions are facilitatory, white regions are inhibitory and gray regions are neutral. b,c) Contrast sensitivity maps inside a circular and elliptical contour, from Kovács and Julesz (1994). d,e) Sensitivity maps from the model based on averaged connection fields of aligned Gabor units.

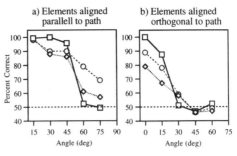

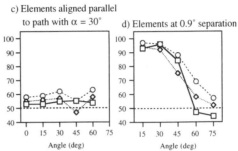

Figure 3. Simulation results are compared to the data from 2 subjects (AH, DJF) in Field *et al.* (1993). Stimuli consisted of 256 randomly oriented Gabor patches with 12 elements aligned to form a contour. Each data point represents results for 500 simulations. AH (diamond); DJF (circle); Model (square).

presentation, the network "chooses" the stimulus containing the contour with the higher salience. Network performance was measured by computing the percentage of correct detection. The network was tested on a range of stimulus variables governing the target contour: 1) the angle, β, between elements on a contour, 2) the angle between elements on a contour but with the elements aligned orthogonal to the contour passing through them, 3) the angle between elements with a random offset angle, $\pm\alpha$, with respect to the contour passing through them, and 4) average separation of the elements. 500 simulations were run at each data point. The results are shown in Figure 3. The model shows good qualitative agreement with the psychophysical data. When the elements are aligned, the performance of the network is mostly modulated by the co-axial connections, whereas when the elements are oriented orthogonal to the contour, the trans-axial connections mediate the performance of the network. The performance of both the model and human subjects are adversely affected as the weights between consecutive elements decrease in strength. This reduces the length of the contour and thus the saliency of the stimulus.

3.4. Effects of Contour Closure (Kovács and Julesz, 1993, 1994)

In a series of experiments using similar stimuli to Field *et al.* (1993), Kovács and Julesz (1993) found that closed contours are much more salient than open contours. They found that when the inter-element spacing between all elements was gradually increased, the maximum inter-element separation for detecting closed contours (Δ_c, defined at 75% performance) is higher than that for open contours (Δ_o). In addition, they showed that when elements spaced at Δ_o are added to a "jagged" (open) contour, the saliency of the contour increases monotonically but when elements spaced at Δ_c are added to a circular contour, the saliency does not change until the last element is added and the contour becomes closed. In fact, at Δ_c, the contour is not salient until it is closed, at which point it suddenly "pops-out" (see Figure 4c). This finding places a strong constraint on the computation of saliency in visual perception.

Interestingly, it has been shown that synchronization in a chain of coupled neural oscillators is enhanced when the chain is closed (Kopell and Ermentrout, 1986; Ermentrout, 1985; Somers and Kopell, 1993). This property seems to be related to the differences in boundary effects on synchronization between open and closed chains and appears to hold across different families of coupled oscillators. It has also been shown that synchronization is dependent on the coupling between oscillators—the stronger the coupling, the better the synchronization, both in terms of speed and coherence (Somers and Kopell, 1993; Wang, 1995). We believe these findings may apply to the psychophysical results.

As in Kovács and Julesz (1993), we generated stimulus arrays containing 2025 elements. Contours were made up of 24 elements. Again the network is presented with two stimuli, one containing a contour and the other made up of all randomly oriented elements. The network picks the stimulus containing the synchronized contour with the higher salience. In separate trials, the threshold separation for open and closed contours

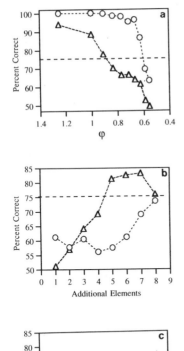

Figure 4. Simulation of the experiments of Kovács and Julesz (1993). Stimuli consisted of 2025 randomly oriented Gabor patches, with 24 elements aligned to form a contour. Each data point represents results from 500 trials. a) Plot of the performance of the model with respect to the ratio of the separation of the background elements to the contour elements. Results show closed contours are salient to a more salient than open contours. b) Changes in salience as additional elements are added to open and closed contours. Results show that the salience of open contours increase monotonically while the salience of closed contours only change with the addition of the last element. Open contours were initially made up of 7 elements while closed contours were made up of 17 elements. c) The data from Kovács and Julesz (1993) are re-plotted for comparison (Circle: closed; triangle: open).

were determined. The ratio of the separation of the background elements to the that of elements on a closed curve, φ_c, was found to be 0.6 (which is similar to the threshold of 0.65 recently reported by Kovács *et al.*, 1996), whereas the ratio for open contours, φ_o, was found to be 0.9. (Δ is the threshold separation of contour elements, φ, at a particular background separation). We then examined the changes in saliency for open and closed contours. The performance of the network was measured as additional elements were added to an initial short contour of elements. The results are shown in Figure 4b. At φ_o, both open and closed contours are synchronized but at φ_c, elements are synchronized only when the chains are closed. If salience can only be computed for synchronized contours, then as additional elements are added to an open chain at φ_o, the salience would increase since the whole chain is synchronized. On the other hand, at φ_c, as long as the last element is missing, the chain is really an open chain, and since φ_c is smaller than φ_o, the elements on the chain will not be able to synchronize and adding elements has no effect on salience. Once the last element is added though, the chain is immediately able to synchronize and the salience of the contour increases dramatically and causes the contour to "pop-out".

4. CONCLUSION

We have presented a cortically-based model that is able to identify perceptually salient contours in images containing high levels of noise. The model is based on the use of long distance intra-cortical connections that facilitate the responses of cells lying along smooth contours. Salience is defined as the combined activity of the synchronized population of cells responding to a particular contour. The model qualitatively accounts for a range of physiological and psychophysical results and can be used in extracting salient contours in real images (Yen and Finkel, 1996).

ACKNOWLEDGMENTS

Supported by the Office of Naval Research (N00014–93–1–0681), The Whitaker Foundation, and the McDonnell-Pew Program in Cognitive Neuroscience.

REFERENCES

Bair, W., Koch, C., Newsome, W. & Britten, K. (1994). Power spectrum analysis of bursting cells in area MT in the behaving monkey. *Journal of Neuroscience, 14*, 2870–2892.

Baldi, P. & Meir, R. (1990). Computing with arrays of coupled oscillators: An application to preattentive texture discrimination. *Neural Computation, 2*, 458–471.

Ermentrout, G. B. (1985). The behavior of rings of coupled oscillators. *Journal of Mathematical Biology, 23*, 55–74.

Field, D. J., Hayes, A. & Hess, R. F. (1993). Contour integration by the human visual system: Evidence for a local "Association Field". *Vision Research, 33*, 173–193.

Fitzpatrick, D. (1996). The functional-organization of local circuits in visual-cortex --insights from the study of tree shrew striate cortex. *Cerebral Cortex, 6*, 329–341.

Freeman, W. T. & Adelson, E. H. (1991). The design and use of steerable filters. *IEEE Transactions on Pattern Analysis and Machine Intelligence, 13*, 891–906.

Gilbert, C. D. (1992). Horizontal integration and cortical dynamics. *Neuron, 9*, 1–20.

Ghose, G. M. & Freeman, R. D. (1992). Oscillatory discharge in the visual system: Does it have a functional role? *Journal of Neurophysiology, 68*, 1558–1574.

Gray, C. M. & McCormick, D. A. (1996). Chattering cells -- superficial pyramidal neurons contributing to the generation of synchronous oscillations in the visual-cortex. *Science, 274,* 109–113.

Kapadia, M. K., Ito, M., Gilbert, C. D. & Westheimer. G. (1995). Improvement in visual sensitivity by changes in local context: Parallel studies in human observers and in V1 of alert monkeys. *Neuron, 15,* 843–856.

Kopell, N. & Ermentrout, G. B. (1986). Symmetry and phaselocking in chains of weakly coupled oscillators. *Communications on Pure and Applied Mathematics, 39,* 623–660.

Kovács, I. & Julesz, B. (1993). A closed curve is much more than an incomplete one: Effect of closure in figure-ground segmentation. *Proceedings of National Academy of Sciences, USA, 90,* 7495–7497.

Kovács, I. & Julesz, B. (1994). Perceptual sensitivity maps within globally defined visual shapes. *Nature, 370,* 644–646.

Kovács, I., Polat, U. & Norcia, A. M. (1996). Breakdown of binding mechanisms in amblyopia. *Investigative Ophthalmology & Visual Science, 37,* 3078.

Lund, J., Fitzpatrick, D. & Humphrey, A. L. (1985). The striate visual cortex of the tree shrew. In Jones, E. G. & Peters, A. (Eds), *Cerebral Cortex* (pp. 157–205). New York: Plenum.

McGuire, B. A., Gilbert, C. D., Rivlin, P. K. & Wiesel, T. N. (1991). Targets of horizontal connections in macaque primary visual cortex. *Journal of Comparative Neurology, 305,* 370–392.

Nelson, J. I. & Frost, B. J. (1985). Intracortical facilitation among co-oriented, co-axially aligned simple cells in cat striate cortex. *Experimental Brain Research, 61,* 54–61.

Parent, P. & Zucker, S. W. (1989). Trace inference, curvature consistency, and curve detection. *IEEE Transactions on Pattern Analysis and Machine Intelligence, 11,* 823–839.

Polat, U. & Sagi, D. (1993). Lateral interactions between spatial channels: Suppression and facilitation revealed by lateral masking experiments. *Vision Research, 33,* 993–999.

Polat, U. & Sagi, D. (1994). The architecture of perceptual spatial interactions. *Vision Research, 34,* 73–78.

Rockland, K. S. & Lund, J. S. (1983). Intrinsic laminar lattice connections in primate visual cortex. *Journal of Comparative Neurology, 216,* 303–318.

Singer, W. & Gray, C. M. (1995). Visual feature integration and the temporal correlation hypothesis. *Annual Review of Neuroscience, 18,* 555–586.

Somers, D. & Kopell, N. (1993). Rapid synchronization through fast threshold modulation. *Biological Cybernetics, 68,* 393–407.

Wang, D. (1995). Emergent synchrony in locally coupled neural oscillators. *IEEE Transactions on Neural Networks, 6,* 941–948.

Weliky M., Kandler, K., Fitzpatrick, D. & Katz, L. C. (1995). Patterns of excitation and inhibition evoked by horizontal connections in visual cortex share a common relationship to orientation columns. *Neuron, 15,* 541–552.

Yen, S-C. & Finkel, L. H. (1996). Salient Contour Extraction by Temporal Binding in a Cortically-Based Network. In Touretzky, D. S., Mozer, M. C. & Hasselmo, M. E. (Eds), *Advances in Neural Information Processing Systems 9.* Massachusetts: MIT Press.

88

TYPE-BASED ANALYSIS OF NEURAL ENSEMBLES*

Lin Yue, Don H. Johnson, and Charlotte M. Gruner

Computer and Information Technology Institute
Department of Electrical and Computer Engineering
Rice University
Houston, TX
E-mail: `linyue@rice.edu`
E-mail: `dhj@rice.edu`
E-mail: `cmkruger@rice.edu`

ABSTRACT

Theoretical considerations indicate that discharges from individual neurons in the auditory pathway cannot adequately represent stimulus features. To assess how well a neural ensemble encodes information, we develop a type-based approach that assumes no *a priori* knowledge of how the component response patterns are inter-related and which naturally accommodates response non-stationarities. From records or simulations of a neural population, we measure neural group activity and determine what aspects of the group response crucially represent sensory variables. The measure thus produced makes optimal use of data, and can assess sensory representation fidelity from both detection and estimation theoretic viewpoints.

1. INTRODUCTION

Theoretical studies indicate that single-neuron discharges found in the auditory pathway cannot represent stimulus variables (loudness, frequency, sound location) to psychophysical accuracy. However, we have shown that the sustained discharges (i.e., the average discharge rates) of individual lateral superior olive (LSO) neurons encode crucial information for sound localization [2, 3, 5]. The issue is that a single LSO neuron's discharge rate is not large enough to yield accurate angle estimates within reasonable time intervals (on the order of tens to a few hundreds of milliseconds). When transient stimuli are used, all LSO neurons exhibit an oscillatory response lasting some 20–30 ms that varies from neuron to neuron [6, 8]. Judging

*Supported by grants from the National Science Foundation and from the National Institute of Mental Health.

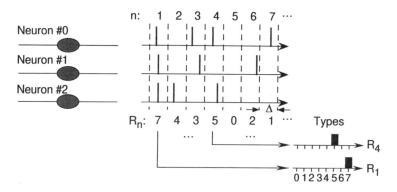

Figure 1. Individual neurons are given an arbitrary identification number. The discharge pattern for each is measured, and individual discharges placed in the nth bin (each bin has width Δ). Using the neuron identification number and the presence of discharges in a binary code, a number denoted by R_n is assigned to each bin to represent which neurons fired during that bin. Because the intensity corresponding to each neuron typically varies with time, we estimate types for each bin separately using multiple stimulus presentations. Only one of such presentations is illustrated in the figure.

how, if at all, this transient represents location information is difficult because the response's nonstationary nature renders standard analysis techniques powerless. Thus, to understand sensory coding, we must have a technique that not only allows us to consider LSO neurons as a functional group and to determine how they cooperatively represent a sound source's location, but also copes with nonstationarities.

We have developed such a technique based on the theory of types. The type-based approach assumes no *a priori* knowledge of how stimulus parameter(s) are represented in a population's response, and it naturally accommodates any non-stationarity in the responses, such as the transient prevalent in all LSO recordings. The technique's properties have a strong mathematical foundation, among them is that it can discriminate response patterns optimally, it can easily handle (if provided with sufficient data) neural ensembles of more than trivial size, and it directly measures how well ensemble responses represent the information.

2.   TYPE-BASED ENSEMBLE RESPONSE MEASURE

To develop a measure of the population's response, we first convert the population's discharge pattern into a convenient representation as depicted in figure 1. Here, a neural population's response during the nth bin is summarized by a single number R_n that equals a binary coding for which neurons, if any, discharged during the bin. From the sequence $\{R_1, \ldots, R_N\}$ (N represents the total number of bins in the stimulus repetition period), the discharge pattern of the entire population can be reconstructed exactly (down to the precision of the binwidth Δ), and any population response measure can be calculated from it. This procedure simply recasts ensemble activity into a convenient form for later signal processing.

"Type" is a word used in information theory for the histogram estimate of a discrete probability distribution [1, Chap. 12]. If $R_1, \ldots, R_N$ comprise a sequence of statistically independent, identically distributed (i.i.d.) random variables that each assume a value drawn

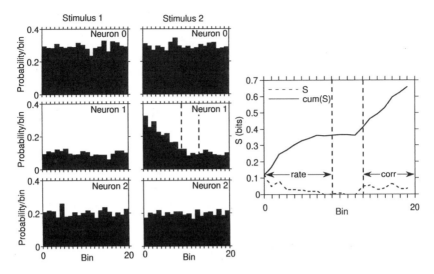

Figure 2. We simulated a three-neuron ensemble responding to two stimulus conditions. The left portion of the display shows PST histograms of each neuron, and these indicate that neuron 1 had a rate response to stimulus 2 (ending at the first vertical dashed line). The right panel shows the result of computing Gutman's statistic to measure the difference between the responses under the two stimulus conditions. The dashed line shows the distance computed in each bin and the solid the cumulative value of these component distances. Not only does the rate response create a difference between the responses, but also a later response difference not evident in the PST histograms (starting at the second vertical dashed line). This difference occurred because of a stimulus-induced correlation between neuron 1 and 2 in the last six bins. Interestingly, the correlation response is nearly as significant for response distinguishability as the rate response: The contribution of each to the total Gutman statistic is about the same.

from a finite set $X_n \in \{a_1, \ldots, a_K\}$, its type is defined to be

$$\widehat{P}_\mathbf{R}(a_k) = \frac{1}{N}\sum_{n=1}^{N} I(R_n = a_k), \ k = 1, \ldots, K$$

where $I(\cdot)$ is the indicator function, equaling one if its argument is true and zero otherwise. This expression equals the fraction of times the value a_k occurred in the observation vector $\{R_1, \ldots, R_N\}$. Because we have assumed that each observation is drawn from a finite set, a type is also a valid probability distribution for the set.

For neuron discharges, however, $R_1, \ldots, R_N$ are generally not i.i.d. because of response nonstationarity. Types are accumulated separately in each bin because we have no *a priori* reason to believe that the response is stationary. We accumulate types in each bin by incorporating responses from multiple stimulus presentations, as in the formation of PST histograms. Each set of discharges occurring in the nth bin forms the vector $\mathbf{R}_n = \{R_n, R_{n+P}, R_{n+2P}, \ldots\}$, where P represents the number of bins for *each* stimulus repetition (i.e., the stimulus period in bins). Thus, the value of L, the number of observations used to compute a type for each bin, equals the number of stimulus repetitions $\lfloor N/P \rfloor$. The type $\widehat{P}_{\mathbf{R}_n}$ — the histogram estimate of the probability distribution of the ensemble's discharges during the nth bin — is accumulated by the length L vector as shown in figure 1. Given these types under stimulus conditions i and j, Gutman's type-based statistic [4] determines the ultimate discriminability of these two datasets when no *a priori* model exists. The statistic solves hypothesis testing problems optimally, with error rates that no other empirically based decision rule could surpass. In other words, it discriminates between datasets optimally while exploiting the amount of data available to the greatest possible extent. The type-based discrimination statistic in the nth bin of responses under stimulus conditions i and j equals

$$S_n^{(i,j)} = D(\widehat{P}_{\mathbf{R}_n^{(i)}} \| \widehat{P}_{\mathbf{R}_n^{(i)}\mathbf{R}_n^{(j)}}) + D(\widehat{P}_{\mathbf{R}_n^{(j)}} \| \widehat{P}_{\mathbf{R}_n^{(i)}\mathbf{R}_n^{(j)}}), \tag{1}$$

where the quantity $D(P\|Q) \equiv \sum_k P(a_k) \log \frac{P(a_k)}{Q(a_k)}$ is known as the *Kullback-L eibler distance* [7] between the probability distributions P and Q, and $\widehat{P}_{\mathbf{R}_n^{(i)}\mathbf{R}_n^{(j)}} = \frac{1}{2}\left(\widehat{P}_{\mathbf{R}_n^{(i)}} + \widehat{P}_{\mathbf{R}_n^{(j)}}\right)$ if the same number of stimulus repetitions are performed under each stimulus condition. If the observations were generated according to the same probabilistic model, which is the case when the two stimulus conditions are the same, the distances in between each type and the linear combination should both be small. If different, the distances should increase, with greater distance meaning greater difference in the response, which would mean that stimulus encoding is occurring. Thus, the quantity $S_n^{(i,j)}$ measures the degree of distinguishability between the responses during bin n. Gutman also showed that the error rate of any empirical decision rule that attempts to determine which stimulus condition was used from the observations $\mathbf{R}_n^{(i)}$ and $\mathbf{R}_n^{(j)}$ cannot exceed $S_n^{(i,j)}$: Formally, the probability of decision errors cannot decrease more rapidly than $\exp\{-LS_n^{(i,j)}\}$ as the amount of data (i.e., the number of stimulus repetitions L) becomes large. Assuming the responses in a bin are statistically independent,* add the statistics $S_n^{(i,j)}$ across bins to obtain the total statistic $S^{(i,j)}$. Thus, it is this type-based quantity that determines how well any empirical decision rule based on training sets can perform.

3. SIMULATION RESULTS

By computing Gutman's statistic from the recorded responses to two different stimuli determined by the experimenter in a controlled way, one can probe how an ensemble processes the stimulus. We can not only reveal which portion of the response encodes the stimulus condition change, but also determine which portion encodes the change most vigorously. Because the calculation of Gutman's statistic makes no assumption about the response, the statistic measures objectively when stimulus coding occurs in the response and how well each response component represents the stimulus. Figure 2 illustrates the application of this approach to a simple population of three neurons. This result indicates the generality of our analysis: Both stimulus-induced rate response and inter-neuron correlation can be detected, and the relative contribution of each response component quantified by considering the detailed behavior of $S_n^{(i,j)}$. We can do more than just understand what portion of the response represents the neural code; we can also determine the exact nature of the neural code temporally and spatially.

As illustrated in figure 2, the type-based calculation quantifies any response changes without distinguishing what induced such change. For that example, changes due to either rate or correlation structure resulted in similar change in Gutman's statistic. We can, however, determine the presence of correlation in an ensemble's response with this technique. Figure 3 illustrates an example. The type that would have been produced by an ensemble if it had statistically independent members can be calculated from the measured response as well as the type for the ensemble response. By calculating Gutman's statistic between the response types

---

*If neural dependence is present from bin to bin, conditional types should be used in this technique.

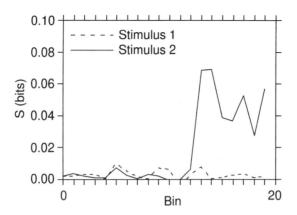

Figure 3. The Gutman statistic between the ensemble response and the forced-independent response for the two stimulus conditions used in figure 2 is shown. As was the case in actuality, the responses to stimulus 1 contained no trans-neural correlation. The second stimulus did induce a correlation in the latter portion of the response, and the Gutman statistic clearly indicates the presence of such correlation.

of the recorded ensemble and the forced-independent ensemble, we can infer when correlated responses are present.

Figure 4 illustrates an example of the simulated two-neuron ensemble responses of LSO fast and slow choppers. We can explicitly measure what response features are more important for stimulus discrimination (transient or sustained, fast chopping or slow chopping, for example). The slow-chopper response type contributes to angle discrimination soon after stimulus onset, while the fast chopper provides increasing discrimination afterwards. Gutman's statistic not only demonstrates this phenomenon, but also quantifies the amount of discrimination.

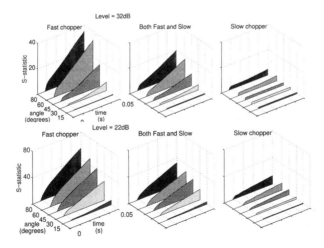

Figure 4. The cumulative values of the Gutman statistic in time are shown for three two-neuron population models: fast chopper alone, slow chopper alone, and both. In each panel, the response to a midline stimulus is compared to responses obtained from stimuli located increasingly more lateral. The top portion corresponds to a louder sound source than the bottom; note the reduced discrimination obtained with the louder stimulus. The statistic must generally increase with time as longer observation times usually result in greater discrimination.

4. RELATION TO ANGLE ESTIMATION

In addition to the detection viewpoint, Gutman statistic (eqn. 1) is related to the mean-squared error in estimating a stimulus parameter, the azimuthal angle of the sound source, for example. Suppose $\widehat{P}_{\mathbf{R}(i)}$ corresponds to responses to the sound source at θ, and $\widehat{P}_{\mathbf{R}(j)}$ responses to $\theta + \Delta\theta$. If $\Delta\theta$ is small, i.e., $\widehat{P}_{\mathbf{R}(i)}$ is a small perturbation of $\widehat{P}_{\mathbf{R}(j)}$, then we have [7]

$$S^{(i,j)} = \frac{F(\theta)}{4}(\Delta\theta)^2,$$

where $F(\theta)$ is the Fisher Information of the parameter θ. By the Cramer-Rao bound, the smallest mean-squared error any unbiased estimator of θ can yield is $F(\theta)^{-1}$. Thus, by obtaining $S^{(i,j)}$ for small perturbations, we can determine the lower bound on estimation error of any unbiased estimator.

5. SUMMARY

The type-based statistic has optimal properties when it comes to assessing a population's response to a stimulus: No other technique can use data more efficiently, and the resulting quantity can be directly related to discrimination performance. This statistic will vary with population composition, and can thus be used to understand what response types are most important. Furthermore, because it assumes so little about the form of the response, we can also assess the relative effectiveness of response components. The local behavior of $S^{(i,j)}$ is directly related to the minimum mean-squared estimation error. Thus, this single statistic can be used to assess information representation in a variety of ways.

REFERENCES

[1] T. M. Cover and J. A. Thomas. *Elements of Information Theory*. John Wiley & Sons, Inc., 1991.

[2] A. Dabak and D.H. Johnson. Function-based modeling of binaural interactions: Interaural phase. *Hearing Res.*, 58:200–212, 1990.

[3] A. Dabak and D.H. Johnson. Function-based modeling of binaural interactions: Level and time cues. *J. Acoust. Soc. Am.*, 94(5):2604–2616, 1993.

[4] M. Gutman. Asymptotically optimal classification for multiple tests with empirically observed statistics. *IEEE Trans. Info. Th.*, 35:401–408, 1989.

[5] D.H. Johnson, A. Dabak, and C. Tsuchitani. Function-based modeling of binaural interactions: Interaural level. *Hearing Res.*, 49:301–320, 1990.

[6] D.H. Johnson, C. Tsuchitani, D.A. Linebarger, and M.J. Johnson. Application of a point process model to responses of cat lateral superior olive units to ipsilateral tones. *Hearing Res.*, 21:135–159, 1986.

[7] S. Kullback. *Information Theory and Statistics*. Wiley, New York, 1959.

[8] M. Zacksenhouse, D.H. Johnson, and C. Tsuchitani. Transient effects during the chopping response of LSO neurons. *J. Acoust. Soc. Am.*, 98:1410–1422, 1995.

89

THE ONTOGENY OF THE NERVOUS CONTROL OF CARDIOVASCULAR CIRCULATION

A Neural Network Model

B. Silvano Zanutto,[1,3] Enrique T. Segura,[2] Bruno Cernuschi Frias,[3] and Alberto E. Dams[4]

[1]Department of Psychology: Experimental
Duke University
Box 90086, Durham, North Carolina 27708
silvano@psych.duke.edu
[2]IBYME (CONICET) Vuelta de Obligado 2490
1428 Buenos Aires, Argentina
etsegura@ibyme.edu.ar.
[3]LIPSIRN
[4]Dto Electronica
Facultad de Ingeniería, Universidad de Buenos Aires
Paseo Colón 850, 1063, Buenos Aires, Argentina
bcf@acm.org, adams@aleph.fi.uba.ar.

ABSTRACT

In adult animals, the nervous system controls the long-term mean blood pressure. The nucleus of the tractus solitarius (NTS) behaves as a comparator between its rostral afferents (the reference) and the cardioreceptor afferents (the feedback). It is hypothesized that, during ontogeny, the nervous system learns the sympathetic efferents from the chemoreceptor feedback. This process is simulated by a neural network model. In a stable metabolism, the model converges to a state in which all the tissues are irrigated with optimal oxygen and carbon dioxide levels.

INTRODUCTION

Cardiovascular variables are regulated by humoral nervous and auto regulatory mechanisms (Guyton 1991). It is generally accepted that the renal output curve regulates the long-term arterial blood pressure, but the role of the nervous system in this regulation is not well established. Guyton (1991, p. 1816) noted that, "Many prominent researchers believe that much, if not most, hypertension in human beings is initiated by nervous stress. But how can stress cause hypertension?" In this paper, we study the role of the nervous system in the long-term regulation of blood pressure. In an adult animal, if the information of the arterial (baro and chemoreceptor) and cardiopulmonary receptors do not feedback to the central nervous system, a permanent hypertension with large fluctuations is observed (Persson et al. 1988; 1989). This drift of the mean arterial pressure due to the lack of feedback is not stabilized by any other system. Thus, the nervous system regulates the long-term mean arterial pressure.

The NTS is the only structure that receives information from the rostral nuclei and from the cardiovascular receptors and sends efferents to nuclei that control the circulatory variables (Galosy et al. 1981). In this way, if the nervous system behaves as a linear system, the NTS will function as a comparator. In this paper, we propose a model where this feature is an emergent property of the NTS, and its rostral afferents provide the reference for the cardiovascular variables.

The local blood flow in each tissue is under sympathetic control, Shepherd (1983, p. 352) noted that "the role of the sympathetic nerves may be to modulate the local dilator to maintain the most economical ratio of blood flow to oxygen extraction." The state of maturation of sympathetic enervation at birth varies among animal species (Bevan et. al. 1980) but, in any case, the regulatory neural mechanisms during this period adapt to environmental and behavioral conditions (Dworkin and Miller 1977, Friedman et al. 1968). During ontogeny, the chemoreceptors discharge when the partial pressure of oxygen and carbon dioxide in the arterial blood values are not normal (95mmHg and 40 mmHg respectively) (Itskovits et al., 1987, Boekkooi et al., 1992). Also, the baroreflex are functional, but it is not involved in the long term regulation (Kunze 1981). The cardiopulmonary receptors during development have not been well studied yet. During this period, because the system is not mature, the feedback information from the receptor will not necessarily produce the right sympathetic discharge in all the tissue.

Some researchers postulate that plasticity during ontogeny is influenced by individual environmental interaction (Cohen and Randall 1984) and instrumental learning has been postulated as a visceral learning mechanism (Dworkin and Miller 1977; Dworkin 1986). In particular, it has been suggested that the cardiovascular system learns to avoid chemoreceptor discharge.

2. SIMULATION AND RESULTS

In this model, during development the nervous system adapts to stabilize the arterial blood pH to a normal value (7.4) using chemoreceptors' information. The nervous regulation of blood flow can be considered optimal when, even despite maximal drift, the local blood flow is compensated by auto-regulation. This condition minimizes the chemoreceptive discharge.

The model simulated is represented in the Fig. 1. The network has "T" nodes where each node "k," have two source inputs, one from the chemoreceptors and another from the rostral nuclei to the NTS. The outputs of the neural network y_k represent the sympathetic discharge frequency that controls the blood flow of each tissue. During learning, of all the possible output of each node (y_k), the system has to learn the values that will cause minimum chemoreceptors discharge.

As it was said, during ontogeny at least the arterial receptors are functional; because of that, we assume that the arterial blood pressure will be above a threshold, in such way

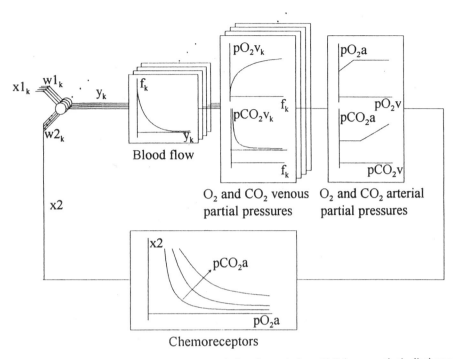

Figure 1. Neural network model of cardiovascular regulation. For each tissue "k," the sympathetic discharge, the blood flow, the venous partial pressure after the tissue irrigation, and the arterial partial pressure after the lung gases diffuse are shown. Finally chemoreceptors discharge "x2" is depicted. $x1_k$ represents the rostral nuclei inputs from the NTS.

that the local blood flow in each tissue can be control by the sympathetic system independent of the other. The blood flow f_k as a function of y_k can be simulated by a hyperbole (Celander 1954),

$$f_k = \frac{0.05}{y_k} \qquad (1)$$

where f_k and y_k take values between 0.05 and 1.

The oxygen and carbon dioxide partial pressure in the venous capillary as a function of the blood flow (Guyton 1986) is given by:

$$pO_2 v_k = 95\left(1 - 0.715\,e^{-2.13 f_k}\right) \qquad (2)$$

for the oxygen and

$$pCO_2 v_k = \frac{0.6}{f_k^{\,0.92}} + 40 \qquad (3)$$

for the carbon dioxide

where pO_2v and pCO_2v are measured in mmHg.

We assume that the partial pressure of the central venous blood results from the random addition of the venous blood pressure of all tissues:

$$pO_2v = \frac{\sum_{k=1}^{T} pO_2v_k r_k}{\sum_{k=1}^{T} r_k} \quad \text{and} \quad pCO_2v = \frac{\sum_{k=1}^{T} pCO_2v_k r_k}{\sum_{k=1}^{T} r_k} \tag{4}$$

where r_k's are random number between 0 and 1.

We suppose that when the pO_2v is lower than 39.55 mmHg and pCO_2v greater than 45.2 mmHg, gas diffusion in the alveolus will not be sufficient. In this case, the arterial partial pressures drift from the optimal value ($pO_2a=95$ mmHg and $pCO_2a=40$ mmHg) following a linear law:

$$pO_2a = pO_2v + 55 \ mmHg \ \text{and} \ pCO_2a = pCO_2v - 5 mmHg \tag{5}$$

The chemoreceptors increase their discharge exponentially if the O_2 concentration decreases or CO_2 concentration increases (Korner 1971). This function is formalized as:

$$x2 = 15\left(c_1 e^{-c_2 pO_2a} - 0.12598142\right) \tag{6}$$

$$c_1 = \frac{72}{pCO_2a} \tag{7}$$

$$c_2 = 0.2175 \, e^{-0.06525 pCO_2a} + 0.012 \tag{8}$$

The output of each node of the network is calculated as the difference between the inputs (a linear approximation of Grossberg 1982.)

$$y_k = F(s_k) \ \text{and} \ S_k = w \, 1_k \times 1_k - w \, 2_k \, x2 \tag{9}$$

where k: is number of the tissue, $x1_k$: is discharge frequency in the dendrite (rostral nuclei input), $x2$: represents chemoreceptor discharge frequency, y_k: is axon discharge frequency, $w1_k$: is the synaptic weight of the rostral input, $w2_k$: synaptic weight of the chemoreceptor input, and F: is a sigmoid function, with the inflection point in S_g. The constants have values between 0 and 1.

$$S_g = \frac{s_k}{d_s} - 1 \tag{10}$$

were d_s is a constant to determine the slope of the function when $S_g=0$. The function has two parts:

$$y_k = \frac{\frac{1}{1+e^{-g_dS_g}} - 0.5}{y_d} + 0.5 \text{ if } S_g < 0, \text{ and } y_k = \frac{\frac{1}{1+e^{-g_uS_g}} - 0.5}{y_u} + 0.5 \text{ if } S_g > 0 \tag{11}$$

where y_d, y_u, g_d and g_u are constants, y_d and y_u are given by:

$$y_d = 2\left(\frac{1}{1+e^{-g_d}} - 0.5\right) \text{ and } y_u = 2\left(\frac{1}{1+e^{-\left(\frac{1}{d_s}-1\right)g_u}} - 0.5\right) \tag{12}$$

W1 and w2 are the synaptic weights. The value of w2=1 and the value of w1 is calculated using Hebb's hypothesis (1949) and assuming that in the neuron there is competition for the synaptic resources (Kohonen 1984; 1987) as follows:

$$w \, l_{t+1k} = w \, l_{tk} + \left[\alpha_t \times l_k - \beta_t \, w \, l_{tk}(1+x2)\right] y_k \tag{13}$$

where α and β are constants greater than 0, to simulate learning in a critical period, they diminish with time following a hyperbolic function:

$$\alpha_{t+1} = \frac{\alpha_t}{1+at} \text{ and } \beta_{t+1} = \frac{\beta_t}{1+bt} \tag{14}$$

The initial values of the outputs y_k and of the synaptic weights $w1_k$ are random (between 0 and 1). The venous blood partial pressure and the synaptic weight are simulated for each tissue, as are the arterial blood partial pressure and chemoreceptor discharge frequency. Each element of the input vector $x1_k$ from the rostral nuclei to the NTS has a random value between 0.6 and 0.8. Over a series of iterations, the chemoreceptor frequency discharges (x2) reach the maximum value, then slowly, decays to zero (Fig 2). After reaching a maximum, the system evolves with fluctuations until optimal partial pressure of gases are achieved in all the tissues.

DISCUSSION

The experimental data strongly support that the nervous system stabilizes the long-term blood pressure and that the nucleus of tractus solitarius acts as a comparator. We propose a neural network model to explain the development of the nervous control of the cardiovascular system. Through the chemoreceptors' information, the model learns the discharge frequency of the sympathetic efferent of the tissues. Their values have to maintain a tissue blood flows so that the partial pressures of oxygen and carbon dioxide are near the normal values (pO_2=95 mmHg and pCO_2=40 mmHg for the artery and pO_2=40 mmHg and pCO_2=45 mmHg for the vein.) Then, auto-regulation can control the tissue blood flow to stabilize the partial pressures at these values. Changes in the renal output curve could be considered in the model by adding a constant to the curve of blood flow as a function of the sympathetic discharge.

In this way, during development the nervous system provides a reference for the feedback loop, it is in the chemoreceptors (perhaps the cardiopulmonary receptors but not

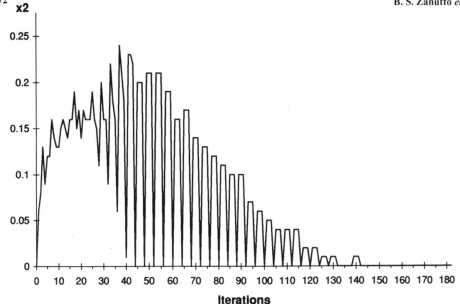

Figure 2. x2 simulates the chemoreceptor's discharge frequency. The following parameters are used: for the learning constant α: initial value $\alpha=0.3$, decrement value $a=0.0008$, for the learning constant β: initial value $\beta=0.23$, decrement value $b=0.0005$; $g_u=2$, $g_d=1.2$, and $d_s=0.25$. The network controls 50 tissues and it converges in 183 iterations.

the baroreceptors). These receptors have information about the partial pressure of CO_2 and O_2. Because the system is not completely developed, some sympathetic efferent frequency could be different from that specified by the chemoreceptors' feedback. When ontogeny is finished, because of the plastic changes in the nervous system, the pH is stabilized through the feedback of the chemoreceptors (and perhaps the cardiopulmonary receptors). In the cardiovascular system the blood flow is controlled by the arterial blood pressure, in this way the pH is regulated by the MABP. The baroreflex would then stabilize the instantaneous pressure value to the prevailing carotid pressure. In this model the reference of the feedback loop is not one value, instead it will be a group of inputs to the NTS that control the cardiovascular variables.

In the adult, the rostral structure to the NTS would provide the reference and would modulate the cardiovascular reflex. In this way the rostral structures play a functional role not only in the steady condition, but also for different behaviors and pathologies like neurogenic hypertension.

ACKNOWLEDGMENTS

We thank Dr. N. Schmajuk, J. Lamoureux, M. Cleaveland and J. Bentes for their comments on an earlier version of this manuscript.

REFERENCES

Bevan JA, Bevan RD, Ductless SP (1980) Adrenergic regulation of vascular smooth muscle. In: Bohr DF, Somlyo AP, Sparks HV (ed)The handbook of physiology: The cardiovascular system. Baltimore, 2 sec 2.

Boekkooi PF, Baan J, Teitel D, Rudolph AM (1992) Chemoreceptor responsiveness in fetal sheep. Am J Physiol 263:H162-H167.

Celander O (1954) The range of control exercised by the sympathico-adrenal system. Act Phys Scand 32:supp 116:45–49.

Cohen DH, Randall DC (1984) Classical Conditioning of cardiovascular responses. Ann Rev Physiol 46:187–197.

Dworkin BR, Miller NE (1977) Visceral learning in the curarized rat. In: Schwartz G, Beatty J (ed) Biofeedback: Theory and Research. Academic Press, New York, pp 221–242.

Dworkin BR (1986) Learning and long-term physiological regulation. In: Davidson RJ, Schwartz GE, Shapiro D (ed) Consciousness and Self-Regulation. Plenum Publishing Corporation, New York, 4 pp163–182.

Friedman WF, Pool PE, Jacobowitz D, Seagreen SC, Braunwald E (1968) Sympathetic enervation of the developing rabbit heart. Circ Res 23:25–32.

Galosy RA, Clarke LK, Vasko MR, Crawford IL (1981) Neurophysiology and neuropharmacology of cardiovascular regulation and stress. Neurosci and Biobehav Rev 5:137-175.

Grossberg (1982) How does a brain build a cognitive code ?. In: Grossberg S (ed) Studies of mind and brain. D. Reidel Publishing Company, Boston, pp 1–52.

Guyton AC (1986). Textbook of medical physiology (7th ed.). Saunders WB, Philadelphia.

Guyton AC (1991). Blood Pressure control-Special role of the kidneys and body fluids. Science 252:1813–1816.

Hebb DO (1949) The Organization of Behavior. Wiley, New York.

Itskovitz J, Rudolph AM (1987). Cardiorepiratory response to cyanide of arterial chemoreceptors in fetal lambs. Am J Physiol 252:H916-H922.

Kohonen T (1984) Self-organization and associative memory. Series in information science. Springer-Verlag, Berlin, Vol. 8.

Kohonen T (1987) Self-organization and associative memory. Springer-Verlag, Berlin.

Korner PI (1971) Integrative neural cardiovascular control. Ann Rev Physiol 51:312–367.

Kunze DL (1981) Rapid resetting of the carotid baroreceptor reflex in the cat. Am J Physiol 241:H802-H806.

Persson P, Ehmke H, Kirchheim H, Seller H (1988) Effect of sino-aortic denervation in comparison to cardiopulmonary deafferentation on long-term blood pressure in conscious dogs. Pfluegers Arch 411:160–166.

Persson P, Ehmke H, Kirchheim H (1989) Cardiopulmonary-arterial baroreceptor interaction in the control of blood pressure. NIPS 4:56–59.

Shepherd JT (1983) Circulation to skeletal muscle. In: Shepherd T, Abboud FM (ed) The handbook of physiology: The cardiovascular system. Williams, Wilkins, Baltimore, 3 sec 2.

SECTION IV

SYSTEMS

90

EFFECTS OF MEDIAL SEPTAL LESIONS

Implications for Models of Hippocampal Function

John J. Boitano,[1,2] Guido Bugmann,[2] Raju S. Bapi,[2] Susan L. McCabe,[2] Carl P. J. Dokla,[1] and Michael J. Denham[2]

[1]Department of Psychology
Fairfield University
Fairfield, Connecticut 06430-7524
[2]Neurodynamics Research Group
School of Computing
University of Plymouth
Plymouth PL4 8AA
United Kingdom

1. INTRODUCTION

The modern reformulation (O'Keefe & Nadel, 1978) of Tolman's (1948) cognitive map theory has largely been the result of two major findings which have provided the impetus for the direction of research over the last 2.5 decades; viz., rats with hippocampal lesions are deficient in negotiating such spatial memory tasks as the radial arm maze (Olton & Samuelson, 1976) and the water maze (Morris, Garrud, Rawlins & O'Keefe, 1982); and the discovery of place cells in the hippocampus which fire when the rat is in a specific location or place field (O'Keefe, 1976). The cognitive map theory of hippocampal functioning suggests that this structure processes spatial information detailing the rat's current position in the environment and provides the necessary computational skills allowing movement to a target goal (O'Keefe, 1989). The major neurotransmitter afferents to the hippocampus from the medial septal area and the associated vertical limb of the diagonal band of Broca (MSDB) include acetylcholine (Amaral & Kurz, 1985) projecting to widely distributed areas, and GABA (Freund & Antal, 1988) which innervate most of the GABA-containing interneurons in the hippocampus. MSDB GABA also projects inhibition to the lateral septal area which in turn receives the output of the hippocampal glutamatergic pyramidal cells. Some of these cells additionally terminate in MSDB. Lateral septal neurons do not, as previously suggested project to MSDB (Leranth, Deller & Buzsaki, 1992) but do project reciprocally to the hypothalamus which returns afferents to the hippocampus and MSDB.

Computational Neuroscience
edited by Bower, Plenum Press, New York, 1997

2. MEDIAL SEPTAL LESION STUDIES

2.1. Effects on Spatial Learning

In two separate experiments, male rats received electrolytic lesions of MSDB. Control animals received either sham-operations or no surgical procedure. In the first experiment (Boitano, Dokla, Parker, Norelli & Fiorini, 1990) animals were tested one week postoperatively in a standard version of the Morris water task which was conducted in blocks of two trials for 16 consecutive days. The results indicated that both the control and lesioned animals were statistically indistinguishable from one another in learning this spatial task. Approximately one month later, the submerged platform was moved to an adjacent quadrant and all subjects (Ss) were retested for four consecutive days. The MSDB Ss had a significantly longer mean escape latency than the control Ss.

This finding was the major impetus for the second study (Boitano, Small, Fiorini, Belanger, Savinelli & Dokla, 1992). An initial five-day adaptation period determined quadrant preference in which each S had to swim to one of two fixed submerged platforms located in either the NE or NW quadrants after starting from an unchanging south position. Platform preference was ascertained on the basis of the five trials on the fifth day. During acquisition lesioned and non-lesioned Ss were trained to locate a fixed submerged platform in either the preferred NE or NW quadrant and avoid a collapsible submerged platform in the non-preferred NW or NE quadrant after starting from the south position. An error was counted whenever any body part touched the collapsible platform. Following criterion performance (10 trials/day with at least 9 out of 10 being correct), the reversal paradigm was introduced. Each animal had to find the stable platform which had been moved to the adjacent NE or NW quadrant and avoid the collapsible platform which was then positioned in the quadrant previously containing the stable platform. Each animal went through 10 reversals. A repeated measures ANOVA revealed that the MSDB Ss took significantly longer in terms of days to acquire criterion than did the control Ss.

In summary, using the standard Morris water task procedure, MSDB animals exhibit no impairment in the acquisition of spatial navigation but when exposed to the reversal paradigm, in which the relation between the cues was changed from what was originally learned during acquisition, severe deficits were observed. Not only do these animals have greater variability in relearning the discrimination, but they also commit more errors by going more often to the previously correct location, and, therefore take longer to unlearn the previous association between the discriminative stimuli and its consequences that was developed during the previous reversal problems. It is suggested that the reversal paradigm is a complex task with both spatial and non-spatial aspects. For example, successful completion is contingent upon estimating the correct distance and angles of one's present position in the maze vis-a-vie the extramaze cues (a spatial component), observing the interrelationships among the cues and their discriminability (a spatial component), exhibiting adequate swimming skills (a non-spatial component, a motor skill), recalling the current goal location and its reward value (a non-spatial component, memory), and overcoming proactive interference from the prior reversal's goal location (a non-spatial component, learning).

3. NEUROPHYSIOLOGY OF THE SEPTOHIPPOCAMPAL SYSTEM

Rhythmical oscillations in the EEG patterns, called theta, recorded from intrahippocampal electrodes were first detailed by Green and Arduini in 1954 in the rabbit, cat and monkey. It was found that lesions of the septum or of the fimbria/fornix completely abolished the synchronous high amplitude slow (5–7 Hz) waves. Petsche, Stumpf and Gogolak (1962) observed rhythmically bursting neurons within the medial septal nuclei which set the hippocampal theta rhythm into resonance. Gogolak, Stumpf, Petsche and Sterc (1968) suggested that excitatory septal projections ended on the soma and basal dendrites of the pyramidal cells of the hippocampus causing activation with a recurrent collateral impinging on an inhibitory interneuron projecting to the apical dendrites of the same pyramidal cell. This alternation between rhythmical excitations and inhibitions was considered the cause of theta, and since no other afferents to MSDB had the same frequency as theta, the originating MSDB cells came to be known as pacemaker neurons.

If the hippocampal field potential is eradicated with lesions of MSDB, what happens to the activity of place cells? Mizumori, McNaughton, Barnes and Fox (1989) reversibly inactivated MSDB by tetracaine injection while the rats were engaged in a spatial working memory problem on a radial arm maze. It was found that: (a) theta was eliminated, (b) choice accuracy in the maze was temporally impaired, (c) place-specific firing of CA1 cells was unaffected, and (d) the firing rates of hilar/CA3 place cells were significantly reduced. The authors speculate that "CA1 pyramidal cell output may have been preserved because of a decline in feedforward inhibition that was proportional to the reduced rates in CA3...(and) that the remaining inputs from CA3 provided adequate spatially relevant information to CA1 pyramidal cells to allow apparently normal place-specific firing to occur" (p. 3925).

One final observation deserves mention. The animals in the Mizumori et al (1989) study had already learned the radial arm maze before medial septal inactivation, and therefore, the place cells were already established. This is radically different from Boitano et al (1990, 1992) in which the MSDB lesions were made before exposure to the spatial water maze procedure.

4. CURRENT MODELS OF HIPPOCAMPAL FUNCTION AND SPATIAL LEARNING

4.1. Spatial Memory Models

Many models of spatial memory found in the literature focus mainly on the learning of places in the environment through the activity of place-specific cells in the hippocampus. A number of these models have concentrated on the autoassociative property of the recurrent network in CA3.

Treves and Rolls (1992) have identified the need for two inputs to CA3. A strong input, probably the mossy fibers projecting from the dentate gyrus to CA3, forces new input patterns onto CA3 which then learn via recurrent connections. A weaker input, probably from the perforant path which projects from the entorhinal cortex to the dentate gyrus and to CA3, is later used to retrieve associatively stored patterns. This model makes no explicit use of the theta rhythm, cholinergic or GABAergic projections. It is unclear how this model may be incorporated into the model of a behaving animal.

In a discussion of Hebb-Marr networks, McNaughton and Nadel (1990) noted the prolonged reverberatory activity caused by associative recall and possible interference between successively recalled memories. To interrupt this reverberatory activity, a resetting mechanism is postulated with theta-induced rhythmic inhibition mediated by medial septal inputs to CA3 accomplishing this task. Although it has not been directly tested in this model, it does suggest an approach for incorporating simulated medial septal lesions. In addition, the model explicitly incorporates a memory-network into a larger system comprising sensory inputs and motor outputs. This model has, therefore, the potential for a behavioral application.

Hasselmo and Schnell (1994) combined computational modeling and brain slice physiology in depicting the role of medial septal inputs in the associative memory function of the hippocampus. The results suggested that the association of the patterns impinging on CA1 from both CA3 and the entorhinal cortex is stored in the form of synaptic modifications of the Schaffer collaterals that connect CA1 and CA3. Learning and recall dynamics are determined by the feedback regulation of the cholinergic input from the medial septum by the level of activity of CA1 cells. In other words, CA1 activity determines if the pattern is novel or familiar which, in turn, regulates the cholinergic input from the medial septal area. The effect of cholinergic modulation is to alter the synaptic connections (weights) at the Schaffer collaterals and CA1 junction. Thus, the medial septum imparts a facilitatory role onto the hippocampus in the learning of new patterns, and also by controlling the degree of interference between the old (familiar) and the new (novel) information. This model has been extended by Hasselmo (1995) to include self-organization of inputs from the entorhinal cortex to both the dentate gyrus and CA1. As in the previous model, novel patterns cause an increased cholinergic input from the medial septum which facilitates the learning of newly self-organized associations. Familiar patterns cause a decrement in the cholinergic modulation leading to recall of the old associated pattern corresponding to the current input. From these models the deficits in reversal learning due to MSDB lesions can be explained as a problem of memory interference. Both these models, however, ignore the role of GABA inputs from the medial septum to the hippocampus. This needs to be investigated in future computational models of septohippocampal interactions.

4.2. Spatial Navigation Models

These models focus mainly on the exploitation of the properties of place cells for navigating towards a goal. There is also a memory component regarding the storage of goal-location and the actions (motor patterns) leading to the goal.

Burgess, Recce and O'Keefe (1994) utilize the properties of place cells in depicting rodent spatial navigation. The specific phase of firing of place cells in relation to the phase of the theta rhythm is used to form representations where cells respond to places ahead of the rat's movement. These cells, which are located in the subiculum, store the position of objects of interest, such as food or obstacles. This model simulates behavior and therefore has the potential for testing the effects of medial septal lesions, via their effect on the theta rhythm. A foreseeable problem with the model is that when the theta phase reference is missing (as through medial septal lesioning), general learning is disrupted while the current data show only that reversal learning is affected.

Touretzky and Redish (1995) present a comprehensive model which includes both the learning of place cell firing in response to the locations of landmarks as well as navigation in an environment using these place cells. They also postulate the existence of a path integrator unencumbered by biological reality/neural substrate. At the start of the simulation, locations are selected at random from the task-environment. Place cells are then recruited and trained to respond to both external and internal state information. External state information constitutes egocentric angles and distances to all the visible landmarks from that location. Internal state information consists of path integrator coordinates giving the animal's position with respect to reference points, such as the location where reinforcement was obtained. Once a specified number of place cells have been established in this way, navigation is accomplished by the use of the difference between external visual information and the bearings obtained by place cell firing. This vector difference drives the system towards the goal location which is linked to the landmarks. The authors did not explicitly use the known anatomy of the hippocampus which, therefore, precludes medial septal involvement in this model.

5. CONSTRAINTS ON A MODEL OF SPATIAL LEARNING

The medial septal lesion experiments described above require that hippocampal models satisfy several criteria as well as overcoming certain problems that may be encountered in designing models of spatial learning.

    i. A model must be part of a control system with a measurable behavior. Ideally, it should simulate the rat swimming in the water maze with the appropriate dependent variables as escape latency and days-to-criterion being measured. This requirement has wide implications. It necessitates simulating perception, action generation, and certain emotional components to account for drive and learning/remembering. All these "accessories" are mediated by very complex brain structures which are technically impossible to simulate in detail. Therefore, gross simplifications must be made.

    ii. On the other hand, medial septal lesions act at a relatively detailed level, by removing certain types of neurotransmitters at the input of a certain class of cells in the hippocampus, or by modifying the temporal organization of spikes produced by place cells. Therefore, at least a part of the behavioral system must be modeled at a sufficient level of complexity. These two general constraints, of coarse and detailed modeling, call for a new design method addressing the problem of multiscale integration. Such a method could be investigated within the framework of spatial learning and the more general design of intelligent systems.

    iii. Medial septal projections reach not only the dentate gyrus, CA3 and CA1, but also such perihippocampal areas as the subiculum and the entorhinal cortex. Consequently, modeling of these regions at some level of detail may also be required. This is further supported by studies showing that hippocampal place cells do not show goal dependent firing (O'Keefe, 1989), which suggests that spatial learning of the goal location also involves these extra-hippocampal regions. This further indicates the need to model at least two spatial representations: (a) place cells in the hippocampus, and (b) some form of goal-location memory represented outside the hippocampus.

    iv. Sutherland, Kolb and Whishaw (1982) have observed deficits in medial frontally lesioned rats in the water maze similar to CA3–4 hippocampal lesions, which were interpreted as impairments of spatial learning. It would be interesting to confirm these results in the reversal paradigm. Nevertheless, there is a strong possibility that the frontal lobe may be involved in reversal tasks. On the one hand, this would extend the number of brain areas which would have to be modeled at some degree of intricacy. On the other hand, the identification of a convergence zone for hippocampal and frontal areas (perhaps the entorhinal cortex) may help restrict the number of brain regions which would have to be considered.

v. According to Hasselmo (1995), the medial septum provides a global control signal determining if learning should take place or not in the hippocampus. In a simulation, this function can be achieved with a single neuron, which reduces the computational load of a comprehensive model. However, in a biological system, a large number of neurons may be required because of the size of the areas to be innervated and because of the limited size of the axonal tree that a single projecting neuron can support. These projection limitations also apply to the afferents to MSDB; for instance, the local (place) cell firing properties of CA1 cells may also be reflected in the activation of MSDB. However, to ensure that the output of all neurons in the medial septum carries the same non-specific signal, one method is to provide them with strong lateral coupling. A side effect of such a coupling is the emergence of oscillations. Thus, there may be a link between the global control hypothesis and the existence of theta rhythm. However, the control hypothesis must be taken with caution, as it does not include a role for the strong GABA inhibitory projections from the medial septum to the hippocampus.

6. CONCLUSION

The experimental data referenced here indicate that medial septal lesions impaired selectively the learning of reversal tasks but not simple learning. No current model has the adequate mix of behavioral capabilities and level of physiological detail to reproduce these results. However, by combining selected features of several models it should be possible to build a comprehensive model to reproduce the lesion experiments. The discussion of septohippocampal interactions and the review of the current models has indicated the need for a review of experimental results involving medial septal projections into perihippocampal areas as well as the characterization of constraints for developing future models of spatial learning. The distinction between learning and relearning constitutes a new target and a very selective test for evaluating any future model of hippocampal function.

7. ACKNOWLEDGMENTS

Part of this research was supported by a visiting fellowship research grant, GR/K79734 to JJB and a postdoctoral research fellowship grant, GR/J42151 to RSB from Engineering and Physical Sciences Research Council, UK.

8. REFERENCES

Amaral, D. G. & Kurz, J. (1985). An analysis of the cholinergic and noncholinergic septal projections to the hippocampal formation of the rat. *Journal of Comparative Neurology*, **240**, 37–59.
Boitano, J. J., Dokla, C. P. J., Parker, S., Stalzer, K., Norelli, N. & Fiorini, M. (1990). Effect of medial septal lesions on activity and water maze performance. *Society for Neuroscience Abstracts*, **16**, 1248.
Boitano, J. J., Small, T., Fiorini, M. M., Belanger, S., Savinelli, T. & Dokla, C. P. J. (1992). Medial septal lesions impair spatial reversal learning. *Society for Neuroscience Abstracts*, **18**, 1421.
Burgess, N., Recce, M. & O'Keefe, J. (1994). A model of hippocampal function. *Neural Networks*, **7**, 1065–1081.

Freund, T. F. & Antal, M. (1988). GABA-containing neurons in the septum control inhibitory interneurons in the hippocampus. *Nature*, **336**, 170–173.

Gogolak, G., Stumpf, C., Petsche, H. & Sterc, J. (1968). The firing pattern of septal neurons and the form of the hippocampal theta wave. *Brain Research*, **7**, 201–207.

Green, J. D. & Arduini, A. A. (1954). Hippocampal electrical activity in arousal. *Journal of Neurophysiology*, **17**, 533–557.

Hasselmo, M. E. & Schnell, E. (1994). Laminar selectivity of the cholinergic suppression of synaptic transmission in rat hippocampal region CA1: Computational modeling and brain slice physiology. *The Journal of Neuroscience*, **14**, 3898–3914.

Hasselmo, M. E. (1995). A network model of hippocampus combining self-organization and associative memory function. *Proceedings of the World Congress on Neural Networks WCNN'95*, **2**, 909–912.

Leranth, C., Deller, T. & Buzsaki, G. (1992). Intraseptal connections redefined: lack of a lateral septum to medial septum path. *Brain Research*, **583**, 1–11.

McNaughton, B. L. & Nadel, L. (1990) Hebb-Marr networks and the neurobiological representation of action in space. In M. A. Gluck & D. E. Rumelhart (Eds.), *Neuroscience and connectionist theory* (pp. 1–63). Hillsdale, NJ, USA: Lawrence Erlbaum Associates.

Mizumori, S. J. Y., McNaughton, B. L., Barnes, C. A. & Fox, K. B. (1989). Preserved spatial coding in hippocampal CA1 pyramidal cells during reversible suppression of CA3c output: evidence for pattern completion in hippocampus. *Journal of Neuroscience*, **9**, 3915–3928.

Morris, R. G. M., Garrud, P., Rawlins, J. N. P. & O'Keefe, J. (1982). Place navigation impaired in rats with hippocampal lesions. *Nature*, **297**, 681–683.

O'Keefe, J. (1976). Place units in the hippocampus of the freely moving rat. *Experimental Neurology*, **51**, 78–109.

O'Keefe, J. (1989). Computations the hippocampus might perform. In L. Nadel, L. A. Cooper, P. Culicover & R. M. Harnish (Eds.), *Neural Connections and mental computation* (pp. 225–284). London, England: MIT Press.

O'Keefe, J. & Nadel, L. (1978). *The hippocampus as a cognitive map.*. Clarendon Press, Oxford.

Olton, D. S. & Samuelson, R. J. (1976). Remembrances of places past: spatial memory in rats. *Journal of Experimental Psychology: Animal Behavior Processes*, **2**, 97–116.

Petsche, H., Stumpf, C. & Gogolak, G. (1962). The significance of the rabbit's septum as a relay station between the midbrain and the hippocampus. 1. The control of hippocampal arousal activity by the septum cells. *Electroencephalography and Clinical Neurophysiology*, **14**, 202–211.

Sutherland, R. J., Kolb, B. & Whishaw, I. Q. (1982). Spatial mapping: Definitive disruption by hippocampal or medial frontal cortical damage in the rat. *Neuroscience Letters*, **31**, 271–276.

Tolman, E. C. (1948). Cognitive maps in rats and men. *Psychological Review*, **55**, 189–208.

Touretzky, D. S. & Redish, A. D. (1995). Landmark arrays and the hippocampal cognitive map. In L. Niklasson & M. Boden (Eds.), *Current trends in connectionism-Proceedings of the 1995 Swedish conference on connectionism* (pp. 1–13). Hillsdale, NJ, USA: Lawrence Erlbaum Associates.

Treves, A. & Rolls, E. T. (1992). Computational constraints suggest the need for two distinct input systems to the hippocampal CA3 network. *Hippocampus*, **2**, 189–200.

91

WEIGHT-SPACE MAPPING OF FMRI MOTOR TASKS

Evidence for Nested Neural Networks

Jeremy B. Caplan, Peter A. Bandettini,[*] and Jeffrey P. Sutton

Neural Systems Group and NMR Center
Massachusetts General Hospital, Harvard University Medical School
Building 149—9th Floor, Thirteenth Street
Charlestown, Massachusetts 02129
jcaplan@nmr.mgh.harvard.edu
pab@post.its.mcw.edu
sutton@nmr.mgh.harvard.edu

ABSTRACT

In this report, we examine the notion that networks of neocortical activity exist simultaneously across different spatial scales. Our investigations concern human motor data obtained from six motor tasks using functional magnetic resonance imaging (fMRI) at 1.5 T (Sutton et al., 1996). Weights for all pairs of fMRI voxels were calculated across tasks without anatomical bias, and weight maps were produced. We observed regions of interest, which averaged 1 cm^3 in size, and a theoretically predicted network two orders of magnitude larger that linked anatomically related regions together. Temporal analysis revealed the persistence of multilevel clustering through time. Our results support the concept that some dynamic networks in the neocortex are nested within other networks.

1. INTRODUCTION

New functional neuroimaging techniques are providing exciting opportunities to study spatiotemporal patterns of brain activity in ways that were unimaginable only a few years ago (Kwong et al., 1992; Ogawa et al., 1992). Spatially localized regions of activity

---

[*] Current address: Biophysics Research Institute, Medical College of Wisconsin, 8701 W. Watertown Plank Rd., Milwaukee, WI 53226.

Computational Neuroscience
edited by Bower, Plenum Press, New York, 1997

are temporally correlated with behavioral tasks, and there is growing evidence that regions modify with learning (e.g., Karni et al., 1995). There is considerable interest in the temporal course of activity within and between circumscribed regions, which themselves appear to form parts of large distributed brain networks. The similarity in problems faced in trying to understand network dynamics at the scale of neuroimaging data and at the scale of multiple unit recordings is striking.

We are beginning to examine an integration between computational neuroscience and neuroimaging as a means to better understand network organization across large brain regions, especially within the neocortex. Our approach is motivated by an analogy with work on small neural networks and by computational models of nested networks which specifically develop theoretical tools for traversing scales of neural organization. Building upon recent mathematical studies (Sutton and Anderson, 1995), we sought to test the hypothesis that networks of regional brain activity in the neocortex exist simultaneously across different orders of spatial magnitude. We further hypothesized that some networks are nested, and that the detection of such networks is not obvious by standard techniques. In an attempt to identify such functional networks and characterize their organization, we transformed functional images across tasks into a weight-space representation. This approach is similar to the way many small artificial neural networks are constructed.

2. METHODS

Human motor data were examined from the performance of six motor tasks using fMRI at 1.5 T (Sutton et al., 1996). The tasks consisted of simple repetitive finger movements, complex alternating finger movements and imagined complex finger movements. Each of these tasks was performed over ten trials for each hand. Each trial lasted 16 seconds and the trials were interspersed with 16 seconds of rest. Data were collected using the T_2^* BOLD contrast method (Bandettini et al., 1993), with a single-shot, blipped echo-planar imaging sequence (initial flip angle = $\pi/2$, TE = 40 ms, TR = 2000 ms). Both static and time-course functional maps were obtained by cross-correlation with a reference function, and time-averaged patterns of activity were generated for each task. Localized regions of interest (ROIs) averaged 1.0 cm^2 x 1.0 cm, with a maximum ROI = 3.5 cm^2 x 1.0 cm. Specific ROIs corresponded anatomically with the behavioral tasks [e.g., M1 activation with contralateral finger movements ($p < 0.005$)]. On the basis of signal intensity differences and cross-correlational analysis, there was no objective evidence for large-scale inter-connected networks among the ROIs, either within or between tasks.

Within the conceptual framework of modeling parallel distributed networks, we looked at how the connections among individual voxels represented functions across tasks. In particular, we devised a simple technique for discerning clusters of activity within and between different scales of organization.

For the static case, we considered the six time-averaged activation maps— one for each task. The activation signal was mapped onto:

$$A = \begin{cases} +1 & r > r_\tau \\ 0 & |r| \leq r_\tau \\ -1 & r < -r_\tau \end{cases}$$

(1)

where r is the Pearson cross-correlation co-efficient and r_τ is a threshold value. Voxels of varying magnitude of r above the threshold were dealt with similarly.

For each pair of voxels i,j within the fMRI slice (64 x 64 voxels), a weight w_{ij} was computed by summing the inner products of transformed signal intensity across all tasks:

$$w_{ij} = \sum_{k=1}^{N} A_i^k A_j^k$$

(2)

where A are the activation values and N is the number of tasks. Weight maps were produced consisting of lines connecting all pairs of voxels satisfying $|w_{ij}| \geq w_\tau$, where w_τ was a threshold weight value. No assumptions were made about anatomy or ROIs.

As each task followed a repetitive off-on paradigm with nine usable trials, time-course information was also available. To examine temporal evolution of the maps, the procedure for the static maps was applied to each of the nine off-on trials.

3. RESULTS

3.1. Functional Maps

The six time-averaged functional maps revealed complex patterns of activation, including the anticipated ROIs [e.g., primary and supplementary motor cortex, sensory cortex; Bandettini et al. (1993)]. No large-scale networks were apparent at this stage of analysis, nor was any nesting evident.

3.2. Static Weight Maps

$W_\tau=6$. At the highest weight value, a single, tight cluster of connections was observed measuring 1.0 cm^2 x 1.0 cm. This corresponded with a ROI in the right S1 (white lines in Figure 1).

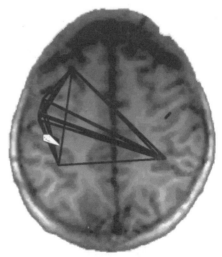

Figure 1. Static Weight Maps. Transverse slice of the human brain at a level through the neocortex. Overlaid upon the anatomical MRI scan, white lines represent connections with w_τ=6 and black lines represent connections with w_τ=5. The front of the brain appears at the top of the image. As per MRI convention, the left side of the brain is shown on the right side of the image and vice-versa.

threshold: w=6

threshold: w=5

$W_\tau=5$. At the next weight threshold, a large, fully-connected network spanning a considerable portion of both hemispheres emerged (black lines in Figure 1), with nodes at bilateral S1 and motor cortex, including the small $w_\tau=6$ cluster as a node. The lack of symmetry may have been due, in part, to the (mis)-alignment of the head within the scanner. This large network was approximately two orders of magnitude larger than the $w_\tau=6$ network nested within it (large network 70 cm^2 x 1.0 cm). It was not observed at the higher threshold.

$W_\tau=4$. At an even lower weight threshold, the $w_\tau=4$ network (not shown) covered the cortex much more densely, including the ROIs that had been identified from the functional maps. However, the lines at this threshold did not cover the whole cortical slice.

3.3. Temporally Varying Weight Maps

Weight values in the static and temporal cases differed in their statistical properties, due to the greater temporal averaging in the static relative to the temporal maps. Neverthe-

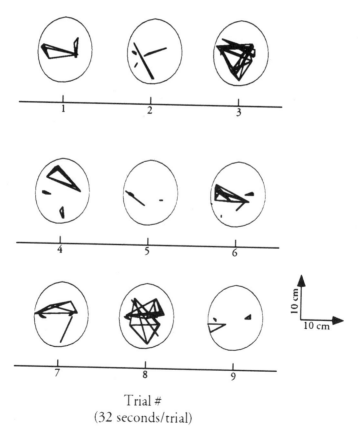

Figure 2. Temporally Varying Weight Maps. Nine images are displayed corresponding to the $w_\tau=6$ weight maps for each of the trials across the six tasks. The time axis refers to the trial number. Each image is oriented the same as in Figure 1, and the outline of the cortical slice appears in grey.

less, fully-connected networks were recovered at different spatial scales throughout the course of the experiment. The tasks showed temporal variability within and between scales of organization (Figure 2).

4. DISCUSSION

We found weight maps to be an interesting and potentially useful way to visualize patterns of functional brain activity across time and across tasks. The fully-connected nature of the clusters observed at $w_\tau=6$ and $w_\tau=5$ suggests that cohesive networks may mediate related brain functions. The relationship between functional clustering of MRI data and behavior was strongly dependent upon thresholding. In particular, thresholding at higher weight values was tantamount to selecting for more strongly coupled brain regions. Alternatively, one could view weight-value thresholding as a method of drawing cluster boundaries at a specific strength. Even at $w_\tau=4$, for the static case, no lines were found connecting voxels in the skull or empty space outside the brain where there was considerable noise.

While it is difficult to interpret patterns from the temporal analysis, it was evident that changes in activity were present during repeated task performance. This type of analysis may be helpful in studies of learning and cortical reorganization.

In conclusion, the appearance of functional clustering at distinctly different spatial scales, and the nesting of these clusters, suggests that some motor tasks activate brain regions that are linked by dynamic networks at different scales. One scale consists of discrete ROIs and a second is manifest by networks linking behaviorally related ROIs. Only by looking *across* tasks was it possible to identify these higher-level functional networks. Finally, networks at both scales appeared to have similar properties.

Supported by NIH grant MH01080 and ONR grant N00014-97-1-0093.

5. REFERENCES

Karni, A., Meyer, G., Jezzard, P., Adams, M. M., Turner, R., Ungerleider, L. G. 1995. Functional MRI evidence for adult motor cortex plasticity during motor skill learning. *Nature:* 377:155–158.

Kwong, K. K., Belliveau, J. W., Chesler, D. A. et al. 1992. Dynamic magnetic resonance imaging of human brain activity during primary sensory stimulation. *Proc. Natl. Acad. Sci. USA:* 89: 5675–5679.

Ogawa, S., Tank, D. W., Menon, R. et al. 1992. Intrinsic signal changes accompanying sensory stimulation: Functional brain mapping with magnetic resonance imaging. *Proc. Natl. Acad. Sci. USA:* 89: 5951–5955.

Sutton, J. P. & Anderson, J. A. 1995. Computational and neurobiological features of a network of networks. In: Bower, J. M. (ed.), *Neurobiology of Computation.* Boston: Kluwer Academic. 317–322.

Sutton, J. P., Caplan, J. B. & Bandettini, P. A. 1996. fMRI evidence of nested networks associated with motor tasks [abstract]. *Human Brain Mapping.* 3:S370.

92

TEMPORAL CODING OF SENSORY INFORMATION

Peter A. Cariani

Eaton Peabody Laboratory
Massachusetts Eye & Ear Infirmary
243 Charles St., Boston, Massachusetts 02114
peter@epl.meei.harvard.edu

1. GENERAL CLASSES OF NEURAL CODES

The neural coding problem—how populations of neurons represent and convey information—is fundamental to understanding how the brain works as an information-processing system. Three independent aspects of any time-varying signal are the physical channel through which the signal is transmitted, its internal temporal structure (e.g. its Fourier spectrum), and its absolute time of arrival. Any of these aspects of a signal can be used to convey what kind of distinction is being conveyed, be it a pitch, or a color, or a smell. Correspondingly, neural spike codes can be classified into "connectivity-based codes" in which the meaning of a spike train depends upon which specific neuron produced it (e.g. "place" or "labelled line" codes) and "temporal codes" in which meaning depends upon specific timings of spikes. "Temporal codes" can be further subdivided into "time-of-arrival codes" (e.g. latency and synchrony codes) vs. "temporal pattern codes" (e.g. interspike interval and higher-order pattern codes).[6,53] Since these aspects of spike codes are complementary, combinations of these primitive coding strategies are also possible (Fig. 1). While some evidence for temporal coding exists in each sensory modality, relatively few attempts have been made to gather together existing physiological and psychophysical evidence to search for common temporal coding strategies.[2,5,6,10,48,49,53,60,67,76]

2. RELATIVE LATENCY CODES IN SENSORY SYSTEMS

The relative time-of-arrival of an external stimulus at different parts of the body is a highly robust cue for the direction of that stimulus. Receptor systems translate relative-times-of-arrival at spatially-dispersed receptor sites into relative spike latencies in their associated sensory pathways, and these in turn can be analyzed more centrally using sets of

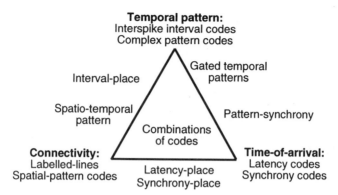

Figure 1. A space of possible pulse-codes.

delay lines and coincidence detectors. Such canonical coding strategies are found over wide phylogenetic spans.[10,25]

In auditory localization, stimuli create azimuth-specific interaural time differences that precisely encode neurally by latency differences of spikes produced in each monaural pathway. Cross-correlation operations using coincidence detectors and tapped delay lines provide a readout of interaural time delays, and consequently, of azimuth estimates.[11,12,33] More central representations of auditory space may entail more complex latency patterns[46] and/or population-latency profiles[4].

Echolocation systems use latency differences between sounds emitted by the animal and their return echoes. In bat and cetacean echolocation systems, precise time-of-arrival of echo-evoked spikes relative to cry-evoked ones permit extremely precise estimation of target ranges and shapes.[64,65] Analogous temporal cross-correlation strategies in electroception represent the space around the body by latency differences of spikes produced at different body locations in response to an emitted, sinusoidally-varying electrical field deformed by the presence of nearby objects.[10,24] These "phase offset" codes are structurally similar to those being contemplated for the hippocampus[26] and other CNS stations,[25] albeit with different time scales.

In the fly visual system, cross-correlations between the latencies of small numbers of stimulus-locked spikes are used to rapidly detect motion (at CNS*96 Strong reported precisions of < 1 ms).[1,18,54,56,60] If such stimulus locking were utilized in human vision, then depth illusions produced by binocular time differences (Pulfrich Effect) could be seen in terms of binocular temporal cross-correlation operations not unlike binaural ones in audition.

Even in the chemical senses with their long response latencies, variances of latencies across neural populations can nevertheless be relatively small, such that small differences of arrival times (~1–2 ms) reportedly can be registered and used to localize.[3,71,72,73,74,75]

3. TEMPORAL PATTERN CODES

While relative latency codes depend upon arrival times of spikes with respect to some external time reference, temporal pattern codes depend upon the internal time structures of spike trains. A very general hypothesis is that latency codes, being independent of the internal structure of a stimulus, are best adapted for localization tasks, while temporal pattern codes, being independent of the relative times of arrival on different receptor surfaces, are better suited to representing the form of the stimulus.

3.1 Phase-Locked Receptor Systems

To the extent that receptors follow the stimulus waveform, the temporal structure of that waveform is impressed upon the discharge patterns that are produced in primary sensory neurons. In simplest terms, the discharge times are threshold-crossing times in an integrate-to-threshold element, such that intervals between spikes reflect time intervals present in the stimulus waveform. In a phase-locked system, the all-order interval distribution of each primary sensory neuron constitutes a direct, temporal neural representation of the short-time autocorrelation function of the stimulus, taking into account how the stimulus has been filtered by the sensory organ.

Thus, in those sensory systems where spikes are locked to stimulus transients and ongoing periodicities (audition, vibration perception, vision), direct temporal representation of the form of the stimulus is possible. The autocorrelation function of the stimulus has the same information as its power spectrum, so many analyses of complex perceptual form that can be implemented by recognizing patterns in power spectra (e.g. pitch, timbre, consonance, phonetic quality, rhythm, vibrotactile frequency and complex pattern) can analogously be carried out by means of an autocorrelation analysis, i.e. by recognizing complex patterns of intervals.

In the auditory system, from physiological studies and computer simulations it appears that such autocorrelation analyses account for a large number of complex pitch phenomena at the level of the auditory nerve and brainstem: the pitch of the "missing fundamental", the perceptual equivalence of spectrally-diverse stimuli, the relative invariance of pitch and pitch strength with sound pressure level, pitches of stimuli with unresolved harmonics, as well as complex pitch shifts that are heard for inharmonic stimuli.[8,9,38,39,44,58,63,66] Characteristic patterns in population interval distributions are also observed for vowels[6,52] and musical intervals[7]. In the somatosensory system, coding of flutter-vibrations and of complex tactile patterns may well be explicable in analogous terms.[47,50,61]

Could visual form be encoded in time intervals in an analogous manner? The eyes are in constant (slow) drift even during fixation, and many central visual neurons are known to discharge with relatively precise latencies when contrast gradients (edges, spatial transients) cross their receptive fields. Under such conditions stimulus-synchronizations down to millisecond resolutions are observed[32] such that the resulting post-stimulus time histograms[55] replicate the spatial structure of the drifting image. Precisely replicating spike triplet patterns similarly suggest that such timing precisions exist[35,36,37] If scanning is achieved by slow drifts of the image across the retina coupled with phase-locking, then spatial intervals are converted into time intervals, and all-order interspike interval distributions (autocorrelation histograms[55]) carry the (running) spatial autocorrelation component in the direction of drift. As in the auditory example, this is a means of carrying out a (spatial) Fourier analysis in the time domain, where frequencies are carried in time intervals and phases carried in relative latencies. Such mechanisms couple directly to autocorrelation-based models of visual form and texture.[51,57,68,69]

There is also the phenomenon of the "missing fundamental" in vision, wherein multiple spatial harmonics (4+5+6) are easily matched to their "missing" spatial fundamental.[15,23] If spatial periods are converted to time intervals by scanning, then the most common time interval in the sensory array will correspond to the spatial fundamental that

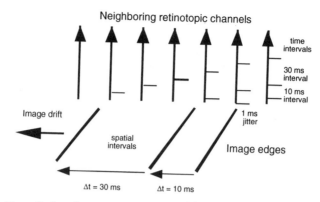

Figure 2. Scanning processes convert spatial intervals into temporal ones.

is seen, just as the most common interval in the auditory nerve corresponds to the "missing fundamental" pitch that is heard.[8,9]

3.2 Non Phase-Locked Receptor Systems

In sensory systems where spikes are not locked to stimulus waveforms (color vision, chemical senses), temporally-structured neural responses characteristic of particular stimulus qualities may nevertheless still be produced through the kinetics of receptor activations and lateral inhibition. A considerable literature exists on color percepts evoked by temporally-structured white light (the Benham Top, "subjective" or "achromatic" colors).[20,29,31,62] Wavelength-dependent interval patterns[30] latency patterns, and characteristic time courses of discharge[59,70] have also been observed. Achromatic visual form and other characteristics of the visual scene also have temporal correlates.[13,59,70]

Interval sequences may subserve the central code for pain.[19] In primary gustatory neurons of the rat, neural responses exhibit tastant-specific temporal discharge patterns.[14,16] In olfaction, odorant-specific time patterns have been observed,[34,40,41,42] but the analysis of temporal codes here has been hampered by complexities caused by concentration effects,[27,28,45] intrinsic neural rhythms, and history dependent responses.[21,22] Temporal coding may provide a solution for the problematic nature of "across-neuron" rate pattern codes given the ideosyncratic, broadband nature of chemical responsiveness. Because temporal patterns are largely independent of each other (transparent, superposable), even a few temporal pattern primitives, are sufficient to construct a rich space of sensory qualities (of dimensionality N-1) from their ratios.

4. PERCEPTS EVOKED BY TEMPORALLY-PATTERNED ELECTRICAL STIMULI

The ability of temporally-structured electrical stimuli to evoke specific percepts given the nonspecific excitation of whole populations of neurons is strongly suggestive that a particular time-pattern plays a functional role in sensory coding. Different pitches are evoked by different stimulation periodicities in single-channel cochlear electrodes.[17,43] Sensations of flutter-vibration are similarly evoked by periodic electrical stimulation of the skin.[50] Temporally-patterned electrical stimulation of the retina using patterns not unlike those of the Benham Top produces correspondingly colored phosphenes.[77] Electrical stimulation of the gustatory systems of rats using time-patterns observed in single units in response to particular tastants elicits oro-facial responses normally associated with that tastant, unlike those seen when the time-patterns are scrambled.[14,16] Similarly, particular time-patterns of central stimulation can either elicit or mask out pain.[19]

5. CONCLUSIONS

Clearly a considerable amount of evidence currently exists for temporal coding of many different kinds of sensory information, even in those modalities, such as in vision and the chemical senses, where one might least expect it. The existence of such widespread time patterns and their close linkages to perception begs the question of what kinds of neural architectures are capable of making effective use of time structure in their inputs.

6. ACKNOWLEDGMENTS

This work was supported by NIH Grant DC03054 from the National Institute for Deafness and Communications Disorders, a part of the National Institutes of Health.

REFERENCES

1. Bialek, W., F. Rieke, R.R. van Stevenink and W. de Ruyter, D., 1991, Reading a neural code, *Science* **252**(28 June): 1854–1856.
2. Boring, E.G. 1942, *Sensation and Perception in the History of Experimental Psychology*, Appleton-Century-Crofts, New York.
3. Bower, T.G.R., 1974, The evolution of sensory systems, in: *Perception: Essays in Honor of James J. Gibson*, (R. B. MacLeod and H. Pick Jr., ed.), Cornell University Press, Ithaca.
4. Brugge, J.F., R.A. Reale and J.E. Hind, 1996, The structure of spatial receptive fields of neurons in primary auditory cortex of the cat, *J. Neurosci.* **16**(14): 4420–4437.
5. Bullock, T.H., 1967, Signals and neural coding, in: *The Neurosciences: A Study Program*, (G. C. Quarton, T. Melnechuck and F. O. Schmitt, ed.), Rockefeller University Press, New York.
6. Cariani, P., 1995, As if time really mattered: temporal strategies for neural coding of sensory information, *Communication and Cognition - Artificial Intelligence (CC-AI)* **12**(1–2): 161–229. Preprinted in: *Origins: Brain and Self-Organization*, (Pribram, K., ed.), Lawrence Erlbaum, Hillsdale, NJ.
7. Cariani, P., 1996, Temporal coding of musical form, *Proceedings, International Music Perception Conference, Montreal, August, 1996.* : 425–430.
8. Cariani, P.A. and B. Delgutte, 1996, Neural correlates of the pitch of complex tones. I. Pitch and pitch salience., *J. Neurophysiol.* **76**(3): 1698–1716.
9. Cariani, P.A. and B. Delgutte, 1996, Neural correlates of the pitch of complex tones. II. Pitch shift, pitch ambiguity, phase-invariance, pitch circularity, and the dominance region for pitch, *J. Neurophysiol.* **76**(3): 1717–1734.
10. Carr, C.E., 1993, Processing of temporal information in the brain, *Annu. Rev. Neurosci.* **16**: 223–243.
11. Carr, C.E. and M. Konishi, 1990, A circuit for detection of interaural time differences in the brain stem of the barn owl, *J. Neuroscience* **10**(10): 3227–3246.
12. Casseday, J.H. and E. Covey, 1987, Central auditory pathways in directional hearing, in: *Directional Hearing*, (W. A. Yost and G. Gourevitch, ed.), Springer Verlag, New York.
13. Chung, S.H., S.A. Raymond and J.Y. Lettvin, 1970, Multiple meaning in single visual units, *Brain Behav Evol* **3**: 72–101.

14. Covey, E. *Temporal Neural Coding in Gustation.* Ph.D., Duke University (1980)

15. de Valois, R.L. and K.K. de Valois. 1990, *Spatial Vision,* Oxford University Press, Oxford.

16. Di Lorenzo, P.M. and G.S. Hecht, 1993, Perceptual consequences of electrical stimulation in the gustatory system, *Behavioral Neuroscience* **107**: 130–138.

17. Eddington, D.K., W.H. Dobelle, D.E. Brackman, M.G. Mladejovsky and J. Parkin, 1978, Place and periodicity pitch by stimulation of multiple scala tympani electrodes in deaf volunteers, *Trans. Am. Soc. Artif. Intern. Organs* **24**: 1–5.

18. Egelhaaf, M. and A. Borst, 1993, A look into the cockpit of the fly: visual orientation, algorithms, and identified neurons, *J. Neurosci.* **13**(11): 4563–4574.

19. Emmers, R. 1981, *Pain: A Spike-Interval Coded Message in the Brain,* Raven Press, New York.

20. Festinger, L., M.R. Allyn and C.W. White, 1971, The perception of color with achromatic stimulation, *Vision Res.* **11**: 591–612.

21. Gesteland, R.C., J.Y. Lettvin and W.H. Pitts, 1965, Chemical transmission in the nose of the frog, *J. Physiol.* **181**: 525–559.

22. Gesteland, R.C., J.Y. Lettvin, W.H. Pitts and S.H. Chung, 1968, A code in the nose, in: *Cybernetic Problems in Bionics,* (H. L. Oestereicher and D. R. Moore, ed.), Gordon and Breach, New York.

23. Hammett, S.T. and A.T. Smith, 1994, Temporal beats in the human visual system, *Vision Res* **34**(21): 2833–2840.

24. Heiligenberg, W., 1994, The coding and processing of temporal information in the electrosensory system of the fish, in: *Temporal coding in the brain,* (G. Buzaki, R. Llinas, W. Singer, A. Berthoz and Y. Christen, ed.), Springer-Verlag, Berlin.

25. Hopfield, J.J., 1995, Pattern recognition computation using action potential timing for stimulus representation, *Nature* **376**: 33–36.

26. Jenson, O. and J.E. Lisman, 1996, Hippocampal CA3 region predicts memory sequences: accounting for the phase precession of place cells, *Learning & Memory* **3**: 279–287.

27. Kauer, J.S., 1974, Response patterns of amphibian olfactory bulb neuones to odour stimulation, *J. Physiol.* **243**: 695–715.

28. Kauer, J.S., 1990, Temporal patterns of membrane potential in the olfactory bulb observed with intracellular recording and voltage-dye imaging: early hyperpolarization, in: *Chemosensory Information Processing,* (D. Schild, ed.), Springer Verlag, Berlin.

29. Kaufman, L. 1974, *Sight and Mind: An Introduction to Visual Perception,* Oxford University Press, New York.

30. Kozak, W.M. and H.J. Reitboeck, 1974, Color-dependent distribution of spikes in single optic tract fibers of the cat, *Vision Research* **14**: 405–419.

31. Kozak, W.M., H.J. Reitboeck and F. Meno, 1989, Subjective color sensations elicited by moving patterns: effect of luminance, in: *Seeing Contour and Colour,* (J. J. Kulikowski, Dickenson, C.M., ed.), Pergamon Press, New York.

32. Kreiter, A. and W. Singer, 1996, Stimulus-dependent synchronization of neuronal responses in the visual cortex of the awake macaque monkey, *J. Neurosci.* **16**(7): 2381–2396.

33. Kuwada, S. and T.C.T. Yin, 1987, Physiological studies of directional hearing, in: *Directional Hearing,* (W. A. Yost and G. Gourevitch, ed.), Springer Verlag, New York.

34. Laurent, G. and H. Davidowitz, 1994, Encoding of olfactory information with oscillating neural assemblies, *Science* **265**(23 Sept): 1872–1875.

35. Lestienne, R., 1996, Determination of the precision of spike timing in the visual cortex of anesthetized cats, *Biol. Cybern.* **74**: 55–61.

36. Lestienne, R., E. Gary-Bobo, J. Przybyslawski, P. Saillour and M. Imbert, 1990, Temporal correlations in modulated evoked responses in the visual cortical cells of the cat, *Biol. Cybern.* **62**: 425–440.

37. Lestienne, R. and B.L. Strehler, 1988, Differences between monkey visual cortex cells in triplet and ghost doublet informational symbols relationships, *Biol. Cybern.* **59**: 337–352.

38. Licklider, J.C.R., 1951, A duplex theory of pitch perception, *Experientia* **VII**(4): 128–134.

39. Licklider, J.C.R., 1959, Three auditory theories, in: *Psychology: A Study of a Science. Study I. Conceptual and Systematic,* (S. Koch, ed.), McGraw-Hill, New York.

40. Macrides, F., 1977, Dynamic aspects of central olfactory processing, in: *Chemical Signals in Vertebrates,* (D. M. Schwartze and M. M. Mozell, ed.), Plenum, New York.

41. Macrides, F. and S.L. Chorover, 1972, Olfactory bulb units: activity correlated with inhalation cycles and odor quality, *Science* **175**(7 January): 84–86.

42. Marion-Poll, F. and T.R. Tobin, 1992, Temporal coding of pheromone pulses and trains in Manduca sexta, *J Comp Physiol A* **171**: 505–512.

43. McKay, C.M., H.J. McDermott and G.M. Clark, 1994, Pitch percepts associated with amplitude-modulated current pulse trains in cochlear implantees, *J. Acoust. Soc. Am.* **96**(5): 2664–2673.

44. Meddis, R. and M.J. Hewitt, 1991, Virtual pitch and phase sensitivity of a computer model of the auditory periphery. I. Pitch identification, *J. Acoust. Soc. Am.* **89**(6): 2866–2882.

45. Meredith, M. and D.G. Moulton, 1978, Patterned response to odor in single neurones of goldfish olfactory bulb: influence of odor quality and other stimulus parameters, *J. Gen. Physiol.* **71**: 615–643.

46. Middlebrooks, J.C., A.E. Clock, L. Xu and D.M. Green, 1994, A panoramic code for sound location by cortical neurons, *Science* **264**: 842–844.

47. Morley, J.W., J.S. Archer, D.G. Ferrington, M.J. Rowe and A.B. Turman, 1990, Neural coding of complex tactile vibration, in: *Information Processing in Mammalian Auditory and Tactile Systems,* ed.), Alan R. Liss, Inc,

48. Morrell, F., 1967, Electrical signs of sensory coding, in: *The Neurosciences: A Study Program,* (G. C. Quarton, T. Melnechuck and F. O. Schmitt, ed.), Rockefeller University Press, New York.

49. Mountcastle, V., 1967, The problem of sensing and the neural coding of sensory events, in: *The Neurosciences: A Study Program,* (G. C. Quarton, Melnechuk, T., and Schmitt, F.O., ed.), Rockefeller University Press, New York.

50. Mountcastle, V., 1993, Temporal order determinants in a somatosthetic frequency discrimination: sequential order coding, *Annals New York Acad. Sci.* **682**: 151–170.

51. Pabst, M., H.J. Reitboeck and R. Eckhorn, 1989, A model of preattentive texture region definition based on texture analysis, in: *Models of Brain Function,* (R. M. J. Cotterill, ed.), Cambridge University Press, Cambridge.

52. Palmer, A.R., 1992, Segregation of the responses to paired vowels in the auditory nerve of the guinea pig using autocorrelation, in: *The Auditory Processing of Speech,* (S. M.E.H., ed.), Mouton de Gruyter, Berlin.

53. Perkell, D.H. and T.H. Bullock. Neural Coding. (1968)

54. Poggio, T. and W. Reichardt, 1976, Visual control of orientation behaviour in the fly. Part II. Towards the underlying neural interactions, *Quart. Rev. Biophys.* **9**(3): 377–438.

55. Pollen, D.A., J.P. Gaska and L.D. Jacobson, 1989, Physiological constraints on models of visual cortical function, in: *Models of Brain Function,* (R. M. J. Cotterill, ed.), Cambridge University Press, Cambridge.

56. Reichardt, W., 1961, Autocorrelation, a principle for the evaluation of sensory information by the central nervous system, in: *Sensory Communication,* (W. A. Rosenblith, ed.), MIT Press/John Wiley, New York.

57. Reitboeck, H.J., M. Pabst and R. Eckhorn, 1988, Texture description in the time domain, in: *Computer Simulation in Brain Science,* (R. M. J. Cotterill, ed.), Cambridge University Press, Cambridge, England.

58. Rhode, W.S., 1995, Interspike intervals as correlates of periodicity pitch in cat cochlear nucleus, *J. Acoust. Soc. Am.* **97**(4): 2414–2429.

59. Richmond, B.J., L.M. Optican and T.J. Gawne, 1989, Neurons use multiple messages encoded in temporally modulated spike trains to represent pictures, in: *Seeing Contour and Colour,* (J. J. Kulikowski and C. M. Dickenson, ed.), Pergamon Press, New York.

60. Rieke, F., D. Warland, R. de Ruyter van Stevenick and W. Bialek. 1997, *Spikes: Exploring the Neural Code,* MIT Press, Cambridge.

61. Rowe, M., 1990, Impulse patterning in central neurons for vibrotactile coding, in: *Information Processing in Mammalian Auditory and Tactile Systems,* ed.), Alan R. Liss, Inc,

62. Sheppard, J.J. 1968, *Human Color Perception: A Critical Study of the Experimental Foundation,* American Elsevier, New York.

63. Simmons, A.M. and M. Ferragamo, 1993, Periodicity extraction in the anuran auditory nerve, *J. Comp. Physiol. A* **172**: 57–69.

64. Simmons, J.A., 1987, Directional hearing and sound localization in echolocating animals, in: *Directional Hearing,* (W. A. Yost and G. Gourevitch, ed.), Springer Verlag, New York.

65. Simmons, J.A., M. Ferragamo, C. Moss, F., S.B. Stevenson and R.A. Altes, 1990, Convergence of temporal and spectral information into acoustic images of complex sonar targets perceived by the echolocating bat, *Eptesicus fuscus.*, *J. Comp. Physiol. A* **167**: 589–616.

66. Slaney, M. and R.F. Lyon, 1993, On the importance of time - a temporal representation of sound, in: *Visual Representations of Speech Signals,* (M. Cooke, S. Beet and M. Crawford, ed.), John Wiley, New York.

67. Uttal, W.R. 1973, *The Psychobiology of Sensory Coding,* Harper and Row, New York.

68. Uttal, W.R. 1975, *An Autocorrelation Theory of Form Detection,* Wiley, New York.

69. Uttal, W.R. 1988, *On Seeing Forms,* Lawrence Erlbaum, Hillsdale, NJ.

70. Victor, J.D., K. Purpura, E. Katz and B. Mao, 1994, Population encoding of spatial frequency, orientation, and color in Macaque VI, *J. Neurophysiol.* **72**(5): 2151–2166.

71. von Bekesy, G., 1955, Human skin perception of travelling waves similar to those on the cochlea, *J. Acoust. Soc. Am.* **27**(5): 830–841.

72. von Bekesy, G., 1962, Synchrony between nervous discharges and periodic stimuli in hearing and on the skin, *Annals of Otology, Rhinology, and Laryngology* **71**(3): 678–692.

73. von Bekesy, G., 1963, Interaction of paired sensory stimuli and conduction in peripheral nerves, *Journal of Applied Physiology* **18**(6): 1276–1284.

74. von Bekesy, G., 1964, Olfactory analogue to directional hearing, *Journal of Applied Physiology* **19**(3): 369–373.

75. von Bekesy, G., 1964, Rythmical variations accompanying gustatory stimulation observed by means of localization phenomena, *Journal of General Physiology* **47**(5): 809–825.

76. Wasserman, G.S., 1992, Isomorphism, task dependence, and the multiple meaning theory of neural coding, *Biol. Signals* **1**: 117–142.

77. Young, R.A., 1977, Some observations on temporal coding of color vision: psychophysical results, *Vision Research* **17**: 957–965.

93

TESTING THE ROLE OF ASSOCIATIVE INTERFERENCE AND COMPOUND CUES IN SEQUENCE MEMORY

Frances S. Chance and Michael J. Kahana

Volen Center for Complex Systems
Brandeis University
Waltham, MA
E-mail: chance@binah.cc.brandeis.edu
E-mail: kahana@fechner.ccs.brandeis.edu

ABSTRACT

A major question facing computational models of human memory concerns the storage and retrieval of sequentially processed information. Many current models assume a chaining of associations. According to this view, each item, or memory pattern, is associated with the preceding item in the sequence. In reproducing the sequence, each recalled item serves as a retrieval cue for the next item. A serious problem facing these chaining models is their susceptibility to associative interference. This paper presents a novel experimental method designed to assess the effects of associative interference in the retrieval of ordered lists of items. Experimental findings presented here suggest that subjects use multiple prior items, as well as context, to overcome the effects of associative interference in list recall.

1. INTRODUCTION

The hypothesis that the pattern of neural activity at time t is associated with the pattern of neural activity at time $t + \tau$ is central to most neural network and abstract mathematical models of sequence memory (Abbott & Blum, 1996; Lewandowsky & Murdock, 1989; Sompolinsky & Kanter, 1986; Kleinfeld, 1986; Riedel, Kuhn & van Hemmen, 1988). This notion of chained associations, in which each recalled item in a sequential list facilitates recall of the next item, formed the cornerstone of Ebbinghaus's (1885) seminal research on human learning. During the 1960's experimental psychologists began to find holes in the notion of pure-chaining and proposed that chained, positional, and compound item associations were all factors in ordered recall (see Young, 1968 for a review). These early studies, based on the "transfer of training" technique, failed to provide consistent evidence for or against a particular associative model.

A basic problem for all associative models is how subjects cope with interference from competing associations. Consider the challenge facing subjects when two lists containing a common element must be reproduced (e.g., A-B-C-D-E-F and P-Q-R-B-S-T). According to a chaining model the common element (B) is involved in two competing associations (B-C and B-S). Consequently, recall of the list-appropriate item following the shared element will be drastically impaired.

In this paper, a direct approach to studying the problem of associative interference in sequence memory is presented. In this new methodology, human subjects learn two sequences of unique items (common words) to a performance criterion. If the sequences share a single item, the two sequences may be thought of as "crossing paths". If the sequences share a sub-sequence of items, the sequences will have joined and then divided. Consider the sequences A-B-C-D-E-F-G-H and U-V-W-X-E-F-Y-Z. These two sequences contain a common subsequence E-F. In a pure-chaining model in which each recalled item facilitates recall of the following item, F is now associated with both G and Y. Thus a decrease in performance would be expected when the subject tries to recall the letter following F. However, in the case of a compound cue model, in which some combination of prior items facilitates recall of the following list item, there is no ambiguity between the cues for G and Y and the decrease in performance should be attenuated. Rather than measuring accuracy, we have chosen to examine a more sensitive measure of performance – namely, the inter-response time (IRT) at the exit transition out of the shared sub-sequence (i.e. F-G or F-Y).

2. METHODS

Twenty four paid volunteers from the Brandeis community participated in one practice session followed by four experimental sessions. In each experimental session, subjects learned two lists of fifteen words until they could recite them perfectly on three successive trials following a short distractor task. The words were randomly sampled from among the 300 highest frequency nouns in the Toronto Word Pool (Friendly, et al., 1982) and displayed, one at a time, on a computer monitor. Lists were chosen so as to "cross paths" for one, two, four, or eight words. To discourage positional coding, the overlapping items always began at a random point in the list such that there were at least three words following and three words preceding which were not included in the other list. At the end of each session subjects recalled both lists three times. IRTs were determined using a computer-assisted scoring program. IRTs deviating 3σ or more from the mean of each condition were excluded (σ represents the standard deviation for a particular position relative to the exit transition).

3. RESULTS

The results of an earlier unpublished study suggest that IRTs are not affected by list position, except possibly for those times at the very beginning and end of the list (results not shown). As discussed in the methods, the overlapping items of the two lists were placed so that the exit transition did not fall onto these positions.

Although all subjects were instructed to recite the list as quickly as possible without losing accuracy, IRT data was still extremely variable, as is typical in studies examining IRTs

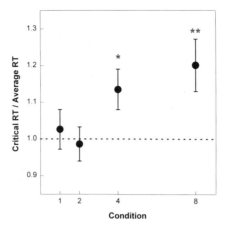

Figure 1. The effects of one, two, four, and eight word "crossed paths". On the vertical axis is the mean exit IRT (normalized). The four and eight word overlap conditions are significantly different from baseline ($p < 0.05$ in the four-word condition and $p < 0.01$ in the eight-word condition).

in recall tasks. To control for this, IRTs were normalized by dividing each IRT by the mean IRT for that list. Thus an unaffected transition should approximately equal the mean IRT and the ratio should equal one. If the IRT is significantly longer than the mean IRT, the ratio will be significantly greater than one.

Figure 1 shows the effect of increasing the number of overlapping words on the ratio of the critical (exit) IRT to the average list IRT. A significant increase in IRTs is observed for the four ($p < 0.05$) and eight ($p < 0.01$) word overlap conditions.

According to any simple associative chaining model of sequence memory, we would expect that even a single shared element should create substantial interference in recalling the item following the shared element in either list. The obtained IRT data fail to demonstrate associative interference when one or two consecutive elements are shared across two lists. These findings are consistent with data on intraserial repetition effects: if an item is repeated within a single list (e.g., A-B-C-D-E-F-C-G-H), IRTs to items following the repeated elements are not elevated (Kahana & Loftus, in press). Taken together with the results reported here, it seems clear that human subjects are relatively immune from associative interference in sequence learning tasks. This finding, coupled with the significant interference observed when large subsequences are shared across two lists, suggest the role of compound cues in sequence recall. The next section presents a preliminary mathematical analysis of the role of context and compound cues in associative models of sequence memory.

4.   THEORETICAL ANALYSIS

In a chaining model, associations are formed between each item and its predecessor. Thus the effective stimulus is the prior item. If we represent item i in lists one and two as the vectors $\mathbf{a}_i$ and $\mathbf{b}_i$, the weight matrix which codes for the associations which make up the two lists is given by:

$$\mathbf{W} = \sum_{i=1}^{L_1} \mathbf{a}_i \mathbf{a}_{i-1}^T + \sum_{i=1}^{L_2} \mathbf{b}_i \mathbf{b}_{i-1}^T.$$

In this equation, vectors are columns, the superscript T above a vector denotes its transpose, and L_1 and L_2 are the lengths of lists one and two respectively.

When cued with an item which has two competing associates in the list (for instance, $\mathbf{a}_n\mathbf{a}_{n-1}^T$ and $\mathbf{b}_m\mathbf{a}_{n-1}^T$), the following information is retrieved:

$$\mathbf{W}\mathbf{a}_{n-1} = \mathbf{a}_n + \mathbf{b}_m.$$

In a non-linear model, both attractors will have equivalent strength and the model will not be able to disambiguate to the desired target. According to chaining models, the size of the overlapping region of the lists should not affect performance at the ambiguous exit point.

Adding context to a chaining model provides a powerful basis for list discrimination. In such a model, features representing item information may be added to features representing contextual information. If the vector features are sparse, addition should not produce many overlapping elements.

Two different patterns of activity, $\mathbf{x}$ and $\mathbf{y}$, represent context in lists one and two respectively:

$$\mathbf{a}' = \mathbf{a} + \mathbf{x}$$
$$\mathbf{b}' = \mathbf{b} + \mathbf{y}.$$

In this model, the effective stimulus is the prior item combined with the appropriate list context. The weight matrix for two lists sharing a common element ($\mathbf{a}_{n-1} = \mathbf{b}_{m-1}$, but $\mathbf{a}'_{n-1} \neq \mathbf{b}'_{m-1}$) is given by:

$$\mathbf{W} = \sum_{i=1}^{L_1} \mathbf{a}'_i \mathbf{a}'^T_{i-1} + \sum_{i=1}^{L_2} \mathbf{b}'_i \mathbf{b}'^T_{i-1}.$$

Consider the information which is retrieved when the memory is cued with $\mathbf{a}'_{n-1} = \mathbf{a}_{n-1} + \mathbf{x}$.

$$\mathbf{W}(\mathbf{a}_{n-1} + \mathbf{x}) = 2\mathbf{a}_n + \sum_{i \neq n}^{L_1} \mathbf{a}_i + \mathbf{b}_m + L_1\mathbf{x}.$$

In this model both associates are retrieved together with all of the $\mathbf{a}$ terms as well as a very strong contextual term ($\mathbf{x}$) list one. If the vectors are partitioned into context and content elements, we can retrieve the desired pattern $\mathbf{a}_n$.

In a compound cue model, associations are stored between the current item and an exponentially weighted sum of all prior items. The strength of the associations are greatest for the immediately preceding item and fall off exponentially:

$$\mathbf{W} = \sum_{i=1}^{L_1} \mathbf{a}'_i \sum_{j=0}^{i-2} e^{-kj} \mathbf{a}'^T_{i-j-1} + \sum_{i=1}^{L_2} \mathbf{b}'_i \sum_{j=0}^{i-2} e^{-kj} \mathbf{b}'^T_{i-j-1}$$

When the memory matrix is cued with $\mathbf{a}_{n-1}$:

$$\mathbf{W}\left(\sum_{j=0}^{n} e^{-kj}[\mathbf{a}'_{n-j-1}]\right) = 2\sum_{j=0}^{n} e^{-2kj}\mathbf{a}_n + \sum_{j=0}^{S} e^{-2kj}\mathbf{b}_m + \dots$$

In this model, the size of the overlapping region, S, increases the interference produced by the major competitor, $\mathbf{b}_m$. Thus the size of the overlapping region affects performance of this model, unlike the chaining model with context. The effect of context is to double the cue strength for the list-appropriate response $\mathbf{a}_n$ – making this model far more robust in the presence of noise. Note: the omission of numerous weakly matching terms in the equation above are indicated by the ellipses.

5. SUMMARY AND CONCLUSIONS

We have presented a novel experimental method for examining the effect of associative interference in recalling sequences of discrete items. Human subjects' performance during ordered recall of word lists is adversely affected when a sequence of several words is shared in two simultaneously learned lists. Contrary to the predictions of simple associative chaining models, the presence of one, or even two, shared items does not result in elevated inter-response times when exiting the shared sequence. However, significant interference is observed following shared sequences of four and eight words. These results support models in which multiple prior items combine to form compound cues used in retrieval of subsequent list items.

REFERENCES

[1] Blum, K. I., & Abbott, L. F. (1996) A Model of Spatial Map Formation in the Hippocampus of the Rat. *Neural Compuatation* 8: 85-93.
[2] Ebbinghaus, H. (1885/1913 [Reprinted by Dover, 1964]). *Memory: A Contribution to Experimental Psychology*. Teachers College, Columbia University, New York.
[3] Friendly, M., Franklin, P. E, Hoffman, D., Rubin, D. C. (1982) Nouns for the Toronto Word Pool. *Behavior Research Methods and Instrumentation* 14: 375-379.
[4] Kahana, M. J. & Loftus, G. (in press). Response Time and Accuracy in Human Cognition: Are They Two Sides of the Same Coin? In R. Sternberg (Ed.) *The Concept of Cognition*. MIT Press.
[5] Kleinfeld, D. (1986) Sequential State Generation by Model Neural Networks. *Proc. Natl. Acad. Sci. USA* 83: 9469-9473.
[6] Lewandowsky, S. & Murdock, B. B. (1989) Memory for Serial Order. *Psychological Review* 96: 25-57.
[7] Riedel, U., Kuhn, R., & van Hemmen, J. L. (1988) Temporal Sequences and Chaos in Neural Nets. *Physical Review A* 38: 1105-1108.
[8] Sompolinsky, H. & Kanter, I. (1986) Temporal Association in Asymmetric Neural Networks. *Physical Review Letters* 61: 259-262.
[9] Young, R. K. (1968) Serial Learning. In T. R. Dixon & D. L. Horton (Eds) *Verbal Behavior and General Behavior Theory*. Prentice Hall, England Cliffs, NJ. pp. 122-148.

94

ANATOMICAL ORGANIZATION OF FEEDFORWARD PATHWAYS IN CORTICAL MICROCIRCUITS

Jeffrey B. Colombe and Philip S. Ulinski

Committee on Neurobiology and
Department of Organismal Biology and Anatomy
The University of Chicago
Chicago, Illinois 60637
jcolombe@midway.uchicago.edu
pulinski@midway.uchicago.edu

ABSTRACT

Cortical microcircuits are defined as stereotyped, repeating patterns of local connectivity in the cortex. Those pathways by which thalamic afferents influence pyramidal neurons, through either direct synaptic connections or through interneurons, constitute feedforward pathways in the microcircuit. The anatomical organization of feedforward pathways in turtle visual cortex was studied by comparing the dendritic geometry and spatial distribution of Golgi-impregnated neurons to the spatial distribution of HRP-filled geniculocortical afferents. Seven morphologically distinct cortical neuron types are positioned to receive geniculocortical synapses. The geniculocortical afferents form parallel, horizontal delay lines that run from lateral to medial across the visual cortex in the coronal plane. The horizontal dimensions of dendrite arbors for populations of each neuron type were thus measured and found to vary between extremes of 70 μm and 1100 μm. The time it takes an afferent volley to cross the arbors of the seven types of neurons should vary and determine the properties of the compound postsynaptic potentials elicited by geniculate inputs to each type of cell.

1. INTRODUCTION

Pyramidal cells in turtle visual cortex have characteristic responses to 1 sec light flashes of varying intensity that must result from a temporal interaction between sequences of neurons carrying information from the lateral geniculate complex to the cortex (feedforward pathways) and those dependent on activity in pyramidal cells (feedback

Computational Neuroscience
edited by Bower, Plenum Press, New York, 1997

pathways). This paper uses a combination of Golgi impregnation of cortical neurons and anterograde labeling of geniculocortical afferents to anatomically characterize the feedforward pathways that converge on pyramidal cells in turtle visual cortex. The analysis suggests that seven morphologically distinct classes of neurons receive geniculate input. Their characterization provides a basis for detailed, functional analyses of microcircuits in visual cortex.

2. METHODS

Eleven brains of freshwater turtles were prepared by the Adams modification of the Golgi-Kopsch method. Detailed camera lucida drawings were made of examples of each type of neuron in visual cortex and their spatial distribution was charted in serial sections. Horizontal dimensions of dendrite arbors were measured for populations of each cell type. Large pressure injections of horseradish peroxidase (HRP) were made in the dorsal lateral geniculate nucleus (dLGN) to anterogradely label geniculocortical afferents [1]. Serial sections of the HRP-labeled material were used to construct a map of the trajectory of geniculocortical afferents with respect to the boundaries of the cortical layers. The dendritic dimensions and spatial distributions of cortical neuron types were compared with the trajectory of the thalamic afferents to identify those neuron types that are positioned to receive geniculocortical synapses.

3. RESULTS

The visual cortex (D) of the pond turtle, *Pseudemys scripta*, receives geniculate afferents and consists of lateral (D_L) and medial (D_M) subfields (Fig. 1). Geniculocortical afferents run horizontally from lateral to medial across the visual cortex [1,2]. They enter the visual cortex at the lateral edge of D_L, perforate through layer 2, then rise to occupy the top half of layer 1 in D_M and finally terminate at the medial edge of D_M. Under the light microscope, HRP-filled geniculocortical axons have a variety of fiber diameters. The thickest are unmyelinated axons approximately 2 μm in diameter; the smallest are at the limit of resolution of the light microscope.

Seven morphologically distinct types of neuron lie within the path of the thalamic afferents through D (Fig. 2). We were able to classify 955 of 1000 impregnated neurons based on their morphology. Measurements of arbor widths were obtained for 94 completely impregnated neurons. *Pyramidal cells* are the primary output neurons of the visual cortex and form the great majority of neurons in the second or intermediate layer of the cortex. The spiny dendritic trees of *medial pyramidal cells* have apical dendrites rising toward the pia and basal dendrites radiating downward toward the ependyma (Fig 2A). *Lateral pyramidal cells* also have spiny apical and basal dendritic arbors near the soma which spread out to give the cell an overall stellate shape (Fig. 2B). The dendritic trees measure 346 ± 100 μm (n=32) in D_M and 487 ± 104 μm (n=16) in D_L. Pyramidal cells project from the visual cortex to other cortical areas and subcortical structures. They are known to use excitatory amino acids as neurotransmitters. *Bowl cells* are found in layer 2 slightly above the pyramidal cell somata. Their dendrites radiate horizontally from the soma and then curve toward the pia, giving their dendritic arbors a characteristic bowl shape that averages 599 ± 172 μm (n=12) in diameter. Their dendrites are spiny, and these neurons are presumed to be excitatory interneurons (Fig. 2D). *Spiny horizontal cells* are found in lay-

Figure 1. Anatomy of turtle visual cortex. The turtle brain and eyes are shown in (A), with the visual cortex outlined *(VC)*. The horizontal lines in [A] indicate the plane of coronal cross-section through the right hemisphere shown in (B). Cytoarchitectonically distinct cortical areas are labeled: medial *(M)*, dorsomedial *(DM)*, medial dorsal *(D_M)*, lateral dorsal *(D_L)* and lateral *(L)*. The visual cortex, shown in coronal cross-section in (C), corresponds to cortical area D, which consists of subareas D_M and D_L. The geniculate afferents run from lateral to medial across D *(genic)*. The cortical layers 1, 2 and 3 are labeled, and typical medial pyramidal and smooth stellate cells are illustrated. All scale bars are 1 mm. Other abbreviations: cerebellum *(CB)*; cortex *(CTX)*; olfactory bulb *(OB)*; optic tectum *(OT)*; dorsal ventricular ridge *(DVR)*; striatum, *(STR)*.

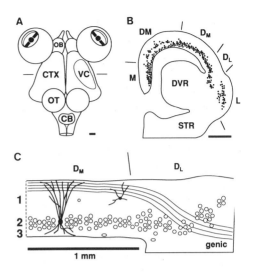

ers 1 and 2. Their spiny dendrites radiate horizontally from the soma in parallel with the geniculate afferents, giving them wide dendritic arbors of 850 ± 87 µm (n=3). They are presumed to be excitatory interneurons (Fig. 2C). *Smooth stellate cells* are found in layers 1 and 2 of D_M and layers 1, 2 and 3 of D_L. Their smooth, sometimes beaded dendritic arbors are stellate in shape. *Medial smooth stellate cell* dendritic arbors measure 180 ± 50 µm (n=11) in diameter (Fig. 2F), while *lateral smooth stellate cell* dendritic arbors measure 318 ± 104 µm (n=16) in diameter (Fig. 2E). These neurons react positively to immunohistochemical labeling for glutamic acid decarboxylase (GAD) and γ-aminobutyric acid (GABA) [3], and are presumed to be inhibitory interneurons. *Smooth hemispheric cells* are quite rare. Four of these were found in upper layer 1 of D_L. Their smooth, beaded dendrites arise from the pial surface of the somata, branch and radiate outward to form a hemispheric-shaped arbor measuring 128 ± 17 µm (n=4) in diameter. These are also presumed to be inhibitory interneurons (Fig. 2G).

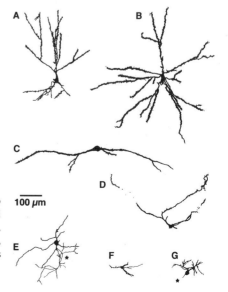

Figure 2. Cell types in visual cortex which are positioned to receive geniculate input. Medial pyramidal cell (A), lateral pyramidal cell (B), spiny horizontal cell (C), bowl cell (D), lateral smooth stellate cell (E), medial smooth stellate cell (F), smooth hemispheric cell (G). Axons are marked by asterisks (*). Scale bar 100 µm. See text for descriptions.

4. SUMMARY AND CONCLUSIONS

Seven distinct cortical neuron types were found in the feedforward pathway of turtle visual cortex. Four of these are presumed to be excitatory cells, due to the moderate to large numbers of spines on their dendrites. The other three types are presumed to be inhibitory interneurons. The horizontally oriented afferents form *en passant* synapses as they course lateromedially across the dendritic arbors of each cortical neuron type. The average horizontal width of dendritic arbors varies by a factor of nearly seven between neuron types. The *spiny horizontal* cells have the largest dendritic arbors, which average 850 μm in diameter and are oriented parallel to the geniculate afferents. The smooth hemispheric cells have the smallest arbors, which average 128 μm and are hemispheric in shape. Since geniculate afferents have slow conduction velocities, on the order of 0.01 to 0.2 mm/msec [4], the time required for a geniculate volley to cross the arbors of neurons receiving geniculate input should vary from 12.8 to 85.0 msec. It is likely that these temporal differences are a major factor in determining the amplitudes and time courses of compound postsynaptic potentials elicited in the neurons of the feedforward pathway. The anatomical organization of the feedforward pathway, thus, may be important in establishing the functional properties of microcircuits in visual cortex.

Aspects of this work were supported by a grant from the National Eye Institute to PSU. JBC is supported by an NIH predoctoral training grant to the Committee on Neurobiology at the University of Chicago.

REFERENCES

1. Heller, S.B. and P.S. Ulinski (1987) Morphology of geniculocortical axons in turtles of the genera *Pseudemys* and *Chrysemys*. *Anat. Embryol.* 175:505–515
2. Mulligan, K.A. and P.S. Ulinski (1990) Organization of Geniculocortical Projections in Turtles: Isoazimuth Lamellae in the Visual Cortex. *J. Comp. Neurol.* 296:531–547
3. Blanton, M.G., J.M. Shen and A.R. Kriegstein (1987) Evidence for the Inhibitory Neurotransmitter γ-Aminobutyric acid in Aspiny and Sparsely Spiny Nonpyramidal Neurons of the Turtle Dorsal Cortex. *J. Comp. Neurol.* 259:277–297
4. Colombe, J.B. and P.S. Ulinski (1996) Temporal integration windows for neurons in the feedforward pathways of visual cortex. *Soc. Neurosci. Abstr.* 22:284

95

A MODEL FOR PERIODICITY CODING IN THE AUDITORY SYSTEM

Socrates Deligeorges and David C. Mountain

Boston University
Department Biomedical Engineering
Boston, Massachusetts

ABSTRACT

The processing of complex sounds is thought to require the analysis of both spectral and temporal features. The initial processing of temporal features is believed to take place in a monaural pathway or pathways from the cochlea to the inferior colliculus (IC). We hypothesize that temporal processing begins with enhancement of temporal features by the cochlea and by cells in the cochlear nucleus and ends with a coincidence detection mechanism in the IC.

Amplitude modulated (AM) stimuli were used in all model simulations for comparison with data from physiological experiments. The peripheral model consists of a basilar membrane, inner-hair cell, and auditory nerve fiber models. The cochlear model output, was compared to auditory nerve experimental data taken by Joris and Yin (1992) with respect to modulation gain. The cochlear nucleus model, likewise, was compared to data collected by Rhode and Greenberg (1994). The IC model was compared to the findings of Langner and Schreiner (1988).

The result from each stage of the model showed good agreement with the physiological data. At the cochlear level, the model was able to reproduce physiological responses to AM over the same stimulus intensity range as well as frequency range. The cochlear nucleus model also performed well duplicating levels of modulation gain and temporal enhancement seen in the physiological experiments. The IC model was able to process the temporal features passed through the first two model stages and produced modulation transfer functions (MTFs) similar to those seen by Langner and Schreiner. The complete model output can be thought of as a cellular matrix whose activity maps the temporal content of spectral components within an acoustic stimulus.

Computational Neuroscience
edited by Bower, Plenum Press, New York, 1997

1. INTRODUCTION

In this work we propose a model for a monaural pathway where the temporal features of sound are processed. This pathway begins at the auditory periphery and then projects to the cochlear nucleus (CN) which in turn projects to the inferior colliculus (IC). The auditory periphery performs an initial stage of spectral processing which separates the various spectral components of a sound. The spectral information is carried to the cochlear nucleus by an array of auditory nerve fibers which project to a particular group of cells that are especially sensitive to temporal variations in the spectral information. These cells in the cochlear nucleus then enhance the temporal variations or what will be referred as the 'highlights' in the temporal information. These collections of temporal highlights will then be processed by an array of cells in the IC. The output of the IC will be a representation of both the temporal and spectral information in the acoustic signal.

2. AUDITORY PERIPHERY MODEL

The auditory periphery model begins with a model of cochlear mechanics based on the linear gamma-tone filter bank implementation (Patterson et al., 1988). Each band-pass filter in the bank has a center frequency (CF) and the model band-pass filter CFs are logarithmically distributed over the range of frequencies. The length of the membrane is represented in the model as a set of parallel channels where each channel is analogous of a position on the basilar membrane. To adequately represent to basilar membrane motion, 64 filters were used to create a smooth representation of the motion in the model response. The total bandwidth of the filterbank was adjustable, but was usually set to cover CFs from 100 to 20,000 Hz with each filter having an approximate Q of 4. The compressive non-linearity of the membrane is not directly incorporated into the band-pass filter model. This problem was addressed indirectly by pre-compressing the acoustic input using a relation taken from the basilar membrane input-output function for the chinchilla (Ruggero and Rich 1991).

The next stage of the peripheral model processing is the inner hair cell model which performs a simplified version of the hair cell's complex physiological transduction process. The model transduction process is non-linear using a sum of two two-state Boltzsmann functions whose parameters were fit to the physiological data from Russell *et al.* 1986. The function as well as the parameters used are shown in **equation 1**. This function allows both saturation at high levels as well as an enhancement of amplitude variations in the mid-range of function. These properties increase the temporal enhancement and allow the enhancement properties extend over approximately 70–100 dB range of function. There is a fourth order stage of low-pass filtering after transduction which causes the model to lose phase-locking to the stimulus with increasing frequency. The filter cutoffs are set so that phase-locking decays markedly at the same approximate frequency as seen in the cat experimental data, roughly between 2 to 3 kHz.

$$\text{IHC}_{Mv} = \frac{A_0}{1 + \exp[-(x - x_0)/Sx_0]} + \frac{1 - A_0}{1 + \exp[-(x - x_1)/Sx_1]} \tag{1}$$

IHC_{Mv} = IHC membrane voltage
$A_0 = 0.6$, $x_0 = 17.0$ nm, $Sx_0 = 85.0$ nm, $x_1 = 17.0$ nm, $Sx_1 = 11.0$ nm

The inner hair cell synapse with the auditory nerve is modeled by a cascade of two generic depletion mechanisms. They were derived from models for the calcium flux and vesicle depletion (see Mountain and Hubbard 1996) and the implementation is described in Singh and Mountain (this volume) with the addition of a low-pass filtering stage between the automatic gain control and a thresholding/spiking function at the output. The equation for the spiking/threshold function used is shown in **equation 2**. The output of the model is a representation of the instantaneous firing rate of a population of auditory nerve fibers with the same CF. The model does not generate individual spikes but instead creates a waveform representing the instantaneous firing rate of the nerve in response to the stimulus.

$$S_{rate} = (V_m - T_{out}) * \frac{\left[\tanh(V_m - T_{out}) + 1\right]}{2}$$

(2)

S_{rate} = Instantaneous spike rate, V_m = Thresholded membrane voltage
T_{out} = Output threshold

Figure 1 shows a comparison of the peripheral model output with the data of Joris and Yin (1992). The model performed well over most of the modulating frequency range with only slight discrepancies at the lower frequencies. The previous modeling work by Hewitt and Meddis(1994) used the physiological data of Joris and Yin(1992) for comparison to determine their model efficacy as well. Their model response is also shown in figure 1 for comparison. The physiological experiment as modeled consisted of presenting a 100% amplitude modulated acoustic signal with a fixed carrier frequency and varying modulation frequency to the auditory periphery.

3. THE COCHLEAR NUCLEUS MODEL

The cochlear nucleus(CN) model tries reproduce the response of a particular cell type in the CN, the On$_L$ cell type. The model begins with a simulation of a broad range of auditory nerve CFs converging on each onset cell which is implemented through a spatial filter which allows a wide range of "adjacent" CF channels to influence each onset cell model frequency channel. In order for the spatial filter to keep feature events across CF in the proper time frame, a special delay bank was implemented. The pre-processing bank of delays for the spatial filtering stage is implemented as a simple set of frequency channel specific time delays. The delays were chosen so that the maximum value of the impulse response of each frequency channel was aligned in time. The delay values were determined explicitly using the model output to an impulse stimulus to determine the delay

Figure 1. A comparison of the peripheral model synchronization to AM stimuli to the experimental data taken by Joris and Yin 1992. The experimental data was taken from a auditory nerve fiber with CF = 21 kHz, SR = 61 s⁻¹, at a stimulus s level of 39 dB SPL. The equivalent model 'fiber' had CF = 20 kHz, SR = 37 s⁻¹, at a stimulus level of 39 dB SPL. The figure also shows the results of the Hewitt and Meddis model for simulation of the same data.

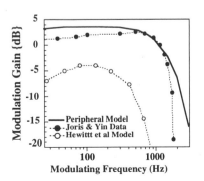

necessary to maintain alignment of the model response across CF. The delay size for each channel was approximately one and quarter cycles of the CF of the frequency channel.

Properties stemming from morphology and arrangement of different types of synaptic connections, which would be difficult to model explicitly, were lumped together as a general intrinsic membrane property of the entire cell. This intrinsic membrane is modeled simply as a filter based on competing excitatory and inhibitory processes with different time constants which reproduces the general frequency response seen in the physiological data. The filter is a fifth order filter with a frequency response fit roughly to the modulated response seen in the experimental data of Rhode and Greenberg (1994) from On_L type cell. The response of the filter represented by the combined input from the excitatory and inhibitory synapses described mathematically by the transfer function shown in **equation 3**.

$$\frac{\left[A * (T_I s + 1)^2 - B * (T_E s + 1)^3 \right]}{\left[(T_E s + 1)^3 (T_I s + 1)^2 \right]} \tag{3}$$

T_E = Excitatory time constant, T_I = Inhibitory time constant
A = Excitatory weight, B = Inhibitory weight

In this implementation T_E and T_I are equal, but the order of the filter representing the excitatory process is one order higher. This arrangement of leading excitatory response reproduces the mildly bandpass shaped response which is seen in the experimental data of Rhode and Greenberg 1994. The net effect is slight attenuation of frequencies below 100 Hz and strong attenuation above 1 kHz with a slight enhancement between 150 and 850 Hz.

The delayed and filtered information was passed to the synaptic model. The synaptic model implementation is identical to that used in the auditory nerve stage of the cochlear model with only slight adjustments to parameters in the calcium activation function (Singh and Mountain this volume). To adjust the model to generate a lower spontaneous rate the value of K_s was doubled and the V_o was adjusted upwards to compensate for the increased tonic input due to both the spontaneous rate of the auditory nerve model and the converging of inputs due to spatial filtering. The adjustment was needed to reduce the model spontaneous rate from the medium to high range used in the peripheral model to a low spontaneous rate more typical of the onset cell type.

In addition, an adaptive threshold was used, modeled here as a slow potassium current, which regulates the average response of the model. This was implemented as a negative feedback loop using an attenuated low-passed version of the signal to threshold the output. The corner frequency of the low-pass filter is on the order of 10 Hz and the output is scaled by a factor 0.1 to create an adaptive threshold of about 10% of the average value of the signal. The output is then passed through a thresholding/spiking function of the same form as was used in the peripheral model (see equation 2).

Figure 2 shows a comparison of the CN model response to the Rhode and Greenberg experimental data. The modulated and average responses from the simulation were compared to the modulated and average responses seen in the experimental data. The average rate responses of the model reproduces the same trends seen in the average rate responses of the experimental data (top). However, the modulated response of the model does not show the same marked increase over a narrow frequency range evident in the Rhode and Greenberg data (bottom). The physiological experiment, as modeled, consisted of presenting the auditory periphery response to a 100% AM acoustic signal with a fixed carrier frequency and varying modulation frequency to the onset cells of the cochlear nucleus.

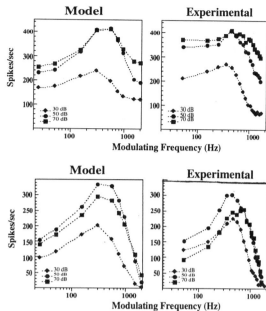

Figure 2. A comparison of the average (top) and modulated (bottom) firing rates of the cochlear nucleus model and the data taken by Rhode and Greenberg(1994) for an On_L type cochlear nucleus cell. The experimental data was taken from a unit with CF = 7.9 kHz at three different stimulus levels (dB SPL).

4. THE INFERIOR COLLICULUS MODEL

For a large number of neurons in the IC, the initial input appears to be dominated by inhibition. Whole cell patch-clamp recordings indicate that inhibition dominates the early part of the response (Casseday et al. 1994) Some earlier work by Grinnell (1963) and Suga (1964) has also shown evidence that units in the IC have intervals during which their response to a second acoustic event or highlight some interval after the first event is either facilitated or suppressed. In work done by Cassedy and Covey (1996) using 50% recovery as a criterion the interval over which suppression occurred ranged from 0.7 msecs to 50.2 msecs. The window of facilitation created by this mechanism could be the functional basis for a temporal filtering scheme in the IC. It is not difficult to see how a system using temporal filters could be generated through something like a delay network that produce similar sequential windows of facilitation and suppression that repeat at intervals.

The structure of the IC is very tonotopically or 'cochleotopically' organized where within and across each laminae of the IC there is a fairly orderly progression of spectral and temporal frequencies to which each area will respond (Langner and Schreiner 1988). The range of best modulating frequencies (BMF) for IC units is roughly from 10 Hz to as high as 1 kHz, where most BMFs fall in the range from 10 to 300 Hz. The BMFs of the IC seem to be distributed logarthmically over the range of modulating frequencies similar to the way CFs of the cochlea are distributed over spectral frequency.

The IC model mimics this tonotopic organization by creating a two dimensional array of units each with an associated spectral and modulating frequency. Each model IC unit receives input from a single spectral channel and performs two stages of processing. The first stage uses a type of filter based delay with explicit excitatory and inhibitory inputs to create a delayed signal of a specific delay interval size. The implemented filters used competing excitatory and inhibitory 'synapses' modeled as filters. The response of the filter represented by the combined the excitatory and inhibitory input can be described mathematically using the transfer function shown in **equation 4.**

$$\frac{A*\left[(T_E s+1)^4-(T_I s+1)^4\right]}{\left[(T_I s+1)^4(T_E s+1)^4\right]}$$

$$(4)$$

T_E = Excitatory time constant, T_I = Inhibitory time constant, A = System gain

The ratio between T_I and T_E is fixed to approximately 0.67 which means that the inhibitory response will lead the excitatory. This arrangement of leading inhibitory response with a slight difference in time constant creates a very bandpass system with a phase change of 360 degrees very near the peak of the bandpass region. The 360 degree phase shift is equivalent to a single cycle delay of a temporal interval size approximately equal to the period associated with the peak frequency of the bandpass region. By changing the value of T_E and maintaining the ratio of T_E to T_I, a set of responses with the desired temporal intervals and constant filter shape can be created. In addition to constant shape characteristics, this implementation will only respond to an interval of specific size that falls with in the bandpass region of the filter response.

This delayed response can be thought of as the 'window' of facilitation where an incoming signal can occur and cause the cell to respond. The number of channels needed to represent a range of temporal intervals associated with modulation frequencies was determined by parameters taken from experimental data. The BMFs of the IC unit filters ranged from approximately 20 to 1000 Hz (the range resolved by human), with a majority of BMFs being distributed between the 30 to 300 Hz range. The model's temporal intervals were logarithmically spaced to match the distribution seen in the experimental data. The total number of BMF channels used in this model was 16.

The second stage of processing is a coincidence detection scheme which combines the delayed and undelayed signal in such a manner that there must be activity in both the delayed and undelayed versions of the signal at the same time to create an output. The coincidence detection stage of the model uses the delayed signal's window of facilitation as a coincidence window where a second temporal highlight in the undelayed signal can cause the unit to respond. The delayed signal is first thresholded and then multiplied with the undelayed signal. The complete model output from periphery to IC is an array of units, each with a CF and a BMF associated with it, for each point in time.

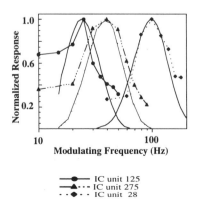

Figure 3. A comparison of the IC model MTFs and the MTFs seen in the experimental data of Schreiner and Langner(1988). The CFs of the IC units in the experimental data were unit 125 =1.3 kHz, unit 275 =0.8kHz, unit 28 = 1.8 kHz ; the stimulus was varied din level from 30 to 60 dB. The model equivalent units were all at CF = 2.0 kHz and the stimulus level was fixed at 50 dB.

Figure 3 shows a comparison of a model unit MTF and the Langner and Schreiner (1988) experimental data. The physiological experiment as simulated consisted of presenting the output for the model of the lower auditory system in response to 100% amplitude modulated acoustic signals with a fixed carrier frequency and varying modulation frequency to the IC model. To reproduce the single physiological unit response in the simulation, the model channel whose center frequency and BMF was closest to the CF and BMF of the unit in the experiment was used. The instantaneous rate waveform of the model unit was used as the period histogram and the same calculations performed on the physiological response were performed on the model response.

5. DISCUSSION

The performance of the collected models presented here, with respect to the experimental data, demonstrates that it is plausible for the proposed temporal pathway to exist in the auditory system and to perform the temporal processing task. The model reproduced the general trends and characteristics of the experimental data for AM stimuli for each stage of the proposed pathway.

REFERENCES

Casseday, J.H., Ehrlich, D. and Covey, E. Neural Tuning for Sound Duration : Role of Inhibitory Mechanisms in the Inferior Colliculus. *Science* 264:847–850, 1994.

Cassedy, J.H. and Covey, E. Mechanisms for Analysis of Auditory Temporal Patterns in The Brainstem of Echolocating Bats. In: *Neural Representations of Temporal Patterns* H.L. Hawkins, T.A. McMullen, R.R. Popper, and R.R. Fay, eds., Springer Verlag, New York, 1996.

Grinell, A.D. The Neurophysiology of Audition in Bats: Intensity and Frequency Parameters. *J. Physiol* 167:38–67, 1963 (a).

Hewitt , M.J. and Meddis, R. A Computer Model of Amplitude Modulation Sensitivity of Single Units in the Inferior Colliculus. *J. Acoust. Soc. Am.* 95:2145–2159, 1994.

Joris, P.X. and Yin, C.T. Responses to Amplitude-modulated Tones in the Auditory Nerve of the Cat. *J. Acoust. Soc. Am.* 91:215–232, 1992.

Langner, G. and Schreiner, C.E. Periodicity Coding in the Inferior Colliculus of the Cat . I. Neuronal Mechanisms. J. Neuro-Physiol. 60:1799–1822, 1988.

Mountain, D.C., and Hubbard, A.E. Computational Analysis of Haircell and Auditory Nerve Processes. In: *Auditory Computation, Springer Handbook of Auditory Research*, H.L. Hawkins, T.A. McMullen, R.R. Popper, and R.R. Fay, eds., Springer Verlag, New York, pp. 121–156, 1996.

Patterson, R.D., Nimmo-Smith, I. Holdsworth, J., and Rice, P. Spiral vos final report, part A: The auditory filterbank. Cambridge Electronic Design. Contract Rep. (APU 2341), 1988.

Rhode, W.S. and Greenberg, S. Encoding of Amplitude Modulation in the Cochlear Nucleus of the Cat. *Journal of Neurophysiology*. 71:1797–1825, 1994.

Ruggero, M.A. and Rich, N.C. Application of a Commercially Manufactured Doppler-shift Laser Velocimeter to the Measurement of Basilar-membrane Vibration. Hear. Res. 51:215–230, 1991.

Russell, I.J., Cody,A.R., and Richardson, G.P. The Responses of Inner and Outer Hair Cells in the Basal Turn of the Guinea-pig cochlea and in the Mouse Cochlea Grown *invitro*. Hear. Res. 22:199–216, 1986.

Singh, S. and Mountain, D.C. A Model for Duration Coding in the Inferior Colliculus. *Computational Neuroscience '96*.

Suga, N. Recovery cycles and Responses to Frequency Modulated Tone Pulses in Auditory Neurones of Echo-locating Bats. J. Physiol 175:50–80, 1964.

96

A MODEL OF CONTEXTUAL INTERACTIONS IN PRIMARY VISUAL CORTEX: EXAMINING THE INFLUENCE OF CORTICOGENICULATE FEEDBACK

Valentin Dragoi

Department of Psychology: Experimental
Duke University
Box 90086
Durham, NC
E-mail: valentin@psych.duke.edu

INTRODUCTION

The current view of visual processing in the mammalian brain is based on the general idea that the information is transmitted from retina to neocortex via a feedforward sequence of hierarchical levels that create a detailed topographic map of the visual world. However, in reality, anatomical data show that the feedforward visual pathways are far outnumbered by corticofugal projections [1] which terminate in the LGN of the thalamus. Furthermore, electrophysiological data [2] show that the massive cortical feedback is able to modulate the processing of sensory information by determining the sensitivity of LGN neurons to visual stimuli. Despite this overwhelming neuroanatomical and neurophysiological evidence, the role of cortical feedback has not been addressed yet in relation to mechanisms of orientation selectivity and context effects. The current views completely neglect the role of recurrent feedback in favor of feedforward and horizontal connections [3, 4]. The aim of this study, therefore, is to investigate whether models that incorporate the effect of cortical feedback, in addition to feedforward and horizontal connections, can explain neurophysiological data reporting context-dependent distortions in the orientation tuning and firing pattern of cells in primary visual cortex (V1).

In an attempt to look for context-sensitivity at the level of single cells in the primary visual cortex (V1), Gilbert & Wiesel [5] showed that if a cell in the superficial layers of V1, that shows preference for 30° stimuli, is presented a surround oriented 30° counterclockwise from the optimal orientation, the cell's tuning curve is shifted 10° in a direction away from the orientation of the surround. The shift in orientation specificity is reversible, and is fully controlled by the situations in which the surround is either present or absent. Furthermore, Knierim and Van Essen [6] showed that if a texture surround is added, the response of a cell in the superficial layers of V1 to a single line segment placed within the center is suppressed. The amount of suppression varies with the mismatch in orientation between the textured surround and the center stimulus: more suppression for a parallel surround, less suppression for a random surround, and even less suppression (or none) for an orthogonal surround.

METHODS

The model examines the role of short and long-range horizontal connections as well as corticogeniculate feedback projections. For simplicity, a monocular patch of the visual field is divided into 21 x 21 locations, where each location can be uniquely mapped onto a full set of 72 orientation columns (2.5^o resolution). The model configures 31,752 LGN neurons arranged on the array of 21 x 21 locations, with 72 cells per each location of the visual patch. Each LGN cell receives a stimulus-specific input which is maximal for the LGN cell that corresponds topographically to the cortical cell whose orientation preference matches that of the input stimulus.

The model simulates the following circuit: retina $\rightarrow$ LGN $\rightarrow$ V1a (layer 4C) $\rightarrow$ V1b (layer 2-3) $\rightarrow$ LGN. For simplicity, a monocular patch of the visual field is divided into 21 x 21 locations, where each location can be uniquely mapped onto a full set of 72 orientation columns. The model configures 31,752 LGN neurons arranged on the array of 21 x 21 locations, with 72 cells per each location of the visual patch. Each LGN cell has a specific orientation bias that establishes the initial orientation preference of each cortical column to which it projects. The spread of geniculate inputs to the cortex ensures that each LGN cell synapses within the same hypercolumn to a group of cortical cells within a broad range of orientations (with a spread of 25°). The thalamocortical synapses comprise about 23% of all synapses received by V1a. LGN cells are modeled as single units in which the mean rate of firing, LGN_{ij}^k, of each cell located at position (i, j) and with orientation bias k is given by the equation

$$\frac{dLGN_{ij}^k}{dt} = -\alpha_1 LGN_{ij}^k + \alpha_2(R_{ij}^k + s)F_{ij}^k(1 - LGN_{ij}^k) - \alpha_3(LGN_{ij}^{k-1} + LGN_{ij}^{k+1})LGN_{ij}^k \quad (1)$$

where α_1, α_2, α_3, are positive constants. R_{ij}^k is the stimulus-specific external input to LGN_{ij}^k ($R_{ij}^k = \max(0, \cos(3\pi(\theta - k)/36))$, where θ is the stimulus orientation. Spontaneous firing results from the background excitation of all inputs outside LGN, where the variable s expresses the global effect of this action. F_{ij}^k is the total feedback that LGN_{ij}^k receives from area V1:

$$F_{ij}^k = h\left(\sum_{i_1, j_1, k_1} f_{k_1} V1b_{i_1 j_1}^{k_1} \right) \quad (2)$$

where $h(x) = \frac{x^8}{x^8 + q^8}$ is a sigmoidal function. $V1b_{i_1 j_1}^{k_1}$ are cells in area V1b, and f_{k_1} are coupling parameters. Corticogeniculate feedback is implemented such that each LGN cell receives feedback projections from outside its RF from a group of iso-oriented cells in V1b within a circular patch of radius 1 at the center. The activity of the presynaptic cells in V1b is integrated via the cortical feedback in order to provide global information used to modulate the rate at which the input stimuli are transmitted through the LGN [2]. It is noteworthy that recent neurophysiological evidence [7] shows that cortical feedback serves to lock or focus the cortical circuitry onto the stimulus feature by providing global information that can be used to adjust local encoding mechanisms to ensure that information transmission through the LGN is improved.

The model configures 63,504 cortical neurons arranged on two layers that correspond to V1a (layer 4C) and V1b (superficial layers). Both V1a and V1b cells develop short-range connections within each hypercolumn, i.e., cross-orientation inhibition in the range of $\pm 45^o$, with the strength of connections decreasing as cortical neurons are more widely separated. Long-range connections are made only onto cells in V1b. Given that the situations discussed in this report refer to high-contrast stimuli, I investigate the effect of the long-range inhibitory connections. They are made onto iso-oriented cells in the range of $\pm 35^o$ within a circular patch of radius 3 at the center, the strength of inhibitory connections decreasing with distance [8, 9].

Cells in V1 are modeled as single units in which the mean rate of firing of each cell located at position (i, j) and with orientation preference k is given by the following equations:
- V1a (layer 4C)

$$\frac{dV1a_{ij}^k}{dt} = -\beta_1 V1a_{ij}^k + \beta_2 INP_{ij}^k(1 - V1a_{ij}^k) - \beta_3 INH_{ij}^k V1a_{ij}^k \tag{3}$$

where β_1, β_2, β_3 are positive constants. INP_{ij}^k is the total input that $V1a_{ij}^k$ receives via the thalamo-cortical afferents. INH_{ij}^k is the local cross-orientation inhibition from cells within the same hypercolumn.
-V1b (layers 2-3)

$$\frac{dV1b_{ij}^k}{dt} = -\beta_1 V1b_{ij}^k + \beta_2 V1a_{ij}^k(1 - V1b_{ij}^k) - \beta_3 (INH_{ij}^k + LRINH_{ij}^k)V1b_{ij}^k \tag{4}$$

The projection from layer 4C to superficial layers is one-to-one. INH_{ij}^k is the local cross-orientation inhibition from cells within the same hypercolumn. $LRINH_{ij}^k$ expresses the total effect of long-range inhibition.

RESULTS

Fig. 1A illustrates the surround-dependent orientation shift effect. Fig. 1B shows tuning curves generated when the model is exposed to the same conditions as in the original experiment. The stimulus which is oriented away from the surround, i.e., 40°, is able to elicit a higher response in the cells whose original tuning curve peaked at 30°. This repulsive shift in orientation specificity is a byproduct of the interplay between surround inhibition, cross-orientation inhibition and cortico-geniculate feedback. When the orientation-dependent surround inhibition is applied, the center cells which receive feedforward input via the spread of the thalamocortical afferents, but prefer orientations different from that of the center stimulus, are released from cross-orientation inhibition in a direction away from the surround. If the center is stimulated with a bar oriented away in a direction clockwise from the orientation of the surround, e.g., 40°, the overall responsiveness of the center cells increases relative to the situation when the center stimulus is oriented nearer the surround, e.g., 30°. The orientation-dependent increase in cortical responsiveness contributes to a stronger corticogeniculate feedback that amplifies the response of the LGN source cells which begin to fire

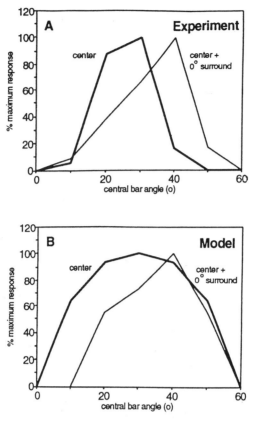

Figure 1. Surround-dependent orientation shift effect. (**A**) Tuning curves for cortical cells tested with different oriented stimuli in the presence of a fixed surround. [adapted from Gilbert & Wiesel (1990)]. (**B**) Theoretical tuning curves obtained in the same conditions as in panel A.

more vigorously eliciting stronger cortical responses that account for the tuning curve profiles depicted in Fig. 1B.

The surround-dependent orientation shift effect is difficult to explain without invoking the presence of the corticogeniculate feedback. How can a stimulus (40° in Fig. 2), which under normal conditions does not elicit a maximal response, provide a level of excitation that, in the presence of a 0° surround, generates a response which is stronger than that measured when the cell is stimulated with its preferred orientation (30° in Fig. 2)? In the absence of the corticogeniculate feedback, whenever the 40° center stimulus is presented the spread of the thalamocortical afferents ensures only a low level of the excitatory feedforward inputs to the 30° cortical cells (the level of excitation increases as the center stimulus approaches 30°). In these conditions, the inhibitory influences exerted by the long-range horizontal connections cannot drive, by local disinhibitory mechanisms, the increase in the response firing rate of the cells tuned to 30°, despite the fact that the cells tuned to 40° do show an increase in their response firing rate when the 40° stimulus is presented (a joined effect of feedforward excitation and lateral disinhibition). However, the orientation-dependent long-range inhibitory connections (LRIC) contribute to the disinhibition, from local cross-orientation inhibition (CI), of the cells oriented away from the orientation of the surround. The level of disinhibition is more effective when the center receives the 40° stimulus, a situation that favors a stronger

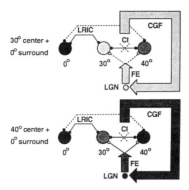

Figure 2. The interaction between representative cells in V1b embedded in their recurrent network. Surround - 0°; center - 30° and 40°. Empty circles - excitatory synapses; filled circles - inhibitory synapses. Dashed lines - weak presynaptic effects; solid lines - strong presynaptic effects. Relative neuron activation patterns are plotted as shading intensity of circles and arrows.

corticogeniculate feedback (CGF) which amplifies the feedforward excitation (FE) to cortical cells. The only way to use the increased responsiveness of the cells tuned to 40° is by means of cortical feedback projections which act directly on the LGN source cells to amplify the feedforward excitatory inputs.

Fig. 3A illustrates experimental results in which normalized cortical responses are obtained for a cell that prefers the vertical orientation, tested in situations when the surround pattern varies in orientation. According to the model (Fig. 3B), at high stimulus contrast the maximum suppression is obtained when the surround is parallel with the center, and, as the surround orientation deviates further from the center, its inhibitory influence diminishes because the strength of the long-range inhibitory connections decreases with the mismatch in orientation between center and surround. When the surround and the center are orthogonal, the strength of long-range inhibition reduces drastically, a situation which favors the perceptual effect of pop-out, i.e., facilitation in the detection of features which are very different from context. When the center stimulus is absent, the cell whose response rate is measured shows only the background level of activity.

In order to analyze alternative mechanisms underlying surround effects, computer simulations were repeated in the absence of corticogeniculate feedback, adding instead long-range

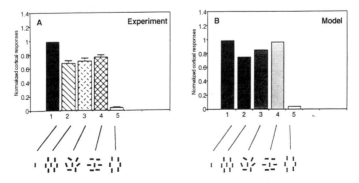

Figure 3. Surround suppressive effect. (**A**) Normalized responses for cortical cells tested with different orientation stimuli [adapted from Knierim & Van Essen (1992)]. (**B**) Model normalized responses

excitatory connections. The ratio between the strengths of long-range excitation and inhibition was varied over a broad range. Computer simulations showed negative results in the case of the surround-dependent orientation shift effect, while the surround suppressive effects were not affected. Furthermore, when the corticogeniculate feedback is restored such that the ratio between the strength of feedback projections and long-range inhibitory connections is subunitary the surround-dependent orientation shift effect decreased in magnitude, thus suggesting that the presence of the corticogeniculate feedback represents a necessary condition for the mechanism of specific surround effects.

CONCLUSIONS

The present model integrates known neuroanatomical and neurophysiological evidence to explore the role of corticogeniculate feedback in shaping the orientation tuning of individual neurons in V1. It is shown that in the presence of an inhibitory oriented surround the long-range horizontal connections contribute to the disinhibition of orientation detectors along a direction away from the surround orientation, an effect amplified by the corticogeniculate feedback. As opposed to the current views on contextual interactions in V1 [6, 7], the present model suggests that all three types of connections, i.e., feedforward, horizontal, and feedback, contribute to generate surround effects.

REFERENCES

[1] P. S. Churchland, V. S. Ramachandran, T. J. Sejnowski, In Large-scale Neuronal Theories of the Brain, C. Koch and J. L. Davis Eds. (MIT Press, Cambridge MA, 1993), pp 23-60.
[2] J. W. McClurkin, L. M. Optican, B. J. Richmond, Visual Neurosci. **11**, 601 (1994).
[3] M. Stemmler, M. Usher, E. Niebur, Science **269**, 1877 (1995).
[4] D. C. Somers, L. J. Toth, E. Todorov, S. C. Rao, D. Kim, S. B. Nelson, A. G. Siapas, and M. Sur, In J. Sirosh, R. Miikkulainen, and Y. Choe Eds. (Electronic book, 1996).
[5] C. D. Gilbert and T. N. Wiesel, Vision Res., **30**, 1689 (1990).
[6] J. Knierim and D. C. Van Essen, J. Neurophys. **67**, 961 (1992).
[7] A. M. Sillito, H. E. Jones, G. L. Gerstein, D. C. West, Nature **369**, 479 (1995).
[8] K. Toyama, M. Kimura, K. Tanaka, J. Neurophys. **46**, 191 (1981).
[9] Y. Hata, T. Tsumoto, H. Sato, K. Hagihara, H. Tamura, Nature **335**, 815 (1988).

97

A MODEL OF THE RODENT HEAD DIRECTION SYSTEM

Adam N. Elga,[1] A. David Redish,[2] and David S. Touretzky[2]

[1] Department of Linguistics and Philosophy
MIT
Cambridge, MA
E-mail: adam@mit.edu
[2] Computer Science Department and CNBC
Carnegie Mellon University
Pittsburgh, PA
E-mail: dredish@cs.cmu.edu, dst@cs.cmu.edu

1. INTRODUCTION

Head direction cells in the postsubiculum (PoS, also known as dorsal presubiculum) were first described by Ranck *et al.* [10]. In subsequent work, Taube *et al.* [14] characterized these cells as having triangular tuning curves: the firing rate drops off linearly from a peak at the *preferred direction* until it reaches a baseline value. Taube *et al.* [15] report that PoS cells typically have baseline-to-baseline tuning curve widths of 100°. Similar cells have been found in the anterior thalamic nuclei (ATN) [4, 6, 13]. See Figure 1(b) for a sample PoS tuning curve. These curves can also be modeled very closely by Gaussians with an average standard deviation of approximately 66° [4, 18].

One way of interpreting the activity of these cells is as a distributed representation of the rat's current head direction. A population of HD cells with preferred directions ϕ_i evenly distributed through 360° represents the direction of the weighted vector sum $\sum_i f_i \vec{v}_i$, where f_i is the normalized firing rate and $\vec{v}_i$ is a unit vector pointing in direction ϕ_i.

Although the two populations in PoS and ATN seem similar, Blair and Sharp [4] and Taube and Muller [16] have recently shown a difference between them: ATN cell activity is best correlated not with current head direction, but with head direction approximately 20-40 ms in the future. PoS head direction cells, on the other hand, are best correlated with the animal's current (or recent) head direction.

The rat's head direction system remains active even in total darkness [8, 13], so nonvisual input must be sufficient to update the heading estimate. In this paper, we sketch a model of the how the head direction representation can be updated by integrating angular head velocity over time. We model the head direction system as a pair of coupled attractor networks

Computational Neuroscience
edited by Bower, Plenum Press, New York, 1997

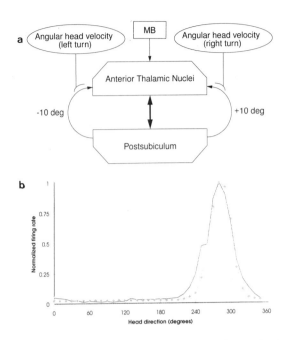

Figure 1. (a) Schematic layout of the model, including postsubiculum (PoS), anterior thalamic nuclei (ATN), and mammillary bodies (MB). Matching connections are drawn as a straight arrow; left and right offset connections as curved arrows. Activity of the *Angular head velocity (left turn)* units is proportional to angular speed during left turns, and zero otherwise. They modulate the strength of the left offset connections. *Angular head velocity (right turn)* behaves analogously. (b) Comparison of tuning curves from a real PoS cell (solid line) and a simulated PoS cell (grey line with diamonds) during tracking of a series of rat head rotations. Single cell recording data courtesy of Tad Blair and Pat Sharp.

and demonstrate accurate tracking of a series of actual rodent head rotations. Full details are available in [11]. Previous models of the head direction system [2, 8, 12, 18] either did not include simulations, or did not report results on realistic trajectories.

2.   THE MODEL

Our model of PoS-ATN interactions, shown in Figure 1(a), consists of two coupled attractor modules and an additional inhibitory gain control population which, like Blair [2], we will identify with the mammillary bodies. All connections between the two attractor modules are excitatory.

To create an attractor module with triangular stable states we follow [5] and create an excitatory pool E and an inhibitory pool I, each composed of units governed by equations (2)–(4) below. See Figure 2(a). For simplicity, we assume that the units in each pool have evenly distributed preferred directions. A unit in the excitatory pool strongly excites those units in both pools whose preferred directions are close to it. A unit in the inhibitory pool weakly inhibits practically all units in both pools, but units close in preferred direction are inhibited slightly more.

In addition to the intrinsic connections within each pool and between the two pools making up an attractor module, which are necessary for the maintenance of the representation, there are two types of connections between the excitatory PoS and ATN pools: *matching* and *offset*. See Figure 2(b).

Matching connections are reciprocal connections between units with corresponding preferred directions. In the absence of any head rotation, matching connections dominate and serve to synchronize the locations of the peaks in the PoS and ATN modules. Offset connections are responsible for changing the represented head direction. They come in two forms: left and right. Each element in the excitatory pool of the PoS structure with preferred direction ϕ has a left offset connection to the unit in the excitatory pool of the ATN structure with preferred direction $\phi - \delta$ and a right offset connection to the unit with preferred direction $\phi + \delta$, where δ is the amount of the offset and is the same for all units. In the simulations presented here, $\delta = 10°$.

All offset connections have strengths modulated by angular head velocity in the following way: While the head is rotating to the right, right offset connections have strength proportional to the speed of rotation and left offset connections have strength zero. The opposite holds true for rotations to the left. During periods when the animal is not turning, both sets of offset connections have zero strength; only the matching connections remain effective, synchronizing the PoS and ATN representations.

We use a neuronal model that is more realistic than standard integrate-and-fire models, but more abstract than compartmental models. We assume for simplicity that action potentials are boolean events with infinitesimal duration. Given an action potential at time s in a neuron j that synapses onto neuron i, we assume (again for simplicity) that the postsynaptic potential (PSP) in neuron i has an instantaneous rise and exponential fall-off with time constant τ. We model this PSP by the product of a synaptic weight w_{ij} and an α function $\alpha_i(t - s) = e^{-(t-s)/\tau_i}$. The voltage $V_i(t)$ of neuron i is taken to be the linear sum of the effect at time t of all PSPs that have ever occurred there. Let $\mathcal{F}_j(s)$ be 1 if cell j fired a spike at time s and 0 otherwise. We write $V_i(t)$ as a tonic inhibition term γ_i plus a sum over all synapses j of the integral of all PSP's ever induced by that synapse:

$$V_i(t) = \gamma_i + \sum_j \left(w_{ij} \int_0^t \alpha_i(t - s) \cdot \mathcal{F}_j(s) \, ds \right) \tag{1}$$

In order to work with a continuous formulation, we replace the spike record $\mathcal{F}_j(s)$ with a probability of firing $F_j(s)$, defined as a sigmoidal function of the voltage $V_j(s)$. This probabilistic approximation of spiking behavior can also be understood as a model of a neuronal population [9]; $F_j(s)$ is then the fraction of neurons in population j firing at time s. Following [9], we rewrite these equations in a form similar to the Wilson-Cowan equations [17] by defining the *Synaptic Drive* $S_i(t)$ supplied by neuron i:

$$V_i(t) = \gamma_i + \sum_j w_{ij} S_j(t) \tag{2}$$

$$F_i(t) = \frac{1 + \tanh(V_i(t))}{2} \tag{3}$$

$$\tau_i \frac{dS_i(t)}{dt} = -S_i(t) + F_i(t) \tag{4}$$

The synaptic drive $S_i(t)$ of neuron i thus approaches the neuron's firing rate $F_i(t)$ with time constant τ_i. Note that this time constant is the same as the time constant of the decay of

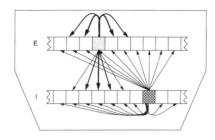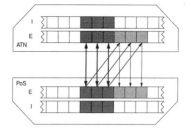

Figure 2. (a) Connection structure of the excitatory (E) and inhibitory (I) pools in one attractor module. Shown are the strong, narrowly focused excitatory connections from an excitatory unit and the weak, diffuse inhibitory connections from an inhibitory unit. (b) Active connections between the PoS and ATN modules during a right turn.

the PSP α function, but in our reformulation the time constant is a function of the presynaptic neuron, not the postsynaptic one. Since each neuron makes only one type of synapse in this model (e.g., we're not modeling both GABA$_A$ and GABA$_B$ synapses), the change is of no consequence.

Synaptic drive can be interpreted as a coarse-grained time-average of the firing rate [9, 17] and is not necessarily a measurable neuronal property. It can be understood as the effect neuron j has on neuron i divided by the synaptic weight between them [9].

The connection weights within a module follow gaussian distributions. Let x be the difference in degrees between the preferred directions of the source and target unit. Then the connection strength is given by a normalized version of the gaussian

$$g_E(x) = \exp\left(\frac{-x^2}{\sigma_E^2}\right) \qquad \text{or} \qquad g_I(x) = \exp\left(\frac{-x^2}{\sigma_I^2}\right) \tag{5}$$

depending on whether the source unit is excitatory or inhibitory. In our simulations, the tuning widths were $\sigma_E = 30°$ and $\sigma_I = 360°$. Inter-module matching and offset connections, which involved only the excitatory pools, used the σ_E value.

3. UPDATING THE REPRESENTATION

The angular velocity modulation of the offset connections to ATN is an empirically determined function which we denote $\xi(\dot{\phi})$. We implement the compensatory gain control signal from the mammillary bodies by setting the tonic inhibition on all ATN:E cells to $\gamma_E - \frac{1}{2}\xi(\dot{\phi})$, where γ_E is the normal tonic inhibition term for an excitatory pool. This nonspecific modulation is applied to all units in the ATN:E population, independent of preferred direction.

This matching-plus-offset connections architecture entails that the locations of Gaussians in the two attractor modules will be synchronized during periods of no rotation, but during rotations ATN will lead PoS. Furthermore, the amount of lead will depend on the angular velocity of the rotation, but due to the gain control mechanism, the shape of the hill in ATN will remain largely unchanged.

Our model is thus compatible with the data presented in the introduction: that PoS and ATN are reciprocally connected, and that cells in PoS are better correlated with current or recent heading, while ATN cells are better correlated with future heading.

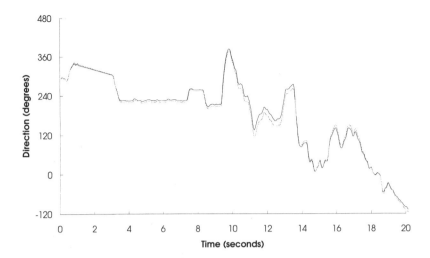

Figure 3. Performance of the model tracking 20 seconds of rodent head rotations. Black line is the animal's actual head direction, grey line is the direction calculated by our PoS:E population by integrating angular velocity. Rodent data courtesy of Tad Blair and Pat Sharp.

4. RESULTS

We show that our model tracks head direction accurately, using cells with tuning curves similar to real cells in postsubiculum and the anterior thalamic nuclei. For our simulations, we used units with evenly spaced preferred directions.

We begin by comparing the model to data recorded by Blair and Sharp from freely moving rats (for details on recording methods, see [4]). This data included the rat's head direction, sampled at 60 Hz. Missing data points were linearly interpolated. To counteract quantization error, we calculated the head direction at every time as the average direction over a 133 ms window centered at that time. This filtering smoothed out fine fluctuations without removing important detail from the angular velocity trace. We then estimated the rat's angular head velocity over the 16 ms time period between two samples ϕ_t and ϕ_{t+1} as the change in head direction divided by 16 ms. These head velocity estimates served as the vestibular input $\dot{\phi}$ for our simulations.

Simulations thus consisted of (1) initializing the units to random states, (2) allowing the two modules to settle to stable attractor states (approx 20-30 ms), (3) identifying the direction represented in PoS:E with the initial head direction sample, and (4) allowing the system to run using the $\dot{\phi}$ sequence calculated as per above, and at each step comparing the direction represented in PoS:E with the measured head direction of the animal.

Figure 3 shows the model's ability to integrate actual rodent head movements. Cumulative HD tracking error fluctuated, but typically did not exceed 20° for simulations shorter than 3 minutes.

Tracking accuracy was largely dependent on four parameters: how strongly vestibular input modulated offset connections $\xi(\dot{\phi})$, the offset amount δ, and the time constants τ_E and τ_I. The offset connection modulation $\xi(\dot{\phi})$ controlled the angular velocity of the Gaussian for a given input head velocity, while the amount of offset ($\delta = 10°$) controlled the lead of the ATN population. The τ parameters controlled the resistance of units to changing activation, determining the inertia of each pool.

Tuning curves of model PoS:E and ATN:E cells (say cell i) were determined by recording $F_i(t)$ at each time t and storing $F_i(t)$ and the actual head direction (not the represented direction) at that point in the simulation. A histogram of F_i was then generated, binned by head direction in $10°$ bins. Tuning curves of real PoS and ATN cells were generated from spike timing data supplied by Blair and Sharp. The model shows an excellent fit to real ATN and PoS cells. A comparison for a sample PoS cell is shown in Figure 1(b).

Given our parameter values, the direction represented by the ATN:E population leads the direction represented by the PoS:E population by approximately 10 ms over a wide range of turning speeds. Although this is smaller than the 20-40 ms discrepancy reported in neurophysiological experiments [4, 16], our model at least replicates the qualitative observation that ATN leads PoS. With further refinement, we hope to come closer to replicating the actual lead of ATN relative to the PoS population.

5. DISCUSSION

We have described a coupled attractor model of the rodent head direction system whose components closely match the tuning curves of cells in PoS and ATN. The model tracks actual head rotations with good accuracy, and the ATN representation of head direction leads the PoS representation during turns, in qualitative agreement with neurophysiological observations. The model makes explicit predictions about the attractor nature of PoS and ATN representations, and the shape of ATN tuning curves during rotations, which we hope to see tested in future experiments.

The model as presented here requires multiplicative connections between PoS and ATN. An alternative was suggested by Zhang (this volume) in which cells are sensitive to both head direction and angular velocity; cells with angular velocity tunings preferring leftward rotations form the left offset connections, while cells with tunings preferring rightward rotations form the right offset connections. Leonhard *et al.* [7] recently presented data showing that cells in the lateral mammillary nuclei (LMN) are tuned to both head direction and angular velocity. This suggests that the representation of head direction in the rodent is maintained by a loop consisting of PoS, LMN, and ATN. PoS sends a strong connection to LMN, LMN sends a projection to ATN (the mammillothalamic tract), and ATN sends a strong projection back to PoS.

Another complication of the model presented here is that we model ATN as an attractor network. The anterior dorsal (AD) nucleus (where most of the HD cells are recorded from [13]) does not have excitatory recurrent connections or inhibitory interneurons [1]. If ATN is not an attractor network, then the tuning curves in ATN distort dramatically when the animal rotates at large angular velocities. Blair *et al.* [3] report that HD tuning curves in ATN distort in exactly the way predicted by this variant of the model.

ACKNOWLEDGMENTS

Supported in part by National Science Foundation grant IBN-9631336. Adam Elga was supported by the Center for the Neural Basis of Cognition summer program 1995. We thank Bard Ermentrout for assistance with attractor dynamics, David Pinto for illuminating discussions, and Tad Blair and Pat Sharp for sharing their rat recording data with us.

REFERENCES

[1] M. Bentivoglio, K. Kultas-Ilinsky, and I. Illinsky. Limbic thalamus: Structure, intrinsic organisation, and connections. In B. A. Vogt and M. Gabriel, editors, *Neurobiology of Cingulate Cortex and Limbic Thalamus: A comprehensive handbook*, pages 71–122. Birkhauser, Boston, 1993.

[2] H. T. Blair. A thalamocortical circuit for computing directional heading in the rat. In D. S. Touretzky, M. C. Mozer, and M. E. Hasselmo, editors, *Advances in Neural Information Processing Systems 8*, pages 152–158. MIT Press, 1996.

[3] H. T. Blair, B. W. Lipscomb, and P. E. Sharp. Anticipatory time intervals of head-direction cells in the anterior thalamus of the rat, implications for path integration in the head-direction circuit. Manuscript, 1996.

[4] H. T. Blair and P. E. Sharp. Anticipatory head direction signals in anterior thalamus: Evidence for a thalamocortical circuit that integrates angular head motion to compute head direction. *Journal of Neuroscience*, 15(9):6260–6270, 1995.

[5] B. Ermentrout and J. Cowan. A mathematical theory of visual hallucination patterns. *Biological Cybernetics*, 34:137–150, 1979.

[6] J. J. Knierim, H. S. Kudrimoti, and B. L. McNaughton. Place cells, head direction cells, and the learning of landmark stability. *Journal of Neuroscience*, 15:1648–59, 1995.

[7] C. L. Leonhard, R. W. Stackman, and J. S. Taube. Head direction cells recorded from the latermal mammillary nucleus in rats. *Society for Neuroscience Abstracts*, 22:1873, 1996.

[8] B. L. McNaughton, L. L. Chen, and E. J. Markus. "Dead reckoning," landmark learning, and the sense of direction: A neurophysiological and computational hypothesis. *Journal of Cognitive Neuroscience*, 3(2):190–202, 1991.

[9] D. J. Pinto, J. C. Brumberg, D. J. Simons, and G. B. Ermentrout. A quantitative population model of whisker barrels: Re-examining the wilson-cowan equations. *Journal of Computational Neuroscience*, 3(3):247ff, 1996.

[10] Ranck, Jr., J. B. Head-direction cells in the deep cell layers of dorsal presubiculum in freely moving rats. *Society for Neuroscience Abstracts*, 10:599, 1984.

[11] A. D. Redish, A. N. Elga, and D. S. Touretzky. A coupled attractor model of the rodent head direction system. *Network*, 7(4):671–685, 1996.

[12] W. E. Skaggs, J. J. Knierim, H. S. Kudrimoti, and B. L. McNaughton. A model of the neural basis of the rat's sense of direction. In G. Tesauro, D. S. Touretzky, and T. K. Leen, editors, *Advances in Neural Information Processing Systems 7*, pages 173–180. MIT Press, 1995.

[13] J. S. Taube. Head direction cells recorded in the anterior thalamic nuclei of freely moving rats. *Journal of Neuroscience*, 15(1):1953–1971, 1995.

[14] J. S. Taube, R. I. Muller, and J. B. Ranck, Jr. Head direction cells recorded from the postsubiculum in freely moving rats. I. Description and quantitative analysis. *Journal of Neuroscience*, 10:420–435, 1990.

[15] J. S. Taube, R. I. Muller, and J. B. Ranck, Jr. Head direction cells recorded from the postsubiculum in freely moving rats. II. Effects of environmental manipulations. *Journal of Neuroscience*, 10:436–447, 1990.

[16] J. S. Taube and R. U. Muller. Head direction cell activity in the anterior thalamic nuclei, but not the postsubiculum, predicts the animal's future directional heading. *Society for Neuroscience Abtracts*, 21:946, 1995.

[17] H. R. Wilson and J. D. Cowan. Excitatory and inhibitory interactions in localized populations of model neurons. *Biophysical Journal*, 12(1):1–24, 1972.

[18] K. Zhang. Representation of spatial orientation by the intrinsic dynamics of the head-direction cell ensemble: A theory. *Journal of Neuroscience*, 16(6):2112–2126, 1996.

98

SEPTOHIPPOCAMPAL CHOLINERGIC MODULATION IN CLASSICAL CONDITIONING

Brandon R. Ermita,[1][*] Catherine E. Myers,[1] Michael Hasselmo,[2] and Mark A. Gluck[1]

[1]Center for Molecular and Behavioral Neuroscience
Rutgers University
Newark, New Jersey
[2]Department of Psychology
Harvard University
Cambridge, Massachusetts

INTRODUCTION

In previous papers we have investigated the functional role of the hippocampal region in learning and memory using connectionist modeling techniques to focus on behavioral processes involved in associative learning (Gluck & Myers, 1993; Myers & Gluck, 1994). Additional work has extended the model by using known neuroanatomical architecture of the hippocampal formation to ascribe specific hippocampal and entorhinal cortical functions (Myers, Gluck & Granger, 1995). These models assume that the hippocampus develops new stimulus representations that enhance the discriminability of differentially predictive cues while the entorhinal cortex develops new stimulus representations that decrease the discriminability of similarly predictive cues. In subsequent work, (Myers, et al., 1996) by incorporating the ideas proposed by Hasselmo and Schnell (1994), the model has been extended to account for the physiologically-based effects of septohippocampal neuromodulation on associative learning. The modulation of hippocampal processing states may be controlled by cholinergic input from the basal forebrain, thus allowing for the hippocampus to be shifted between states of information storage and recall. By manipulating the learning rate of the hippocampal component, the model is useful in understanding the role of disruption of hippocampal processes by cholinergic blockade. Hippocampal computation may be more directly assessed through the application of the generalized Gluck & Myers model on both hippocampal-region mediated (delay conditioning) and hippocampal-region dependent (latent inhibition) associative learning tasks.

[*]**Contact information:** Brandon R. Ermita, Center for Molecular and Behavioral Neuroscience, Rutgers University, 197 University Ave, Newark, NJ 07102, (201) 648–1080 ext 3255, (201) 648–1272 (fax), ermita@pavlov.rutgers.edu

REVIEW OF THE MODEL

In the Gluck & Myers (1993) theory of cortico-hippocampal interaction, the model consists of two parts. The hippocampal-region component is able to monitor predictive and correlational stimuli that constructs new stimulus representations by processes of compressing redundant information and differentiating predictive information. The cortical component is thought to be the final repository of long-term stores. While capable of learning some simple (stimulus-response) mappings, it is unable to modify its own representations.

Computationally, this framework may be instantiated with a hippocampal-region autoencoder network which constructs new representations in an internal layer which is able to compress redundancies while differentiating predictive information. A cortical network acquires these hippocampal representations and is able to map from them to behavioral responses. Simulations of the effect of hippocampal-region lesion may be done by disabling the hippocampal-network. While the cortical network can map from existing (fixed) representations to new behavioral responses, the lesioned model, cannot modify its own internal-layer representations (Figure 1).

REVIEW OF CLASSICAL CONDITIONING

One impediment to a fuller understanding of hippocampal-region function is the large number of species, preparations and paradigms that have been adopted by different researchers. In an effort to simplify matters, classical conditioning has been used an elementary paradigm for which the neural bases are well understood. Moreover, several detailed mathematical and computational models exist which account for a wide-range of conditioning behaviors (Pearce & Hall, 1980; Rescolra & Wagner, 1972). The most researched type of motor-reflex conditioning has been classical conditioning of the eyeblink response. Subjects are repeatedly presented with a corneal airpuff (US), which evokes a

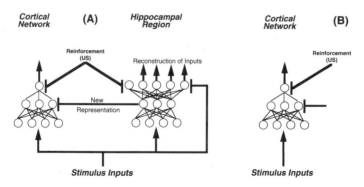

Figure 1. The cortico-hippocampal model (Gluck & Myers, 1993). (A) The intact model. A hippocampal-region network learns to reconstruct its inputs, plus a prediction of US arrival, while forming new stimulus representations in its internal layer which compress redundant information but differentiate predicative information. These new representations are acquired by a cortical network, which learns to map from them to a prediction of a behavioral response, and which is the site of long term memory. (B) The lesioned model. Disabling the hippocampal network is assumed to result in the cortical network no longer being able to acquire new representations, although it can still learn to map from existing representations to new behavioral responses. Figure reprinted from Gluck & Myers, 1993, Figure 5.

protective and reflexive eyeblink (UR); this airpuff is always immediately preceded by a tone stimulus (CS). After repeated CS-US pairings, subjects learn to generate an anticipatory blink response when they hear the tone so that eyelid closure occurs at the expected airpuff times. This simple paradigm has been the subject of detailed behavioral studies and theoretical analysis that has been well studied in both animals (Gormezano, et al., 1983; Thompson, 1986) and in humans (Solomon, et al., 1989). Considerable amounts of data suggest that the circuitry for acquisition of the CR is within the brainstem and cerebellum (Thompson, 1986) whereas the hippocampal region plays a modulatory role (Berger, et al., 1976; Solomon, 1981; Solomon et al., 1983). It is widely accepted that the cerebellar interpositus nucleus is the essential site of learning of the conditioned response (Thompson, 1986). However, if stimulus contingencies are more complex, learning may become hippocampal dependent. For example, hippocampal region damage can attenuate the ability to extinguish learned CSs (extinction), to learn about multiple CSs (discrimination), compound stimuli (blocking), unreinforced stimuli (latent inhibition). However, even under simple conditions which are not hippocampal-dependent, hippocampal activity does change to reflect learning, as evidenced through neurophysiological recordings (Berger & Orr, 1983). Thus the hippocampal region is performing some function during simple CS-US associational learning, and this function becomes critical for cases where complex stimulus-stimulus relationships must be mastered.

The Gluck & Myers (1993) model has been used to model the acquisition of a simple conditioned eyeblink response. Empirical data from hippocampal lesion studies repeatedly show that conditioned eyeblink acquisition in humans (Gabrieli, et al., 1995; Woodruff-Pak, et al., 1993) and rabbits (Solomon & Moore, 1975, Solomon, 1977) is not impaired under simple CS-US delay condition pairings. The model assumes that hippocampal region damage eliminates the ability to form new stimulus representations, but not the ability to learn new stimulus-response mapping based on pre-existing representation. Therefore, the lesioned model shows no particular deficit in acquisition of a simple CS-US association (Figure 2).

In contrast with simple CS-US acquisition, the model predicts that the hippocampal region should be critical for more complex conditioning paradigms, which require learning information about stimulus-stimulus relationships. For example, in sensory preconditioning, prior exposure to a CS1-CS2 compound increases the amount by which learning of the CS1 component generalizes to the other CS2 component (Thompson, 1972). The model predicts that this enhanced generalization arises from representational compression of the two components during the pre-exposure phase. Thus, the intact, but not lesioned

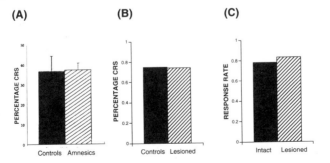

(A) **(B)** **(C)**

Figure 2. Hippocampal-region damage does not impair acquisition of conditioned eyeblink responding in (A) humans (Gabrieli et al., 1995) or (B) rabbits (Solomon & Moore, 1975), measured as percent of trials generating conditioned responses after equal amounts of training. (C) The cortico-hippocampal model similarly predicts no deleterious effects of hippocampal-region damage. Reprinted from Myers et al. (1996), Figure 4.

cortico-hippocampal model shows sensory preconditioning (Gluck & Myers, 1993); this is consistent with data showing that hippocampal region damage eliminates sensory precon-ditioning in rabbit eyeblink conditioning (Port & Patterson, 1984).

ROLE OF CHOLINERGIC NEUROMODULATION ON CONDITIONING

Functionally, removal of the hippocampal region results in a characteristic inability to learn and remember new information. In addition to hippocampal lesion, disruption of the cholinergic projections to the hippocampus results in profound memory impairments. The hippocampus appears as a curved cortical structure in the floor and medial wall of the temporal lobe. Information from the various primary and associational cortices converges on the entorhinal cortex which in turn projects to the hippocampus though GABAergic and glutamatergic neurons. The cortical input to the hippocampus is mirrored by efferent projections from the hippocampus to entorhinal cortex and returning to cerebral cortex. Another known bi-directional hippocampal pathway is via the fornix. Through the fornix, information from the hippocampus can reach a variety of subcortical structures including the anterior thalamic nuclei, hypothalamus and the septal nuclei. The septal nuclei of the basal forebrain cholinergic system consists of a variety of cholinergic (and GABAergic) neurons which send afferent projections to the hippocampus. Several animal studies have examined the role of the septo-hippocampal cholinergic projection in conditioning. As aforementioned in simple CS-US conditioning, an intact cerebellum is necessary for learn-ing the conditioned response, while lesions of the hippocampus do not affect simple learn-ing. However surgical lesions of the medial septum result in a characteristic slowing of the acquisition of the conditioned response. Pharmacological lesions using the cholinergic blocker scopolamine, injected into the medial septum (Berry & Thompson, 1979) or ad-ministered systemically in both rabbits (Solomon, et al., 1983) and humans (Solomon, et al., 1993) likewise impairs hippocampal function as demonstrated by a slowed acquisition of the relevance of the CS-US pairing. It thus appears that incomplete functional disrup-tion of the hippocampus via dysfunction of the basal forebrain cholinergic system is be-haviorally more devastating than complete hippocampal removal (Figure 3).

The effects of acetylcholine are well known as a primary neurotransmitter within pe-ripheral neural structures, such as at the neuromuscular junction and within ganglia of the autonomic nervous system; however it does not appear to be involved in the primary transfer of information-rich signals within cortical structures. Hasselmo and Schnell (1994) have suggested that the cholinergic input from the medial septum can be used to switch the hippocampus between two processing states: in the presence of acetylcholine (ACh), the hippocampus stores new information and suppresses pattern reconstruction via the recurrent collaterals in the CA3 hippocampal field; in the absence of ACh, storage is suppressed, and pattern reconstruction is allowed to occur to retrieve stored patterns for transfer to long-term cortical or cerebellar storage. This hypothesis is consistent with physiological studies (Hasselmo, Schnell & Barkai, 1995) showing that ACh both en-hances synaptic plasticity and suppresses activity in hippocampal stratum radiatum (the site of synapses from recurrent collaterals) more than activity in stratum lacunosum-moleculare (the site of synapses from extrinsic inputs).

We have approximated Hasselmo's storage-recall switching hypothesis within the original Gluck & Myers's (1993) cortico-hippocampal model by noting the tendency of the hippocampal-region network to store new information, as opposed to simply process-

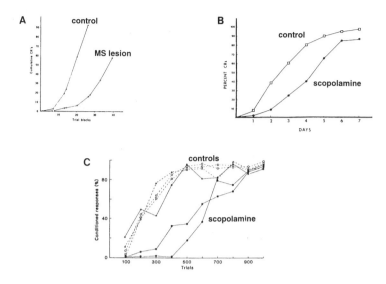

Figure 3. Acquisition of conditioned eyeblink responding is impaired by (A) medial septal lesion in rabbits (Solomon & Gottfried, 1981), and also by systemic administration of the anticholinergic scopolamine in (B) rabbits (Solomon et al., 1983) and (C) humans (Solomon et al., 1993). Reprinted from Myers et al. (1996), Figure 5.

ing it and recalling old information, is determined by the hippocampal network's learning rate (Myers et al., 1996). Disrupting septal input can therefore by approximated by lowering the learning rate – although not the rate at which information is transferred to the cortical network, nor the rate at which cortical associations develop. The consequence of this depression of hippocampal learning rates is to strongly retard classical conditioning in the model, proportional to the amount of depression. The anticholinergic drug scopolamine causes a similar dose-dependent effect in both rabbits (Solomon et al., 1983) and humans (Solomon, et al., 1993).

EFFECT OF CHOLINERGIC NEUROMODULATION ON SIMPLE ACQUISITION AND LATENT INHIBITION

The complete details of the simulation appear in Myers, et al., 1996. The model parameters were optimized to produce a system in which both intact and lesioned models learn a simple conditioned acquisition quickly and at approximately the same speed. The model was not optimized for any other tasks. In summary, 18 binary external inputs are used: 3 "CS"s and 15-element "context". One binary external reinforcement signal specifies the US. The hippocampal network is a fully connected autoencoder (Hinton, 1989) with 19 input nodes, 10 internal-layer nodes, and 19 output nodes. This network is trained by the standard backpropogation algorithm (Rumelhart, Hinton & Williams, 1986). The hippocampal learning rate is 0.02, except when adjusted to simulate the effects of cholinergic drugs, and is increased 10-fold on trials where the US is present.

The cortical network is a fully connected (18-60-1) network. Lower layer weights are trained by LMS (learning rate 0.005) to reproduce random fixed recoding of hippocampal region hidden layer activations. Upper layer weights are trained by LMS (learning rate 0.001) to predict US (taken as CR). Hippocampal lesions are simulated by disabling the hippocampal-region network resulting in cortical lower layer weights then remaining fixed. The effect of the cholinergic antagonist, scopolamine, is stimulated by selectively reducing the hippocampal network learning rate to either LR = 0.01 or 0.005, without changing any other parameters.

SIMPLE ACQUISITION

In humans and animals, disruption of septohippocampal acetylcholine with scopolamine results in impaired conditioning. This effect is dose-dependent with increasing concentrations of scopolamine resulting in increasing cholinergic blockade, and further hippocampal disruption resulting in greater impairments in conditioned learning behavior. In the model, reductions in the hippocampal learning rate result in disrupted hippocampal storage processes. As such the cortical network is unable to correctly map to correct responses, until the hippocampal output is stable.

It is important to note that disruption of septohippocampal acetylcholine impairs learning by delaying its onset, while not affecting the eventual rate of learning or its asymptote. In the model, disrupting hippocampal storage shows the similar pattern of delayed onset, but once the hippocampal output representation is stable then the cortex can learn at a normal rate.

LATENT INHIBITION

As aforementioned, the involvement of hippocampal computation in classical conditioning becomes more evident with more complex forms of associative learning. Latent inhibition is one example of complex stimulus-stimulus learning that has been well studied in animals and humans. During this two-phase learning paradigm, unreinforced pre-exposure to a cue (phase 1) slows learning to respond to subsequent CS-US pairings (phase 2) (Lubow, 1973). In the Gluck & Myers intact model, latent inhibition is caused by compression of the pre-exposed cue with co-occurring and equally non-reinforced contextual cues. The subsequent increase in learning time results because in phase 2 the model must

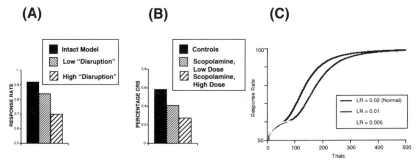

Figure 4. Disrupting septohippocampal acetylcholine via scopolamine results in dose-dependent impairments in conditioning in both (A) humans (Solomon et al., 1993) and in the (B) model. (C)The effect of lowering hippocampal-region learning rate in this system is to delay the onset of conditioned responding, but not its rate or eventual asymptote, much like the effects of the anticholinergic scopolamine in vivo. Figure C reprinted from Myers et al. (1996), Figure 7A.

first differentiate the cue from the context before a response can be selectively associated with the cue. Because this effect is assumed to depend primarily on hippocampal-region representational changes, the effect is correctly absent in the lesioned model (Myers & Gluck, 1994). Consistent with this model behavior, animals with large hippocampal region lesions also show this impairment (Solomon & Moore, 1975; Kaye & Pearce, 1987).

While scopolamine disrupts CS-US acquisition (simple delay conditioning), the detrimental effect of scopolamine neither eliminates nor significantly disrupts the effect of CS pre-exposure (latent inhibition). Although disruption of hippocampal computation by lowering the hippocampal learning rate slows acquisition, the hippocampal network is still able to construct compressed representations. Thus in the "scopolamine" model, phase 2 learning is slowed in forming CS-US associations, as a combined result of the non-reinforced stimulus pre-exposure in phase 1 and disrupted hippocampal learning processes of compression and differentiation.

CONCLUSION

Through gradual extensions and improvements of the original Gluck & Myers (1993) connectionist model to include behavioral functions of hippocampal computation (Myers & Gluck, 1994), known anatomical architecture of the hippocampal-region (Myers, Gluck & Granger, 1995), as well as principles of cholinergic neuromodulation (Myers, et al., 1996), the computational results obtained more completely test the model and hypothesis against existing empirical conditioning data. The combination of functional, behavioral tests with informed anatomical and selective lesion studies suggest further empirical and modeling studies for the examination of the processes and mechanisms involved in hippocampal function.

The additional integration of the role of other neuromodulators and other corticopetal activating systems (e.g. locus coerulues-noradrenergic system, substantia nigra-dopaminergic system, raphe nuclei-serotonergic system) will allow for future

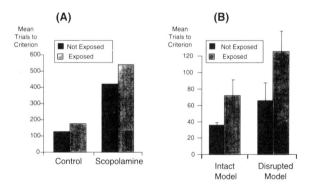

Figure 5. (A) Latent inhibition, the slowing in speed of learning a CS-US association following prior exposure to CS alone, is shown in normal rabbits; the effect appears to be spared under systemic scopolamine. Figure plotted from data presented in Moore et al. (1976). (B) Latent inhibition occurs in the "normal" intact model, as the hippocampal-region network compresses the representations of unreinforced CS and co-occurring contextual cues during the exposure phase (150 trials); this must be undone to allow later responding to the CS but not context alone, slowing conditioning (Myers & Gluck, 1994). The model correctly expects latent inhibition under "scopolamine": although the drug is assumed to retard hippocampal-region processing, it does not eliminate it, and so compressed representations are eventually formed which retard learning still further beyond the basic effects of the drug. Figure reprinted from Myers, et al., 1997/submitted, Figure 7.

understanding of the role of functional neuromodulation on affecting GABAergic and glutamatergic neurotransmission that may allow for a more complete understanding of the effects of neuropharmaceuticals on cortical processes and on human cognition.

REFERENCES

Berger, T., Alger, B., & Thompson, R. (1976). Neuronal substrate of classical conditioning in the hippocampus. *Science, 192*(4238), 483–485.

Berger, T., & Orr, W. (1983). Hippocampectomy selectively disrupts discrimination reversal learning of the rabbit nictitating membrane response. *Behavioral Brain Research, 8,* 49–68.

Berry, S., & Thompson, R. (1979). Medial septal lesions retard classical conditioning of the nictitating membrane response in rabbits. *Science, 205,* 209–211.

Gabrieli, J., McGlinchey-Berroth, R., Carrillo, M., Gluck, M., Cermack, L., & Disterhoft, J. (1995). Intact delay-eyeblink classical conditioning in amnesia. *Behavioral Neuroscience, 109*(5), 819–827.

Gluck, M., & Myers, C. (1993). Hippocampal mediation of stimulus representation: A computational theory. *Hippocampus, 3,* 491–516.

Gluck, M., Myers, C., & Goebel, J. (1994). A computational perspective on dissociating hippocampal and entorhinal function (Response to Eichenbaum, et al.). *Behavioral and Brain Sciences, 17,* 478–479.

Gormezano, I., Kehoe, E. J., & Marshall, B. S. (1983). Twenty years of classical conditioning research with the rabbit. *Progress in Psychobiology and Physiological Psychology, 10,* 197–275.

Hasselmo, M., & Schnell, E. (1994). Feedback regulation of cholinergic modulation and hippocampal memory function. In *Proceedings of World Congress on Neural Networks,* 2 (pp. 729–734). San Diego: INNS Press.

Hasselmo, M., Schnell, E., & Barkai, E. (1995). Learning and recall at excitatory recurrent synapses and cholinergic modulation in hippocampal region CA3. *Journal of Neuroscience, 15*(7), 5249–5262.

Hinton, G. (1989). Connectionist learning procedures. *Artificial Intelligence, 40,* 185–234.

Kaye, H., & Pearce, J. (1987). Hippocampal lesions attenuate latent inhibition and the decline of the orienting response in rats. *Quarterly Journal of Experimental Psychology, 39B,* 107–125.

Lubow, R. (1973). Latent Inhibition. *Psychological Bulletin, 79,* 398–407.

Moore, J., Goodell, N., & Solomon, P. (1986). Central cholinergic blockade by scopolamine and habituation, classical conditioning, and latent inhibition of the rabbit's nictitating membrane response. *Physiological Psychology, 4*(3), 395–399.

Myers, C., Ermita, B., Harris, K., Gluck, M., Hasselmo, M., Solomon, P., Gluck, M. (1996) A computational model of the effects of septohippocampal disruption on classical eyeblink conditioning. *Neurobiology of Learning and Memory,* 51–66.

Myers, C., Hasselmo, M., & Gluck, M. (1997/submitted). Further applications of a computational model of septohippocampal cholinergic modulation. submitted to *Psychobiology*

Myers, C., & Gluck, M. (1994). Context, conditioning and hippocampal re-representation. *Behavioral Neuroscience, 108*(5), 835–847.

Myers, C., Gluck, M., & Granger, R. (1995). Dissociation of hippocampal and entorhinal function in associative learning: A computational approach. *Psychobiology, 23*(2), 116–138.

Port, R., & Patterson, M. (1984). Fimbrial lesions and sensory preconditioning. *Behavioral Neuroscience, 98,* 584–589.

Rescorla, R., & Wagner, A. (1972). A theory of Pavlovian conditioning: Variations in the effectiveness of reinforcement and non-reinforcement. In A. Black & W. Prokasy (Eds.), *Classical Conditioning II: Current Research and Theory* (pp. 64–99). New York: Appleton-Century-Crofts.

Rumelhart, D., Hinton, G., & Williams, R. (1986). Learning internal representations by error propagation. In D. Rumelhart & J. McClelland (Eds.), *Parallel Distributed Processing: Explorations in the Microstructure of Cognition* (pp. 318–362). Cambridge, MA: MIT Press.

Solomon, P. (1977). Role of the hippocampus in blocking and conditioned inhibition of the rabbit's nictitating membrane. *Journal of Comparative and Physiological Psychology, 91*(2), 407–417.

Solomon, P., & Gottfried, K. (1981). The septohippocampal cholinergic system and classical conditioning of the rabbit's nictitating membrane response. *Journal of Comparative and Physiological Psychology, 95*(2), 322–330.

Solomon, P., Groccia-Ellison, M., Flynn, D., Mirak, J., Edwards, K., Dunehew, A., & Stanton, M. (1993). Disruption of human eyeblink conditioning after central cholinergic blockade with scopolamine. *Behavioral Neuroscience, 107*(2), 271–279.

Solomon, P., & Moore, J. (1975). Latent inhibition and stimulus generalization of the classically conditioned nictitating membrane response in rabbits (Oryctolagus cuniculus) following dorsal hippocampal ablation. *Journal of Comparative and Physiological Psychology, 89,* 1192–1203.

Solomon, P., Pomerleau, D., Bennett, L., James, J., & Morse, D. (1989). Acquisition of the classically conditioned eyeblink response in humans over the life span. *Psychology and Aging, 4*(1), 34–41.

Solomon, P., Solomon, S., Van der Schaaf, E., & Perry, H. (1983). Altered activity in the hippocampus is more detrimental to classical conditioning than removing the structure. *Science, 220,* 329–331.

Thompson, R. (1972). Sensory preconditioning. In R. Thompson & J. Voss (Eds.), *Topics in Learning and Performance* (pp. 105–129). New York: Academic Press.

Thompson, R. (1986). The neurobiology of learning and memory. *Science, 233,* 941–947.

Woodruff-Pak, D. (1993). Eyeblink classical conditioning in H.M.: Delay and trace paradigms. *Behavioral Neuroscience, 107*(6), 911–925.

99

MECHANISM FOR TEMPORAL ENCODING IN FEAR CONDITIONING

Delay Lines in Perirhinal Cortex and Lateral Amygdala

Billie Faulkner,[1] Kinh H. Tieu,[2] and Thomas H. Brown[1,2]

[1]Interdepartmental Neuroscience Program
[2]Departments of Psychology and Cellular and Molecular Physiology
Yale University
New Haven, Connecticut 06520
billie.faulkner@yale.edu
thomas.brown@yale.edu
kinh.tieu@yale.edu

1. ABSTRACT

Our neurobiological studies of the lateral amygdala (LA) and perirhinal cortex (PR) suggest a novel and testable hypothesis regarding encoding the time interval between the conditioned stimulus (CS) and unconditioned stimulus (US) in fear conditioning. In such conditioning, the interval between the onset of the CS and the onset of the US can be several seconds or longer. Here we outline our working hypothesis in qualitative terms, showing how the functional anatomy and neurophysiology furnish a platform for encoding the CS-US interval by mapping time onto space and then relying upon a conventional Hebbian learning rule in which the modifications require temporal contiguity of pre- and post-synaptic activity.

2. INTRODUCTION

Pavlovian conditioning furnishes a productive tool and framework for exploring the neurobiology of learning and memory.[4,8,12–16] One important application involves understanding the manner in which emotional responses are acquired.[1,8,9,13–15] In a typical fear conditioning paradigm, the conditioned stimulus (CS) is commonly a tone or light that signals the onset of an aversive unconditioned stimulus (US) such as a mild, electric footshock. Usually a long-delay paradigm is used, meaning that the CS and US co-terminate but the CS onset precedes the US onset by durations that can be several seconds. In a long-delay paradigm, the animal not only learns to associate the CS and US but also acquires temporal information about the CS-US interval[9].

This raises the question of how the circuitry encodes durations of such a magnitude. One might postulate the existence of a clock-like mechanism, although there is no cellular evidence for this in the particular brain regions that seem critically involved in fear conditioning. Alternatively, one could invoke some variation of the notion of reverberating activity around a polysynaptic circuit. To account for intervals of several seconds, this mechanism would require huge numbers of neurons or else huge numbers of reverberations around a smaller group of neurons. Such a mechanism would seem unstable and/or easily disrupted.

3. THEORY AND BACKGROUND

We pursued a third possibility—that there is a way, using a small number of neurons, to create a suitable array of delay lines. In the present context, the idea is that the CS is represented and made available to neurons in LA at various times (t, t+δ, t+2δ, ... t+Nδ) after its onset (at time t). Time is thus mapped onto space[5,6] and stored via a standard Hebbian mechanism,[2,11] whereby modifications in the CS pathway occur in just those synaptic inputs that are active at the same time as US-generated activity in other synaptic inputs to the same set of postsynaptic neurons.

How might such a mechanism be built from characteristics of the neurons in the critical circuitry? Both the amygdala and perirhinal cortex (PR) appear to be essential for some aspects of fear conditioning. In studies of fear conditioning in the rat, it is known that PR is importantly involved in processing the CS; that PR projects to the lateral amygdala (LA); that both the CS and US can drive single unit activity in the lateral amygdala; and that long-term synaptic potentiation occurs in LA.[3,7,8,14,15] Here we report that PR and LA cell physiology and morphology furnish a plausible platform for creating an appropriate spectrum of delay lines and corresponding activity windows representing the CS input to LA.

4. METHODS

Horizontal brain slices (400 mm thick) containing PR and LA were prepared and whole-cell recordings were made from visually preselected neurons.[10] The patch pipettes were conventional and usually contained 0.5% biocytin. Input-output relationships were determined under current-clamp conditions at the start of the whole-cell recording and again 30–60 min later. Subthreshold and spike firing characteristics were analyzed and quantified. After recording from a cell, slices were fixed and resectioned at 75 mm. Following histological processing of the biocytin-filled cell, a camera lucida was used for serial reconstruction of the neuron.

5. RESULTS

Perirhinal layer I contains sparsely-distributed fast-firing (FF) neurons, presumed to be inhibitory, with processes contained within the layer. FF cells occur throughout all layers of PR and LA. Regular-firing (RF) and late-firing (LF) neurons in layers II/III are rela-

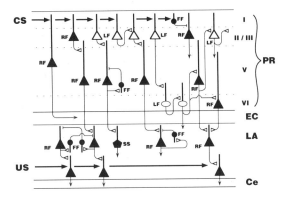

Figure 1. Schematic representation of the working hypothesis. Regions illustrated include perirhinal cortex (PR), the external capsule (EC), the lateral nucleus of the amygdala (LA), and the central nucleus of the amygdala (Ce). Open triangles or ovals represent late-firing (LF) cells, filled triangles are regular firing (RF) neurons, and filled circles show examples of various types of fast-firing (FF) inhibitory interneurons. Feed-forward inhibition in PR layer I mediates vector normalization; recurrent inhibition in PR layers II-VI help create activity windows; lateral inhibition in LA is part of a competitive learning scheme to facilitate sparse representations.

tively small pyramids whose apical dendrites extend to the edge of cortex (Fig. 1). Their axons collateralize throughout PR, but are most extensive within layers II/III.

There is no layer IV. Layer Va contains large pyramids, either RF or burst-firing (BF), whose dendrites extend through layer I and whose primary axon travels in the external capsule. In contrast, large pyramids in Vb often have bifurcating apicals that do not extend beyond layer V and whose primary axons cross the external capsule and project across LA. Layer VI is replete with RF stellates and horizontally-elongate LF cells. LA contains all of the physiological heterogeneity seen in PR as well as cells that exhibit extreme accommodation (SS), being resistant to spiking more than once or twice during a one second current step.

Pyramidal neurons in layers II/III and Va, whose spiny dendrites branch extensively in layer I, are optimally positioned to receive CS-generated input activity from axons conveying sensory information that courses through layer I. Some of these neurons can be expected to be recipients of CS-generated input activity. In response to a just-threshold current step, these cells commonly delayed firing for more than a second (Fig 2). The firing onset latency de-

Figure 2. Whole-cell recording from a LF neuron in PR. Characteristically, in response to a long current step, the cell fires with a delay greater than one second. Even longer delays have been observed. Application of a longer current step produces repetitive firing following the first spike. Once repetitive firing begins, the cells often exhibit anti-accommodation, meaning that the successive interspike intervals shorten. This is in contrast to the RF cells, which do not exhibit this delayed firing and show pronounced accommodation, meaning that the successive interspike intervals lengthen until the cell ceases to fire. The FF cells fire with little delay and are non-accommodating, meaning that the interspike interval is relatively constant.

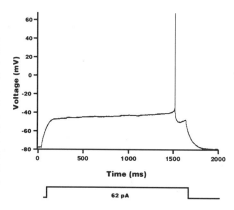

pended on both the particular cell and the strength of the depolarizing current. The axons of these layer II/III cells were frequently traced to layers Va, Vb, and VI.

Interestingly, some neurons in layer V projected back to layers II/III and also to LA. This raises the possibility of a reverberatory loop between the superficial and deep layers. Because the firing of a single cell can be delayed by a second or more, long loops or huge numbers of reverberations would not be required to account for a considerable amount of time. Similarly, layer VI and LA neurons often exhibited delayed firing in response to an outward current step; whereas neurons of layers I and V did not evidence delayed firing. Presumptive inhibitory (FF) interneurons were present in layers I–VI. The architecture of PR makes it easy to see how a spectrum of delays can emerge from this structure. LA might also participate in this process, but its internal spatial organization could make this possibility less evident.

6. SUMMARY AND CONCLUSIONS

The goal was to discover how to create a wide spectrum of delays in the CS pathway, avoiding the need to resort to clocks or seemingly implausible reverberations. The experimental results suggest a mechanism for creating appropriate delay lines.[10,17] These delayed representations of the CS can emerge naturally from the firing patterns and local circuitry. The firing delay was found to be a consequence of both the cell type, which varies by layer, and the intensity of the input to the neuron. The cell properties and local circuitry seem to enable a suitably wide spectrum of delays, from tens of milliseconds to several seconds. Our more complete proposed mechanism for mapping time onto space adds inhibitory interneurons to create restricted temporal windows of activity in the delay lines and also to control overall excitability. Computer simulations of this conceptual model indicate that it is in fact capable of accurately encoding a wide span of CS-US intervals.[17]

Cellular properties have previously been invoked to build delay lines to account for circuit-level computations. Action-potential propagation times were proposed to account for delays in the sub-millisecond range in studies of sound localization in the owl.[6] A more recent effort to explore longer durations incorporated a representation of synaptic facilitation into a conventional artificial neural network model, which was capable of temporal encoding in the sub-second range.[5] Here we proposed a novel mechanism that extends temporal encoding to the range of several seconds. This proposal emerged naturally from the neurobiology[10] of the specific brain regions that are critically involved in CS processing during fear conditioning. However, the principles of temporal encoding may well be applicable to many brain regions, perhaps in concert with additional timing mechanisms, such as those proposed for aspects of cerebellar function.[4]

7. REFERENCES

1. Brandon, S. E., Bombace, J. C., Falls, W. A., & Wagner, A. R. 1991. Modulation of unconditioned defensive reflexes by a putative emotive Pavlovian conditioned stimulus. *J. Exp. Psych.: Animal Behavior Processes,* 17, 312–322.
2. Brown, T. H., Ganong, A. H., Kairiss, E. W., & Keenan, C. L. 1990. Hebbian synapses: Biophysical mechanisms and algorithms. *Ann. Rev. Neurosci.,* 13, 475–512.
3. Burwell, R. D., Witter, M. P. & Amaral, D. G. 1995. The perirhinal and postrhinal cortices of the rat: A review of the neuroanatomical literature and comparison with findings from the monkey brain. *Hippocampus,* 5, 390–408.

4. Buonomano, D. V. & Mauk, M. D. 1994. Neural network model of the cerebellum: Temporal discrimination and the timing of motor responses. *Neural Computation*, 6, 38–55.

5. Buonomano, D. V. & Merzenich, M. M. 1995. Temporal Information transformed into a spatial code by a neural network with realistic properties. *Science*, 267, 1028–1030.

6. Carr, C. E. & Konishi, M. 1988. Axonal delay lines for time measurement in the owl's brainstem. *Proc. Natl. Acad. Sci.*, 85, 8311–8316.

7. Chapman, P. F., Kairiss, E. W., Keenan C. L. & Brown, T. H. 1990. Long-term synaptic potentiation in the amygdala. *Synapse*, 6, 271–278.

8. Corodimas, K. P. & LeDoux, J. E. 1996. Disruptive effects of post-training perirhinal cortex lesions on conditioned fear: Contributions of contextual cues. *Behav. Neurosci.*, 109(4), 613–619.

9. Davis, M., Schlesenger, L. S. & Sorenson, C. A., 1989. Temporal specificity of fear conditioning: effects of different conditioned stimulus—unconditioned stimulus intervals on the fear-potentiated startle effect. *J. Exp. Psych.: Animal Behavior Processes*, 15, 295–310.

10. Faulkner, B & Brown, T. H. 1996. Spatial organization and diversity of neurons in the rat perirhinal-lateral amygdala region. *Soc. Neurosci. Abstr*, 22, 1742.

11. Kelso, S. R., Ganong, A. & Brown, T. H. 1986. Hebbian synapses in hippocampus. *Proc. Natl Acad. Sci.*, 83, 5326–5330.

12. Moore, J. W. & Desmond, J. E. 1988. Adaptive Timing in Neural Networks: The Conditioned Response. *Biol Cybern.* 58, 405–515.

13. Lam, Y. W., Wong, A., Canli, T. & Brown, T. H. 1996. Conditioned enhancement of the early component of the rat eyeblink reflex. *Neurobiol. Learn. Mem.*, in press.

14. LeDoux, J. E. 1993. Emotional memory systems in the brain. *Behav. Brain Res.* 58, 69–79.

15. Quirk, G. J., Repa, J. C., & LeDoux, J. E. 1995. Fear conditioning enhances short latency auditory responses of neurons in the lateral nucleus of the amygdala: Parallel recordings in freely behaving rats. *Neuron*, 15(5), 1029–1039.

16. Thompson, R. F. & Krupa, D. J. 1994. Organization of memory traces in the mammalian brain. *Ann. Rev. Neurosci.*, 17, 519- 549.

17. Tieu, K. H., Faulkner, B. & Brown, T. H. 1996. Perirhinal-Amygdala neural network model of temporal encoding in fear conditioning. *Soc. Neurosci. Abstr*, 22, 1742.

100

NEURAL DYNAMICS FOR THE PERCEPTION OF ILLUSORY CONTOURS

Linking Gestalt-Laws and Synchronized Oscillations

Winfried A. Fellenz

Universität-GH Paderborn
FB 14, Pohlweg 47–49, 33098 Paderborn, Germany
fellenz@cs.tu-berlin.de

1. INTRODUCTION

The unrevealing of neuronal mechanisms responsible for the generation of illusory contours is important both for a better understanding of the human visual system and the improvement of an image segmentation method trying to imitate human performance. Studies of the human visual system support the conjecture that this segmentation process is purely preattentive [2,23,14] occurring at early visual stages [26,5,12,27]. We have extended existing schemes for boundary perception and completion [24,6,7,22,15,8,17,21, 28] by a relaxation labeling mechanism [3] which groups the phases of parametric neural oscillators into perceptual objects exploiting simple geometric constraints between the basic features. The proposed model consists of three successive processing stages which model some of the mechanisms found in visual cortex.

1.1. Model Overview

The network for the perception of illusory contours is inspired from neurophysiological and psychophysical findings, incorporating a relaxation phase labeling and diffusion process. The network groups visual features into perceptual entities by (de)synchronizing parametric phase labels of stochastic neuronal oscillators using a constraint satisfaction mechanism. The local constraints between features, which model the Gestaltist grouping principles proximity and good continuation [16], act horizontally in, and vertically between feature dimensions to allow for the emergent segregation of globally salient illusory contours. An initial nonlinear edge detection stage extracts two saliency maps for oriented and orthogonal cooperation between edges and endpoints. The energy maps are rectified into ON- and OFF-channels which cooperate for nearby orientations, but compete for orthogonal and opposing directions in phase space, suppressing false responses generated from the edge detection stage and forcing the global percept to appear.

1.2. Implementation Details

To model the intrinsic synchronization mechanism, an efficient modeling scheme based on parametric phase labels was chosen for which stability has been proved [25]. To reduce the computational burden of directly propagating the interaction constraints between all feature detecting neurons, some biologically motivated simplifications were made. Neurons responsive for all directions at each spatial position were combined into a hypercolumn, resembling the cortical microstructure in V1 [9]. The columns were tagged with a parametric phase label which was updated according to the combined interaction response calculated from the contribution of all feature dimensions (directions) of the model hyper-column. The propagation of constraints was decomposed into two separate stages: the spatial (horizontal) interaction between different sites within a specific feature map and the subsequent local (vertical) interaction between primitive features within each column. To allow the spread of synchronized phase labels into the interior of contour defined regions, a small activity is added at each location, resembling the probabilistic activity found with real neurons in visual cortex. The interacting feature maps were initialized using two extraction schemes shown in figure 2b for the direction specific detection of illusory contours (para-grouping) and figure 3b for the extraction of curved illusory contours at the endpoints of lines.

2. TYPES OF ILLUSORY CONTOURS

As can be seen from Figure 1, illusory contours can arise in various stimulus configurations and appear to induce sharp perceptual contours, which are most of the time accompanied by a variation of the perceived brightness of the emerging illusory figure. Although human expectation and attentional factors can influence the perceived illusory form [20], a fast bottom-up component prewired in biological visual systems may preattentively reveal situations of object occlusion in the visual world. The question concerning the origin of the perceived illusions can not be solved without incorporating multiple sources, like perceived depth, stereopsis [1], expectation, and knowledge [10,18,19].

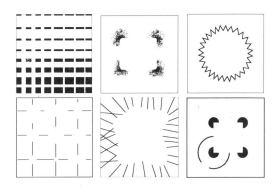

Figure 1. Different types of illusory contours at edges and line-ends.

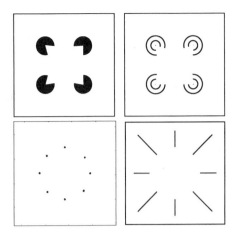

Figure 2. Perceptual grouping and illusory contours.

These sources can modify a perceived illusory surface but are not causal factors for its generation.

Figure 2 shows that both oriented edges and line-ends can produce stable illusory contours indicating two different mechanisms responsible for their generation. Opposite contrast polarity of the inducing edges or isoluminant stimulus conditions do not override the illusory contour but can produce contours without brightness contrast [13,23].This findings and studies concerning the time constraints of illusory contours suggest a decoupling of the process generating illusory contours from an illumination process, which eventually fills in perceived brightness and form. Some theories argue that illusory contours arise due to a process of amodal completion [11] thereby requiring amodal completing inducing elements. However, the completion of the occluded background is not necessary which can be seen in figure 1e, even if it is still possible to find the corresponding lines in the image. Figure 2c-d depicts a comparison of perceptual grouping and illusory contours showing that the sharp circular contour which appears in figure 2d is not visible in a similar stimulus without a specified orientation of the endpoints (Fig.2c). Both grouping mechanisms seem to be independent from each other, leading to oriented cooperation between lines and points or the cooperation between endpoints orthogonal to the inducing line.

2.1. Grouping of Parallel and Orthogonal Features

To specify a prewired architecture for the autonomous generation of illusory contours two mechanisms were used. A first nonlinear stage extracts oriented edge energy followed by the grouping of aligned oriented edges. Figure 3a shows the processing stages for the extraction of direction specific contours by rectifying oriented edge energy into ON- and OFF-channels. Opposing the edge grouping parallel to the direction of the contour [3], a second processing scheme for the detection of curved contours generated at the endpoints of lines was used. This scheme, depicted in figure 3b, takes directed endpoints extracted from the edge energy maps and sums the results of convoluting them with rotated versions of the function $F(x,y)=\exp(-a * x^2 - (b * y^2/(abs(x)+1)))$ depicted in figure 4. The parameters $a=0.008$ and $b=0.07$ were chosen empirically using a dicretization

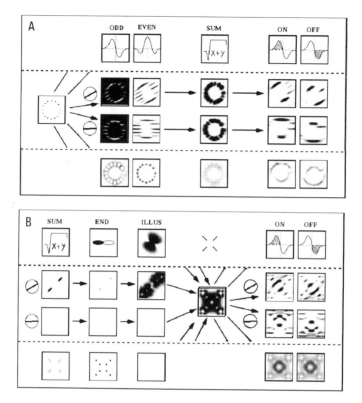

Figure 3. a) Extraction of direction specific illusory contours; b) Extraction of illusory contours orthogonal to endpoints. Upper row shows applied filter operation, middle rows shows two sample orientation maps out of six, lower row shows the summed responses of all maps.

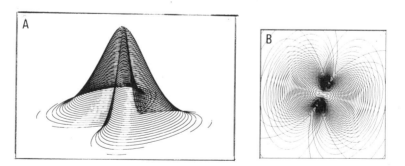

Figure 4. a) 3D-plot of IC-filter for the detection of illusory contours orthogonal to line ends. b) Contour plot of two IC-filters with 30 degrees orientation difference summing up to a 15 degree IC-filter.

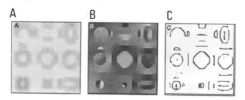

Figure 5. Circular illusory contours (Ehrenstein-Illusion); a) contour saliency (ortho-grouping); b) phase image after 20 iteration steps; c) binarized gradient of b.

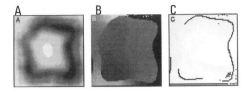

Figure 6. Artificial illusory 'object'; a) contour saliency (ortho-grouping); b) phase image after 28 iteration steps; c) binarized gradient of b.

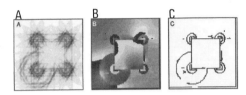

Figure 7. Kaniza-Square; a) contour saliency (para-grouping); b) phase image after 400 iteration steps; c) binarized gradient of b.

of size 71 x 71. The generated saliency-map is again rectified into opposite orientation channels to allow the emergent forming of contours independent of illumination.

3. RESULTS

Figure 5–7 show the results of relaxing phase labels corresponding to the discrete positions of the hypercolumnar array. Using the ortho-grouping mechanism, the blob-like illusory disks in figure 1d elicit a saliency map depicted in fig. 5a, which was used to extract sharp illusory contours in phase space (fig. 5b). The binarized gradient of the phase map (fig. 5c) shows the extracted illusory contours which resemble the perceived illusory disks. The simulation result using one complex illusory object (fig. 1e) is depicted in figure 6, showing global synchronization after 28 iteration steps with a sparse connectivity scheme between the parametric oscillators. Figure 7 shows a modified Kaniza square and contour lines extracted by the para-grouping mechanism. The illusory contours of the central square in figure 1f which are initially hidden in the long range directional responses (Fig. 7a) are almost completely extracted, and numerous false responses are suppressed, as can be evaluated from the phase gradient in Figure 7c.

REFERENCES

1. B. L. Anderson and B. Julesz. A theoretical analysis of illusory contour formation in stereopsis. Psychological Review, 24:681–684, 1995.
2. J. Driver, G. C. Baylis, and R. D. Rafal. Preserved figure-ground segregation and symmetry perception in visual neglect. Nature, 360:73–75, 1992.
3. W. A. Fellenz and G. Hartmann. Preattentive grouping and attentive selection for early visual computation. 13th IAPR Int. Conference on Pattern Recognition, August 25–29, 1996, TU Vienna, Austria, IV:340–345, IEEE, 1996.
4. C. D. Gilbert, A. Das, M. Ito, M. Kapadia, and G. Westheimer. Spatial integration and cortical dynamics. Proceedings of the National Academy of Science (USA), 93:615–622, 1996.

5. D. H. Grosof, R. M. Shapley, and M. J. Hawken. Macaque V1 neurons can signal 'illusory' contours. Nature 365:550–552, 1993.

6. S. Grossberg and E. Mingolla. Neural dynamics of perceptual grouping: Textures, boundaries, and emergent segmentations. Perception and Psychophysics, 38 (2):141–171, 1985.

7. G. Guy and G. Medioni. Perceptual grouping using global saliency-enhancing operators. Proc. 11th IAPR Int. Conf. on Pattern Recognition, The Hague, The Netherlands, I:99–103, IEEE, 1992.

8. F. Heitger and R. von der Heydt. A computational model of neural contour processing: Figure-ground segregation and illusory contours. Proc. Fourth Int. Conference on Computer Vision, pages 32–40, Berlin, IEEE, 1993.

9. D. Hubel and T. Wiesel. Functional architecture of macaque visual cortex. Proc. Royal Society London B, 198:1–59, 1977.

10. G. Kaniza. Organization in Vision. Praeger, New York, 1979.

11. P. J. Kellman and T. F. Shipley. A theory of visual interpolation in object perception. Cognitive Psychology, 23:141–221, 1991.

12. V. A. F. Lamme, B. W. van Dijk, and H. Spekreijse. Contour from motion processing occurs in primary visual cortex. Nature, 363:541–543, 1993.

13. G. W. Lesher and E. Mingolla. The role of edges and line-ends in illusory contour formation. Vision Research, 33(16):2253–2270, 1993.

14. D. S. Levine and S. Grossberg. Visual illusions in neural networks: line neutralization, tilt after effect, and angle expansion. J. Theor. Biol., 61:477–504, 1976.

15. B. S. Manjunath and R. Chellappa. A unified approach to boundary perception: edges, textures and illusory contours. IEEE Transactions on Neural Networks, 4(1):96–107, 1993.

16. W. Metzger. Gesetze des Sehens. Frankfurt/M: W. Kramer, 1975.

17. M. Nitzberg, D. Mumford, and T. Shiota. Filtering, Segmentation, and Depth. Springer, Berlin, 1993.

18. S. Petry and G. E. Meyer (Eds). The Perception of Illusory Contours. Springer, Berlin, 1987.

19. F. Purghé and S. Coren, Subjective contours 1900–1990: Research trends and bibliography, Perception and Psychophysics, 51:291–304, 1992.

20. I. Rock and R. Anson. Illusory contours as the solution to a problem. Perception, 8:665–681, 1979.

21. P. Sadja and L. H. Finkel. Intermediate-level visual representations and the construction of surface perception. Journal of Cognitive Neuroscience, 7(2):267–291, 1995.

22. J. Skrzypek and B. Ringer. Neural network models for illusory contour perception. IEEE Proc. Comput. Soc. Conf. Computer Vison and Pattern Recognition, pages 681–683, 1992

23. L. Spillmann and B. Dresp. Phenomena of illusory form: can we bridge the gap between levels of explanation. Perception, 24:1333–1364, 1995.

24. S. Ullman. Filling-in the gaps: the shape of subjective contours and a model for their generation. Biological Cybernetics, 25:1–6, 1976.

25. J. L. van Hemmen and W. F. Wrezinski. Lyapunov function for the Kuramoto model nonlinearly coupled oscillators. J. Stat. Phys., 72:149–166, 1993.

26. R. von der Heydt and E. Peterhans, Mechanisms of contour perception in monkey visual cortex. I. Lines of pattern discontinuity. II. Contours bridging gaps, Journal of Neuroscience, 5(9):1731–1763, 1989.

27. G. Westheimer. Illusory figures and real neurons. Nature, 371:745–746, 1994.

28. L. R. Williams and D. W. Jacobs. Stochastic completion fields: a neural model of illusory contour shape and salience. Proc. Fifth Int. Conference on Computer Vision, pages 408–415. IEEE, Washington, 1995.

101

NONLINEAR FUNCTIONS INTERRELATING NEURAL ACTIVITY RECORDED SIMULTANEOUSLY FROM OLFACTORY BULB, SOMATOMOTOR, AUDITORY, VISUAL AND ENTORHINAL CORTICES OF AWAKE, BEHAVING CATS

G. Gaál[1] and W. J. Freeman[2]

129 Life Sciences Addition
Department of Molecular and Cell Biology
University of California
Berkeley, California 94720
gaal@violet.berkeley.edu
wfreeman@garnet.berkeley.edu

1. INTRODUCTION

Population coding algorithms have been designed to retrodict sensory stimuli or predict motor behavior from neuronal responses (Georgopoulos et al., 1986). These include calculation of Jacobian matrices in nonlinear systems (Gaál, 1995), which made it possible to model the visuomotor hand movement task of reaching in a plane, in adaptive feedback control while updating the joint angles of a three-joint arm (Lee and Kil, 1994). The control signal was the dot product between the visual error signal and the transpose of the Jacobian matrix of the direct kinematic equation of hand movement. The trajectories of the hand were synchronized with the x and y time series outputs of coupled nonlinear equations. The equations used to calculate the adaptive feedback signal were similar to those used by Kocarev et al. (1993) to show that two different nonlinear systems could synchronize, when the difference between the goal (Lorenz system) and target signals (Chua system) was added as an adaptive feedback signal to modify the equations of the entrained (Chua) system. In robotics, the Jacobian matrix was defined by the makers of the robots. In biological systems, the matrix needs to be derived from observed time series. Experimental control and synchronization of chaos in nonlinear dynamical systems by self-controlling feedback have already been demonstrated (Pyragas, 1992; Pecora and Carroll, 1990; McKenna et al., 1994; Kelso and Ding, 1992), with applications in neurobiological control, prediction and synchronization.

Computational Neuroscience
edited by Bower, Plenum Press, New York, 1997

The sensorimotor system controlling arm movement with visual feedback is especially promising to investigate/model synchronization. Widespread oscillatory synchrony has been observed at different levels of the sensory and motor system between central action potentials and muscle activity (e.g. Kreiter and Singer, 1992; Murthy and Fetz, 1992; Sanes and Donoghue, 1993; Donoghue et al., 1996). In the model (Gaál, 1995) used to simulate results observed in neurophysiological experiments with primates (Georgopoulos et al. 1986; Schwartz 1992; 1993), the equations combined the transpose of Jacobian matrices with visual error signals to reproduce reaching trajectories of visually guided hand movement. Movement trajectories could also be synchronized with visual moving targets. The Jacobian matrix estimation method was first introduced to map neural signals, muscle activity and behavioral signals recorded in behaving primates engaged in a visuomotor task with instructed delay (Sanes and Donoghue, 1993; Donoghue et al., 1996). In the present study, the method is used to map nonlinear functions of neural activity observed in five brain areas of behaving cats trained to release bars on visual cues. The matrices were derived from simultaneously observed time series by numerical estimation of their partial derivatives with respect to each other.

For arm movement simulations, the elements of the Jacobian matrix were estimated from correlations between observed changes in joint angles and Cartesian coordinate values of the hand and compared to elements calculated as partial derivatives. One thousand small random movements at a given reference location were used to estimate the location-dependent Jacobian matrix (Gaál, 1995). A new algorithm was introduced to estimate the Jacobian matrix by measuring changes in joint angles and Cartesian coordinate values of the hand from continuous simulated (or experimentally observed) arm movement time series and computing the correlation matrices of the joint angle and hand velocities (Figure 1). This step gave Jacobian matrices also for other nonlinear functions, e.g. Lorenz attractors, and mapping between muscle activity and neural activity.

These gave erratic, approximate estimates for simulated arm movement using 15 trials. 1000 trials gave exact estimates compared to analytic expressions of the Jacobian matrices.. The numerical implementation of the algorithm was evaluated by estimation of the Jacobian matrices at various reference locations along Lorenz (Figure 2) and Rössler attractors.

Related calculations have originally been used to determine Lyapunov exponents from experimental time series (Sano and Sawada, 1985; Eckmann and Ruelle, 1985).

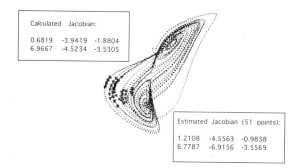

Figure 1. Calculated and estimated Jacobian matrices at a single spot on the elbow angle–shoulder angle attractor while the arm is tracing the Lorenz pattern (the equations used to obtain the simulated data of Figure 1 are presented in Gaál, 1995). The calculated values were derived from the equations, the estimated values were obtained as combinations of correlation matrices at 51 points in phase space.

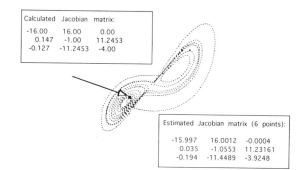

Figure 2. Calculated and estimated Jacobian matrices at a single location on the Lorenz attractor. Six nearest neighbours around the reference point (white arrowhead) were used for estimation.

2. ANALYSIS AND RESULTS

The Jacobian matrices for nonlinear functions of neural activity in the present work were derived from multiple cortical EEGs simultaneously recorded in operant behavior in cats engaged in a sensorimotor task. The cats were trained to release bars on visual (Figure 3) or auditory cues after a brief instructed delay period for food reward in a task with a

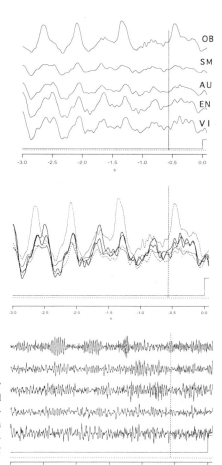

Figure 3. EEG recordings from five brain areas (OB—olfactory bulb, SM—somatomotor, AU—auditory, EN—entorhinal and VI—visual cortices, respectively, from top to bottom) filtered between 1–12 Hz (upper and middle insets) and 10–100 Hz during a single trial in a visuomotor task from subject S2. Vertical bar indicates onset of light conditioned stimulus (CS). Lowest trace is DC code for bar release signal (CR).

fixed intertrial interval and fixed waiting period. The recordings were obtained from five areas. Two depth electrodes were placed in olfactory bulb (first trace in upper and lower Figure of Figure 3; thin dotted line in middle). Square electrode arrays of 4x4 electrodes were placed over the somatomotor cortex (second trace in upper and lower Figures of Figure 3, thin continuous line in middle), auditory cortex (third trace, thin dashed line) and visual cortex (fifth trace and thick dotted line), respectively. A flat electrode array was placed on the surface of the entorhinal cortex containing 8+8 wires (fourth trace, thick continuous line).

Anatomical studies show pathways by which the primary sensory cortex interact with each other, and by which the entorhinal cortex interacts with all sensory areas and the hippocampus. On this basis, the relations between the EEGs from the several structures during performance of the operant were examined by visual inspection. The vertical dotted line in the Figures labels the onset of the sensory stimulus; the solid line is the bar release signal. The relationships in the high frequency range are illustrated in a single visual trial from all four subjects.

The upper Figure of Figure 3 shows five EEG traces (one from each area) band-pass filtered in the frequency range of 1–12 Hz. The middle Figure shows the same traces superimposed on each other to illustrate the relationships along the time axis 3 sec before bar release. The lower Figure shows the same EEG recordings as the two other Figures, but the traces were band-pass filtered between 10 and 100 Hz to reveal activity in the gamma range.

The middle part of Figure 3 illustrates relationships in the low frequency range between recordings from different areas: in the early stages of this particular 6 sec trial, approximately 3 sec before bar release, the visual recording led all the others by several milliseconds in this visuomotor task. The entorhinal cortex led in the time period immediately preceding the appearance of the visual stimulus. The somatomotor cortex led immediately before bar release.

The times of the appearance and disappearance of oscillations in the gamma range (35–60Hz in cats) in all areas were not precisely locked to any of the cues used in the task (beginning of the trial, appearance of the sensory stimulus, beginning of reward period, end of reward period, end of sensory stimulus period as well as self-initiated bar release). A high frequency olfactory bulb burst spindle often appeared 200–800 ms before the sen-

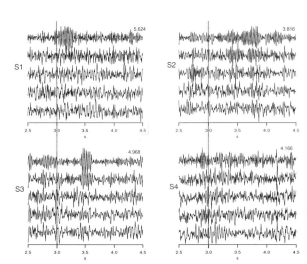

Figure 4. Spatial averages of EEG recordings from five brain areas (OB, SM, AU, EN and VI cortices, respectively, from top to bottom). EEGs recorded from four subjects, S1, S2, S3 and S4, during a single trial in a visuomotor task, were filtered between 10–100 Hz. Numbers indicate timing of bar release; the onset of the visual stimulus is at 3 sec.

sory cue or immediately preceding bar release. Both were preceded or accompanied by a somatomotor gamma burst, but never in all five areas simultaneously. The lead and lag relationships observed at different time points and different frequencies, tended to average out in sliding spectrograms, coherence plots, Joint Perievent Time Histogram calculations or estimations of delays from maxima of cross-correlation functions between EEGs.

The motor output of the cats was manifested in the EEGs in two ways. First, the low frequency wave in the olfactory bulb EEG correlates closely with the animals' respiratory cycle (Davis and Freeman, 1982). Second, the somatomotor EEG manifests changes associated with onset of conditioned responses. Therefore, Jacobian matrices were derived on the assumption that the olfactory bulb and somatomotor EEGs were functions of neural recordings from the other three EEGs. Correlation matrices were estimated for the three areas (auditory, entorhinal and visual) at each point in phase space visited by the normalized EEG signals multiple times. Locally linearized correlation matrices between putative input (auditory, entorhinal and visual) and output (olfactory bulb and somatomotor) neural activities were also calculated. The Jacobian matrix estimates were then obtained by matrix multiplication of the pseudo inverse of the first correlation matrix with the second correlation matrix. The hypothesis was that errors between Jacobian matrices estimated for a sufficiently high enough number of sequential trials would be zero, when the cortical areas were interacting strongly. The observed error functions calculated for trial sets consisting of nine trials each, aligned for sensory stimulus or bar release, respectively, showed segments of low error (Figure 5). Peaks in the error functions revealed episodic uncoupling between the cortices.

The peaks for the OB-AU, OB-EN, OB-VI, SS-AU, SS-EC, SS-VI nonlinear functions were related with behavioral events in the 6 sec trials: anticipation immediately following the trial-start cue; expectation before the onset of the sensory stimulus, which could be a CS+ or a CS-(depending on the location of the light or sound frequency); sensory identification; expectation of the beginning of the reward period (although a variable delay could be imposed between the appearance of the sensory cue and the beginning of the reward period, the delay was generally fixed in most training/recording sessions and set to 350 ms); preparation for movement; execution of the movement (bar release); and consuming the food reward. For the reasons already given, pairwise linear methods, such as peaks of cross-correlation functions in sliding 512, 128 or 64 msec windows, locations

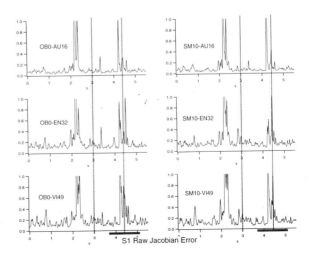

Figure 5. Errors between Jacobian matrices estimated for two sets of nine trials for subject S1 (presented along a single reference trial, not on a phase plane plot). By assumption the olfactory bulb and the somatomotor cortex activity are nonlinear functions of the neural activity in auditory, visual and entorhinal cortices. Note the strong time segmentation of the EEGs that was correlated with the behavior of the subject, and was found in all four subjects. Such strong segmentation was not present in EEGs recorded during auditory mapping at rest, without performance of an operant. First vertical bar at 3 indicates onset of light stimulus, second bar indicates mean bar release with standard deviation.

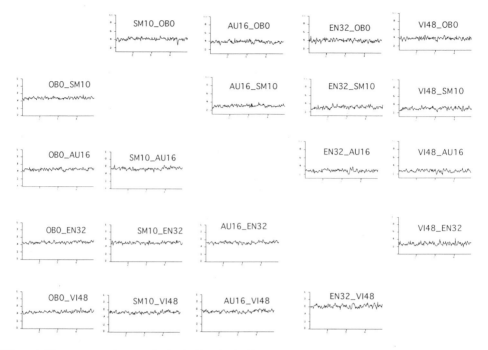

Figure 6. Time segmentation was obtained with pairwise linear methods for the same EEG signals of subject S1. Peaks of pairwise cross-correlation functions calculated in 64 msec windows stepped at 20 msec and averaged for 20 trials are shown in lower left corner. Peaks of cross-frequency spectra derived for the same time windows are shown in upper right corner. Although the linear method shows subtle changes at various stages in the task for various pairs of brain areas, it does not produce a time segmentation as robust as the one revealed by the locally linearized nonlinear method of estimating errors between Jacobian matrices. Note that both the strength of correlation and the segmentation effect were strongest for the entorhinal–visual cortex pair in this visuomotor task. This effect was also reproducible for four subjects.

of peaks of cross-frequency spectra, strengths of cross-coherence, have not provided such robust time segmentation so far for the EEGs in the visuomotor tasks (Figure 6).

3. CONCLUSIONS

The Jacobian matrix estimations used only correlations between observed signals. They might be more useful for interpreting physiological observations than the methods of calculating correlation dimensions, Lyapunov exponents or Kolmogorov entropies in nonlinear dynamics. The time segmentation of the EEGs in the sensorimotor task appears to provide a method to reveal uncoupling among brain structures, and conversely, the coupling that must occur to mediate the behaviors in operant conditioning. Synchronization is more difficult to detect by means of sliding windows and Fourier decomposition, owing to the nonlinearities manifested in the broad spectra of the EEGs. The Jacobian matrix estimation method relates the nonlinear functions between CS, CR, and neural activity, which unites two routine analytic methods (calculating effective connectivity in neural networks, and mapping receptive field weighting functions in sensory areas or preferred directions in motor areas) in a single multidimensional framework. The Jacobian matrices derived from observed biological time series might be useful as control signals and in treating movement disorders or neurohormonal diseases. We are now investigating various frequency ranges during the same and different sets of training sessions within and between subjects.

ACKNOWLEDGMENTS

This work was supported by NIMH 06686 and ONR N00014-93-1-0938 research grants awarded to W. J. Freeman.

REFERENCES

D. Bullock, S. Grossberg and F.H. Guenther, A self-organizing neural model of motor equivalent reaching and tool use by a multijoint arm, J. Cogn. Neurosci. 5 (1993) 408–435.

Y. Burnod, P. Grandguillaume, I. Otto, S. Ferraina, P.B. Johnson and R. Caminiti, Visuomotor transformations underlying arm movements toward visual targets: a neural network model of cerebral cortical operations, J. Neurosci. 12 (1992) 1435–1453.

G.W. Davis and W.J. Freeman, On-line detection of respiratory events applied to behavioral conditioning in rabbits. IEEE Trans. on Biomed. Eng. 29 (1982) 453–456.

J.N. Donoghue, J.N. Sanes, N. Hatsopoulos, and G. Gaál, Oscillations in neural discharge and local field potentials in primate motor cortex during voluntary movement. J. Neurophys. (1996) pending.

J.P. Eckmann and D. Ruelle, Ergodic theory of chaos and strange attractors, Rev. Modern Phys. 57 (1985) 617–656.

G. Gaál, Relationship of calculating the Jacobian matrices of nonlinear systems and population coding algorithms in neurobiology, Physica D 84, (1995) 582–600.

A.P. Georgopoulos, A. Schwartz and R.E. Kettner, Neuronal population coding of movement direction. Science 233 (1986) 1416–1419.

D. Hebb, The Organization of Behavior (New York, Wiley, 1949).

J.F. Kalaska and D.J. Crammond, Cerebral cortical mechanisms of reaching movements, Science 255 (1992) 1517–1523.

J.A.S. Kelso and M.Z. Ding, Fluctuations, intermittency and controllable chaos in motor control. In: K. Newell and D. Corcos (eds.) Variability in Motor Control. Human Kinetics. Champaign (1992).

L. Kocarev, A. Shang and L.O. Chua, Transitions in dynamical regimes by driving: A unified method of control and synchronization of chaos, Int. J. of Bifurcation and Chaos 3 (1993) 479–483.

A.K. Kreiter and W. Singer, Oscillatory neuronal responses in the visual cortex of the awake macaque monkey, Eur. J. Neurosci. 4: (1992) 369–375.

M. Kuperstein, Neural model of adaptive hand-eye coordination for single postures, Science, 289 (1988) 1308–1311.

S. Lee and R.M. Kil, Redundant arm kinematic control with recurrent loop, Neural Networks 7 (1994) 643–659.

T.M. McKenna, T.A. McMullen and M.F. Shlesinger, The brain as a dynamic physical system. Neurosci. 60 (1994) 587–605.

V.N. Murthy and E.E. Fetz, Coherent 25–35 Hz oscillations in the sensorimotor cortex of the awake behaving monkey, Proc. Natl. Acad. Sci. 89 (1992) 5670–5674.

L.M. Pecora and T.L. Carroll, Synchronization in chaotic systems, Phys. Rev. Lett. 64 (1990) 821–824.

K. Pyragas, Continuous control of chaos by self-controlling feedback, Phys. Lett. A170 (1992) 421–428.

J.N. Sanes and J.P. Donoghue, Oscillations in local field potentials of the primate motor cortex during voluntary movement, Proc. Natl. Acad. Sci. 90 (1993) 4470–4474.

M. Sano and Y. Sawada, Measurement of the Lyapunov spectrum from a chaotic time series, Phys. Rev. Lett. 55 (1985) 1082–1085.

A.B. Schwartz, Motor cortical activity during drawing movements: single-unit activity during sinusoidal tracing, J. Neurophys. 68 (1992) 528–541.

A.B. Schwartz, Motor cortical activity during drawing movements: population representation during sinusoidal tracing, J. Neurophys. 70 (1993) 28–36.

102

POPULATION CODING, POPULATION REGULARITY AND LEARNING

Pierre Germain[1]* and Yves Burnod[2]

[1] DRET/ETCA/CREA/SP
16 bis, Av Prieur de la Côte d'Or
F-94114 ARCUEIL cedex
E-mail: germain@etca.fr
fax: 33 (1) 42 31 99 64
[2] INSERM-CREARE
Neurosciences et Modélisation
9 quai St Bernard
F-75252 PARIS
E-mail: ybteam@snv.jussieu.fr

ABSTRACT

The validity of population coding for movement within the motor cortex has now been confirmed by many experimental studies. The preferred direction of the population units seemed to be uniformly distributed. Previous studies have highlighted two points: (1) a regularity condition of the population activity distribution called H-regularity is a necessary and sufficient condition for learning with a class of biologically plausible correlation-based rules of synaptic modification without any bias. (2) H-regularity can result from a self-organization driven by the difference between feedforward and lateral inputs, whatever the distribution of inputs is. The set of learning rules is extended here to include a Hebb-like rule.

1. INTRODUCTION

Many experimental data are now available on the activity of cortical cells of the monkey parietal, premotor and motor areas recorded during reaching movement in selected directions (see [3] for a review). These data point out first that almost every cell the activity of which is correlated with reaching movements responds maximally for a particular direction called "the preferred direction". Tuning to this direction is broad and well approximated by the cosine $M.C_i$

---

*correspondence should be sent to this address

Computational Neuroscience
edited by Bower, Plenum Press, New York, 1997

between the preferred direction C_i and the direction of movement M. The fact that many cells are involved in each movement leads to the concept of "Population Coding". One possible code for the direction of movement proposed in [4] is the "population vector" (P) which is the vector sum of the preferred direction C_i weighted by the change $o_i(M)$ in cell activity associated with a particular movement direction M:

$$P(M) = \sum_i o_i(M)C_i \qquad (1)$$

The population vector has been shown to be a reasonably good predictor of the movement direction. Another characteristic of such populations is that the preferred direction seems to be uniformly distributed across the 3D space. A weaker form of regularity we called HL-regularity happens when the inertia matrix of the preferred direction is an homothety:

$$Q = \sum C_i C_i^\top = \lambda I \qquad (2)$$

HL-regularity has been shown to be a necessary and sufficient condition for the population vector to be an unbiased estimator of the direction of movement.

The next step is to sketch how the cerebral cortex could use such a coding to learn correct behavior and in particular to learn to perform reaching movements toward a visual target. This point has been discussed in [2]; a model consistent with the known anatomical and experimental data has been developed. Here again, H-regularity of the population is necessary to learn the appropriate behavior.

Our research aims at investigating the impact of uniformity and more generally regularity of the population activities on the learning properties.

Two key ideas set up our research framework :

1. **The regularity of the population allows to use a wider set of learning mechanisms so as to include more biologically plausible ones.**

2. **The population could be self-organized to become regular using lateral connections. Theses connections link the units inside the population and are learned with the same learning rule as the feedforward connections.**

These two ideas highlight the general principle telling that adaptation and redundancy mechanisms are used by the brain so as to correct the bias of the obviously limited computation mechanisms provided by the biological bulk. In previous papers [5, 6] we have shown that the set of learning mechanisms include a class of correlation-based rules of synaptic modification.

The work described in this paper extends the set of possible learning mechanisms so as to include a classical Hebb-like learning rule.

We now exposed more technically our main results.

2. LEARNING MODEL

The model used is drawn in fig 1. The bottom line of units represents the input population. o_{ff} is a linear combination of the activities o_i of the population units through weights w_i. o_{ref} is a reference signal. o_{ref} and the o_i's are functions of a variable x denoting "the state of the world" i.e. the concatenation of all the variables that could be represented in the cortex. The learning goal is to predict this reference signal with the "feedforward" signal o_{ff}. The upper

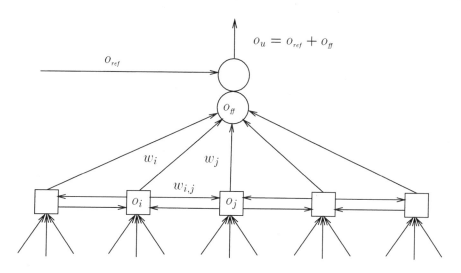

Figure 1. General model used in this paper

unit u has been divided in two parts to separate the reference and feedforward inputs. Only a few sample units and connections are shown. Lateral connections $w_{i,j}$ allow the population to reach self-organization.

Unit u output activity is the sum of the feedforward and reference signals:

$$o_u = o_{ref} + o_{ff} \tag{3}$$

3. LEARNING MECHANISM

As a learning rule we use a classical Hebb rule with a multiplicative weight decay to keep the weights bounded :

$$\delta w_i = \alpha o_u o_i - |W|^2 w_i \tag{4}$$

where $|W|^2 = \sum w_i^2$. This rule is well known and has been widely studied (see [7, 8] for reviews). However, we have adopted a different point of view since the reference signal entry is not adapted and we look in which conditions the feedforward signal is the best approximation of the reference signal.

The following **"best approximation property" has been shown : the feedforward signal is proportional to the orthogonal projection of the reference signal onto the linear subspace spanned by the population unit activities *if and only if* the population unit activities fulfill the following H-regularity condition:**

$$\sum_i C_i^* C_i^{*\top} = \lambda I \tag{5}$$

where C_i^* is the the vector of components of the unit activity o_i in an orthonormal basis for the scalar product $< a, b >= E_\xi[ab]$, E_ξ being the mean value over the distribution of the examples presented during learning.

4. APPLICATION TO THE MOTOR CORTEX

We applied the previous results in a model of learning motor cortex cell command for reaching movements consistent with the experimental results mentioned in the introduction. We restrict first to a fixed arm position. u is a motor cortex cell and the population units are cells of the parietal cortex tuned to this arm position. Their activity is supposed to be a linear function of the desired movement direction $o_i = M.C_i$. As in [2], we modeled the contribution of a motor cortex cell to the movement by the product $o_u M_{eff}^u$ of its activity o_u and a 3D vector M_{eff}^u; o_{ref} is a proprioceptive feedback that has an activity $o_{ref} = M_{eff}^u.M$ which is maximum for $M = M_{eff}^u$.

Then the "best approximation property" means that the motor cell preferred direction C_u is proportional to the unit contribution M_{eff}^u which itself is a condition of movement efficiency.

In this case, H-regularity (5) is exactly equivalent to HL-regularity (2).

It can be shown that the overall effect of the motor cell activation is also equal to the intended movement due the regularity of the motor cell population. Using several populations of parietal neurons tuned to different initial positions allows to generalize to a moving position of the hand.

It should be noted that our basic result applied even without a linear coding hypothesis in the parietal and motor cortex.

REFERENCES

[1] Michael A.Arbib. *The Handbook of Brain Theory and Neural Networks*. MIT Press, Cambridge, Massachusetts, 1995.

[2] Yves Burnod, Philippe Grandguillaume, Isabelle Otto, Stefano Ferraina, Paul B. Johnson, and Roberto Caminiti. Visuomotor transformations underlying arm movements toward visual targets: A neural network model of cerebral cortical operations. *The Journal of Neuroscience*, 12(4):1435–1453, April 1992.

[3] Apostolos P. Georgopoulos. *Reaching: Coding in Motor Cortex*, chapter Part III: Articles, pages 783–787. In A.Arbib [1], 1995.

[4] Apostolos P. Georgopoulos, Andrew B. Schwartz, and Ronald E. Kettner. Neuronal population coding of movement direction. *Science*, 233:1416–1419, August 1986.

[5] Pierre Germain and Yves Burnod. Population coding and hebbian learning. In Prof José Mira-Mira, editor, *Proceedings of the Internationnal Conference on Brain Processes, Theories and Models*. ICBP'95, MIT Press, 1995. also on the web: http://www.etca.fr/Users/Pierre%20Germain/english/index.html.

[6] Pierre Germain and Yves Burnod. Computational properties and auto-organization of a population of cortical neurons. ICNN'96, 1996.

[7] Richard G. Palmer John Hertz, Anders Kroph. *Introduction to the Theory of Neural Computation*. A Lecture Notes Volume in the Santa Fe Institute Studies in the Sciences of Complexity. Addison Wesley Publishing Company, 1991.

[8] Kenneth D. Miller and David J.C. MacKay. The role of constraint in hebbian learning. *Neural Computation*, 6:100–126, 1994.

103

A DEVELOPMENTAL LEARNING RULE FOR COINCIDENCE TUNING IN THE *BARN OWL* AUDITORY SYSTEM

Wulfram Gerstner,[*][1] Richard Kempter,[1] J.Leo van Hemmen,[1] and Hermann Wagner[2]

[1] Institut für Theoretische Physik
Physik-Department der TU München
D-85748 Garching bei München
Germany
[2] Lehrstuhl für Zoologie/Tierphysiologie
Institut für Biologie II, RWTH Aachen
Kopernikusstr. 16, D-52074 Aachen, Germany

Binaural coincidence detection is essential for the localization of external sounds and requires auditory signal processing with high temporal precision. We present an integrate-and-fire model of spike processing in the third order nucleus laminaris of the auditory pathway of the barn owl. Each input spike generates an excitatory postsynaptic potential with a width of 250 μs. Output spikes occur with a tenfold enhanced temporal precision. This is possible since neuronal connections are fine tuned during a critical period of development as has been suggested by recent experiments. This rule does not only explain the temporal presicion in the output, but causes also a tuning to interaural time difference. The learning rule is of the Hebbian type: A synaptic weight is increased if a presynaptic spike arrives at about the same time as or slightly before postsynaptic firing. A presynaptic spike arriving after postsynaptic firing leads to a decrease of the synaptic efficacy.

1. INTRODUCTION

Owls are able to locate acoustic signals based on the extraction of interaural time difference (ITD) by coincidence detection [1, 2, 3]. The spatial resolution of sound localization found in behavioral experiments corresponds to a temporal resolution of auditory signal processing in the range of a few μs. It follows that both the firing of spikes and their transmission along the so-called time pathway of the auditory system must occur with high temporal precision.

---

[*]Present address: Swiss Federal Institute of Technology, Center for Neuromimetic Systems, EPFL-DI, CH-1015 Lausanne, E-mail: wgerst@di.epfl.ch

Computational Neuroscience
edited by Bower, Plenum Press, New York, 1997

Each neuron in the nucleus laminaris, the third processing stage in the ascending auditory pathway, responds to signals in a narrow frequency range. Its spikes are phase locked to a stimulation tone for frequencies up to 8 kHz [4, 5]. The nucleus laminaris is of particular interest, because it is the first processing stage where signals from the right and left ear converge. Owls use the interaural phase difference (that is, the interaural time difference modulo the period T of the stimulus) for azimuthal sound localization. For a successful localization, the temporal precision of spike encoding and transmission must be at least in the range of some 10 μs.

This poses at least two severe problems. First, the neural architecture has to be adapted to operating with high temporal precision. Considering the fact that the total delay from the ear to the nucleus magnocellularis is approximately 2-3 ms [4], a temporal precision of some 10 μs requires some fine tuning, possibly based on learning. Recent experiments have shown that the tuning to ITDs evolves during a critical period of the development [6]. Here we suggest that Hebbian learning is an appropriate mechanism. Second, neurons must operate with the necessary temporal precision. A firing precision of some 10 μs seems truly remarkable considering the fact that neuronal time constants are probably in the range of 100-300 μs [7, 8, 9]. It is shown below that neuronal spikes can be transmitted with a temporal precision of 25 μs despite the width of a single excitatory postsynaptic potential (EPSP) of 250 μs. This is possible because spikes are always triggered during the rise time of an EPSP.

2. NEURON MODEL

We concentrate on a single frequency channel of the auditory pathway and model a neuron of the nucleus laminaris. Since synapses are directly located on the *soma* or very short dendrites, the spatial structure of the neuron can be reduced to a single compartment. In order to simplify the dynamics, we take an integrate-and-fire unit. Its membrane potential changes according to

$$\frac{d}{dt}u = -\frac{u}{\tau_0} + I(t) \tag{1}$$

where $I(t)$ is some input and τ_0 is the membrane time constant. The neuron fires, if $u(t)$ crosses a threshold ϑ. This defines a firing time t_{out}. After firing u is reset to an initial value $u_0 = 0$. Since auditory neurons are known to be fast, we assume a membrane time constant of 100 μs.

The laminaris neuron receives input from several presynaptic neurons $1 \leq k \leq K$. Each input spike at time t_k^f generates a current pulse which decays exponentially with a time constant $\tau_r = 100$ μs. The magnitude of the current pulse depends on the coupling strength J_k. The total input is

$$I(t) = \sum_{k=1}^{K} \sum_{f=1}^{F_k} J_k \exp(\frac{t - t_k^f}{\tau_r}) \theta(t - t_k^f) \tag{2}$$

where $\theta(x)$ is the unit step function and the sum runs over all synapses $1 \leq k \leq K$ and all input spikes $1 \leq f \leq F_k$. Combining Eqs. (1) and (2) we find that each input evokes an EPSP with a width of 250 μs at half maximum. This EPSP is twice as fast as in chicken [7, 8, 9], but seems reasonable for an expert auditory-cue based hunter like the barn owl.

We focus on a laminaris neuron which is tuned to a 5 kHz signal (period $T = 200 \mu s$) and stimulate both ears with a tone of the optimal frequency and vanishing interaural time difference (ITD=0). Input spikes from presynaptic terminals are phase locked to the stimulus tone with a jitter of 40 μs. The mean phase of spikes arriving at a synapse k is different from the mean phase at another synapse k' because the transmission lines from the ear to the nucleus laminaris have different delays. It has been estimated from experimental data that the signal transmission delay Δ_k from the ear to the nucleus laminaris varies between 2 and 3 ms [4]. Note that a delay as small as 0.2 ms shifts the signal by a full period. Thus, if the delays Δ_k for the synapses k with $1 \leq k \leq K$ are chosen at random between 2 and 3 ms, then the periodicity of the signal is lost after integrating over different presynaptic inputs. We therefore need a tuning process which selects transmission lines with matching delays.

3.   TEMPORAL TUNING THROUGH LEARNING

We assume a developmental period of unsupervised learning during which a fine tuning of the temporal characteristics of signal transmission takes place. Before learning the laminaris neuron receives many inputs ($K = 600$, 300 from each the left and the right ear) with weak coupling ($J_k = 1$). Due to the broad distribution of delays the total input (2) has, apart from fluctuations, no temporal structure. The output is not sensitive to the interaural time difference (ITD); cf. Fig. 1, top. During learning some connections increase their strength to a maximum of $J_{max} = 3$, others are removed. A slight sensitivity to the ITD develops as seen in experiments [6]; cf. Fig. 1, middle. After learning, the laminaris neuron receives 75 inputs from each ear [5]. The connections to those neurons have become very effective; cf. Fig. 1, bottom. The surviving inputs all have nearly the same time delay modulo the period T of the stimulus. Thus, after learning the number of synapses is reduced and only inputs with matching delays survive. On the other hand, with 75 inputs remaining from each ear there is a high degree of convergence or 'pooling' which leads to a significant increase in temporal precision. To see this, we compare the phase jitter of the output spikes with that of the presynaptic input lines. The phase jitter of the presynaptic spike trains corresponds to a temporal precision of 40 μs. In contrast, after learning the output spikes have a temporal precision of 25 μs. Thus our learning rule leads to the fine tuning of the neuronal connections necessary for precise temporal coding.

We use a variant of Hebbian learning. In standard Hebbian learning, synaptic weights are changed if pre- and postsynaptic activity occurs simultaneously. In the context of temporal coding by spikes, the concept of 'simultaneous activity' has to be refined. We assume that a synapse k is changed, if a presynaptic spike t_k^f and a postsynaptic spike t_{out} occur within a *learning window* $W(t_k^f - t_{out})$. More precisely, each pair of presynaptic and postsynaptic spikes changes a synapse J_k by an amount

$$\Delta J_k = W(t_k^f - t_{out}). \qquad (3)$$

Depending on the sign of $W(x)$, a contact to a presynaptic neuron is either increased or decreased, in agreement with recent experimental results [10]. A decrease below $J_k = 0$ and an increase above J_{max} is not allowed. In our model, we assume a function $W(x)$ with two phases. For $x \approx 0$, the function $W(x)$ is positive. This leads to a strengthening (potentiation) of the contact with a presynaptic neuron k which is active shortly before or after a postsynaptic spike. Synaptic contacts which become active more than 3 ms later than the postsynaptic spike are decreased. Finally, an active synapse is always increased by a small amount γ, even if no

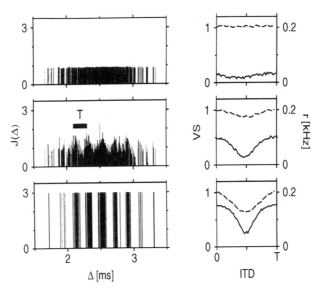

Figure 1. *Development of tuning to a 5kHz tone.* In the left column, we show the strength of synaptic contacts as a function of the delay Δ. On the right-hand side, we show the vector strength (vs, solid line) and the output firing rate (r, dashed) as a function of the interaural time delay (ITD). **Top.** Before learning, there are 600 synapses (300 from each ear) with different delays, chosen randomly from a Gaussian distribution with mean 2.5 ms and variance 0.3 ms. The output is not phase-locked ($vs \approx 0.1$) and shows no dependence upon the ITD. **Middle.** During learning, some synapses are strengthened others decreased. Those synapses which increase have delays which are similar to each other or differ by multiples of the period $T = 0.2$ ms of the stimulating tone. The vector strength of the output increases and starts to depend on the ITD [6]. **Bottom.** After learning, only 150 synapses (75 from each ear) survive. Both the firing rate r and the vector strength vs show the characteristic dependence upon the ITD as seen in experiments with adult owls [5]. The neuron has the maximal response ($r = 200$ Hz) for ITD=0 which is the stimulus used during the learning session of the model neuron. The vector strength at ITD=0 is $vs \approx 0.8$ which corresponds to a temporal precision of 25 μs. By definition, the vector strength is proportional to the amplitude of the first Fourier component of the distribution of output phases, normalized so that $vs = 1$ indicates perfect phase locking (infinite temporal precision).

output spike occurs. The effect of learning on a synapse k after F_k input spikes and N_{out} output spikes is

$$\Delta J_k = F_k \gamma + \sum_{n=1}^{N_{\text{out}}} \sum_{f=1}^{F_k} W(t_k^f - t_{\text{out}}^n). \tag{4}$$

We require $\int W(s)ds = -W_0 < 0$. The combination of decrease and increase then balances the average effects of potentiation and depression and leads to a normalization of the number and weight of synapses.

Learning occurs in three phases. At the beginning, input and output spikes are not correlated. In particular, output spikes are not phase locked to the stimulation tone. During this phase, all weights J_k approach a transitory equilibrium (defined by $\langle \Delta J_k \rangle = 0$) which leads to a mean output firing rate $r = \gamma/W_0$. We recall that each synapse k corresponds to a delay line with a delay Δ_k. In Fig. 1 we have plotted the value J for each synapse as a function of the delay Δ. The initial distribution is flat.

Shot noise due to the spike nature of the signals leads to spontaneous symmetry breaking and starts the second phase of learning (Fig. 1, middle). As a result, $J(\Delta)$ develops a T-periodic modulation. In the third and final stage of learning, all values $J(\Delta)$ approach a limiting value, either 0 or $J_{max} = 3$. The output spikes now show a high degree of phase locking measured by the vector strength. Moreover, the output firing rate r depends on the interaural time difference (ITD), as found in experiments with adult owls.

ACKNOWLEDGMENTS

R.K. holds scholarship of the state of Bavaria. W.G. has been supported by the Deutsche Forschungsgemeinschaft (DFG) under grant number He 1729/2-2, H.W. by a Heisenberg fellowship of the DFG.

REFERENCES

[1] L. A. Jeffress, J. Comp. Physiol. Psychol. **41**, 35 (1948).
[2] M. Konishi, Trends in Neurosciences **9**, 163 (1986).
[3] C. E. Carr, Annual Rev. Neurosci. **16**, 223 (1993).
[4] W. E. Sullivan and M. Konishi, J. Neurosci. **4**, 1787 (1984).
[5] C. E. Carr and M. Konishi, J. Neurosci. **10**, 3227 (1990).
[6] C. E. Carr, in *Proceedings of the 10th International Symposium on Hearing* (Irsee, Bavaria, 1994), pp. 21–26.
[7] E. M. Overholt, E. W. Rubel, and R. L. Hyson, J. Neurosci. **12**, 1698 (1992).
[8] A. D. Reyes, E. W. Rubel, and W. J. Spain, J. Neurosci. **14**, 5352 (1994).
[9] A. D. Reyes, E. W. Rubel, and W. J. Spain, J. Neurosci. **16**, 993 (1996).
[10] H. Markram, J. Lübke, M. Frotscher, and B. Sakmann, preprint (1996).

104

EXPERIENCE-BASED AUDITORY MAP FORMATION AND THE PERCEPTUAL MAGNET EFFECT

Marin N. Gjaja and Frank H. Guenther

Department of Cognitive and Neural Systems
Boston University
Boston, MA

1. INTRODUCTION

The perceptual magnet effect (e.g., Kuhl, 1991, 1995) is one of the most actively discussed topics in the recent speech perception literature. The effect is characterized by a warping of perceptual space such that acoustic patterns near phonemic category centers are perceived as closer together than equally spaced acoustic patterns that are further away from phonemic category centers. This language-specific effect is evident in infants by six months of age and is maintained through adulthood (Kuhl, 1991). In this paper we propose an explanation for the perceptual magnet effect using a self-organizing feature map neural network (e.g., von der Malsburg, 1973; Kohonen, 1982). This model is a component of a larger computational modeling framework of speech development, perception, and production called DIVA (Guenther, 1995).

In this account, the magnet effect arises as a natural consequence of the formation of neural maps in the auditory system as an infant is exposed to the sounds of his/her native language. These maps are assumed to develop according to the same principles believed to be involved in map formation for other modalities such as vision and somatic sense. This explanation requires only two fundamental hypotheses: (1) sensory experience leads to language-specific nonuniformities in the distribution of the firing preferences of cells in an auditory map, and (2) a population vector analysis (e.g., Georgopoulos et al., 1984) can be used to predict psychological phenomena based on the pattern of cell activities in this map. Both of these hypotheses receive significant support from various studies in the computational neuroscience literature. The hypothesis that linguistic experience results in nonuniformities in the firing preferences of cells in an auditory map is supported by a large number of neurophysiological studies of auditory, somatosensory, and visual cortices, and many neural models similar to the one posited here have been used to explain this phenomenon. The hypothesized utility of the population vector for predicting psychological phenomena from the ensemble of cell activities in a neural map is also supported by several neurophysiological and modeling studies of different tasks in different modalities.

2. MODEL DESCRIPTION

The model uses two layers of neurons, referred to as the formant representation and the auditory map, connected by a set of adaptive weights. All formant representation cells project to all auditory map cells.

2.1. Formant Representation

The model assumes that peripheral auditory processing yields a neural representation of the formant frequency values of speech sounds. Simulations used either formants in Mels or in Hertz; nearly identical results were obtained in the two cases. Only two formants were used in each simulation, either F1 and F2 or F2 and F3. A normalized, agonist-antagonist neural representation was used for each formant. Similar results would be expected for many different neural representations of formant values. It is further expected that the results reported herein will hold with input representations that include more than the first two formants; this limited representation was used here to simplify the simulations and graphical presentation of simulation results.

2.2. Auditory Map

In keeping with most models in the neural network literature, the input to each cell in the auditory map is calculated as the dot product of the input vector (i.e., the neural representation of formant values) and the vector of weights projecting to the map cell. When an input is received, the L nodes in the map with the largest input were allowed to remain active, and the activities of all other nodes in the map were set to zero. This process approximates the effects of competitive interactions between map cells (Grossberg, 1976; Kohonen, 1982; von der Malsburg, 1973), and the inhibitory synaptic connections believed to mediate this type of competition are seen in primary sensory areas of cortex, including auditory cortex, somatosensory cortex, and visual cortex. During training, L is a monotonically decreasing function of time, as implemented in the SOFM of Kohonen (1982). When testing the network, a fixed value of L is used. The activity levels of the L nodes with the largest input are assumed to be proportional to the sizes of their inputs.

The learning process used to adjust the synaptic weights between formant representation cells and cells in the auditory map is defined by the following equation:

$$\frac{dz_{ij}}{dt} = \varepsilon m_j (x_i - z_{ij}) \tag{1}$$

where z_{ij} is the weight projecting from the ith formant representation cell to the jth auditory map cell, x_i is the activity of the ith formant representation cell, m_j is the activity of the jth auditory map cell, and ε is a learning rate parameter. This is essentially the same learning law that is used in many other models of map formation, including the models of von der Malsburg (1973), Grossberg (1976), and Kohonen (1982). Nearly identical learning laws have also been put forth by neurophysiologists to explain observed synaptic strength changes in areas such as visual cortex.

One key property of this model is that the distribution of firing preferences of auditory map cells comes to reflect the distribution of the inputs used to train the model (see Kohonen, 1982), which in turn reflects the distribution of sounds in a particular language. This represents one of the two main hypotheses embodied by the model. Such experience-based nonuniformities in the distribution of cell firing preferences have been identified by neurophysiologists studying sensory maps in several different modalities.

2.3. Population Vector

After the model described in the preceding paragraphs is trained by exposure to a set of sounds approximating the distribution of sounds in a particular language, we are left with the problem of interpreting map cell activities in perceptual terms. The second major hypothesis of the current model is that a population vector (e.g., Georgopoulos et al., 1984) can be used to predict psychological phenomena from the pattern of neural activities in this map. This population vector is calculated as follows. Each cell in the current model's auditory map will be maximally activated by a particular vector in formant space (e.g., a particular F1-F2 pair). This formant vector is the one whose neural representation at the first stage of the model is parallel to the vector of weights projecting to the auditory map cell. This is clear when one considers that the input to each map cell is the dot product of the neural representation vector and the cell's afferent weight vector, and the dot product of two vectors is maximal when the angle between them is zero. The formant vector that maximally activates the jth auditory map cell will be referred to as its preferred stimulus and denoted as F_j. It is assumed that the perceived sound can be derived from the pattern of cell activities at the auditory map using the following population vector equation:

$$F_{perceived} = \frac{\sum_j m_j F_j}{\sum_k m_k} \qquad (2)$$

Population vectors of this form have been used by neuroscientists to predict a wide range of behavioral and psychological phenomena from the ensemble of cell firing rates in neural maps from many different modalities.

3. SIMULATION RESULTS

Numerical simulations were run to illustrate the model's ability to explain general characteristics of the magnet effect and accurately reproduce specific psychophysical data. Random initial weight vectors chosen from a uniform distribution over the input space were used in all simulations. The model was then trained with formant input vectors chosen from Gaussian distributions centered on phonetic categories. The auditory map self-organizes given only the input vectors without any information regarding their category membership. Between 300 and 1500 inputs from each category were used in the simulations.

The results of one simulation are shown in Figure 1. This figure compares model and experimental generalization scores (i.e., the percentage of time two stimuli were identified as the same when they were different) for test stimuli spaced 1, 2, 3, and 4 steps away from the referent stimulus. Results are shown for a prototypical /i/ referent stimulus (top curves) and

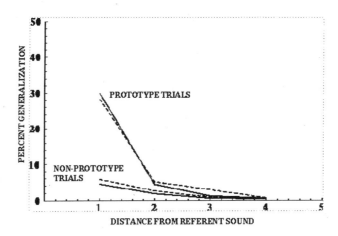

Figure 1. Comparison of the model's generalization properties after learning (solid lines) to the results of Kuhl (1991) for adult subjects (dashed lines) for prototypical and nonprototypical /i/ stimuli.

a nonprototypical /i/ referent stimulus (bottom curves). The solid lines indicate the model's performance, and dashed lines indicate the performance of subjects in the Kuhl (1991) study. The model accurately reproduces the Kuhl (1991) experimental data.

4. SUMMARY

Although the current model contains a relatively large number of cells and synaptic weights, only three free parameters were available to fit the data in these simulations: a learning rate, a neighborhood size in the self-organizing map, and the variance of the Gaussian distributions of the training inputs. Despite the model's simplicity, it captures the salient known aspects of the magnet effect: a shrinking of perceptual space near phonemic category centers, an expansion of perceptual space away from centers, and language-specificity in this warping. Furthermore, the model provides very close fits to specific psychophysical results reported in Kuhl (1991) and Kuhl (1995). The model also requires no assumptions about the abilities of infants to identify sounds as members of linguistic categories and therefore has no trouble explaining why the magnet effect is evident in six month old infants.

Previous attempts to explain the magnet effect have typically been formulated within the theoretical framework of cognitive psychology. Kuhl and others have described the magnet effect as involving a stored representation of a category prototype, i.e., an exemplar whose perceptually measured goodness is maximal for that category. This prototype serves as a sort of anchor whose functional role as a perceptual magnet serves to strengthen category cohesiveness (e.g. Kuhl, 1991, p. 99). In contrast, the explanation provided here uses tools from computational neuroscience to account for the effect, thereby allowing the interpretation of psychological descriptions of the magnet effect in terms of the properties of neural systems. In this account, the "stored representation" of vowels that leads to the magnet effect is simply the set of synaptic weights projecting to the cells in an auditory map. "Category prototypes" are stimuli located at the peaks in the distribution of map cell firing preferences, which in turn reflect peaks in the distribution of sounds in a particular language. The "magnet effect" itself is a warping of perceptual space resulting from a nonuniform distribution of the preferred stimuli of auditory map cells, and "category cohesiveness" describes the psychological result of the multiply peaked form of this nonuniformity: inputs near the peaks of the distribution are perceived as closer together than inputs near the valleys.

ACKNOWLEDGMENTS

Marin Gjaja supported in part by the National Science Foundation (NSF IRI 94-01659) and the Office of Naval Research (ONR N00014-95-1-0409 and ONR N00014-95-0657). Frank Guenther supported in part by the Alfred P. Sloan Foundation and the National Institutes of Health (1 R29 DC02852-01).

REFERENCES

[1] Georgopoulos, A.P., Kalaska, J.F., Crutcher, M.D., Caminiti, R., and Massey, J.T. (1984). The representation of movement direction in the motor cortex: Single cell and population studies. In Edelman, G.M., Gall, W.E., and Cowan, W.M. (eds.): *Dynamic aspects of cortical function*. New York: Wiley.

[2] Guenther, F.H. (1995). Speech sound acquisition, coarticulation, and rate effects in a neural network model of speech production. *Psychological Review*, **102**, pp. 594-621.

[3] Grossberg, S. (1976). Adaptive pattern classification and universal recoding: I. Parallel development and coding of neural feature detectors. *Biological Cybernetics*, **23**, pp. 121-134.

[4] Kohonen, T. (1982). Self-organized formation of topologically correct feature maps. *Biological Cybernetics*, **43**, pp. 59-69.

[5] Kuhl, P.K. (1991). Human adults and human infants show a 'perceptual magnet effect' for the prototypes of speech categories, monkeys do not. *Perception & Psychophysics*, **50**, pp. 93-107.

[6] Kuhl, P.K. (1995). Mechanisms of developmental change in speech and language. In Elenius, K., and Branderud, P. (eds.): *Proceedings of the XIIIth International Congress of Phonetic Sciences*, vol. 2, pp. 132-139. Stockholm: KTH and Stockholm University.

[7] von der Malsburg, C. (1973). Self-organization of orientation sensitive cells in the striata cortex. *Kybernetik*, **14**, pp. 85-100.

105

SPATIOTEMPORAL PATTERNS OF LIGHT STIMULATION ARE CORRELATED WITH LARGE SCALE DYNAMICAL PATTERNS OF SYNCHRONISED OSCILLATIONS

Rowshanak Hashemiyoon and John K. Chapin

Allegheny University of the Health Sciences
Department of Neurobiology and Anatomy
3200 Henry Avenue, Philadelphia, Pennsylvania 19129

1. ABSTRACT

Although there have been many reports of fast frequency (10–50 Hz) oscillations in the vertebrate visual system, much debate persists about their functional significance to visual processing.[1,2,3,4,5] In previous reports, we have demonstrated the existence of widespread oscillatory discharges throughout multiple retinorecipient nuclei in the subcortical visual system of the anaesthetised and awake rat.[6] The results of our studies indicate these oscillations may play a role in a form of dynamic-distributed processing of visual information which has yet to be described in the lower levels of the visual system. In this paper, we suggest the visual system may incorporate two modes when making representations of visual attributes. The local feature mode is linked to the theory of circumscribed receptive fields associated with individual neurons.[1] The global feature mode makes use of a periodic oscillatory signal emerging from a network distributed across the retina.

2. INTRODUCTION

The prevailing doctrine of the mechanism of visual processing is that elaborate visual processing takes place only through the multi-synaptic serial transfer of information from lower levels in the system (retina) to higher ones (cortex)[7]. The drawback of this model is that it reduces the retina and non-thalamic subcortical nuclei to little more than pixel transfer sites. However, the retina is not a simple sheet of cells whose capability is restricted to parcelling information. It is in fact a complexly organised network equipped with the capacity to perform elaborate processing functions. Furthermore, the retina itself is a component of a larger network consisting of the components of the subcortical visual nuclei.

Computational Neuroscience
edited by Bower, Plenum Press, New York, 1997

Since synchronous oscillations by their nature are distributed across networks, they imply some sort of distributed processing wherever they are found. While any hypothesis about the functional significance of oscillations may be premature, their level and distribution in the visual system are crucial to advancing knowledge of this complex system and its behavior.

3. METHODS

Multiple single unit recordings were simultaneously obtained from 16-electrode arrays implanted in both the LGN and SC in anaesthetised and awake rats. In these chronic recordings, arrays consisted of two rows of eight 50 micron electrodes (NB Labs, Denison, TX), spaced at 0.2 mm intervals, with 0.5 mm between the rows. Discrimination of multi-neuron data was made using a 64-channel unit recording system from Spectrum Scientific (Dallas, TX). Light stimuli were provided by a spatially circumscribed light source, repetitively occluded and revealed by a computer controlled shutter. Stimuli were presented at a range of frequencies (0.5–20 Hz) and light-ON durations (0.02–1.0 sec).

4. RESULTS

We have reported that retinally derived subcortical visual system oscillations exist throughout the medial pretectal area (MPA), stratum griseum superficiale of the superior colliculus (SGS-SC), and dorsal lateral geniculate nucleus of the thalamus (DLGN)[6]. The frequency and magnitude of these spontaneous oscillations were different in the dark vs. during light stimulation. Although these oscillations responded to light stimuli in virtually any part of the contralateral visual field of the rat, oscillatory response patterns to light changed dramatically depending upon the frequency of the light stimulus. In figure 1, the oscillations became very strong after the onset of the dark period. Though the oscillations were also observed during the light period, they were slightly weaker (lower amplitude) and of a lower frequency. At a slightly higher frequency of light flash, transitions between light and dark produced sudden alterations in the frequency and phase structure of the spontaneous oscillations. The joint peri-stimulus time histograms (JPSTHs) of these neurons' joint activities revealed marked changes in their dynamical properties during light/dark transitions, as produced by rapid movement of the shutter over the visual field. The oscillatory phases shifted suddenly at the onset and also the offset of such transitions. In the peri-event histograms (PEHs), these transition periods appeared as depressions in firing, and appeared as long latency effects of the light stimulation, which in this case was 100ms in duration. Figure 2 reveals oscillatory responses at a 5Hz stimulus frequency. The spontaneous oscillations were confined within narrow envelopes corresponding roughly to the light-OFF rebound response. Within these envelopes, the diagonal bands were curved, indicating transient changes in the joint oscillation frequencies. In this case, the joint histogram (the diagonal) was similar to the PEH, indicating that most peri-event neural activity revolved around stimulus induced phase-resetting of the oscillations. At stimulus frequencies harmonic with the endogenous oscillation, the JPSTH (figure 3) again revealed smooth periodicity, but complex structures between the diagonal ridges still revealed patterns often associated with rhythmic forcing of intrinsically oscillatory systems. Supra-harmonic stimulus frequencies produced complexly patterned frequency dependent responses.

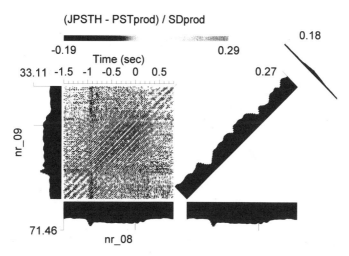

Figure 1. Frequency and magnitude of spontaneous oscillations are different in the dark vs. during light stimulation. Joint peri-stimulus time histograms (JPSTHs) allow one to view the influence of stimulus driving on the correlated activity between two simultaneously recorded neurons. Each pixel in the 2D gray scaled display indicates the correlation coefficient between two neurons in the LGN (neurons 8 and 9), displayed as a function of time shifts before and after the light stimulus. The magnitudes of the correlation coefficients within each pixel are depicted in the gray scale above. The peri-event histograms (PEHs) of neurons 8 and 9, centered around the onset of Light-ON, are shown against the horizontal and vertical axes, respectively. The magnitude of the *joint* activity (i.e. correlation) of these two neurons, characterised by a time window around the diagonal in the JPSTH, is shown in the larger diagonal histogram. The time-shifted cross correlogram between these neurons is shown in the smaller histogram to the upper right corner. The frequency of visual stimulation was 0.5 Hz. Parallel diagonal lines in the JPSTH reflect ongoing oscillatory activity. These oscillations become very strong after the onset of the dark period (-1.0s). Bin size=20ms; diagonal window=24ms.

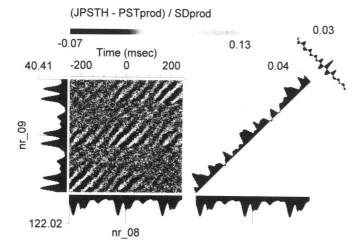

Figure 2. Stimulus induced phase resetting and frequency alteration of oscillations. Increasing the frequency of light flash results in dynamical changes in both the frequency and phase of the oscillations. Duration of the light-ON was 50 ms, stimulus frequency was 5Hz. Bin size=5ms; diagonal window=20ms.

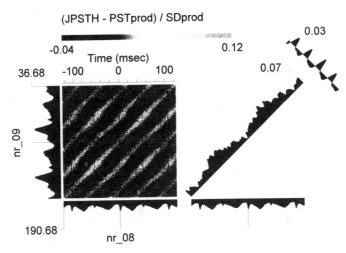

Figure 3. At near harmonic stimulus frequencies JPSTHs reveal smooth periodicity with overlying complexity. Intrinsic oscillations are beginning to be more influenced by neuron 9 as revealed by the slight bending of the diagonal lines. The continuity of these lines reveals that the same dynamical *structure* is alternately influenced by intrinsic and extrinsic factors. Duration of the light-ON was again 50 ms, stimulus frequency was increased to 9.4 Hz. Bin Size=1ms; diagonal window=2ms.

5. SUMMARY AND CONCLUSIONS

Feature processing of a visual percept has been suggested to occur via the abstraction of attributes by the specific receptive fields associated with specific neurons[7]. Reports have been made associating the binding of this neuronal operation with synchronised oscillatory discharges in the visual cortex.[1,8] These oscillations are thought to be cortically-derived and serve simply to dynamically link the processing of otherwise independent neuronal function.

Our results suggest that information in the visual field need not be processed exclusively by this mechanism. We suggest that representations of features, particularly global features, of the visual scene may be distributed across the retina and, therefore, the subcortical visual system. We have previously reported that properties of a stimulus, such as locus, strength, and timing, could be reflected in the spatiotemporal patterning of retinally-derived, fast frequency, spontaneous subcortical oscillations.[9] We have also reported oscillatory responses to visual stimuli which are *distinct* from those elicited from the highly circumscribed receptive fields of the neurons under study.[3,6,10]

In this study, we demonstrate that light stimulation has marked effects on the amplitude, frequency, and phase of spontaneous oscillations in the subcortical visual system. Though the intrinsic oscillatory phenomena can exhibit clear resonance with harmonically compatible forcing stimuli, even small deviations from the intrinsic frequency produce unique, distinct spatiotemporal dynamical patterning. The curved joint activity surfaces in the JPSTHs suggest a dynamic linkage between the intrinsic vs. the stimulus dependent components of this activity. Thus, these results provide more clear evidence that spatiotemporal patterns of light stimulation are represented not just in the point-to-point transmission of pixel data by the retina, but also in unique and highly structured large scale dynamical patterns in the ascending retinofugal visual system.

Therefore, we conclude that visual processing occurs through at least two modes: local feature processing and global feature processing. Local feature processing would incorporate the prevailing, or classical, theory of feature abstraction through circumscribed receptive fields. An additional, more "global" mode of feature processing, could be mediated through global network interactions in the retina, as reflected in the wide-field dynamical phenomena reported here. Distributed representations of time-varying features of the visual image could be contained within complex patterns of phase-shifted, amplitude modulated neural activity waves across the retinal network. More specifically, our results suggest that light intensity may be inversely related to the amplitude of the oscillations, and information about the position of the feature in space-time might be carried in the phase.

REFERENCES

1. Eckhorn R., Bauer R., Jordan W., Brosch M., Kruse W., Munk M., Reitboeck H.J. Coherent Oscillations: A Mechanism of Feature Linking in Visual Cortex? *Biol Cybern* (1988) 60:121–130.
2. Gray C.M., Konig P., Engel A.K., Singer W. Oscillatory Responses in Cat Visual Cortex Exhibit Inter-columnar Synchronization Which Reflects Global Stimulus Properties. *Nature* (1989) 338:334–337.
3. Hashemiyoon, R. Synchronous Oscillations in the Subcortical Visual System. Thesis (1996).
4. Ghose G.M., Freeman R.D. Oscillatory Discharge in the Visual System: Does it Have a Functional Role? *J Neurophys* (1992) 68:1558–1574.
5. Bullock T.H., Hofmann M.H., Nahm F.K., New J.G., Prechtl J.C. Event-Related Potentials in the Retina and Optic Tectum in Fish. *J Neurophys* (1990) 64:903–914.
6. Hashemiyoon, R. and Chapin, J. Retinally Derived Fast Oscillations Coding for Global Stimulus Properties Synchronise Multiple Visual System Structures. *Soc. Neurosci. Abstr.* (1993) 19: 528.
7. Hubel, D.W, Wiesel, T.N. Receptive Fields, Binocular Interaction, and Functional Architecture in the Cat's Visual Cortex. *J Physiol* (1962) 160:106–154.
8. Gray C.M., Engel A.K., Konig P., Singer W. Stimulus-Dependent Neuronal Oscillations in Cat Visual Cortex: Receptive Field Properties and Feature Dependence. *Eur J Neurosci* (1990) 2:607–619.
9. Hashemiyoon, R. and Chapin, J. *Computational Neuroscience: Trends in Research* James M. Bower, ed. N.Y., N.Y.: Academic Press Inc., 1996.
10. Hashemiyoon, R., and Chapin, J. Spatiotemporal Patterns of Visual Stimuli Modify Subcortical Visual System Oscillations. *Assoc. Res. Vis. and Ophthal. Abstr.* (1996) 37: 1059.

106

THE IMPORTANCE OF HIPPOCAMPAL GAMMA OSCILLATION FOR PLACE CELLS

A Model That Accounts for Phase Precession and Spatial Shift

Ole Jensen and John E. Lisman

Volen Center for Complex Systems
Brandeis University, Waltham, Massachusetts 02254
jensen@volen.brandeis.edu
lisman@binah.cc.brandeis.edu

ABSTRACT

The firing of rat hippocampal place cells advances to earlier phases of the theta cycle as the rat moves through a place field. It has also been observed that when a rat repeatedly traverses a track, the place field shifts slowly in a direction opposite to the direction of movement. We have constructed a physiologically realistic network model of the CA3 region, based on nested theta and gamma oscillations. This model accounts for the phase precession and place-field shift. From this work, a set of further predictions follows which are testable if both theta and gamma oscillations are measured in conjunction with place cell activity.

INTRODUCTION

The activity of pyramidal cells in CA3 and CA1 regions of the rat hippocampus are dependent on the position of the rat within its environment. These cells are termed place cells.[12] It is often hypothesized that the CA3 region of the hippocampus serves as an autoassociative memory[10,16] which recalls information about spatial location. However, Muller et al.[11] experimentally showed that place cells in CA3 and CA1 predict upcoming locations rather than code for actual position. Theoretical work has suggested that the temporal asymmetry of NMDA-mediated LTP allows the encoding of sequences[1,4,5,7,17] and the prediction of future position. We have analyzed such predicting circuits with the added constraints that theta and gamma frequency oscillations occur in the hippocampus. The fact that there are about 7 gamma cycles within a theta cycle points to a model in which stored sequence information is read out at gamma frequency, allowing the prediction of information 7 items ahead in the sequence. In this paper we apply this framework to two experimental findings: 1) When a rat enters a place field, its place cells fire in the late phase of the theta cycle, and then advance systematically in phase as the rat continues.[12,14] 2) Mehta et al.[8,9] recently found that the location of a place cell slowly shifts in the direction opposite to the movement of the rat during repeated runs on a familiar linear track.

METHODS

The network representing a simplified model of CA3 is shown in Figure 1. The details of the model are described in previous papers.[2,3,6] The network is implemented using firing model neurons. A memory representing a location is constituted by a subset of the pyramidal neurons firing in synchrony within a gamma cycle. The cholinergic and GABAergic projections from the medial septum provide the phasic theta drive to the pyramidal cells. Each time a set of pyramidal neurons fires, the interneuron network is activated and produces feedback inhibition, thereby producing gamma oscillations. The pyramidal neurons in CA3 have abundant recurrent collaterals with modifiable synapses. In two previous papers[4,5] we argued that hippocampal LTP will inevitably lead to encoding of sequences. This is because sequential items are represented in different gamma cycles and because the associational mechanism that governs synaptic modification in the recurrent connections, the NMDA channels, have kinetics that are slow (>100msec) relative to the period of a gamma cycle. We further showed that synaptic excitation mediated by NMDA channels provide a potential mechanism by which memory sequences could be read out at the rate of one memory per gamma cycle. In this paper we will expand the conclusions and predictions made in these previous papers.[5] We will not show simulations, but restrict ourselves to schematic illustrations explaining the function and predictions of the model.

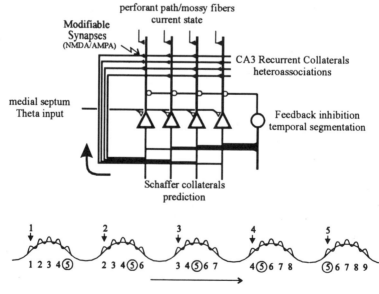

Figure 1. The architecture of the CA3 hippocampal network model. A memory coding for a place is represented by a subset of pyramidal neurons firing in synchrony within a gamma cycle. The modifiable synapses allow encoding and recall of sequences. The circular cell is an inhibitory interneuron that is excited by pyramidal cells.

locations:

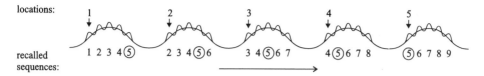

recalled
sequences:
| 1 2 3 4⑤ | 2 3 4⑤6 | 3 4⑤6 7 | 4⑤6 7 8 | ⑤6 7 8 9 |

Figure 2. As a rat moves trough a place field in a well-known environment, here enumerated 1 to 9, sequences of upcoming locations are recalled. Number at top is current location. Numbers at bottom show positions recalled within a given theta cycle. Each memory (location) is recalled in a gamma subcycle of the theta period. As can be seen, the firing of cells representing location 5 advances within the theta cycle as the rat passes the place field.

RESULTS AND DISCUSSION

O'Keefe and Recce[13] showed that the firing of place cells advances systematically in the phase of the theta cycle as the rat passes through a place field. The following example explains how the theta/gamma network can account for this finding. Let us assume that a rat has been traversing a linear maze enough times to have memorized a representation of the environment as a sequence of locations.

Each location in the sequence is represented by a number (Figure 2). The physical distance between the encoded locations in the maze is determined by the average velocity of the rat during learning (v_{learn}) multiplied by the average period of the theta oscillations. When the rat traverses a well-learned track, sensory or path integration information about the current position provides an input to the CA3 region at the beginning of each theta cycle, triggering the recall of subsequent memories (positions) through the internal synaptic dynamics of the network. Each memory is recalled in subsequent gamma subcycles of a theta period. In the example in Figure 2, input 1 triggers the recall of the predicted upcoming path (2 to 5). The ratio of the theta to gamma period approximately equals the number of memories recalled per theta cycle. One theta cycle later the rat has moved ahead to position 2, which then triggers the recall of the path 3 to 6. Consider now the firing of place cells coding for location 5. Since the phase advances one gamma cycle per theta cycle, we can derive an analytical expression for the slope of the phase advance (as a function of location):

$$\alpha = 360\frac{1}{width_{place\,field}} = 360\frac{1}{v_{learn}\frac{T_\theta}{T_\gamma}T_\theta} = 360\frac{1}{v_{learn}}\frac{f_\theta^2}{f_\gamma} \tag{1}$$

where $f\theta$ and $f\gamma$ denotes the theta and gamma frequency. The phase advance is independent of the rat's velocity when measuring the phase of firing.[5,13] No experiment has yet been published in which these three parameters were simultaneously measured. Using the estimated value's $f\theta$ = 8 Hz, f_γ = 50 Hz and v_{learn} = 30 cm/sec, we find the phase advance to be 15.3 o/cm. This is in accord with the experimentally measured slope.[13,14] One would expect that the later a place cell fires in the theta cycle the less spatially specific they are, since late firing cells are a consequence of predictions about upcoming locations. This expectation is consistent with experimental results.[14] The model furthermore implies that the number of gamma cycles per theta cycle determines the number of theta cycles during which a place cell is active when the rat passes through a place field with a certain speed, v. In the above example where the rat's velocity is the same as the velocity during learning, this number is 5. If we take a higher value of gamma, the number could reasonably be 7–8. In general the number is

$$N = \frac{v_{learn}}{v} \frac{f_\gamma}{f_\theta}$$

(2)

Skaggs et al.[14] have experimentally found this number to be about 8–10. A related model based on sequence readout has been proposed by Tsodyks et al.[17] Their model however does not involve oscillations and predicts the phase advance only in a qualitative way.

Mehta et al.[8,9] have found that place fields on a linear track are not stable, but expand in the direction opposite to the rats movement when the rat repeatedly traverses the track (shown schematically in Figure 3A). Importantly, these shifts occur even though the track is highly familiar to the animal as a result of many days of previous exposure. Not only do the place fields expand, but the position of the place field shifts from trial to trial. By assuming learning during recall of known sequences, our model can account for this effect. As mentioned earlier, the NMDA channels of the CA3 collaterals are assumed to be responsible for the readout of sequences. Because NMDA channels are used in readout, a use that leads to synaptic strengthening, the EPSP's between CA3 cells will become stronger with increasing number of trials on the familiar track. The effect of synaptic strength on gamma frequency can be understood in terms of the synaptic interactions. In the model, gamma frequency emerges as a result of successive excitation, inhibition cycles: the excitation of pyramidal cells triggers feedback inhibition, which must then partially decline before the next round of firing occurs. The time it takes before the next round of excitation can occur will depend on the excitatory drive, a drive that is affected by potentiation of the synapses involved. Thus, each time a synapse is activated, its efficacy is enhanced by LTP and there will be a speedup of gamma frequency. The higher the gamma frequency, the more gamma cycles can fit into a theta cycle and the further ahead sequences can be predicted. This mechanism has successfully been implemented in our network model.[5] The outcome is schematically illustrated in Figure 3B: after repeated laps the sequence of locations is recalled at a faster rate within each theta cycle, resulting in place cells firing earlier on the track. We can quantitatively predict the spatial advance on the track by considering the increase in the number of gamma cycles per theta cycle:

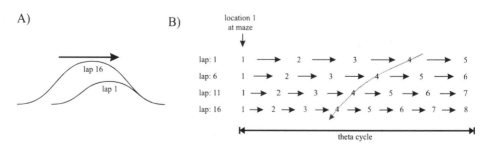

Figure 3. A) Firing as a function of position on a linear track. As a rat makes repeated traversals of a linear track, the place field expands in the direction opposite to the rat's movement (arrow). B) When the rat arrives at location 1, a sequence of upcoming locations is read out. The readout of sequences during repeated runs causes stronger encoding, resulting in an increased gamma frequency (see text). Hence firing of recalled place cell activity, e.g., 4, progresses opposite to the rat's direction of movement and the place field expands. This is essentially because the rat can predict further ahead within a theta cycle if gamma is faster.

$$\Delta X = (n_N - n_0) T_\theta v_{learn} = \left(\frac{T_\theta}{T_{\gamma.N}} - \frac{T_\theta}{T_{\gamma.0}} \right) T_\theta v_{learn} = \left(f_{\gamma.N} - f_{\gamma.0} \right) \frac{1}{f_\theta^2} v_{learn} \tag{3}$$

where $f_{\gamma.0}$ denotes the gamma frequency at the first lap and $f_{\gamma.N}$ the frequency after N laps. Assuming that the initial gamma frequency of 50 Hz (6 gamma cycles per theta cycle) increases to 70 Hz (9 gamma cycles per theta cycle) the firing of place cells will advance 9.4 cm (the rest of the parameters are as in Eq. 1). Note that, when measured, the advance is independent of the rat's velocity. As predicted by Eq. 1, the increase in gamma frequency will result in a decrease in the slope of the phase advance with respect to the theta oscillation. In the current example the slope will decrease from 15.3 o/cm to 10.9 o/cm.

Our last point relates to the phase precession in bidirectional and multidirectional place fields. The example in Figure 4 illustrates how the phase relation of bidirectional place cells depends on the direction of movement of the rat. When the rat enters from the *left*, the sequence 2 to 5 is recalled as soon as the rat arrives at location 1. Cells coding for location 5 will at first fire late in the theta cycle. When the rat enters from the *right*, cells coding for location 5 will fire late in the theta cycle as soon as the rat arrives at location 9 and the sequence 8 to 5 is recalled. Consequently bidirectional place cells will show opposite phase profiles depending on which side the rat enters the place field, and the phase of firing will have a bimodal profile. This principle can be extended to two dimensions: place cells firing late in the theta cycle will form an annular ring in the two-dimensional domain. This prediction has been confirmed by Skaggs et al.[14] (their Figure 8B). By combining this principle with the finding of Mehta et al.,[8,9] another testable prediction becomes evident: when a rat re-explores a two-dimensional environment repeatedly, not only will the place field grow in size, but the annular ring formed by cells firing late in the theta cycle will increase in diameter.

A fundamental problem regarding the multidirectional place fields observed in a two-dimensional environment needs to be considered: when the rat arrives at a location, say 5, how does CA3 know to recall 5 6 7 8 9 or 5 4 3 2 1? One solution is to cue the sequence with more than one memory, i.e., both the memory for the current and previous location. For the first sequence, the cue would be 4 5 and for the second 6 5 (Figure 5). In a previous paper,[4] we showed that the cumulative build up of synaptic excitation mediated by the slowly inactivating NMDA channels provides a mechanism for integrating inputs that occur at different times. This provides a way of breaking symmetry. Another possibility is that head direction cells[15] are part of the cue.

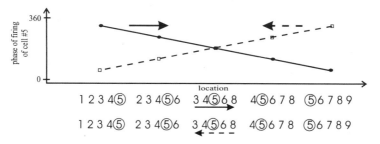

Figure 4. The phase of firing of place cells coding a *bidirectional* field will have two peaks, one for each direction of motion. This property emerges because CA3 predicts upcoming locations. In a two-dimensional place field the phase relationship would form an annular ring.

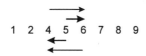

Figure 5. The NMDA conductive integrates multiple inputs, thereby, providing a mechanism for symmetry breaking. Say that the rat arrives at location 5 from the left. Location 6 (and *not* 4) is then recalled since synaptic input from cells coding for both location 4 and 5 help to bring cells coding for location 6 to fire.

CONCLUSION

We have suggested a physiologically realistic network model that can account for important aspects of place cell data. The model makes testable predictions:

1. The number of theta cycles in which place cells remain active when the rat passes trough a place field is determined by Eq. 2.
2. The model predicts the slope of the phase precession with respect to location (Eq. 1). The slope is determined by the theta and gamma frequency and the average velocity of the rat during learning.
3. The spatial shift of place cells with repeated runs is related quantitatively (Eq. 3) to predicted shifts in gamma frequency. The advance is independent of the rat's velocity during the measurement.
4. The slope of theta phase advance (Eq. 1) will decrease with repeated runs due to the increase in gamma frequency.
5. The annular ring emerging when considering place cells firing late in the theta cycle will increase in diameter as a rat repeatedly re-explores a known two-dimensional environment.

Checking these predictions will require the simultaneous measurement of place fields, gamma oscillations and theta oscillations. These variables have not been measured simultaneously in previous work. We hope these predictions will encourage the acquisition of such data.

ACKNOWLEDGMENTS

This work was supported by the Alfred P. Sloan Foundation (94-10-1).

REFERENCES

1. Abbott L.F. and K. Blum 1996. Cerebral Cortex 6:406–416.
2. Jensen O., M.A.P. Idiart and J.E. Lisman 1996. Learning and Memory 3:243–256.
3. Jensen O. and J.E. Lisman 1996a. Learning and Memory 3:257–263.
4. Jensen O. and J.E. Lisman 1996b. Learning and Memory 3:264–278.
5. Jensen O. and J.E. Lisman 1996c. Learning and Memory 3:279–287.
6. Lisman J.E. and M.A.P. Idiart 1995. Science 267:1512–1515.
7. Manai, A.A. and Levy, W.B. 1993. The dynamics of sparse random network. Biol.Cybernet. 70:177–187.
8. Mehta M.R., B.L. McNaughton, C.A. Barnes, M.S. Suster, K.L. Weaver, and J.L. Gerrard 1996. Soc. Neurosci. Abs. 22:1872.
9. Mehta M.R. and B.L. McNaughton 1997. Computational Neuroscience. (this issue)
10. McNaughton B.L. and R.G.M. Morris 1987 Trends in Neurosci. 10:408–415.

11. Muller R.U. and Kubie J.L. 1989. J. Neurosci. 9: 4101–4110.
12. O'Keefe J. and J. Dostrovsky 1971. Brain Res. 34:171–175.
13. O'Keefe J. and M.L. Recce. 1993. Hippocampus 3:317–330.
14. Skaggs W.E., B.L. McNaughton, M.A. Wilson and C.A. Barnes. 1996. Hippocampus 6:149–172.
15. Taube J.S., R.U. Muller, and J.B. Ranck 1990. J. Neurosci. 10:420–435.
16. Treves A. and E.T. Rolls 1994. Hippocampus 4:374–391.
17. Tsodyks, M.V., W.E. Skaggs, T.J. Sejnowski, and B.L. McNaughton. 1996. Hippocampus, 6:271–280.

107

A STATISTICAL LINK BETWEEN LEARNING AND EVOLUTION

Brendan Kitts[*]

Center for Complex Systems
Brandeis University
Waltham, MA
E-mail: brendy@cs.brandeis.edu

ABSTRACT

Widely replicated experimental findings show that faced with multiple variable reward outcomes, animals sample each proportional to their mean payoff. This finding is explained computationally using the Holland theorem, in which this style of sampling is optimal given certain assumptions about the distribution. Other adaptive phenomena which seem to be consistent with this law are discussed, and it is suggested that the 'adaptive problem' may be broadly similar across different domains.

1. RESPONSE MATCHING: A CURIOUS BEHAVIORAL CONSTANT

The *matching law* was first uncovered by Richard Herrnstein in operant conditioning experiments in the 1950s. Herrnstein was examining how pigeons behave on a concurrent reinforcement schedule pecking at two response keys, k_1 and k_2. Reinforcement was supplied at a different rate for each key, given by $r_1 (k_1)$ and $r_2 (k_2)$. Herrnstein found that the relative frequency of responding, b_1 to k_1 and b_2 to k_2 was equal to the relation:

$$\frac{b_1}{b_2} \propto \frac{r_1}{r_2} \tag{1}$$

Soon after this result research into the area exploded, with the same finding replicated with only small deviations across a number of labs, preparations and species (for a good survey see Baum, 1979). For many years however, the finding remained puzzling, as given two rewards it would seem that a higher overall reward could be achieved just by sampling

[*]Supported by ONR grant no. N00014-95-1-0759

the higher-payoff alternative. Why do animals continue to sample from *both* at these specific frequencies? The answer may be found in a sampling theorem derived from the genetic algorithms literature some 25 years later.

2. THE HOLLAND THEOREM

The problem of maximizing payoff from two or more stochastic processes is a classic problem, which may be formalized as a *K-armed bandit* problem. Imagine a person walks into a room with K slot machines, each of which may be biased differentially. The problem is to maximize this person's expected payoff from these machines, from an outlay of only N trials (or coins).

One strategy might be to simply allocate all the trials to the machine with the highest observed payoff, say after t trials. Unfortunately, if after this time, the machine's payoff was the result of some "lucky pulls", then by settling on this machine our person can expect to incur a loss over the following trials equal to $N(u_1 - u_2)$, i.e., the difference between the mean of the true best and selected choice, multiplied by every trial the erroneous sampling is performed.

Holland (1975) investigated this scenario in the context of trying to develop a search strategy for genetic algorithms (where different individuals had various payoffs, and the algorithm had to decide which individuals should be retained as good solutions). The theoretical result he derived would show that payoff could be maximized by sampling at an exponentially higher rate between an observed higher and lower payoff alternative, and this can be approximated by sampling proportional to fitness.

3. OPTIMAL ALLOCATION OF TRIALS

Assume we have N trials to allocate between two random variables. Let $K_{(1)}$ be the observed highest payoff, and $K_{(2)}$ be the observed second highest, with observed means $u_1 > u_2$ and variances σ_1^2 and σ_2^2. We now want to choose some value of n trials to allocate to $K_{(2)}$ and $N - n$ trials to $K_{(1)}$ so as to minimize future expected loss.

3.1. Theorem

Expected loss is minimized when $N - n = \sqrt{8\pi} \cdot b \cdot \exp\left[\frac{b^{-2}n + \lg n}{2}\right]$, which is approximated by setting $N - n \propto$ observed payoff $K_{(1)}$.

3.2. Proof

The loss equation which expresses how much we loose from dividing $N - n$ and n trials between the two random variables, is equal to

$$L(N-n, n) = [q(N-n, n)(N-n) + (1 - q(N-n, n))n]|u_1 - u_2| \tag{2}$$

where q is the probability that the observed best so far is actually second best. This follows from the standard formulation in decision theory. The loss equation states that our expected loss will be equal to (trials expended on best * probability that its second best) + (trials expended on second best * probability that its the best), multiplied by the difference between the means of the best and second best. Our task is to select n which minimizes L. We will do this as follows: First we will determine q, which involves calculating it using Bayes rule and getting its value over repeated samples using the Central Limit Theorem. We will then put this value into our loss equation, L, set $\frac{dL}{dn} = 0$, and solve to find the value of n which minimizes L.

Derive Bayes expression for $q(N - n, n)$ We first break $q(N - n, n)$ into a Bayes probability density, which involves introducing the following quantities:

- $q' = Pr(K' = K_{(2)}|K' = K_1)$ This is the probability that K' is observed to be the second highest, given that its actually the best. This probability can also be restated as $q' = Pr(\frac{1}{n}\sum_1^n K_1 < \frac{1}{N-n}\sum_1^{N-n} K_2)$ which means that the mean payoffs for K_1 are less than the mean payoffs for K_2.

- $q'' = Pr(K' = K_{(2)}|K' = K_2)$ This is the probability that K' is observed to be the second highest, given that it is second highest.

- $p = Pr(K' = K_1)$ The prior probability (*a priori* bias) that the variable K' is the best. Without additional information we will eventually just write this as 0.5.

- $1 - p = Pr(K' = K_2)$ The prior probability that K' is second best. The value for $1 - p = 0.5$.

By Bayes theorem, we write the probability for q as

$$q(N - n, n) = \frac{q'p}{q'p + q''(1 - p)} \tag{3}$$

Use the Central Limit Theorem to show that over series of trials, q' **approaches normal** By the central limit theorem, $\frac{1}{N-n}\sum_1^{N-n} K_2$ (payoff from actual second highest) approaches a normal distribution with mean u_2 and variance $\frac{\sigma_2^2}{N-n}$. $\frac{1}{n}\sum_1^n K_1$ (payoff from actual highest) approaches normal with mean u_1 and variance $\frac{\sigma_1^2}{n}$. The difference between these two normal distributions (giving us the probability that the mean payoffs from $K_1 < K_2$ or $Pr(\frac{1}{n}\sum_1^n K_1 < \frac{1}{N-n}\sum_1^{N-n} K_2)$, is equal to a normal with mean $u_1 - u_2$ and variance $\frac{\sigma_1^2}{n} + \frac{\sigma_2^2}{N-n}$. Thus q' or $Pr(\frac{1}{n}\sum_1^n K_1 - \frac{1}{N-n}\sum_1^{N-n} K_2 < 0)$ the probability that $K_{(1)}$ achieves an average payoff less than $K_{(2)}$ is equal to $N[(u_1 - u_2), (\frac{\sigma_1^2}{n} + \frac{\sigma_2^2}{N-n})]$, or when normalized to mean 0 and unit variance (z-score),

$$q' = N\left[\frac{u_1 - u_2}{\sqrt{\frac{\sigma_1^2}{n} + \frac{\sigma_2^2}{N-n}}}\right] \tag{4}$$

Introduce equation for normal and drop small terms The tail of a normal is equal to $\frac{1}{\sqrt{2\pi}} \cdot \frac{e^{-x^2/2}}{x}$. Therefore, substituting in the parameters of our normal distribution for q' (4) we derive the expansion

$$q' = \frac{1}{\sqrt{2\pi}} \cdot \frac{\sqrt{\frac{\sigma_1^2}{n} + \frac{\sigma_2^2}{N-n}}}{u_1 - u_2} \cdot \exp\frac{-(u_1 - u_2)^2}{2\sqrt{\frac{\sigma_1^2}{n} + \frac{\sigma_2^2}{N-n}}}.$$

Analogous to the above derivation of q', we can also derive q'', obtaining:

$$q'' = \frac{1}{\sqrt{2\pi}} \cdot \frac{\sqrt{\frac{\sigma_1^2}{N-n} + \frac{\sigma_2^2}{n}}}{u_1 - u_2} \cdot \exp \frac{-(u_1 - u_2)^2}{2\sqrt{\frac{\sigma_1^2}{N-n} + \frac{\sigma_2^2}{n}}} \cdot$$

By inspection, as n increases, both q' and $1 - q''$ decrease exponentially, becoming very small. Therefore we can approximate the Bayes expression for q (3) as $\frac{q'p}{q'p + q''(1-p)} \approx q'(\frac{p}{1-p})$. With the assumption that $p = 0.5$ (our prior knowledge about each), we get

$$q(N - n, n) \approx q' \tag{5}$$

Find n at $\frac{dL}{dt} = 0$ We want to know how the expected loss changes with respect to an increase or decrease in n. By setting this to zero, we can find the value of n which maximizes L.

$$\frac{dL}{dn} = |u_1 - u_2|[-q + (N-n)\frac{d}{dn}q + 1 - q - n\frac{dq}{dn}]$$

$$= |u_1 - u_2|[(1 - 2q) + (N - 2n)\frac{dq}{dn}]$$

$$= |u_1 - u_2|[(1 - 2q) + (N - 2n)q\frac{x^2 + 1}{2n}]$$

when $\frac{dL}{dt} = 0$ we get $0 = (1 - 2q) - \frac{N-2n}{2n}q(x^2 + 1)$, or $\frac{N-2n}{2n} = \frac{1-2q}{q(x^2+1)}$. Expanding using our value for q, (5) ($q \approx q'$) this becomes $\frac{N-2n}{n} = \frac{2\sigma_1\sqrt{2\pi}}{u_1 - u_2} \cdot \frac{1}{\sqrt{n}} \cdot \exp \frac{(u_1-u_2)^2 n}{2\sigma_1^2}$. Introducing $b = \sigma_1(u_1 - u_2)$ for simplification, we obtain our solution $N - n = \sqrt{8\pi} \cdot b \cdot \exp \frac{b^{-2}n + \lg n}{2}$. This verifies the first part of the theorem. It can be observed that since the cost of the RHS (trials allocated to the observed poorer alternative) grows exponentially faster than LHS, then $N - n$, the trials allocated to the better observed variable, should be increased exponentially compared to n in order to maintain the equality. Without worrying too much about the precise order of growth, this will be the case when $\frac{d}{dt}N - n = cK_{(1)}$ since $K_{(1)} > K_{(2)}$. Thus, a sampling regime in which $N - n$ is sampled proportional to the observed payoff of $K_{(1)}$ will approximately minimize loss.

4. BIOLOGICAL APPLICATIONS OF THE THEOREM

The Hebbian neural learning rule (or "outstar") can provide a neurally plausible implementation of the matching algorithm. Consider elements of a set of behaviors, $b \in B$ each with an associated probability of being selected, $P(b)$. After sampling, a reward $R(b)$ is calculated. Probability of selection can be increased as follows (Grossberg, 1975; Levy and Desmond, 1985):

$$\frac{dP}{dt} = [R(b) - P(b)]\alpha \tag{6}$$

where α is an arbitrary constant $1 \geq \alpha \geq 0$. In the limit, the probabilities $P(b)$ of response approach the true reward of the environment $R(b)$. At any time therefore, sampling of behavior is proportional to reward.

The theorem also crops up in some intriguing biological areas. According to modern evolutionary theory, evolution works by organisms reproducing proportional to their fitness. If this is true, then there may be an important similarity between ethology and evolution: both use the same sampling method.

5. CONCLUSION

Its heartening that a significant behaviorist finding from the 1950s, which spurred research and then lay almost forgotten in the intervening paradigm shifts, could 40 years later be explained in terms of sampling optimality. Indeed, the only mention of Herrnstein's (widely replicated) data today that the author has been aware of, was in passing as an example of the brain's "puzzling" disposition towards contiguity in a modern neuroscience text (Thompson, 1993; pp. 343). The contemporary work of John Holland and others in the neural networks and statistics communities provides an excellent example of the importance of basic mathematical research, and its potential to cross-fertilize with other disciplines and lead to "deep understandings" of seemingly disparate processes.

REFERENCES

[1] Baum, W. (1979), "Matching, undermatching, and overmatching in studies of choice", *Journal of the Experimental Analysis of Behavior*, Vol. 32, No. 2, pp. 269-281.

[2] Grossberg, S. (1976), "On the Development of Feature Detectors in the Visual Cortex with Applications to Learning and Reaction-Diffusion Systems", *Biological Cybernetics*, Vol. 21, pp. 145-159.

[3] Herrnstein, R. (1958), "Some factors influencing behavior in a two-response situation", *Transactions of the New York Academy of Science*, Vol. 21, pp. 35-45.

[4] Holland, J. (1992), *Adaptation in Natural and Artificial Systems*, University of Michigan Press, Ann-Arbor.

[5] Levy, W. and Desmond, N. (1985), "The Rules of Elemental Synaptic Plasticity", from Levy, W., Anderson, J. and Lehmkuhle, S. (eds), *Synaptic Modification, Neuron Selectivity, and Nervous System Organisation*, Lawrence Erlbaum Associates.

[6] Thompson, R. (1993), *The Brain*, W.H.Freeman and Co., New York.

108

THE ROLE OF V1 IN SHAPE REPRESENTATION

Tai Sing Lee,[1] David Mumford,[2] Song Chun Zhu,[2] and Victor A. F. Lamme[3]

[1] Center for the Neural Basis of Cognition
and Department of Computer Science
Carnegie-Mellon University
Pittsburgh, PA
[2] Department of Applied Mathematics
Brown University
Providence, RI
[3] Department of Medical Physics
University of Amsterdam
The Netherlands Ophthalmic Research Institute
1100 AC Amsterdam, The Netherlands

ABSTRACT

The analysis of higher level global visual attributes such as geometric shapes is thought to be a function of specialist modules in the extrastriate cortices. However, V1 is unique because only in this area can one find cells tuned to orientation and other local features with the spatial precision needed for representing high resolution and geometric aspects of an image. Geometric computation that demands such resolution may need to involve V1 via feedback. V1 neurons are known to exhibit very different types of responses in the short latency (40-80

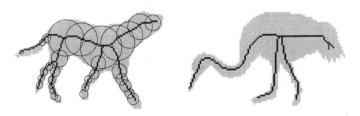

Figure 1. The skeletons of biological forms can be computed by tracing the locus of the centers of the set of maximally inscribing disks inside the figures[10]. Hence, the skeletal or the medial axis transform actually include information about the location of the skeleton and the diameter of the inscribed disk. This representation leads directly to methods for calculating salient shape properties of a figure such as relationship between boundaries, aspect ratio, parts and symmetries[9–11]. In contrast to contour which is inherently local, the medial axis transform is a robust semi-global descriptor for region.

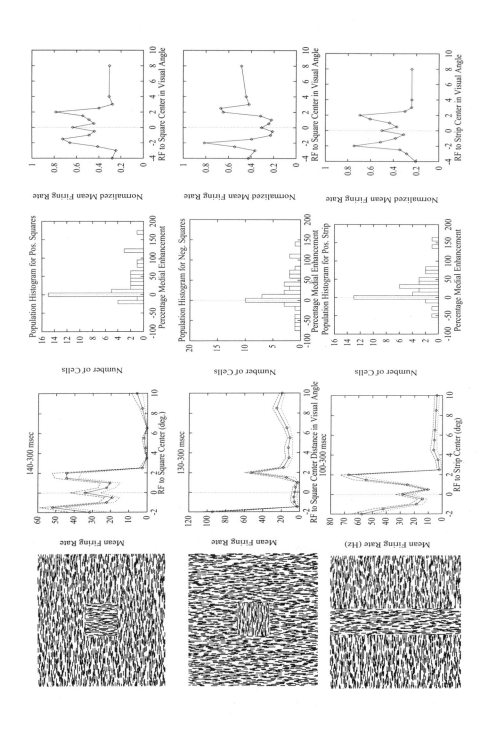

milliseconds post-stimulus time window) versus longer latencies (80-200 milliseconds)[1-5]. Here we report that the later part of V1 neurons' responses is sensitive to geometric attributes of globally defined shapes. We propose a computational interpretation of these results.

The outline of an object defined by luminance contrast or by discontinuity in surface qualities such as texture, color, disparity and optical flow can easily be recognized by *local filters*. But the computation of shape requires *global methods*: one of these is to label all the 'pixels' that belong to the same surface, in what Ullman[6] referred to as 'coloring'. Coloring a region takes time: Paradiso and Nakayama[7] showed that the percept of a white disk forms in stages over 50-100 milliseconds, propagating in from the edges and that it can be interrupted by masks of different shapes. Recent neurophysiological data suggested two possible mechanisms could be used to 'color' a region. One is to link activity of already active cells, as Gray and Singer's data[8] suggested, using synchrony of spikes to represent the presence of a single large percept. The other is to dedicate a cell or part of a cell's activity to signalling that specific elementary regions are part of a single figural surface that are not cut up by boundaries, as suggested by Lamme's finding that the response of V1 neurons are enhanced when their receptive fields are located within the figure than when they are outside the figure [2].

Only when a figure is 'popped out', its surface 'colored' and its boundaries detected, can one begin to compute properties of its shape. These properties may suggest the identity of the visible object or they may need to be modified if the object is partly occluded or in shadow. Therefore, we propose that *the processes of scene segmentation, shape computation and object recognition must happen both concurrently and interactively in feedforward/feedback loops* that involve the whole hierarchical circuit in the visual system at the same time. In the presence of feedback, any perceptual reasoning such as shape computation that requires high resolution and spatial precision may be expected to show up in the single cell recording in V1.

What would be the strategies employed by the visual system to encode figural shapes? Shapes in nature are complex and do not have easy geometry. Biological shapes, in particular, with flexible joints, vary drastically in movement and different postures. Blum[9], in the 70s, observed that with view point changes and motion, the edge-based description can change dramatically while the region-based description such as the skeleton remain relatively stable.

Figure 2. The square and the strip figures were defined by texture contrast. Both *positive* and *negative* stimuli were tested: a *positive* stimulus in which the figure was composed of the texture of the preferred orientation of the cell and a *negative* stimulus in which the figure's texture was of the orthogonal orientation. The width of the figure was selected to be 4-6 times that of the classical receptive field. Typically, the width of the figure was 4° at about $3 - 4^o$ eccentricity, where the RF were about $0.7 - 1.2^o$ in diameter. The monkeys fixated within a 0.3^o fixation window while the stimulus was presented on the screen for 350 msec. The texture defined figure was presented in a randomized series of sampling positions relative to the cell's receptive field (see reference 2 and 14 for details in method). The responses of a vertically oriented V1 neuron to the different part of the figures 100-300 msec after stimulus onset were shown in column 2. The abscissa is the distance in visual angle from the RF center to the center of the figure. The solid lines in these graphs indicate the mean firing rate within the 100-300 msec time window, and the dashed lines depict the envelopes of the standard errors. The dots on the solid lines are the data points. To examine the prevalence of these central peaks, we analyzed a population of 41 vertically oriented cells in response to the positive and negative square figures along a horizontal cross-section-section in the middle of the figure. Column 3: The histograms of the percentage response enhancement at the center of the figure relative to other interior points within the figure showed that there was a general bias toward extra enhancement at the center. A percentage medial enhancement was computed from $(R_0 - R_a)/R_a$, where R_0 is the response at the center (i.e. position=0) and $R_a = [\min(R_1,R_2) + \min(R_{-1},R_{-2})]/2$ is the averaged valley response at the adjacent positions. 13 out of these 41 cells showed at least 30 percent extra enhancement at the center of the figure. Column 4: The average normalized response of these 13 cells, each normalized by its maximum firing rate, revealed response peaks at the center of both the positive and negative squares and strips (fig 2 column 4).

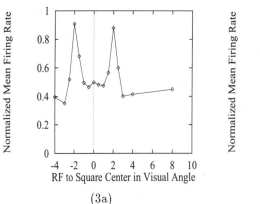
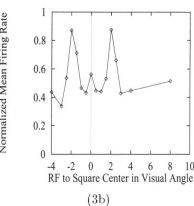

(3a)   (3b)

Figure 3. (a) The population average of the net normalized responses of the 37 vertically oriented neurons to square figures showed a considerable enhancement at the boundary and 25 percent enhancement within the figure, with a hint of extra central enhancement. This was computed by summing the response of each individual V1 neurons to *positive* and *negative* squares at the corresponding positions and then normalized the resulting net response by its maximum. (b) The population average of the net normalized responses of the 13 neurons that showed the extra central enhancement. Note that the average response of the rest of the population, apart from the enhancement at the boundary, was generally flat within the figure.

He therefore proposed to describe the 'infinite' geometrical variability using the skeleton and a small finite set of shape primitives. These ideas have been further developed and beautifully utilized by Zhu and Yuille[10] in their *flexible object recognition and modeling system* (FORM).

Recent psychophysical evidence suggested that the skeleton transform might play a role in the biological vision[12-13]. In particular, Kovacs and Julesz[12] studied the contrast-sensitivity threshold to the perception of Gabor filter patches on a gray background with a surround of many randomly oriented similar patches. They arranged some of the Gabor patches so as to form closed contours of various shapes. This caused a marked increase in sensitivity at the points of the medial axis of their closed contours. In neurophysiological studies, Lee et al[4,14] had found the figure-ground enhancement had certain spatial structure: the responses were particularly accentuated at the figural boundaries for most V1 neurons, and at the medial axis of the figure for a subset of the neurons, as shown in fig 2.

Since the responses within the positive figure was partly due to the tuning properties of cells, we can compute a texture-independent net response by summing together the responses of each cell to the positive and the negative figures, and then normalizing it by its observed maximum firing rate. The computed net response (shown in fig 3a) showed that for this population of cells, on the average, the cells responded 25 % better when they were inside the figure than when they were outside the figure, with only a hint of extra enhancement at the figural center. Some individual cells showed very significant response peaks at the center of the figures (fig 2, col 2). The population histograms of the percentage response enhancement at the center of the figure relative to other interior points within the figure showed that there was a general bias toward extra enhancement at the center. The average net response of the subset of neurons (N=13/43 cells) that were positive for medial axis (as described in fig 2) showed that the figure-ground enhancement was concentrated at the center for this subset of neurons (fig 3b).

For texture strips, the figure-ground enhancement was significantly less. The central enhancement histogram still showed a bias toward positive extra enhancement at the center.

We tested the responses of the cells to strips of different widths and found that collectively, at each eccentricity, there were cells sensitive to the center of strips of different widths. However, individual neurons only showed central peaks for a narrow range of widths (fig 4).

In another experiment (Lee et al 1997), we observed that the central enhancement at the center of uniformly colored disks, e.g. a black disk in a white background or vice versa. This suggests that the central enhancement effect is of a more general nature. It is not unique to texture figures nor can it be simply an effect of intracortical lateral inhibition or disinhibition. These data suggest that V1 might play a role in the computation of figure-ground (figural enhancement) and the medial axis (the central peaks). Abstractly, the medial axis could be produced by a conjunction of three features: two bounding border segments with a homogeneous surface in between (fig 5). One mechanism for computing the medial axis of a figure was the grass fire algorithm originally proposed by Blum[9], which mimics a *grass fire* propagating in from the boundary of the figure; the points of collision form the skeleton of the figure. However, the grass fire spreads in both direction (in and out of the figure). To ensure only the medial axis of the figure is computed, only the borders belonging to the surface of the figure (intrinsic border) is allowed to ignite the fire within the surface it belongs to. von de Heydt et al's finding[16] of border belongingness signal in V2 and Zipser's[5] finding on amodal surface completion effect lend further support to these ideas.

These data, together with other works[1-5,16-17], inspire a possible reinterpretation of the classical paradigm: V1 is not just the first stage of visual processing, but might be considered to be a *high-resolution buffer* that participates in high as well as low level visual algorithms. Any computation that requires high acuity, resolution, and spatial specificity such as finding object boundaries and medial axes would refer back to V1 for image details and spatial registration. Individual cells in V1 are therefore involved in the computation and representation of many kinds of global perceptual structures rather than being a dedicated detector of a particular local feature. It is possible that different types of information coexist or are multiplexed in the spike trains of individual neurons through synchronized firings with different neuronal assemblies. For V1 to play the role we are sketching, it is essential that feedback from extra-striate areas play a major role in driving V1. Consistent with this hypothesis, psychophysical [18] and PET/MRI studies[19] have shown that V1 activity can be caused or influenced by mental imagery. Recent neurophysiological evidence[20] also suggests some context-sensitive surround effects can be eliminated by deactivating V2. Furthermore, the observation that these higher order effects were observed at long latencies suggested they might might be a consequence of the continuous interaction between V1 and the extrastriate cortices.

ACKNOWLEDGMENTS

We are grateful to P.H. Schiller, W. Slocum, V. Lamme, K. Zipser, D. Pollen A. Tolias, J. Mazer, T. Moore for insightful discussion and technical assistance. This research is supported by a McDonnell-Pew grant to T.S. Lee, a NSF grant DMS-93-21266 to D. Mumford, and a NIH grant EY00676 to P.H. Schiller.

REFERENCES

[1] Knierim, J.J. & Van Essen D.C. (1992). Neuronal responses to static texture patterns in area V1 of the alert macaque monkey. *J. Neurophysiology, 67*, 961-980.

Width = 3 deg (*N* = 3/11)

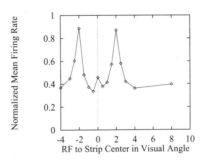

Width = 4 deg (*N* = 16/36)

Width = 6 deg (*N* = 2/10)

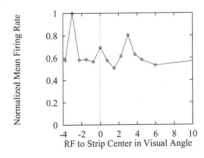

Figure 4. Individual neurons at V1 were sensitive to center of the strips of a narrow range of width. Collectively, the neurons were sensitive to the medial axis of strips of different widths. Shown here are the average normalized net responses of neurons which show central peaks for strips of 3 different widths. 3 out of 11 cells tested for 3° width strip, 16 out of 36 cells tested for 4° width strip, and 2 out 10 cells tested for 6° width strip show 30 percent extra enhancement at the medial axis of the strip. Because of the difficulty in maintaining a neuron for more than a thousand trials, we usually could test the neurons' response to strips of only one or two widths.

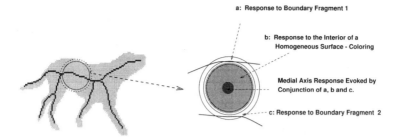

Figure 5. The figure illustrates how a cell may be constructed so that it fires when located on the medial axis of an object. The conjunction of 3 properties has to be present: at least 2 distinct boundary points on a disk of a certain radius and the homogeneity of surface qualities within an inscribing disk. Such a response is highly nonlinear, but can be robustly computed[9-13].

[2] Lamme, V.A.F. (1995). The neurophysiology of figure-ground segregation in primary visual cortex. *J. Neuroscience, 10*, 649-669.

[3] Zipser, K. Lee, T.S. Lamme, V.A.F. & Schiller, P. (1994) The spatial extent of contextual modulation in macaque V1. *Soc of Neuros cience Abstract*, **20** 608.7.

[4] Lee, T.S., Mumford, D. & Schiller, P.H. (1995). Neuronal correlates of boundary and medial axis representation in primate striate cortex. *Invest. Opth. Vis. Sci, 36*, 477.

[5] Zipser, K. Lamme, V.A.F. & Schiller, P.H. (1996) Contextual modulation in primary visual cortex. *J. Neuroscience* **16**, 22, 7376-7389.

[6] Ullman, S. (1984). Visual routines. *Cognition*, **18**, 97-159.

[7] Paradiso, M.A. & Nakayama, K., (1991) Brightness perception and filling-in *Vision Research*, **31**, 1221-1236.

[8] Gray , C.M. & Singer, W. (1989). Stimulus-specific neuronal oscillations in orientation columns of cat visual cortex. *Proc. Nat. Acad. Sci. USA, 86*, 1698-1702.

[9] Blum, H. (1973). Biological shape and visual science. *J. Theoretical Biology, 38*, 205-287.

[10] Zhu, S.C. & Yuille, A.L. (1995) FORMS: a flexible object recognition and modelling system. *Int. J. of Comp. Vis.* 187-212.

[11] Ogniewicz, R. (1994) Skeleton-space: a multiscale shape descritpion combining region and boundary information. *Proc. Conf. On Computer Vision and Pattern Recognition* 746-751.

[12] Kovacs, I. & Julesz, B. (1994). Perceptual sensitivity maps within globally defined visual shapes. *Nature, 370*, 644.

[13] Burbeck, C.A. & Pizer, S.M. (1995). Object representation by cores: identifying and representing primitive spatial regions. *Vision Research, 35*, 13, 1917-1930.

[14] Lee, T.S. (1996). Neurophysiological evidence for image segmentation and medial axis computation in primate V1. *Computational Neuroscience*, Bower, J. Eds., 373-378, Academic Press.

[15] Li, C.Y. & Li, W. (1994). Extensive integration field beyond the classical receptive field of cat's striate cortical neurons - classification and tuning properties. *Vision Research, 34*, 2337-2355.

[16] Zhou, H. Friedman, H. von de Heydt, R. (1996) Edge assignment in cells of monkey area V2. *Invest. Ophthalmol. Vis. Sci.* **37**, 904.

[17] Gilbert, C.D. Das, A. Ito, M. Kapadia, M. & Westheimer, G. (1996). Spatial integration and cortical dynamics. *Proc. Nat. Acad. Sci. USA. 93*, 615-622.

[18] Ishai, A. & Sagi, D. (1995). Common Mechanisms of Visual Imagery and Perception. *Science, 268*, 1772-1774.

[19] Kosslyn, S., Thompson, W.L., Kim, I.J. & Alpert, N.M. (1995). Topographical representations of mental images in primary visual cortex. *Nature, 378*, 496-498.

[20] James, A.C., Hupe, J.M., Lomber, S.L., Payne, B., Girard, P. & Bullier, J., (1995). Feedback connections contribute to center surround interactions in neurons of monkey area V1 and V2. *Soc. Neuroscience Abstract*, **21**, 359.10.

109

PREDICTING NOVEL PATHS TO GOALS BY A SIMPLE, BIOLOGICALLY INSPIRED NEURAL NETWORK

William B. Levy and Xiangbao Wu

Department of Neurological Surgery
University of Virginia Health Sciences Center
Charlottesville, Virginia 22908
wbl@virginia.edu
xw3f@virginia.edu

1. ABSTRACT

Although recurrent networks can be used as content addressable memories, they can also be used as sequence prediction systems. Because problem solving can often be viewed as a sequence prediction problem, we hypothesize that such networks can be used as problem solvers. There are many aspects to problem solving. Here we concentrate on a single but important aspect, goal finding without search. Using a highly simplified, clearly prototypical version of this problem, a sparsely connected recurrent network successfully predicts novel paths to reach a goal.

2. INTRODUCTION

Problem solving is often recognized as a sequence completion or sequence prediction problem (e.g., Newell & Simon 1972). In the artificial intelligence approach to problem solving, logical sequences to reach goals are typically investigated goal-by-goal and sequence-by-sequence. This approach becomes rapidly computationally infeasible. Here, we present a neural network model that uses attractors to find goals and even novel paths, without search.

3. THE NETWORK

The network (see Fig. 1a and 1b), inspired by a small piece of hippocampal region CA3, has a sparse (10%) recurrent connectivity, c_{ij}, between its 512 excitatory neurons

a.   b.

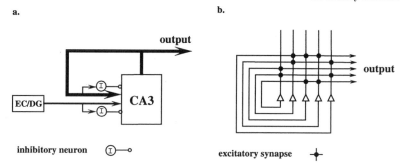

Figure 1. The Model. a) Simplified Hippocampal model. In the model the input layer is a combination of the entorhinal cortex and dentate gyrus. Accompanying this feedforward excitation is a proportional feedforward inhibition. The strong excitation of the network results from the recurrent connections, which is also accompanied by a feedback inhibition. The output of the network is the state of the excitatory CA3 cells themselves, and this is decoded by a simple cosine comparison. b) The recurrent excitatory synapses are sparse and randomly placed.

and a paucity (two) inhibitory neurons that maintain control over net activity levels in a crude, imprecise way (Minai & Levy 1993) via fixed parameters K_I and K_R. The simultaneously updated McCulloch-Pitts neurons transmit $z_i(t) \in \{0,1\}$ every increment of time, t through synaptic weights w_{ij}.

The excitation y_j of CA3 neuron j is given by:

$$y_j(t) = \frac{\sum_i w_{ij} c_{ij} z_i(t-1)}{\sum_i w_{ij} c_{ij} z_i(t-1) + K_I \sum_i x_i(t) + K_R \sum_i z_i(t-1)}$$

and its output of neuron j is

$$z_j(t) = \begin{cases} 1 & \text{if } y_j(t) \geq \theta \text{ or if } x_j(t) = 1; \\ 0 & \text{otherwise} \end{cases}$$

where each neuron j receives a single external input $x_j(t) \in \{0,1\}$.

Synaptic modification proceeds adaptively via a self-supervised, local modified Hebbian process of the form $w_{ij}(t+1) = w_{ij}(t) + 0.01 \cdot z_j(t) (z_i(t-1) - w_{ij}(t))$. The time delay, similar to a proposal of Amari (1972), is qualitatively consistent with physiological observations (Levy & Steward 1979, 1983).

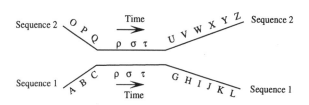

Figure 2. Pictorial representation of the enviroment. There are two sequences of 12 input patterns to learn. Each sequence contains three orthogonal segments. The two sequences share a subsequence of three patterns.

The cosine function for comparing two vectors is used to understand the output of the network during testing. Specifically, after learning, the network states that evolve in response to the test input are compared with the states produced by a complete sequence of input patterns. By definition, the largest value at each time step is the decoded network state at that moment. Although other techniques (e.g., maximum likelihood) would produce superior decoding, such methods would obscure our object of our study, i.e., what the CA3-like network is actually doing.

4. RESULTS

4.1. Sequence Completion under Ambiguous Conditions (The Disambiguation Problem)

Here we look at an analog of a useful skill in problem solving — the ability to combine separately learned inferences. In this problem the network learns two deterministic sequences of 12 patterns (see Fig. 2). Each sequence contains three orthogonal segments. Within each segment, the successive external inputs share 7 of 8 active neurons. The two sequences share a common subsequence $[\rho,\sigma,\tau]$ that is three patterns long. We initially studied this problem of shared subsequences (Minai et al 1994 and again in Levy et al. 1995, Wu et al. 1996) to prove that the network, as a model of the hippocampus, can learn and can use context. Here, we start by studying the disambiguation problem. As before, the network successfully disambiguates two related sequences based only on context. After learning, when the network is given pattern O, it produces a sequence culminating in pattern Z. Similarly (see Fig. 3a), giving the network pattern A produces a sequence culminating in pattern L. Once again, we emphasize that there is no explicit spectrum of delays (e.g. Kleinfeld 1986) nor is there any microscopic time spanning properties greater than one step so that the solution derives from the dynamic code created by the sparsely interconnected network itself (Levy 1989).

4.2. Goal Finding Via Novel Sequence Generation

In addition to solving sequence prediction under ambiguous conditions, we now want to show goal finding by the network. This is a compositional problem involving a recomposition of subsequences. Therefore, in addition to giving the network an initial starting point as an input, we also give it a partial description of a goal that requires a novel path. Thus, the input is initially pattern A as before, but now two neurons of pattern Z are continuously activated — as if the model has some partial idea about what the goal looks like. In this case the network hypothesizes the path $[A,B,C,\rho,\sigma,\tau,U,V,W,X,Y,Z]$ which reaches the appropriate goal pattern. Figure 3b helps show how the network succeeds. Thus, the model decomposed the two learned sequences and recomposed the pieces as appropriate to solve the problem.

To avoid unrepresentative results, here we present only robust results where robust is defined as replicable in 4 out of 5 randomly constructed networks. For each network there are four test cases, i.e. given pattern A, given pattern O, given pattern A and two neurons of pattern Z, and given pattern O and two neurons of pattern L. Out of the 5 networks we tried, the problems were correctly solved between 80–90% of the time with an average of 85%.

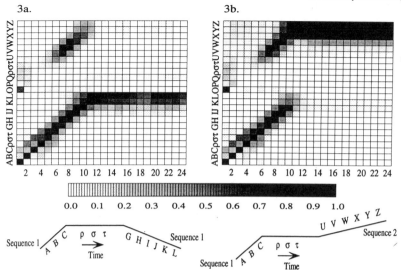

Figure 3. Sequence prediction after learning the two partially overlapping sequences illustrated in Fig. 2. These similarity matrices use the cosine function. They compare the network states generated by each full pattern sequence after learning (ordinate) with the network states generated over time during testing (abscissa). Decode the output during testing by finding the darkest square in each column; the letter on the ordinate is the decoded answer. 3a. Sequence completion in the disambiguation problem is made difficult by the shared subsequence [ρστ]. Here we show the similarity values (used for decoding) when the network is transiently given pattern A. Appropriately enough for the learning and for the starting point, the states go to pattern L, a pattern essentially orthogonal to the representation of the other learned goal pattern, Pattern Z. 3b. When the same network is given the same transient input but two neurons of goal Z are also turned on, the network produces a sequence of representations leading to this partially specified goal. To create this path, the network must produce a novel sequence that appropriately combines its knowledge of the two separately learned sequences.

5. DISCUSSION

Because the network solves the context-dependent disambiguation problem, representations of ρ through τ must differ depending on where the network starts its sequence of representations. Even so, there is enough similarity between each set of the ρ-τ states for the network to follow either path depending on the relative strength of the two attractors at the end of each sequence. In addition to the network's immediate representation of τ, the winning attractor is a function of experience and of external biasing.

Hopfield's research (Hopfield 1982, 1984) has inspired a great amount of analytical work on the properties of recurrent networks. As a result, we now know that even networks with a time spanning Hebb rule can form, at the very least, transient attractors (e.g. Sompolinsky & Kanter 1986, Wu & Liljenström 1994, Liljenström & Wu 1995). However, using the dynamics of such networks to solve sequence prediction problems with novel goals seems a new use of such networks.

In sum, the hippocampal model studied here is capable of goal finding, albeit a highly simplified form of this problem.

6. ACKNOWLEDGMENTS

This work was supported by NIH MH48161 and MH00622, by EPRI RP8030-08, and Pittsburgh Supercomputing Center Grant BNS950001 to WBL, and by the Department of Neurosurgery, Dr. John A. Jane, Chairman.

7. REFERENCES

Amari, S.I. (1972) *IEEE Trans. on Computers* **C-21**, 1197–1206.

Hopfield, J. J. (1982) *Proc. Natl. Acad. Sci. USA* **79**, 2554–2558.

Hopfield, J. J. (1984) *Proc. Natl. Acad. Sci. USA* **81**, 3088–3092.

Kleinfeld, D. (1986) *Proc. Natl. Acad. Sci. USA* **83**, 9469–9473.

Levy, W. B (1989) in *Computational Models of Learning in Simple Neural Systems*, eds. Hawkins, R. D. & Bower, G. H. (Academic Press, New York), pp. 243–305.

Levy, W. B & Steward, O. (1979) *Brain Res.* **175**, 233–245.

Levy, W. B & Steward, O. (1983) *Neurosci.* **8**, 791–797.

Levy, W. B, Wu, X. B., & Baxter, R. A. (1995) *Intl. J. Neural Syst.* **6 (Supp.)**, 71–80.

Liljenström, H. & Wu, X. B. (1995) *Intl. J. Neural Syst.* **6**, 19–29.

Minai, A. A., Barrows, G., & Levy, W. B (1994) *World Congress on Neural Networks* **IV**-176–181

Minai, A. A. & Levy, W. B (1993) In: *Advances in Neural Information Processing Systems*, eds.

Giles, C. L., Hanson, S. J., & Cowen, J. D. (Morgan Kaufmann, San Mateo, CA), pp. 556–563

Newell, A. & Simon H. (1972) *Human Problem Solving*. Englewood Cliffs, NJ: Prentice-Hall.

Sompolinsky, H. & Kanter, I. (1986) *Phys. Rev. Ltrs.* **57**, 2861–2864.

Wu, X. B., Baxter, R. A., & Levy, W. B (1996) *Biol. Cybern.* **74**, 159–165.

Wu, X. B. & Liljenström, H. (1994) *Network: Comput. Neural Syst.* **5**, 47–60.

110

INVESTIGATING AMPLIFYING AND CONTROLLING MECHANISMS FOR RANDOM EVENTS IN NEURAL SYSTEMS

Hans Liljenström[1] and Peter Århem[2]

[1]Department of Physics
Royal Institute of Technology
S-100 44 Stockholm, Sweden
hali@theophys.kth.se
[2]Nobel Institute for Neurophysiology
Karolinska Institutet
S-171 77 Stockholm, Sweden
peter.arhem@neuro.ki.se

ABSTRACT

Microscopic and individual events in the nervous system, such as single channel openings and individual action potentials, often drown in the summed activity of the surround. Yet, under certain circumstances such events can be amplified and have meso– and macroscopic effects. We use experimental as well as computational methods to investigate mechanisms by which neural systems can amplify weak signals and individual events, and control the system at a larger scale. By this multilevel approach we try specify the rules and constraints for the resulting state transitions.

1. INTRODUCTION

The effects of microscopic and mesoscopic events in the nervous system, e.g. single channel openings and individual action potentials, are normally assumed to be negligible on a macroscopic scale; they often drown in the summed activity of the surround. Yet, under certain circumstances such events can be amplified and indeed have macroscopic effects.

For example, we have in previous experimental studies [7] shown that random single channel activity can induce action potentials in small isolated hippocampal interneurons (see Fig. 1a and 1b). This finding suggests that a subset of neurons can function as random generators in the hippocampus. Other reports show that single inhibi-

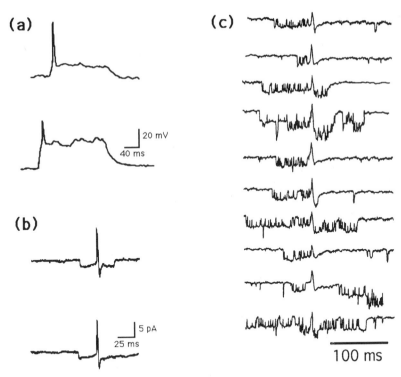

Figure 1. Spontaneous activity in small hippocampal interneurons, induced by single channel openings. (a) Action potentials associated with plateau potentials, caused by single channel openings. (b) Spontaneous action currents associated with single channel openings in isolated small hippocampal neurons. Pipette potential was −26 mV. (c) Action currents and single channel openings in small interneurons of a hippocampal slice. Pipette resistance was 5 MΩ and presented data filtered by a Bessel filter with a corner frequency of 3 kHz. The neurons in (a) and (b) were prepared from embryonic (day 18–21) rat hippocampi and cultured 4–9 days as described in [7]. The preparation in (c) is described in Methods.

tory hippocampal interneurons are capable of regulating/synchronizing the activity of more than a thousand pyramidal cells [4].

In a computational study we have shown that the complex behavior of cortical neural networks can be regulated by additive and multiplicative noise [10]. By changing the noise level in a neural network model of the olfactory cortex we demonstrated that the system dynamics could switch from one attractor state to another. These states can be point attractor states or limit cycle or strange attractor states, similar to those obtained from EEG readouts. It was also shown that noise can improve system performance by reducing recall time in associative memory tasks. At an optimal noise level recall time reaches a minimum.

In the present study we analyze these problems by a combination of experimental and computational techniques. The specific aims are: (i) to experimentally investigate spontaneous activity of small interneurons in hippocampal slices, and (ii) to computationally investigate the functional role that micro- and mesoscopic random events, might play for the macroactivity of cortical networks. A more general aim is to relate the findings to previously proposed mechanisms, such as amplification by oscillations and enhancement of neuronal excitability [15,9], stochastic resonance [14], control of chaos [2], and computation by action potential timing [6].

2. METHODS

The experimental data come from patch clamp measurements of small neurons (diameter between 10 and 20 µm) in hippocampal slices (thickness 200 µm) from young rats (8–10 days). The recordings were made under visual control according to the technique described by Edwards et al [5]. The neurons recorded from were selected mainly on basis of size. Neurons from the dentate gyrus, CA3 and CA1 are included in the material.

The simulations are made on a cortical neural network model, having network units corresponding to populations of neurons with a continuous input-output function, and which are connected according to the architecture of hippocampus or the olfactory cortex. The model, which is modified from a previous olfactory cortex model [8], consists of a three layered structure; two layers of inhibitory units and one layer of excitatory units (see Fig. 2). Noise enters as additive and multiplicative noise with various characteristics imposed on each network unit activity. The cortical model is simulated on a parallel Connection Machine, with 3x32x32 network units used in the current simulation.

3. RESULTS

Figure 1c shows ten consecutive action current impulses and channel openings recorded in a small interneuron of a hippocampus slice, lined up to show the relation between the occurrence of impulses and channel openings. In this case, eight impulses out of ten occur while the channel is in open state, whereas the fraction open state is about 35%, suggesting a causal relationship. More detailed statistical analysis revealed a clear correlation between single channel openings and impulses for a subset of interneurons, implying that single channel openings can induce regenerative action potentials in neurons in relatively intact conditions. This finding strengthens the conclusion from a study of cultured neurons that a subset of hippocampal interneurons can function as cellular random generators [7].

To investigate the possible role of cellular random generators, we simulated spontaneous activity as noise of Gaussian distribution in both excitatory and inhibitory units of the hippocampus model. We showed that global oscillatory behavior can be induced by a

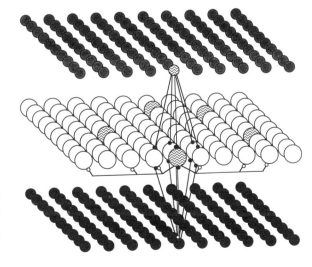

Figure 2. The hippocampal network model used. The top layer consists of "fast inhibitory interneurons", the middle layer of excitatory "pyramidal cells", and the bottom layer consists of "slow inhibitory interneurons". The two sets of inhibitory units are characterized by two different time constants and somewhat different connections to the excitatory units. All connections are modeled with time delays for signal propagation, corresponding to the geometry and fiber characteristics of rat cortex. Spontaneously active units are indicated. See figure 3.

few noisy units in certain configurations. Fluctuations that occasionally sum up to surpass a threshold can eventually lead to an oscillatory spatiotemporal activity that is sustained for several hundreds of milliseconds. (There is no external input to the network in these simulations). The character of the transition, as well as the transition point, depends on noise characteristics, such as the noise level and the spatial distribution of the noisy units. For low noise levels, no state transition occurred within the observed time period. At higher noise levels, pseudo-chaotic behavior can arise. For a different random seed, the onset of the oscillatory state may occur at a quite different point in time. In Fig. 3 we show how global oscillatory behavior can be induced by noise in five excitatory units, distributed in two different configurations (low and high density), and how the transition depends on the spatial distribution of the noisy units. We have also shown that global

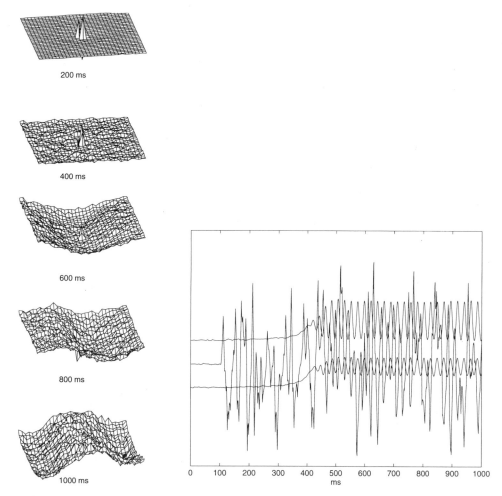

Figure 3. Synchronous oscillatory activity is induced by five noisy units in a network with 32 by 32 network units in each of the three layers of the hippocampal model, corresponding to a 10 by 10 mm square of the real cortex. The noise is turned on at t=100ms. (a) The onset of the network oscillations appears at approx. t=500ms when the noisy units are densely packed. (b) The time course for one noisy and two regular excitatory units is shown for the same simulation as in (a). (c) and (d) show simulation results when the five noisy units are spread out. Here, the onset of network oscillations appears at approx. t=800ms.

oscillatory or more complex coherent behavior can be induced in the excitatory network layer by noise in a single (feedforward) inhibitory unit. This suggests that spontaneous disinhibition (simulating the action of inhibitory neuronal random generators) can result in synchronized oscillatory activity in the excitatory layer. These results should be seen in relation to the abovementioned finding that noise can improve system performance in cortical networks.

4. CONCLUSIONS

In conclusion, the present study supports the view that stochastic activity in neurons plays a functional role in CNS information processing [11]. More precisely the study suggests: (i) that a subset of small hippocampal neurons have intrinsic spontaneous activity, (ii) that single ion channels elicit action potentials in a subset of hippocampal neurons, that consequently function as cellular random generators, (iii) that neuronal random generators can induce synchronous oscillations in cortical networks, and (iv) that neuronal ran-

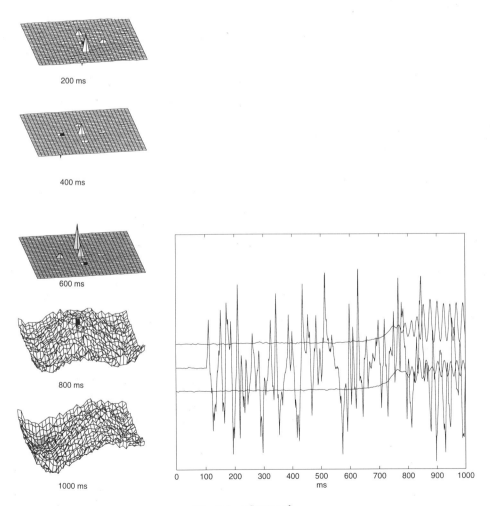

Figure 3. (*Continued*)

dom generators may have significance for cognitive functions in giving rise to transitions between different dynamical states, representing for instance oscillatory or pseudochaotic attractor memory states.

Furthermore, the discussed findings illustrate a form of stochastic resonance and the more general phenomenon of amplifying microevents to global events in biological systems [3, 14]. They may also suggest an explanation to the origin of the randomly changing activity patterns, recently observed in visual cortex [1].

At present, we are trying to specify in more detail the constraints under which fluctuations in the hippocampal system can lead to more ordered behavior, and how chaotic processes in cortical circuits can be controlled [2]. We are also trying to relate the findings of the oscillatory behavior to the recently proposed computation mechanism by action potential timing [6].

REFERENCES

1. Arieli, A., Sterkin, A., Grinvald, A. & Aertsen, A. (1996) Science 273:1868–1870.
2. Braiman, Y., Lindner, J.F., & Ditto, W.L. (1995) Nature 378:465–467.
3. Buhmann, J. & Schulten, K. (1987) Biol. Cybern. 56:313–327.
4. Cobb, S.R., Buhl, E.H., Halasy, K., Paulsen, O. & Somogyi, P. (1995) Nature 378:75–78.
5. Edwards, F.A., Konnerth, A., Sakmann, B. & Takahashi, T. (1989) Pluegers Arch. 414:600–612.
6. Hopfield, J.J. (1995) Nature 376:33–36.
7. Johansson, S. & Århem, P. (1994) Proc. natl. Acad. Sci. U.S.A. 91:1761–1765.
8. Liljenström, H. (1991) Int. J. Neural Syst. 2:1–15.
9. Liljenström, H. & Hasselmo, M.E. (1995) J. Neurophysiol. 74:288–297.
10. Liljenström, H. & Wu, X. (1995) Int. J. Neural Syst. 6:19–29.
11. Smetters, D.K. & Zador, A. (1996) Current Biology 6:1217–1218.
12. Softky, W.R. & Koch, C. (1993) J. Neurosci. 13:334–350.
13. Strassberg, F.A. & DeFelice, L.J. (1993) Neural Computation 5:843–855.
14. Wiesenfeld, K. & Moss, F. (1995) Nature 373:33–36.
15. Wu X. & Liljenström, H. (1994) Network 5:47–60.

111

PERCEPTUAL COMPLETION ACROSS THE FOVEA UNDER SCOTOPIC/MESOPIC VIEWING CONDITIONS

Jennifer Linden

Computation and Neural Systems Program
Caltech 216-76
Pasadena, CA
E-mail: linden@vis.caltech.edu

ABSTRACT

Perceptual completion has been shown to occur across the blind spot, across the blue-blind region of the fovea, and across blind fields created by retinal scotomas. The purpose of this study was to determine whether perceptual completion can also occur across the dark-adapted fovea. The central fovea, which contains only cones, is blinded at low light levels, when rod vision is still possible in the periphery. Does the visual system "fill in" features across the center of gaze at low light levels? To answer this question, three dark-adapted subjects were asked to report the presence or absence of a circular gap in a field, line, or checkerboard pattern viewed under high scotopic or low mesopic conditions. The gap in the field, line, or pattern appeared on half the trials, positioned either at the fovea or 4° in the periphery. All three subjects were significantly more likely to report "no gap" on foveal gap trials than on peripheral gap trials for field and line stimuli; results for the patterned stimuli were inconclusive. These results indicate that perceptual completion of both fields and lines occurs across the fovea under high scotopic/low mesopic viewing conditions. The completion of lines suggests mechanisms involving cortical circuitry.

1. INTRODUCTION

Perceptual completion, or "filling in," has been shown to occur across blind regions of the visual field corresponding to gaps in the photoreceptor mosaic. The blind spot in the monocular visual field, corresponding to the optic disk, completes fields, lines, patterns, and even illusory contours (Ramachandran, 1992). Similarly, the blue-blind zone in the central 0.4° of the visual field, corresponding to the region of the retina which lacks blue cones, has

Computational Neuroscience
edited by Bower, Plenum Press, New York, 1997

been shown to complete blue fields (Williams, 1981; Curcio *et al.*, 1991). In patients with retinal lesions, blind areas corresponding to retinal scotomas have also been shown to complete fields (Ramachandran, 1993).

Another blind region corresponding to a gap in the photoreceptor mosaic is the night-blind fovea. The central 1°-2° of the visual field is represented by a part of the foveal retina which contains only cones (Curcio *et al.*, 1990). Under scotopic and low mesopic viewing conditions, when rods dominate vision, this area is blinded (Jayle *et al.*, 1959; Crawford, 1977; Stabell and Stabell, 1993). Why are we not aware of a hole at the center of vision at night? Does perceptual completion occur across the fovea?

2. METHODS

Three subjects (including the author), all with vision corrected to normal, participated in this study. Subjects dark-adapted for 30 minutes before the experiments. During the experiments, subjects wore goggles made from pairs of polarizing filters (Prinz), cross-polarized to block 95% of incident light. With these goggles in place, average apparent luminance of all experimental stimuli was within high scotopic to low mesopic range (< 0.05 cd/m^2) for all subjects. A chin rest held subjects' heads 50 cm from the monitor screen used to display experimental stimuli. Eye position was not monitored.

Experimental design is described in Figure 1. Open pinwheel figures were used as a fixation guide instead of a fixation dot or crosshair to avoid interference with foveal vision. The upper pinwheel figure was centered on the peripheral gap position, to ensure that any local afterimages created by the fixation guide would be identical at the fovea and in the periphery. The Field Completion (Figure 1A), Line Completion (Figure 1C), and Pattern Completion (Figure 1D) experiments were designed to determine which visual features, if any, could be completed across the fovea under scotopic/mesopic viewing conditions. Trials for these three experiments were completely interleaved. The Extent of Field Completion experiment (Figure 1B) was conducted separately from the other experiments.

At the end of each trial, subjects were asked to choose one of two alternative descriptions of the stimulus: "gap" or "no gap." To reduce the likelihood of a "no gap" response bias, subjects were informed that half the trials would be gap trials. Furthermore, they were instructed to report "gap" if they saw **any** distortion in the stimulus at either of the two possible gap locations.

3. RESULTS

Results for the four experiments are summarized in Figure 2. For field and line stimuli, subjects were significantly more likely to report "no gap" on foveal gap trials than on peripheral gap (Figures 2A, 2B, and 2C). The response bias for fields was evident over all gap diameters tested.

For patterned stimuli, there was no significant difference in frequency of "no gap" responses for the foveal and peripheral gap trials (Figure 2D). Subjects performed poorly on both the peripheral gap detection task and the foveal gap detection task. A control experiment (not shown) ruled out the possibility that the pattern was simply not visible. All subjects were able to discriminate perfectly between the patterned stimulus and an equiluminant gray field, even with checks half the size of those used in this experiment.

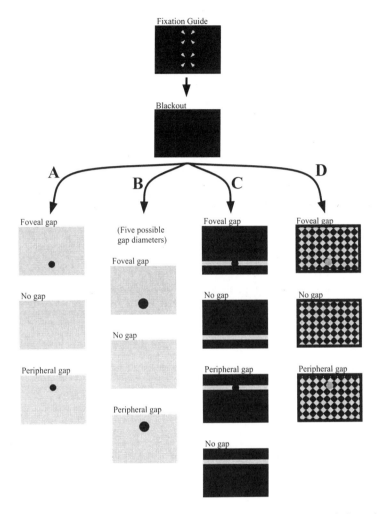

Figure 1. Each trial began with presentation of a fixation guide for 1000 ms. Subjects were asked to maintain fixation throughout the entire trial on the location indicated by the center of the lower pinwheel figure. Experimental stimuli were presented for 200 ms after a 500 ms blackout period. **A. Field Completion.** The stimulus could be either a gray field (40 trials), a gray field with a black circular 1°-diameter gap centered at the fovea (20 trials), or a gray field with a gap centered 4° in the periphery (20 trials). **B. Extent of Field Completion.** Stimuli were similar to those used in the Field Completion experiment, but now the diameter of the circular gap was varied as well as its location. Five different gap diameters were tested: 0.75°, 1.00°, 1.25°, 1.50°, and 1.75°. On half of the trials, the stimulus was a gray field (200 trials). A black gap of varying diameter appeared in the fovea (20 trials per gap diameter) or in the periphery (20 trials per gap diameter) on the remaining trials. **C. Line Completion.** In this experiment, four different stimuli were used: a gray horizontal line 0.5° thick, extending unbroken across the fixation location (20 trials); the same line with a 1°-diameter gap centered over the fixation location (20 trials); an unbroken line extending across a point 4° above the fovea (20 trials); and the same line with a gap centered 4° above the fovea. **D. Pattern Completion.** The stimulus could be either a complete gray-and-black checkerboard pattern with 0.4°-square checks (40 trials); the same checkerboard pattern with an equiluminant gray 1°-diameter gap centered on the fixation location (20 trials); or the checkerboard pattern with a gap centered 4° above the fovea (20 trials).

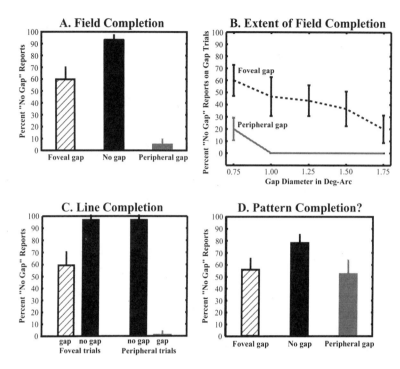

Figure 2. Data from all three subjects are averaged; error bars represent propagated error in the mean assuming binomial statistics for each subject. **A. Field Completion.** Subjects were significantly more likely to report "no gap" on foveal gap trials than on peripheral gap trials. The hypothesis that "no gap" response frequency was equal on foveal and peripheral trials could be rejected at a significance level of 0.001 (Fischer-Irwin test). **B. Extent of Field Completion.** A significant bias toward "no gap" responses on foveal gap trials persisted over all gap diameters tested, up to 1.75°. **C. Line Completion.** Again, subjects were more likely to report "no gap" on foveal gap trials than on peripheral gap trials. The difference in response frequency for foveal vs. peripheral gap trial conditions was highly significant ($p < 0.001$, Fischer-Irwin test). **D. Pattern Completion?** For the checkerboard stimulus, there was no significant difference between foveal and peripheral gap trials in the frequency of "no gap" responses.

4. DISCUSSION

4.1. Field Completion Experiments

The results of the two field completion experiments demonstrate that field completion does occur in the fovea under scotopic/mesopic viewing conditions, over an area at least 1.75° in diameter. Subjects were much more likely to report "no gap" on foveal gap trials than on peripheral gap trials. This response bias was highly significant, despite the fact that two features of the experimental design could have obscured any differences in foveal and peripheral gap detection performance. First, subjects were told to report "no gap" only on trials in which they were very confident that the stimulus was complete and unbroken; any sense of distortion in the field at either possible gap location should have prompted a "gap" response. This response criterion would counter any bias toward "no gap" responses. Second, fixation was not monitored. Many of the presumed "foveal" gap trials may in fact have been more accurately classified as peripheral gap trials, due to loss of fixation. Loss of fixation would obscure differences in gap detection performance between foveal and peripheral gap trials.

The quantitative results of these two experiments were confirmed by the qualitative impressions of subjects. Subjects said that while the peripheral gap in the field was often obvious, the foveal gap was never distinct; the region at the fixation point almost always appeared just as gray as the surrounding field.

The foveal field completion demonstrated here might be explained entirely in terms of retinal mechanisms. Under scotopic viewing conditions, rod signals are transmitted by AII amacrine cells to cone bipolar and ganglion cells (Wässle and Boycott, 1991). Under low mesopic conditions, direct communication between rods and cones takes precedence over the indirect amacrine-cell circuit (Wässle and Boycott, 1991). Such circuits might link rods near the fovea to cone ganglion cells in the rod-free zone when light levels are low.

4.2. Line Completion Experiment

Results of the Line Completion experiment indicate that lines as well as fields are completed across a 1° region of the fovea under scotopic/mesopic viewing conditions. Subjects were significantly more likely to report "no gap" on foveal gap trials than on peripheral gap trials, despite the fact that the response instructions and the lack of fixation control could have obscured any such bias (as discussed above). Subjects said they rarely, if ever, perceived a gap in the line at the fixation location; at most, they saw slight distortions such as narrowing and dimming of the line in the center of the visual field. All three subjects also noticed an unexpected manifestation of line completion: the wedges forming the pinwheel fixation guides sometimes seemed to extend across the fixation location, creating an X.

While sufficient to explain field completion, retinal mechanisms alone probably cannot explain the line completion observed in this experiment. The circuits connecting rod bipolar cells to cone ganglion cells, and those linking rods to cones, are likely to be circularly symmetric. Such circuits would not support completion of oriented lines. The completion of lines across the fovea suggests mechanisms involving cortical as well as retinal circuitry.

4.3. Pattern Completion Experiment

The response bias observed in the previous experiments was not evident in the Pattern Completion experiment. Subjects were almost equally likely to report "no gap" on foveal gap trials and on peripheral gap trials. Although subjects could easily distinguish the global checkerboard pattern from an equiluminant gray field, they described the pattern as locally indistinct. Small equiluminant gray gaps in the pattern were perceived neither at the fovea nor in the periphery. Therefore, while patterns may be completed across the fovea, the present experiment does not conclusively demonstrate whether or not foveal pattern completion occurs. Future experiments will attempt to resolve this question by modifying stimulus conditions to make the peripheral gap easier to see.

4.4. Implications

These experiments demonstrate that fields and lines are completed across the fovea under high scotopic/low mesopic viewing conditions. Line completion suggests a mechanism involving cortical as well as retinal circuits. If cortical completion mechanisms do exist, perceptual completion across the fovea may involve an enormous area of cortex. Almost 10%

of primary visual cortex, 100 mm^2 of cortical surface, is dedicated to the central 1° of the visual field (Wässle and Boycott, 1991). Under photopic conditions, the fovea is the most important source of visual information available to the higher visual system; under scotopic or low mesopic conditions, the fovea is blinded and will complete fields and lines. Like completion across the blind spot, the blue-blind zone, and regions corresponding to retinal scotomas, completion across the fovea requires the visual system to adapt to a gap in the photoreceptor mosaic; unlike the other examples, it also requires the visual system to do so dynamically.

ACKNOWLEDGMENTS

I would like to thank Maneesh Sahani, for providing many helpful suggestions, for assisting me with the poster, and for participating in the experiment; Grace Chang, for participating in the experiment; Dr. Pietro Perona, for offering advice on experimental design; and Dr. David Williams, for sharing the unpublished results of his own investigations. This research was supported by a Howard Hughes Medical Institute Predoctoral Fellowship to the author.

REFERENCES

[1] Crawford M.L.J. (1977) Central vision of man and macaque: cone and rod sensitivity. *Brain Research* 119: 345-356.
[2] Curcio C.A., Sloan K.R., Kalina R.E., and Hendrickson A.E. (1990) Human photoreceptor topography. *J. Comp. Neurol.* 292: 497-523.
[3] Curcio C.A., Allen K.A., Sloan K.R., Lerea C.L., Hurley J.B., Klock I.B., and Milam A.H. (1991) Distribution and morphology of human cone photoreceptors stained with anti-blue opsin. *J. Comp. Neurol.* 312: 610-624.
[4] Jayle G.E., Ourgaud A.G., Baisinger L.F., and Holmes W.J. (1959) *Night Vision.* Springfield, Illinois: Charles C. Thomas, Publisher.
[5] Ramachandran V.S. (1992) Blind spots. *Scientific American* 266(5): 86-91.
[6] Ramachandran V.S. (1993) Filling in gaps in perception: Part II. Scotomas and phantom limbs. *Curr. Dir. Psychol. Sci.* 2(2): 56-65.
[7] Stabell B. and Stabell U. (1993) Rod-cone interaction in form detection. *Vision Research* 33(2): 195-201.
[8] Wässle H. and Boycott B.B. (1991) Functional architecture of the mammalian retina. *Physiol. Reviews* 71(2): 447-480.
[9] Williams D.R., MacLeod D.I.A., and Hayhoe M.M. (1981) Foveal tritanopia. *Vision Research* 21: 1341-1356.

112

NEURONS AND NETWORKS WITH ACTIVITY-DEPENDENT CONDUCTANCES

Zheng Liu, Mike Casey, Eve Marder, and L. F. Abbott

Volen Center for Complex Systems
Brandeis University
Waltham, Massachusetts 02254

ABSTRACT

Cultured lobster stomatogastric ganglion (STG) neurons appear to regulate their intrinsic properties in an attempt to maintain stable activity patterns. Previous theoretical descriptions of activity-dependent regulation resulted in highly restricted sets of maximal conductances that involved a number of free parameters. We first explore a single-compartment model with a conductance regulation scheme that is guided by calcium-dependent feedback pathways that span multiple time scales. The exploration of maximal conductance parameters in the model is completely unrestricted and the number of parameters is greatly reduced. We then explore the interplay between network interactions and single cell regulation in a network model of simple cells each having a single calcium feedback mechanism. It is shown that when the cells are properly coupled such a network is able to produce an oscillatory pattern that requires interaction between the elements, but that when the cells are uncoupled, they become capable of oscillating independently.

Activity-dependent changes in neuronal properties are involved in functionally significant phenomena ranging from ion channel development to synaptic modification.[1,2] Experimental work on cultured STG neurons of the spiny lobster, *Panulirus Interruptus*, demonstrated that the intrinsic properties of these neurons are affected by their recent history of activity.[3] These neurons spontaneously change their intrinsic activity from tonic firing to bursting after long-term isolation in culture. When stimulated for a prolonged time, they revert to tonic firing behavior. The role of activity-dependent phenomena has not yet been experimentally explored on the network level in the STG. We have built networks with activity-dependent cells to explore the issues that arise in networks of these cells and to generate experimentally verifiable hypotheses. We have explored the interplay between the slow time scale regulation of system properties and system behavior by focusing on two questions. First, how does a single STG neuron regulate its conductances to stabilize different activity patterns? Second, what role do network interactions play in determining and coordinating the intrinsic properties of each element to produce the pyloric rhythm?

Computational Neuroscience
edited by Bower, Plenum Press, New York, 1997

Guided by the experimental data, we first built a conductance-based single compartment model with dynamically regulated conductances that accurately reproduces the sequences of excitability changes seen in cultured STG neurons. Previous models with activity-dependent conductances used the intracellular Ca^{2+} concentration to control modification of the maximal conductance parameters that determine the strengths of membrane currents.[4,5] These models suffer from a rather severe limitation; the equilibrium maximal conductances always end up lying on a single line in the space of maximal conductance values. We have now reformulated the dynamics describing the activity-dependent maximal conductances so that the equilibrium solutions are completely unrestricted. This increased freedom required us to construct a more elaborate feedback signal, although one still based on Ca^{2+} dynamics.

The intracellular Ca^{2+} concentration does not always provide an unambiguous indication of the activity of a neuron. For instance, a fast tonic firing pattern and a burst firing pattern with a long inter-burst interval can produce the same mean Ca^{2+} concentration. To specify uniquely the temporal properties of an activity pattern, information about both its high and low frequency components is needed. The feedback signal we used before, the Ca^{2+} concentration, is essentially a low frequency filter and does not give much information about the high frequency component of the activity.

Presumably, intracellular Ca^{2+} does not affect conductances directly but rather through a molecular cascade. This cascade may act as either a low or a high pass filter. In the present model, we use three feedback signals based on Ca^{2+} that span multiple time scales, a high frequency (or fast) sensor F, a low frequency (or slow) sensor S and a DC sensor D. These signals can be constructed in biophysically plausible ways. We represent these cascades as sensors acting on the total Ca^{2+} current i_{Ca}, normalized to total membrane capacitance. The equations for the sensors have similar forms (Fig 1). The value of S depends on the slow wave envelope of the voltage oscillation. The value of F reflects fast events, like the Na^+ spikes on top of the slow wave oscillation. The value of D reflects the DC component of Ca^{2+} current, and indirectly, the average level of depolarization of the voltage oscillation. These non-linear sensors are constructed to distinguish reliably the activity features that occur on different time scales.

The currents of the model are modified from previous work[6] based on voltage-clamp data from cultured STG neurons. The model has a single compartment and seven active

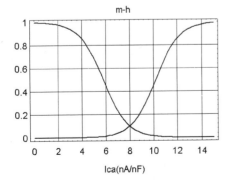

$$S = 3m_S^2 h_S \,, \tau_m=50\text{ms}, \tau_h=60\text{ms}$$

$$F = 10m_F^2 h_F \,, \tau_m=0.5\text{ms}, \tau_h=1.5\text{ms}$$

$$D = m_D^2 \,, \tau_m=500\text{ms}$$

$$\tau_m \frac{dm}{dt} = m_\infty - m \,, \tau_h \frac{dh}{dt} = h_\infty - h$$

Figure 1. Sensors F, S and D depend on the total calcium current and have forms similar to Hodgkin-Huxley equations. The equations for the sensors are shown on the right side of the figure. The figure shows m_∞ and h_∞ for the fast sensor. The crossing point of m_∞ and h_∞ for the slow sensor is shifted to (5, 0.1) and m_∞ for the DC sensor is shifted 2nA/nF further to the left.

currents: fast Na$^+$(I_{Na}), delayed rectifier K$^+$ (I_{Kd}), high-threshold Ca^{2+} (I_{CaP}), transient Ca^{2+} (I_{CaT}), Ca^{2+}-dependent K$^+$ (I_{KCa}), fast transient K$^+$ (I_A), hyperpolarization-activated inward (I_h), as well as a passive leak current with G_{leak} = 0.01nS/nF and E_{leak} = -50mV. Calcium enters the cell through membrane Ca^{2+} currents and is removed by a buffering system modeled as exponential decay (decay time constant is 20ms).

The seven maximal conductances are regulated according to the equation:

$$\tau \frac{dg_i}{dt} = \left[\pm(F - F_T) \pm (S - S_T) \pm (D - D_T) \right] \overline{g}_i$$

(1)

F_T, S_T and D_T are the target average value of the fast and slow sensors. τ is the time constant for conductance changes and is longer than any other dynamic time scale in the model.

The signs in equation (1) are determined by the function of the individual currents in producing burst firing pattern and are given in Table 1. When no sign is given, the corresponding term is absent. Equation (1) reaches equilibrium when F = F_T, S = S_T and D = D_T. The target average values (F_T = S_T = D_T = 0.1) are set so that when this occurs, the neuron displays the desired type of activity, namely burst firing. Note that this, in no way, limits the values of the maximal conductances and that all seven conductances are controlled by only these three parameters.

We tested the model against the results from experimental work on cultured STG neurons of the spiny lobster, *Panulirus interruptus*.[3,6] These neurons spontaneously changed their intrinsic activities after long-term isolation in culture. Voltage clamp data showed that the changes were accompanied by a decrease in outward current densities and an increase in inward current densities.[6] The model with the new regulation rule accurately reproduces the sequences of excitability changes seen in cultured STG neurons (Fig. 2). The new regulation rule eliminates a large number of parameters that existed in the original models and provides the model with the freedom to explore a much larger range of conductance strengths. Fig 3 shows the conductance distributions of model neurons that exhibit bursting behavior at equilibrium. Note that these models can achieve the same approximate activity pattern using a fairly wide range of conductances.

To explore the roles that cellular regulation plays in a network, we have built a three cell network of model neurons with activity dependent conductances. Slow time-scale regulatory processes will, in general, undermine the ability of a network to act as a functional unit (by making the cells in the network more autonomous and potentially destabilizing otherwise stable behaviors), but they provide an adaptive advantage by allowing partial recovery from unusual environmental perturbations. Our network was built around behavioral constraints loosely given by the pyloric network of the STG, and therefore was required to produce a stable triphasic rhythm. Previous experimental results further informed the construction of the model by requiring that only one of the network's three cells bursts intrinsically (the AB/PD cell), but that all three cells eventually become bur-

Table 1.

Current	I_{Na}	I_{CaP}	I_{CaT}	I_{Kd}	I_{KCa}	I_A	I_h
Sign for F	+			+			
Sign for S		+	+	−	−	−	+
Sign for D					−	−	+

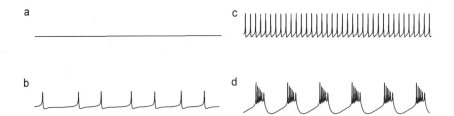

Figure 2. Voltage traces of a typical sequence of model activity that reproduces the spontaneous intrinsic activity changes seen in cultured STG neurons. Trace a) shows the initial behavior of the neuron. Traces b) and c) show the behavior as the neuron evolves. Trace d) shows the final equilibrium behavior.

sters when isolated.[3] The other two cells (the LP and PY cells) were required to act as followers in the network (and therefore to fire tonically for a significant period of time when removed from the network), but to burst at equilibrium when isolated. Together we will say that a three cell network is a *robust conditional burster network* (RCBN) if it satisfies all of these constraints. Note that being a RCBN implies the existence of at least one slower timescale process.

To focus on the issues related to network properties, we returned to a simple, one sensor version of a dynamically regulated maximal conductance cell. The cells that we use in the network model have two compartments – a soma/neurite and a spike generating zone. The main compartment has an activating/inactivating calcium current and a large potassium current. The maximal conductances of these currents are regulated by the calcium sensor. The second compartment contains an activating/inactivating sodium current and a delayed rectifier, neither of which is regulated by the calcium. All cells were identical, except for their coupling patterns, coupling strengths and their target calcium levels. The coupling patterns and strengths were chosen to reflect the connectivity of the pyloric network, while the target calcium levels were chosen to give each cell a different response to synaptic perturbations. Limiting the differences between cells allowed us to focus on network properties, and to explore the amount of asymmetry required in the network parameters to accommodate the asymmetric behavioral constraints.

Such a network was capable of producing a RCBN in two ways. The aspect of the network that is critical for being a RCBN is that the cells which are to be followers in the network and bursters out of the network must be entrained to a frequency that is much higher or lower than their intrinsic frequencies. Consider the case where a cell is entrained to a frequency much higher than its intrinsic frequency. Such a cell will be bursting more than it would at equilibrium, and since the slow-wave component of its burst will be pri-

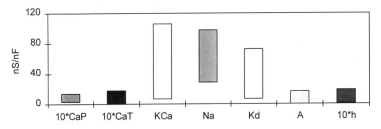

Figure 3. The range of equilibrium maximal conductances (less top and bottom 5%) of 1000 model neurons starting with random conductances.

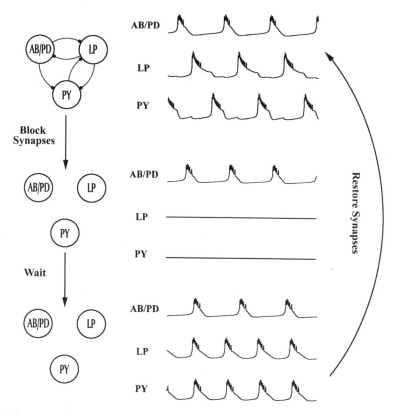

Figure 4. Voltage traces from the somata in the coupled and uncoupled network models showing that the model is a robust conditional oscillator network.

marily due to its calcium current, it will be getting more intracellular calcium than it would at equilibrium, and therefore will down regulate its maximal calcium conductance. It is then possible that it will down regulate its maximal conductance until it no longer bursts autonomously, but will burst due to post-inhibitory rebound when periodically inhibited (i.e. when in the network). Hence when removed from the network it will initially be silent, but then will not get enough intracellular calcium and will, therefore, up regulate its maximal conductance until it begins to burst autonomously. A symmetric case exists where the cell is entrained to a frequency that is much lower than its intrinsic frequency, in which case it will potentially up regulate its maximal calcium conductance until it no longer bursts. Figure 4 shows such a network.

In this paper we have explored several properties of models of cells with activity regulated maximal conductances. First, we have examined the sufficiency of previously proposed calcium feedback schemes for regulating maximal conductances and have proposed new feedback schemes that more accurately accommodate experimental results and allow for more flexible recovery from a wider number of environmental perturbations. Second, we have begun exploring the interplay between cellular regulation and network interactions in producing a fault-tolerant network oscillator. It is our hope that these theoretical investigations will lead to interesting, empirically verifiable hypotheses and a deeper understanding of cellular regulation in general.

ACKNOWLEDGMENTS

Research supported by the Sloan Center for Theoretical Neurobiology and Brandeis University, MH 46742 and the W.M. Keck Foundation.

REFERENCES

1. P. Linsdell and W.J. Moody, 1994, Na$^+$ Channel Mis-expression Accelerates K$^+$ Channel Development in Embryonic *Xenopus laevis* Skeletal Muscle. *J. Physiol.* **480.3**:405–410.
2. J.H. Byrne and W.O. Berry, 1989, *Neural Models of Plasticity*, Academic Press, San Diego.
3. G. Turrigiano, L. F. Abbott and E. Marder, 1994, Activity-dependent Changes in the Intrinsic Properties of Cultured Neurons. *Science* **264**, 974–977.
4. G. LeMasson, E. Marder and L. F. Abbott, 1993, Activity-dependent Regulation of Conductances in Model Neurons, *Science* **259**, 1915–1917.
5. L. F. Abbott and G. LeMasson, 1993, Analysis of Neuron Models with Dynamically Regulated Conductances, *Neural Comp.* **5**, 823–842.
6. G. Turrigiano, G. LeMasson and E. Marder, 1995, Selective Regulation of Current Densities Underlies Spontaneous Changes in the Activity of Cultured Neurons, *J. Neurosci.* **15**, 3640–3652.

113

ESTIMATION OF SIGNAL CHARACTERISTICS DURING ELECTROLOCATION FROM VIDEO ANALYSIS OF PREY CAPTURE BEHAVIOR IN WEAKLY ELECTRIC FISH

Malcolm A. MacIver, John L. Lin, and Mark E. Nelson

The Neuroscience Program and
Beckman Institute for Advanced Science and Technology
University of Illinois
Urbana-Champaign Urbana, Illinois 61801

INTRODUCTION

Weakly electric fish can actively influence the strength and spatiotemporal patterns of incoming electrosensory signals by controlling the velocity and orientation of their body and by adjusting the gain and filtering properties of neurons in the electrosensory lateral line lobe (ELL) via descending control (review: Bastian 1995). To better understand the signal conditions under which the active electric sense normally operates, we have undertaken a set of behavioral studies aimed at characterizing how electric fish use their electrosensory system to locate and capture small prey. Key questions we are pursuing include the range of detectability for small prey; the typical signal magnitude of prey within this range; the typical velocity of the prey relative to the receptor array; the movement strategies of the fish during prey search, localization, and strike phases of behavior; and the spatiotemporal patterns of receptor activation during prey capture behavior and how these patterns relate to the filtering properties of sensory neurons in the ELL. Answers to these questions will be used to guide and constrain our electrophysiological and neural modeling studies.

In general, we find that under natural conditions the electrosensory system operates at signal levels that are much lower than those typically used in electrophysiological studies to characterize the system. We hope that by focusing our attention on the signal processing required to extract small natural signals, we can gain a better understanding of the active components of sensory acquisition and of the adaptive filtering task of the ELL. In this paper we describe how we record and analyze the sensory acquisition behavior of a South American weakly electric fish (*Apteronotus albifrons*) during capture of small aquatic prey *(Daphnia magna)*, and how we estimate the spatiotemporal pattern of transdermal potential and afferent firing rate changes during this challenging electrosensory task.

Computational Neuroscience
edited by Bower, Plenum Press, New York, 1997

METHODS

These studies were conducted with adult *Apteronotus albifrons* (black ghost) weakly electric knife fish, 12–16 cm in length, maintained in water of 250 ± 25 mS conductivity at $27 \pm 1.0°C$, and pH 6.9 ± 0.2, on a 12-hour light/dark cycle. For prey we used mature (2–3 mm) *Daphnia magna* (water fleas) that were cultured in our laboratory for this purpose.

Estimation of Prey Detection Distance

Behavioral Data Acquisition. Prey search and prey capture behavior was recorded within an infrared videotaping setup (see Figure 1). Fish were housed in a rectangular Plexiglas aquarium (77 x 30 x 20 cm) with a central area partitioned from the rest of the aquarium to form an arena (40 x 30 x 20 cm) which constrained fish-prey interactions to the region imaged by our video cameras. Four fish were held in individual holding bays that were electrically insulated from the arena in which behavior was observed; prior to recording, one fish was allowed into the recording arena through a Plexiglas door. Prey were introduced into the aquarium one at a time using a narrow tube with minimal mechanical disturbance and without introduction of visible light.

Infrared Videotaping. For this study we developed an infrared videotaping setup (Figure 1A) to record behavior that has no visually-mediated component. Two black and white cameras provided top and side views. The two video cameras were coupled and their output sent to a commercial video splitter to merge the top and side views of the tank. The merged images were recorded on videotape. Infrared illumination was provided by two custom-fabricated IR light sources, each consisting of 100 high-intensity infrared diodes (radiant power 35 mW, 880 nm). The illuminators, cameras, and aquarium were within a light-tight enclosure.

Video Digitizing and Trajectory Reconstruction. Videotaped records of prey capture behavior were visually scanned to identify segments to be digitized. Segment selection was based on several criteria: 1) the segment included an orienting response toward the

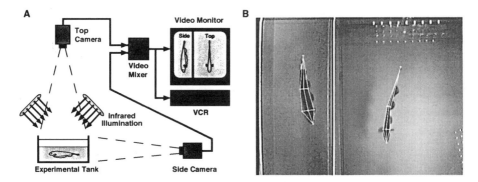

Figure 1. Behavioral recording setup and video reconstruction. (A) Schematic diagram of two-camera infrared video setup. The experimental tank and cameras were housed within a light-tight enclosure. (B) One picture of raw video data with an illustrative wireframe model overlaying the fish. The actual wireframe mesh used has 7 nodes per (top and side) view to represent the fish, and 1 node to represent the prey.

prey, either a lunge or a change in swimming pattern near the prey 2) the prey had to be located on the sides of the fish closest to the top and side view cameras, as otherwise our view of the prey was blocked; 3) the trajectory of the fish must have been at least 4 cm away from the bottom and sides of the tank, to minimize distortions in the fish's electric field that occur near boundaries. We used a Peak Performance Technologies motion measurement system for digitizing videotape of behavior. The sampling rate was 60 pictures per second, where a picture is one video field with the alternate scan lines interpolated. For each digitized image, a wireframe mesh was constructed to indicate the fish body position. The fish mesh had 7 nodes in each view (top and side), corresponding to the tip of the snout, the right and left pectoral fins at their anterior insertion points, a ventral and dorsal surface point midway between the pectoral fins, the midpoint of the fish on the central long axis, and the tip of the caudal fin. Figure 1B shows a typical picture of the raw video data used during digitization, overlaid with a sample wireframe mesh. A single point was used in each view to mark the location of the prey. The coordinates of the wireframe nodes were then used to reconstruct the 3-D trajectory of fish and prey during prey search, detection, and capture phases of behavior.

Data Analysis and Interpretation. Data analysis was carried out using MATLAB running on Sun workstations. Reconstructed node trajectories were digitally low-pass filtered to attenuate jitter due to small variations in manual point placement across pictures, then combined to reconstruct the 3-D trajectories of the fish and prey. The trajectories were analyzed to extract prey position, prey velocity, fish position, fish velocity, fish acceleration, the distance between the prey and the closest point on the fish body, and the distance between the prey and the fish mouth. The putative time of detection was identified using two criteria: 1) a longitudinal acceleration peak exceeding a threshold value, and 2) an orienting response which was terminated with a lunge towards the prey. Analysis of post-detection events was carried out using observations of the raw videotape as well as quantitative measures obtained from the digitized trajectories.

Reconstruction of Spatiotemporal Patterns of Transdermal Potential and Afferent Activity. When visualizing changes in transdermal potential we represented the unfolded bilateral electroreceptor surface of the fish as an idealized rectangular grid (Figure 3). The projected coordinates of the *Daphnia* were transformed into this representation. When estimating changes in transdermal potential, we treat the *Daphnia* as a 2 mm spherical insulating object. Based on the physics of electrosensory image formation (Rasnow, 1996) we then estimated spatial spread and signal intensity of the electrosensory "shadow" cast upon the fish at each time frame and calculated the transdermal potential change at each point on the idealized electroreceptor surface. We then used the estimated transdermal potential as the input to a linear-nonlinear cascade model of primary electrosensory afferent response dynamics (Xu et al. 1994) to estimate the associated changes in P-type tuberous electrosensory afferent activity during the prey-capture sequence.

RESULTS AND DISCUSSION

We reconstructed the 3-D trajectories of 81 prey capture sequences and extracted estimates of acceleration, velocity and distance of closest approach from each sequence. Figure 2A show a sample of how these parameters vary during a typical sequence. Initially the fish is swimming with a relatively constant forward velocity of about 10 cm/s,

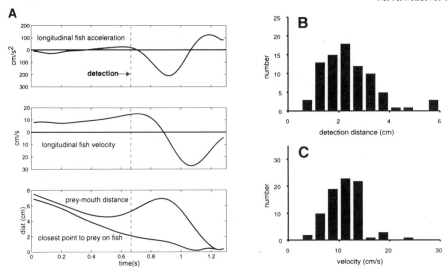

Figure 2. (A) The longitudinal acceleration, velocity, and distance between prey and fish during a typical prey capture. The estimated time of prey detection is indicated by the dashed vertical line at t = 0.7 s (B) Distance between prey and closest point on fish body surface at time of detection (mean 2.4 cm, s.d. 1.1 cm, N=81). (C) Relative velocity between fish and prey at time of detection (mean 11 cm/s, s.d. 3.6 cm/s, N=81).

then there is an abrupt longitudinal deceleration (t = 0.7–1.0 s) and longitudinal velocity reversal (t = 0.9 s) as the fish backs up to capture a *Daphnia* that has been detected by electroreceptors on the trunk. The abrupt deceleration and velocity reversal are characteristic of most prey strike sequences we have recorded. Based on analysis of the longitudinal acceleration profile we estimate that prey detection occurred near the onset of the deceleration at t = 0.7 s (dashed vertical line). At the time of detection, the fish was moving with a forward velocity of 14.6 cm/s and the prey was approximately 2.1 cm from the closest point of the fish's body, which was in the mid-trunk region for this detection event. Preliminary analysis of 81 such sequences indicates that the typical detection distance for small prey (*Daphnia*, 2–3mm diameter) is approximately 2.4 cm (Figure 2B); the typical fish-prey velocity at the time of detection is about 11 cm/s (Figure 2C).

From the reconstructed trajectories we then estimate the change in transdermal potential at 100 ms intervals as shown in Figure 3A. Based on the physics of electric image formation for spherical objects (Rasnow 1996) and approximating the *Daphnia* as a perfect insulator due to its non-conducting carapace, we compute the peak amplitude and full-width at half maximum for the electric image the Daphnia casts on the electroreceptor array at each time point. Figure 3A shows the resulting pattern of transdermal potential change for the final half (0.6–1.2 s) of the prey capture sequence shown in Figure 2A. Note that the electric image is weak and diffuse at the beginning of the sequence and becomes both more intense and more tightly focused as the *Daphnia* gets closer to the electroreceptor array. At the time of detection (t = 0.7 s), the estimated peak transdermal potential modulation is about 0.1 μV RMS and just before prey capture (t = 1.2 s) the peak modulation is about 5.0 μV RMS.

Based on the estimated transdermal potential change, we then compute the corresponding change in afferent firing rate based on a model of P-type electrosensory afferent response dynamics that we have developed (Xu et al., 1994). The afferent model is a linear-nonlinear cascade model consisting of a second order linear model that describes the

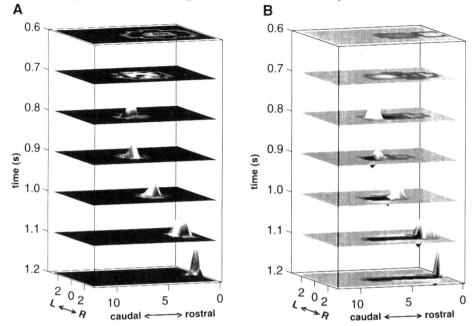

Figure 3. (A) Spatiotemporal pattern of the change in transdermal potential during the last segment (0.6–1.2 s) of the prey capture event shown in Figure 2. (B) Corresponding pattern of P-type afferent firing rate changes. Note the trailing depression in firing rate in planes from t=0.8 s to the end of the sequence (dark patches). This is due to the depression in afferent firing rate that occurs at the offset of a stimulus.

frequency dependence of the gain and phase of the response, in series with a static non-linearity that incorporates the effects of firing rate rectification and saturation. Figure 3B shows the change in afferent firing rate corresponding to the change in transdermal potential shown in Figure 3A. At the time of detection (t = 0.7 s), the estimated peak rate modulation on an individual afferent is only 0.2 spike/s, but since the electric image is spatially diffuse, approximately 600 P-type electrosensory afferents are influenced simultaneously. One unexpected result from this analysis is the trailing suppression of afferent firing rate (dark regions in Figure 3B) as the stimulus moves across the surface of the fish's skin. This trail of suppressed activity, which arises from the response dynamics of the afferents, provides a short-term spatial memory of the trajectory history which could possibly be used centrally to aid in target tracking computations.

Using natural prey-capture sequences to estimate electrosensory signal characteristics has provided us with a much better understanding of the conditions under which the electrosensory system normally operates. While *Daphnia* are probably not a prey item in the native habitat of *A. albifrons*, they are similar in size to prey found in stomach-content analyses of this species in their natural environment (M. Hagedorn 1996, unpublished data). We have not yet demonstrated that prey detection is mediated by the active electric sense alone, since we do not know the extent to which sensory modalities other than the tuberous (active) electrosensory system, most importantly the mechanosensory lateral line and low-frequency ampullary (passive) electrosensory system, may contribute to prey detection and localization. In many cases, the apparent detection event was followed by backward scanning, dorsal edge scanning, and tail-bending behaviors, which are suggestive of active electrolocation. We are about to undertake a study that will assess the relative contributions of the active electrosense, passive electrosense and mechanosensory lateral line to prey capture behavior. We are confident that there is no visually-mediated component, since we use a wavelength of illumination (880 nm) which is well beyond the range of teleost photoreceptors (Fernald, 1988) and our fish show no startle response when the infrared illuminators are switched on. The long term aim of behavioral studies is to provide quantitative measurements of the spatiotemporal patterns of activation during active electrolocation, which we will compare with the spatial and temporal filtering properties of primary afferents and sensory neurons in the ELL.

This research is supported by grant R29MH49242 from NIMH.

REFERENCES

Bastian, J. (1995) Electrolocation. In: *The Handbook of Brain Theory and Neural Networks*, (Arbib M, ed.), pp 352–356. Cambridge: MIT Press.

Carr, C. E., L. Maler, and E. Sas (1982). Peripheral organization and central projections of the electrosensory nerves in Gymnotiform fish. *Journal of Comparative Neurology* (2): 139–153.

Fernald R.D. (1988) Aquatic adaptations in fish eyes. In: *Sensory Biology of Aquatic Animals* (Atema J, Fay R.R., Popper A.N., Tarolga W.N., eds), pp 433–466. New York: Springer-Verlag.

Rasnow B. (1996) The effects of simple objects on the electric field of *Apteronotus*. *Journal of Comparative Physiology A* 178:397–411.

Xu, Z. and Payne, J.R. and Nelson, M.E. (1994). System identification and modeling of primary electrosensory afferent response dynamics. *Computation in Neurons and Neural Systems*, Eeckman, F., ed. Kluwer Academic Press, 197–202.

114

THE INTEGRATION OF PARALLEL PROCESSING STREAMS IN THE SOUND LOCALIZATION SYSTEM OF THE BARN OWL

James A. Mazer

Division of Biology, 216-76
California Institute of Technology
Pasadena, CA
Current Address:
MIT, E25-634
45 Carleton St.
Cambridge, MA
Phone: 617-253-5792
FAX: 617-253-8943
E-mail: mazer@ladyday.mit.edu

ABSTRACT

The barn owl, *Tyto alba*, exhibits both peripheral and central nervous system specializations that enable it to accurately localize sounds in space. Interaural time difference (ITD), for azimuthal localization, and interaural level difference (ILD), for vertical localization, are computed in parallel by two independent pathways. These pathways remain anatomically and physiologically segregated up to the level of the lateral shell of the central nucleus of the inferior colliculus (LS). Electrophysiological mapping studies of the LS were performed to characterize the mechanisms underlying the convergence of the ITD and ILD processing streams. These studies suggest that the space specificity observed in the owl's auditory midbrain is an emergent property of a hierarchically organized feed-forward network arranged anatomically across the mediolateral extent of the LS.

1. REPORT

Behavioral studies of sound localization ability in the barn owl, *Tyto alba*, have demonstrated that the owl uses interaural time difference (ITD) cues for auditory localization in the horizontal place and interaural level difference (ILD) for localization in the vertical plane [3,

]. ITDs arise when sound sources are displaced horizontally from the midline, leading to a difference in path length, and corresponding travel time, between the source and each ear. In the barn owl, ILDs occur when a sound source is located above or below the horizontal meridian. The external ears of the barn owl exhibit a distinct left-right asymmetry, the result of which is that the right ear is maximally sensitive to sounds from high elevations and the left ear to sounds from low elevations. This asymmetry means that that interaural level cues encode information about sound source elevation. Anatomical and physiological studies have demonstrated that ITD and ILD cues are processed by independent processing streams originating at the level of the cochlear nuclei [4, 5, 6, 1,]. These two processing streams are both anatomically and physiologically segregated up to the level of the inferior colliculus (IC) in the auditory midbrain.

"Space specific" neurons have spatially restricted auditory receptive fields; these neurons first appear in the external nucleus of the IC [2,]. The formation of these space specific neurons requires convergence of the time and intensity processing streams at the level of the single neuron. Space specific neurons perform an additional computation, namely, the elimination of phase ambiguity. Phase ambiguity arises as a consequence of the narrow band frequency tuning observed in the coincidence detection neurons located in the nucleus laminaris (NL). Single NL neurons are sensitive only to interaural phase difference and are therefore unable to distinguish between ITDs differing only by multiples of the period of their best frequencies. This ambiguity is eliminated at higher levels in the auditory system by integrating across multiple frequency channels to transform a representation of interaural phase difference into one of true interaural time difference.

The first site at which the time and intensity processing streams converge anatomically is the lateral shell (LS) subdivision of the central nucleus of the inferior colliculus [5, 6,]. In order to characterize the mechanisms underlying the integration of the parallel time, intensity and spectral processing channels, I quantified auditory tuning properties related to sound localization in single LS neurons and correlated the distribution of these tuning properties with anatomical position within the LS.

Three possible models for pathway convergence in the lateral shell are depicted in figures 1-3. Each of these models makes specific predictions about the distribution of tuning properties one would expect to observe in the LS. The single step convergence model shown in figure 1 predicts that neurons would exhibit either space specific or "primary-like" tuning, but nothing intermediate. In contrast, the separate processing model, shown in figure 2, predicts intermediate tuning properties, but single neurons sensitive to only time or only level cues, and no neurons sensitive to both. Finally, figure 3 shows a hierarchical model which predicts both intermediate tuning properties and a large percentage of neurons sensitive to both ITD and ILD.

1.1. Methods

I recorded from 423 well isolated single neurons located in the LS region of the inferior colliculus in eight adult barn owls using conventional single neuron recording techniques in Ketamine anesthetized animals (25mg/kg; Ketaset, Fort Dodge). LS neurons were identified based on a combination of stereotaxic coordinates, auditory tuning properties and histological reconstruction of electrode tracks. Auditory stimuli were presented dichotically and tuning curves were measured for each localization cue by systematically varying the cue value, while holding all other stimulus parameters constant. Each LS neuron was characterized by measuring frequency tuning (best frequency and half-maximal tuning width), ITD tuning (best

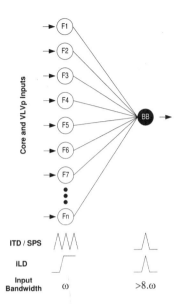

Figure 1. Single Step Convergence: One of several possible models for the integration of ITD, ILD and spectral cues in the LS. The schematic tuning curves indicate ITD and ILD tuning for neurons at particular levels in the model and ω indicates the input bandwidth of the narrow band coincidence detectors located in nucleus laminaris. In this particular model, all processing channels converge in a single step onto single target LS neurons.

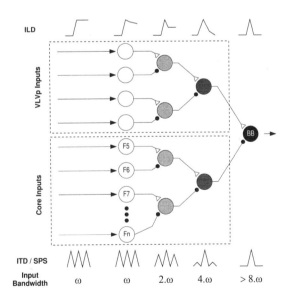

Figure 2. Separate Processing: ITD and ILD cues are processed by separate populations of LS neurons. Though the figure indicates anatomical segregation of the time and intensity streams, this need not necessarily be the case.

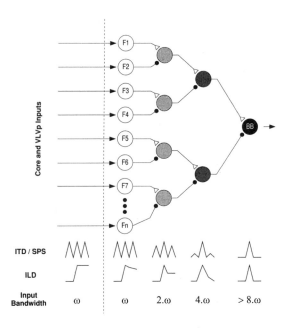

Figure 3. Hierarchical Convergence: The site of ITD and ILD convergence lies "early" in the LS, but generation of space specific neurons (BB, broad band), requires a refinement of tuning properties. In this model, the refinement is the result of a feed-forward network.

ITD, width of ITD tuning and degree of phase ambiguity) and ILD tuning (best ILD, width of ILD tuning). These tuning properties where then correlated with anatomical position along the mediolateral (ML) axis of the LS.

1.2. Results

The distributions of the various tuning properties generally associated with space specificity were found to be highly correlated with cell position along the ML axis of the LS. Response latencies in the LS varied systematically with the anatomical location of the neurons along ML extent of the shell ($r = 0.49$, $p < 0.0001$, n=420), with medially located neurons having latencies of around 5ms and lateral neurons latencies of >10ms. Frequency tuning widths were also positively correlated with ML position ($r = 0.32$, $p < 0.0001$, n=368). In contrast, both ITD and ILD tuning widths were negatively correlated with ML position ($r = -0.230$, $p < 0.0001$, n=357 and $r = -0.144$, $p = 0.010$, n=317, respectively). The degree of phase ambiguity was correlated with both ML position ($r = 0.399$, $p < 0.0001$, n=354) and response latency ($r = 0.457$, $p = 0.0001$, n=323), consistent with the idea that phase ambiguity is eliminated over the course of several synaptic interactions distributed systematically along the ML axis. Finally, 92% (389/423) of LS neurons exhibited some degree of sensitivity to both time and intensity cues. Of the remaining neurons 6% (25/393) were sensitive to ILD only and 2% (9/393) to ITD only.

1.3. Conclusions

These data are most consistent with the model presented in figure 3, in which there exists a feed-forward network arranged along the mediolateral axis of the LS responsible for smoothly refining the auditory tuning properties of LS neurons. This network results in the formation of space specific neurons with spatially restricted auditory receptive fields. The smooth continuum of tuning properties related to sound localization reported here argues strongly against a model based on one-step convergence model (see figure 1). Likewise, the fact that more than 90% of LS neurons appear to be sensitive to both time and intensity cues is inconsistent with a model in which ITD and ILD are processed separately within the shell (see figure 3).

REFERENCES

[1] Adolphs, R. (1993). Acetylcholinesterase staining differentiates functionally distinct auditory pathways in the barn owl. *J. Comp. Neurology*, 329(3):365–377.
[2] Knudsen, E. I. and Konishi, M. (1978). A neural map of auditory space in the owl. *Science*, 200:795–797.
[3] Moiseff, A. (1989). Bi-coordinate sound localization by the barn owl. *J. Comp. Physiol. A*, 164(5):637–644.
[4] Takahashi, T., Moiseff, A., and Konishi, M. (1984). Time and intensity cues are processed independently in the auditory-system of the owl. *J. Neurosci.*, 4(7):1781–1786.
[5] Takahashi, T. T. and Konishi, M. (1988a). Projections of nucleus angularis and nucleus laminaris to the lateral lemniscal nuclear-complex of the barn owl. *J. Comp. Neurology*, 274(2):212–238.
[6] Takahashi, T. T. and Konishi, M. (1988b). Projections of the cochlear nuclei and nucleus laminaris to the inferior colliculus of the barn owl. *J. Comp. Neurology*, 274(2):190–211.

115

EXPANSION AND SHIFT OF HIPPOCAMPAL PLACE FIELDS: EVIDENCE FOR SYNAPTIC POTENTIATION DURING BEHAVIOR

Mayank R. Mehta and Bruce L. McNaughton

ARL Division of Neural Systems, Memory and Aging
Department of Psychology
University of Arizona
Tucson AZ
E-mail: Mayank@NAMA.Arizona.EDU
E-mail: Bruce@NSMA.Arizona.EDU

ABSTRACT

Rat hippocampal neurons fire in a spatially selective fashion [1]. We show that place fields enlarge (by 75%) and shift (by $1.4cm$) in a direction opposite to the direction of movement of the rat, within a few traverses of a route, even if the environment has been experienced extensively on previous days. The expansion is not a result of locomotion or neural activity per se because it reoccurs when the rat runs on a different track immediately after running on the first one. This provides an evidence for systematic changes in neuronal firing properties due to and during experience. The results are consistent with the predictions of models [2, 3] of learning of sequences via Hebbian [4] synaptic potentiation. Thus, these data provide an evidence for Hebbian synaptic enhancement during behavior, and show that such learning occurs even when a rat enters a highly familiar environment after a day's absence.

Although it is a common belief that learning occurs via Hebbian long term potentiation (LTP) of the synapses, there is little evidence to suggest that such changes indeed occur during behavior. With the present technology, it is not possible to measure the strength of individual synapses during behavior. Most of the experiments on LTP involve artificial stimulation of brain tissue. These experiments show that the NMDA receptor mediated LTP is associative [5] and temporally asymmetric [6] –i.e. it occurs only if the postsynaptic neuron is depolarized within a short duration ($\sim 100ms$) *after* the spiking of the presynaptic neuron. It is not obvious that the anatomical connections between the neurons and the pattern of neuronal firing during behavior is indeed such that the pre and the post synaptic neurons have the required order of activation so as to strengthen the synapse. Thus it is important to investigate whether Hebbian LTP occurs during behavior. Given the present technological constraints, one has to devise

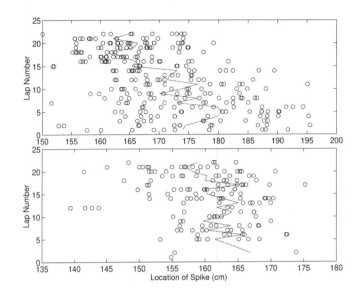

Figure 1. The location of the rat when two isolated cells (top and bottom panels) fired a spike is indicated by open circles, as a function of the lap number. The mean value of the locations of all the spikes fired by a cell during a lap corresponds to the location of the place field at that time. The place field locations during successive laps are joined by a solid line. The average value of all the spikes across *all the laps* corresponds to the place field center and is indicated by a light vertical line. The amount of firing per lap increased with time. In fact the cell in the bottom panel did not fire at all during some of the early laps. Further, the lap specific place field center shifted backwards.

indirect measures which would allow us to detect the consequences of changes in synaptic strengths. We have therefore looked for systematic changes in neuronal firing properties during behavior, which could arise due to Hebbian LTP.

The majority of experiments on LTP have been carried out on neurons of the Hippocampus. When a rat explores an environment, the neurons in the CA1 and CA3 regions of the hippocampus are known to fire in a spatially selective fashion [1] and are often called place cells. The region of space where a neuron fires selectively is called it's place field. A large number of neurons were recorded simultaneously [7] from two rats as they ran repeatedly in a counterclockwise direction on the perimeter of an elevated, rectangular track.

The closed linear track was mapped onto a straight line for the purpose of analyses, with the rats running in the direction of increasing distance. Fig. 1 shows the location of each spike, from two different cells, as a function of the lap number. As can be seen, the amount of firing increased with experience and the location of the center of mass of the lap specific place field moved in a direction opposite to the direction of movement of the rat.

However, these data are noisy. Few spikes (~ 10) were emitted by each cell during each lap. In order to do a statistically meaningful analysis of changes in place field properties, the 'ensemble averaged place field profile' of a population of cells was computed for each lap, as follows. For each place field, the location of the place field center (averaged over all the laps) was subtracted from the location of each spike, thereby obtaining the relative location of each spike. The firing rate as a function of the distance from the center, averaged across all the cells, gives the average place field profile of a population of cells. This profile was computed separately for each lap. Such an analysis corresponds to evaluating the ensemble average over all the place fields for each lap as opposed to the usual, time average over all the laps for each cell. The area under the 'ensemble averaged place field' curve is called the size of the place

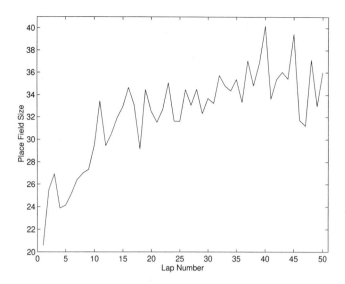

Figure 2. The size of the place field, averaged across 43 place fields is shown as a function of the lap number. The size increases by about 75% in fifty laps in a saturating way, with that most of the increase occurring within the first ten laps.

field for the given lap and the center of mass of the profile is called the location of the place field for that lap.

Figure 2 shows the average size of the ensemble averaged place field as a function of the lap number (i.e. time). The size of the place field increases by about 75% in a saturating way with experience. Further, the firing rate distribution shifts to the left (Fig. 3) such that the place field center in the last lap is shifted by 1.4*cm* with respect to that in the first lap in a direction opposite to the direction of motion of the rat.

These changes could not arise due to changes in rat's behavior because when some rats were run on a second maze immediately after running the first maze, the size of the place fields again increased and their locations again shifted backwards [8].

These data were recorded from the hippocampi of rats which were highly familiar with the maze. Therefore these changes must occur every time the rat enters a familiar environment after a day's absence.

Place cells are activated in the same temporal sequence, every time the rat follows the same route. Artificially induced, NMDA mediated LTP is temporally asymmetric, occurring only if the postsynaptic neuron is depolarized after and within a short duration of depolarization of the presynaptic neuron. Thus synapses between place cells will be strengthened asymmetrically, i.e. the synapses from the cells which fire earlier on the maze, onto those that fire later on the maze, would become stronger; however the reciprocal connections would be relatively unaffected. Such an asymmetric strengthening would cause the post-synaptic neurons to fire earlier, and more vigorously, thereby expanding the place fields and shifting their locations backwards [2].

Hippocampal neurons fire in a noisy fashion in a small region of the environment, thus providing a sparse spatial code. For such a network, expanded place fields would result in an improved signal to noise ratio and hence the accuracy of the hippocampal population code would increase, as was observed by by Wilson and McNaughton [9] in data from rats exploring

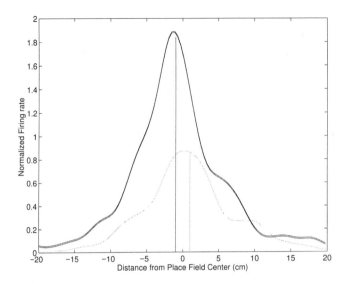

Figure 3. The ensemble averaged place field profile for the first(lower, light curve) and the last (upper, bold curve) laps are shown. The area under the curve is larger for the last lap than for the first lap and the firing rate distribution is shifted to the left in the last lap compared to that in the first lap. The center of mass of the two firing rate distributions (vertical lines) are 1.4*cm* apart.

novel environments. However, we observe place field expansion in the familiar environments too.

The backward shift in the neuronal firing means that the cell commences firing before the animal reaches the location that was originally coded for by that neuron. This could enable the animal to learn to navigate in an environment [2].

To conclude, we have shown that place fields are rapidly modified by experience, even when a rat enters a familiar environment after a day's absence. This supports the hypothesis that the hippocampus is involved in short term memory. Further, the present results provide an indirect evidence for asymmetric Hebbian synaptic strengthening, which may underlie learning of sequences during behavior.

ACKNOWLEDGMENTS

This work was supported by AG12609, MH01227, and MH46823. MRM was supported by a McDonnell-Pew Center for Cognitive Neuroscience Grant and by a fellowship from the Human Frontier in Science Program. We thank W.E. Skaggs, M.S. Suster, K.L. Weaver, J.L. Gerrard, and R. D'Monte for assistance with recording, and K. Stengel for engineering support.

REFERENCES

[1] O'Keefe, J., & Dostrovsky, *J. Brain Res.* 34, 171–175 (1971).
[2] Blum, K. & Abbott, L., *Int. J. Neural Systems* 6, 25–32 (1995); Abbott, L., & Blum, K., *Cerebral Cortex*, in press.
[3] Tsodyks, M. & Sejnowski, T., *Int. J. Neural Systems* 6, 81–86 (1995); Tsodyks, M., Skaggs, W.E., Sejnowski, T.J. & McNaughton, B.L., *Hippocampus*, in press.

[4] Hebb, D.O., *The Organization of Behavior*, Wiley, New York, 1986.

[5] McNaughton, B.L., Douglas, R.M. & Goddard, G.V. *Brain Res.* 157, 227–293 (1978).

[6] Levy, W.B. & Steward, O., *Soc. for Neurosci. Abs.* 8, 791–797 (1983); Gustaffson B. & Wigstrom, H., *J. Neurosci.* 6, 1575–1582 (1986); Markram H. & Sakmann, B., *Soc. Neurosci. Abs.* 21, 2007 (1995).

[7] For details about recording procedure see: Gothard, K., Skaggs, W.E. & McNaughton, B.L. *J. Neurosci.* 16, 832–836 (1996).

[8] Mehta, M. R. & McNaughton, B. L., Society for Cognitive Neuroscience abstract, March '96; Society for Neuroscience abstract # 734.15, 1872, Nov '96; submitted for publication.

[9] Wilson, M.A. & McNaughton, B.L., *Science* 261, 1055–1058 (1993).

116

FOURIER ANALYSIS OF INTERSEGMENTAL COORDINATION DURING FICTIVE SWIMMING IN THE LAMPREY

W. L. Miller and K. A. Sigvardt

Center for Neuroscience
University of California
Davis, California 95616
miller@itd.ucdavis.edu
kasigvardt@ucdavis.edu.

1. INTRODUCTION

The control of rhythmic behavior by the nervous system is a subject of common interest to neuroscientists, clinicians, and engineers. One of the primary models of the neurobiology of rhythmic movement is the system of spinal neurons responsible for locomotion in the lamprey, a primitive fish. The lamprey swims by generating lateral undulatory waves of muscle contraction along the body. Smooth and efficient progression of the bending wave from one part of the body to the next requires precise coordination of the segment-to-segment phase delay in neural activity in the spinal cord. In addition, the lamprey maintains approximately one wavelength of curvature on its body at all times when swimming, independent of its swimming speed. Thus, the intersegmental phase lag must be maintained at a constant value of about 1% of the full wave between each neighboring pair of segments in the 100-segment animal.

Our present goal is to determine the important features of neuronal coupling that lead to this precise intersegmental coordination. Our approach has been to investigate coupling between segments at the system-level using an *in vitro* preparation of lamprey spinal cord, and to compare our experimental results with predictions generated by our collaborators from analysis of a coupled-oscillator model of the system (e.g., Kopell and Ermentrout, 1986). To analyze the results of our recent experiments we have utilized Fourier analysis to characterize the range of experimentally induced effects on the performance of the locomotor system.

The isolated lamprey spinal cord can be induced to generate a stable oscillatory pattern of neural activity throughout its length by bathing the preparation in physiological saline containing excitatory amino acid (Figure 1a). The output motor rhythm, recorded with suction electrodes placed near motor nerves in selected segments, consists of regular bursts of action potentials. In the intact animal, each burst in the motor nerve triggers a contraction of the associated muscle fibers. This isolated rhythmic activity has been termed 'fictive swimming' since its analysis yields values of frequency, intersegmental phase and other characteristics of locomotion similar to those measured in the intact animal (Wallén and Williams, 1984). For instance, the characteristic phase delay between segments can be seen in Figure 1a by reference to the vertical gray line. Since fictive swimming takes place in the absence of descending or sensory input, the system of propriospinal neurons involved in controlling locomotion has been termed the 'lamprey locomotor central pattern generator (CPG)'.

We have recently performed a range of experiments on the isolated lamprey spinal cord in which the intersegmental coupling available to the locomotor CPG is systematically reduced. For example, Figure 1b illustrates an experiment in which the local coupling pathways in the middle of the preparation are blocked with a low calcium, high manganese saline solution confined with partitions placed over and around the spinal cord, while the ends of the preparation are bathed in saline containing the excitatory amino acid agonist D-glutamate (cf. Rovainen, 1985; Cohen, 1987). The low calcium solution blocks only local synaptic activity while still allowing passage of action potentials in any fibers which are long enough to span the blocked region. Thus, this experiment tests whether multi-segmental propriospinal fibers are by themselves capable of maintaining appropriate intersegmental coordination. By systematically moving the partitions to increase the number of blocked segments, we can also assess the collective functional length of the longer fibers.

2. EVALUATING FICTIVE LOCOMOTION AND INTERSEGMENTAL COUPLING

The traditional method of analyzing fictive locomotion has been to define the event times of the individual bursts in each recorded nerve (e.g., start, end, midpoint, etc.), and use these to estimate the rhythm cycle period, bursting duty cycle and intersegmental phase as, respectively, the average time between bursts, the average burst duration over the period, and the delay between bursts in different segments within the same cycle divided by the period (e.g. Wallén and Williams, 1984). However, we found that the burst event method was inadequate for our present analyses, for several reasons. First, our systematic reduction of coupling often degraded the fictive rhythm in complex ways (for example, greatly escalating the occurrence of spikes outside of bursts), which caused highly increased uncertainty and subjectivity in estimating burst start and end times. In essence, the act of imposing a 'square-wave' representation onto the degraded motor output pattern became itself an untenable source of measurement error. Second, we additionally needed a robust measure of the relative amount of rostro-caudal coupling, a value that is difficult to obtain confidently with the burst analysis method.

For our recent studies we have instead employed Fourier (frequency-domain) time-series analysis to estimate characteristics of the motor rhythm, in particular following the point-process (spike time) method of Brillinger (Rosenberg, et al. 1989). Fourier analysis seemed a natural choice given that frequency and phase are the basic variables under study, and that the squared coherence, a normalized frequency-domain correlation measure, is easily calculated and interpreted in terms of relative coupling. Furthermore, few interventions or subjective decisions are required during the analysis process, so that the potential for user-imposed bias is minimal.

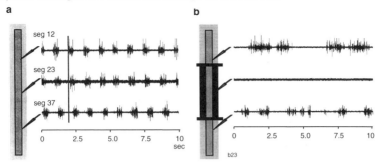

Figure 1. Fictive locomotion in one animal during the experiment described in the text. a. Fictive locomotion in the control trial. b. The rostral and caudal motor rhythms during block of the synaptic activity in 20 middle segments of the preparation.

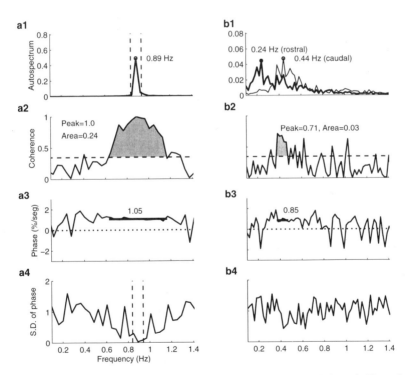

Figure 2. Frequency-domain analysis of the motor rhythms in segments 12 and 37 shown in Figure 1. a. Control trial. b. Blocked trial. a1,b1. Rostral (thick line) and caudal (thin line) autospectra, averaged over 8 non-overlapping time windows; (the rostral control autospectrum obscures the caudal one); the autospectra are normalized to the squared number of spikes to facilitate comparisons. a2,b2. Squared coherence spectra; dashed lines indicate 95% confidence level. a3,b3. Phase spectra; gray bars indicate the range used in calculating mean phase (values above bars). a4,b4. Spectra of standard deviation of the phase; gray dashed lines in a4 and a1 delimit the frequency bandwidth of the autospectra in a1. Refer to the text for details.

The steps involved in analyzing fictive swimming with the Fourier spike-time approach are outlined in the subsections below, together with examples from the above experiment.

2.1. Estimating Rhythm Frequency from the Spike Train Autospectrum

Spike times are first extracted via thresholding the raw data. Generally between 150 to 200 cycles of fictive locomotion are digitized at 5 KHz. Using spike times rather than real-time series requires much lower storage and memory requirements during analysis. Since we record from motor nerve roots containing many tens of axons, a portion of the captured spikes are compound.

As discussed in detail in the Rosenberg paper, the empirical Fourier transform of a signal X consisting of spike times is:

$$FT_X(f) = \sum_{(i = 1 \text{ to } N)} \{\exp(-i\, 2\pi\, f\, t_{Xi})\}$$

where f is frequency in Hz , N is the total number of spikes and t_{Xi} are the spike times. Note that in the usual FT for real-time series the exponential term is multiplied by the signal amplitude; here that signal amplitude is always equal to 1 since the algorithm only considers spike occurrences and ignores all other times. The autospectrum (i.e., power spectrum for a single signal) is then obtained by multiplying the FT by its complex conjugate and dividing by the total time of the signal. To reduce the variance of the spectral estimate (and for later calculation of the coherence), the spike train is divided into several (e.g., 8–10) successive time windows, so that the final autospectrum is an average over all windows:

$$P_X(f) = 1/(LT) * \sum_{(j = 1 \text{ to } L)} \{FT_{Xj}(f) * \text{conj}[FT_{Xj}(f)]\}$$

where L is the number of windows, T is the duration of each window, and 'conj' denotes the complex conjugate.

Figures 2a1 and 2b1 show the autospectra of the fictive rhythms represented in Figures 1a and 1b, respectively, for the blocking experiment described above. In the control trial, the rostral and caudal autospectral peaks are relatively narrow and are tightly co-aligned. When 20 middle segments are blocked with the low calcium saline, the two ends of the spinal cord take on different preferred frequencies, both lower than for the control, and the bandwidths of expressed frequencies at each end are significantly broadened.

2.2. Estimating the Relative Amount of Coupling from the Squared Coherence Spectrum

The squared coherence is can be understood as the linear predictability of signal Y given X, and is defined by:

$$C_{XY}(f) = |P_{XY}(f)|^2 / \{P_X(f) * P_Y(f)\}$$

where P_{XY} is the window-averaged cross-spectrum, which is calculated like the autospectrum except that FT_X is multiplied by the complex conjugate of FT_Y. The squared coherence spectrum is bounded in amplitude between 0 (Y is not predictable from X) and 1 (Y is perfectly predictable from X).

Figures 2a2 and 2b2 display the coherence spectra for the control and blocked trials, respectively. The dashed lines in the figures represent a level of confidence of 95% (Rosenberg et al. 1989). The peak control coherence almost always takes on the maximal value of 1, and range of frequencies over which the control coherence is significant (i.e. above the dashed line) is typically broader than the apparent bandwidth of the control autospectra. When 20 middle segments are blocked, most of the earlier coherence is lost, which can be quantified in a relative way by comparing the heights of the peaks, or the integrated significant coherence (gray areas in the figures). The residual significant coherence indicates that there are still pro-priospinal fibers long enough to span the blocked segments, although this residual coupling is not strong enough to entrain the two ends to a common frequency (Figure 2b2).

2.3. Estimating the Phase Lag from the Phase Spectrum

The phase spectrum (normalized per segment) between signals X and Y is defined as:

$$\text{Phase}(f) = 1/(2\pi d) * \arctan(\text{Im}[P_{XY}(f)] / \text{Re}[P_{XY}(f)])$$

where Im[ ] and Re[ ] are the imaginary and real parts of the cross-spectrum, and d is the number of segments between the electrodes. As discussed above, the locomotor CPG maintains a fixed intersegmental phase independent of swimming frequency. It is indeed interesting to observe from the control phase spectrum in Figure 2a3 that frequency-independence is maintained over the whole range of frequencies for which the coherence is significant. During blocking (Figure 2b3), the flatness of the phase spectrum is mostly replaced by rapid jumps above and below zero, except in a small region of frequencies for which the coherence is still significant and for which the mean phase lag is reduced relative to the appropriate control value.

There are a number of choices available for estimating the 'typical' phase lag in the frequency regime of interest. One option is to simply use the value of the phase spectrum at the autospectral (or cross-spectral) peak frequency; this generally will work for controls; however, as is evident in Figure 2b1, it is often not possible to decide on a single peak frequency in the blocked trials. A second option is to take the mean value of the phase spectrum in a neighborhood of frequencies defined by the bandwidth (dashed vertical lines in Figures 2a2 and 2a4) or half-maximal width of the cross-spectrum (not plotted), or alternatively as defined by the contiguous region of significant coherence (e.g., the gray regions in Figures 2a2 and 2b-2). This last method is employed for this example.

We found another interesting and useful tool for determining the frequencies over which phase is controlled by inspecting the spectrum of standard deviation of phase (sdPhase), which is obtained by calculating individual phase spectra for each of the time windows. Figure 2a4 shows the typically deep 'well' of the sdPhase in the neighborhood of the autospectral peak for controls. The 'well' shape in the control trial all but disappears (Figure 2b4) when the coupling between the two ends of the spinal cord is severely reduced, and the minimal value of sdPhase is significantly increased.

3. CONCLUSIONS

The Fourier approach provides a convenient and powerful means of quantifying changes in the important parameters of fictive locomotion, particularly when the experimental method degrades the regular bursting. The real gain of this method over burst analysis is that it allows straightforward display of these parameters over a range of frequencies, instead of at a single average value. Change in the relative amount of coupling between two points on the spinal cord due to experimental manipulations, and the effect of that change on phase, are also easily gauged with the spectra of coherence, phase, and standard deviation of phase. In the experiment shown here, for example, significant residual coupling and some residual control of phase can still be demonstrated despite block of synaptic activity over 20 segments disrupts rhythmic bursting.

REFERENCES

Cohen, A. H. (1987). Intersegmental coordinating system of the lamprey central pattern generator for locomotion. Journal of Comparative Physiology A, 160: 181–193.

Kopell, N. and G. B. Ermentrout (1986). Symmetry and phase coupling in chains of weakly coupled oscillators. Comm. Pure and Appl. Mathe. 39: 623–660.

Rosenberg, J. R., A. M. Amjad, P. Breeze, D. R. Brillinger and D. M. Halliday (1989). The Fourier approach to the identification of functional coupling between neuronal spike trains. Progress in Biophysics and Molecular Biology, 1989, 53(1):1–31.

Rovainen, C. M. (1985). Effects of groups of propriospinal interneurons on fictive swimming in the isolated spinal cord of the lamprey. Journal of Neurophysiology 54:299–317.

Wallén, P. and T. L. Williams (1984). Fictive locomotion in the lamprey spinal cord *in vitro* compared with swimming in the intact and spinal animal., Journal of Physiology 347: 225–239.

117

ADAPTIVE RESONANCE IN V1–V2 INTERACTION

Grouping, Illusory Contours, and RF-Organization

Heiko Neumann, Wolfgang Sepp, and Petra Mössner

Universität Ulm
Fakultät für Informatik
Abt. Neuroinformatik
Oberer Eselsberg
D-89069 Ulm, Germany
hneumann@neuro.informatik.uni-ulm.de

1. INTRODUCTION

A model for visual cortical boundary detection and contour grouping is proposed that takes into account the structure and functionality of the primate visual system. The architecture relates to visual cortical areas V1 and V2 which are bidirectionally interconnected via feedforward as well as feedback projections. It is suggested that their functionality is primarily determined by the measurement and integration of signal features that are continuously matched against neural codes of expectancies generated on the basis of long-range integration of compatible arrangements of initial measurements. Feedforward signal detection and the generation of feedback expectances is dedicated to different visual layers or areas. Thus, the bidirectional interaction between cortical areas can be understood as an active and continuing mechanism for the prediction and selection of elements in the visual input data stream. The net effect produces contour grouping and illusory contour completion as well as context-sensitive shaping in the tuning of orientation selective cells.

In the remainder of the paper we briefly describe the elements of the functional architecture and their contribution to the overall functionality. For details of the mathematical denotation we refer to (Neumann&Mössner96).

2. DESCRIPTION: COMPETENCE, MECHANISMS AND ARCHITECTURE

The functional architecture has been inspired by the principles of adaptive resonance theory (Grossberg80). The model consists of two major stages of feedforward and feedback processing each of which having different sublayers. The input to the recurrent network of resonant processing is generated by orientation selective units for local contrast detection. In order to demonstrate the capability of the network interactions alone, we omitted this initial stage of input measurement. Instead, at each spatial location equal magnitude of input activation is fed for all orientations.

Subsequent processing is based on a layered bidirectionally connected structure of model cortical areas V1 and V2, respectively. These layers are implemented as input feature activation (F_1) and a code layer (F_2), the latter of which is used to test the distribution of input activity against "expected" activity in a much broader visual context (Grossberg80, Mumford91). Activity generated at V1 feeds forward to V2 where it is integrated from different spatial branches in the visual field. Conjunctive arrangement of oriented input activity in turn generates an activation that is fed back to gate activities of orientation selective units in V1. The net effect of gating and positional and orientational competition shapes the orientation selectivity of cells in a context-sensitive manner (Gilbert&Wiesel90). Furthermore, inconsistent activity that fails to match the V2 feedback prediction (expectancy) will be suppressed.

The different layers implement the primary mechanisms for adaptive input processing. We identified the following key processing principles that contribute to the predicted functionality. These are

- top-down matching of higher level activation against activity distributions in the preceeding layer (generates a context-sensitive receptive field selectivity),
- contrast enhancement in *orientation* domain,
- contrast enhancement in *spatial* domain and end-stop generation, and
- spatial integration of activity and completion (grouping).

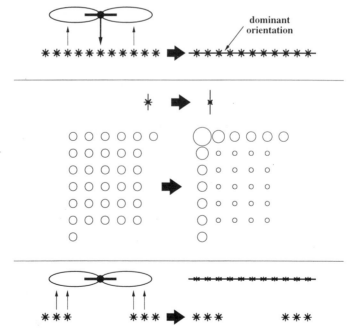

Figure 1. Mechanisms and their contribution to functionality: Top-down generated predictions based on long-range integration of activities (top), contrast enhancement in orientation and spatial domain (with generation of end-stopping) (center, two graphs), spatial integration of activity and completion via curvilinear grouping (bottom).

Figure 1 shows graphical illustrations to denote the desired transformations.

Based on these principles, a neural architecture has been proposed that consists of processing stages with analog model neurons. Each one of the principle layers consists of sub-layers, where each one is identified by two major stages of processing. In the feature layer F_1 we have

1. local competition (on-center/off-surround feedback processing) generates activity based on resonant matching between feedforward and feedback streams, and
2. local competition of 'match activity' to enhance spatial and orientation contrast.

Activity in the different sublayers is therefore generated in response to the combination of the bottom-up feedforward and the top-down feedback stream. It is subsequently followed by a spatial and orientational competition between localized activation in the parametrized representation of oriented contrast.

In the code layer F_2 curvilinear arrangements of oriented contrast activation will be integrated based on the evaluation of a much broader spatial context. Processing again consists of two hierarchically organized stages:

1. Long-range cooperative interaction integrates contrast activation in a curvilinear arrangement, and
2. local intra-orientational on-center/off-surround competition (between spatial locations) and activity normalization over orientations to generate a locally contrast-enhanced activation distribution.

The integration mechanism realizes a spatial *relatability* measurement (Kellman&Shipley91). Activity is integrated from opposite branches in the visual field (vonderHeydt&Peterhans89) utilizing bipole cells of different sizes (viz. spatial scales; see e.g. Grossberg&Mingolla85, vonderHeydt95). The support of a given activation in an orientation field is determined on the basis of the evaluation of a spatial compatibility function (Parent&Zucker89). Cooperative support for oriented activity at a target location is determined primarily by local circular arrangements of oriented contrast whereas incompatible arrangments contribute an inhibitory influence. The input from both branches is combined in a non-linear feedforward processing micro-circuit of disinhibitory interaction of conjunctive input activation. Therefore, a bipole generates a net response only if both subfields receive positive input activation. The activity generated at one spatial position for one orientation is normalized w.r.t. the spatially blurred activation of orientation columns and then fed back to layer F_1 to match oriented contrast measures. Figure 2 sketches the primary elements of the architecture.

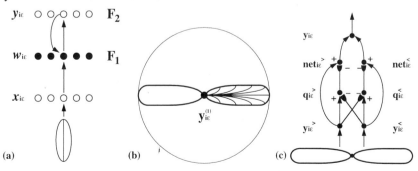

Figure 2. Overview of the architecture: (a) principle layers of the model with bidirectional coupling of F_1 and F_2, (b) geometry of a bipole cell, (c) non-linear circuit for combination of bipole branches.

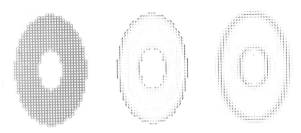

Figure 3. Processing results for an elliptic shape with a gap: Input activation for model V1 hypercolumns (left), activity generated at model V1 (F_1-layer) with cells tuned to orientation (middle), activity generated at model V2 (F_2-layer) with averaged contributions from two differently scaled bipole cells (right).

3. EVALUATION: TEST DATA SETS AND MODEL SIMULATIONS

We tested the functional behavior generated for input activity distributions with no initial orientation bias at the stage of individual hypercolumns. The spatial arrangements of synthetically generated activity distributions correspond to figural patterns such as square and elliptic shape outlines as well as Kanizsa square pattern and phase shifted bar patterns. The latter two cases generate illusory percepts of shape outline. These patterns of input activation have been processed with two different spatial integration widths for the bipole. The activities of each processing scale have been summed up for display purposed. Parameters have been kept identical in all simulations. Figure 3 shows the processing of the network for an elliptic shape, Figure 4 displays the result of processing for a Kanizsa square pattern. The net effect produces local curvilinear grouping and long-range illusory contour completion. In particular, the results of model simulations demonstrate that

1. the orientation selectivity of V1 cells is sharpened depending on the spatial arrangement and visual context,
2. orientation selective contrast measurements are enhanced at boundaries of spatially homogeneous stimulus arrangements and suppressed in the interiors,
3. end-stop behavior is generated at locations of abrupt termination or sharp corners,
4. oriented cells at model V2 stage only respond to curvilinear input arrangements of activity at both branches of bipole cells, and
5. model V2 cells show fine tuning for orientation selectivity and generate subjective contours to bridge gaps in arrangements of oriented contrast.

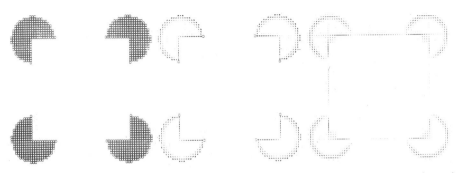

Figure 4. Processing results for a Kanizsa square pattern (different stages as in Figure 3, top, center, bottom).

4. SUMMARY

A new model architecture has been proposed for visual boundary detection and grouping. The functional organization has been motivated by a computational theory based on the principles of adaptive resonance. We postulate that the role of V2 to V1 feedback processing can be understood as a process of dynamic code testing and prediction. The net effect produces context sensitive shaping of orientation selectivity, suppression of activation in homogeneous arrangements, as well as grouping and illusory contour completion.

5. REFERENCES

C.D. Gilbert and T.N. Wiesel. The influence of contextual stimuli on the orientation selectivity of cells in primary visual cortex of the cat. *Vision Research*, 30(11):1689–1701, 1990

S. Grossberg. How does a brain build a cognitive code? *Psychological Review*, 87:1–51, 1980.

S. Grossberg and E. Mingolla. Neural dynamics of perceptual grouping: Textures, boundaries, and emergent segmentation. *Perception and Psychophysics*, 38(2):141–171, 1985

R. von der Heydt. Form analysis in visual cortex. In M.S. Gazzaniga, editor, *The Cognitive Neurosciences*, chapter 23, pages 365–382. MIT Press (Bradford Book), Cambridge (MA/USA), 1995

R. von der Heydt and E. Peterhans. Mechanisms of contour perception in monkey visual cortex. I. Lines of pattern discontinuity. *The Jorunal of Neuroscience*, 9(5):1731–1748, 1989

P.J. Kellman and T.F. Shipley. A theory of visual interpolation in object perception. *Cognitive Psychology*, 23(2):141–221, 1991

D. Mumford. On the computational architecture of the neocortex II: The role of cortico-cortical loops. *Biological Cybernetics*, 65:241–251, 1991

H. Neumann and P. Mössner. Neural model of cortical dynamics in resonant boundary detection and grouping. In C. von der Malsburg, W. von Seelen, J.C. Vorbrüggen, B. Sendhoff, editors, Lecture Notes in Computer Science 1112 "*Artificial Neural Networks* - ICANN 96", (Proc. Int. Conf. on Artificial Neural Networks, Bochum, Germany, July 16–19, 1996) Springer, Berlin, 1996

P. Parent and S.W. Zucker. Trace inference, curvature consistency, and curve detection. *IEEE Transactions on Pattern Analysis and Machine Intelligence*. 11(8):823–839, 1989

118

A HOLISTIC MODEL OF HUMAN TOUCH

Dianne T. V. Pawluk and Robert D. Howe

Division of Engineering and Applied Sciences
Harvard University
Cambridge, Massachusetts 02138

INTRODUCTION

The peripheral mechanoreceptive system in the skin consists of several functional components: the mechanical response of the skin, the mechanical response of the end organ, the creation of the generator potential, the initiation of an action potential, and (for some units) the branching structure of the afferent fibers[5]. Experimentally, only the stimulus applied to the surface of the skin and the final afferent nerve fiber response (as the signal propagates toward the central nervous system) can be measured. The system must therefore be treated as a series of black boxes for which we only have access to the first input and the last output. Previous research has focused on one or another of these boxes and related them to the final output, despite the fact that the components act together to produce the response and cannot be treated in isolation. Here we examine the system as a whole, with the goal of attributing different aspects of the final nerve fiber response to the various components of the system. Our approach is to determine the components which are necessary and sufficient to describe this overall system response.

1.1. Background

There are many sensors which respond to mechanical stimuli throughout the body. We will focus on the four primary types of mechanoreceptive units found in the nonhairy (glabrous) skin of the human hand, as they are the most important for tactile exploration and manipulation. These receptors are classified in terms of speed of adaptation, either 'Fast Adapting' (FA, no static response) or 'Slowly Adapting' (SA, static response present), and by the size of their receptive fields, either type I (small, sharp borders) or type II (large, diffuse borders). Previous models of mechanoreception include static models of the skin mechanics[1,7] and dynamic models of the generator potential and impulse initiation.[2,6] Our models additionally consider the end organ mechanics and the issue of consistency between these contributing factors.

The models that we present are primarily based on the frequency response of the four types of mechanoreceptors experimentally obtained by Johansson and colleagues[4].

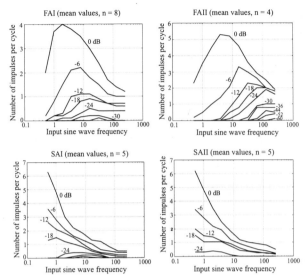

Figure 1. Frequency response functions of the mechanoreceptors. Adapted from Johansson, Landström and Lundström (82). Note that the vertical axis units are in terms of nerve impulses per input frequency cycle and that 0 dB corresponds to an amplitude of 1 mm.

The nerve impulses of individual afferent fibers were measured using microneurography in alert human subjects. The test stimuli were sinusoidal displacements applied perpendicular to the skin varying in amplitude (0.002–1.0 mm) and frequency (0.5 - 1024 Hz). The measured frequency response for each receptor type is given in Figure 1. As shown in the figure, the responses of the mechanoreceptive units are highly nonlinear, with the frequency at which the peak output occurs shifting to higher frequencies at lower stimulus amplitudes.

PRELIMINARY MODELS AND SIMULATION RESULTS

Because of uncertainty about the contribution of the highly nonlinear neural impulse initiation component to the response characteristic of any mechanoreceptor model, very simple models were initially examined. The initial approach uses one-dimensional, lumped parameter models. The goal of these models is to capture the experimental results shown in Figure 1 with physically plausible representations of each of the constituent elements.[*]

The first such model is shown in Figure 2a. It represents the skin, the end organ and the nerve membrane as simple springs, the generator potential as a simple proportionality to the nerve membrane displacement, and the conduction of the nerve fiber by the Hodgkin-Huxley equations[3]. The input to the model is the displacement of the skin surface and the output is the time history of the nerve impulses initiated. Note that in this model the mechanical components contribute essentially nothing to the form of the output signal. This is because they form a simple proportionality factor which only affects the input range to which the model responds.

---

[*] The effect of the nerve fiber branching structure, which exists for some types of units, is assumed negligible in these initial models as the stimulus probe for the given experimental data was a relatively flat indentor covering all branches of each afferent unit.

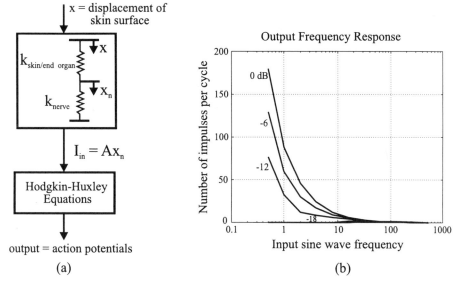

Figure 2. (a) Spring model of the skin, end organ and nerve membrane proportionally coupled to the Hodgkin-Huxley equations. (b) Frequency response function of the 'spring' model. 0 dB corresponds to the maximum input level.

The model was simulated for sinusoidal displacement inputs varying in amplitude (over the entire range which produced a nerve impulse train) and frequency (0.5 -512 Hz). The frequency response is shown in Figure 2b. The amplitude levels are presented in decibels, where 0 dB corresponds to the maximum input level; this non-dimensionalizes the input range and enables the model to be viewed independently of the proportionality constants. Note that the output is in terms of nerve impulses per input frequency cycle to facilitate the comparison with the experimental results in Figure 1.

The shape of the frequency response function (Figure 2b) is similar to those experimentally obtained for both SA type units: at large amplitudes the response is hyperbolic and at lower amplitudes it is an inverted U-shape. In addition, the variation of the response as a function of the input amplitude (not shown) is also similar, being logarithmic in both cases. The results also suggest that the shifting of the 'peak' of the frequency response with amplitude, most apparent in the responses of the FA type units, is an inherent property of the nerve membrane rather than due to nonlinearities in the skin mechanics.

However, there are two major discrepancies between this model and the SA type units: (1) the nerve impulse rate in the model is over an order of magnitude larger than the afferent units, and (2) the input range to which the model responds is smaller than the range for all the receptors except the SAII units.

The addition of a simple mechanical high-pass filter (i.e., a single zero in the transfer function of the mechanical components) alleviates both of these discrepancies. In this second model (Figure 3a), the simple springs used to model the skin, end organ and nerve membrane are replaced with dampers, and the generator potential becomes proportional to the derivative of the nerve membrane displacement. This emphasizes the viscous properties of these viscoelastic materials. The model results in a derivative relationship between the displacement of the skin surface and the input current of the nerve fiber equations.

The model was simulated for sinusoidal displacement inputs of varying amplitude and frequency, as above for the 'spring' model. The frequency response is shown in Fig-

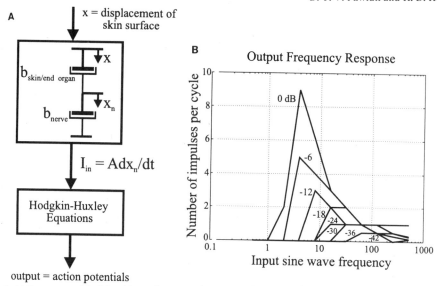

A

x = displacement of
skin surface

B

Output Frequency Response

output = action potentials

Figure 3. (a) Damper model of the skin, end organ and nerve membrane coupled by its derivative to the Hodgkin-Huxley equations. (b) Frequency response function of the 'damper' model. 0 dB corresponds to the maximum input level.

ure 3b. Both the magnitude of the nerve impulse rate and the decibel input range are comparable to the mechanoreceptive units. More specifically, the results are also similar to the FA type units in their general form and exhibition of 'shifting peaks' with amplitude.

The results from both the spring model and the damper model suggest that modeling the mechanical components with carefully placed simple zeros and poles coupled with the Hodgkin-Huxley equations will explain the frequency responses of the mechanoreceptive units. A further significant aspect of the experimental results which is important to consider is the portion of the sinusoidal indentation cycle to which the different types of mechanoreceptive units respond (i.e., the phase response). Qualitatively, the SAI and SAII units respond principally to the indentation portion; the FAI units respond to both the indentation and the retraction, but much less to the retraction portion; and the FAII units also respond to both portions, but more to the retraction[4]. These experimental results can be compared to simulations of models using simple mechanical components, as above.

The simulation results showed that for a mechanical component consisting of: (1) a simple gain (e.g., the spring model), the response occurs over the entire input cycle; (2) a simple derivative dx/dt (e.g., the damper model), the response occurs only on the indentation; and (3) a second derivative d^2x/dt^2, the response occurs primarily on the retraction, but to some degree on the indentation. These results suggest that the SAI and SAII units can be modeled by first order systems, and the FAI and FAII units by second order systems (with the poles placed at much lower frequencies for the FAI units than for the FAII units).

3. PROPOSED MODELS

From the insight gained in examining these initial models, we are developing more elaborate models that: (1) produce the desired overall shape of the frequency responses, including phase characteristics, (2) are morphologically plausible based on the known structure of the individual components and their connectivity and (3) meet the additional

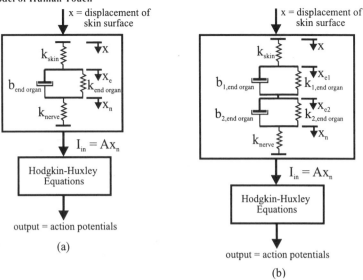

Figure 4. Proposed models. (a) Model of the SAI and SAII units. (b) Model of the FAI and FAII units. Same parameters for the skin and nerve fiber for all models. Different parameters for the end organs of each of the four different types of mechanoreceptors.

constraint that all units must share the same model of skin.[†] The proposed models are given in Figure 4.

The first and third criteria were used in examining many different first order and second order models which could potentially describe the mechanoreceptor responses. The resulting models were then compared to the known morphology to determine their plausibility. For example, the end organs of the SAI units can be described as fluid-filled sacs, with the afferent nerve terminal entering along the bottom surface. The fluid in the sac corresponds to the damper in our model, whereas the membrane of the sac corresponds to its parallel spring. Both presumably apply forces on the nerve ending to cause mechanosensitive channels to open, represented by the spring and proportionality constant in the generator current. The end organ of the FAII units, in contrast, can be described as a series of fluid filled lamellae in the form of an onion, with weak interconnections between the layers. Each of the lamellar spaces can plausibly be modeled by a parallel damper (for the fluid) and spring (for the interconnections) between rigid lamellae. These series of spring damper pairs can then be represented by a single equivalent spring damper pair on either side of the end organ; mathematically this is equivalent to the representation shown in Figure 4b. Subsequent work will be directed at verifying these models.

4. CONCLUSIONS

We have shown that the Hodgkin-Huxley equations coupled with simple mechanical components capture the essential properties of the experimental frequency responses.

[†] Although the type I and type II units are at different depths, statistically there is very little variation due to this parameter in the experimental data used to develop the models.

Based on these results, we have proposed more complex models to explain the responses of the four types of mechanoreceptors in the human hand. The simplicity of the models should facilitate further examination of mechanoreception, including models of branching afferent fibers and of population responses.

5. REFERENCES

1. Dandekar, K. and Srinivasan, M.A. (1995). A Three Dimensional Finite Element Model of the Monkey Fingertip for Predicting Responses of Slowly Adapting Mechanoreceptors. *Summer Bioengineering Conference of the ASME*, Beaver Creek, CO.
2. Freeman, A.W. and Johnson, K.O. (1982). Cutaneous Mechnoreceptors in Macaque Monkey: Temporal Discharge Patterns Evoked by Vibration, and a Receptor Model. *J. Physiology*, 323, 21–41.
3. Hodgkin, A.L. and Huxley, A.F. (1952). A Quantitative Description of Membrane Current and Its Application to Conduction and Excitation in Nerve. *J. Physiology*, 117, 500–544.
4. Johansson, R.S., Landström, U. and Lundström, R. (1982). Response of Mechanoreceptive Afferent Units in the Glabrous Skin of the Human Hand to Sinusoidal Skin Displacements. *Brain Research*, 244, 17–25.
5. Johnson, K.O., Phillips, J.R. and Freeman, A.W. (1985). Mechanisms Underlying the Spatiotemporal Response Properties of Cutaneous Mechanoreceptive Afferents. In *Development, Organization, and Processing in Somatosensory Pathways*. Alan R. Liss, Inc., 111–122.
6. Nemoto, I., Miyazaki, S., Saito, M. and Utsunomiya, T. (1975). Behavior of Solutions of the Hodgkin-Huxley Equations and Its Relation to Properties of Mechanoreceptors. *Biophysical Journal*, 15, 469–479.
7. Phillips, J.R. and Johnson, K.O. (1981). Tactile Spatial Resolution III. A Continuum Mechanics Model of Skin Predicting Mechanoreceptor Responses to Bars, Edges, and Gratings. *J. Neurophysiology*, 46, 1204–1225.

119

POST SYNAPTIC DENSITY (PSD) COMPUTATIONAL OBJECTS: ABSTRACTIONS OF PLASTICITY MECHANISMS FROM NEUROBIOLOGICAL SUBSTRATES

James K. Peterson*

Department of Mathematical Sciences
Clemson University
Clemson, SC
E-mail: peterson@math.clemson.edu

1. INTRODUCTION

We focus in this work on the development of a software analog of the **Post Synaptic Density** (PSD) structure and its use as a fundamental building block in the construction of general connectionist architectures of computational modules. Each generic computational object, hereafter referred to as a *neural object*, interacts with other neural objects via a PSD object which mediates information transfer between axonal and dendritic pathways. By deriving appropriate children from core classes of NEUROTRANSMITTER, RECEPTOR and PSD objects, it is possible to model very arbitrary connectionist architectures of quite varied computational nodes.

2. THE PSD OBJECT

The Post Synaptic Density structure (PSD) is an agent that mediates the interaction between dendritic and axonal pathways. (see [Black, [1] and also Dudai, [2], Hall, [3], Hardie, [4] and Hille, [5]]). There are a variety of biochemical mechanisms for altering the structure of both the pre- and post-sides of the PSD structure and we wish our software PSD object to be endowed with sufficient plasticity to alter architectural structure as well.

We can abstract from neurobiological structure a number of guiding principles for the development of a useful object-oriented infrastructure using neuromodulatory agents. Following [Black, [1]], we note that the principle of multiple function implies that there is

*Research partially supported by NASA Johnson Space Center Grant NGT-70386 and by NSF Grant ECS-9412430

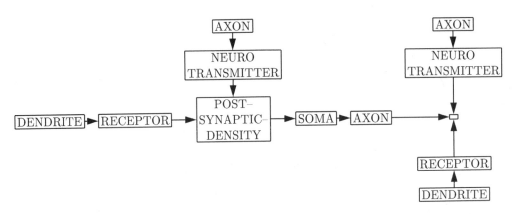

Figure 1. Abstract Neuronal Process

no clear distinction among the processes of cellular metabolism, intercellular communication and symbolic function in the nervous system. **Hence our software infrastructure should possess the same blurring of responsibility.** In addition, the argument can be made that software and hardware are the same in the nervous system implying the basic building blocks of our software system should be able to alter the software architecture itself. This capability can be implemented using **dynamically bound** computational objects and strategies. Finally, ongoing function, the very fact of communication, alters nervous system structure. Hence, the structure of our software architecture should be mutable in response to the external input environment.

2.1. The Neurotransmitter Abstraction

Consider Figure 1 as an illustration of a prototypical abstract neuronal object. Each DENDRITE object contains a number of RECEPTOR objects, while an AXON object contains NEUROTRANSMITTER objects. A DENDRITE–AXON object pair interact via the POST-SYNAPTIC-DENSITY object which plays a role that is similar to its biological function The output of the POST-SYNAPTIC-DENSITY is sent to the computational body of the neuronal object, the SOMA object. The output of the SOMA is collected into an AXON object containing the previously mentioned NEUROTRANSMITTER objects. There could be many different types of NEUROTRANSMITTERs. We indicate the interaction pathways in Figure 1 on the dendritic and axonal side of the SOMA. More detail is shown in the close up view given by Figure 2. Here, the DENDRITE object contains five types of NEUROTRANSMITTERs and the AXON uses the associated RECEPTOR objects for these NEUROTRANSMITTERs. The result of the dendritic–axonal interaction is computed by the POST-SYNAPTIC-DENSITY object. Now, how should we handle the intricacies of the neurotransmitter–receptor interactions? For our purposes, we will concentrate on a few salient characteristics. The probability of neurotransmitter efficacy will be denoted by the scalar parameter p. The variable p models neurotransmitter efficacy in a lumped parameter manner. Each neurotransmitter will also have its own reabsorption rate denoted by q with a corresponding intrinsic time interval of action which can be modelled crudely by controlling the (p,q) interactions. Finally, each neurotransmitter has an associated locality that sets the scope of its interaction with multiple dendrites and axons.

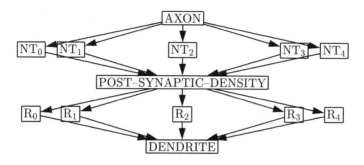

Figure 2. Dendrite–Axonal Interaction Pathway

2.2. PSD Computation

Let's denote the PSD computation by the symbol •. To model the process by which we obtain the value of a prototypical interaction, I, we need to introduce some notational conventions. Let $\mathcal{N}^i$ be the ith neural object in our system and $\mathcal{A}^i$ and $\mathcal{D}^i$ the associated axon and dendrite, respectively. Each axon and dendrite use neurotransmitters and receptors in the PSD calculation and we will let p_t^i and q_t^i be the efficacy and the reabsorption rate of the neurotransmitter σ_t for the axon of neural object $\mathcal{N}^i$. The values carried by the axon and dendrite will be denoted by v^i and w^i, respectively. Finally, since more than one axon and dendrite may be involved in these interactions, we let $\mathcal{L}_t^i$ denote the **axonal locality** and $\mathcal{M}_t^i$ the **dendritic locality** of the neurotransmitter σ_t in the ith neuron's axon.

Then, in the case of $T + 1$ different transmitters $\{\sigma_0, \sigma_1, \ldots, \sigma_T\}$ at one PSD junction, we find that the interaction value I is given by

$$I = \sum_{t=0}^{T} \sum_{m \in \mathcal{M}_t^j} \sum_{k \in \mathcal{L}_t^i} (p_t^k - q_t^k)\, v^i \bullet w^m \tag{1}$$

Software Implementation A rough picture of the PSD object must therefore allow for a DENDRITE object entering the PSD computational unit. The receptors required for evaluation are embedded in the DENDRITE object's instantiation. In order to access all required NEURON objects, we will need to pass in the starting address of the full NEURON vector for the chain architecture, `IOMAP **iN`. We also need the index of the dendritic object (`int dendritic_object`) and of the particular dendrite we use in the PSD computation (`int dendrite_index`). In addition, we need the index of the AXON object entering the PSD unit. (this information has already been supplied through the argument `IOMAP **iN`) and the index of the axonal object, (`int axonal_object`). Finally, we need to supply the index of the particular axon we use in the PSD computation as the axonal NEURON will potentially have many axons, (`int axonal_index`). Since the actual computational model used to handle the PSD interactions will also be plastic, we also provide a computational engine object (modeled as a pointer to an ENGINE object, `ENGINE *S`) to handle the computation of the value I.

The PSD object is connected to axonal and dendritic pairs via LINK objects. Each dendritic–axonal interaction that is mediated through the PSD object requires an associated LINK object. The ith IOMAP object has associated with it a set of other IOMAP objects that connect some of their axons to i^{th} IOMAP dendrites. We need to specify the indices of both the connecting IOMAP objects and the required dendrite–axon pairs. Hence, a LINK object must be able to specify both the particular IOMAP objects that are connected and also the particular dendrite and axon of these objects which are to be used in the PSD computations. The LINK and PSD object can be specified through the class descriptions below:

```
class LINK{ class PSD{
public: public:
LINK(PSD *iPsd); PSD(IOMAP **iN,ENGINE *iE,
LINK(void); int dendritic_object,
LINK(const LINK&); int dendritic_index,
~LINK(); int axonal_object,
LINK& LINK::operator=(const LINK&); int axonal_index);
protected: PSD(void);
//pre_neuron PSD(const PSD&);
int pre_neuron; ~PSD();
//PSD interaction axon PSD& PSD::operator=(const PSD&);
int axon_index; float compute_value();
//post_neuron protected:
int post_neuron; //dynamically bound NEURON vector
//PSD interaction dendrite //of core class IOMAP
int dendrite_index; IOMAP **N;
//value of PSD computation value //dynamically bound
float value; //computational engine
//LINK to LINK address ENGINE *E;
int ptr; //dendritic object
//dynamically bound PSD object int dobject;
PSD *Psd; //dendritic index
}; int dindex;
//axonal object
int aobject;
//axonal index
int aindex;
};
```

To shed insight into the PSD computational engine, we will consider the following code fragment for the PSD::compute_value() agent.

```
float PSD::compute_value()
{
int t, number_neurotransmitters;
int number_local_dendrites;
int number_local_axons;
int den_object,den_index;
int axon_object,axon_index;
float sum;

number_neurotransmitters
= N[aobject]->axon[aindex]->get_number_neurotransmitters();
number_local_dendrites
= N[dobject]->dendrite[dindex]->get_number_local_dendrites();
number_local_axons = N[aobject]->axon[aindex]->get_number_local_axons();

sum = 0.0;
for(t=0;t<number_neurotransmitters;++t){
for(m=0;m<number_local_dendrites;++m){
for(k=0;k<number_local_axons;++k){
axon_object = N[aobject]->axon[aindex]->get_axon_object(t,k);
axon_index = N[aobject]->axon[aindex]->get_axon_index(t,k);
den_object
= N[dobject]->dendrite[dindex]->get_dendrite_object(t,m);
```

```
den_index
= N[dobject]->dendrite[dindex]->get_dendrite_index(t,m);
sum +=
( N[axon_object]->axon[axon_index]->get_efficacy(t)
-N[axon_object]->axon[axon_index]->get_reabsorption(t) )
*E->compute_value(N[aobject]->axon[aindex]->get_value(),
N[den_object]->dendrite[den_index]->get_in());
}// local axon loop
}// local dendrites loop
}// local neurotransmitter loop
return(sum);
};
```

As you can see, there are many unexplained terms in this code fragment. For example, in the above code the functions `get_axon_object()`, `get_axon_index()` and so forth are public agents designed to retrieve indexing information from the protected areas of the LINK object structure. However, these details are unimportant for our present purpose. Each neural object will need its own evaluation agent and there will be an rich interplay between them. There will be the low-level evaluation agents at the PSD level whose structure is mutable via the neurotransmitter, receptor, axon and dendrite interactions; midlevel agents to handle the evaluation at the single neuron level and high level agents for the evaluation of clusters and other complicated ensembles of neural objects. Using dynamic binding and polymorphism, all of these agents will have the same syntactical form, `object->compute_value()`. For example, a neural object can be built to model a memory subsystem or an I/O loop and inserted and used in a very transparent way.

3. CONCLUSION

Through the use of dynamic binding and polymorphism, we are able to look at connectionist models at varying levels of detail, but whose linking and evaluation agents share a common interface. The low-level details of the computations required for neural objects interaction are all buried in the appropriate class agent code of particular neural object children, yet the "look" of the calling interface is the same. This architecture provides a simple and effective mechanism for the building of many connectionist architectures that include subsystems of neurobiological meaning that are useful in our quest for artificial systems capable of interesting behavior. We discuss this in more detail in [Peterson, [6]]. We are also hopeful that this modeling technique will be of some use in constructing high level models of reasonable plasticity for neurobiologically motivated subsystems.

REFERENCES

[1] Black, I. 1991. *Information in the Brain: A Molecular Perspective*, A Bradford Book, MIT Press.

[2] Dudai, Y. 1989. *The Neurobiology of Memory: Concepts, Findings, Trends*, Oxford University Press.

[3] Hall, Z. 1992. *An Introduction to Molecular Neurobiology*, Sinauer Associates Inc., Sunderland, MA.

[4] Hardie, D. G. 1991. *Biochemical Messengers: Hormones, Neurotransmitters and Growth Factors*, Chapman & Hall, London.

[5] Hille, B. 1992. *Ionic Channels of Excitable Membranes*, Second Edition, Sinauer Associates Inc., Sunderland, MA.

[6] Peterson, J. 1995. *A White Paper on Neural Object Design: Draft 1.0*, Technical Report 644 Department of Mathematical Sciences, Clemson University.

120

V1 RECEPTIVE FIELDS REFLECT THE STATISTICAL STRUCTURE OF NATURAL SCENES

A Projection Pursuit Analysis

William A. Press[1,2] and Christopher W. Lee[1]

[1]Washington University School of Medicine
St. Louis, Missouri 63110
[2]California Institute of Technology
Pasadena, California 91125

1. ABSTRACT

The strategy by which the visual system encodes our environment has long been a topic of debate. One compelling hypothesis is that the tuning characteristics of cells in the visual system provide a sparse representation of natural scenes (Barlow, 1972; Field, 1994). While recent work by Olshausen and Field (1996) supports this hypothesis, other studies suggest alternate explanations (Law and Cooper, 1994; Fyfe and Baddeley, 1995). To address this question, we employ exploratory projection pursuit to investigate the statistical structure of natural scenes—the images the visual system evolved to represent.

Applying projection pursuit to over 130,000 natural image patches, we find that searching for sparse and other non-normal structure results in oriented, band-pass, localized projections. Our results suggest that V1 simple cell receptive fields directly reflect the statistical structure in natural scenes, consistent with Field's hypothesis (Field, 1994). We relate our technique to that of Olshausen and Field (1996), as well as compare our results to those of similar efforts (Law and Cooper, 1994; Fyfe and Baddeley, 1995). In addition, we demostrate how projection pursuit can be used to investigate non-linear processing found in the visual system, such as on-off channel segregation and receptive fields derived from feed-forward projections.

2. INTRODUCTION

The natural world is made up of complex visual scenes, rich in structure and information. Evolutionarily, our survival has depended upon the visual system's ability to represent and interpret these scenes quickly and accurately. The space of all images is enormous: an 8-by-8 image patch, for example, represented by 256 gray levels subtends 256^{16}, or approximately 10^{154}, possible images. Natural images form a minute subspace of all possible images, and much recent work has focused on how the visual system might represent this subspace efficiently (Barlow, 1972; Atick and Li, 1992; Field, 1994; Baddeley, 1996).

Field (1994) suggested that receptive fields in primary visual cortex (V1) are optimally tuned to sparsely represent natural scenes—that is, tuned so that each cell responds to only a small subset of possible images. Recent work by Olshausen and Field (1996) demonstrated that a set of functions constrained to sparsely represent natural images converge to oriented, localized, band-pass wavelets—similar to receptive fields in V1.

Motivated by these results, we have investigated the statistical structure of natural scenes directly. To this end, we employed exploratory projection pursuit (see Huber, 1985, for a review), a technique specifically designed to explore high-dimension data sets for structure. In addition, we used projection pursuit to investigate known non-linearities in visual processing, such as on-off channel segregation, and suggest a principled way to investigate response properties of cells in higher cortical areas.

3. METHODS

Projection pursuit (Huber, 1985; Friedman, 1987; Intrator, 1992) searches for structure in a high-dimensional data set by finding the linear projection whose distribution maximizes some function Q, where Q measures the "interestingness" of a one-dimensional distribution. To be specific, let the d-dimensional X denote a random variable representing the data set (possibly after pre-processing), and let the d-dimensional w define a direction, or *projection*, onto which the data are projected via the standard inner product, $w^T X$ ($\Sigma_i w_i x_i$). Exploratory projection pursuit maximizes $Q[w^T X]$ over all possible projections w.

We perform projection pursuit via conjugant gradient ascent (Press *et al.*, 1992). The initial conditions of w are determined by the coarse stepping method described by Friedman (1987). A potential problem arises if the data set remains unchanged with each pursuit: the deterministic coarse stepping algorithm results in identical initial conditions, and thus identical projections, every time. To remedy this, we remove each projection's structure from the data set before beginning the next iteration of coarse stepping and projection pursuit (Friedman, 1987).

Structure removal entails redistributing the data along the last projection w so as to minimize Q, while leaving orthogonal directions unchanged. To achieve this, the data are initially rotated so that the first coordinate lies in the same direction as w; that is, $Y = UX$, where U is an orthonormal basis whose first row is w. The first row of Y is therefore the projected data, $w^T X$. The values of this row are then adjusted so that, while their relative rankings remain unchanged, the values form a standard normal distribution—a distribution of arguably minimal structure (Huber, 1985). The resulting Y' is then rotated back into the data's original coordinate system—$X' = U^T Y'$—whereupon the next projection is sought.

To form our standard data set, we extracted a large number (typically 131,000) of 8-by-8 or 12-by-12 image patches from eight 512-by-512 images of natural scenes (generously provided by David Field). The means by which these images were obtained is described in detail by Field (1994). Briefly, photographs of natural scenes were taken on

Ilford XP1 film and sampled at twice their nyquist frequency using a 512-by-512 8-bit Barneyscan digitizer. The images were calibrated to account for both the optics of the camera and the intensity compression of the film, and then logarithmically transformed so that pixel value differences reflect the original images' intensity ratios, or contrast. From these 512-by-512 images we extracted our data set of smaller image patches. These image patches were then low-pass filtered (to reduce sampling artifacts) and sphered, or whitened, (to remove second-order correlations) as putatively performed in the retina (Atick and Redlich, 1990; Atick and Redlich, 1992).

All image patches can be mapped onto $\mathbf{R}^n$, so as to be suitable substrates for projection pursuit. For example, an 8-by-8 image patch is mapped onto a point in $\mathbf{R}^{64}$; the value along each dimension is given by the intensity of a particular pixel. A set of image patches, then, becomes a set of points, X, and projection pursuit can look for structure in their distribution. Inversely, a given projection, w, can be represented by a two-dimensional distribution of intensities. As each projection is defined by its inner product with the data, its corresponding image is equivalent to a filter, or receptive field, whose output over the image data set maximizes the projection pursuit objective function Q.

4. RESULTS

We begin by searching the space of natural images for sparse distributions, as suggested by Field (1994) and Olshausen and Field (1996). This is achieved by applying projection pursuit to the 12-by-12 standard data set, using the objective function

$$Q[w] = E[(w^T X - E[w^T X])^4] / E^2[(w^T X)^2] \qquad (1)$$

This function corresponds to the projected data's kurtosis. A kurtotic distribution has most of its values near zero, with the balance likely to assume high values; it thus represents an effective measure of sparseness. The resulting projections (Figure 1, left) resemble V1 receptive fields in that they are band-pass, oriented, and sometimes localized (that is, spatially restricted in both dimensions); however, their structure degenerates abruptly after the 70th iteration. This corresponds, approximately, to the effective dimen-

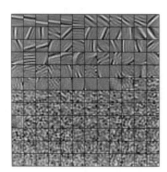
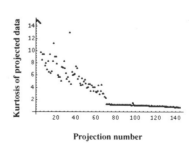

Figure 1. (*left*) 144 projections of kurtotic structure (Eq. 1), found with exploratory projection pursuit and structure removal in ~130,000 12-by-12 natural image patches. The first projection is in the upper left, with subsequent projections shown to the right, by row. (*right*) The kurtosis of each projection drops precipitously after ~70 projections. This corresponds to the effective dimensionality of the data, and to when the projections' oriented, band-pass structure degenerates.

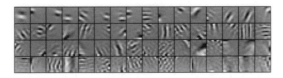

Figure 2. A set of projections resulting from a more robust measure of kurtosis (Eq. 2.) These projections show greater spatial localization than those found without a sigmoid (Eq. 1 and Fig. 1.)

sionality of the image patches after low-pass filtering (where the dimensions containing the smallest eigenvalues were removed). The degeneration is probably a result of the structure removal algorithm, as the kurtosis of subsequent projections also drops precipitously after 70 iterations (Figure 1, right).

While some of the projections in Figure 1 are localized, the majority are substantially elongated. This might be due to the kurtosis index's sensitivity to outliers, where rarer elongated contours become over-emphasized. To test this hypothesis, we made our measure of kurtosis more robust against outliers by adding a sigmoid. Figure 2 shows the projections found with the objective function

$$Q[w] = E[\sigma((w^T X - E[w^T X])^4)] / E^2[\sigma(w^T X)^2] \qquad (2)$$

where $\sigma(y) = tanh(y)$. These projections are significantly more localized than those found without the sigmoid.

In addition to searching for kurtotic, or sparse, structure in natural scenes, we can investigate alternate definitions of structure, as well. The BCM model of visual cortical plasticity is a biologically plausible developmental model akin to projection pursuit (Intrator and Cooper, 1992). Single cell simulations of this model have been shown to develop band-pass, oriented (primarily horizontal and vertical) receptive fields (Law and Cooper, 1994). We compared their results with our technique by performing projection pursuit using a variant of their objective function,

$$Q[w] = E[\sigma((w^T X - E[w^T X])^3)] / 3 - E^2[\sigma(w^T X)^2] / 4 \qquad (3)$$

The sigmoid, in this case, performs positive half-rectification. This was done because the third moment ($E[(w^T X)^3]$) is an odd function, and symmetric values of $w^T x$ would otherwise cancel each other out. Projections found with this objective function, shown in Figure 3, exhibit randomly oriented, band-pass structure, similar to those projections found with the kurtosis index.

We also applied projection pursuit to investigate non-linear processing found in the visual system. As early as the retina, separate processing channels, on- and off-channels, represent increases and decreases in central receptive field visual stimulation, respec-

Figure 3. A set of projections found using the BCM objective function (Eq. 3.) These show oriented, band-pass structure, similar to those found with the kurtosis index (Eq. 1 and Fig. 1.)

Figure 4. A set of projections found with on-off segregated inputs. Every projection pair is displayed as two separate, adjacent projections. Segregated inputs result in quadrature pairs of oriented, band-pass, and sometimes localized structure.

tively. We modeled these separate channels using a modified data set: each 8-by-8 image patch was duplicated and placed beside itself to form an 8-by-16 image patch; the left side was negative-half-rectified to form an off channel, while the right side was positive-half-rectified to form an on-channel. Projections capturing the kurtotic structure in these data are shown in Figure 4. Each projection pair shows the oriented, band-pass, and sometimes localized structure found in the standard data set. In addition, each filter pair tends to comprise quadrature pairs—simple-cell-like receptive fields that share the same location, orientation, and band-pass characteristics, but are 90° out of phase with one another.

5. DISCUSSION

Our results demonstrate that a search for sparse structure in natural scenes leads to a set of oriented, band-pass, localized projections. Field (1994) has shown that, when comparing the outputs of three sets of filters applied to natural scenes, oriented wavelet filters provided the sparsest responses. Our methods extend the range of this comparison across all possible linear filters, and yield results consistent with Field's.

Motivated by Field's results, Olshausen and Field (1996) have demonstrated that a finite set of filters constrained to sparsely reconstruct natural scenes develop oriented, band-pass, localized structure. Our method, though significantly different, shows a strong relationship to that of Olshausen and Field.

The objective function Olshausen and Field maximize is given by

$$Q[a,w] = -\Sigma_{x,y}[I(x,y) - \Sigma_i a_i w_i(x,y)]^2 - \beta \Sigma_i S(a_i - \mu_i / \sigma_i) \tag{4}$$

The first term measures the error in image ($I(x,y)$) reconstruction by coefficient-weighted (a_i) filters (w_i), and the second term (S) requires that the reconstruction coefficients be sparse. This differs fundamentally from our objective function, as we do not constrain our projections to reconstruct the data set. Projections resulting from projection pursuit, as we have described it, directly reflect the statistical structure of natural scenes alone. Filters resulting from Olshausen and Field's method, however, reflect both the image statistics and the reconstruction constraint.

Despite this difference, projections found with projection pursuit can satisfy Olshausen and Field's reconstruction constraint. This term is minimized whenever the filters *wi* are linearly independent; structure removal fulfills this constraint by forcing projections to be different from one another. When the *wi* form a tight frame over the image space (Daubechies, 1990) and the $w_i^T X$ are equated with the coefficients a_i, our method and that of Olshausen and Field approach equivalence.

A caveat to the above analysis draws attention to the limitations imposed by structure removal, where the data set is modified after each projection. First, only the first projection is derived from the original data; subsequent projections result from an increasingly modified data set. This limits the degree to which these subsequent projections represent the original natural scene statistics. This is in contrast to Olshausen and Field, where filters are derived in parallel from the original data set. Second, for the w_i to form a tight frame, they must span the image space. As shown in Figure 1, though, the number of meaningful projections obtainable with structure removal is fewer than the full image dimensionality. Thus, a portion of projections that contribute to their forming a tight frame will have little relevance to the image statistics. To relieve these limitations, projection pursuit can be performed exclusively on the original unmodified data set. Instead of determining each iteration's initial conditions by coarse stepping, they can be determined by the projections obtained with structure removal. Then, all projections would reflect the unmodified statistics of natural scenes, and their number can be unlimited.

In addition to finding V1-like sparse (kurtotic) structure in natural images, similar structure was found when using the BCM model's skew objective function. This suggests that the structure we see may have more to do with deviations from normality than with the projection index's precise formulation.

A fundamental difference between our results and those of the original BCM model is that BCM-derived projections are primarily of either horizontal or vertical orientation, while our projections are randomly oriented. This bias was also reported by Fyfe and Baddeley (1995), who trained a neural network on natural images using a kurtosis objective function and found exclusively horizontal and vertical receptive fields. Our not finding a bias may be, in part, due to our use of structure removal. Once a projection was found, structure removal forced subsequent projections of similar location and band-pass characteristics to be of different orientation.

The horizontal and vertical bias reported by Law and Cooper (1992) and Fyfe and Baddeley (1995) may, alternatively, be related to how the images used as data were acquired. When a natural image is sampled with a regular cartesian grid, its circularly symmetric (approximately $1/f$) fourier amplitude spectrum is replicated with a spacing inversely proportional to the density of the sampling lattice. When the lattice is sufficiently dense, these replicated fourier amplitude spectra do not interfere with one another. When an image is undersampled, though, these circularly symmetric functions overlap, preferentially in the direction of the axes. Thus, undersampling images would result in significantly increased horizontal and vertical power, and, consequently, horizontally- and vertically-biased results.

In addition to searching for linear projections, we have shown that representing images with two parallel streams, one reflecting image intensity increases and the other reflecting image intensity decreases, results in a projection pair whose corresponding regions are in quadrature. These results suggest that these more complicated response properties may arise from simple transformations, such as on-off channel segregation, of natural scene statistics.

How receptive fields, or projections, are combined to form downstream responses can also be examined with projection pursuit. For example, the output from first-stage projections can be passed through a non-linearity and then used as a data set for our projection pursuit algorithm. Each pixel in a second-stage projection would correspond to a first-stage projection, and each pixel's intensity would indicate the degree to which that first-stage projection contributes to the second-stage projection's structure. Thus, it is possible to apply projection pursuit iteratively to investigate hierarchical processing, such as that seen in the visual system.

6. CONCLUSION

Exploratory projection pursuit provides a way to examine directly the statistical structure of natural images. Searching for sparse (kurtotic) and skew structure results in oriented, band-pass, localized projections. This suggests that receptive fields in V1, which share these characteristics, directly reflect the statistical structure of the world they represent. Encoding visual scenes more similarly to the retina and geniculate, with on- and off-channels, results in projections that capture additional structure in the visual environment. Projection pursuit will provide a fertile ground for future research, including the study of additional non-linear processing found in the visual system.

7. REFERENCES

Atick JJ and Li Z (1992) Towards a theory of the striate cortex. *Neural Comput.* **6**, 127–146.

Atick JJ and Redlich AN (1990) Towards a theory of early visual processing. *Neur. Comput.* **2**, 308–320.

Atick JJ and Redlich AN (1992) What does the retina know about natural scenes? *Neur. Comput.* **4**, 196–210.

Baddeley R (1996) Searching for filters with "interesting" output distributions: an uninteresting direction to explore? *Network* **7**, 409–421.

Barlow HB (1972) Single units and sensation: A neuron doctrine for perceptual psychology? *Perception* **1**, 371–394.

Daubechies I (1990) The wavelet transform, time-frequency localization and signal analysis. *IEEE Trans. Inform. Theory* **36**(5), 961–1005.

Field DJ (1994) What is the goal of sensory coding? *Neur. Comput.* **6**, 559–601.

Friedman JH (1987) Exploratory projectin pursuit. *J. Amer. Statis. Assoc.* **82**(397), 249–266.

Fyfe C and Baddeley R (1995) Finding compact and sparse-distributed representations of visual images. *Network* **6**, 333–344.

Huber PJ (1985) Projection pursuit. *Annals Statis.* **13**(2), 435–475.

Intrator N (1992) Feature extraction using an unsupervised neural network. *Neur. Comput.* **4**, 98–107.

Intrator N and Cooper LN (1992) Objective function formulation of the BCM theory of visual cortical plasticity: statistical connections, stability conditions. *Neural Networks* **5**, 3–17.

Law CC and Cooper LN (1994) Formation of receptive fields in realistic visual environments according to the Bienenstock, Cooper, and Munro (BCM) theory. *Proc. Natn. Acad. Sci. U.S.A.* **91**, 7797–7801.

Olshausen BA and Field DJ (1996) Emergence of simple-cell receptive field properties by learning a sparse code for natural images. *Nature* **381**, 607–609.

Olshausen BA and Field DJ (1996) Natural scene statistics and efficient coding. *Network* **7**, 333–339.

Press WH, Teukolsky SA, Vetterling WT, and Flannery BP (1992) Minimization or maximization of functions. *Numerical recipes in C, 2nd ed.*, Cambridge University Press.

121

A COMPUTATIONAL MODEL OF SPATIAL REPRESENTATIONS THAT EXPLAINS OBJECT-CENTERED NEGLECT IN PARIETAL PATIENTS

Rajesh P. N. Rao* and Dana H. Ballard

Department of Computer Science
University of Rochester
Rochester, NY
E-mail: {rao,dana}@cs.rochester.edu

1. INTRODUCTION

Patients with parietal cortex lesions exhibit unusual visual deficits, typically involving the neglect of a part of their visual space. Without being consciously aware of it, they act as if that part of space is not visible. At first, it was thought that this neglected space coincided with the visual hemifield contralateral to the lesioned hemisphere, but more recently, an increasing number of experiments [1, 2, 5, 6, 7, 8, 11] have shown that in many cases, the neglected part of space is related to a reference object of immediate interest. For example, in a recent experiment by Behrmann et al. [3], a patient with a right parietal lobe lesion was asked to count the number of instances of the letter "A" in a field of letters on a TV screen (see Figure 1). The eye movements recorded from the subjects showed that they typically neglected to look at most of the "A"s in the left side of the TV screen. Note that the neglect cannot be explained as a visual hemifield neglect because, *as the patient makes eye movements, the letters that appear in each hemifield change.*

We know that the observed behavior pertains to the object of immediate interest owing to another experiment by Behrmann *et al.* [2]. In this case, the patient gazes at a dumbbell-shaped object consisting of two circles (one red and one blue) joined by a horizontal bar. The patient is asked to press a button when a target (a small white circle) appears within either the left circle or the right. As expected, the response to targets in the contralesional circle (opposite side of the lesion) is typically much slower. Now the patient sees the dumbbell slowly rotate 180 degrees clockwise about its midpoint so that the colored circles have now exchanged positions. When the same test is repeated using the rotated dumbbell, the patient is now much

---

*To whom correspondence should be addressed.

Computational Neuroscience
edited by Bower, Plenum Press, New York, 1997

Figure 1. The Visual Counting Task. Subjects were asked to count the number of occurrences of the letter "A" on the current display screen.

slower with respect the ipsilesional circle (same side as the lesion). This result is explained if the patient assigns an object centered frame to the dumbbell initially and maintains that frame throughout the experiment. The neglect is thus consistently in object-centered coordinates. Other experiments [6, 7] have demonstrated similar results.

While neglect in object-centered coordinates is easy to understand abstractly, it is much more difficult to explain how the brain's object recognition machinery could be organized to explain such results. The purpose of this paper is to describe a systems-level model of spatial representations in the parietal cortex, and show by simulations that it can explain the above experimental data in a concise manner.

2. REFERENCE FRAMES

A central problem in object recognition is that of determining the *pose* of an object, where pose characterizes the transformation between an object-centered reference frame and the current view frame (retinal reference frame). For humans, the retinal frame is determined by the current fixation point. However, it is easy to demonstrate the usefulness of a third frame. In reading, the position of letters with respect to the retina is unimportant compared to their position in the encompassing word. In driving, the position of the car with respect to the fixation point is unimportant compared to its position with respect to the edge of the road. In both of these examples, the crucial information is contained in the transformation between an object-centered frame and a *scene* frame [9].

Figure 2 shows the relationships between the three reference frames for the image of the letter "A" (the object) depicted on a TV screen (the scene). The transformations between the frames are denoted by T_{os} (object-scene), T_{sr} (scene-retina), and T_{or} (object-retina). The position of remembered objects with respect to a scene, T_{os}, together with the scene-to-retina transformation T_{sr}, determines the current position of objects T_{or} in retinotopic space.

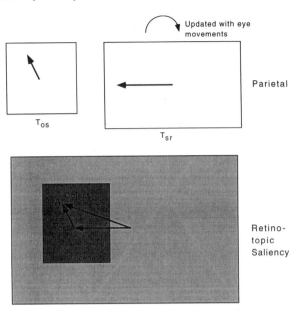

Figure 2. **Three Fundamental Spatial Transformations**. To represent the geometric relations of visual features, three transformations are fundamental. The first (T_{sr}) describes how a particular depiction of the world, or scene, is related to the retinal coordinate system with respect to the current fixation. The second (T_{os}) describes how objects can be related to the scene. The third (T_{or}), which is the composition of the other two, describes how objects are located with respect to the retina.

3. THE MODEL

The reference frames discussed in the previous section provide a basis for constructing a systems-level model of spatial representations in the parietal cortex. The model uses "iconic representations" for recognizing and searching for targets in natural scenes. These iconic object representations are obtained by filtering the scene with a large number of oriented spatial filters at multiple scales [15]. This allows each location in the scene to be characterized by a vector of filter responses that serves as an effective signature of the photometric intensity variations in the region surrounding the given scene location. We refer the reader to [4, 12, 15, 19, 20] for more information regarding recognition methods based on such iconic representations.

In order to search for possible locations containing a target (for example, an "A") in a given scene, the remembered filter response vector for the target is correlated with those for all scene locations. This results in a retinotopic *saliency map* (see Figure 3; brighter spots indicate higher correlations with the target, in this case, an "A"). Given a saliency map denoting possible target locations in the scene, the number of targets in the scene can be counted by approximately fixating on each candidate target location in succession. Unfortunately, this strategy changes the location of the targets in retinal coordinates. Thus, the central issue is *how to keep track of the already counted locations, which shift in a retinotopic frame after each eye movement.* One possible solution is to inhibit the counted locations in the retinal frame, but this requires elaborate circuitry to keep track of the inhibited locations across eye movements. A more elegant solution is to use a separate representation for the scene that describes the location of the scene with respect to retinal coordinates. In the model, the transform of the scene frame with respect to the retina is activated in a separate area T_{sr} in parietal cortex. The transform T_{sr} is continually updated using posture signals ("efference copies" or "corollary" discharges) as derived from eye, head, and body movements. The use of the scene frame allows the relations of the parts of the scene (for example, letters on a TV screen) to be depicted in a separate object-centered frame denoted by T_{os}, which is assumed to be represented bilaterally in parietal cortex, with each half in a separate hemisphere. In addition, the scene frame is assumed to be task-dependent. Once initialized (at the beginning of the task), it is maintained until the task is deemed completed.

Consider the situation where objects are being counted sequentially by approximately foveating them. After an eye movement is made to an object, its corresponding location in T_{os} is inhibited. Note that these locations do not change with eye movements, thereby avoiding the problem of shifting inhibitory markers. After an object has been foveated and counted, the saliency map is recomputed. The scene frame and the object-centered frame then combine to inhibit previously visited locations in the saliency map, yielding a new location for the next eye movement in retinal coordinates. Note that this location is represented in exactly the same coordinate system as that used by the oculomotor system, which allows a saccade to be executed to the desired target location ([14] suggests a possible method for learning such saccadic eye movements). Object centered neglect can be easily explained as unilateral damage to T_{os}. This damage does not allow parts of the scene on the damaged side to be represented. Thus, regardless of eye position, they can never be accessed.

4. EXPERIMENTAL RESULTS

The model was tested using the visual search task of Behrmann *et al.* [3] involving the counting of the number of occurrences of the letter "A" on a display. Figure 3 (left) shows the case without damage, where the search is successful for the display shown in Figure 1. The images depict the alternating sequence of saliency maps (in retinal coordinates) that are used to initiate the next saccade, followed by the resulting retinal image after the saccade (the fixation point is always the center of the image and the fovea is denoted by a circle). The correlation peaks in the saliency map (bright regions in (a)) depict possible "A" locations. An eye movement to the highest peak allows the corresponding "A" to be foveated and counted as shown in (b). The foveated location is then inhibited in object-centered coordinates. This inhibition shows up in retinal coordinates in the saliency map in (c) by virtue of the fact that the scene frame is continuously translating T_{os} locations to T_{or} locations. A subsequent eye movement is shown in (d), and the final state of the model at the end of the counting process is shown in (e).

Figure 3 (right) shows the model with right parietal lobe damage. The damage is assumed to lie in the area of parietal cortex responsible for the T_{os} transformation. Since T_{os} is required for computing the current saliency map, its damage prevents the appearance of any task-relevant saliency peaks in the contralesional side of the object-centered frame. As a result, any targets (such as the "A" locations) on the contralesional side of the current scene (the TV screen) fail to be noticed and are not attended to or foveated during the course of the task. Note that the "damage" (dark region) shows up in different parts of the retinal frame as a result of eye movements and the updating of the scene frame.

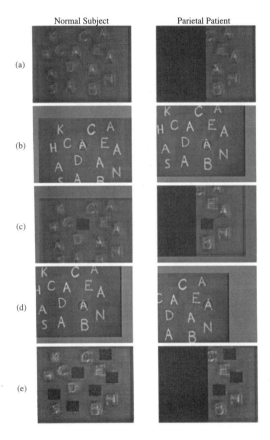

Figure 3. Simulation Results. The left side of the figure shows the model without a lesion ("normal subject") counting the number of occurrences of the letter "A" in the display shown in Figure 1. The right side shows the model with a right hemispheric lesion ("parietal patient"). The panels in (a), (c), and (e) depict the retinotopic saliency map of Figure 2. (a) Correlation peaks in retinal coordinates. (b) The first "A" foveated with an eye movement and counted. (c) Inhibition of the location of the counted "A". (d) The second "A" is foveated and counted. The process repeats until the correlation peaks fall below a preset threshold for detecting the presence of an "A". (e) The final state of the model. Note that the model without a lesion ("normal subject") is able to count all 6 occurrences of the letter "A" while the lesioned model ("parietal patient") counts only 3.

5. DISCUSSION

Observations from related experiments such as Behrmann et al.'s rotating dumbbell task [2] can be succinctly explained in the context of the present model by allowing the scene transformation to be sufficiently general so as to allow for rotation. In this case, the explanation for the neglect is the same. The targets are represented in T_{os}, the movement of the display is interpreted by the subject as a change in T_{sr}, and the neglect follows by the same mechanism.

Some object-centered neglect effects can be obtained by using only retinal frames, but such models cannot explain complicated effects such as those elicited by the dumbbell display. Previous models of unilateral neglect [10, 13] have either relied on highly abstract interpretations of experimental observations or have concentrated on deriving relatively low-level implementations that can explain a given set of experimental observations. The model presented herein attempts to bridge this gap by suggesting a systems-level mechanism for explaining a wide variety of neglect-related phenomena while simultaneously retaining the possibility of a neural implementation. Ongoing work involves integration of the present model with the Kalman filter-based neural model of the cortex [17, 18]. Preliminary results in this direction have been encouraging [16].

REFERENCES

[1] M. Behrmann and M. Moscovitch. Object-centered neglect in patients with unilateral neglect: Effects of left-right coordinates of objects. *Journal of Cognitive Neuroscience*, 6(1):1–16, 1994.

[2] M. Behrmann and S.P. Tipper. Object-based attentional mechanisms: Evidence from patients with unilateral neglect. In C. Umilta and M. Moscovitch, editors, *Attention and Performance XV: Conscious and Nonconscious Information Processing*. MIT Press, Cambridge, MA, 1994.

[3] M. Behrmann, S. Watt, S.E. Black, and J.J.S. Barton. Impaired visual search in patients with unilateral neglect: An oculographic analysis. Submitted, 1996.

[4] J.M. Buhmann, M. Lades, and C.v.d. Malsburg. Size and distortion invariant object recognition by hierarchical graph matching. In *Proc. IEEE IJCNN, San Diego (Vol. II)*, pages 411–416, 1990.

[5] R. Calvanio, P. N. Petrone, and D. Levine. Left visual spatial neglect is both environment-centered and body-centered. *Neurology*, 37:1179–1183, 1987.

[6] A. Caramazza and A.E. Hillis. Spatial representation of words in the brain implied by studies of a unilateral neglect patient. *Nature*, 346:267–269, 1990.

[7] M. J. Farah, J. L. Brunn, A. B. Wong, M. Wallace, and P. Carpenter. Frames of reference for the allocation of spatial attention: Evidence from the neglect syndrome. *Neuropsychologia*, 28:335–347, 1990.

[8] M. Gazzaniga and E. Ladavas. Disturbances in spatial attention following lesion or disconnection of the right parietal lobe. In M. Jeannerod, editor, *Neurophysiological and Neuropsychological Aspects of Spatial Neglect*. North-Holland, Amsterdam, 1987.

[9] G.E. Hinton. A parallel computation that assigns canonical object-based frames of reference. In *7th International Joint Conference on Artificial Intelligence*, pages 683–685, 1981.

[10] J. Beng-Hee Ho, M. Behrmann, and D.C. Plaut. The interaction of spatial reference frames and hierarchical object representations: A computational investigation of drawing in hemispatial neglect. In *Proc. 17th Annual Conf. of the Cognitive Science Society*, pages 148–153, 1995.

[11] E. Ladavas. Is hemispatial deficit produced by right parietal damage associated with retinal or gravitational coordinates. *Brain*, 110:167–180, 1987.

[12] B. Mel. A neurally-inspired approach to 3-D visual object recognition. Presentation at Telluride Workshop on Neuromorphic Engineering, Telluride, Colorado, July 1994.

[13] A. Pouget and T.J. Sejnowski. A model of spatial representations in parietal cortex explains hemineglect. In D. Touretzky, M. Mozer, and M. Hasselmo, editors, *Advances in Neural Information Processing Systems 8*, pages 10–16. Cambridge, MA: MIT Press, 1996.

[14] R.P.N. Rao and D.H. Ballard. Learning saccadic eye movements using multiscale spatial filters. In G. Tesauro, D.S. Touretzky, and T.K. Leen, editors, *Advances in Neural Information Processing Systems 7*, pages 893–900. Cambridge, MA: MIT Press, 1995.

[15] R.P.N. Rao and D.H. Ballard. An active vision architecture based on iconic representations. *Artificial Intelligence (Special Issue on Vision)*, 78:461–505, 1995.

[16] R.P.N. Rao and D.H. Ballard. A class of stochastic models for invariant recognition, motion, and stereo. Technical Report 96.1, National Resource Laboratory for the Study of Brain and Behavior, Department of Computer Science, University of Rochester, June 1996.

[17] R.P.N. Rao and D.H. Ballard. Dynamic model of visual recognition predicts neural response properties in the visual cortex. *Neural Computation* (in press). Also, Technical Report 96.2, National Resource Laboratory for the Study of Brain and Behavior, Department of Computer Science, University of Rochester, 1996.

[18] R.P.N. Rao and D.H. Ballard. Cortico-cortical dynamics and learning during visual recognition: A computational model. In J. Bower, editor, *Computation Neuroscience 1996*. New York, NY: Plenum Press, 1997.

[19] R.P.N. Rao, G.J. Zelinsky, M.M. Hayhoe, and D.H. Ballard. Modeling saccadic targeting in visual search. In D. Touretzky, M. Mozer, and M. Hasselmo, editors, *Advances in Neural Information Processing Systems 8*, pages 830–836. Cambridge, MA: MIT Press, 1996.

[20] P. Viola. Feature-based recognition of objects. In *AAAI Fall Symposium on Learning and Computer Vision*, 1993.

122

CORTICO-CORTICAL DYNAMICS AND LEARNING DURING VISUAL RECOGNITION: A COMPUTATIONAL MODEL

Rajesh P. N. Rao* and Dana H. Ballard

Department of Computer Science
University of Rochester
Rochester, NY
E-mail: {rao,dana}@cs.rochester.edu

1. INTRODUCTION

A ubiquitous feature of the neocortex is the reciprocity of connections between its many distinct areas: if area A projects to area B, then area B almost invariably projects to area A [5, 22]. While the role of the feedforward projections in facilitating tasks such as visual recognition is generally well-acknowledged, the precise computational role of the corresponding feedback projections has remained relatively unclear (cf. [1] p. 23).

In this paper, we describe a model of visual recognition wherein feedback pathways play as important a role in guiding perception and recognition as the feedforward pathways. The visual cortex is treated as a neural network that implements a hierarchical form of the Kalman filter [10, 11] from optimal control theory [13]. At each hierarchical level, top-down signals from a higher, more abstract area are combined with bottom-up signals from a lower area for optimal estimation of current recognition state. The Minimum Description Length (MDL) principle [21, 27] is used for deriving the local Kalman filter dynamics of the network as well as the synaptic weight modification rules necessary for learning objects that need to be recognized. Preliminary simulation results using realistic objects indicate that the model can successfully recognize objects even in the presence of partial occlusions and can maintain fine discrimination ability between highly similar visual stimuli.

2. ARCHITECTURE OF THE MODEL

The model employs a hierarchical neural network as shown in Figure 1. A canonical module of this network at a given hierarchical level is comprised of three sets of synaptic

---

*To whom correspondence should be addressed.

Computational Neuroscience
edited by Bower, Plenum Press, New York, 1997

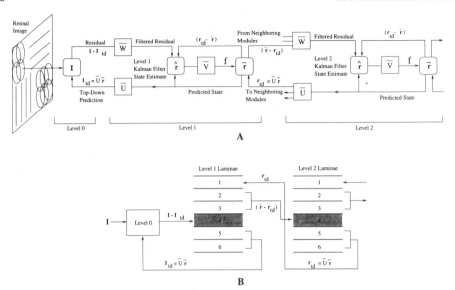

Figure 1. Architecture of the Model. (A) The feedforward pathways for the first two levels are shown in the top half of the figure while the bottom half represents the feedback pathways. The feedback pathways carry top-down predictions $\mathbf{r}_{td}$ (or $\mathbf{I}_{td}$ at the first level) and the feedforward pathways carry residuals between the state estimate $\bar{\mathbf{r}}$ and the top-down predictions. These residuals are filtered through the feedforward weights $\overline{W}$ (a learned estimate of W) at the next level. Each feedforward matrix $\overline{W}$ is approximately the transpose of its corresponding feedback matrix $\overline{U}$ (which is a learned estimate of U). The current state estimate $\widehat{\mathbf{r}}(t)$ at each level is continually updated using the top-down and bottom-up residuals according to the Kalman filter update equation. The next state prediction $\bar{\mathbf{r}}(t+1)$ is computed from the estimate $\widehat{\mathbf{r}}(t)$ using the prediction weights $\overline{V}$ and the activation function f. The figure illustrates the architecture for the simple case of three level 1 modules feeding into a level 2 module. However, this arrangement easily generalizes in a recursive manner to the case of an arbitrary number of lower level modules feeding into a higher level module. **(B)** shows how the model can be implemented in a laminar structure such as the cortex.

weights: feedforward weights (denoted by the matrix W), prediction weights (V), and feedback weights (U). At each time instant t, each module maintains an optimal estimate $\widehat{\mathbf{r}}(t)$ of the current state $\mathbf{r}(t)$ of the observed visual process. The estimate $\widehat{\mathbf{r}}(t)$ is calculated by linearly filtering the "bottom-up" input $\mathbf{I}(t)$ through the feedforward weights W and integrating the result with a "top-down" prediction $\mathbf{r}_{td}$ from a higher level (see below). A prediction $\bar{\mathbf{r}}(t+1)$ for the next time instant is then generated from $\widehat{\mathbf{r}}(t)$ using the prediction weights V and a possibly nonlinear activation function f. This prediction is then conveyed as output to the lower level by multiplying it with the feedback matrix U which translates the prediction to the lower abstraction level (see Figure 1).

The outputs $\bar{\mathbf{r}}$ of spatially adjacent modules at each level are also conjoined and fed as input to the next higher level, whose outputs are in turn conjoined with those of its neighbors and fed into yet another higher level, until the entire visual field has been accounted for. As a result, the receptive fields of units become progressively larger as one ascends the hierarchical network in a manner similar to that observed in the occipitotemporal pathway [3, 25]. At the same time, the feedback connections allow top-down influences to modulate the output of lower level modules. From a computational perspective, such an arrangement allows a hierarchy of abstract internal representations to be learned while simultaneously endowing the system with properties essential for dynamic recognition such as perceptual completions using higher level context. From a neuroanatomical perspective, such an architecture falls well within the known complexity of cortico-cortical and intracortical connections in the mammalian visual cortex [4, 5, 22].

3. NETWORK DYNAMICS AND SYNAPTIC LEARNING RULES

At each intermediate level, bottom-up input $\mathbf{I}$ from the preceding level is optimally combined with the top-down prediction $\mathbf{r}_{td}$ from a higher level using the following Kalman filter update equation for the optimal state estimate $\widehat{\mathbf{r}}$ at time t:

$$\widehat{\mathbf{r}}(t) = \overline{\mathbf{r}}(t) + K_1(\mathbf{I} - U\overline{\mathbf{r}}(t)) + K_2(\mathbf{r}_{td} - \overline{\mathbf{r}}(t)) - K_3 g(\overline{\mathbf{r}}(t)) \tag{1}$$

The vector $\overline{\mathbf{r}}(t)$ denotes the neural responses comprising the prediction of the state at time t computed from the state estimate at time $t - 1$:

$$\overline{\mathbf{r}}(t) = f(V\widehat{\mathbf{r}}(t - 1)) + \overline{\mathbf{n}}(t - 1) \tag{2}$$

where $\overline{\mathbf{n}}$ is an additive noise term.

The update equation 1 is derived from a minimum description length (MDL) based optimization function that penalizes deviations from a hierarchical stochastic model of the image generation process [18]. The update equation corrects the prediction $\overline{\mathbf{r}}$ using the sensory residual $(\mathbf{I} - U\overline{\mathbf{r}}(t))$ and the top-down residual $(\mathbf{r}_{td} - \overline{\mathbf{r}}(t))$. Both residuals are weighted by their respective gain matrices $K_1 = PU^T\Sigma_{bu}^{-1}$ and $K_2 = P\Sigma_{td}^{-1}$ which take into account the uncertainties in the prediction process, the input process, and the top-down feedback process. The quantities P, Σ_{bu} and Σ_{td} denote the state, bottom-up and top-down covariance matrices respectively (T denotes matrix transpose).

An attractive feature of the dynamics embodied in Equation 1 is that it implements an efficient trade-off between information from three different sources: the system prediction $\overline{\mathbf{r}}(t)$, the top-down prediction $\mathbf{r}_{td}$, and the bottom-up data $\mathbf{I}$. Intuitively, the bottom-up and top-down gain matrices K_1 and K_2 can be interpreted as *signal-to-noise ratios*. Thus, when the bottom-up noise variance $\widehat{\Sigma}_{bu}$ is high, the bottom-up term $(\mathbf{I} - U\overline{\mathbf{r}}(t))$ is given less weight (due to a lower gain matrix K_1). On the other hand, when the top-down noise variance $\widehat{\Sigma}_{td}$ is high, the estimate relies more heavily on the bottom-up term and the state prediction. Finally, if the state prediction $\overline{\mathbf{r}}(t)$ has a large noise variance, the matrix P assumes larger values which implies that the system relies more heavily on the top-down and bottom-up input data rather than on its noisy prediction. The dynamics of the network at each level thus strives to achieve a delicate balance between the current prediction and the actual inputs from various cortical sources by exploiting the signal-to-noise characteristics of the corresponding input channels. By keeping track of the degree of correlations between units at any given level, the covariance matrices also dictate the degree of *lateral interactions* between the units as determined by the gain terms K_i in Equation 1. The possibly nonlinear decay term $-K_3 g(\overline{\mathbf{r}}(t))$ arises from the MDL based optimization function [18] and allows the network to seek higher-order statistical correlations in the input data, as suggested by Olshausen and Field [15] in their sparse maximization approach to learning localized receptive fields.

Synaptic learning rules analogous to Equation 1 above can be derived for adapting the weights U, V and W [18]. For example, a Kalman filter-based Hebbian learning rule for computing an estimate $\overline{U}$ of the actual weights U can be obtained by collapsing the rows of $\overline{U}$ into a vector $\overline{\mathbf{u}}$ and deriving an update equation similar to Equation 1 for the optimal weight vector estimate $\widehat{\mathbf{u}}$ [18]:

$$\widehat{\mathbf{u}}(t) = \overline{\mathbf{u}}(t) + C_1(\mathbf{I} - \overline{U}\overline{\mathbf{r}}(t)) - C_2\overline{\mathbf{u}}(t) \tag{3}$$

Once again, the above learning rule involves adding a correction term $(\mathbf{I} - \overline{U}\overline{\mathbf{r}}(t))$ to the prior estimate $\overline{\mathbf{u}}(t)$ (which is computed as $\widehat{\mathbf{u}}(t - 1)$ plus an additive noise term). The correction term

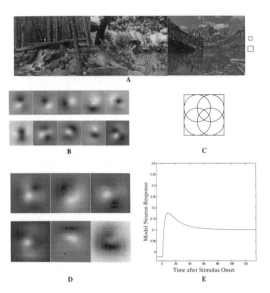

Figure 2. Receptive Fields from Natural Images. Training the hierarchical network on natural images resulted in synaptic weight vectors that closely resembled the spatial receptive field profiles of visual cortical cells (see also [8, 15]). (A) Three of the eight images used in training the weights. The boxes on the right show the size of the first and second-level receptive fields respectively on the same scale as the images. (B) Ten of the twenty feedforward receptive fields (rows of $\overline{W}$) in a module at the first level (black is negative and white is positive). These resemble classical oriented edge/bar detectors, which have been previously modeled as difference of offset Gaussians or Gabor functions. (C) The effective second level receptive field as spanned by four overlapping first level receptive fields. (D) Six of twenty receptive fields at the second level. These appear to be tuned towards more complex visual features such as corners and curves in addition to localized edges or bars. (E) A typical response of a model neuron in level 1 (equivalent to V1) computing the residual $(\hat{\mathbf{r}} - \mathbf{r}_{td})$, which is the feedforward signal transmitted to level 2 (see Figure 1). The neuron exhibits an initial transient response that is subsequently suppressed by the inhibitory effect of appropriate top-down signals from level 2.

is again weighted by a gain matrix C_1 computed from the state and bottom-up error covariances. The linear weight decay term $-C_2\overline{\mathbf{u}}(t)$ serves two purposes. First, it acts as a regularizer by penalizing overfitting, thereby increasing the potential for generalization. Second, by using a similar learning rule for the feedforward weights $\overline{W}$, one can show that these weights converge to the transpose of the feedback weights $\overline{U}$ i.e. $\overline{W} \cong \overline{U}^T$. This allows the network of Figure 1 to implement the Kalman filter update equation 1 by using $\overline{W}$ in the place of U^T when filtering the correction term $(\mathbf{I} - U\overline{\mathbf{r}}(t))$ with the gain term K_1. A learning rule similar to the one above may be formulated for the prediction weights $\overline{V}$ [18].

4.   EXPERIMENTAL RESULTS

In order to validate the model, we trained a two-level hierarchical network on a sample of natural images. Image patches of size 16×16, windowed by a Gaussian, were used for training the network (the need for windowing can be eliminated by using non-linear decay functions g in the Kalman filter update equation above [15]). Since only static images were used, an identity matrix was used as the prediction matrix V, making $\overline{\mathbf{r}}(t) = \widehat{\mathbf{r}}(t-1)$, and the Kalman filter network was allowed to stabilize for each static input before updating the estimate of U (and W). The spatial receptive field (RF) profiles of the feedforward model

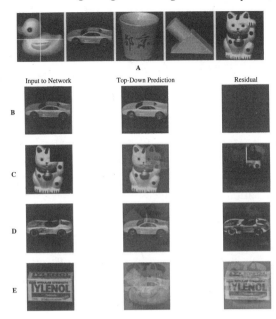

Figure 3. Simulation Results. (**A**) The objects used for training the hierarchical network. The four first level modules each contained 10 feedforward units (rows of W) while the single second level module contained 25. Each input image was of size 128×128 and was split into four equal subimages that were fed to the four lowest level modules. (**B**) through (**E**) show the responses of the hierarchical network to various input stimuli. When a training image is input as in (**B**), the network predicts an almost perfect reconstructed image resulting in a residual that is almost everywhere zero, which indicates correct recognition. If a partially occluded object from the training set is input as in (**C**), the unoccluded portions of the image together contribute to predict and fill-in the missing portions of the input. When the network is presented with an object that is highly similar to a trained object as in (**D**), the prediction is that of the closest resembling object (the car in the training set). However, the large residual (at the right) allows the network to judge the input as a new object rather than classifying it as the training object and incurring a false positive error, an occurrence that is common in most purely feedforward systems. (**E**) shows that a completely novel object results in a prediction image that resembles an arbitrary mixture of the training images and as such, generates large residuals. The system can then choose to either learn the new object (see Section 3), or ignore the object if it happens to be behaviorally irrelevant.

neurons after training were found to be comparable to those known to exist in the mammalian visual cortex (see Figure 2).

In a second experiment, the two-level network was trained on a set of man-made objects. After training, the network was tested for recognition performance. All objects that were in the training set were correctly recognized (close to zero residual errors at all pixels). In addition, the network was able to correctly predict missing portions of the input in the presence of partial occlusions (Figure 3 (C)) and was able to discriminate between objects that were highly similar in appearance (Figure 3 (D)). An additional property that accrues to the Kalman filter network is the ability to function as a novelty detector. Large prediction residuals imply that the object is possibly novel and worthy of further attention (Figure 3 (E)).

5. DISCUSSION AND CONCLUSION

From a computational perspective, the Kalman filter model possesses several properties essential for dynamic recognition such as the ability to perform pattern completions during occlusions and simultaneous top-down segmentation during recognition, in addition to facilitating top-down priming for tasks such as visual search [20]. While several models of the visual cortex have previously been proposed [2, 7, 9, 12, 14, 16, 23, 24, 26], the hierarchical application of Kalman filter theory in conjunction with the Minimum Description Length principle is new to this model. Furthermore, the model views the entire visual cortex as a hierarchical network, the lower levels coding for objects spanning smaller spatial extents (such as edges and bars) and the higher levels coding for increasingly more complex objects. A reassuring feature of the model is that it does not explicitly depend on the input signals being visual. Thus, given that the neocortex is structured along similar laminar input-output principles regardless of cortical area, it is reasonable to assume that the general framework proposed herein may very well be uniformly applicable to other cortical areas such as the parietal (see [17]), auditory, or motor cortex as well.

The laminar structure of the cortex is well-suited to implement a hierarchical Kalman filter such as the one suggested in this paper. When implemented in a laminar structure (Figure 1 (B)), the Kalman filter model provides succinct explanations of complex "extra-classical" phenomena such as the suppression of neural responses due to stimuli from beyond the classical RF (see [19] and response suppression during free-viewing of natural scenes [6] (see [18]). It is known that supragranular pyramidal cells (e.g. layers 2 and 3) project to layer 4 of the next higher visual cortical area while infragranular pyramidals (e.g. layer 6) send their axons back to layer 1 (and other layers) of the preceding lower area [5]. The architecture of the model (Figure 1) thus suggests that the supragranular cells carry the residual $(\bar{\mathbf{r}} - \mathbf{r}_{td})$ to the next level while the infragranular cell axons convey the current top-down prediction $\mathbf{r}_{td}$ to the lower level. The filtered residuals would then be computed by layer 4 cells. In such a setting, the Kalman filter dynamics emerges as a systems-level property of the cortical network, sculpted and structured to function in this manner by the ever-tinkering hands of evolution.

REFERENCES

[1] P. Churchland and T. Sejnowski. *The Computational Brain*. Cambridge, MA: MIT Press, 1992.

[2] P. Dayan, G.E. Hinton, R.M. Neal, and R.S. Zemel. The Helmholtz machine. *Neural Computation*, 7:889–904, 1995.

[3] R. Desimone and L.G. Ungerleider. Neural mechanisms of visual processing in monkeys. In F. Boller and J. Grafman, editors, *Handbook of Neuropsychology*, volume 2, chapter 14, pages 267–299. New York: Elsevier, 1989.

[4] R.J. Douglas, K.A.C. Martin, and D. Whitteridge. A canonical microcircuit for neocortex. *Neural Computation*, 1:480–488, 1989.

[5] D.J. Felleman and D.C. Van Essen. Distributed hierarchical processing in the primate cerebral cortex. *Cerebral Cortex*, 1:1–47, 1991.

[6] J.L. Gallant, C.E. Connor, H. Drury, and D.C. Van Essen. Neural responses in monkey visual cortex during free viewing of natural scenes: Mechanisms of response suppression. *Invest. Ophthalmol. Vis. Sci.*, 36:1052, 1995.

[7] S. Grossberg. How does the brain build a cognitive code? *Psychological Review*, 87, 1980.

[8] G.F. Harpur and R.W. Prager. Development of low-entropy coding in a recurrent network. *Network*, 7:277–284, 1996.

[9] E. Harth, K.P. Unnikrishnan, and A.S. Pandya. The inversion of sensory processing by feedback pathways: A model of visual cognitive functions. *Science*, 237:184–187, 1987.

[10] R.E. Kalman. A new approach to linear filtering and prediction theory. *Trans. ASME J. Basic Eng.*, 82:35–45, 1960.

[11] R.E. Kalman and R.S. Bucy. New results in linear filtering and prediction theory. *Trans. ASME J. Basic Eng.*, 83:95–108, 1961.

[12] M. Kawato, H. Hayakawa, and T. Inui. A forward-inverse optics model of reciprocal connections between visual cortical areas. *Network*, 4:415–422, 1993.

[13] P.S. Maybeck. *Stochastic Models, Estimation, and Control*. New York: Academic Press, 1979.

[14] D. Mumford. On the computational architecture of the neocortex. II. The role of cortico-cortical loops. *Biological Cybernetics*, 66:241–251, 1992.

[15] B.A. Olshausen and D.J. Field. Emergence of simple-cell receptive field properties by learning a sparse code for natural images. *Nature*, 381:607–609, 1996.

[16] T. Poggio. A theory of how the brain might work. In *Cold Spring Harbor Symposia on Quantitative Biology*, pages 899–910. Cold Spring Harbor Laboratory Press, 1990.

[17] R.P.N. Rao and D.H. Ballard. A class of stochastic models for invariant recognition, motion, and stereo. Technical Report 96.1, National Resource Laboratory for the Study of Brain and Behavior, Department of Computer Science, University of Rochester, 1996.

[18] R.P.N. Rao and D.H. Ballard. Dynamic model of visual recognition predicts neural response properties in the visual cortex. *Neural Computation* (in press). Also, Technical Report 96.2, National Resource Laboratory for the Study of Brain and Behavior, Department of Computer Science, University of Rochester, 1996.

[19] R.P.N. Rao and D.H. Ballard. The visual cortex as a hierarchical predictor. Technical Report 96.4, National Resource Laboratory for the Study of Brain and Behavior, Department of Computer Science, University of Rochester, September 1996.

[20] R.P.N. Rao, G.J. Zelinsky, M.M. Hayhoe, and D.H. Ballard. Modeling saccadic targeting in visual search. In D. Touretzky, M. Mozer, and M. Hasselmo, editors, *Advances in Neural Information Processing Systems 8*, pages 830–836. Cambridge, MA: MIT Press, 1996.

[21] J. Rissanen. *Stochastic Complexity in Statistical Inquiry*. Singapore: World Scientific, 1989.

[22] K.S. Rockland and D.N. Pandya. Laminar origins and terminations of cortical connections of the occipital lobe in the rhesus monkey. *Brain Research*, 179:3–20, 1979.

[23] W. Softky. Could time-series prediction assist visual processing? *Soc. Neuro. Abstr.*, 21:1499, 1995.

[24] S. Ullman. Sequence seeking and counterstreams: A model for bidirectional information flow in the cortex. In C. Koch and J.L. Davis, editors, *Large-Scale Neuronal Theories of the Brain*, pages 257–270. Cambridge, MA: MIT Press, 1994.

[25] D.C. Van Essen. Functional organization of primate visual cortex. In A. Peters and E.G. Jones, editors, *Cerebral Cortex*, volume 3, pages 259–329. Plenum, 1985.

[26] D.C. Van Essen, C.H. Anderson, and B.A. Olshausen. Dynamic routing strategies in sensory, motor, and cognitive processing. In C. Koch and J.L. Davis, editors, *Large-Scale Neuronal Theories of the Brain*, pages 271–299. Cambridge, MA: MIT Press, 1994.

[27] R.S. Zemel. *A Minimum Description Length Framework for Unsupervised Learning*. PhD thesis, Department of Computer Science, University of Toronto, 1994.

123

IMAGING WITH ELECTRICITY

How Weakly Electric Fish Might Perceive Objects

Brian Rasnow and James M. Bower

Caltech
Division of Biology, 216-76
Pasadena, California 91125

1. INTRODUCTION

The study of exotic sensory systems, such as electroreception in fish, echolocation in bats, and sound localization in owls, has revealed general principles of neuronal organization that are frequently present but more difficult to discern in other animals and humans. Weakly electric fish are an exceptional model system to study sensory acquisition, neuronal information processing, and sensory-motor integration. These animals detect nearby objects by sensing object-induced distortions in their electric organ discharge (EOD) electric field (reviewed in Bastian 1994; Carr 1990; Bullock and Heiligenberg 1986). Sensory electroreceptor organs, distributed across the fish's body, are acutely sensitive to small changes in transdermal voltage, which constitute an "electric image" of the object. We have investigated how electric fish might identify object features, such as size, shape, location, and impedance, from the object's electric images (Fig. 1). For example, how might a fish differentiate between a large, distant object and a small, nearby one; or a large object with impedance similar to water, and a small object with greater impedance difference? To resolve these questions, we constructed detailed and accurate simulations of the electric images of spheres and ellipsoids placed in EOD fields (Rasnow 1996). Electric images were computed analytically by assuming the measured EOD field was uniform around the object. Measured electric images of large metal spheres verified the simulations, and revealed their robustness to this assumption. In this paper, we summarize the algorithms for electrolocation presented by Rasnow (1996) and propose a plausible neural implementation of these algorithms in the fish's hind and midbrain.

2. ALGORITHMS FOR IDENTIFICATION OF SMALL ISOLATED OBJECTS

Electric images of small objects are broad 1–3 phase "bumps" in the transdermal potential across the body and contain only low spatial frequencies (Rasnow 1996; Bacher

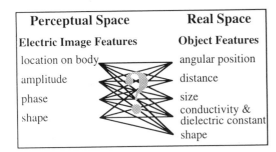

Perceptual Space	Real Space
Electric Image Features	Object Features
location on body	angular position
amplitude	distance
phase	size
shape	conductivity & dielectric constant
	shape

Figure 1. Understanding how fish electrolocate objects can be thought of as a problem of determining a mapping between two abstract feature spaces: a sensory electric image space internal to the animal, and the external environment. Listed here are what we consider some of the most important features or dimensions of these spaces. In the abstract language of this figure, our first goal is to constrain and sort out the nature of this mapping.

1983; Heiligenberg 1975). As such, they can be described by just a few parameters. The location of the main image peak approximately coincides with the latitude and longitude of the object, and thus unambiguously reveals two of the object's 3-dimensional coordinates. Determining object distance from an electric image is not as simple because it not related to single image parameters listed in Fig. 1. For example, the magnitudes of electric images are strongly dependent on object distance because of the rapid attenuation of the fish's electric field with distance from its body. But image magnitude also depends on other parameters such as object size, shape, and impedance. The first hint of a simple solution to disentangle these variables came from analyzing electric images of conducting spheres. We found that the relative width of the image peak (i.e., a parameter like the standard deviation of a Gaussian function) depends solely, and linearly, on the distance from the skin to the sphere's center. Object size can now unambiguously be solved for using the distance and peak amplitude of the image, which is proportional to the sphere's volume.

What if the sphere isn't a conductor? A sphere of radius a perturbs the potential at position $\mathbf{r}$ from its center by (Rasnow 1996):

$$\delta\varphi(\mathbf{r}) = \mathbf{E} \cdot \mathbf{r} \left(\frac{a}{r}\right)^3 \frac{\rho_w - \rho_o + i\omega\rho_w\rho_o(\varepsilon_o - \varepsilon_w)}{2\rho_o + \rho_w + i\omega\rho_w\rho_o(2\varepsilon_w + \varepsilon_o)} \tag{1}$$

where ρ and e are the resistivity and dielectric constants of the object and water (subscripts o and w respectively); $\omega = 2\pi f$ is the angular frequency of the unperturbed EOD field at the object, $\mathbf{E}$; and $i = \sqrt{-1}$. This is the equation of a dipole with a complex amplitude and phase shift. The right-hand term reduces to unity for ideal conductors and to $-1/2$ for ideal insulators. If the EOD field is oriented normal to the fish's skin, then conductors increase the current and transdermal potential directly below them, and insulating spheres decrease the transdermal potential by half the amount of a conductor of the same size. Objects with intermediate resistivities and dielectric strengths produce images with phase shifts between conductors and insulators and lesser magnitudes than conductors.

Since object and water impedances affect electric images globally (i.e., by multiplying them by a complex constant that is independent of position), the distance to the object's center is still proportional to the relative width of the image and thus can be determined unambiguously. However, object size and impedance are confounded unless the fish uses polarity and phase information to separate a^3 from the magnitude of the right-hand term in the above equation. Electric fish are extremely sensitive to EOD phase and possess specialized electroreceptors that encode minute phase shifts (Kawasaki et al 1988; Emde 1992).

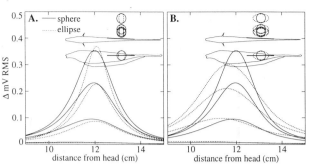

Figure 2. Electric images of conducting spheres and ellipsoids in the midplane of *Apteronotus leptorhynchus*, for 4 object distances (insets). Object centers are 1.2, 1.4, 1.9, and 4.2 cm lateral of the skin, and the spheres have 1 cm radii. The electric images have been averaged (RMS) over the EOD cycle.

What if the object isn't spherical? Figure 2 shows the simulated electric images of spheres and two orientations of ellipsoids with eccentricity of 2. The relative width of the images increases with object distance, however the rostrocaudally compressed ellipsoids have narrower images and the rostrocaudally expanded ellipsoids have wider images than corresponding spheres. Interpreting the first ellipsoid images (Fig. 2A) according to our algorithm for spheres results in the perception of a smaller and nearer sphere (because the image is narrower and has larger peak amplitude). Likewise, the second ellipsoid images (Fig. 2B) correspond to larger and more distant spheres. Although globally incorrect, these inferences are consistent with spheres whose proximal surfaces to the fish approximately correspond with the proximal surfaces of the actual ellipsoids.

The EOD field attenuates rapidly with distance from the fish, causing the electric image amplitude to attenuate with distance to the negative third to fifth power (Rasnow 1996). Therefore, the image of an object will be dominated by the nearest or proximal parts of the object. The resulting distortion, somewhat analogous to perspective distortion inherent in wide–angle optical lenses, might make it difficult for a fish to discriminate between the ellipsoid in Fig. 2A and a smaller, nearer sphere. However, phase information might resolve this ambiguity, and provide object shape information from electric images. We have shown that the electric field vector changes direction during the EOD cycle (Fig. 3; Rasnow & Bower 1996). The electric image of a conducting ellipsoid will be larg-

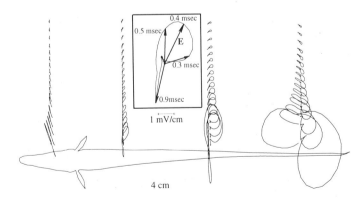

Figure 3. The electric field vector is shown here along four lateral lines in the midplane of *Apteronotus leptorhynchus*. The initial phase of the field vector is shown as a line from each measurement point (dots). At subsequent times, just the tip of each field vector is traced. At any phase, the field is represented by a vector from the measurement point to the curve (inset shows 4 example phases). The field vectors rotate counterclockwise in the caudal part of the body, whereas rostral of the gill, only the magnitudes and sign, but not the direction, changes during the EOD cycle.

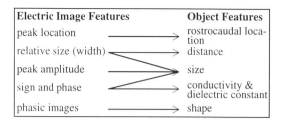

Electric Image Features	Object Features
peak location	rostrocaudal location
relative size (width)	distance
peak amplitude	size
sign and phase	conductivity & dielectric constant
phasic images	shape

Figure 4. This constrained mapping between sensory electric images and the external environment is sufficient to locate and identify small homogeneous objects. It remains to be tested whether and how electric fish might implement these algorithms with their neural hardware.

est when the EOD field is parallel to the major axis, because the ellipsoid short circuits a larger region of water. The variation of electric image amplitude with EOD phase could thus reveal object asymmetry, crudely analogous to how an object's shadow shape depends on illumination angle. Exploratory behaviors such as tail bending could also provide shape information by changing the field orientation. Finally, differences in the ELL maps could also convey information about object shape and eccentricity (see below).

The proposed mappings between electric image and object feature spaces, summarized in Fig. 4, are only a first guess and working hypothesis of how weakly electric fish might process electrosensory information to perceive objects. The corresponding algorithms are capable of identifying the major features of small, isolated, stationary objects near a stationary fish. We have only begun to explore their robustness in the fish's more complex environment, where potentially confounding factors include multiple active and passive sources; large and heterogeneous, non isotropic objects; and EOD envelope modulations from relative motion of the fish and objects. Although some of these confounding factors can be isolated by simple temporal or spatial filtering, more substantial modifications of the algorithms may be necessary to resolve other ambiguities. Note that these simple algorithms ignore a vast amount of electrosensory information, for example, correlations within temporal sequences of electric images generated as the fish moves. Given the complex repertoire of behaviors electric fish use for exploration, it is evident that electrolocation involves additional algorithms to those presented here.

3. NEURAL IMPLEMENTATION

For the proposed algorithms to be involved in electrolocation, they must have a plausible neural implementation in the fish's nervous system. We propose here one such mapping of the computations discussed above onto the neural network in the electric fish brain. Figure 5 summarizes the Gymnotiform fish's electrosensory pathways and central processing structures (Carr & Maler 1986). The electrosensory lateral line lobe (ELL) receives the raw peripheral field encoded by the transdermal electroreceptors. The first processing step we believe may be to extract the object's perturbation or image. This could be done in ELL by the descending feedback and gain control (labeled 1), which is capable of subtracting out the expected EOD (Bell et al. 1997; Bastian 1996). The algorithms suggest that relative image size should be the next calculation. Cells in the ELL maps have center-surround type receptive fields. Convolving the object's image with center-surround spatial filters, and thresholding, results in an area of activity proportional to the image size (labeled 2). The object distance can be calculated by integrating over this active area. Such integration could be achieved within the convergent projections from ELL onto higher areas (labeled 3). Although ELL projects to both the dorsal preeminential nucleus (PEd) and the torus semicircularis (TS), the former is part of the feedback loop to compute object images in ELL. Therefore this scenario predicts

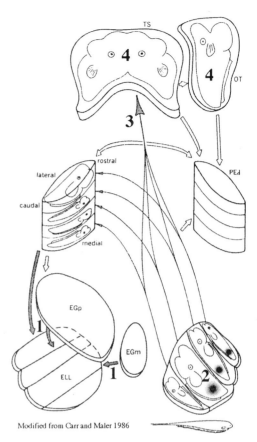

Figure 5. On this figure of electrosensory pathways in the Gymnotiform fish, we have labeled regions where the proposed computations for high frequency electrolocation might be implemented. 1. Extraction of the object image, by subtracting expectation conveyed through descending feedback to ELL. 2. Convolution of the electric image with center-surround receptive fields and thresholding activates a region of ELL proportional to the image's relative width. 3. Integrating over the ELL surface, in the convergent projections to the torus semicircularis (TS), measures the image size, which is proportional to object's distance. 4. Object size, shape, and higher order features could be computed in the torus, optic tectum (OT), and higher areas.

Modified from Carr and Maler 1986

that object distance may first be represented unambiguously in the amplitude pathway input layers of the torus. In particular, input layer neurons might respond similarly to large and small spheres centered at the same positions (and perhaps even ellipsoids at the corresponding locations), even though the electroreceptor responses in these cases would be quite different. Bastian (1986) found neurons in the optic tectum that selectively responded to particular object distances. Finally, knowing object distance is a prerequisite (or corequisite) in the model for deconfounding size, impedance, and shape, so these features would first appear in torus and higher areas.

4. FUTURE DIRECTIONS

Although somewhat speculative, we believe the above proposals may be a useful working hypothesis for interpreting and further exploring parts of the electrosensory nervous system. In addition to providing a functional model for electrolocation, they lead to several testable predictions, especially in higher and more complex parts of the nervous system where object features may begin to emerge unambiguously. For example, tuning to object-center distance should first emerge in the torus semicircularis. Distance preferences have been found in tectal neurons (Bastian 1986), but how the preferred distance related to the object's size, velocity, etc. was not investigated. In ELL, object information is more distributed within the somatotopic maps, making interpretation of single unit encoding of electric images more difficult to interpret. The three lateral ELL maps differ in spatial frequency of their center-surround filters. We are currently exploring whether each map may compute image width at different heights, thereby permitting object eccentricity estimation without relying on phase information.

At a more qualitative level, the models suggests that identifying small objects using high frequency electric sense may be simpler algorithmically and neurocomputationally than may have previously been thought. Some of this simplicity results from our ignoring many factors (especially those involving motion and the time domain) that could confound real electrolocation tasks. We have also ignored many electrosensory circuits in the fish's brain, most notably, the cerebellum. Perhaps it is not just coincidental that conventional ideas of cerebellar function are also intimately associated with both motion and time domain processing. Direct evidence for its role in recognizing and/or tracking approaching stimuli has been reported (Bombardieri and Feng 1977), and Meek (1992) suggested specific roles of cerebellum in timing. Future studies of electric image sequences during exploratory behaviors may clarify the role of the cerebellum in sensory acquisition and behavior.

5. ACKNOWLEDGMENTS

Christopher Assad provided valuable input to this study, and in particular has contributed to the neural algorithms. The work was funded by NSF IBN 9319968.

6. REFERENCES

Bacher M (1983) A new method for the simulation of electric fields generated by electric fish, and their distortions by objects. Biol Cybern 47:51–58.

Bastian J (1986) Electrolocation: behavior, anatomy and physiology. In: Bullock TH and Heiligenberg W (eds). Electroreception. Wiley, New York, pp 577–612

Bastian J (1994) Electrosensory organisms. Physics Today 47(2):30–37

Bastian J (1996) Plasticity in an electrosensory system. II. Postsynaptic events associated with a dynamic sensory filter. J Neurphys 76:2497–2507

Bell C, Bodsnick D, Montgomery J, Bastian J (1997) The generation and subtraction of sensory expectations within cerebellum-like structures. Brain Behav Evol (in press)

Bombardieri RA and Feng AS (1977) Deficit in object detection (electrolocation) following interruption of cerebellar function in the weakly electric fish Apteronotus albifrons. Brain Res 130:343–347

Bullock TH and Heiligenberg W (1986) Electroreception. Wiley & Sons, New York.

Carr CE and Maler L (1986) Electroreception in gymnotiform fish. In: Bullock TH and Heiligenberg W (Eds) Electroreception. Wiley, New York, pp 319–373

Carr CE (1990) Neuroethology of electric fish. Bioscience 40:259–267

Emde G von der (1992) Extreme phase sensitivity of afferents which innervate mormyromast electroreceptors. Naturwissenschaften 79:131–133

Heiligenberg W (1975) Theoretical and experimental approaches to spatial aspects of electrolocation. J Comp Physiol 103:247–272

Kawasaki M, Rose G, Heiligenberg, W (1988) Temporal hyperacuity in single neurons of electric fish. Nature 336:173–176

Meek J (1992) Why run parallel fibers parallel? Teleostean purkinje cells as possible coincidence detetors, in a timing device subserving spatial coding of temporal differences. Neuroscience 48:249–283

Rasnow B and Bower JM (1996) The electric organ discharges of the gymnotiform fishes: I. Apteronotus leptorhynchus. J Comp Physiol 178:383–396

Rasnow B (1996) The effects of simple objects on the electric field of Apteronotus. J Comp Physiol 178:397–411

Shumway, CA (1989) Multiple electrosensory maps in the medulla of weakly electric gymnotiform fish. I. Physiological differences. J Neurosci 9:4388–4399

124

POISSON-LIKE NEURONAL FIRING DUE TO MULTIPLE SYNFIRE CHAINS IN SIMULTANEOUS ACTION

Raphael Ritz,[1] Wulfram Gerstner,[2] René Gaudoin,[3] and J. Leo van Hemmen[3]

[1]CNL, The Salk Institute
La Jolla, CA
[2]Centre for Neuro-mimetic Systems
Mantra-LAMI EPFL, IN-J,
1015 Lausanne, Switzerland
[3]Physik-Department der TU München, T35
D-85747 Garching bei München, Germany

1. INTRODUCTION

The irregularity of neuronal firing times is commonly interpreted as being due to noise[6]. Here, an alternative approach is taken to show that even in a completely deterministic model — without any noise — neuronal firing times might appear random. This can be achieved in an attractor model with spiking neurons where the limit cycles are complex spatio–temporal spiking patterns also called synfire chains[1]. Simultaneous activation of several synfire chains can lead to arbitrarily complex-looking spike patterns at the single neuron level. In addition, a learning rule is presented that allows to store general spatio–temporal spiking patterns.

2. SPIKE RESPONSE MODEL

The activity of a neuron $i, 1 \leq i \leq N$ is described by a two state variable $S_i(t) \in \{0, 1\}$ in discrete time t according to

$$S_i(t + \Delta t) = \frac{1}{2}\{1 + \text{sgn}[h_i(t)]\} ,$$

where the local field $h_i(t)$ summarizes synaptic interactions, refractory behavior, and external enervation

$$h_i(t) = h_i^{\text{syn}}(t) + h_i^{\text{refr}}(t) + h_i^{\text{ext}}(t) .$$

Computational Neuroscience
edited by Bower, Plenum Press, New York, 1997

Whereas h_i^{ext} is chosen as needed, h_i^{refr} reflects the spiking history of the given cell

$$h_i^{\text{refr}}(t) = \sum_{\tau=0}^{\infty} \eta(\tau) S_i(t - \tau)$$

with a suitable response kernel[2,3] $\eta(\tau)$.

All synaptic interactions enter through

$$h_i^{\text{syn}}(t) = \sum_{\tau=0}^{\infty} \sum_{j=1}^{N} \sum_{\Delta^{\text{ax}}=\Delta^{\min}}^{\Delta^{\max}} J_{ij}^{\Delta^{\text{ax}}} \varepsilon(\tau) S_j(t - \tau - \Delta^{\text{ax}}) \, .$$

Note that in principle several axonal delays Δ^{ax} are allowed. The synaptic response kernel is usually taken to be an alpha function. There might be some global inhibition as well to control the overall activity. More details are presented elsewhere[2,3].

3.   RESULTS

3.1.   Three different kinds of patterns

First, we show that it is possible to evoke three different kinds of patterns within the proposed framework: Fig. 1-top: different neurons or groups of neurons fire with different mean firing rates but without coordinating their precise spiking times. This corresponds to the standard notion of a neuronal firing pattern and can be detected through temporal averaging of the neuron's activity. Fig. 1-middle: now, all active cells fire synchronously. This additional structure can be revealed by spatial averaging. Fig. 1-bottom: here, all cells fire with the same mean firing rate. Nevertheless, there is a well defined structure hidden in the precise spiking times of all neurons as made obvious for neurons 1 to 10 which are labeled such that they fire one after the other. This we call a spatio–temporal spiking pattern. It is a generalization of the synfire chain concept as introduced by Abeles[1]. Neither a temporal nor a spatial average is able to reflect the existing structure calling for more sophisticated methods to analyze neuronal data. Using this kind of patterns we now turn to the problem of high variability.

3.2.   High variability — Basic idea

To keep things simple, we assume cyclic patterns, i.e., after a certain time, the pattern repeats. This time, called period of a pattern, may vary from pattern to pattern. As long as there is only one pattern active in a given network, a single cell's activity is perfectly periodic if it contributes exactly one spike to the pattern resulting in a delta-like inter-spike interval distribution; cf. Fig. 2-top, left.

As soon as there are more patterns active simultaneously, things become more complicated. Now, there may be neurons participating in two or more patterns with different periods. This, in turn, results in broader inter-spike interval distributions as shown in Fig. 2-top, right and bottom for 2, 3 and 4 patterns respectively. For as few as three patterns, the coefficient of variation (cv) defined as the standard deviation divided by the mean of the inter-spike interval distribution can be greater than a half and the inter-spike interval distributions come close to experimentally observed ones.

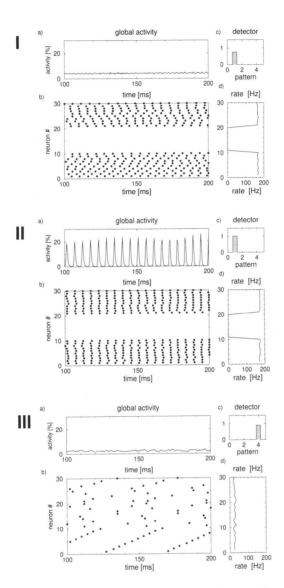

Figure 1. Three different kinds of patterns. (I) incoherent activity, (II) coherent activity, and (III) spatio–temporal spiking patterns. In (a) the spatially averaged activity is shown; (b) gives a spike raster of 30 neurons; (c) a formal classifier; (d) mean firing rate of the 30 neurons along the abscissa. These data have been produced using a network of 4000 cells.

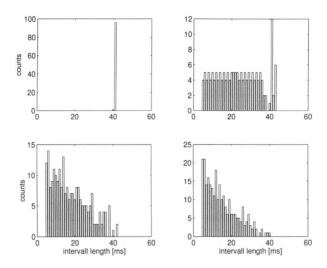

Figure 2. Inter-spike interval distributions of a cell participating in 1 to 4 cyclic patterns. Due to refractoriness, short intervals are suppressed. The periods ranged from 41 ms to 53 ms. The coefficient of variation is about 0.6 for the lower two plots.

3.3.   Example of high variability

To show that the proposed superposition of patterns is possible in practice we performed a simulation implementing a network of 400 cells that learned several patterns with differing periods. Then we stimulated the network for a short period of time with a stimulus adequate to trigger three different patterns which are then repeated by the network simultaneously after the stimulus has been turned off. For illustration Fig. 3 shows in **a** the activity of a subset of 25 cells. Cells 1 to 4 are chosen such that cells 1 to 3 participate in one of each pattern respectively whereas cell 4 takes part in all three of them. Note, that cell 4 therefore should always fire synchronously with cell 1, 2, and 3 but due to refractoriness it sometimes misses a spike or fires a little delayed; cf. Fig. 3**b**.

So far, we have shown that the proposed explanation for high variability works in principle and in practice but we did not yet discuss learning procedures allowing to entrain a suitable network. This is done next.

3.4.   Learning

Let $\xi_j^\mu(t)$ and $\xi_i^\mu(t)$ denote the pre- and postsynaptic spike trains of cell j and i in pattern μ respectively.

According to Hebb[4], a synapse is strengthened whenever both cells are simultaneously active. At the site of the synapse, this can be implemented as

$$\delta J_{ij}^{\Delta^{ax}}(\mu) \propto \sum_{\tau=0}^{\infty} \nu(\tau)\, \xi_i^\mu(t - \Delta^{\text{dent}})\, \xi_j^\mu(t - \Delta^{ax} - \tau)\ .$$

Here, $\nu(\tau)$ defines a time window during which the pre- and postsynaptic spikes have to occur[5]. In addition, there is a need for the postsynaptic spike to be detectable at the synapse. This could be the functional role of action potentials travelling back into the dendritic tree[7].

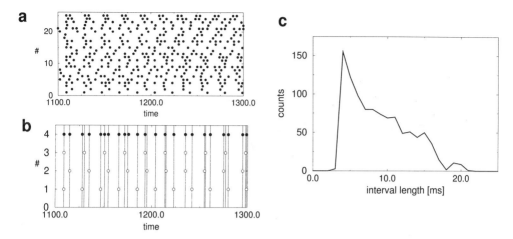

Figure 3. Sample simulation showing three different patterns evolving simultaneously (periods: 19 ms, 20 ms, and 21 ms). **a**: Spike raster of 25 cells out of 400. **b**: Whereas cells 1, 2, and 3 belong each to a different pattern, cell 4 belongs to all three of them. Therefor, this cell fires together with all the other three whenever possible. This leads to a irregular firing pattern as exemplified by the inter-spike interval histogram of cell 4 shown in **c**.

The time it takes the signal to travel back to the synapse on the postsynaptic side is denoted by Δ^{dent}.

To increase the stability of the learned patterns so as to improve the performance of the network, a supervised, iterative learning procedure has been evoked. That way, the network is more robust and better able to deal with noise. During learning a *supervisor* notes the state of any cell at every single time step and induces synaptic changes in the weights toward the regarded cell whenever its output differs from the required one. This is just the well known perceptron learning algorithm adapted to arbitrary spiking patterns.

Of course, there are problems with the biological plausibility of this learning scheme but we do not claim that the brain uses perceptron learning but still it is possible that different parts of the brain teach each other. In addition, we postulate a specific increase as well as decrease of synaptic efficacies due to learning. This would require a precise interplay of LTP and LTD for instance.

As we start out with a fully connected network with even multiple connections between any given pair of cells implementing different delays one might suspect that we end up with a tremendous amount of connections. But it turns out that only 5% of all possible connections are needed to store a reasonable amount of patterns.

4. SUMMARY AND CONCLUSIONS

There are to basic statements to be made from this study: First, irregular firing at the single cell level does not necessarily reflect noise. Second, almost arbitrary spatio–temporal spiking patterns can in principle be learned in a Hebbian way but other mechanisms could provide a better performance. To test the biological relevance of these claims more multi-unit recordings as well as more precise data on the temporal structure of synaptic plasticity are needed.

ACKNOWLEDGMENTS

Supported by DFG (He 1729/2-2 and Ri 821/1-1)

REFERENCES

[1] Abeles, Moshe, *Local cortical circuits.* Springer-Verlag, Berlin Heidelberg New York, 1982.
[2] Gerstner, W. 1995 Time structure of the activity in neural network models. Phys. Rev. E, **51**:738–758.
[3] Gerstner, W., Ritz, R., and van Hemmen, J.L. 1993 Why spikes? Hebbian learning and retrieval of time-resolved excitation patterns. Biol. Cybern., **69**:503–515.
[4] Hebb, Donald O., *The organization of behavior.* Wiley, New York, 1949.
[5] Markram, H. and Sakmann, B. 1995 Action potentials propagating back into dendrites triggers changes in efficacy of single-axon synapses between layer V pyramidal neurons. Soc. Neurosci. Abstr., **21**:2007.
[6] Shadlen, M.N. and Newsome, W.T. 1994 Noise, neural codes and cortical organization. Curr. Opin. Neurobiol., **4**:569–579.
[7] Stuart, G.J. and Sakmann, B. 1994 Active propagation of somatic action potentials into neocortical pyramidal cell dendrites. Nature, **367**:69–72.

125

ATTENTIONAL GAIN MODULATION AS A BASIS FOR TRANSLATION INVARIANCE

Emilio Salinas[1,2] and L. F. Abbott[2]

[1] Instituto de Fisiología Celular, UNAM
México D. F., México
E-mail: esalinas@ifcsun1.ifisiol.unam.edu
[2] Volen Center
Brandeis University
Waltham MA
E-mail: abbott@volen.brandeis.edu

ABSTRACT

Inferotemporal (IT) neurons in monkeys can respond to visual stimuli in a translation- and scale-invariant manner. In a neural circuit model based on a recently reported form of gain modulation by attention in area V4, the modulated visual responses of model V4 neurons produce object-centered receptive fields further down the visual processing stream, accounting for invariances exhibited by IT neurons.

1. INTRODUCTION

The responses of inferotemporal (IT) neurons in monkeys are highly selective for certain types of visual images but, to a large degree, are insensitive to the exact location and size of those images[1,2]. These invariances to translation and scale at the neuronal level parallel our ability to recognize objects independently of the location and size of their images on the retina. Although some ideas have been put forward regarding possible mechanisms for generating invariances[3,4], these have not been confirmed neurophysiologically.

Monkeys with lesions in area V4 cannot recognize images when they are subject to a variety of visual transformations (see Ref. [5]). Other experiments reveal that attention affects the responses of many cells in V4[6,7]. In particular, recent findings by Connor et al.[8] indicate that the visual responses of many neurons in V4 are modulated by the location where attention is directed. Here we show, through analytic calculations and computer simulations, that V4 can give rise to invariant visual responses further downstream if the modulatory effect of attention is considered.

Computational Neuroscience
edited by Bower, Plenum Press, New York, 1997

2.   THE MODEL

Our model consists of an array of V4 neurons that project, through feedforward synaptic connections, to a single model IT neuron. For simplicity, we consider the case in which an image corresponds to a one-dimensional luminance distribution I that varies with position x along a single axis. The firing rate of cell i in the V4 array is denoted by r_i, and is equal to the product of two terms: the output of a visual filter that represents the cell's receptive field, and a modulatory factor that corresponds to the gain field. We denote the output of the visual filter by $F_i(a_i; I)$, where a_i is the receptive field center and I is the image being presented. We model the gain field as a Gaussian function of the distance between the locus of attention and a neuron-specific quantity we call the preferred attentional locus, denoted by b_i. This is in agreement with the reported observations[8] indicating that the gain field of a cell is close to its maximum value when attention is directed near the preferred attentional locus of the cell, decreasing as attention moves away from it. A key requirement in the model is that receptive field centers and preferred attentional loci are not aligned or directionally correlated. The experimental data collected support this assumption[8]. The gain factor for cell i is denoted by $G(y - b_i)$, where y is the currently attended location and G indicates a Gaussian function. Since the gain field is multiplicative, the firing rate of cell i in the V4 array in response to an image I when attention is focused at the point y is given by

$$r_i = F_i(a_i; I)\, G(y - b_i).\tag{1}$$

The response of the model IT neuron, denoted by R, is computed by rectifying a synaptically weighted sum of V4 responses minus a threshold θ:

$$R = [\sum_i W_i r_i - \theta]_+,\tag{2}$$

where $[x]_+ = \max\{x, 0\}$.

3.   METHODS

In the simulations, the visual responses of V4 cells are modeled according to an 'energy' mechanism[9], which depends on the outputs S_i and C_i of two localized linear filters:

$$S_i = \int dx\, I(x)\, \mathrm{hcos}(\pi(x - a_i)/\lambda)\, \sin(k\pi(x - a_i))$$
$$C_i = \int dx\, I(x)\, \mathrm{hcos}(\pi(x - a_i)/\lambda)\, \cos(k\pi(x - a_i)).\tag{3}$$

The half-cosine function $\mathrm{hcos}(x)$ that determines the filter envelopes is equal to $\cos(x)$ if $-\pi/2 < x < \pi/2$ and is zero otherwise. Parameters λ and k determine the receptive field width and frequency selectivity, respectively. To produce the final visual response, the above terms are squared and added: $F_i(a_i; I) = (S_i)^2 + (C_i)^2$. This expression is substituted into Eq. (1) to compute the V4 firing rates. Finally, all firing rates are divided by $\sum_i (r_i)^2$ to account for the effects of contrast normalization[9].

The simulated one-dimensional visual field spans 64 degrees; 65 values for a_i and 33 values for b_i were uniformly distributed throughout it. Neurons with frequency selectivities equal to 1/8, 2/8 and 3/8 cycles per degree were included. All combinations of a_i and b_i were

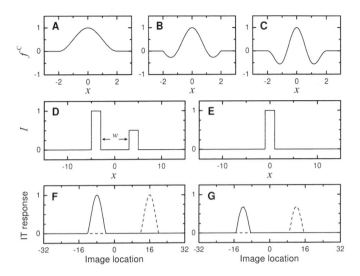

Figure 1. Computer simulations of the model network. A, B and C, localized, even filters f^C used to compute the term C_i; they correspond to $k=1$, $k=2$ and $k=3$ in Eq. (3). D, image used during training: two bars separated by six degrees. E, test image used after training. F, normalized model IT responses when the training image is shown at different locations. G, response when the image in E is shown at different locations. In F and G the continuous lines result when attention is focused at $y=-8°$; the broken lines result when attention is focused at $y=16°$. The IT receptive field moves with the locus of attention.

present, giving a total of $65 \times 33 \times 3$ neurons. The baseline width of the visual filters was $\lambda=4°$, and for the Gaussian gain fields $\sigma=2°$. The threshold θ was set to one half the maximum IT response when $\theta=0$. One-dimensional images consisted of square patches or 'bars' of varying intensities and widths. The locus of attention y was provided to the program as an external parameter that could be changed arbitrarily.

4. ANALYTIC RESULTS

The crucial elements in Eq. (2) are the synaptic weights. The model V4 responses give rise to IT receptive fields that shift with attention if the synaptic weight W_i depends on the receptive field center and the preferred attentional locus only through their difference; that is, $W_i = W(a_i - b_i)$. Consider the simple case in which the visual responses are given by linear filters: $F(a_i;I) = \int dx\, I(x)\, f(x - a_i)$, where $f(x - a_i)$ is the filter characterizing the receptive field of cell i. The sum over cells that determines the IT response (Eq. (2)) can be approximated by a multiple integral if we assume uniformity, independence and high density,

$$\sum_i W_i r_i \propto \int da\,db\,dx\, W(a - b)\, G(y - b)\, I(x)\, f(x - a). \tag{4}$$

Using the substitutions $a \to a+y$ and $b \to b+y$ the integral takes the form $\int dx\, I(x)\, \bar{F}(x-y)$, which is a filtered version of I that shifts with y. Thus, the IT receptive field is not fixed to a retinal location, but rather is attached to the attentional locus. For clarity, this result was derived using one-dimensional linear filters. It should be stressed that analogous results can be obtained for two-dimensional nonlinear filters as well.

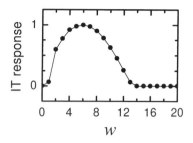

Figure 2. Firing rate of the model IT neuron when variations of the training image are presented. Each point represents the response to one of the test images, which consisted of the same two bars in the original image separated by a variable amount w. For each point attention was directed to the image center. The curve peaks at the original separation $w=6°$, indicating selectivity for the training image.

5. SIMULATION RESULTS

Computer simulations support the above analysis and provide some insight into the mechanism that gives rise to shifting receptive fields. In the simulations, synaptic weights satisfying the condition discussed above are generated using a simple Hebbian learning scheme. During a learning period, a particular image was shown at all possible locations, while the model IT neuron was set to an active state and the synaptic weights W_i changed by an amount proportional to Rr_i. During this process attention was always focused at the center of the training image. This simple procedure generates weights that give rise to shifting receptive fields —this can be shown analytically— but it is meant mainly as an example of a plausible mechanism; synaptic weights with this property are not unique.

In the simulations, the visual filters of model V4 neurons were nonlinear (see METHODS). Panels A, B and C of Fig. 1 show examples of localized, even filters of different frequencies used to generate the C_i terms in Eq. (3). Analogous odd filters were used for the S_i terms. Panel F shows the response of the model IT neuron, after training, when the training image shown in D is presented at different locations. The image location is taken as the midpoint between the two bars. The continuous and dashed curves correspond to runs in which the attentional locus was directed at points $y=-8°$ and $y=16°$. The responses are maximal when the image center coincides with the locus of attention. Panel G shows the results of another run in which the image shown in E, a single bar, was presented at different positions, the image location corresponding to the center of the bar. The firing rate produced when the bar appears exactly where attention is directed ($-8°$ or $16°$) is zero. In contrast, a response is seen when the bar appears to the left of the attentional locus, where the image partly matches the one used during training. These results demonstrate that the model IT neuron is selective for the training image and that its firing is determined by the relation of the image location to the attentional locus, no matter where the image appears retinotopically.

Fig. 2 shows the results of another set of simulations in which variations of the training image were presented to the network. These consisted of the same two bars, but separated by a variable amount w indicated in Fig. 1 D. The responses shown were obtained with attention directed to the center of the image, *i.e.* to the midpoint between the two bars. The maximum response occurs for $w=6°$, which is the original separation. As w deviates from this value the responses fall off to zero. This provides further evidence of the selectivity acquired by the model IT neuron.

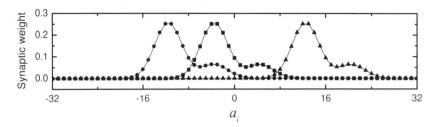

Figure 3. Distribution of synaptic weights after training, arranged according to the receptive field centers a_i of the cells they project from. Weights from three sub-populations are shown: from cells with $b_i=-8°$ (circles), $b_i=0°$ (squares) and $b_i=16°$ (triangles).

The distribution of synaptic weights produced after training clarifies the relationship between gain modulation and the condition required for translation invariance. Some synaptic weight values are plotted in Fig. 3, ordered according to the receptive field centers of the neurons they project from. This is done for all of the neurons with a frequency selectivity of 3/8 cycles per degree and preferred attentional loci $b_i=-8°$, $b_i=0°$ and $b_i=16°$. The weight patterns are translated versions of one another, which is precisely what the condition $W_i = W_i(a_i - b_i)$ means. When attention is directed at a particular point, say $y=0°$, only the neurons with b_i close to $0°$ are active, and the system behaves as if only those weights existed. As shown in Fig. 3, the weight patterns can be considered as templates that are retinotopically ordered. There are a variety of templates, centered at different points, and when attention is focused on a particular location the corresponding template centered on that location is activated by attentional modulation.

6. SUMMARY

The model circuit described reproduces the characteristic features of neurons in IT cortex: substantial translation invariance coupled to high selectivity for certain complex images. A similar gain-modulatory mechanism depending on a currently attended scale gives rise to receptive fields that are scale-invariant (results not shown). Although a single model IT neuron was considered, others driven by the same array of V4 cells could be added to the network. Through the same Hebbian mechanism described, these could acquire different selectivities but, more importantly, all would have receptive fields that moved with the locus of attention. Thus the gain-modulated V4 neurons work as a basis for arbitrary receptive fields that shift.

The model is akin to others developed for parietal cortex[10-12], where gain modulation is thought to be responsible for transformations from retinal to head- and body-centered coordinates useful for sensory/motor coordination. Thus, similar mechanisms for coordinate transformations may be present in the dorsal-where and ventral-what visual pathways.

REFERENCES

[1] M.E. Hasselmo, E.T. Rolls, G.C. Baylis and V. Nalwa, 1989, Object-centered encoding by face-selective neurons in the cortex in the superior temporal sulcus of the monkey, *Exp. Brain Res.* **75**:417-429.
[2] M. Tovee, E.T. Rolls and P. Azzopardi, 1994, Translation invariance in the responses to faces of single neurons in the temporal visual cortical areas of the alert macaque, *J. Neurophysiol.* **72**:1049-1060.

[3] G.E. Hinton, 1981, Shape recognition and illusory conjunctions, in: *Proceedings of the Seventh International Joint Conference on Artificial Intelligence* Vol. II, p. 683-685.

[4] B.A. Olshausen, C.H. Anderson and D.C. Van Essen, 1993, A neurobiological model of visual attention and invariant pattern recognition based on dynamical routing of information, *J. Neurosci.* **13**:4700-4719.

[5] P.H. Schiller, 1995, The role of the primate extrastriate area V4 in vision, *Nature* **376**:342-344.

[6] J. Moran and R. Desimone, 1985, Selective attention gates visual processing in the extrastriate cortex, *Science* **229**:782-784.

[7] B.J. Motter, 1993, Focal attention produces spatially selective processing in visual cortical areas V1, V2, and V4 in the presence of competing stimuli, *J. Neurophysiol.* **70**:1-11.

[8] C.E. Connor, J.L. Gallant, D.C. Preddie and D.C. Van Essen, 1996, Responses in area V4 depend on the spatial relationship between stimulus and attention, *J. Neurophysiol.* **75**:1306-1308.

[9] Heeger, 1991, Nonlinear model of neural responses in cat visual cortex, in: *Computational Models of Visual Processing* (M. Landy and J.A. Movshon ed), p. 119-133, MIT Press, Cambridge, MA.

[10] D. Zipser and R.A. Andersen, 1988, Nonlinear model of neural responses in cat visual cortex. *Nature* **331**:679-684.

[11] E. Salinas and L.F. Abbott, 1995, Transfer of coded information from sensory to motor networks. *J. Neurosci.* **15**:6461-6474.

[12] A. Pouget and T.J. Sejnowski, 1996, Spatial transformation in the parietal cortex using basis functions. *J. Cogn. Neurosci.* (in press).

[13] Research was supported by the Sloan Center for Theoretical Neurobiology at Brandeis University, the National Science Foundation (DMS-9503261), the W.M. Keck Foundation and by the Conacyt-Fulbright-IIE program.

126

MODELLING RAT BEHAVIOR IN AN ELEVATED PLUS-MAZE CONFRONTED WITH EXPERIMENTAL DATA

Cristiane Salum, Silvio Morato, and Antônio C. Roque-da-Silva-Filho

Faculdade de Filosofia, Ciências e Letras de Ribeirão Preto
Universidade de São Paulo
Avenida Bandeirantes, 3900
14040-901–Ribeirão Preto–SP, Brazil
crisalum@usp.br

1. ABSTRACT

This work describes a neural network model of rat exploratory behavior in the elevated plus-maze, a test used to study anxiety. It involves three parameters: drive to explore, drive to avoid aversive stimuli and spontaneous locomotive activity. Competitive learning is used to generate a sequence of network states each corresponding to a place in the maze. The work also presents experiments made with real rats providing data to be compared with the simulation results. The simulations are consistent with the experimental evidence, and may provide an efficient way of describing anxiety-like behaviors of rats in the elevated plus-maze.

2. INTRODUCTION

There have been several attempts at using neural networks to understand emotional psychological disorders such as depression, schizophrenia, juvenile hyperactivity and Parkinson's disease.[1,2] However, anxiety, an emotional state involving the same neural mechanisms as fear, has not been approached by such models.

The animal model most recently and widely used for the study of anxiety is the elevated plus-maze which consists of a plus-shaped maze elevated from the floor, with two open and two closed arms. This model, inspired in Montgomery,[6] is based on the rat spontaneous aversion to open and high spaces and was validated as a measure of anxiety from the pharmacological, biochemical and behavioral standpoints.[7] A rat will typically explore the closed arms more than the open arms, both in terms of frequency of entries and of time spent in the arms. One explanation for this, which was first advanced by Montgomery,[6] postulates a conflict between the motivation to explore a new environment and its incompatible motivation to stay in a safe place. The animal's reluctance to explore the open arms of the maze probably results from a combination of rodent aversion to open spaces and the elevation of the maze.[4] Anxiolytic drugs increase the number of entries and the time spent into the open arms whereas anxiogenic agents do the opposite.[3,7] Besides, fear reduces exploratory behavior which also decreases with time of direct exposure to novel stimulation.[5]

The aim of the present work is to construct a neural network simulation of a three-parameter (spontaneous activity, exploratory drive and safety seeking drive) mathematical model for rat exploratory behavior, and compare the simulation results with experimental data by real rats.

3. THE NETWORK MODEL

The network model is based on the structure of a plus-maze divided into squares of equal size, five per arm plus a central one (Figure 1). Each network unit corresponds to a specific place in the maze and the connections, only between closest neighbors, represent the possible adjacent places where a rat could go. At each time step, the connection weights are modified according to the function

$$w_{ij} = R_{ij} + M_{ij} - A_{ij} \tag{1}$$

where w_{ij} is the connection weight between units i and j, R_{ij} corresponds to the spontaneous activity of the rat and is generated randomly; M_{ij} is the exploratory motivation of the animal and A_{ij}, is its aversion.

The exploratory motivation is inversely proportional to the number of times a given connection is visited by the rat,

$$M_{ij} = M_{max} / (N_{ij} + \delta)^{\alpha}, \tag{2}$$

where M_{max} is a constant representing the basal motivation of the rat to explore; N_{ij} is the number of times a rat explores a connection between units i and j; α is a constant that determines the motivation decay to explore and δ is a variability factor for the motivation decay. This means that each time the virtual rat explores one connection (from place j in the maze to place i) its motivation to explore this place decays according to the above function.

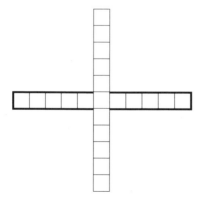

Figure 1. The virtual plus-maze. Two open and two closed arms.

The value of the aversion changes according to the equation:

$$Aij = C \left(1 \pm \beta \, Nij \right) \tag{3}$$

where C is a constant representing the animal's initial aversion for a new environment; N_{ij} is the number of times the animal explores the connection between i and j; β is a factor that determines the decay or the increase of aversion, depending on the place that is been explored, i.e. if the animal is exploring a unit (or square) belonging to the closed arms the plus sign is assumed, otherwise the minus sign is assumed. This means that aversion increases when the unit which is being explored refers to a square in the open arms of the maze and decreases when the unit refers to a square in the closed arms. There is evidence that this happens to a real rat probably due to its fear of height and open spaces (negative thigmotaxis).[8]

Only one unit of the network can be active at a moment, and it represents the place that is being explored by the rat. There is a competition among the activities of neighboring units and the one with larger activity wins.

Virtual sessions were performed with a computational implementation of the mathematical model described above. The measures scored for virtual rats were statistically indistinguishable from the ones recorded for real rats.

4. THE ANIMAL MODEL

4.1. Subjects

Male Wistar rats weighting between 180 and 230 g were used. They were housed in groups of six per cage for at least 72 hours (habituation) before testing with food and water freely available throughout the whole experiment.

4.2. Apparatus

The plus-maze consisted of two open arms, 50 x 10 cm with acrylic borders (1 cm) and two enclosed arms, 50 x 10 x 40 cm, without roof, arranged in such a way that like arms were opposite to each other. The maze was elevated 50 cm from the floor. An arena was also used. It was a wooden square box (60 x 60 x 30 cm) with the floor divided in 10-cm squares. It was used to improve the locomotor activity of the rats in the maze.[7]

4.3. Procedure

Each animal was placed individually in the arena and allowed to explore for 5 minutes. Immediately after this procedure, the rat was transferred to the maze and placed gently in the central square for a 5-minute test. The sessions were recorded with a video camera so that number of entries and time spent into the arms could be analyzed later.

5. RESULTS

A series of virtual sessions were performed using different values for the parameters α, β and C. In addition, the random parameter R_{ij}, which bias the directions the rat can

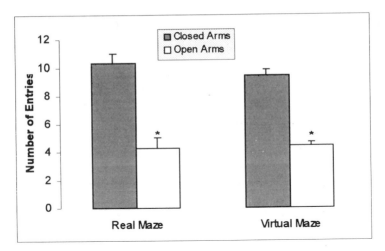

Figure 2. Frequency of entries in the arms (open and closed) of the real and virtual mazes. Bars represent the mean (± SEM). * p < 0.05 compared entries in the open and closed arms within the same maze.

take, was given different probabilities to each possible direction in relation to the animal's head: 0.4 for forward movement, 0.2 for each lateral movement, and 0.1 for backward or no movement.

Simulation and experimental results were statistically compared via two-way Anova followed by multiple comparison tests or t-tests. Anova showed that there was not a significant difference between real and virtual rats (Figure 2) in the number of entries in both open and closed arms ($F_{[1,47]}$ = 0.385, P = 0.538). The number of entries in the open arms was significantly smaller than in closed arms both for real rats ($t_{[22]}$ = −6.333, P < 0.001) and for virtual rats ($t_{[22]}$ = −9.876, P <0.001). Anova also showed that the percentage of time spent in the arms of the virtual maze (Figure 3) was not significantly different from the real one ($F_{[1,47]}$ = 0.000, P = 1.000). The percentage of time spent in the open arms was

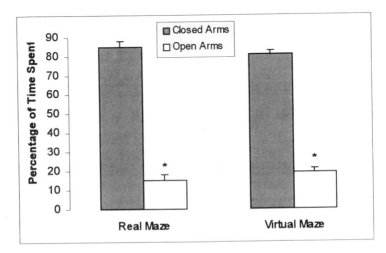

Figure 3. Percentage of time spent in the arms (open and closed) of the real and virtual mazes. Bars represent the mean (± SEM). * p < 0.05 compared % of time in the open and closed arms within the same maze.

significantly smaller than that in the closed arms both in the real maze ($t_{[22]} = -18.822$, $P <$ 0.001) and in the virtual maze ($t_{[22]} = -19.887$, $P < 0.001$).

6. DISCUSSION

We finally reached a set of parameter values which led the virtual rat to exhibit similar measures of exploratory behavior as real rats, both in terms of frequency of entries and of percentage of time spent in the open and closed arms. The network model also presented the typical result shown by real rats, that is, massive exploration of the arms with walls (closed) and less exploration of the arms without walls (open), which is considered a manifestation of anxiety.

According to these results, virtual animals produced data similar to the real ones indicating that the mathematical basis of the model is consistent with real rat exploratory behavior suggesting that the former offers a reasonable representation for reality.

REFERENCES

1. Cohen, J. D. and Servan-Schreiber, D. (1992). Context, cortex, and dopamine: A connectionist approach to behavior and biology in schizophrenia. *Psychological Review*, 99: 45–77.
2. Grossberg, S. (1984). Some normal and abnormal behavioral syndromes due to transmitter of opponent process. *Biological Psychiatry*, 19: 1075–1117.
3. Handley, S. L. and Mithani, S. (1984). Effects of alpha-adrenoceptor agonists and antagonists in a maze-exploration model of 'fear'-motivated behaviour. *Archives of Pharmacology*, 327: 1–5.
4. Lister, R. G. (1990). Ethologically-based animal models of anxiety disorders, *Pharmacology Therapeutics*, 46: 321–340.
5. Montgomery, K. C. (1951). The relation between exploratory behavior and spontaneous alternation in the white rat. *Journal of Comparative Physiological Psychology*, 44: 582–589.
6. Montgomery, K. C. (1955). The relation between fear induced by novel stimulation and exploratory behavior. *Journal of Comparative Physiological Psychology*, 48: 254–260.
7. Pellow, S., Chopin, P., File, S. E. and Briley, M. (1985). Validation of open:closed arm entries in an elevated plus-maze as a measure of anxiety in the rat. *Journal of Neuroscience Methods, 14*: 149–167.
8. Treit, D., Menard, J. and Royan, C. (1993). Anxiogenic stimuli in the elevated plus-maze. *Pharmacology, Biochemistry and Behavior, 44*: 463–469.

127

MODELING THE PRECEDENCE EFFECT FOR SPEECH

Odelia Schwartz,[1]* John G. Harris,[2] and Jose C. Principe[2]

[1]New York University
  Center for Neural Science
[2]University of Florida
  Electrical and Computer Engineering

ABSTRACT

The human auditory system is capable of perceiving the direction of a sound source despite reflections off of walls, ceiling and other interfering objects. This phenomenon, the precedence effect, has been previously modeled only for synthetic click stimuli. This paper presents a biologically plausible computer simulation for speech, that combines a cross-correlation model with an onset enhancement scheme. As part of the complete model, a novel onset enhancement method is described and implemented, based on adaptive prediction. The adaptive method is promising in its ability to enhance onsets and reduce the steady state portion of the speech signal. The model can serve as a tool to understand the precedence effect for complex signals and should be advantageous to engineering applications such as preprocessors for speech recognition systems.

1. INTRODUCTION

The ability of the human auditory system to localize the direction of a sound source in a reverberant environment extends from simple clicks to complex speech signals. The perceived location of the auditory event is dominated by the direct sound, for small enough time intervals between the direct sound and reflection–a phenomenon known as the precedence effect [1, 3, 12, 13]. For time differences that exceed a so-called echo threshold, we perceive two distinct auditory events. Most models formulated to explain the precedence effect have been tested only on synthetic click stimuli [6]. With more realistic signals such as speech, the direct sound and reflection overlap in time, creating a complex and variable spectrum. The benefits in creating a computer model of this effect on complex signals is twofold:

---

*This work was done while a M.S. student in Computer Science and Engineering at the University of Florida.

Computational Neuroscience
edited by Bower, Plenum Press, New York, 1997

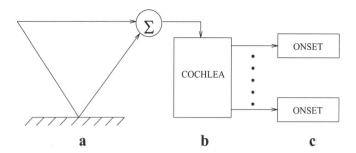

Figure 1. Preprocessing of signal to enhance onsets. (a) Delay and add. (b) cochlear model. For each band produced by the cochlea: (c) Onset enhancement and peaks of slope of onset.

1. Theoretically, to comprise part of our more general understanding of how complex acoustic stimuli arriving at our ears develop into a sensible auditory scene [2].

2. As a basis for engineering applications, such as a preprocessor to a more robust speech recognition system.

Traditional cross-correlation models for horizontal localization do not perform well in reverberant environments, because they lack a mechanism to detect early peaks corresponding only to the direct signal. Lindenmann [6] solves this problem for click stimuli, by initiating an inhibition mechanism from the primary cross-correlation peak, causing the suppression of subsequent peaks of the echo. The model lacks an explicit onset mechanism that could provide a more robust measure of when to trigger the inhibition mechanism for complex signals. The role of onsets as a trigger for the precedence effect is established by psychoacoustic experiments [4, 8]. In addition, from the signal processing perspective, onsets enable us to isolate portions of the signal that correspond mostly to the direct sound or mostly to the reflection. This is particularly crucial for the localization of complex speech signals which overlap in time. This study demonstrates a computer model of the precedence effect for speech, that incorporates an onset enhancement mechanism.

2. THE PRECEDENCE MODEL

The precedence model is implemented using Matlab and is tested on speech signals from the TIMIT database. The computer simulation attempts to follow the signal processing of the peripheral auditory system. The input is first preprocessed at each ear by a cochlear model and onset enhancer. Then, the preprocessed input from the two ears are fed into a cross-correlation inhibition model. These two stages, namely preprocessing and cross-correlation, are discussed in detail in the following subsections.

2.1. Preprocessing of signal to enhance onsets–an adaptive approach

The preprocessing stages of the model are depicted in Figure 1. We assume that the raw input signal to each ear consists of the summation of the direct signal and a delayed version of the direct signal (corresponding to the reflection). The horizontal locations of the direct sound and reflection are simulated by introducing interaural time differences. The raw input signal is then entered into a cochlear model consisting of a bandpass filter bank (16 filters) and full wave rectification. The bandpass filter bank [10] models the basilar membrane. The center frequency is determined based on psychoacoustic experiments. The rectification models a population of inner hair cells. Each frequency band produced by the cochlea undergoes onset enhancement. Finally, we consider only the areas in which the onset enhanced signal changes maximally. This is in line with psychoacoustic studies that demonstrate that the rate of onset is important in triggering the precedence effect [8].

We will analyze the onset enhancement technique with respect to a single frequency band. An onset is usually defined as an increase in signal energy. Most onset enhancement strategies compute an energy function using fixed low pass filters [11]. Here we utilize an adaptive filter approach, which does not compute energy per se, but rather offers an alternative definition to onsets as changes in the input signal the system cannot predict. The adaptive approach is advantageous in reducing steady state portions of the signal that are repetitive and rather predictable in nature [9].

In the adaptive approach, we utilize a one-step adaptive predictor. The adaptive predictor attempts to predict the current value of the input based on its previous values. The adaptive filter is designed as an FIR filter with adjustable weights. The number of taps corresponds to the number of past values of the signal, used to predict the current value. The weights of the filter are adjusted, such that the squared error is minimized. According to the Exponential weighted recursive least squares algorithm (EWRLS), we define the squared error at time n over an exponential window [2]:

$$J(n) = \sum_{\ell=-\infty}^{n} \alpha^{n-\ell} \varepsilon(\ell)^2 \qquad (1)$$

where $0 < \alpha < 1$ and $\varepsilon(\ell)$ is the prediction error for time ℓ. The goal then is to adjust the weights such that $J(n)$ is minimized:

$$\nabla_w J(n) = 0 \qquad (2)$$

We interpret high error readings as an onset, since we cannot predict the current input based on previous input values. Also, higher error levels can only be detected if there is a high signal energy. Therefore, this definition of onset is in effect a combination of signal energy and predictability. The two parameters of interest in the adaptive filter approach are the number of taps and α. Increasing the number of taps increases the performance but also increases the computational complexity. As α increases, the algorithm is more stable but the filter becomes less sensitive to changes in the error [2].

Figure 2 depicts the 4 stages of preprocessing of the model for a single ear, with the word "teeth". It includes: (a) The delay and add to produce a 10 millisecond echo, (b) A single band produced by the cochlea, (c) The adaptive filter for onset enhancement and (d) The peaks of the slope of the onset enhanced signal. The number of taps and α for the EWRLS are taken as 40 and .993, respectively. Note that the steady state portions of the signal following the onset at "t" are significantly reduced by the adaptive filter. Also, we expect that the initial peaks in (d) belongs to the direct sound. The peaks thereafter might belong to the echo but they will be inhibited. This process is described in detail in the following section.

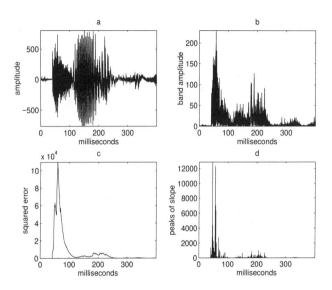

Figure 2. Complete preprocessing of "teeth": (a) Delay and add to produce 10 millisecond echo. (b) One rectified band – 2nd band (6000-4000 Hz). (c) Adaptive filter onset operator. (d) Peaks of slope.

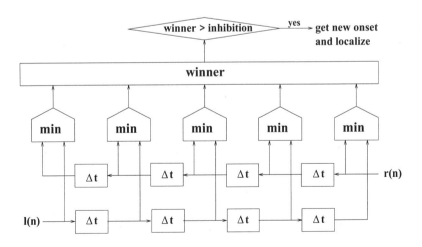

Figure 3. Cross-correlation model with inhibition.

2.2. Cross-correlation model with inhibition

The precedence model is presented in Figure 3. The preprocessed inputs, $l(n)$ and $r(n)$, are fed into two delay lines. The cross-correlation value at each tap is computed as the minimum of the values coming in from the left and right ears. This is to ensure that an onset will be detected only if the signal values at both ears are sufficiently high. If the value of the winning tap, W, is high enough to signify an onset, it will encode the interaural delay of the sound source. This interaural delay value is used to indicate the horizontal location of the stimulus, such that a value of zero represents the midline and large values indicate sources toward one side.

The inhibition mechanism sets a threshold as to when a cross-correlation peak should be considered an onset. The inhibition function I is defined as follows:

$$I(t) = kV_{ons}e^{-(t-t_0)/\tau_f} + V_{ons}e^{-(t-t_0)/\tau_s} + V_{thr} \qquad (3)$$

I is initially some minimal threshold, V_{thr}. If an onset is detected, I rises by a constant $k+1$ times V_{ons}, and then decreases exponentially based on a fast and a slow time constant, τ_f and τ_s, respectively. τ_f models the precedence effect. It is set such that after echo threshold time the function returns to the value of the onset. τ_s causes the function to return to the default threshold if no onset is detected for a long time.

The model consists of 16 delays in the delay lines, representing 17 horizontal locations of the sound source. The time delay between each tap is .0625 milliseconds, based on the observation that the total time for sound to travel from one ear to the other is at most one millisecond [5]. τ_f is 50 milliseconds corresponding to the echo threshold for speech [5], and τ_s is 250 milliseconds. k is chosen as 20 and V_{thr} as 500 (equivalent to about 4 percent of the highest onset values that were experienced in this study). In the following examples the direct sound is presented to the right ear .5 milliseconds after to the left, corresponding to a correlation at tap 13. The echo is presented simultaneously to both ears, corresponding to a correlation at tap 9.

Figures 4 and 5 demonstrate the results of the cross-correlation inhibition model. Figure 4 depicts the winners and the inhibition function for the second band of "teeth" (6000-4000 Hz) with a 10 millisecond echo, i.e., below the echo threshold. The winning tap at each time step, is the tap in which the minimum of the onset information from the two ears is highest. This winning tap is considered an onset, only if the value of the tap is higher than the inhibition function. The dotted line represents the inhibition function. In this example, a single auditory event is perceived, corresponding to the direct sound. Since the echo delay is within the echo threshold, the perception of the direct sound and the suppression of the directional information thereafter is in line with our expectations from psychoacoustics. The directional information is first sampled at 42.4 milliseconds, with the maximum at tap number 12 (i.e., one tap off). However, a second higher peak of the direct sound is able to push through the inhibition at 47.4 milliseconds. The maximum this time corresponds to tap number 13 (i.e., the correct tap). As such, if the original default threshold is too low, the system is able to adjust itself and correct this localization error.

Figure 5 contains the second band of "teeth" (6000-4000 Hz) with a 100 millisecond echo, i.e., above the echo threshold. This time, two auditory events are perceived at 42.4 and 145.6, respectively. The first event corresponds to the direct sound at tap number 13 and the second to the echo at tap number 8 (i.e., the localization is off by one tap). Since the echo delay is above the echo threshold, the onset of the echo is not suppressed and two distinct events are determined—as expected by psychoacoustic theory.

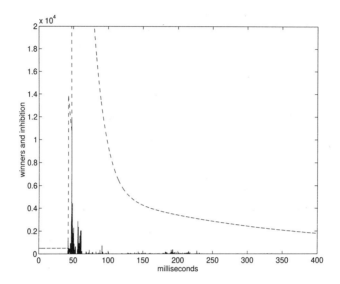

Figure 4. Precedence effect on 2nd band of "teeth" (6000-4000 Hz) with 10 millisecond echo. A single auditory event is perceived, in the direction of the direct sound at 42.4 milliseconds and 47.4 milliseconds.

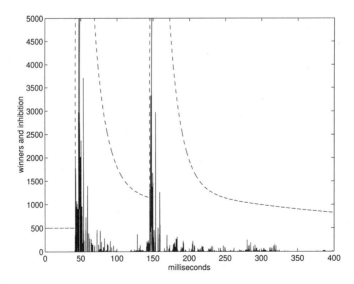

Figure 5. Precedence effect on 2nd band of "teeth" (6000-4000 Hz) with 100 millisecond echo. Two separate auditory events are detected at 42.4 and 145.6 milliseconds.

3. CONCLUSIONS

The results signify that the model is able to localize a band of speech containing abrupt onsets quite accurately. Moreover, the model is in line with our expectations from psychoacoustics. For small time differences between the direct sound and echo, a single auditory event is perceived in the direction of the direct sound. For time differences that exceed the echo threshold, two distinct auditory events are perceived. Therefore, this model demonstrates the plausibility of onsets as a low-level cue for the precedence effect. The model has been tested on a variety of words by different speakers. A more comprehensive review of the model and its results is provided in [9].

This work is a first step in deciphering the relation between precedence, speech and onset enhancement. The precedence model can serve as a tool to understand the precedence effect for complex signals and should be advantageous to engineering applications such as preprocessors for speech recognition systems. The adaptive model offers a novel approach to onset enhancement that deserves further investigation. In our current research, we have been comparing between fixed methods and adaptive methods for onset enhancement, and using the gamma filter [7] as a unifying approach. Both lines of research provide a promising platform for the development of more sophisticated models in auditory scene analysis, that involve real speech signals.

ACKNOWLEDGMENTS

The authors would like to acknowledge Office of Naval Research contract #N00014-94-1-0858.

REFERENCES

[1] Blauert J. *Spatial Hearing + supplement*. MIT Press, Cambridge, MA, 1983.

[2] Clarkson P. *Optimal and Adaptive Signal Processing*. CRC Press, Boca Raton, FL, 1993.

[3] Haas H. The influence of a single echo on the audibility of speech. *Acoustica*, 1:49–58, 1951.

[4] Hartman W.M. and Rakerd B. Localization of sound in rooms, 4: The franssen effect. *Journal Acoust. Soc. Am.*, 86:1366–1373, 1989.

[5] Jeffress L.A. A place theory of sound localization. *J. Comp Physiol. Psychol.*, 61:468–486, 1947.

[6] Lindenmann W. Extension of a binaural cross-correlation model by contralateral inhibition, part 1 and part 2. *Journal Acoust. Soc. Am.*, 80:1608–1630, 1986.

[7] Principe J.C., De Vries B., and De Oliveira P.G. The gamma filter-a new class of adaptive iir filters with restricted feedback. *IEEE Transactions on Signal Processing*, 41:649–656, 1993.

[8] Rakerd B. and Hartman W.M. Localization of sound in rooms, 3: Onset and duration effects. *Journal Acoust. Soc. Am.*, 80:1695–1706, 1986.

[9] Schwartz O. Modeling the precedence effect for speech. Master's thesis, University of Florida, Gainesville, FL, 1996.

[10] Slaney M. An efficient implementation of the patterson and holdworth auditory filter bank. *Apple Technical Report 35*, 1993.

[11] Smith L.M. Onset based sound segmentation. *Journal of New Music Research*, November 1993.

[12] Wallach H., Newman, and Rosenzweig. The precedence effect in sound localization. *Amer. J. Psychol.*, 57:315–336, 1949.

[13] Zurek P. M. *Directional Hearing*. Springer-Verlag, New York, 1987.

128

ADAPTABILITY AND VARIABILITY IN THE OCULOMOTOR SYSTEM

David F. Scollan,[1] Beau K. Nakamoto,[2] and Mark Shelhamer[3]

[1]Department of Biomedical Engineering
Johns Hopkins University School of Medicine
411 Traylor Bldg., 720 Rutland Ave.
Baltimore, Maryland 21205
[2]University of Hawaii
School of Medicine
Honolulu, Hawaii 96822
[3]Departments of Otolaryngology–Head and Neck Surgery, and Biomedical
    Engineering
Johns Hopkins University School of Medicine
210 Pathology Bldg., 600 N. Wolfe St.
Baltimore, Maryland 21287

1. INTRODUCTION

We have been investigating the adaptive capabilities of a simple oculomotor system: saccadic eye movements. Saccades are rapid eye movements that take the line of gaze from one point to another very quickly, as when reading. Saccades are highly ballistic — once a saccade to a visual target has been programmed, the eyes will go to that target (or very nearly so) even if the target subsequently changes position.[1] A further corrective saccade must then be made to bring the eyes to the target.

This open-loop property implies that the saccadic system must be under parametric adaptive control,[1] since real-time visual feedback cannot be relied upon to keep saccades accurate. Yet they are very accurate; saccade endpoints typically have a standard deviation on the order of 1 deg when made to visual targets at 10 deg. The adaptive capability has been demonstrated in several experiments.[2] Typically, a visual target is presented in a dark room. When the subject's eyes begin to move toward the target, it is displaced several degrees from its original location. When the eyes arrive at the original target location, they must therefore make another (secondary) saccade to the new target location. Eventually (usually within a few minutes) the saccadic system becomes recalibrated so that a target at the original location (e.g. 15 deg to the right) triggers a single saccade to the altered location (e.g. 10 deg).

We were interested in predicting the adaptive capabilities of a group of subjects — that is, predict who would adapt the fastest or the most completely — by using measures of saccade variability and/or accuracy. We hypothesized that the normal (and usually small) scatter in the endpoints of saccades made by subjects would be related to their adaptive capabilities, such that large scatter would indicate high flexibility and the ability to make neural parameter adjustments easily, and hence to adapt more readily. This hypothesis was motivated by the recent interest in "dynamical diseases", and the related notion that too-small variability in a physiologic system (such as cardiac interbeat intervals) might be a sign of pathology.[3]

2. METHODS

Eight human subjects were tested over three years. Each subject was tested from one to four times; multiple test sessions on a given subject were at least one month apart. All subjects gave their informed consent to participate in the experiment after the procedures were fully explained. Subjects sat in a dark room facing an array of visual targets (red LEDs). Horizontal eye movements were recorded with a scleral search coil in a magnetic field.

Each test session had two parts, one to assess variability and one to assess adaptability. First, the subject made saccades to visual targets that were presented alternately 10 deg to the right and 10 deg to the left, with 750 msec between target motions, for a total of 40 saccades of 20 deg amplitude each. The subject then continued to make saccades to the remembered locations of the targets, with no targets lit, while maintaining the previous pace with the aid of timing tones. The velocities of these saccades, and their ending positions, were measured to assess variability and accuracy.

The second part of the experiment, in which saccades amplitudes were adaptively altered, followed immediately. A target at 5 deg to the right was lit, and the subject looked at it. This target was extinguished, and a target at 10 deg to the left was lit, requiring a 15 deg leftward saccade. When the subject's eyes were halfway to the 15 deg left target, its position was changed to 5 deg left, requiring a secondary, corrective, saccade of 5 deg to the right to bring the eyes to the target. This was repeated in the other direction, and the entire sequence was re-

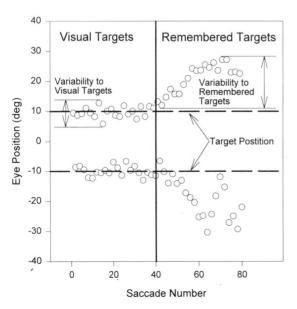

Figure 1. Measures of saccade variability and accuracy used in this study. The first 40 saccades were made to visual targets and have little variability in their endpoints. The second 40 saccades were made to the remembered locations of the visual targets; their variability is greater and the accuracy is also much worse.

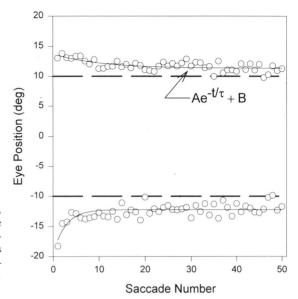

Figure 2. Example of saccade adaptation, showing rate of adaptation (τ) and asymptote (B). With each adaptation trial, the eye position at the end of the primary saccade moves steadily toward 10 deg, which is the final target position and the desired saccade endpoint.

peated 100 times. After several trials, the subject adaptively altered the amplitudes (gains) of his or her saccades so that saccades were made directly to the 5 deg position when a target at 10 deg was presented. We measured the amplitude of the first saccade in each trial – a saccade which should take the eyes from 5 deg on one side to 5 deg on the other. Initially these saccades were on the order of 15 deg in amplitude (requiring a secondary saccade to the target at 5 deg); as the adaptation progressed the amplitudes approached 10 deg. A decaying exponential ($e^{-t/\tau}$) was fit by least squares techniques to the series of saccade amplitudes. The decay time constant τ was used as a measure of adaptation rate.

From the first part of the experiment, we calculated saccade variability – the standard deviations of the amplitudes (STD_A) and the velocities (STD_V) of the saccades made to remembered target positions—and saccade accuracy—the distance of the average saccade amplitudes from the target position (AVG_A). From the second part of the experiment, time constants for adaptation to the right and to the left were calculated.

3. RESULTS

The amplitude variability of saccades to remembered positions (STD_A) was positively correlated with slower rates of adaptation (longer time constants) (Spearman R=0.65, P=0.003). This is a key result, and indicates that to a large extent the higher the variability, the slower the adaptation.

One might attribute both increased variability and slowed adaptation to fatigue or inattention – fatigue might cause a spurious correlation between these quantities. Fatigue is often manifest as a decrease in saccade velocity.[1] However, adaptation rate and fatigue (velocity) were not correlated in this study.

4. MATHEMATICAL MODELING

Using combinations of the variability and accuracy measures, we constructed linear regression models that accurately predicted adaptation rates ($R^2 > 0.8$, adjusted for number

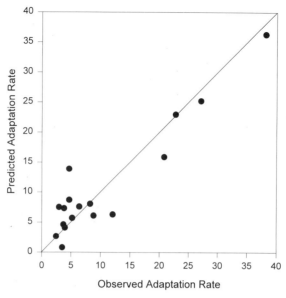

Figure 3. Linear regression model that predicts adaptation rate based on a set of four measures of variability and accuracy (see text).

of variables). These models passed the required statistical tests of normality of the residuals, homoscedasticity (variance of dependent variable constant regardless of value of independent variable), and absence of multicollinearity.[4] One such model is:

$$\tau_L = -51.0 + 3.7\ STD_{A,L} + 4.1\ AVG_{A,R} + 0.6\ STD_{V,R} - 0.2\ STD_{V,L}$$

where τ_L is the time constant for adaptation of leftward saccades, $STD_{A,L}$ is the variability (standard deviation) of leftward saccade amplitudes to remembered positions, $AVG_{A,R}$ is the mean of rightward saccade amplitudes to visual targets, $STD_{V,R}$ is the variability of rightward saccade velocities to remembered positions, and $STD_{V,L}$ is the variability of leftward saccade velocities to visual targets.

In this model, all variables were significant ($P < 0.03$), and adjusted $R^2 = 0.83$. Models using more variables achieved adjusted R^2 values as high as 0.97. All legitimate regression models required at least one measure of velocity variability (STD_V).

5. DISCUSSION AND CONCLUSIONS

We have developed mathematical models by which we are able to relate adaptive ability to variability and accuracy. There is rising support, stemming from the application of nonlinear dynamics to physiology, for the notion that variability is a hallmark of a healthy physiologic system, and that a decrease in variability may be indicative of pathology[5]. In cardiology, for example, reduced variability (or "complexity") of beat-to-beat intervals may be correlated with myocardial infarction.[3] Thus we originally hypothesized that saccade variability and saccade adaptability might be positively correlated – high variability might confer increased adaptability. Variability might then reflect an underlying neural flexibility, and ability of the system to react to environmental changes in an adaptive manner. One might make an analogy to artificial neural networks, which are initialized with a set of random synaptic weights; too little variation in the initial weights might preclude the network from exploring fully the solution space of the particular problem, leading to slowed or absent convergence. We conjecture that there may be an optimal amount of variation in such situations.

Our results contradict this hypothesis. Those subjects with low variability in their saccades have a greater ability to alter adaptively the gain of their saccades. This is consistent with a view of the oculomotor system as one that is continuously adapting. In this view, the saccadic system is not primarily fixed, with adaptive mechanisms coming into play only secondarily. Instead, adaptive processes are always in operation, constantly correcting for changes such as those due to aging. Thus, those subjects who have accurate and consistent saccades have them because of an active and strong adaptive capability that keeps them that way all the time. Such adaptive processes in oculomotor control are well known; however, the intimate relationship of the adaptive mechanisms to the normal operation of the system is one that has not been explored. It remains to be seen to what extent these results can be generalized to other sensorimotor tasks.

6. ACKNOWLEDGMENTS

Supported by the Whitaker Foundation and NASA grant NAGW-2874. We thank Olga Telgarska for her participation in this work.

7. REFERENCES

1. W Becker (1989) Metrics. In: Review of Oculomotor Research, vol. 3, The Neurobiology of Saccadic Eye Movements, R Wurtz, M Goldberg (eds) New York: Elsevier, pp. 13–67.
2. LM Optican (1985) Adaptive properties of the saccadic system. In: Adaptive Mechanisms in Gaze Control, A Berthoz & G Melvill Jones (eds) New York: Elsevier, pp. 71–79.
3. JE Skinner, CM Pratt, T Vybiral (1993) A reduction in the correlation dimension of heartbeat intervals precedes imminent ventricular fibrillation in human subjects. Amer Heart J 125: 731–743.
4. RA Johnson, DW Wichern, Applied Multivariate Statistical Analysis. 3rd ed. (1992) Prentice Hall, Englewood Cliffs, NJ.
5. DT Kaplan, MI Furman, SM Pincus, SM Ryan, LA Lipsitz, AL Goldberger (1991) Aging and the complexity of cardiovascular dynamics. Biophysical Journal 59:945–949.

129

USING MEASURES OF NONLINEAR DYNAMICS TO TEST A MATHEMATICAL MODEL OF THE OCULOMOTOR SYSTEM

Mark Shelhamer[1] and Nabeel Azar[2]

[1]Departments of Otolaryngology–Head and Neck Surgery, and Biomedical
    Engineering
Johns Hopkins University School of Medicine
210 Pathology Bldg., 600 N. Wolfe St., Baltimore, Maryland 21287
[2]Department of Biomedical Engineering
Johns Hopkins University
210 Pathology Bldg., 600 N. Wolfe St., Baltimore, Maryland 21287

1. INTRODUCTION

Mathematical models are widely used and well-established in studies of the oculomotor system. Some of the newer techniques of nonlinear systems analysis also are now beginning to be applied to data from eye movement experiments. This convergence of established models and new methods allows us to address two issues in computational neuroscience. First, many published models are seldom subject to verification and further study by independent researchers. Second, some of the newly-developed measures of nonlinear dynamics are not easily interpreted in a way that is physiologically meaningful. We set out to address these issues by applying nonlinear dynamical measures to a specific model of oculomotor control, and also to the animal data upon which the model was based, to verify the model. We then made systematic alterations to the model to see how the dynamic measures changed, to help provide some basis for interpretation of these measures.

2. MATHEMATICAL MODEL

The model that we chose for our analysis simulates optokinetic nystagmus (OKN) in the red-eared turtle.[1] OKN is the reflexive eye movement that results when an animal is presented with a homogeneously-moving wide-field visual scene.[2] The eyes track the scene to stabilize the visual field on the retinas, making a slow phase eye movement. Occasionally, the eyes will reset themselves to a new position in order to pick a new part of

Computational Neuroscience
edited by Bower, Plenum Press, New York, 1997

the visual scene and track it; to do this they make a fast phase eye movement in the opposite direction from the slow phase. The alternation of slow and fast phases resembles a shaky sawtooth waveform, and is called "nystagmus".

The central concept of the model is the generation of fast phases at appropriate intervals. For each fast phase, this interval is generated by multiplying the previous fast phase interval by a random number; the random number is drawn from a truncated normal distribution (truncated on the low end so that fast phases do not occur too close to each other). Thus the fast phase timing is essentially generated by a noisy clock. If this process generates an interval that is an outlier (based on actual OKN statistics), then the clock is reset to a default initial interval value and the process continues to run. There are also occasions, based on eye position and timing, when the clock "skips a beat" – the clock is recycled and a new interval is generated without a fast phase occurring.

To test this model, it was changed systematically, and simulated data sets were created from each modified model. In one case, the variance of the clock was set to zero, so that fast phases (except for an occasional skipped beat) occurred periodically. (Since the velocity of the slow phases could still vary somewhat, this did not create perfectly periodic OKN.) In another case, the variance of the clock was increased by a factor of two. Thus four different OKN data sets were available for analysis: original turtle OKN, standard (published) model, model with no clock variance, model with increased clock variance.

3. DYNAMIC MEASURES

Five measures of nonlinear dynamics were applied to the animal data and to the corresponding model data. All five measures begin with a time-delay embedding of the OKN time series. This is a means of reconstructing the state space trajectories of a dynamic system, in an artificial state space. Instead of using, as each dimension of the state space, the actual state variables of the system (which may be unknown or inaccessible), time-delayed values of a single variable are used instead. Thus one "axis" may be $x(t)$, the next "axis" $x(t-T)$, then $x(t-2T)$, and so on to $x(t-nT)$. The value of T is the delay time of the reconstruction, and n is the "embedding dimension". Each OKN time series was used to reconstruct such state space trajectories, upon which the following procedures were applied. (This approach is also referred to as "attractor" reconstruction, since in many cases the dynamic system under study has trajectories that converge to (are attracted to) a specific region of the state space. The term "attractor" will be used in the discussion that follows.)

3.1. Correlation Dimension[3]

This common measure is an estimate of the "fractal" dimension of a system's state space trajectories (its attractor). It quantifies the spatial structure of an attractor in phase space, and has been widely applied in physiology, with mixed results. An earlier dimensional study[4] showed some evidence for nonlinear dynamics in human OKN.

3.2. Nonlinear Prediction[5]

On the state space attractor, two points, consecutive in time, are found and a linear regression made to predict subsequent points (1 to 10 time steps into the future). The error in this prediction, averaged over the entire attractor, behaves in a systematic fashion depending on the nature of the system under study. A random system has poor prediction for all time steps, a classic deterministic system has uniformly good prediction, and a chaotic system has good short-term prediction that rapidly drops off. The relative predictability of different systems can be compared by this method.

3.3. Exceptional Events[6]

Nearby points on the state-space attractor are projected into the future, and the projections examined to see how close together they are. For a deterministic system, as the initial points get closer together, the projected points should also get closer together. For a stochastic system, the projected points will get closer together as the original points get closer together, but will not converge toward zero in the limit because of the random component. This provides one method to distinguish between deterministic and stochastic dynamics in a time series.

3.4. Recurrence Analysis[7]

A recurrence plot is a graphical display of the time correlation in a signal. A ball of radius r is taken around a reference point x(i) on the attractor. If another point on the attractor, say x(j), is within distance r of the point x(i), then a dot on the recurrence plot is placed at coordinates (i,j). The is repeated for all points x(j), and the entire procedure is repeated for successive reference points x(i). The resulting recurrence plot has dimensions N×N (N is the number of points on the attractor). The coordinates represent time indices of points on the attractor.

3.5. Mutual Recurrence

We have been developing a method of applying recurrence analysis to two time series simultaneously, to look for dynamics that are associated between the two. This was done by overlapping the individual recurrence matrix from each signal, and taking the difference between them (i.e. the element-by-element difference of the two recurrence matrices, with each matrix entry (i,j) representing the distance between points x(i) and x(j)). This new matrix then had a distance threshold applied to it, and if a point (i,j) exceeded the threshold a dot was plotted in the mutual recurrence plot at location (i,j). A denser mutual recurrence plot then indicates that similar recurrences occur at the same times in the two signals. (Some problems remain with this method, which is still in its preliminary stages. The most critical is the issue of phase delay between the signals, which would cause a very low density plot even if the two time series were otherwise identical. We are developing methods to take this factor into account.)

For the correlation dimension, the entire OKN time series was used in the computations. In the other cases, parameters derived from the OKN were used. These parameters are: time interval between fast phases, position of the eye at the start of each fast phase, position of the eye at the end of each fast phase, amplitude of each fast phase, and velocity of each slow phase. Each of these parameters was extracted from the OKN sample under study, and time-delay embedding was performed on the parameter values. For standard single-parameter recurrence analysis, both the eye position time series data and the derived parameters were used.

4. RESULTS

4.1. Dimension, Prediction, Exceptional Events

Most of the dynamic measures revealed little or no difference between the actual turtle OKN data and the simulated data generated by the mathematical model in its original form. Even though the model contains several random processes, the correlation dimensions were finite values (noise is infinite dimensional and can in some cases lead to very high dimensions). The dimensions of the simulated data were less than those of the real data (e.g. real data: 3.2; standard model data: 1.86, 1.62). The dimensions were mostly unaffected (remaining < 2.0) by changes to the model. Prediction of fast phase intervals, starting positions, ending positions, and amplitudes generally indicated random behavior (little predictability) for both real and simulated OKN. The single exception to this was the presence of some slight predictability of fast phase ending positions in the standard model (which has biological significance in terms of the point at which an organism thinks it important to begin subsequent slow phase tracking[8]). Predictability of all parameters was unaffected by changes to the model except for the loss of this slight predictability of fast phase ending positions. Similarly, the method of exceptional events showed essentially no convergence toward zero at small distances, indicative of stochastic dynamics, in all cases except one – there was some indication of deterministic structure in the sequence of slow phase velocities in all models (less so in the original OKN). Overall, these results are highly suggestive of significant random behavior in the original and various simulated OKN samples.

4.2. Recurrence Analysis

Recurrence plots for the eye position data of the original OKN (Fig. 1, top left) show some clustering of points, with relatively constant overall density as one moves away from the main diagonal. These observations indicate that the OKN time series is nearly periodic and highly stationary. Recurrence plots for the standard model (Fig. 1, top right) show a similar general pattern, but the points are grouped more tightly and the regions of non-recurrence (white areas) are more pronounced; this indicates a similar overall structure to the data, but more tightly grouped time epochs in which the behavior is highly periodic interspersed with non-periodic epochs. The recurrence plots from the modified models, with

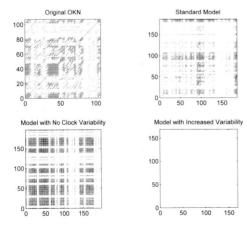

Figure 1. Recurrence plots of OKN time series data. Actual turtle OKN (top left) has a recurrence structure that is slightly different from that of data from the standard model (top right). Modifications were made to the model to decrease (lower left) or to increase (lower right) variability in the clock that generates fast phase intervals; in each case only minor changes in recurrence are seen.

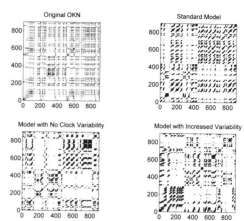

Figure 2. Recurrence plots of the time interval between OKN fast phases. Data from the standard model (top right) are overall more homogeneous in recurrence structure than actual turtle OKN data (top left). With no clock variability (lower left) the recurrence structure is highly periodic, while the recurrence decreases overall when the clock becomes more variable (lower right).

either zero or increased clock variability (Fig. 1, bottom), are surprisingly similar to the plot from the standard model.

Similar analysis was carried out on the parameters derived from each OKN time series. Results for one of these parameters, the interval between fast phases, are shown in Fig. 2. In this case, the standard model has a somewhat more uniform recurrence than does the original OKN, indicating that perhaps the clock variation in the model is higher than necessary. The model with no clock variability, as expected, shows very high periodicity (gaps in the recurrence plot in this case are likely due to the skipping of a beat, as described previously). The model with increased clock variability again has a similar overall structure to the standard model, but the plot density is lower which indicates overall lower repeatability (recurrence).

4.3. Mutual Recurrence

Mutual recurrence plots were created to compare several pairs of OKN parameters. In most cases, the coupling between parameters is similar for both the original OKN and the model OKN, which can be quantified as the density of each mutual recurrence plot (a measure of percent recurrence for a given distance criterion on the attractor[9]). In the case of fast intervals and slow phase amplitudes, however, the model data exhibit a much stronger dynamic coupling than does the original OKN data set, although this coupling decreases partially when the clock variability is increased (Fig. 3). This, and the slightly altered periodicity, are the two essential difference between the original OKN and the model OKN, as determined by our methods.

5. CONCLUSIONS

The results largely confirm the fidelity of the particular model under study, especially in terms of the random nature of many of the properties of the simulated OKN that it produces. Model data have dimensions less than that of the actual data, however, indicating that perhaps some increase in the level of the random processes used in the model should be made. This might presumably have the added advantage of decreasing the overly-high coupling between some of the OKN parameters. The application of several computational tools to this model has given us some hints as to how the model might be changed to increase its physiologic fidelity.

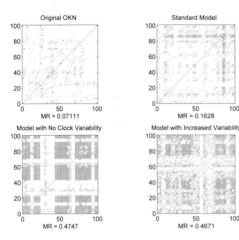

Figure 3. Mutual recurrence plots. Each mutual recurrence plot is a measure of the overlap of the individual recurrence plots of slow phase amplitude and fast phase interval. Strong dynamic coupling between the two parameters would be expected to produce a dense mutual recurrence plot. The measure of "percent mutual recurrence" (i.e. the density of the plot) is shown below each graph (labeled as "MR"). All models show an overly-high coupling between these two parameters.

6. ACKNOWLEDGMENTS

Supported by the Whitaker Foundation.

7. REFERENCES

1. CD Balaban, M Ariel (1992) A "beat-to-beat" interval generator for optokinetic nystagmus. Biol Cybern 66:203–216.
2. RJ Leigh, DS Zee (1991) The Neurology of Eye Movements. Edition 2. Philadelphia: F. A. Davis.
3. J Theiler (1990) Estimating fractal dimension. J Opt Soc Am A 7: 1055–1073.
4. M Shelhamer (in press) On the correlation dimension of optokinetic nystagmus eye movements: computational parameters, filtering, nonstationarity, and surrogate data. Biol Cybern.
5. M Casdagli (1989) Nonlinear prediction of chaotic time series. Physica D 35:335–356.
6. DT Kaplan (1994) Exceptional events as evidence for determinism. Physica D 73:38–48.
7. J-P Eckmann, S Oliffson Kamphorst, D Ruelle (1987) Recurrence plots of dynamical systems. Europhys Lett 4: 973–977
8. K-S Chun, DA Robinson (1978) A model of quick phase generation in the vestibuloocular reflex. Biol Cyber 28:209–221.
9. CL Webber, JP Zbilut (1994) Dynamical assessment of physiological systems and states using recurrence plot strategies. J Appl Physiol 76:965–973.

130

MATHEMATICAL MODELS OF THE CRAYFISH SWIMMERET SYSTEM

Frances K. Skinner,[1] Nancy Kopell,[2] and Brian Mulloney[1]

[1] Section of Neurobiology
Physiology and Behavior
University of California
Davis, CA
[2] Department of Mathematics
Boston University
Boston, MA

1. INTRODUCTION

The neural basis of intersegmental coordination in animals that have been studied lies in the oscillatory output of neural networks or central pattern generators (Delcomyn, 1980). Mathematical models which describe such circuits have the potential to provide insight into neural mechanisms and design principles that operate in them. In this paper we look at intersegmental coordination in the crayfish swimmeret system, and consider two different mathematical models. In the first model, the system is represented by phase-coupled oscillators and no cellular details are present. In the second model, a simple cellular framework is formulated and preliminary results are presented.

2. THE CRAYFISH SWIMMERET SYSTEM

Swimmerets are paired, jointed limbs on each segment of the crayfish abdomen (4 segments: A2-A5) that can move rhythmically through cycles of power-strokes (PS) and return-strokes (RS) that propel the animal forward in the water. The rhythmic movements of the swimmerets are known to be driven by neural networks in their segmental ganglia. Recent results show that each swimmeret is controlled by its own pattern-generating module, whose function is to drive alternating PS and RS movements (Murchison et al., 1993). Movements of swimmerets on neighboring segments have the same period and a constant phase relationship. Each cycle begins with a PS by the most posterior pair. Swimmerets on the same segment move synchronously, but more anterior pairs lag behind their nearest neighbor by about 25%

Computational Neuroscience
edited by Bower, Plenum Press, New York, 1997

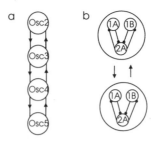

Figure 1. Model Schematics of the Crayfish Swimmeret System. Each of the 4 ganglia shown is capable of oscillation and is represented by a circle numbered from anterior to posterior in the same way as it is in the crayfish abdomen. Arrows indicate the bidirectional coupling. (a) Schematic for the PCO model. Each oscillator has an intrinsic frequency and is described by a single variable, its phase. (b) Schematic for the simple cellular model. Each module generates oscillatory output via a reciprocally inhibitory circuit of 2A/1A/1B non-spiking interneurons. Black circles represent inhibitory synapses.

(or $\pi/2$) of the cycle period. A separate, bidirectional coordinating network organizes the activity of these modules in different segments of the body (Mulloney et al., 1993).

In recent experimental work, several properties of the swimmeret system have been determined: (i) There is no significant difference in the intrinsic frequencies of the different swimmeret modules. (ii) The constant phase lag of about $\pi/2$ down the abdominal cord remains the same with changes in frequency. These frequency changes can be elicited with nicotinic analogs of acetylcholine which excite the system in a dose-dependent fashion. (iii) When anterior or posterior ganglia are selectively excited, a significant change in the phase lag at the excitation boundary is obtained; anterior excitation shortens the phase lag and posterior excitation lengthens it. Phase lag changes occur between segments other than at the excitation boundary, but the maximum change always occurs at this boundary. (iv) The periods of the swimmeret motor patterns decrease as the number of excited ganglia increases, regardless of whether posterior or anterior modules are excited (Braun and Mulloney, 1993, 1995; Mulloney and Hall, 1995).

3. A PHASE COUPLED OSCILLATOR (PCO) MODEL

The four coordinated ganglia of the swimmeret system (A2-A5) are represented by a chain of four oscillators (Osc2-Osc5) bidirectionally coupled to their nearest neighbors (Fig.11a). Each oscillator has an intrinsic frequency and is described by a single variable, its phase, with interactions occurring via phase differences. This model approach is based on a formal mathematical framework developed by Kopell and Ermentrout (1988) who examined intersegmental coordination in the lamprey spinal cord. Details of the PCO model in the swimmeret system can be found in Skinner et al. (1997).

4. RESULTS

The model had the ability to reproduce the experimentally observed changes in intersegmental phases caused by uniform and non-uniform excitation of the abdominal ganglia. Some aspects are illustrated in Figs.2 and 3. Fig.2 shows that the system frequency increases in a

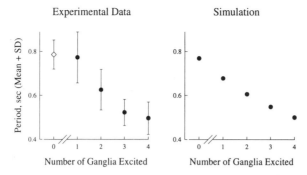

Figure 2. Combined Anterior and Posterior Periods. The entrained period decreases monotonically as the number of ganglia that are excited increases and this decrease is independent of whether posterior or anterior modules are excited.

monotonic way as the number of excited ganglia increases, and that this increase is independent of whether anterior or posterior ganglia are excited. Fig.3 illustrates how phase lags change when three anterior or posterior ganglia are excited. The maximal change occurs across the excitation boundary, shortening if anterior modules are excited and lengthening if posterior modules are excited.

The parameter choices needed to match the experimental data led to the following predictions: (i) The ascending and descending coupling are approximately equal in strength and both contribute to the production of phase lags. (ii) The ascending and descending coupling are each separately constructed to create the observed lag of $\pi/2$. Since the coupling goes in opposite directions, inducing lags in the same direction requires the coupling to be asymmetric. (iii) The frequency of the coupled system is not significantly affected by the intersegmental coupling. (iv) Excitation affects either the intrinsic frequencies alone or the intrinsic frequencies and the strength of the intersegmental coupling.

Even though the PCO model (Skinner et al., 1997) was successful in revealing certain features of the swimmeret system, the abstract nature of the model does not allow us to consider other features such as burst duration or the type of coupling that may exist within and between segments. For example, the PCO model tells us that excitation must change the intrinsic properties of the particular modules, but we do not know whether it is the properties of the individual neurons or the properties of the connections within the particular ganglion that should change.

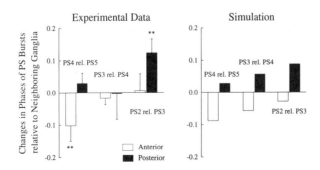

Figure 3. Anterior versus Posterior Excitation of 3 Ganglia. Excitation of selected ganglia causes a maximal change in phase lags across the excitation boundary. It shortens if anterior modules are excited and lengthens if posterior modules are excited.

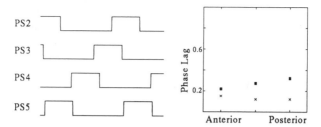

Figure 4. Phase Lags down the Abdominal Cord. (a) The burst duration for the 2A cell (representing PS) is shown for the 4 ganglia. Cell is "on" and "off" at upper and lower values respectively and the activity is shown for 1.5 sec. (b) If the frequency is increased by increasing the rate constant of the outward current (open squares), then the phase lags do not change significantly (solid and open squares overlap). However, if the frequency is increased by increasing the injected current (crosses), the phase lags change dramatically (compare crosses and solid squares). The solid squares correspond to part (a).

5. A SIMPLE CELLULAR MODEL

To try to understand the swimmeret system in a bit more detail, we formulated a simple cellular model using what is currently known about the system. Each module contains a small set of non-spiking local interneurons which depolarize in phase with the PS motor neurons (Type 2A) and in phase with the RS motor neurons (Type 1A/1B) (Paul and Mulloney, 1985). We assume that these interneurons form a reciprocally inhibitory circuit to produce the observed, antiphasic PS and RS behavior. It is also known that for the bidirectional coupling, most of the ascending axons fire in PS time and the descending axons fire in RS time (Stein, 1971).

The cellular model is schematically shown in Fig.1b. Each module consists of four local interneurons; two electrically coupled 2A neurons (these neurons are assumed to be tightly coupled so that they can be described as one cell), one 1A neuron and one 1B neuron. Graded reciprocally inhibitory synapses exist between them as shown in Fig.1b. Each interneuron is modelled using Morris-Lecar like equations (potassium, calcium and leak currents), and the graded synapses are modelled such that they activate above a given synaptic threshold.

The four modules are connected by coordinating interneurons which are themselves driven by the local interneurons described above. These coordinating interneurons are not specifically modelled. Instead, a "spike-mediated transmission threshold" is used. If the presynaptic voltage rises above this level, spikes of a given duration and frequency are transmitted to the next level, causing synaptic currents in the target neurons.

6. PRELIMINARY RESULTS

Using the PCO model predictions and what is known experimentally, we obtain phase lags of about $\pi/2$ down the cord with: (i) an ascending inhibitory connection from cell 2A to cell 1A, (ii) an ascending excitatory connection from cell 2A to cell 1B, (iii) a descending inhibitory connection from cell 1A to cell 1A, and (iv) a descending inhibitory connection from cell 1A to cell 2A. This is shown in Fig.4a. As the cord is uniformly excited, these phase lags are maintained. How is the frequency changed with excitation? Fig.4b shows that if the frequency is changed via the rate constant of the outward current, then the phase lags are not affected. However, changing other parameters such as injected current (also shown in Fig.4b), graded synaptic and intrinsic conductances, and synaptic thresholds, either do not produce at least a two-fold frequency change (as observed in the animal) and/or do not give phase lags close to the $\pi/2$ value. This suggests that modulating the rate constant of the intrinsic currents may be a way in which the system maintains phase lags across different frequencies.

7. SUMMARY

Using a PCO model and matching experimentally observed changes in intersegmental phases and period caused by non-uniform excitation of selected ganglia, we have predicted several features of the crayfish swimmeret system: coupling between ganglia is asymmetric; the ascending and descending coupling have approximately equal strengths; intersegmental coupling does not significantly affect the system frequency; excitation must affect the intrinsic frequencies.

Using a simple cellular model and reproducing the experimental result that phase lags of about $\pi/2$ are maintained down the cord with frequency changes, we have shown that modulating the rate constant of the outward currents allows this to be achieved.

ACKNOWLEDGMENTS

This work was supported by NSF IBN92-22470, IBN95-14889 (B.M.), NIMH-47150 (N.K.) and by a UC President's Postdoc Fellowship (F.K.S.).

REFERENCES

[1] Braun G, Mulloney B (1993) Cholinergic modulation of the swimmeret system in crayfish. *J Neurophysiol* 70:2391-2398.

[2] Braun G, Mulloney B (1995) Coordination in the crayfish swimmeret system: Differential excitation causes changes in intersegmental phase. *J Neurophysiol* 73:880-885.

[3] Delcomyn F (1980) Neural basis of rhythmic behavior in animals. *Science* 210:492-498.

[4] Kopell N, Ermentrout GB (1988) Coupled oscillators and the design of central pattern generators. *Math Biosci* 90:87-109.

[5] Mulloney B, Murchison D, Chrachri A (1993) Modular organization of pattern-generating circuits in a segmental motor system: the swimmerets of crayfish. *Semin Neurosci* 5:49-57.

[6] Paul DH, Mulloney B (1985) Local interneurons in the swimmeret system of the crayfish. *J Comp Physiol A* 156:489-502.

[7] Murchison D, Chrachri A, Mulloney B (1993) A separate local pattern-generating circuit controls the movements of each swimmeret in crayfish. *J Neurophysiol* 70:2620-2631.

[8] Skinner FK, Kopell N, Mulloney B (1997) How does the crayfish swimmeret system work? Insights from nearest neighbor coupled oscillator models. *J Comput Neurosci* (In Press).

[9] Stein PSG (1971) Intersegmental coordination of swimmeret motor neuron activity in crayfish. *J Neurophysiol* 34:310-318.

131

A MODEL OF CHANGES IN INFEROTEMPORAL ACTIVITY DURING A DELAYED MATCH-TO-SAMPLE TASK

Vikaas S. Sohal and Michael E. Hasselmo[*]

Department of Psychology
Harvard University
Rm. 1440, 33 Kirkland St., Cambridge, Massachusetts 02138
sohal@katla.harvard.edu
hasselmo@katla.harvard.edu

1. ABSTRACT

Neurons in inferior temporal (IT) cortex exhibit selectivity for complex visual stimuli and can maintain activity during the delay following the presentation of a stimulus in delayed match to sample (DMS) tasks. Experimental work in awake monkeys has shown that the responses of IT neurons decline during presentation of stimuli which have been seen recently (within the past several seconds). In addition, experiments have found that the responses of IT neurons to visual stimuli also decline as the stimuli become familiar, independent of recency. Here a biologically based neural network simulation is used to model these effects primarily through two processes. The recency effects are caused by adaptation due to a calcium-dependent potassium current, and the familiarity effects are caused by competitive self-organization of modifiable feedforward synapses terminating on IT cortex neurons.

2. INTRODUCTION

Inferior temporal (IT) cortex has been implicated in both working memory,[2,18] and object recognition.[6,17] Experimental studies have shown that for short interstimulus intervals, the responses of IT neurons decline during repeated presentation of visual stimuli.[7,14,19] Recent studies have focused on the responses of IT neurons to visual stimuli in two variants of delayed match to sample (DMS) tasks. In both variants, a sample stimulus is followed by a variable number (between 1 and 4) of nonmatching stimuli, and the trial ends when a match stimulus (identical to the sample) is presented.

---

[*] Address correspondence to Dr. Hasselmo.

Computational Neuroscience
edited by Bower, Plenum Press, New York, 1997

The first variant used all familiar stimuli and found the following recency effects: (1) the responses of IT neurons to match stimuli are suppressed relative to both nonmatch and sample responses; (2) the difference between nonmatch and match responses decreased as the number of intervening stimuli (measured from the sample presentation) increased; (3) nonmatch responses were suppressed relative to sample responses.[15,16]

The second variant used initially novel sample/match stimuli and familiar nonmatch stimuli and found the following effects as the novel stimuli became familiar: (1) the responses of a subset (approx. one-third) of studied IT neurons to initially novel stimuli declined an average of 40 percent after 12 presentations; (2) the decline in response was stimulus-specific and endured over presentations of >150 other stimulus; (3) the decline in response to successive presentations of a stimulus was greater when there were fewer intervening stimuli; (4) responses to sample and match stimuli declined in parallel, and the decline in response with increasing stimulus familiarity appeared to summate with the match suppression described earlier.[10,15]

These recency and familiarity effects appear to summate in IT cortex.[10] Here a biologically based neural network simulation will be used to model these short-term recency and long-lasting familiarity effects observed in IT cortex primarily through two processes: adaptation caused by a calcium-dependent potassium current, and competitive self-organization of modifiable feedforward synapses terminating on IT cortex neurons.

3. METHODS (COMPUTATIONAL MODELING)

This network simulated the flow of visual information from input areas into IT cortex. A basal forebrain region controlled cholinergic modulation of both the input region and IT cortex. Stimuli consisted of patterns of afferent input applied to the neurons in the input region. Activity spread from these units to the IT region via self-organizing feedforward connections. Synaptic modification proceeded according to a local Hebbian-type learning rule followed by pre-and post-synaptic normalization of connection strengths.

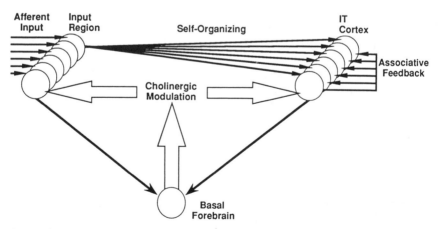

Figure 1. Connectivity between regions in the neural network model of IT cortex. Afferent input enters the network in the input region, and spreads into the IT cortex region via self-organizing feedforward connections. IT cortex contains feedback connections representing recurrent excitation and some units mediating feedback inhibition. The cholinergic neuron in the basal forebrain region is driven by the presence of a stimulus in region 1 and suppressed by activity in IT cortex. Activity of the cholinergic neuron determines the level of cholinergic modulation in IT cortex and the input regions.

IT neurons have been shown to maintain activity in the absence of a stimulus.[3,4,5] The simulation maintained activity using a combination of recurrent excitation and feedback inhibition to produce attractor dynamics in the IT region.

The model also included the phenomenon of adaptation, or slow afterhyperpolarization (ahp) observed in pyramidal cells of the hippocampus,[9] neocortex[20,21] and piriform cortex,[1] but not in inhibitory interneurons.[11] Adaptation in model pyramidal cells was represented by an exponentially decaying hyperpolarization with a time constant of 1 second, based on *in vitro* observations of slow ahp.[21]

IT cortex receives cholinergic innervation from the nucleus basalis of the substantia innominata region in the basal forebrain.[13] As in previous models, cholinergic suppression of recurrent synapses during learning prevented recurrent synapses from interfering with self-organization.[8] Studies have found feedback connections from IT cortex to the particular sector (Ch4i) of substantia innominata from which most of the connections to IT cortex originate.[12] In the model IT activity influenced activity in the basal forebrain region so that cholinergic modulation was high during learning of novel stimuli but low during presentation of familiar stimuli.

4. RESULTS

4.1. Cholinergic Modulation Prevents Recurrent Excitation from Interfering with Self-Organization

Self-organization normally proceeds via competition between patterns. However, in the presence of recurrent excitation, when different patterns of input contain one or more

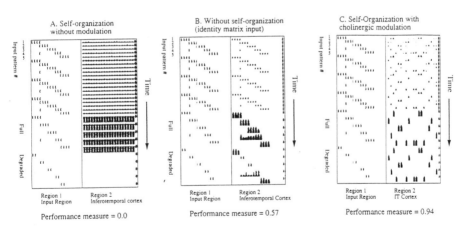

Figure 2. Shows self-organization in the presence and absence of cholinergic modulation. After the set of input patterns is presented for learning, each network is tested using degraded versions of those patterns and performance is measured using a normalized dot product to compare activity in the IT cortex region during presentation of the degraded and complete patterns. *A.* In the absence of cholinergic modulation, recurrent excitation interferes with self-organization and stable, non-overlapping attractor states do not form. *B.* Shows performance when cholinergic modulation is present, but there is an identity mapping rather than self-organizing synapses from input areas to IT cortex. In the absence of self-organization, representations for the input patterns form in IT cortex, but are not non-overlapping. This causes interference between patterns. *C.* When cholinergic modulation is present, self-organization proceeds normally and non-overlapping representations of input patterns form in the IT cortex region. Degraded versions of input patterns recall the complete IT cortex representation, indicating that stable attractor states have formed to represent input patterns.

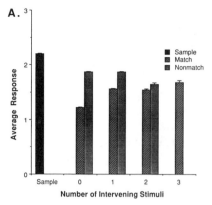

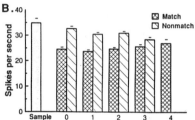

Figure 3. A. Shows the average response of model IT neurons during a DMS task to sample, match and nonmatch stimuli as function of the number of intervening stimuli. B. Shows the average response of IT neurons recorded extracellularly from awake monkeys during performance of a similar DMS task.[15]

common elements, excitation spreads along recurrent excitatory connections during self-organization, activating the overlapping patterns and causing runaway excitation. This problem can be avoided with cholinergic suppression of recurrent connections learning.

4.2. Recency Effects (Match and Nonmatch Suppression)

During DMS trials using all familiar stimuli, adaptation caused IT neurons to respond more strongly the first time that a stimulus appeared on a trial, as a sample, than when it appeared again, as a match. Stimuli occasionally activated overlapping subsets of IT neurons. Therefore, some of the same neurons responded to both nonmatch and sample stimuli on a trial. Adaptation suppressed the responses of these neurons to nonmatch stimuli so that the average response to nonmatch stimuli was weaker than the average sample response but higher than the average match response. As in experiment, match responses recovered and the difference in match and nonmatch responses decreased with increasing numbers of intervening stimuli.

4.3. Familiarity Effects

In the other DMS variant, responses of a subset (45%) of IT neurons to initially novel sample/match stimuli declined an average of 58% after 12 presentations (six as samples and six as matches). In addition to this decline in response with increasing stimulus familiarity, match responses were suppressed relative to sample responses by an average of 44%. Comparison of responses in networks with and without adaptation currents shows that this match suppression is due to adaptation, not just an increase in stimulus familiarity from sample to match.

The number of trials intervening between trials containing the particular sample/match stimulus alternated between 3 and 35. Responses fell an average of 16% when there were only three intervening trials between successive presentations of a stimulus, but fell by only 10% when 35 trials intervened. Earlier experimental studies have similarly found larger response decrements over fewer intervening trials.

5. DISCUSSION

The results demonstrate that in a biologically based neural network model of IT cortex, adaptation with a time constant determined by *in vitro* studies of slow afterhyperpolarization causes responses to a stimulus to decrease depending on its recency. They also show how self-organization causes responses to a stimulus to decrease depending on its familiarity. These results provide mechanisms for similar response decrements observed experimentally.[7,10,14,15,16,20]

These results also make predictions that may help guide future experiments. One consequence of the adaptation used in the model is afterhyperpolarization of the neurons. The adaptation current of neurons which had been "match-suppressed" should continue to exert a hyperpolarizing effect on these neurons, reducing their baseline firing rates for a few seconds after the sample or match presentation. Another consequence of the adaptation of individual neurons is that for a short time after the presentation of one stimulus, responses to any subsequent stimulus should be suppressed if the stimuli activate overlapping groups of IT neurons. This model also predicts that blocking cholinergic innervation should interfere with effective self-organization in IT cortex. One final prediction is that cholinergic agonists or acetylcholinesterase blockers should reduce the amount of match suppression by blocking adaptation currents.

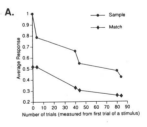

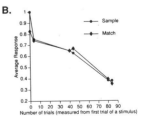

Figure 4. A. Shows the average response of model IT neurons to sample and match stimuli as they became familiar during repeated DMS trials. The average response was computed for the 45% of IT neurons whose responses declined over the course of each simulation. The number of trials intervening between successive presentations of a particular sample/match stimulus alternated between 3 and 35. B. Shows the average responses to sample and match stimuli from an identical network in which adaptation currents were reset to zero after each stimulus presentation, eliminating any recency effects caused by adaptation. C. Shows the experimentally observed decline in the responses of IT neurons to sample and match stimuli as they become familiar.[10,15] Here the number of trials intervening between successive presentations of a particular stimulus alternated between 3 and 35. Again, the average included only those neurons whose responses declined significantly over the course of the recording session (approximately one-third of the total number of IT neurons studied).

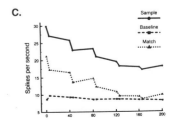

6. REFERENCES

1. Barkai E and Hasselmo ME (1994) Modulation of the input/output function of rat piriform cortex pyramidal cells. *J Neurophysiol* **72**: 644–658.
2. Delacour J (1977) Cortex inférotemporal et mémoire visuelle à court terme chez le singe. Nouvelles domnnées. *Exp Brain Res* **28**: 301–310.
3. Fuster JM (1990) Inferotemporal units in selective visual attention and short-term memory. *J Neurophysiol* **64**: 681–697.
4. Fuster JM, Jervey JP (1981) Inferotemporal neurons distinguish and retain behaviorally relevant features of visual stimuli. *Science* **212**: 952–955.
5. Fuster JM, Jervey JP (1982) Neuronal firing in the inferotemporal cortex of the monkey in a visual memory task. *J Neurosci* **2**: 361–375.
6. Gaffan D, Weiskrantz L (1980) Recency effects and lesion effects in delayed nonmatching to randomly baited samples by monkeys, *Macaca mulatta*. *Brain Res* **196**: 373–386.
7. Hasselmo ME (1988) The representation and storage of visual information in the temporal lobe. Oxford, UK: Oxford University. Unpublished PhD Thesis.
8. Hasselmo ME and Cekic M (1996) Cholinergic suppression of synaptic transmission may allow combination of associative feedback and self-organizing feedforward connections in the neocortex. *Behav Brain Res*: in press.
9. Lancaster B and Adams PR (1986) Calcium dependent current generating the afterhyperpolarization of of hippocampal neurons. *J Physiol (London)* **387**: 519–548.
10. Lin, L, Miller, EK, and Desimone R (1993) The representation of stimulus familiarity in anterior inferior temporal cortex. *J Neurophysiol* **69**: 1918–1929.
11. McCormick DA, Connors BA, Lighthall JW, Prince DA (1985) Comparative electrophysiology of pyramidal and sparsely spiny stellate neurons of the neocortex. *J Neurophysiol* **54**: 782–806.
12. Mesulam MM and Mufson EJ (1984) Neural inputs into the nucleus basalis of the substantia innominata (Ch4) in the rhesus monkey. *Brain* **104**: 253–274.
13. Mesulam MM, Mufson EJ, Wainer BH, and Levey AI (1983) Central cholinergic pathways in the rat: an overview based on an alternative nomenclature. *Neuroscience* **10**: 1185–1201.
14. Miller EK, Gochin PM, and Gross CG (1991a) Habituation-like decrease in the responses of neurons in inferior temporal cortex of the macaque. *Visual Neurscience* **7**: 357–362.
15. Miller EK, Lin L, and Desimone R (1991b) A neural mechanism for working and recognition memory in inferior temporal cortex. *Science* **254**: 1377–1379.
16. Miller EK, Lin L, and Desimone R (1993) Activity of neurons in anterior inferior temporal cortex during a short-term memory task. *J Neurosci* **13**: 1460–1478.
17. Mishkin M (1982) A memory system in the monkey. *Philos Trans R Soc Lond B Biol Sci* **298**: 83–95.
18. Mishkin M and Delacour J (1975) An analysis of short-term visual memory in the monkey. *J Exp Psychol* **1**: 326–334.
19. Rolls ET, Baylis GC, Hasselmo ME, Nalwa V (1989) The effect of learning on the face selective responses of neurons in the cortex in the superior temproal sulcus of the monkey. *Exp Brain Res* **76**, 153–164.
20. Schwindt PC, Spain WJ and Crill WE (1992) Calcium-dependent potassium currents in neurons from cat sensorimotor cortex. *J Neurophysiol* **67**: 216–226.
21. Schwindt PC, Spain WJ, Foehring RC, Stafstrom CE, Chubb MC and Crill WE (1988) Slow conductances in neurons from cat sensorimotor cortex and their role in slow excitability changes. *J Neurophysiol* **59**: 450–467.

132

A NEW APPROACH TO LEARNING VIA SELF-ORGANIZATION

Dimitris Stassinopoulos[1] and Per Bak[2]

[1]Center for Complex Systems
Florida Atlantic University
Boca Raton, FL
[2]Department of Physics
Brookhaven National Laboratory
Upton, NY

ABSTRACT

Recently, we have introduced a simple "toy" brain model to address the problem of learning in the absence of external intelligence.[1] Our model departs from the traditional gradient-descent based approaches to learning by operating at a highly susceptible "critical" state with low activity and sparse connections between firing neurons. Here, quantitative studies of the performance of our model in a simple association task show that tuning our system close to this critical state results in dramatic gains in performance.

1. DEMOCRATIC REINFORCEMENT

One of the most remarkable properties of biological neural networks is the ability to learn via *self-organization*. Simply put, this means that animals acquire experience and make sense of their environment without the aid of a knowledgeable agent to tune relevant parameters (such as the synaptic weights) to appropriate values. To any layperson that has ever seen a toddler acquiring an uncommon skill, such as operating the remote control with no guidance whatsoever, or a bear riding a bicycle at the circus, the significance of self-organization in learning would seem obvious. Yet for all its simplicity and common sense this idea has long remained on the fringes of experimental and theoretical research concerning brain function, in all likelihood because of the severity of the constraints it imposes on brain modelling.

The term self-organization has been used in many different disciplines such as physics, chemistry, biology, and psychology, and often to convey different underlying mechanisms. Here, however, we will discuss self-organization solely in the context of learning.

Computational Neuroscience
edited by Bower, Plenum Press, New York, 1997

The oldest and perhaps still most dominant approach to understanding the brain is what we call the "engineered brain" paradigm. According to this paradigm, brain function emerges because nature, in the role of the engineer, has created all the necessary mechanisms by establishing an intricate web that brings billions of pieces together. But how can evolution achieve such an engineering feat? We do not deny the role of evolution in many aspects of brain function – the very fact that our brain is different from a lobster's brain has to be attributed to evolution. Nevertheless, it cannot possibly account for the brain's ability to deal with unforeseen situations that are specific to an individual's experience, or for novel ones that evolution had never had the opportunity to confront.

In providing an alternative to this view, the field of artificial neural networks (ANN) offered the first evidence that principles for brain function can be captured with models that have simple structure (For reviews see Hertz *et al*[2]). Despite all the important insights ANN offered, however, they have not eliminated the need for an external agency. In the widely used *supervised learning* paradigm this takes the form of a "teacher" providing the system with a detailed scheme for the update of the synaptic weights based on knowledge of the goal to be achieved. Furthermore, most models for learning use gradient-based update rules, such as back-propagation, which are biologically implausible because they impose strong constraints on the architecture and they require computation that cannot be performed by the neural network itself. Thus, once again, we are faced with a network formed by design rather than by self-organization.

The issues of self-organization have been addressed in the context of reinforcement learning models[2]. These models are more realistic in the sense that there is no teacher explaining how to modify the synaptic weights, but only a "critic" telling the system whether its performance is successful or not. Most reinforcement learning models, however, still rely on back-propagation[3], or some other overseeing agency possessing prior knowledge of the problem, for the update of the synaptic weights. There is one exception, however. Barto, in one of the first variants of his *Associative Reward-Penalty*[4] (A_{R-P}) algorithm, discusses the idea of "self-interested" elements which do not have access to information other than a feedback signal from the environment broadcast simultaneously to all elements. We very much agree with his view that there are fundamental difficulties in solving the problem of learning under the severe constraints imposed by self-organization.

Recently, we have proposed *Democratic Reinforcement* (DR) as a new approach to the long-standing issues of learning via self-organization[1]. A similar approach was originally used to solve a non-trivial tracking problem by a continuous modification of its synaptic weights[5]. The model consists of a set of randomly connected or layered *all-or-none* neurons. Each neural unit receives input and sends output to a small fraction of the total number of units (Fig. 1). The update rule affects only connections between firing ($n_i = 1$) neurons, $J_{ij} \rightarrow J_{ij} + [rJ_{ij} + h_{ij}]n'_i n_j$, where n'_i denotes the state of the i'th neuron at the next time step, r denotes the evaluative feedback that is sent democratically to all neurons simultaneously, and h_{ij} is a random noise between $-h_0$ and h_0. The outgoing weights are normalized, $J_{ij} \rightarrow J_{ij}/\sum_i J_{ij}$. The rule differs from standard gradient-descent based update rules in one crucial aspect: When $r > 0$ the system operates in a "learning" mode in which connections are being strengthened and the performance improves, but when $r < 0$ the system operates in an "exploratory" mode in which strong connections are weakened and weaker connections are strengthened. Typically, during this phase the performance deteriorates. In contrast, standard reinforcement schemes, such as A_{R-P}, rely on an improvement of the performance both for positive and negative r and perform the exploration stochastically. The novel feature of our model is that the threshold for firing,

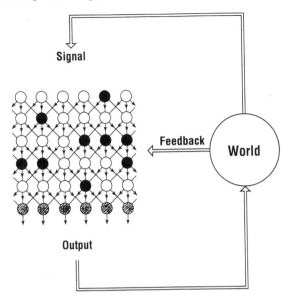

Figure 1. Block diagram of the model, here shown for the layered architecture. The input sites that receive signals from the environment are shown as dark discs. The output sites are shown as shaded disks. Periodic boundary conditions are assumed for the layers.

T, is regulated in order to keep the output activity minimal: $T \rightarrow T + \delta(r)$, where δ assumes a positive value, δ_+, if $r > 0$, and a negative value, δ_-, if $r < 0$.

To the best of our knowledge, the A_{R-P} and the DR represent the only attempts to address the problem of learning via self-organization. However, the two algorithms are fundamentally different. Our studies[1,6] indicate that while the A_{R-P} is a gradient-descent based algorithm, DR solves problems by operating at or near a highly-susceptible "critical" state in which the system becomes very sensitive to modification of the synaptic weights. In such a state the system can establish efficiently causal relationships between changes in the synapses and the output. To achieve such a critical state the system: i) assigns credit and blame only to connections between active neurons, and ii) keeps the activity low by means of the global regulation of the threshold and the local learning rules. By combination of these two mechanisms the system attributes credit and blame selectively by driving the system to the interface between success and failure. To make the point more concrete we shall discuss the two algorithms in the context of a specific association task.

2. SELF-ORGANIZATION IN A SIMPLE ASSOCIATION TASK

In an association task we ask the system to generate a certain input/output pattern. The insets in Figs. 2a,b offer examples of a simple association task. The system accomplishes the task by "carving" paths between the input sites and the output sites. In previous work[1,5] we have investigated the performance of DR in a variety of situations: multiple input/output patterns, recovery from "damage", tracking, conditioning, and so on. Here we will be concerned with the question of degradation of performance as the size of the association task grows.

We consider an $L_1 \times L_2$ layered network (Fig. 2a, inset). The number of layers in this network is kept fixed, $L_1 = 16$, while the lateral dimension is varied, $L_2 = 16, 32, 64, \ldots$

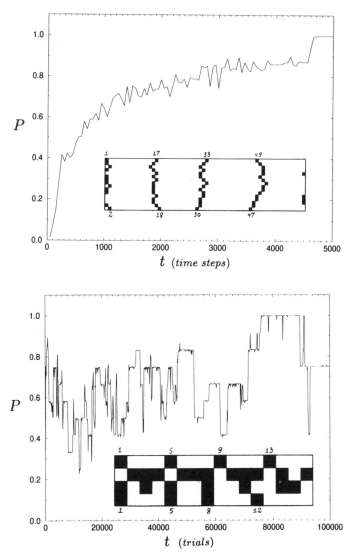

Figure 2. a) *DR*: Performance vs. time, for a 16×64 system and, for the input/output pattern shown in the inset. Light sites denote quiescent units and dark sites denote firing ones. The firing sites connect the input and output sites by effectively forming wires. The parameters of the algorithm have been set to: $r_0 = 0.1, \delta_+ = 0.01/16, \delta_- = -0.05/16$, and $h_0 = 0.01$. The performance is obtained by averaging over 50 time steps. b) A_{R-P}: Performance vs. time (measured in 'trials'), for a 4×16 system, and for the input/output pattern shown in the inset. The connections between input and output sites are more complicated. The central element of the algorithm is the update rule for the synaptic weights, $J_{ij} \rightarrow J_{ij} + \eta(r[n_i - <n_i>] + \lambda(1-r)[-n_i - <n_i>])n_j$, where η is the 'learning' coefficient, λ is the 'penalty learning rate factor', $<n_i> = \tanh(\beta \sum_j J_{ij} n_j)$ is the average firing state, and r is the evaluation feedback (for details see Hertz *et al* in Ref. 2). The parameters have been set to: $\eta = 0.5, \lambda = 0.001$, and $\beta = 0.5$. The performance is obtained by averaging over 100 trials. *Insets:* Typical successful ($P = 1$) activity patterns for *DR* and A_{R-P}.

The number, c, of input and output sites in the input/output pattern is varied accordingly, $c = L_2/16$. Here the input is confined to the top row and the output to the bottom row. To minimize crossover between paths we keep the input and output sites in pairs, well separated from each other. Specifically, each of the c columns of a given network contains a single input and a single desired output site.

In what sense does DR and A_{R-P} differ from one another? We were not as interested in the absolute performances of the two algorithms, which tend to be sensitive to the tuning of the various parameters, as we were in the scaling of the performance with the size of the network. For our simulations with A_{R-P} the same layered architecture was chosen (Fig. 1) but the number of layers was set to $L_1 = 4$ (more layers would degrade the performance of A_{R-P} too much). The lateral dimension is varied, $L_2 = 4, 8, 12, \ldots$ The input/output patterns were chosen with similar considerations as in the DR case (see insets in Figs. 2a,b for a comparison). Figures 2a and 2b show examples of the performance, P, for DR and A_{R-P} as a function of time. P is defined as the temporal average of the activity at the selected output sites minus the activity at the rest of the output sites. Appropriate normalization assures that best performance ($P = 1$) corresponds to persistent firing at the selected output sites only, whereas worst performance ($P = -1$) corresponds to persistent firing everywhere except at the selected sites.

The DR is characterized by intervals of rapid improvement in performance, interrupted by sudden dips. This behavior is a signature of the dual mode of operation of the algorithm: i) the learning or exploitation mode in which the system strengthens connections and "weeds out" irrelevant paths, and ii) the exploration mode in which the system tends to spread out the activity in an attempt to explore new possibilities with subsequent decrease in the performance P. The A_{R-P} performance versus time, although also highly irregular, seems to have a very different structure. It is dominated by very fast fluctuations at the smallest time scale (not seen here due to averaging of P). At longer time scales it seems to be dominated by long periods during which the system seems trapped at a certain level of performance. Once the system escapes this barrier the transition to a new performance level appears to occur very quickly. We would like to point out that at the individual level, and with the limited information available to it, each neuron *always* opts for the change that it expects will increase the collective performance. In its decisions, however, it cannot take into account the positive or negative contributions of the other neurons. Therefore, it is only in a statistical sense that the system senses the gradient towards a better performance and can tune its synapses accordingly. The stochastic nature of the A_{R-P} can also be witnessed in Fig. 3, ($\triangle$). Here we depict the time to completion of the task, $\bar{t}_s$ (averaged over many runs obtained with different initialization) as a function of the number, c, of input/output pairs. In a first order approximation, it seems that $\bar{t}_s$ scales exponentially with c, $\bar{t}_s \sim e^{\alpha c}$, with $\alpha \simeq 1.6$.

DR (Fig. 3, $\bigcirc$) has a significantly better scaling behavior. When plotted in a log-log plot (inset of Fig. 3) $\bar{t}_s$ might follow a power law, $\bar{t}_s \sim c^{\gamma}$, which subsequently breaks down around $c = 8$. If this is true it would not be inconsistent with our suggestion that the algorithm operates near a "critical", highly susceptible regime. Although evidence of such a critical regime have previously been seen in the dynamics of our system[1], this scaling behavior offers the first quantitative evidence. Clearly this initial data seems to be amenable to more than one interpretation, therefore it is imperative that the direct consequences of our critical-state hypothesis be tested further.

The convergence toward the critical state is accomplished by ensuring that the patterns of activity for different input signals do not overlap while not being too sparse to connect inputs with desired outputs. In "sand" models of self-organized criticality,[7] overlap of events

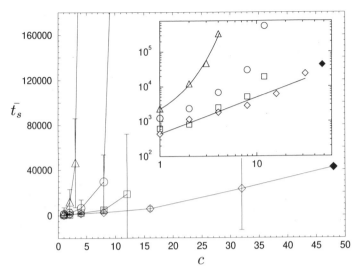

Figure 3. Average time elapsed, $\bar{t}_s$, to completion of an association task vs. number of input/output pairs, c. ($\triangle$) A_{R-P}: systems of size $L_1 \times L_2$, $L_1 = 4$ and $L_2 = 4, 8, 12$, and 16 have been considered. For each case twenty runs were performed (with the exception of the 4×16 system for which we conducted five runs only, due to computing time limitations) for the same association task but with different initialization; ($\bigcirc$) DR: systems of size $L_1 \times L_2$, $L_1 = 16$ and $L_2 = 16, 32, 64, 128$, and 192 have been considered; ($\sqcap$) DR with variable δ_+ ($a = 1.05, \delta_- = -0.05/16$): systems of size $L_1 \times L_2$, $L_1 = 16$ and $L_2 = 16, 32, 64, 128$, and 192 have been considered; ($\Diamond$) DR with variable δ_+ ($a = 1.005, \delta_- = -0.01/16$): systems of size $L_1 \times L_2$, $L_1 = 16$ and $L_2 = 16, 32, 64, 128, 256, 512$ and 768 have been considered. For the same association task, and for all three DR variants, fifty runs with differing initializations, were considered. Vertical bars denote standard deviation. *Inset:* Same data in log-log plot. The solid lines represent least-squares fits. ($\triangle$): $\sim e^{1.6c}$; ($\Diamond$): $\sim c^{1.0}$; the filled diamond denotes failure (for the 16×768 system 48% of all runs failed to complete successfully within the allotted time of two million time steps) thus, it has been excluded from the least-squares fit analysis.

("avalanches") is avoided by keeping the input rate low. Here, criticality is achieved by keeping the output low.

It turns out that one can improve efficiency and carry the system closer to the critical point by taking steps to ensure that changes in activity, due to threshold modulation, do not overlap in time, while not happening too infrequently with respect to the synaptic modification rate. We do so by allowing a variable rate δ_+ for increasing the threshold T. More precisely the rise of the threshold is governed by $\delta_+(t, r)$, $\delta_+ \to a\delta_+$, where $a > 1$. Notice that now δ_+ is time dependent in the sense that, while $r > 0$, it is constantly increased and r dependent in the sense that it is reset to a small value whenever r becomes negative, $\delta_+ \to a^{-L_1}\delta_+$. The rate of decrease of T, δ_- is kept constant as before. The modified algorithm leads to a significant improvement of the performance (Fig. 3, ($\sqcap, \Diamond$)). Furthermore, the new curve, $\bar{t}_s(c)$, gives a stronger indication for the existence of a power law with exponent $\gamma \simeq 1.0$. The modification was chosen for its simplicity rather than its performance and, based on our experience, it appears to be straightforward to obtain further improvements.

3. CONCLUSIONS

In this paper we have been concerned with the issues of self-organization which must play a central role in brain function. The central element of the mechanism we propose is a building-up process that allows the system to operate at a "critical" state, characterized by high sensitivity to small modifications of the synaptic weights and low output activity. The combination of these features allows the system to establish strong cause-effect relationships that allow the coexistence of many input/output patterns.

The work done at Florida Atlantic University was supported by the National Science Foundation, program No. 1151-058-02.

REFERENCES

[1] D. Stassinopoulos & P. Bak, Phys. Rev. E **51**, 5033 (1995).

[2] J. Hertz, A. Krogh, & R. G. Palmer, *Introduction to the Theory of Neural Computation* (Addison-Wesley, Redwood, 1991).

[3] Based on the partial information provided by the critic a target pattern is determined and the output-weight errors computed. The rest of the weights can then be updated by back-propagating this error-signal through the network (see Hertz *et al* in Ref. 2).

[4] A. G. Barto, Human Neurobiology **4**, 229 (1985).

[5] P. Alstrøm & D. Stassinopoulos, Phys. Rev. E **51** 5027 (1995).

[6] D. Stassinopoulos & P. Bak, in *Proceedings of the Fourth Appalachian Conference on Behavioral Neurodynamics - Radford*, edited by K. Pribram, (Lawrence Erlbaum, New Jersey, 1996).

[7] P. Bak, C. Tang, & K. Wiesenfeld, Phys. Rev. Lett. **59** 381 (1987).

133

FUNCTIONAL MAGNETIC RESONANCE IMAGING AND COMPUTATIONAL MODELING

An Integrated Study of Hippocampal Function

Chantal E. Stern[1] and Michael E. Hasselmo[2]

[1]MGH-NMR Center
Harvard Medical School
Bldg 149, 13th St., Charlestown, Massachusetts 02129
chantal@nmr.mgh.harvard.edu
[2]Department of Psychology
Harvard University
33 Kirkland St., Cambridge, Massachusetts 02138
hasselmo@katla.harvard.edu

ABSTRACT

A hippocampal model is utilized to simulate activation in humans detected using functional magnetic resonance imaging (fMRI). During a picture encoding task, fMRI measurements demonstrate increased signal intensity changes in the hippocampal formation and parahippocampal gyrus during sequential presentation of novel pictures, as compared to activation levels during repeated presentation of a single picture. In the hippocampal model, sequential presentation of novel patterns activates separate populations of neurons in the dentate gyrus, region CA3 and region CA1. In contrast, repeated presentation of a single pattern activates the same subpopulation of neurons. In a model of region CA3, repeated activation of the same subpopulation results in decreased activity due to activation of a calcium-dependent potassium current. These results suggest that the changes in fMRI signal intensity during the presentation of novel vs. repeating pictures may be related to neuronal adaptation mediated by calcium-dependent potassium currents in hippocampal neurons.

1. INTRODUCTION

Functional magnetic resonance imaging (fMRI) techniques can be used to localize patterns of activity in the human brain during the performance of specific cognitive tasks.

A complete understanding of the relationship between these patterns of activity and the underlying neuronal processes will require linking our knowledge of the computational processes occuring at the cellular level to these systems level changes. Computational modeling of cortical function provides a technique by which the cellular properties can be linked to network activation—allowing development of theories relating cellular physiology to cognitive and systems level processes in the human.

Advances in magnetic resonance imaging techniques have resulted in fMRI methods which can track changes in blood oxygenation and blood flow non-invasively in the human brain.[9,12] Recently, fMRI techniques have demonstrated robust activation differences in the human hippocampal formation and parahippocampal gyrus during performance of a picture encoding task.[16,17] In these studies, activity is stronger during viewing of different complex novel images than during the viewing of repeated presentations of a single image. Understanding the physiological basis for this difference in activation levels requires understanding the network dynamics of hippocampal function. Here we utilize a network simulation of the hippocampal formation to link the fMRI data to specific physiological phenomena at a cellular level.

Theories of the function of specific subregions of the hippocampal formation have been discussed extensively in the literature.[2,10–13] Previously published simulations of region CA1[5] and region CA3[6] have been combined in a network simulation of the hippocampal formation.[7,8] This simulation can store and recall highly overlapping patterns of activity presented sequentially to an input region representing entorhinal cortex layers II and III. The function of the model corresponds to the picture encoding task which activates the hippocampal formation in the fMRI study described above. Here we analyze how sequential presentation of different novel patterns can elicit greater activity levels than the repeated presentation of a single pattern.

2. METHODS

2.1. Simplified Representation of Hippocampal Neurons

The structure of the simulation was based on physiological and anatomical data from the hippocampus. This simulation allows the level of activity within the full network to be related to the physiological characteristics of individual neurons. In place of sigmoid input-output functions, the network utilized threshold linear neurons with adaptation,

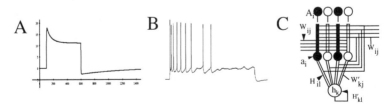

Figure 1. A. Response of single simulated neuron to current injection, showing adaptation and afterhyperpolarization due to simplified representation of calcium-dependent potassium current. Output of this model is a continuous variable proportional to how much membrane potential exceeds threshold. B. Intracellular recording from a cortical pyramidal cell, demonstrating the phenomenon of adaptation. C. Local connectivity of individual subregions, showing population of excitatory neurons receiving feedback from single inhibitory interneuron.

which more closely resemble the input-output function of real cortical neurons.[1] Total network activity was kept bounded by feedback from inhibitory interneurons and by adaptation due to intracellular calcium concentration. Separate variables represent pyramidal cell membrane potential a, intracellular calcium concentration c, and the membrane potential of inhibitory interneurons h. As shown in Figure 1, these equations can account for the spike frequency adaptation and afterhyperpolarization which can be observed during intracellular recording from cortical pyramidal cells.[1,4,6]

2.2. Network Connectivity

As shown in Figure 1C, local subregions of the simulation contained populations of excitatory neurons receiving feedback inhibition from a single inhibitory interneuron. This provided bounded excitatory activity regulated by the strength of connections to and from the inhibitory interneuron. This connectivity contrasts with many network models in which individual units make both excitatory and inhibitory connections with other units. The components of this local circuitry can be related directly to the activity of pyramidal cells and inhibitory interneurons in cortical structures. As shown in Figure 2, these local circuits were connected in a network based on the connectivity of the hippocampal formation. A learning rule of the Hebbian type was utilized at all synaptic connections, with the exception of the mossy fibers from the dentate gyrus to region CA3, and the connections to and from the medial septum. Self-organization of perforant path synapses was obtained through decay of synapses with only pre or post-synaptic activity, and growth of synapses with combined activity. Associative memory function at synapses arising from region CA3 and between CA1 and entorhinal cortex layer IV was obtained through synaptic modification during cholinergic suppression of synaptic transmission. The total output from region CA1 activated an inhibitory interneuron in the medial septum, which suppressed the activity of a cholinergic neuron providing modulation to the full network (which was active in the absence of inhibition). When levels of cholinergic modulation were high, there was strong suppression of synaptic transmission at the excitatory recurrent synapses in CA3[6] and the Schaffer collaterals projecting from region CA3 to CA1.[5] Cholinergic modulation also increased the rate of synaptic modification and depolarized neurons.

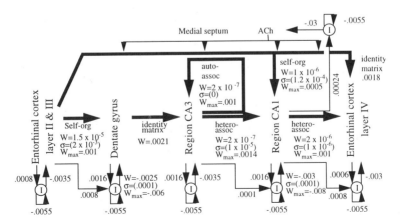

Figure 2. Hippocampal circuitry represented in the model, including strength of excitatory connections between regions, and interaction of excitation and inhibition in local circuits.

3. RESULTS

3.1. Network Simulation of Hippocampal Memory Function

The network simulation of the hippocampal formation can store highly overlapping patterns of activity presented sequentially to entorhinal cortex layers II and III, and can recall these patterns given incomplete cues for recall. In response to each pattern, the network rapidly undergoes self-organization of the afferent input to the dentate gyrus and region CA1. This self-organization results in sparse, nonoverlapping representations of each novel input pattern, which are passed from the dentate gyrus to region CA3, where they undergo autoassociative storage. CA3 activity is associated with CA1 activity and CA1 activity is associated with activity in entorhinal cortex layer IV. Subsequent presentation of incomplete cues activates the sparse representations, allowing effective recall, as shown in Figure 3.

3.2. Simulation of the fMRI Picture Encoding Task

For analysis of the cellular basis for changes in fMRI activation, the simulation was presented with stimuli in a pattern designed to simulate the presentation of stimuli during the picture encoding task utilized in the fMRI study. The picture encoding task alternated between two different conditions. In one condition, novel pictures were presented sequentially for 3 seconds each for a period of 1 minute. In a separate condition, a single picture was presented repeatedly for 1 minute. This allowed comparison of the activation induced by novel stimuli with activation induced by a single stimulus which rapidly became famil-

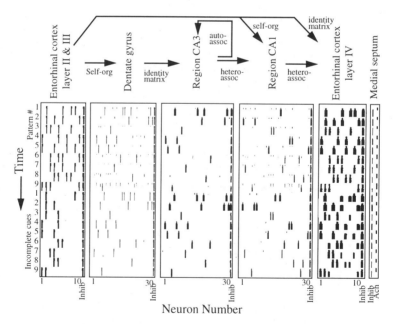

Figure 3. Sequential storage and recall of nine patterns of activity in the full hippocampal simulation. Width of black lines represents the activity level of individual neurons within each region during each step (Time is plotted vertically). Input to the network consists of nine different patterns presented sequentially to entorhinal cortex layer II & III, followed by degraded pattern cues for these nine patterns. Output consists of activity in entorhinal cortex layer IV (right side), which shows full recall of each of the nine stored patterns in response to the incomplete cues.

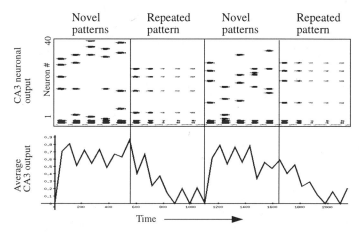

Figure 4. Activity differences in the hippocampal simulation correspond to activity differences in the fMRI experiment. Left: Effect of different stimulus presentation patterns on the time course of summed neuronal activation in the hippocampal simulation (normalized total activity of all simulated pyramidal cells). Presentation of different novel images results in greater activation than repeated presentation of the same image. Right: the time course of fMRI signal intensity changes is shown for a single region of interest. The four time blocks represent the alternating one minute novel (N) and repeating (R) conditions. Horizontal lines represent the average normalized signal intensity for each one minute block. Images were collected every 2.5 seconds.

iar. The changes in fMRI signal within the hippocampal formation in a single subject is shown on the right hand side of Figure 4. In the model, this presentation pattern was replicated by alternating between sequential presentation of novel patterns and repeated presentation of a single pattern. The summed neuronal activation in the model is shown on the left side of Figure 4. As can be seen from the figure, summed activation in the model shows the same differences in activation seen in the fMRI data.

The activation differences observed in the simulation result from repeated activation of the same population of neurons. This is illustrated in a simplified simulation focusing on just region CA3 of the hippocampus, as shown in Figure 5. During sequential presenta-

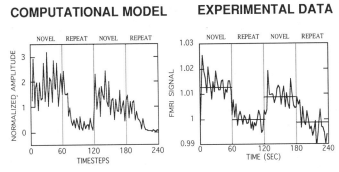

Figure 5. Top. Activity pattern in a representation of region CA3 during alternating phases of sequential presentation of novel patterns and repeated presentation of a single pattern. Width of black lines represents level of activity in individual simulated pyramidal cells. Note that with repeated activation of the same population of neurons, activity in those neurons becomes weaker due to adaptation. (Lines become thinner). Bottom. Average output of neurons in the network computed over 55 step intervals within the simulation. Note the decrease in activity during repeated presentation.

tion of novel patterns to the hippocampal network, separate subsets of hippocampal neurons are sequentially activated. In the CA3 model, this results in strong activity. During repeated presentation of a single pattern, a single group of hippocampal neurons is repeatedly activated. This results in a gradual decrease in response due to adaptation. Each response causes an influx of calcium into the neurons, increasing the total activation of the calcium-dependent potassium current. This potassium current acts to reduce membrane depolarization and decrease neuronal output.

4. DISCUSSION

The activity levels in this model of the hippocampus correspond to the activity levels observed with fMRI during different conditions of a picture encoding task. Interpretation of the fMRI results in the context of the model links the pattern of signal intensity changes in the fMRI data to specific cellular properties of hippocampal neurons. In particular, it appears that the adaptation mediated by calcium-dependent potassium currents in cortical pyramidal cells may underlie these differences in activation. Because the repeated presentation of a single picture activates the same population of hippocampal neurons repeatedly, these neurons show progressively greater adaptation (less activity). In contrast, sequential presentation of novel patterns activates different subpopulations of hippocampal neurons. The initial response of each population of these neurons is stronger, resulting in greater overall activity. This phenomena could also result in differences in overall activity in higher order visual association cortices, which could contribute to differences in hippocampal activity. Indeed, fMRI data shows differences in total activation of the lingual and fusiform gyrii in the paradigm described here [16,17], and recordings from monkey inferotemporal cortex show differences in single unit activity during viewing of familiar versus novel stimuli. A separate chapter in this volume presents a model of these differences in inferotemporal cortex activity [15].

The use of a model to link systems level activation to cellular properties can be used to generate predictions to guide future fMRI studies. In particular, the computational model described here suggests that the decrease in activation should follow the time constant of the calcium-dependent potassium current. Thus, if novel pictures are shown for progressively longer periods, the activation levels observed should decrease with an exponential time course corresponding to the exponential time course of the calcium-dependent potassium current.

ACKNOWLEDGMENTS

This work was supported by an NIMH FIRST award MH52732-01 and by the McDonnell-Pew Program in Cognitive Neuroscience.

REFERENCES

1. Barkai E, Hasselmo ME (1994) Modulation of the input/output function of rat piriform cortex pyramidal cells. J. Neurophysiol. 72: 644–658.
2. Eichenbaum, H. and Buckingham, J. (1990) Studies on hippocampal processing: experiment, theory and model. In: Learning and computational neuroscience: foundations of adaptive networks, M. Gabriel and J. Moore, eds., Cambridge, MA: MIT Press.

3. Hasselmo ME (1995) Neuromodulation and cortical function. Behav. Brain Res.67:1–27.

4. Hasselmo ME, Barkai E, Horwitz G, Bergman RE (1993) Modulation of neuronal adaptation and cortical associative memory function. In: Computation and Neural Systems II (Eeckman F, Bower JM, ed). Norwell, MA: Kluwer Academic Publishers.

5. Hasselmo ME, Schnell E (1994) Laminar selectivity of the cholinergic suppression of synaptic transmission in rat hippocampal region CA1: Computational modeling and brain slice physiology. J. Neurosci. 14: 3898–3914.

6. Hasselmo ME, Schnell E, Barkai E (1995) Learning and recall at excitatory recurrent synapses and cholinergic modulation in hippocampal region CA3. J. Neurosci. 15: 5249–5262.

7. Hasselmo ME and Stern CE (1996) Linking LTP to network function: A simulation of episodic memory in the hippocampal formation. In: Baudry M and Davis J (eds) Long-Term Potentiation, Vol. 3. MIT Press: Cambridge, MA.

8. Hasselmo ME, Wyble BP and Stern CE (1996) A model of human memory based on the cellular physiology of the hippocampal formation. In: Parks R and Levine D (eds.) Neural Networks for Neuropsychologists. MIT Press: Cambridge, MA.

9. Kwong KK, Belliveau JW, Chesler DA, et al. (1992) Dynamic magnetic resonance imaging of human brain activity during primary sensory stimulation. Proc Natl Acad Sci 89:5675–5679

10. Levy WB (1989) A computational approach to hippocampal function. In: Computational models of learning in simple neural systems (Hawkins RD, Bower GH, ed), pp. 243–305. Orlando, FL: Academic Press.

11. Marr D (1971) Simple memory: A theory for archicortex. Phil. Trans. Roy. Soc. B B262:23–81

12. McNaughton BL (1991) Associative pattern completion in hippocampal circuits: New evidence and new questions. Brain Res. Rev. 16:193–220.

13. McNaughton BL, Morris RGM (1987) Hippocampal synaptic enhancement and information storage within a distributed memory system. Trends Neurosci. 10:408–415.

14. Ogawa S, Tank DW, Menon R., et al. (1992) Intrinsic signal changes accompanying sensory stimulation: Functional brain mapping with magnetic resonance imaging. Proc Natl Acad Sci USA 89:5951–5955

15. Sohal V and Hasselmo ME (1997) A model of changes in inferotemporal activity during a delayed match to sample task. In: Bower JM (ed) Computational Neuroscience, Plenum Press: New York, in press.

16. Stern CE, Corkin S, Guimaraes AR, Sugiura R, Carr CA, Baker JB, Jennings PJ, Gonzalez RG and Rosen BR (1994) A functional MRI study of long-term explicit memory in humans. Soc. Neurosci. Abstr. 20: 530.8.

17. Stern CE, Corkin S, Gonzalez R, Guimaraes AR, Baker JR, Jennings PJ, Carr CA, Sugiura RM, Vedantham V and Rosen BR (1996) The hippocampal formation participates in novel picture encoding: Evidence from functional magnetic resonance imaging. Proc. Natl. Acad. Sci. USA 93: 8660–8665.

134

ANALYZING THE VIEW DEPENDENCE OF POPULATION CODES IN INFERIOR TEMPORAL CORTEX

Joshua B. Tenenbaum[*][1] and Emanuela Bricolo[2]

[1] Department of Brain and Cognitive Sciences, E10-124
Massachusetts Institute of Technology
Cambridge, MA
E-mail: jbt@psyche.mit.edu
[2] International School of Advanced Studies
Via Beirut 4, Trieste, Italy
E-mail: bricolo@sissa.it

ABSTRACT

The recent discovery of cells in inferior temporal (IT) cortex that show responses selective for specific views of learned wire-like objects [4] supports a computational model of object recognition built on two-dimensional view-based representations [7] over models based on three-dimensional structural descriptions [1], in line with many psychophysical demonstrations of view-dependent recognition performance [2]. However, many more IT cells show significantly view-sensitive responses without strict view-tuning, which suggests that an analysis of population encodings may be necessary to fully understand the nature of object representation in IT. By analyzing how the similarity of IT population responses to different views of two objects rotated in depth depends on the orientation difference between the views, we provide further evidence that neural representations of object shape depend on abstract two-dimensional views (possibly built from collections of image features), and demonstrate a clear correspondence between the view dependence of IT population responses and the view dependence of human observers' similarity judgments.

INTRODUCTION

The primate visual system can effortlessly recognize an innumerable variety of objects. Recognition performance is robust to common imaging transformations that preserve object

---

[*]To whom correspondence should be addressed.

Computational Neuroscience
edited by Bower, Plenum Press, New York, 1997

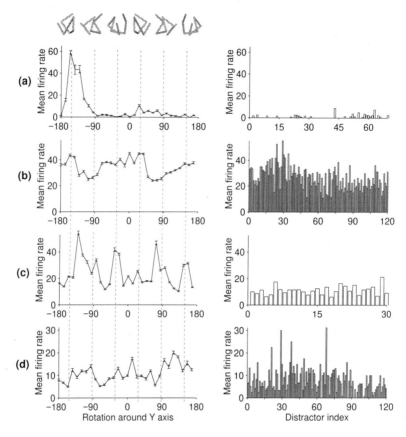

Figure 1. The response (mean firing rate) of four different IT cells to 30 views of a wire-like object rotated in depth (first column), and to various views of other distractor objects (second column).

identity: a particular chair can still be identified easily when viewed from different viewpoints, at different distances, in different positions and under different illumination conditions. In the last two decades, lesion experiments and electrophysiological recordings have established the crucial role played by the inferior temporal (IT) cortex in object recognition [3]. In particular, the recent discovery by Logothetis and colleagues [4] that a significant number of IT cells show responses selective for specific views of wire-like objects learned while performing a recognition task (see Figure 1a) has provoked strong interest in computation and psychophysics circles [2]. This finding offers some support for computational models of recognition built on two-dimensional view-based representations [7] over models based on three-dimensional structural descriptions [1], and thus complements the substantial body of psychophysical evidence indicating that human object recognition is significantly view-dependent [2].

However, two considerations should keep us from drawing any definitive conclusions about the nature of object representations in IT from the mere presence of view-selective cells. First, only 7.7% of the IT cells analyzed in [4] responded significantly better around one or two views of a target object, relative to other views of the target and the other tested distractor objects. Second, much more common than view-tuned cells (Figure 1a) are cells like those shown in Figure 1b-d, which like the view-tuned cells show significantly larger responses to some views of the target object than to other views, but do not have one or two sharply preferred object orientations, and do not necessarily respond significantly better to the target object than to all distractors. Evidently, much of the IT cell population finds the variation in views of the target object produced by rotation in depth meaningful, but only the minority of these view-sensitive cells that can fairly be called view-tuned seem interpretable under the standard paradigm of view-based representations. In this paper, we analyze the object representations encoded in IT population responses, including mostly cells that are more or less view-sensitive but not strictly view-tuned, using the recordings originally reported and analyzed at the single-cell level in [4]. We show that the population representations are most consistent with theories of object recognition based on two-dimensional abstract views, paralleling the findings of recent computational and psychophysical work [7, 2]. We also demonstrate a clear correspondence between the view dependence of IT population responses and the view dependence of human observers' similarity judgments.

ANALYTICAL METHODS AND THEORETICAL PREDICTIONS

The activity of IT cells was recorded while the monkey performed a visual discrimination task (see [4] for details). For two of the wire objects used as target stimuli in this task, more than 60 cells were recorded with significant responses. We chose these two objects and the corresponding cell populations ($N_1 = 63$ cells and $N_2 = 70$ cells) for study because they provide the most reasonable approximations to the complete IT population code.

We have focused on analytic techniques for population codes with the potential to discriminate directly between competing theories of object representation for recognition. In this paper, we compare the similarity of the neural population responses to the 30 views of each object generated as it rotates in depth about a central vertical axis with the predicted similarities of the representations of these 30 views, as given by four candidate theories of object representation [2, 5]: (i) three-dimensional orientation-invariant (which ignores the object's 3-D pose); (ii) three-dimensional orientation-dependent (which explicitly encodes the object's 3-D pose); (iii) two-dimensional feature-based (composed of combinations of image features); (iv) two-dimensional template-based (composed of essentially image-like arrays).

Roughly, these four representational schemes are ordered by their degree of geometric abstraction and consequent invariance under the nonlinear imaging transformation of rotation in depth, and thus make different predictions for how the similarity of two views depends on the orientation difference between them. The 3-D orientation-invariant scheme predicts no effect of orientation difference on view similarity, because information about the object's 3-D pose has already been abstracted away from this representation. The 3-D orientation-dependent scheme predicts that view similarity is simply a monotonically decreasing function of orientation difference. Both 2-D schemes also predict that view similarity generally decreases with orientation difference. However, the 2-D template-based scheme predicts that orientation difference alone can account for only a small percentage of the variance in view similarity, because rotation in depth produces very nonlinear image deformations. The 2-D feature-based scheme predicts that orientation difference can account for much of the variance in view similarity, because key image features, such as vertex angle or segment length in the case of wire-like objects, change smoothly as the object rotates in depth, but under this scheme, view similarity may not always decrease monotonically with increasing orientation difference if the key features are visibly indistinguishable from more than one viewpoint (e.g. left and right, front and back).

Following previous similarity analyses of neural population codes [10, 9], we treat the population response to each view as a vector in a high dimensional "cell space", with each component dimension corresponding to the mean firing rate of one cell in the population recorded while the monkey is presented with that view. Firing rates are normalized for each cell by the overall mean firing rate of that cell to all 30 views of the same object. We then measure the population similarity of two views by taking the negative Euclidean distance in cell space, with the contribution of each cell's response weighted by its standard error, and scale these values uniformly to fill the interval $[0, 1]$. Below we present the results of two analytical procedures applied to the computed view similarities. First, in order to assess the level of representational abstraction in the population code for each object, we used multidimensional scaling (MDS) to represent the view similarities as distances in a Euclidean space of two dimensions (the minimum dimensionality needed to accommodate the "circle" of views produced by rotation in depth) and calculated the percentage of variance in the complete view similarities accounted for by these abstract, reduced-dimensionality view spaces [6, 10, 9]. Second, in order to uncover the specific relation between view similarity and the orientation difference between views, we averaged the similarity of all pairs of views separated by each orientation difference, to obtain a single similarity gradient for each object (Figure 2).

RESULTS AND DISCUSSION

For the two objects, two-dimensional configurations of points produced by MDS could account for (respectively) 72.7% and 68.4% of the variance in the view similarities originally computed from the IT populations in (respectively) 63-dimensional and 70-dimensional cell spaces. Thus, the variation in population responses due to rotation in depth is not strictly reducible to two Euclidean dimensions, as would be expected with a 3-D orientation-dependent representational scheme. On the other hand, the variances accounted for by these two-dimensional configurations are much greater than the values of 39.7% and 32.8% predicted by a 2-D template-based representational scheme in which view similarity is computed as normalized image pixel correlation. These MDS results thus suggest that IT populations may encode 2-D feature-based representations of objects, at an intermediate level of abstraction between 2-D image template-based representations and 3-D orientation-dependent representations. As shown below, a more detailed analysis of the relation between view similarity and the orientation difference between views confirms this hypothesis.

The lefthand panels of Figure 2a,b show the empirically obtained relation between view similarity and orientation difference, as expressed by IT population codes for the two target objects. Both curves show significant violations of monotonicity: view similarity decreases with increasing orientation difference only for differences up to 90°, and then similarity increases again as orientation difference increases to a second peak at 180°. Note that the shape of these curves is not significantly altered by excluding from the analysis those few cells that are distinctly tuned for two views 180° apart.

For the sake of comparison, the central panels of Figure 2 show the relation between view similarity and orientation difference as predicted by a 2-D template-based representational scheme. Each point on these curves represents the normalized image pixel correlation of all pairs of views separated by the indicated orientation difference. As expected, there is no hint of a significant departure from monotonicity. Clearly, a 3-D orientation-dependent representation would give the same monotonically decreasing relation, and a 3-D orientation-invariant representation would generate completely flat curves! Thus, of the four representational schemes

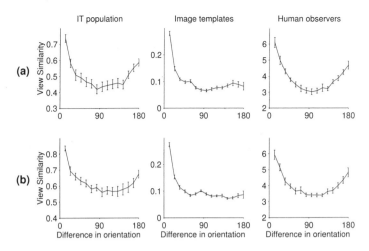

Figure 2. The relation between view similarity and orientation difference for two objects, as calculated from IT population codes, an image template correlation model, and the judgments of human observers.

discussed in the previous section, the empirical U-shaped curves are most consistent with abstract 2-D view-based representations built from collections of image features, such as vertex angles or segment length, that appear indistinguishable in two views separated by 180°. Single view-tuned cells may be selective for combinations of these features and other image features, such as segment direction or ordinal segment occlusion relations, that are not invariant under rotation by 180° and hence may account for why the peak at 180° is still much lower than the maximum.

Of course we don't always expect to obtain such dramatic U-shaped curves for natural objects, with features that are often occluded by large rotations in depth. However, natural objects are often bilaterally symmetric, and both computational considerations [7] and psychophysical findings [8] lead us to expect a nonmonotonic relation between view similarity and orientation difference to hold for these objects as well.

The righthand panels of Figure 2 show the relation between view similarity and orientation difference according to the judgments of human observers. We asked 10 subjects to rate the similarity of each pair of views of the same two wire objects on a 1-7 scale, and averaged their judgments to obtain these curves. These curves clearly exhibit the same U-shape obtained from the IT population responses, which, given that object recognition depends so heavily on shape similarity, provides further support for the hypothesis that view-dependent recognition performance in psychophysical tasks results from the particular view-based representations of objects encoded in IT cortex.

REFERENCES

[1] I. Biederman. Recognition-by-components: A theory of human image understanding. *Psychological Review*, 94:115–147, 1987.
[2] H.H. Bülthoff, S. Edelman, and M.J. Tarr. How are three-dimensional objects represented in the brain? *Cerebral Cortex*, 3:247–260, 1995.
[3] C.G. Gross. How inferior temporal cortex became a visual area. *Cerebral Cortex*, 4(5):455–469, 1994.
[4] N.K. Logothetis, J. Pauls, and T. Poggio. Shape representation in the inferior temporal cortex of monkeys. *Current Biology*, 5(5):552–563, 1995.

[5] S. Pinker. Visual cognition: An introduction. *Cognition*, 18:1–63, 1984.

[6] R.N. Shepard and J.E. Farrell. Representation of the orientations of shapes. *Acta Psychologica*, 59:104–121, 1985.

[7] T. Vetter and T. Hurlbert, A.Poggio. View-based models of 3d object recognition: Invariance to imaging transformations. *Cerebral Cortex*, 3(261–269), 1995.

[8] T. Vetter, T. Poggio, and H.H. Bülthoff. The importance of symmetry and virtual views in three-dimensional object recognition. *Current Biology*, 4(1):18–23, 1994.

[9] Y. Weiss and S. Edelman. Representation of similarity as a goal of early visual processing. *Network: computation in neural systems*, 6(1):19–41, 1995.

[10] M.P. Young and S. Yamane. Sparse population coding of faces in the inferotemporal cortex. *Science*, 256:1327–1331, 1992.

135

PHASE SETTING AND LOCOMOTION

R. E. L. Turner

Department of Mathematics
University of Wisconsin
Madison, WI

1. INTRODUCTION

The nematode *Ascaris suum* has been the object of intensive study by the group working under the direction of Antony Stretton at Madison and there is an extensive picture of the motorneuronal types of synapses in the worm, as well as the neural architecture (see Stretton, 1978, 1985 for further references). However, the way that the elements in the circuit combine to produce motion is not understood. The present report examines the viability of a mechanism for control of locomotion. *Ascaris* has approximately 300 neurons and, of these, 80 or so, including interneurons, control locomotion. The neural architecture can be thought of as consisting of five repetitions of a basic pattern of eleven motor neurons, some excitatory and some inhibitory. There is a dorsal nerve cord and a ventral nerve cord, which synapse onto dorsal and ventral muscle cells, respectively. Here we focus on the propagation of a wave in a single chain of muscles which could represent either the dorsal or ventral chain, and do not address the question of coordination of the two cords.

The muscle cells are geometrically complicated and appear to be capable of spontaneous oscillation (Weisblatt et al, 1976). The capacity for oscillation is one of the features of the muscle that we exploit in our model.

As a simplification of the highly convoluted anatomy, we envision the muscular component in the dorsal or ventral half of the worm to consist of a linear array of cells. We number the muscle cells in the chain $k = 1, 2, 3, ... N$. The anatomy of the excitatory and inhibitory neurons which synapse onto muscle is such that, the amplitude at site k of excitation, $ex(k)$, and of inhibition, $in(k)$, vary with k. This is due to the branching of neurons (Stretton et al, 1978) and to the variation in the density of synapses along a nerve cord. A further feature of *Ascaris* is that an excitatory motorneuron receives its input from an interneuron and the excitatory neuron, in turn, synapses onto muscle and onto an inhibitory motorneuron. As a consequence, the excitatory input to muscle precedes the inhibitory input.

Experiments (Walrond and Stretton, 1985) show that stimulation of single identified motorneurons produces a delayed spiking response in postsynaptic muscle bellies. The signal arrives at successive bellies(imagined as linearly arrayed along the worm) with a speed of 12 to 28 cm. per second. The speed of these electrotonic waves is much higher than the speed of behavioral waves. Measurements done on intact worms (Mead, 1991) showed behavioral waves propagating at 1 cm. to 3 cm. per second. In semi-intact preparations it was observed that recording in muscle cells and identified neurons showed periodic bouts of activity with superimposed, higher frequency activity. It was the period of the bouts, however, and not the higher frequency component that corresponded to the behavioral period. In intact worms, forward propagating behavioral waves have a temporal period of about 7 seconds. It is this slower oscillation that will be the focus of our modeling.

We will assume that at a time T of initiation of locomotion, some sensory system in the worm sends signals 'very rapidly' to elicit excitatory and inhibitory transmitter release. Release will, in fact, be assumed to start at time T as well, so that in the model the 'very rapid' signals (analogous to those which travel 12 to 28 cm. per second) will travel 'infinitely' fast and effect the release of neurotransmitter. The amplitudes, $ex(k)$, $in(k)$ chosen in the model produce a wave of depolarization which propagates along the chain of muscle cells.

2. A MODEL MUSCLE CELL

To study the viability of a wave generation model we have merely taken a muscle cell model which has the principal ionic currents of calcium and potassium, present in *Ascaris* muscle (see Martin, 1992, for references) While there is a small IA current, we have used the predominant, rectifying potassium current of Hodgkin-Huxley type for repolarizing current together with an activating and deactivating calcium current. A leak current has also been included. The model of Chay (1990) provided a starting point. The parameters have been changed, so as to have a cell resting potential near -30 millivolts, a well-established Figure for *Ascaris* (de Bell et al, 1963) and to produce temporal periods like those observed in *Ascaris*. These currents give a simple model for a muscle cell which is excitable and which has several of the properties observed in *Ascaris* muscle cells. However, our aim here is to describe a mechanism which is not highly dependent on the particulars of the excitable cell.

We treat each model cell as an isopotential entity. Its state is described by the variables (V, f, n), where V is the cytoplasmic voltage in millivolts (relative to an extracellular voltage of zero), f is an inactivation variable for the calcium current, and n is an activation variable for the potassium current. The evolution is governed by the three equations:

$$\frac{dV}{dt} = \bar{g}_{Ca}d(V)f \cdot (V - V_{Ca}) + \bar{g}_K n \cdot (V - V_K) + g_L \cdot (V - V_L) \tag{1}$$

$$\frac{df}{dt} = \frac{f_\infty(V) - f}{\tau_f(V)} \tag{2}$$

$$\frac{dn}{dt} = \frac{n_\infty(V) - n}{\tau_n(V)} \tag{3}$$

The first equation gives the current balance with $\bar{g}_{Ca}$, the maximum conductance for calcium; $d(V)$, a unitless activation parameter; and and V_{Ca}, the calcium reversal potential. The other currents are similarly described and the the the cell membrane capacitance is taken to be one microfarad cm^{-2}. The kinetics for f and n are given in the remaining equations. Using the Boltzmann type kinetics from (Chay 1990, p. 307) and taking calcium for the 'slow' current we use parameters: $\bar{g}_{Ca} = 500\ \mu Scm^{-2}$, $\bar{g}_K = 250\ \mu Scm^{-2}$, $g_L = 20\ \mu Scm^{-2}$, $s_f = -10\ mV$, $s_n = 6.5\ mV$, $s_d = 5\ mV$, $\bar{\tau}_f = 22$ sec., $\bar{\tau}_n = 0.005$ sec., $v_f = -45\ mV$ for time < 10 sec.,

$v_f = -39\,mV$ for time > 10 sec., $v_n = 3.5\,mV$, $v_d = -10\,mV$, $V_{Ca} = 34\,mV$, $V_K = -53\,mV$, and $V_L = -33\,mV$. The initial data for each cell is: $V = -30.0$ mV, $f = 0.15$, and $n = 0.0$.

We assume that each cell starts with $v_f = -45mV$, producing an an equilibrium mode and at time $T = 10$ sec., by some physiological mechanism, switches to $v_f = -39mV$ putting it in an oscillatory mode. This mode is without bursts and similar to that in Chay (1990) for a 3 variable model with $\overline{\tau}_n = 0$. It has a period of 6424 milliseconds, a minimum of about -31 mv, and a well defined maximum of about -6 mv. As with Chay's model, for diverse values of the parameters there are other modes of oscillation, some with bursts. Here we are interested not in resolving the spike behavior, but in the phase of the orbits having a period of several seconds, and so have chosen parameters accordingly.

To simulate the synaptic input to the muscle we add a term

$$ex(k) * \text{p1}(t) - in(k) * \text{p2}(t) \tag{4}$$

to the current equation (1) for cell number k. The 'pulse' functions $\text{p1}(t)$ and $\text{p2}(t)$ are step functions which are equal to unity for $10.0 < t < 10.5$ and for $10.5 < t < 11.0$, respectively, and are zero otherwise. We choose $T = 10$ seconds so that prior to that time the cell has reached an equilibrium. The amplitudes $ex(k)$ and $in(k)$ are at our disposal.

Without excitatory and inhibitory input and with no linkage between cells, all cells have identical orbits and asymptotic phases. By delivering varying excitatory and inhibitory pulses to different cells along a chain, an occurrence consistent with the architecture of synapses, one can have different locations in different asymptotic phases in a monotone, steplike fashion. In choosing the inputs we group them in three's to be able later to show the effects of the gap junction linkage and to achieve a representative distribution of asymptotic phases over the period of 6.424 seconds.

Consider a chain of N=18 cells. Suppose the pair $(ex(k), in(k))$ is equal to $(0.0, 1.0)$ for $k = 1, 2, 3$; equal to $(0.0, 0.0)$ for $k = 4, 5, 6$; and equal to $(0.0, 0.016)$, $(0.1, 0.0)$, $(0.1, 0.11)$, and $(0.1, 0.2)$ for subsequent triples. The resulting phases are shown in Figure 1 by the open diamonds, corresponding to the time of a voltage peak. Of course, each cell repeats its cycle indefinitely.

3.   LINKING CELLS

The muscle cells in *Ascaris* are linked by gap junctions near the nerve cord. If the cells were not linked, and had input dependent on position, the asymptotic phases would vary, as exhibited. To simulate gap junctions we link the voltage, $V(k)$, at site k to the voltages at adjacent sites by appending another term:

$$S * (V(k+1) - V(k)) + S * (V(k-1) - V(k)) \tag{5}$$

to equation (1) for cell k, S being the strength of the connection (a cell at an end of the chain has only one link).

With $S > 0$ the junctions have a smoothing effect on the voltages and, if persistent, drive the 18 voltages to a synchronized oscillation, so that eventually all cells would peak at the same times. It is known, however, that gap junctions close when calcium concentration increases. and that the time scale for closure can be on the order of seconds (Rose and Loewenstein, 1976). If the gap junctions close, the phase relations become 'frozen' at the time of closure and the time of peaking of cell k will vary with k. While one can link S to the calcium influx

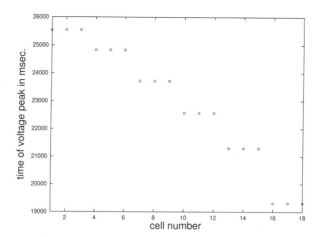

Figure 1. Time of voltage peak without linkage(diamonds); with linkage, s=0.02 (line)

and get similar results, here we merely assume S is a function of the form $s \cdot c(t)$ where the function $c(t)$ is 1 for time t less than 11 seconds, decays linearly to zero between 11 and 12 seconds, and then remains zero.

The effect of having linkage with $s = 0.02$ and subsequent closure, is shown by the connected graph in Figure 1. The asymptotic phases are qualitatively similar for a range of values of s and for small random perturbations of the terms in (4).

If one views the time of maximum depolarization to coincide with the time of maximum contraction of a muscle, then the data in Figure 1. corresponds to a fictive, progressing wave of contraction which occurs first at site 18 and progresses to site 1, taking about 6 seconds to make the passage. The speed in an animal would depend, of course, on the spacing of the cells and that aspect is not addressed here beyond the comment that having the 18 cells spread over 6 centimeters would give a contraction wave traveling at about 1 cm/sec. The number of cells is not important, but rather the way the changes in excitatory and inhibitory inputs are spread over the length of the chain. One sees that this crude model produces a depolarization wave which progresses along a chain of muscle cells. Obviously, refinements of the model to include more of the physiology are needed, but we conclude that a system of excitable cells linked by gap junctions with variable conductance and modulated by excitatory and inhibitory inputs can generate progressing waves of depolarization.

ACKNOWLEDGMENTS

Many thanks are due to the 'Stretton Lab' for their help in introducing me to *Ascaris* and providing a large part of my education in regard to that nematode.

REFERENCES

[1] Chay, T. R. (1990), Bursting excitable cell models by a slow Ca^{2+} current, J. Theor. Biol. 142:305-315.
[2] de Bell J. T., J. del Castillo, and V. Sanchez. 1963. Electrophysiology of the muscle cells of *Ascaris lumbricoides*. J. Cell Comp. Physiol. 62:159-178.

[3] Martin, R. J., P. Thorn, K. A. F. Gration, and I. D. Harrow. 1992. Voltage-activated currents in somatic muscle of the nematode parasitic *Ascaris suum*, J. Exp. Biol., 173:75-90.

[4] Meade, J. A., 1991. Intracellular recordings from neurons and muscle cells in a semi-intact preparation of the nematode *Ascaris suum*: Implications for *Ascaris* locomotion. Thesis, University of Wisconsin, Madison.

[5] Rose, B. and Loewenstein, W., 1976. Permeability of a cell junction and the local cytoplasmic free ionized calcium concentration: a study with aequorin, J. Membrane Biol. 28:87-119.

[6] Stretton A. O. W., R. M. Fishpool, E. Southgate, J. E. Donmoyer, J. P. Walrond, J. E. R. Moses, and I. S. Kass. 1978. Structure and physiological activity of the motorneurons of the nematode *Ascaris*. Proc. Nat. Acad. Sci. USA 75:3493-3497.

[7] Stretton A. O. W., R. E. Davis, J. D. Angstadt, J. E. Donmoyer, and C. D. Johnson. 1985. Neural control of behaviour in *Ascaris*. Trends in Neurosci. 8:294-300.

[8] Walrond J. P. and A. O. W. Stretton. 1985c. Excitatory and inhibitory activity in the dorsal musculature of the nematode *Ascaris* evoked by single dorsal excitatory motoneurons. J. Neurosci. 5:16-22.

[9] Weisblat, D. A., L. Byerly, and R. L. Russell. 1976. Ionic mechanisms of electrical activity in somatic muscle of the nematode *Ascaris lumbricoides*, J. Comp. Physiol. 111:93-113.

136

ACTIVITY-DEPENDENT SELF-ORGANIZATION OF ORIENTATION PREFERENCE PREDICTS A TRANSIENT OVERPRODUCTION OF PINWHEELS DURING VISUAL DEVELOPMENT

F. Wolf and T. Geisel

Max-Planck Institut für Strömungsforschung
Göttingen
SFB Nichtlineare Dynamik
Universität Frankfurt, Germany

ABSTRACT

The pinwheel-like arrangement of iso-orientation domains around orientation centers is a ubiquitous structural element of orientation preference maps in primary visual cortex. Here we investigate how activity-dependent mechanisms constrain the way in which orientation centers can form during visual development. We consider the dynamics of a large class of models for the activity-dependent self-organization of orientation preference maps. We prove for this class of models that the density of orientation centers which proliferate as the map arises from a homogeneous state exhibits a universal lower bound. Due to topological constraints the density of orientation centers can only change by discrete creation and annihilation events. Consequently activity-dependent self-organization of orientation preference implies that low densities of orientation centers develop through an initial overproduction and subsequent annihilation of pinwheels. Monitoring the density of orientation centers during development therefore offers a powerful novel approach to test whether orientation preference arises by activity-dependent mechanisms or is genetically predetermined.

THE LAYOUT OF ISO-ORIENTATION DOMAINS

The orientation preference of neurons in the visual cortex exhibits a columnar organization. Neurons in a cylinder of tissue extending radially across the cortex are selectively activated by stimuli of the same orientation. Determining the orientation preferences of these columns as a function of tangential position $\mathbf{x}$ in the cortical sheet requires knowledge of a

Computational Neuroscience
edited by Bower, Plenum Press, New York, 1997

set of spatial patterns of activity $E_k(\mathbf{x})$ in response to stimuli k of orientations θ_k, which can be measured by optical techniques [1]. This set can be used to define a single complex order parameter field $z(\mathbf{x}) = \sum_k e^{i2\theta_k} E_k(\mathbf{x})$, which captures the layout of iso-orientation domains completely [2]. Reflecting the roughly repetitive arrangement of orientation columns the Fourier representation of $E_k(\mathbf{x})$ and consequently of $z(\mathbf{x})$ exhibits a single finite band centered around modes of characteristic wavelength Λ. Given $z(\mathbf{x})$ the orientation preference map is obtained as $\vartheta(\mathbf{x}) = \frac{1}{2} arg(z)$. In any species investigated so far $\vartheta(\mathbf{x})$ contains topological point defects [3, 4, 5, 6, 7, 8, 9]. Their positions $\mathbf{x}_j$, the pinwheel centers, are the zeros of $z(\mathbf{x})$. They are generically of first order and come in two varieties distinguished by their topological charges $q_j = \frac{1}{2\pi} \oint_{C_j} \nabla\vartheta(\mathbf{x})d\mathbf{s} = \pm\frac{1}{2}$. Results from tangential electrode penetrations [10] as well as an electrophysiological investigation of identified pinwheels [4] indicate that on the cellular level orientation centers are real discontinuities in the pattern of orientation preferences.

For elementary topological reasons, continuous changes of $z(\mathbf{x})$ that might occur during development can affect the defect configuration $\{\mathbf{x}_j, q_j\}$ only in three ways: (1) *motion* of defects, (2) *creation* and (3) *annihilation* of pairs of defects of opposite charge. Therefore the number of pinwheels can only change by discrete creation and annihilation events. In principle there can be orientation maps without orientation centers, the simplest of which is a plane wave $z(\mathbf{x}) = e^{i2\pi/\Lambda \hat{\mathbf{k}}\mathbf{x}}$ ($|\hat{\mathbf{k}}| = 1$). The density of orientation centers ρ quantifies the deviation of a map from a defect-free state. Λ and ρ are independent quantities characterizing a given orientation map. The scaled density $\hat{\rho} = \rho\Lambda^2$ gives the mean number of orientation centers in a region of size Λ^2 and measures the relative abundance of pinwheels in a particular system.

Recent results indicate that the relative abundance of pinwheels varies considerably between species. Fitzpatrick and coworkers report that in tree shrew striate cortex iso-orientation domains are predominantly organized as a system of parallel stripes lacking orientation centers in large regions[9]. In other species as e.g. macaques orientation maps exhibit a fairly high density of orientation centers ($\hat{\rho}_m \approx 3.8$, [11]).

EMERGENCE OF ORIENTATION COLUMNS

To analyze how such differences in the scaled pinwheel density might emerge, we investigated the dynamics of the field $z(\mathbf{x})$ during development under most general assumptions. On the level of individual neurons and synapses, the self-organization hypothesis assumes that afferent and cortical activity patterns guide a continuous refinement of geniculocortical connections. Many theoretical studies (for reviews see [2, 12]) demonstrated that Hebb-type remodeling can lead to the emergence of orientation columns by amplifying weak orientation biases induced by random fluctuations in connectivity or activity patterns. During this process intracortical interactions force neurons to develop orientation preferences similar to those of their neighbors. Consequently the way in which the pattern of orientation selectivities and preferences is modified at a particular point in time depends strongly on the pattern of orientation selectivities and preferences already present.

As an idealization of this picture we assume that the effective modification of $z(\mathbf{x})$ under the stream of sensory activity can be predicted to a large extend from the knowledge of $z(\mathbf{x})$

itself. I.e. it is assumed that $z(\mathbf{x})$ follows a dynamics

$$\frac{\partial}{\partial t} z(\mathbf{x}) = F[z(\cdot)] + \xi, \tag{1}$$

where $F[\cdot]$ is an unknown nonlinear operator and $\xi(\mathbf{x},t)$ is a spatio–temporal random process modeling intrinsic and stimulus induced fluctuations. To ensure that there are no other sources of spatial structure but random fluctuations Eq.(1) must exhibit three basic symmetries.
(i) Invariance under orientation shifts: $F[e^{i\phi}z] = e^{i\phi}F[z]$.
(ii) Invariance under translations of the cortical sheet:
$F[\check{T}_y z] = \check{T}_y F[z]$ with $\check{T}_y z(\mathbf{x}) = z(\mathbf{x}+\mathbf{y})$.
(iii) Invariance under rotations of the cortical sheet:
$F[\check{R}_\beta z] = \check{R}_\beta F[z]$ with $\check{R}_\beta z(\mathbf{x}) = z\left(\begin{bmatrix} \cos(\beta) & \sin(\beta) \\ -\sin(\beta) & \cos(\beta) \end{bmatrix}\mathbf{x}\right)$.
(i) enforces that all orientations are treated equally. (ii) and (iii) ensure that the cortex does not prefer a particular pattern of iso-orientation domains.

Many properties of the dynamics (1) are determined by this set of symmetries. The state $z(\mathbf{x}) = 0$ is a stationary solution as long as the dynamics is invariant under orientation shifts $(F[0] = e^{i\phi}F[0] \Rightarrow F[0] = 0)$. In its vicinity the dynamics is governed by

$$\frac{\partial}{\partial t} z(\mathbf{x}) = \check{L}z(\mathbf{x}) + \xi(\mathbf{x},t), \tag{2}$$

where $\check{L}$ is a linear operator. Like $F[\cdot]$, the operator $\check{L}$ must be invariant under translations and rotations of the cortical plane. This implies that $\check{L}$ is diagonal in Fourier representation and its eigenvalues $\lambda(k)$ depend only on the modulus of the wavevector $k = |\mathbf{k}|$. Eq.(1) and (2) give rise to a columnar pattern if there is one interval (k_{min}, k_{max}) of wavenumbers with positive eigenvalues. While Eq.(2) only describes the primary emergence of an orientation map, Eq.(1) also captures the saturation of orientation selectivity and possible rearrangements of orientation columns in a subsequent nonlinear phase of development.

INITIAL PINWHEEL DENSITY

Most importantly these properties already impose a lower bound on the density of pinwheels which form as the orientation map arises from the homogeneous state $z = 0$. Eq.(2) is solved in terms of the Green's function $G(\mathbf{x},t) = \frac{1}{2\pi}\int d^2k\, e^{-i\mathbf{k}\mathbf{x}+\lambda(|\mathbf{k}|)t}$ by

$$z(\mathbf{x},t) = \int d^2y \int_0^t dt'\, G(\mathbf{y}-\mathbf{x}, t-t')\, \xi(\mathbf{y},t').$$

$z(\mathbf{x})$ thus is the mean of a set of random variables. According to the central limit theorem the field $z(\mathbf{x})$, established during the early (linear) phase of development, therefore realizes a Gaussian random field(GRF) [13]. The density of zeros in a GRF is determined by its power spectrum $P(|\mathbf{k}|) = |\int d^2x\, z(\mathbf{x})\, e^{i\mathbf{k}\mathbf{x}}|^2$

$$\rho = \frac{1}{4\pi} \frac{\int d^2k\, |\mathbf{k}|^2 P(|\mathbf{k}|)}{\int d^2k\, P(|\mathbf{k}|)}$$

[14]. Without loss of generality, we assume $\int_0^\infty dk P(k) = 1$. For convenience the characteristic wavelength Λ is defined as $\Lambda \equiv 2\pi/\bar{k}$ where $\bar{k} = \int_0^\infty dk\, k P(k)$ is the mean wavenumber. The density of zeros can then be rewritten as

$$
\begin{aligned}
\rho &= \frac{1}{4\pi\bar{k}} \int_0^\infty dk\, k^3 P(k) \\
&= \frac{1}{4\pi\bar{k}} \int_0^\infty dk\, (\bar{k} + (k - \bar{k}))^3 P(k) \\
&= \frac{1}{4\pi} \left(\bar{k}^2 + 3 \int_0^\infty dk\, (k - \bar{k})^2 P(k) + \int_0^\infty dk\, \frac{(k - \bar{k})^3}{\bar{k}} P(k) \right) \\
&= \frac{\pi}{\Lambda^2} \left(1 + 3 \int_0^\infty dk\, \frac{(k - \bar{k})^2}{\bar{k}^2} P(k) + \int_0^\infty dk\, \frac{(k - \bar{k})^3}{\bar{k}^3} P(k) \right)
\end{aligned}
$$

Only the third term of ρ can become negative. If one assumes that $P(k)$ vanishes in the range $k > 4\bar{k}$, which holds for spectra containing a single peak centered at $\bar{k}$, the modulus of the third term is bounded by the second

$$
\begin{aligned}
\left| \int_0^\infty dk\, (k - \bar{k})^3 P(k) \right| &\leq \int_0^\infty dk\, \left| (k - \bar{k})^3 \right| P(k) \\
&\leq 3\bar{k} \int_0^\infty dk\, \left| (k - \bar{k})^2 \right| P(k) \\
&= \left| 3\bar{k} \int_0^\infty dk\, (k - \bar{k})^2 P(k) \right|
\end{aligned}
$$

so that

$$
\rho = \frac{\pi}{\Lambda^2} (1 + \alpha) \tag{3}
$$

with $\alpha \geq 0$. Consequently π/Λ^2 forms a lower bound for the density of pinwheels established in the initial phase of development. Therefore the first prediction of the self-organization hypothesis is that at least π/Λ^2 pinwheels must proliferate during the primary establishment of orientation selectivity. Even if no further rearrangement would occur, the observation of more than π/Λ^2 pinwheels early in development would reveal an important fingerprint of the activity-dependent self-organization of orientation preference.

SPONTANEOUS PINWHEEL-ANNIHILATION

The development of lower pinwheel densities is not prohibited by Eq.(1). The invariance of Eq.(1) under translations and orientation shifts even guarantees the existence of pinwheel free stationary solutions

$$
\begin{aligned}
F[\check{T}_y | z_{PW} | e^{i\mathbf{kx}}] &= F[e^{i\phi'} | z_{PW} | e^{i\mathbf{kx}}] \\
&= e^{i\phi'} F[| z_{PW} | e^{i\mathbf{kx}}] \\
&= \check{T}_y F[| z_{PW} | e^{i\mathbf{kx}}] \\
\Rightarrow F[| z_{PW} | e^{i\mathbf{kx}}] &= f(| z_{PW} |) e^{i\mathbf{kx}},
\end{aligned}
$$

that will often be stable attractors of the dynamics of $z(\mathbf{x})$. Such low density states, however, can only form through an early high density state and subsequent annihilation of pinwheel pairs during the nonlinear phase of development. Thus a second prediction of the self-organization hypothesis is that spontaneous pinwheel annihilation must occur in any species in which a pinwheel density lower than π/Λ^2 is observed in the adult, as e.g. in the tree shrew.

To determine whether pinwheel annihilation is indeed an intrinsic property of the activity-dependent refinement of orientation maps we have investigated different biologically plausible models for the development of orientation columns [15]. In these models pinwheels initially proliferate by pairwise creation. In agreement with the theory developed above, pinwheel creation leads to the formation of at least π/Λ^2 orientation centers. All investigated models exhibit spontaneous pinwheel annihilation in the nonlinear phase of development. The degree and the velocity of pinwheel annihilation vary for different initial patterns, models, and choice of parameters. For every individual pattern, however, pinwheel annihilation proceeds fastest close to threshold and for strong lateral interactions. This behavior conforms with the intuitive notion that lateral cooperation favors smooth mappings and hence should tend to remove discontinuities from the orientation preference map. Pinwheel-annihilation breaks the statistical isotropy of the initial pattern and leads to growing regions of stripe-like iso-orientation domains.

Discussion

The second prediction provides a highly significant means to test the self-organization hypothesis. The preferred orientation of the set of columns between annihilating orientation centers changes by 90° during pinwheel-annihilation. Since this requires a radical reshaping of receptive fields it would be incompatible with the assumption that orientation preference is genetically prespecified [16, 17] as well as with the notion that activity-dependent mechanisms only mediate the "selective stabilization" of a subset of exuberant initial connections[18]. Pinwheel-annihilation is only possible if the functional architecture of visual cortex remains in a state of flux beyond the primary establishment of orientation selectivity. Direct observation of pinwheel-annihilation would therefore provide unequivocal evidence for the self-organization of orientation preference.

REFERENCES

[1] Grinvald, A., Frostig, R. D., Lieke, E., and Hildesheimer, R. *Physiological Reviews* **68**, 1285–1365 (1988).

[2] Swindale, N. *Network: Computation in Neural Systems* 7, 161–247 (1996).

[3] Ts'o, D. Y., Frostig, R. D., Lieke, E. E., and Grinvald, A. *Science* **249**, 417–249 (1990).

[4] Bonhoeffer, T. and Grinvald, A. *Nature* **353**, 429–431 (1991).

[5] Bonhoeffer, T., Kim, D.-S., Malonek, D., Shoham, D., and Grinvald, A. *Europ. J. Neuroscience* 7, 1973–1988 (1995).

[6] Blasdel, G. G. *J. Neuroscience* **12**, 3139–3161 (1992).

[7] Blasdel, G., Livingstone, M., and Hubel, D. In *Soc. Neurosci. Abstracts*, volume 12, 1500. Society for Neuroscience, (1993).

[8] Weliky, M. and Katz, L. C. *J. Neuroscience* **14**, 7291–7305 (1994).

[9] Fitzpatrick, D., Schofield, B. R., and Strote, J. In *Soc. Neurosci. Abstracts*, 837. Society for Neuroscience, (1994).

[10] Albus, K. In *Models of the Visual Cortex*, Rose, D. and Dobson, V., editors, chapter 51, 485–491. Wiley (1985).

[11] Obermayer, K. and Blasdel, G. G. *J. Neuroscience* **13**, 4114–4129 (1993).

[12] Miller, K. D. In *Models of Neural Networks III*, Domany, E., van Hemmen, J., and Schulten, K., editors. Springer-Verlag, NY (1995).

[13] Adler, R. J. *The Geometry of Random Fields*. John Wiley, NY, (1981).

[14] Halperin, B. I. In *Physics of Defects, Les Houches, Session XXXV, 1980,* Balian, R., Kléman, M., and Poirier, J.-P., editors (North-Holland, Amsterdam, 1981).

[15] Erwin, E., Obermayer, K., and Schulten, K. *Neural Computation* **7**, 425–468 (1995).

[16] Wiesel, T. N. and Hubel, D. H. *J. Comp. Neurol.* **158**, 307–318 (1974).

[17] Stryker, M. P. In *Neuroscience Res. Prog. Bull., Vol. 15, No.3*, chapter VII, 454–462. MIT Press (1977).

[18] Changeux, J.-P. and Danchin, A. *Nature* **264**, 705–712 (1976).

A SPECIAL ROLE FOR INPUT CODES IN SOLVING THE TRANSVERSE PATTERNING PROBLEM

Xiangbao Wu, Joanna M. Tyrcha, and William B. Levy

Department of Neurological Surgery
University of Virginia Health Sciences Center
Charlottesville, Virginia 22908
xw3f@virginia.edu
joan@sans.kth.se
wbl@virginia.edu

1. ABSTRACT

Rats require a hippocampus to solve the transverse patterning problem. Here, a hippocampal model also solves this configural learning problem. The problem is hard: A learning paradigm, called progressive learning, is required. It is required by rats, humans, and the model. Second, input patterns within a sequence must be repeated. Such repetition increases the statistical dependence, a surprising observation if you assume statistical dependence is undesirable. Such repetition of the same patterns in a sequence facilitates the formation of local context neuronal firings. These neuronal firings are critical, and we hypothesize that they are analogous to place cells found in behaving animals.

2. INTRODUCTION

2.1. Configural Learning and the Problem of Transverse Patterning

Recently, Alvarado and Rudy (1995) demonstrated that the hippocampus is necessary for a configural learning problem called transverse patterning. In this problem, the meaning of a particular stimulus depends upon the context supplied by the other stimuli that are simultaneously present.

Consider three atomic stimuli W, X, and Y. Let these stimuli be presented as pairs including (WX), (XY), and (WY). Then, reinforce behavior in the following manner: for the (WX) pair, W is the correct answer; for the (XY) pair, X is the correct answer; and for the (YW) pair, Y is the correct answer. Thus, each individual stimulus is equally rewarded and punished, and the only way to solve such a problem is to consider the stimulus complex. In this sense the transverse patterning problem requires a system that can learn and use the context provided by the configuration of stimuli themselves. To say it another way, meaning and predictability come by virtue of the context that arises from the pairing of stimuli.

Because an intact hippocampus is necessary for a rat to learn this problem, it is important that a hypothesized model of the hippocampus also solve the problem. Because our hippocampal model is a sequence learning system, we turned the configural learning problem into a very simple sequence. The first pattern in the sequence is the configured stimulus such as we have just described above (e.g., XY). The second pattern in the sequence is a randomly selected motor response that represents a response that chooses one of the two atomic stimuli (e.g., response x chooses X), and the third pattern in the sequence would be either the positive or negative reinforcement as appropriate (e.g., reward =+). Figure 1a shows one such input sequence and some of the recurrently activated neurons. The system is tested as suggested by the method of goal finding (Levy et al. 1995; Wu & Levy 1996; Levy & Wu, 1997). That is, the desired outcome (reward =+) is partially turned on along with a test configuration of paired stimuli.

3. THE NETWORK

The hippocampal model is essentially a model of region CA3 (see Fig. 1 in Levy & Wu, 1997). The input layer corresponds to a combination of the entorhinal cortex and dentate gyrus. To make the system's operation as transparent as possible, decoding is performed by similarity comparisons rather than a CA1-subiculum-entorhinal decoding system. The CA3 model is a sparsely (10%) interconnected feedback network of 512 neu-

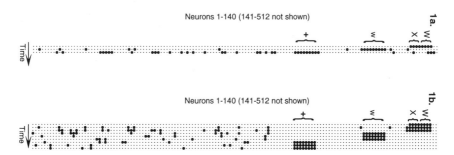

Figure 1. Configural learning requires the development of local context units. CA3 activities at the end of training for two types of inputs. The simple orthogonal input sequence of 1a does not produce local context neurons or learning while the same input sequence "stuttered," in 1b, produces local context neural firings and the appropriate learned "behavior." The two sequences of neural firing illustrated here are for a sequence that is part of the configural learning problem. Here we illustrate an WX trial (W = neurons 3–6, X = 7–10, the response of choice W is represented by neurons 19–26, and the + reinforcement given for this correct response is represented by externally activating neurons 43–50). All other neuronal firings are driven by recurrent connections. In 1b, the externally driven inputs are repeated three times while in 1a just a single pattern of each is given. As a result, in 1b, some of the recurrently driven neurons fire repetitively and selectively in time. Such repetitive firing can be asynchronous relative to all the externally driven input patterns. Note also that these local context neurons overlap with one another so that they can efficiently pass their information on from one time step to another. Only 140 neurons of the 512 are illustrated due to space limitations. Neuron 1 is at the top of the page, neuron 140 is at the bottom.

rons where all direct connections are excitatory and the network elements are McCulloch-Pitts neurons. Inhibition is of the divisive form, but the system is not purely competitive because of a slight delay. Synaptic modification is a postsynaptic rule that includes both potentiation and depression aspects (Levy & Steward 1979; Levy 1982). For details see equations in Levy & Wu, 1997.)

4. RESULTS

We found that this model of the hippocampus could not learn the transverse patterning problem when transverse patterning is coded as described above and when each input just activated neurons orthogonal to all other inputs and patterns. However, when we repeated inputs (as in Fig. 1b) so that instead of giving the input WX, then the response "w", then +, we gave inputs WX, WX, WX, response "w", response "w", response "w", followed by +, +, +, then the network was able to correctly perform the transverse patterning problem with one proviso — the training paradigm effects learnability.

Alvarado and Rudy (1992) noted a controversy — results between labs are contradictory on the learnability of this problem — and they explained its basis. When learning trials totally intermix all stimuli pairs, then even college sophomores, as well as rats, are largely incapable of discovering the correct solution to the transverse patterning problem. However, when a special learning paradigm called, progressive learning, is used, then humans and rats are able to learn and solve the transverse patterning problem. This is exactly the same observation for our network. When the network is taught all three pairs totally intermixed, then the system fails. But when the task is learned gradually, progressively (see Table 1), the network performs the task as successfully as rats do. In the progressive learning paradigm, there is the progression from training on one pair, to training on two pairs, to training on all three pairs. Because our model reproduces this requirement for progressive learning, i.e., it fails when learning is randomized, we are further encouraged to believe in the usefulness of this computational model of the hippocampus. We return now to consider the importance of the input code for controlling the development of local context neurons.

4.1. Local Context Neurons

In studying the transverse patterning problem the network sees inputs that are only sequences made up of orthogonal patterns. Apparently, without capacitive elements and with the narrow time span of synaptic associative modification of this study, such orthogonal sequences do not allow the formation of local context neurons. (Presumably, these

Table 1. The progressive learning paradigm of transverse patterning

	Stage 1.	
		Learn W is correct when WX is the stimulus pair
then	Stage 2.	
		Learn W is correct when WX is the stimulus pair and
		Learn X is correct when XY is the stimulus pair
then	Stage 3.	
		Learn W is correct when WX is the stimulus pair and
		Learn X is correct when XY is the stimulus pair and
		Learn Y is correct when WY is the stimulus pair

neurons are an analogy of place cells that have been discovered in hippocampus (O'Keefe & Nadel 1978). Like place cells, local context neurons fire in response to places in a sequence.)

4.2. Sequence Completion and Local Context Firing with Stuttered Inputs

In general, orthogonal input sequences do not promote local context firing unless patterns are repeated. We have systematically investigated this "preprocessing" and quantified the average length of local context neuron firing and the amount of "stuttering" (i.e., repetition of elements) within the input sequence. In fact, the network can produce local context neuron firing time spans that exceed the length of input stuttering. It is this greater time spanning ability of some recurrently activated neurons that accounts for the appropriate performance by the network in the transverse patterning problem.

We studied the effect of repeating inputs directly on simple sequence completions. For example, we compared CA3 codes for the sequences wxyz and the sequence wwwwxxxxyyyyzzzz. With an activity level of 15%, for instance, such a four-fold stuttering gives an average context neuron firing time span of seven to eight time steps. Such stuttering of the input and creation of context neurons is not without its problems. Sometimes stuttering reduces the sequence length memory capacity (when this capacity is a count of the number of different patterns). For example, at 7.5% activity and a quadrupling of inputs, sequence length capacity goes down approximately 25%.

5. DISCUSSION

Preprocessing an input by repetitively presenting it to the network increases the statistical dependency of the input. In contrast to a prevalent theme of neural computing concerning the importance of lowering statistical dependency — a theme which we support in general — there are certain, very clear situations where statistical dependency should be retained or even increased in order to improve performance (see Levy & Adelsberger-Mangan 1995 for example). Here we have another example where added redundancy helps a neural computation.

6. ACKNOWLEDGMENTS

This work was supported by NIH MH48161 and MH00622, by EPRI RP8030-08, and Pittsburgh Supercomputing Center Grant BNS950001 to WBL, and by the Department of Neurosurgery, Dr. John A. Jane, Chairman. The authors also thank Dr. A. Lansner of the SANS Research Group, Department of Numerical Analysis and Computing Science, Royal Institute of Technology, Stockholm, Sweden for his graciousness and generosity in allowing us access to the SANS computing facilities.

7. REFERENCES

Alvarado, M.C. & Rudy, J. W. Some properties of configural learning: An investigation of the transverse-patterning problem. J. Exp. Psychol.:Animal Behav. Processes 18, 1992, 145–153.

Alvarado, M. C. & Rudy, J. W. Rats with damage to the hippocampal-formation are impaired on the transverse-patterning problem but not on elemental discriminations. Behav. Neurosci. 109, 1995, 204–211.

Levy, W. B Associative encoding at synapses. Proceedings of the Fourth Annual Conference of Cognitive Science Society, 1982, 135–136.

Levy, W. B & Adelsberger-Mangan, D. M. Is statistical independence a proper goal for neural network preprocessors? INNS World Congress on Neural Networks, 1995, I-527–531.

Levy, W. B & Steward, O. Synapses as associative memory elements in the hippocampal formation. Brain Res. 175, 1979, 233–245.

Levy, W. B & Wu, X. B. Predicting novel paths to goals by a simple, biologically inspired neural network. CNS*96, 1997, this proceedings.

Levy, W. B, Wu, X. B. & Baxter R. A. Unification of hippocampal function via computational/encoding considerations. In: Proceedings of the Third Workshop: Neural Networks: from Biology to High Energy Physics. International J. of Neural Sys. 6 (Supp.), 1995, 71–80.

O'Keefe, J & Nadel, L. The Hippocampus as a Cognitive Map. Oxford: Oxford Univ. Press, 1978.

Wu, X. B. & Levy W. B, Goal Finding in a Simple, Biologically Inspired Neural Network. INNS World Congress on Neural Networks, 1996, 1279–1282.

138

A MODEL OF THE EFFECTS OF SCOPOLAMINE ON HUMAN MEMORY PERFORMANCE

Bradley P. Wyble and Michael E. Hasselmo

Department of Psychology
Harvard University
33 Kirkland St. Cambridge, Massachusetts 02138
wyble@berg.harvard.edu
hasselmo@berg.harvard.edu

1. ABSTRACT

The effects of scopolamine on human memory function are simulated in a network model of the hippocampal formation, incorporating simplified simulations of neurons and synaptic connections. In this model, the dentate gyrus, CA1 and CA3 maintain sparse representations of list items and their associated context allowing simulation of recognition and free recall. Blockade of cholinergic effects within this computational representation models the effects of scopolamine. This blockade of cholinergic effects impairs the encoding of new input patterns (as measured by delayed free recall), but not retrieval of previously learned patterns. Additionally, this cholinergic blockade has no significant effect on recognition tasks. The model also accomodates qualitative aspects of both the List Strength Effect and the List Length Effect.

2. INTRODUCTION

Many models of have focused on behavioral data about human memory function (Atkinson and Shiffrin, 1968; Raajimakers and Shiffrin, 1980; Murdock 1982; Gillund and Shiffrin, 1984; Hintzman, 1988; Ratcliff, 1990; Metcalfe, 1993; Chappell and Humphreys, 1994). However, most models of human memory do not attempt to address constraints from physiological data. These models can be seen as interpretive models—they help us understand behavioral data and guide behavioral experiments—but their link to the underlying function is less clear—they do not directly address the actual mechanisms of memory function because they do not link the behavioral data to physiological substrates.

Computational Neuroscience
edited by Bower, Plenum Press, New York, 1997

In contrast, the work presented here concerns a functional model of human memory—this model attempts to directly address the physiological and anatomical substrates of human memory, allowing it to address data on the effect of the cholinergic antagonist scopolamine on human memory function.

Acetylcholine has a number of effects on neuronal function. Among these are enhancement of synaptic modification, direct depolarization of cells, suppression of inhibition, suppression of adaptation and suppression of intrinsic excitatory connectivity. The combination of these different effects appears to set appropriate dynamics for the encoding of new information (Hasselmo et al., 1992; Hasselmo, 1993; 1995)

In this model, we show, at a functional level, how the attenuation of these effects due to scopolamine during encoding of a list impairs the ability of the model to correctly recall items from that list while recognition of them is virtually unimpaired(Ghoneim and Mewaldt, 1975, 1977). Additionally, this simulation demonstrates how the presence of scopolamine does not impair the retrieval process by correctly recalling words encoded prior to its administration (Ghoneim and Mewaldt, 1975; Peterson, 1977). Another relevant aspect of scopolamine effects is the sparing of recognition (Ghoneim and Mewaldt, 1975) which is also modeled here.

Another criteria used to evaluate this model is its ability to mimic certain functional characteristics found in many human memory paradigms apart from pharmacological studies. The List Length Effect (LLE) and the List Strength Effect (LSE)were both chosen as such tests of the model's validity.

3. METHOD

The model included representations of entorhinal cortices II, III, and IV, the dentate gyrus, CA1, CA3 and a septal region used to mediate cholinergic modulation (see figure 1). This network simulation used simplified representations of individual neurons designed to mimic basic properties of hippocampal pyramidal cells and interneurons (Hasselmo et al., 1995). Each region consisted of an array of pyramidal cells connected to a feedback inhibitory interneuron. These cells were then connected directly to excitatory and inhibitory cells within other regions. LTP formed gradually through simulated accumulation of NMDA activation providing hebbian learning. LTD also occurred from mis-

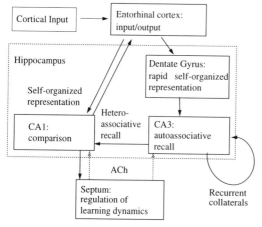

Figure 1. A block diagram of the model's components. Information is presented to the system through the entorhinal cortex and flows through the system as described in the text. Notice that modulation is applied to the entire system from the septum.

matched pre and post synaptic activity. No explicit normalization or winner take all mechanisms existed within the model, and all activity levels derived from continuous firing rate equations similar to Hasselmo and Stern (1996).

Patterns representing list elements were sequentially provided to the entorhinal cortex, layers II and III. These representations consisted of several active elements and were allowed to overlap by up to 50%. While passing through the hippocampus, this information first formed orthogonal and sparse representations in the dentate gyrus via self organization. These representations were next projected onto and encoded within the autoassociative CA3 region through a one to one mapping that is consistent with the restricted connectivity of the mossy fibers. There, attractor dynamics controlled the activity of these representations which were heteroassociated with other self-organized representations in CA1 that had formed simultaneously from direct perforant path projections . These CA1 activity patterns were then associated with a reconstruction of the original pattern on entorhinal cortex IV provided from external cortical input.

Activity in CA1 also served as a mechanism for cholinergic regulation. In the initial state of the model, cholinergic modulation is present in all structures, providing a bias towards learning operations. Based on observed connections within the brain, activity in CA1 inhibited the septum so as to reduce the amount of modulation. This reduction of modulation relaxed suppression on CA3 recurrent collaterals which allowed recall to occur. In this way, the modeled hippocampus was capable of dynamically controlling its own state. As a result, the system could compare the activity in CA3 evoked by cues to that evoked in CA1 by the perforant path projection. If they matched, CA1 would become very active and shut down cholinergic modulation, allowing recall to commence within CA3.

The incorporation of context cues played a crucial role in the function of the model. A static set of cues were provided for each list that remained separated from the information about each of the individual items. These context cues represented an internal designation that a set of items or words constituted a list. This context information was associated within CA3 to each of the items during presentation of the list. Later, recall of a list could be evoked by an input of its context alone. The associative connections between context and the individual items would sequentially activate some subset of the list items (Figure 2) Attractor dynamics mandated that only a single item be could be recalled at a time within CA3 and the order of recall depended on random variations in encoding strengths during learning. Endogenous adaptation currents prevented a given item from being repeatedly recalled.

In a similar fashion, recognition could be accomplished by cueing the model with a representation of a given list item(figure 3). The bi-directional associations between context and item then reconstructed the context of the list that the item belonged to. If context was correctly recalled for a given item, that item was recognized. Sometimes a novel item incorrectly activated a context cue, this was designated a false alarm and was treated accordingly. It should be noted that this sort of recognition addresses specific recall of an episode, and not a perceptual familiarity with an item.

4. RESULTS

The model was able to exhibit all of the aforementioned effects qualitatively: impaired encoding with spared retrieval (Figure 4) and impaired free recall with spared recognition (Figure 5). In some cases the effects in the model were not quantitatively

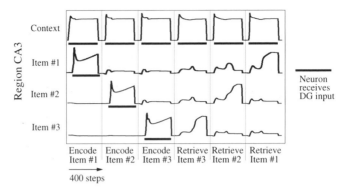

Figure 2. The temporal dynamics of free recall are illustrated by the activity of several excitatory neurons involved in this example. In this figure each horizontal trace represents a neuron's average firing rate over time which flows to the right (unused neurons are omitted for clarity). Black bars underneath the trace indicate time blocks during which powerful excitation from the dentate gyrus is occurring. In the first three columns, a list of three items is presented simultaneously with a constant context. Immediately following this period, recall of this list commences with the afferent activation of the context representation only. A competition between attractors occurs with one emerging in each case, corresponding to successful recall. Intrinsic adaptation currents in each neurons prevent items from being repeatedly recalled.

congruous with the documented effects found in humans but the general trends were definitely present. In Figure 6, a comparison of the model's results to those of Ghoneim and Mewaldt (1975) can be seen.

The cholinergic effects which were primarily responsible for the performance degradation in these simulations of scopolamine impairments were the depolarization of all neurons and the enhancement of synaptic modification. With less depolarization, the CA1 representations did not always form with enough strength with the result that several pat-

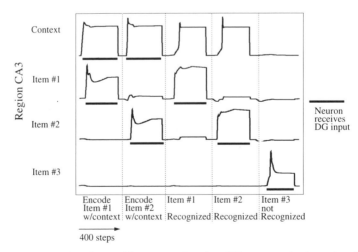

Figure 3. The temporal dynamics of recognition are seen here in a different example. A two item list is first presented for encoding. As in the recall example, context and items are presented simultaneously. However, during testing, the items are presented alone, without context. If the context is reactivated, the item is recognized as having been seen before. In the last column a novel item is presented. Here, no context activation is seen, a correct response.

Figure 4. In this example, the model demonstrates impaired encoding but spared retrieval. Here we are only showing entire layers of the network, but only those of the entorhinal cortex. Time flows to the right as before but active neurons are represented by black bars. The letters C and I to the right of the box illustrate which regions of these layers are context and which are item information storage while the neuron at the bottom of each layer is the inhibitory interneuron for that region. Two lists of four items each are presented to the system each with different contexts. Scopolamine is presented where shown at the bottom. Note that the list learned without scopolamine is recalled successfully while under scopolamine, but the list encoded under the influence is not recalled (recall consists of reconstructing the original patterns on layer IV in any order).

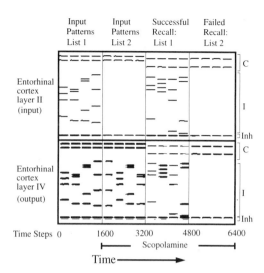

terns successfully recalled by CA3 could not return this information to EC IV. More importantly, the lack of learning enhancement during scopolamine injection retarded the strength of CA3 patterns so that they were less likely to become active during free recall. However, because context was presented repeatedly during a list, the weights within contextual representations were still strong enough to permit attractor dynamics within context, allowing recognition to occur.

The aforementioned LLE and LSE were present but were exaggerated relative to human data. Increasing list length decreased the chance that any item would be recalled and strong items were more frequently recalled at the expense of weak items. Additionally, the LSE was only present during recall, a facet of memory that many models have difficulty accommodating (Ratcliff, Clark and Shiffrin 1990). This distinction results from the fact that competition between items is involved during recall, while the recognition process involves an item recalling only a single context cue.

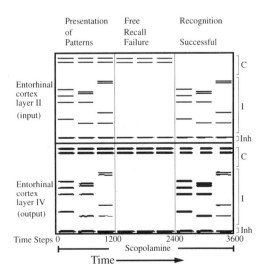

Figure 5. Here we see impaired free recall with spared recognition. The representation is the same as in Figure 4, except that only a single three item list is presented during encoding. In this example, scopolamine is present during the entire trial, learning, recall, and recognition. Notice that recall is greatly impaired while recognition occurs (successful recognition is indicated by the reconstruction of context in layer IV by each of the items).

Free Recall of 8 list of 16 words

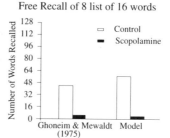

Recognition of 8 list of 16 words

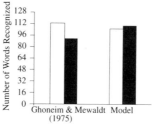

Figure 6. Comparison of model data with human subjects under scopolamine from Ghonheim and Mewaldt (1975).

5. DISCUSSION

Blockade of cholinergic effects within the network simulation of the hippocampus effectively simulates the effects of scopolamine on human memory function, showing selectivity which resembles that observed in psychopharmacological experiments (Ghonheim and Mewaldt, 1975; 1977; Peterson, 1977). In particular, simulation of scopolamine effects impairs the encoding but not the retrieval of words in a test of free recall. In addi-

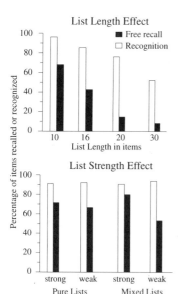

Figure 7. Model data on List Strength and List Length paradigms. In this figure, the model is being compared against itself to exhibit the qualitative effects of these manipulations. Note that in the case of recognition accuracy the percentage correct has been decremented by false alarms committed by the model.

tion, the modeled scopolamine effects significantly impair free recall of a list without significantly impairing recognition of its elements. These impairments in memory function are caused by blockade of cholinergic effects which have been demonstrated in neurophysiological studies of the hippocampus (see Hasselmo, 1995 for review).

Further experimental work could explore in more detail which effects of scopolamine are particularly important for the impairment of memory function. An experiment to analyze the effect of longer or repeated presentations would be most effective at determining the importance of the loss of learning facilitation under scopolamine. On the other hand, a task that provides additional cues for recall may facilitate CA1 pattern formation under scopolamine conditions, alleviating the deficit and providing evidence for the importance of the lack of tonic depolarization the drug effect.

6. REFERENCES

Atkinson, R. and Shiffrin, R.M., Human memory: A proposed system and its control processes. In K. Spence and J. Spence (eds.) The Psychology of Learning and Motivation Vol. 2. Academic Press: New York. pp. 90–195, 1968.

Chappell, M, Humphreys, MS (1994) An auto-associative neural network for sparse representations: Analysis and application to models of recognition and cued recall. Psych. Rev. 101, 103–128.

Ghoneim, M.M. and Mewaldt, S.P. (1975) Effects of diazepam and scopolamine on storage, retrieval and organization processes in memory. Psychopharmacologia 44, 257–262.

Ghoneim, M.M. and Mewaldt, S.P. (1977) Studies on human memory: The interactions of diazepam, scopolamine and physostigmine. Psychopharm. 52, 1–6.

Gillund, G. and Shiffrin, R.M. (1984) A retrieval model of both recognition and recall. Psych. Rev. 91, 1–67.

Hasselmo M.E. (1995) Neuromodulation and cortical function: Modeling the physiological basis of behavior. Behav. Brain Res. 65, 1–27.

Hasselmo M.E. (1993) Acetylcholine and learning in a cortical associative memory. Neural Comp. 5, 32–44.

Hasselmo M.E., Anderson BP, Bower JM (1992) Cholinergic modulation of cortical associative memory function. J. Neurophysiol. 67, 1230–1246.

Hasselmo, M.E. and Stern, C.E. (1996) Linking LTP to network function: A simulation of episodic memory in the hippocampal formation. In: Baudry, M. and Davis, J. (eds.) Long-Term Potentiation, Vol. 3 MIT Press: Cambridge, MA.

Hasselmo, M.E., Wyble, B. and Stern, C.E. (1995b) A model of human memory function based on the cellular physiology of the hippocampal formation. In: Parks, R. and Levine, D. (eds.) Neural Networks for Neuropsychologists. MIT Press: Cambridge, MA.

Hintzman, D.L. (1988) Judgments of frequency and recognition memory in a multiple-trace memory model. Psych. Rev. 95, 528–551.

Metcalfe, J. (1993) Novelty monitoring, metacognition, and control in a composite holographic associative recall model: Implications for Korsakoff amnesia. Psych. Rev. 100, 3–22.

Murdock, B.B. (1982) A theory for the storage and retrieval of item and associative information. Psych. Rev. 89, 609–626.

Peterson, R.C. (1977) Scopolamine induced learning failures in man. Psychopharm. 52, 283–289.

Raajimakers, J. G.W., and Shiffrin, R.M. (1980) SAM: A theory of probabilistic search of associative memory . in G.H. Bower(Eds.) The psychology of learning and motivation (Vol. 14, pp 207–262). New York: Academic Press.

Ratcliff, R. (1990) Connectionist models of recognition memory: Constraints imposed by learning and forgetting functions. Psych. Rev. 97, 285–308.

Ratcliff, R., Clark, S.E., and Shiffrin, R.M. (1990) List Strength Effect: I. Data and Discussion. Journal of Experimental Psychology: Learning Memory and Cognition, 16, 163–178.

139

SPATIAL ORIENTATION AND DYNAMICS OF PAPEZ CIRCUIT

Kechen Zhang

Department of Cognitive Science
University of California, San Diego
La Jolla, California
E-mail: kzhang@cogsci.ucsd.edu

ABSTRACT

Although Papez circuit was initially proposed as a neural substrate for emotions, recent findings of head-direction cells demonstrate that some parts of the structure are involved with representation of spatial orientation. We discuss several problems related to the dynamics of the spatial representation, including (1) reduction of the dynamics of coupled cell groups to that of a simple cell group, in which the formulation of the shift mechanism has been determined uniquely, (2) the relation between dynamic behaviors of coupled cell groups and lesion results, and (3) the limitation of one-dimensional spatial representation in geometric phase problem.

1. BACKGROUND

Head-direction cells have been found in several brain regions close to the classic Papez circuit, which involves the limbic thalamus, subicular complex, cingulate cortices and the mammillary body. For reviews, see [4, 8].

2. REDUCED DYNAMICS

Several authors have considered the possibility that the directional tuning of head-direction cells might be an emergent property of an attractor network [2, 5-7, 9]. Within the framework of the standard simplified dynamics

$$\tau \frac{du}{dt} = -u + \text{"input"}$$

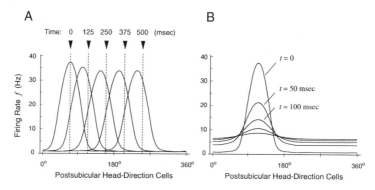

Figure 1. **A.** Snapshots of postsubicular activity when the thalamic activity (peak positions indicated by arrowheads) suddenly starts to move at time zero at the speed 360°/sec. **B.** Without thalamic input, postsubicular activity cannot sustain itself. In this example 75% inputs come from lateral connections within postsubiculum, while thalamus contributes the remaining 25%.

for each individual unit, the preservation of the shape of the activity profile during shift requires that the the antisymmetric component of the synaptic weight distribution be proportional to the derivative of the symmetric component. This weight change can be induced *effectively* by modulating the firing rates of two coupled cell groups, whose derivative weight components have opposite signs [9]. Without external inputs, any differences of activity between the two groups will vanish exponentially. External inputs can induce approximately rigid shift of activity profile at the angular speed

$$\omega \approx \frac{\eta}{\tau} \frac{\bar{f}_1 - \bar{f}_2}{\bar{f}_1 + \bar{f}_2},$$

where $\bar{f}_1$ and $\bar{f}_2$ are the average firing rates of the two groups, and η is a proportional constant for the derivative weight components. The average dynamics of the double cell groups can be identical to that of the standard model, in which the shift mechanism is determined uniquely [9]. Because of space constraint, more details will be published elsewhere.

3. LESIONS

The anterior thalamus might be more closely involved with the updating process during head turn than is the postsubiculum, because of the differential lesion results [8] (but see also [3]). Although postsubiculum projects to the thalamus, its functional influences might be so weak that postsubicular lesions cannot abolish thalamic head-direction cells. Although weak inputs are not efficient to drive activity shift, they can still be sufficient for the calibration process as long as they can offer persistent local-view specific information. In Fig. 1, we only focus on the properties of interacting cells groups in the scenario where postsubiculum passively receive inputs from the thalamus. Here the patterns of synaptic weight distribution from thalamus to postsubiculum and within postsubiculum itself are assumed to be the same, but they have different strength.

The time lag of postsubicular activity relative to anterior thalamic activity can be estimated by the formula

$$\text{Time lag} \approx \Delta t + \frac{\tau}{P},$$

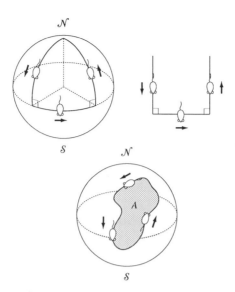

Figure 2. Problem of geometric phase for head-direction cells. Because the internal direction represented by head-direction cells is updated only by yaw of the head, but not by roll or pitch, the system may be fooled when the animal moves in 3-D space instead of on a flat 2-D surface.

where τ is the time constant of the postsubicular network, P is the percentage of the input received from the thalamus, and Δt is conduction delay, which is probably only a few msecs, If $\tau \sim 10$ msec, then a time lag of 40 msec would imply that $P \sim 25\%$, that is, about a quarter of the inputs received by postsubicular cells come from thalamus.

In this example, when the thalamic inputs are removed (corresponding to a thalamic lesion), the remaining weights within postsubicular network cannot support the activity hill, even though they are three times as strong as the thalamic input. *The result is not just a smaller hill of activity, but a totally flat activity profile because now the flat state is the attractor state.* The condition for the stability of a single hill of activity as the attractor state is that the first Fourier component of the weight distribution must exceed certain threshold, whereas all other higher Fourier components must be below the threshold. As a consequence, reducing the overall weights by a small percentage can make the flat state stable. Finally, in Fig. 1A, the amplitude of the postsubicular activity profile is somewhat reduced during traveling, assuming that the thalamic activity profile is constant. In reality, the activities of many head-direction cells in thalamus increase slightly during movement. As a result, the amplitude reduction may be compensated by the increased inputs from thalamus so that the actual amplitude of postsubicular activity could be preserved.

4.   GEOMETRIC PHASE

The following consideration was inspired by the NSMA projects of McNaughton and colleagues. Consider the example in Fig. 2. Parallel transport either on the sphere or on a flat surface will *not* affect the internal direction maintained by head-direction cells because they are affected only by the yaw of the head. In general, the angular speed of internal direction is the dot product

$$\omega = \vec{n} \cdot \vec{\Omega},$$

where $\vec{\Omega}$ is the true angular velocity of the head, and unit vector $\vec{n}$ describes the orientation of the head and is defined as vertical in normal posture.

In the spherical situation in Fig. 2, animal starts from the north pole, makes a loop and returns. The animal's final heading becomes different from the initial one whereas the internal direction is the same. In general, when an animal returns to the starting point following an arbitrary closed path on the sphere, the discrepancy between the internal direction and the actual heading is equal to the solid angle span by the area A enclosed by the path. Only the geometry of the path is important; the animal's turning at any local region does not contribute to the final discrepancy. In other words, the angular discrepancy

$$\Delta\theta = A/r^2,$$

where r is the radius of the sphere. In particular, if an arbitrary path covers half of the total area of the sphere ($A = 2\pi r^2$), there will be no discrepancy. This discrepancy is sometimes called as geometric phase, because it depends only on the geometry of the path [1]. Similarly, the head-direction system might also be fooled by aerobatics such as a half-loop followed by a barrel roll. The internal direction should stay the same during the whole process although the final physical heading becomes opposite to the initial one.

REFERENCES

[1] J. Anandan, Nature 360: 307 (1992).
[2] H. T. Blair, NIPS 8: 152 (1996).
[3] H. T. Blair, B. W. Lipscomb and P. E. Sharp, submitted.
[4] R. U. Muller, J. B. Ranck, Jr. and J. S. Taube, Curr. Opinion Neurobiol. 6: 196 (1996).
[5] A. D. Redish, A. N. Elga and D. S. Touretzky, Network 7: 671 (1996); CNS*96, this volume.
[6] A. Samsonovich and B. L. McNaughton, submitted.
[7] W. E. Skaggs, J. J. Knierim, H. S. Kudrimoti and B. L. McNaughton, NIPS 7: 173 (1995).
[8] J. S. Taube, J. P. Goodridge, E. J. Golob, P. A. Dudchenko, Brain Res. Bull. 40: 477 (1996).
[9] K. Zhang, J. Neurosci. 16: 2112 (1996); CNS*95: 415 (1996).

SECTION V

METHODOLOGY

140

GRID GENERATION FOR BRAIN VISUALIZATION AT THE CELLULAR AND TISSUE LEVEL

David A. Batte[1] and Bruce H. McCormick[2]

[1] Scientific Visualization Laboratory
Texas A&M University, College Station, TX
E-mail: dbatte@cs.tamu.edu
[2] Scientific Visualization Laboratory
Texas A&M University, College Station, TX
E-mail: mccormick@cs.tamu.edu

ABSTRACT

Numerical grid generation is used to provide a framework for brain and neuron visualization. Smoothing spline surfaces are fit to contour data to generate 3D solid model reconstruction of brain tissue. Finite element methods are then used to subdivide the solid models into biologically-consistent finite elements. Numerical grid generation is employed to provide a curvilinear coordinate system within the finite elements. Synthetic and manually traced neurons are mapped into the gridded solid model using the curvilinear coordinate system. To this end grid generation tools, neuron mapping tools, and visualization tools have been implemented.

1. INTRODUCTION

The goal of this work is to provide effective techniques for the visualization and interpretation of neuron data sets within a volume of reconstructed neural tissue. The interpretation and modeling of these neuron data sets rests on having a biologically-consistent framework in which the neuron populations can be visualized. For brain visualization at the cellular and tissue level, a coordinate system must be generated within a brain nucleus or cortical area to produce a biologically-consistent model of the volume filled with a sparse representative population of neurons. Numerical grid generation [1] is a well-established method for producing boundary-conforming curvilinear coordinate systems in irregular 2D and 3D regions. Its development arose from the need to solve partial differential equations within physical regions with complex geometry. Due to the complex, folded nature of the human cerebral cortex we use numerical grid generation to embed a 3D coordinate system inside cortical tissue. Neurons, either individually traced or synthetically generated, are then embedded inside the gridded solid model of the cortical tissue and visualized in this environment. More generally, similar considerations apply to brain nuclei.

1.1. Objectives

Biologically-Consistent Finite Element Decomposition and Grid Generation of Brain Tissue Our primary objective is to embed boundary-conforming grids within solid model reconstructions of cortical areas and brain nuclei. Numerical grid generation techniques rely on boundary information to produce the grids. We first decompose the solid model into finite elements to provide the boundary information for the numerical grid generation methods. To establish finite elements in a manner consistent with developmental neurobiology, the elements are chosen to follow the natural symmetries of the tissue, as defined by its primary native neuron type and thinking how a neuroanatomist would cut tissue locally to make successive sections look as alike as possible. Numerical grid generation can then be employed to establish a curvilinear coordinate system within each finite element, and hence throughout the solid model. This coordinate system provides a means of orienting neurons and specifying physical barriers and chemical gradients within the volume.

Biologically-Consistent Mapping of Neurons into the Gridded Solid Model Our second objective is to embed the gridded solid model with a population of neurons. Sparse populations of neurons can be drawn from databases of measured neurons, or stochastically generated using L-system modeling as described by McCormick and Mulchandani [2, 3, 4]. The grids provide a means of positioning the neurons in the solid model of reconstructed tissue, and specifying positional forces within the volume. The embedded neurons are chosen to be statistically consistent with those found in the actual tissue.

Visualization of Neuron Populations within the Gridded Solid Model Our third objective is to visualize a population of neurons within reconstructed tissue using computer graphics in conjunction with the grid generation and neuron mapping methods. An interactive approach allows the user to freely explore the environment.

2. RESULTS

In this section results generated with the software developed for this project are presented.

2.1. Reconstruction

A small section of human cerebral cortex is reconstructed from a data set containing images collected from the brain of a 76 year old normal female human cadaver. The images were obtained from the UCLA Laboratory of Neuro Imaging World Wide Web site (http://www.loni.ucla.edu)[5]. The brain was cytosectioned through the horizontal plane in 100 μm increments on a heavy duty cytomacrotome. The cytomacrotome was equipped with a high resolution camera for digital image capture of the serial images ($1024^2, 24 - bit$). The images were scaled down $(512^2, 24 - bit)$ for distribution over the Internet. Forty-two serial slices were used for the reconstruction.

Contours were collected from each of the images using *Elastic Reality* (Avid Technologies). *Elastic Reality*, a special effects tool for 2D and 3D animation allows contour construction using piecewise cubic bezier curves and facilitates comparison of consecutive contours. Contours were generated for the outer and inner cortical boundaries. The contours are resampled and the resulting points used for 3D reconstruction.

2.2. Reconstruction with Smoothing Splines

The contours were sampled, and smoothing spline surfaces were fit to the samples using the FORTRAN surface fitting routines written by Dierckx [6]. Interpolating surfaces are first fit to the data. The reconstruction using interpolating spline surfaces is very rough. This most likely demonstrates the "wiggly" noise that is picked up in the surfaces when no smoothing is used. Next, smooth approximating surfaces are fit to the data. A smoothing factor $S = 1$ is used. Figures 1 and 2 show two views of the reconstructed tissue using the smooth approximating surfaces. The resulting reconstruction is much better than the interpolated surfaces.

2.3. Finite Element Decomposition

The finite element decomposition which was used here is a simple sampling of the parameter domain. A grid was constructed for both surfaces in parameter space by choosing lines of constant u and v at even parametric increments. The grid lines in parameter space define u, v grid curves on the surface. The w lines are linear segments connecting the u, v intersection points on the outer surface to the u, v intersection points on the inner surface. Figures 1 and 2 show the finite element grid lines on the surface of the reconstructed tissue. There was no attempt to construct elements which follow the crest lines in the tissue. The finite element lines, which follow the crest and valley lines in the two figures, were purely accidental. Techniques to follow crest and valley lines are described in Batte [8]

2.4. Grid Generation

The 3D transfinite interpolation grid generator was used to generate grids in the reconstructed human cerebral cortex and rat hippocampus. Figure 4 shows the boundary edges of a single finite element from the human cerebral cortex.

A slice of rat hippocampus was reconstructed from an illustration taken from *The Rat Nervous System* [7] using the same reconstruction techniques as above. Figure 3 shows the boundaries of the finite elements.

2.5. Mapping Neurons

Finally, a sparse network of neurons is mapped into the reconstructed tissues. Figure 4 shows a set of pyramidal cells mapped into the the cerebral cortex finite element. Figure 5 shows a set of pyramidal cells mapped into the rat hippocampus finite elements.

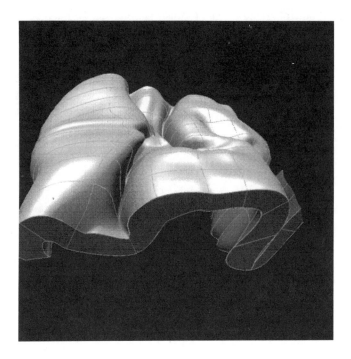

Figure 1. Finite element decomposition of the solid model (outer view).

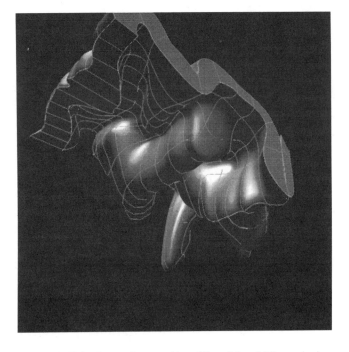

Figure 2. Finite element decomposition of the solid model (inner view).

Figure 3. Hippocampus finite element boundaries.

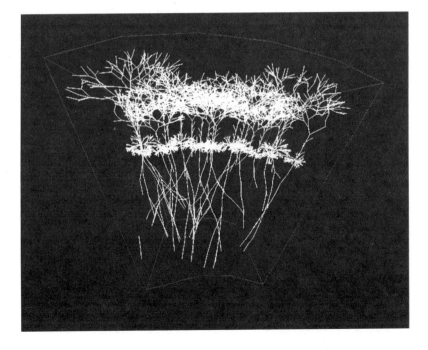

Figure 4. A population of pyramidal cells embedded in a cortical finite element.

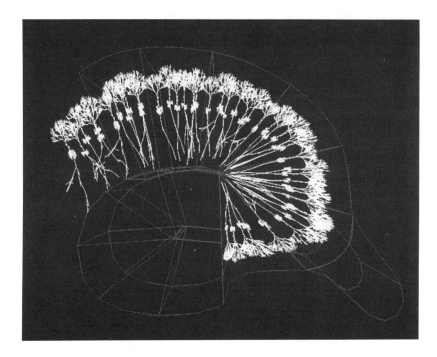

Figure 5. A population of pyramidal cells embedded in hippocampal finite elements.

3. SUMMARY

In this work, smooth approximating B-spline surfaces are used to reconstruct brain tissue from scanned section data. A proposed technique decomposes the reconstructed tissue into finite elements in such a way that preserve symmetry within the tissue. Grid generation methods are used to provide a curvilinear coordinate system within the finite elements. The grids and mappings provided by the grid generators allow either synthetic and manually traced neurons to be embedded inside the finite element model of the tissue. Finally, tools which visualize these grids and neuron data sets were implemented.

REFERENCES

[1] P. Knupp and S. Steinberg, *Fundamentals of Grid Generation*. Boca Raton, FL: CRC Press, 1994.
[2] B. McCormick and K. Mulchandani, "L-system Modeling of Neurons," in *Proc. Visualization in Biomedical Computing*. SPIE, 2359 pp. 693-705, 1994.
[3] K. Mulchandani and B. McCormick, "A Framework for Modeling Neuron Morphology," *Computational Neuroscience: Trends in Research 1995*, J. Bower ed. San Diego, CA: Academic Press, Inc, 1996. pp. 453-458.
[4] K. Mulchandani, "Morphological Modeling of Neurons," Masters thesis, Department of Computer Science, Texas A&M University, 1995.
[5] Toga, A, Ambach, K, and Schluender, S, "High-resolution anatomy from in situ human brain," *NeuroImage*, vol. 1, pp. 334-344, 1994.
[6] P. Dierckx, *Curve and Surface Fitting with Splines,* Oxford, England: Oxford University Press, 1995.
[7] D. Amaral and M. Witter, "Hippocampal Formation," *The Rat Nervous System*, G. Paxinos ed., San Diego, CA: Academic Press, Inc, 1995.
[8] D. Batte, "Finite Element Decomposition and Grid Generation for Brain Modeling and Visualization," Masters thesis, Department of Computer Science, Texas A&M University, May, 1997.

141

PARALLEL GENESIS FOR LARGE-SCALE MODELING

Nigel H. Goddard and Greg Hood

Pittsburgh Supercomputing Center
Pittsburgh, PA
E-mail: ngoddard@psc.edu
E-mail: ghood@psc.edu

1. INTRODUCTION

Simulations of computational models are limited in size and speed by the power and capacity of the computational platform. We have parallelized the GENESIS object-oriented neural simulator [1] for networked workstations, multiprocessors and massively parallel supercomputers. These can provide two orders of magnitude increase in the size of the models that can be effectively simulated. As larger models are partitioned across many processors, interprocessor communication can limit the effective speedup obtainable. This suggests two classes of problems that may benefit most from parallel simulation: parameter searching and network models.

In parameter searching, a simulation (often of a single cell) is repeatedly performed in order to determine the model parameters (e.g. channel conductances) that best fit experimental data. This search through parameter space can involve thousands of individual simulations. In these "embarrassingly parallel" computations the ratio of computation to communication is high: a set of "worker" processors perform the individual simulations, while a "master" processor directs the search. The stereotypical algorithm is an iteration of the following steps: 1) **scatter**: distribute a small amount of data (e.g., parameters) to each worker, 2) **compute**: the workers execute some intensive computation (e.g., a simulation run), 3) **gather**: the workers report back results (e.g., goodness of fit), and 4) **update**: the master uses the reported results to compute the next set of data (parameters) to be distributed.

A second class of models, multicell spiking networks, exhibit two features that can render them suitable for parallelization. First, the frequency of spike generation is often much lower than the frequency of state update in the simulation of a single cell, so that lower bandwidth between processors is acceptable. Second, axonal and synaptic delays are often much greater than the update timestep for a single cell, so that spikes generated by a presynaptic cell on one processor do not have to be delivered to a postsynaptic cell on another processor for several

Computational Neuroscience
edited by Bower, Plenum Press, New York, 1997

timesteps of the simulation. During this delay the destination processor can continue updating its resident cells, thus overlapping communication with computation.

While parallel GENESIS (henceforth PGENESIS) is most effective in simulating these two classes of models, it is a general purpose parallelization of GENESIS and as such is fully capable of simulating large electrotonically connected models (e.g., a single cell) which is distributed across multiple processors. We will not discuss this class of models further except to say that, because of their heavy communication demands (both in bandwidth and latency), shared memory multiprocessors are the most effective platform for their parallelization.

2.   SIMULATION SPECIFICATION AND CONTROL

A PGENESIS simulation consists of N independent serial GENESIS simulations called **nodes** in which objects (e.g., cells) can send and receive data (e.g., spikes) from other objects on any of the N nodes. The nodes are automatically synchronized to ensure data transmitted between them is received before it is needed.

Serial GENESIS simulations are specified and controlled from a scripting language. In PGENESIS this script language is extended with parallel programming constructs so that existing serial scripts generally need only minor modifications. The new constructs allow specification and control of a parallel simulation. PGENESIS assumes nodes can read and write a common file system. In most simulations the nodes are controlled by a small set of script files read simultaneously by all the nodes.

For specifying the structure of distributed network models, the script commands which establish and manipulate connections between cells (e.g., **addmsg** and **volumeconnect** which make axonal connections) have been extended to connect cells residing on different nodes. PGENESIS ensures correct transmission of spikes across these inter-node connections.

For controlling the setup of the model and its simulation, several parallel programming constructs are provided. Remote function calls allow a script on one node to execute a function on another node and obtain the result. Asynchronous function calls with an associated completion construct allow user-controlled optimizations based on script parallelism. Barrier calls allow nodes to synchronize: when a node reaches a barrier, it waits until all other nodes have reached the corresponding barrier. User-specified **zones** can partition the machine to run several independent multi-node simulations simultaneously, e.g., for parameter searching a network model.

3.   USER-LEVEL OPTIMIZATIONS

During model setup, when the structure of the model is being specified by script commands which create compartments, cells and connections, the most important optimizations involve the use of asynchronous execution. For example, establishing remote connections involves the following sequence of events: 1) initial actions at the source node, 2) request to the destination node, 3) actions at the destination node, 4) response to the source node, and 5) completion actions at the source node. When this sequence is executed synchronously, the source node must wait after (2) until the response (4) is received. In addition to the communication latencies and the time taken to execute the necessary actions at the destination node, the destination node may be busy with other uninterruptible tasks when the request arrives. Asynchronous execution of this sequence allows the source node to perform other tasks as soon as the request (2) has been sent, significantly reducing the time spent waiting for other nodes to process and reply to requests.

While PGENESIS ensures that these and all other asynchronous operations, including asynchronous remote function call, have completed before any simulation is run, "future" tags returned to the user script allow it to suspend until asynchronous operations complete. The "future" is a promise of a result (an actual value or simply the fact of completion) at some future time. The completion construct ("waiton") is used in a script to demand the fulfillment of that promise.

During simulation of a network model, PGENESIS transmits data between nodes and synchronizes them as necessary. The decomposition of the model (how cells are allocated to processors) can significantly affect the overhead introduced by data transmission and node synchronization. If the model specifies a delay in data transmission (e.g., axonal and synaptic delays), then spikes can be in transit over the physical network while the destination cell continues to update. At each time step, data from all objects on one node destined for objects on another node are bundled into a single PVM message, reducing communication overhead. If the model incorporates delays, spikes can be buffered across multiple time steps to further reduce the number of individual communication events required of the underlying hardware.

Highly parallelized parameter search simulations can exhibit fall-off in speedup due to poor selection of search strategy. In order to keep all processors busy, desirable characteristics of a search strategy are that (a) a pool of candidate parameterizations be maintained at all times, (b) a candidate be dispatched to a worker nodes as soon as it finishes the previous candidate assigned to it, without waiting for the result of that candidate, and (c) incorporation of simulation results and selection of candidates be sufficiently simple that it does not become a bottleneck - or if it does, that the search strategy itself can be parallelized.

4.   PERFORMANCE RESULTS

4.1.   Parameter Searching Example

In this example we ran a single-cell parameter fitting model [4] which used a genetic algorithm to search parameter space. The individuals in each generation were run in parallel on the Cray T3D. We used 30, 60 and 240 individuals per generation. 100 individuals per generation were used in the original serial simulation [4], the algorithm taking 16 hours to converge on a single processor. On a parallel machine with 16 equivalent processors, this time could be reduced to one hour.

Figure 1 shows linear speedup up to 15 processors for the cases involving 30 and 60 individuals. Analysis of the sub-linear performance for more processors indicates that speedup is increasingly dominated by the sequential gather/update/scatter component and uneven load due to small numbers of individuals being run on each processor. This is confirmed in the performance for 240 individuals per generation. We believe that networked workstations would provide similar speedups for small processor counts (e.g., 6-8). More processors could be effectively utilized with this search strategy if 1) there were more individuals per generation or 2) the sequential component were reduced in complexity or 3) if the cell model were substantially more computationally complex. A better search strategy, the criteria for which were given in section 3, would allow greater speedup without increasing model complexity or the number of individuals per generation.

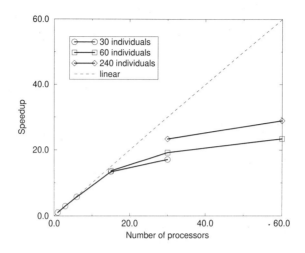

Figure 1. Parameter searching performance

4.2.   Multicell Spiking Network Example

In this example we ran a parallelized version of a simulation of orientation selectivity [3]. The model consists of a group of "retinal" cells projecting to a group of "V1" cells. The retinal cells make contact with the V1 horizontal and vertical cells and generate spikes stochastically under the control of the simulation.

The strategy in allocating cells to processors in network models (decomposition) is important in achieving good parallel performance. The goals are to 1) minimize the number of connections (synapses) between nodes, 2) minimize the average firing frequency experienced by inter-node connections, and 3) maximize the minimum axonal delay over all connections between each pair of nodes. The first two goals aim to reduce the amount of data transferred between processors, while the third goal maximizes the opportunity for the simulation system to exploit interleaved communication and computation, as discussed in Section 1. For this example, in which retinal cells project to "nearby" V1 cells, retina and V1 were partitioned in stripes (Figure 2). Most synapses were within one stripe. Interprocessor communication was only required for synapses between stripes.

We ran the model in three sizes on the Cray T3D (Figure 3). Note the super-linear speedup at low processor counts! This is due to quadratic complexity in the time to setup the fraction of the model assigned to each processor: breaking the model into more pieces produces quadratic speedup in this setup phase. Speedup of the simulation phase alone is approximately linear up to 6, 9 and 10 processors for the small, medium and large cases respectively.

These results show that network models can gain substantial speedup, and that the number of processors that can be effectively utilized grows, although sub-linearly, with the size of the model. Due to the highly simplified cell models used in this simulation, spike communication was a relatively large part of the computational load. Thus these numbers understate what is likely to be achieved in more realistic models.

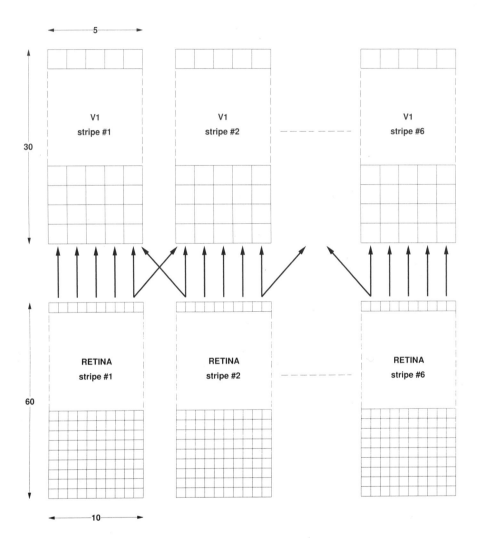

Figure 2. Network model decomposition

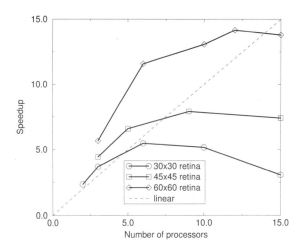

Figure 3. Network model performance

5. PORTABILITY

Using the widely available Parallel Virtual Machine (PVM) message passing library [2], we have achieved portability of PGENESIS code across UNIX-based platforms from networked workstations to parallel supercomputers. Currently PGENESIS runs on the Cray T3D, Cray T3E, Intel Paragon, SUN, SGI and DEC shared-memory multiprocessors, and SGI, SUN and Alpha single processor workstations. Ports to Linux and HP machines are underway.

At the user level, scripts can be written in a way that is independent of the number of processors available in the underlying hardware, with the special case of a single processor resulting in serial simulation. This allows scripts to be run unmodified on widely different platforms, a critical facility in scaling up research simulations.

6. FURTHER GOALS

Many neural models incorporate a degree of randomness e.g., in the structure of synaptic connections. It is desirable that simulation of these models be numerically repeatable, requiring that at each point a random number is needed, the same number be generated in subsequent runs as in the first run. For a parallel simulation, in which the sequence of demands for random numbers can vary depending on exact timing of different processors and data transmission between them, we have developed a method of duplicating a parallel stochastic simulation. Each remote operation carries with it state information for the random number generator to produce a new random number sequence. This results in a repeatable tree of random number sequences, rather than the single sequence required in a serial simulation.

Two avenues to ease parallel scripting are envisaged. First, a global namespace would remove the need for specifying node numbers explicitly in the script. Currently each node in a parallel simulation is cognizant only of the objects simulated on that node, requiring that references to objects on remote nodes have the remote node number explicitly specified in the script. Second, canonical decompositions would allow modelers to prototype their parallel models quickly, and tools which profile and analyze parallel execution will facilitate tuning of decompositions. Once a global namespace is implemented, it will be possible to provide some standard decomposition options, such as random or regular placement. Static and dynamic analysis of the model will enable advice on how to improve decomposition. Dynamic analysis may allow dynamic migration of objects to lightly loaded nodes.

7. CONCLUSIONS

We have extended the GENESIS simulation system to allow it to run on the full range of parallel platforms, from networked workstations to supercomputers. Turnaround time for simulations of large problems in the parameter searching and network modeling classes can be reduced dramatically on parallel platforms - from months to days or from days to hours. Parallel platforms, including the NSF Supercomputing Centers, are increasingly accessible to computational neuroscientists. Large scale computational modeling increasingly requires parallel computing. PGENESIS provides facilities which insulate the modeler from details of the underlying parallel hardware platform, and which provide an upgrade path from small scale prototype models to large scale production simulations.

REFERENCES

[1] J.M. Bower and D. Beeman. *The Book of Genesis*. Springer-Verlag, Santa Clara, CA, 1994.

[2] A. Geist, A. Beguelin, J. Dongarra, W. Jiang, R. Manchek, and V.S. Sunderam. *Parallel Virtual Machine*. MIT Press, Cambridge, MA, 1994.

[3] M. Vanier and D. Beeman. Constructing neural circuits and networks. In J.M. Bower and D. Beeman, editors, *The Book of Genesis*, chapter 17. Springer-Verlag, Santa Clara, CA, 1994.

[4] M. Vanier and J.M. Bower. A comparison of automated parameter-searching methods for neural models. In J.M. Bower, editor, *Computational Neuroscience: Trends in Research 1995*, Monterey, CA, 1995. Academic Press.

142

LARGE NEURAL NETWORK SIMULATIONS ON MULTIPLE HARDWARE PLATFORMS

Per Hammarlund, Örjan Ekeberg, Tomas Wilhelmsson, and Anders Lansner

SANS — Studies of Artificial Neural Systems
Department of Numerical Analysis and Computing Science
Royal Institute of Technology
S-100 44 Stockholm, Sweden
E-mail: {perham,orjan,towil,ala}@nada.kth.se
http://www.nada.kth.se/sans/

ABSTRACT

Efficient simulations of very large networks of interconnected neurons require particular consideration of the computer architecture being used. Techniques for implementing simulators on a number of different computer architectures are presented.

The experience gained from adapting an existing simulator, SWIM, to two very different architectures, vector computers and multiprocessor workstations, is analyzed. This work led to the implementation of a new simulation library, SPLIT, designed to allow efficient simulation of large networks on several different architectures. SPLIT hides from the user most of the architecture dependent details, that is, the particular data structures and computational organization actually utilized.

INTRODUCTION

Simulations of large networks of biologically realistic model neurons are possible utilizing state-of-the-art computing resources in the form of high performance computers. To get the most out of such machines it is generally important to take into account the specific properties of both the machine architecture and the task at hand. How this can be done when the task is to simulate large and detailed neuronal networks is the topic of this presentation.

We focus on neuronal models where the neurons are discretized into a moderate number of compartments and where relevant ion-channels are represented by Hodgkin-Huxley-like differential equations. Communication between such model neurons, in the form of action potentials reaching the terminal after a delay, is modeled as discrete events. Our aim in this work is to allow efficient simulation of networks comprised of up to thousands or even tens of thousands of such neurons, with the number of synapses possibly reaching millions. Simulations made within our group are approaching such sizes [3, 5].

Compared to traditional large-scale scientific computations, these simulations are more irregular. The complex data-sharing leads to unavoidable tradeoffs in optimizing the different sub-computations. For example, a highly optimized code for solving the potential spread in a neuron may lead to inefficient access of the state variables when updating the ion-channel states. Since many state variables are used in several different sub-computations, there is no single optimal layout.

SWIM AND SPLIT

SWIM is a simulator for networks of moderate size [1]. It was primarily targeted towards single processor workstations and is written in ANSI C, making it easily portable to most workstations running standard versions of UNIX. The SWIM program offers an integral object-oriented specification language.

We have adapted SWIM to run efficiently on multiprocessor workstations as well as vector computers. Based on our experiences from this work we created a new simulation library, SPLIT [2]. Simulation networks for SPLIT are specified in terms of neuronal *populations* and synaptic *projections*. This allows SPLIT to exploit the regularity inherent in a simulation network for optimizing performance and memory consumption.

The SPLIT library is designed to run efficiently on several hardware platforms: workstations, multiprocessor workstations, clusters, and vector computers. SPLIT hides from the user most of the architecture dependent details, i.e. the highly optimized data structures and computational organization actually utilized with each computer architecture. Object-oriented programming techniques and conditional compilation are used to allow only parts of the code, both data structures and computational organization, to be optimized for specific architectures while keeping most of the code general.

ARCHITECTURES AND OPTIMIZATIONS

A straightforward implementation of a simulator may often turn out to be surprisingly inefficient. Each architecture comes with its own benefits and drawbacks which must be taken into account in order to fully utilize the potentially available processing power.

Workstations

Workstations are single processor systems with cache, a simple memory system, and low-cost I/O. We have used Sun, Digital, and IBM workstations. An efficient use of cache read-ahead, and reuse, is crucial for good performance. This implies a data structure where data used in the same computation are located close in memory. For instance, all parameters for a channel or a synapse should be allocated together. This property was reasonably well fulfilled in the original SWIM code. In SPLIT we have taken this one step further and made sure that all intensively used data structures are cache-optimal.

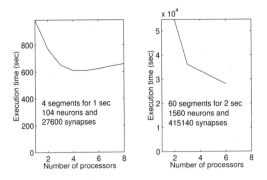

Figure 1. Performance of the threaded SWIM program running on a Sun SPARC Center 2000. Two different sizes of a model of swimming rhythm generation in the lamprey spinal cord is simulated. The timings were done on a system with other users present. The smaller 4 segment simulation can use up to 4 processors before parallelization overheads cancel the benefits of an extra processor. The larger 60 segment simulation makes good use of at least 6 processors.

Multiprocessor Workstations

Many computer vendors now offer systems with multiple processors. We have used Sun, SGI, and Digital multiprocessors. Typically each processor has a local cache and access to a central shared memory.

Threading is a convenient way to use multiple CPUs. Threads can be used to allow concurrent computation of independent parts of the network, e.g. sets of synapses or neurons. In fact, the static structure of the cellular morphology as well as connectivity allows for an optimal scheduling of the computations of each time-step. It is, however, important that this parallelization does not congest the main bottleneck of these systems: the central memory. Figure 1 displays the performance of threaded SWIM on a Sun SPARC Center 2000.

Clusters of Workstations

With the increasing performance of communication networks, it is possible to use multiple workstations as a *cluster*. We have adapted our software for clusters of Suns and for the IBM SP-2.

The simulation must be mapped (i.e. distributed) onto the processors. Typically there is a trade-off between finding a good load balance and minimizing the overhead from interprocessor communication. Communication between processors is needed for: updating state variables computed on one processor and used on others; for synaptic communication; and for collecting result data. Communication is performed with the standard *message passing* libraries PVM and MPI, which ensures portability. Spike messages are buffered and delays in axons are used to hide latencies. We utilize both the sparsity and delays of synaptic connections to find an efficient mapping of the neurons onto the processors. Neurons having many short-delayed synaptic interconnections are kept together on the same processor. Spectral bisection is one useful algorithm for doing this.

SPLIT for clustered systems parallelizes well and the speedup is quite good, see Figure 2. To date, the largest simulations performed included 58 000 neurons and 4.4 million synapses. Simulating this system for 100 ms with a time-step of 50 μs on 8 wide IBM SP-2 processors took about 100 minutes while dynamic initialization took another 100 minutes.

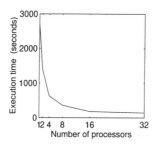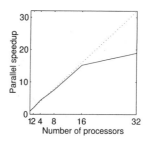

Figure 2. The performance of SPLIT on an IBM SP-2 when simulating a small network of 2535 neurons and 127 000 synapses. This particular simulation uses up to 16 processors efficiently.

Vector Computers

Vector computers are typically regarded as the generic supercomputers. For large neural simulations they offer superb performance. We use CRAY vector computers, with multiprocessors and high-performance shared memory.

The computational organization used for the threaded parallelization also performs well on these systems. Further, it is important to choose data structures so that computational loops can access relevant data sequentially. This implies a data organization quite different from the cache-optimal one. Thus, for multiple neurons, compartments, or synapses, each parameter is collected and allocated in a separate vector.

EFFICIENT SYNAPSE MODELS

Computations on a large number of synapses can totally dominate the execution. Here design decisions may have a large impact on the over-all performance. Delayed communication of the discrete synaptic events between the continuous pre- and post-synaptic neuronal processes calls for different solutions on different architectures.

Synapse models where most of the computation can be associated with each post-synaptic compartment may significantly improve the efficiency. In particular for simulation networks where the number of synapses grows as the square of the number of neurons, e.g. cortical models. We use a method by William Lytton [4] which has significantly cut down execution time for synapses, see Figure 3.

CONCLUSIONS

It is possible to simulate very large networks of synaptically interconnected neurons efficiently on modern high-end computer systems. The same simulation techniques also improve the performance on desktop computers for networks of moderate size. We have analyzed various simulations on different classes of computer architectures and found a number of critical bottlenecks. By careful data structuring and computational organization, these bottlenecks can be overcome. Different architectures call for different solutions and the use of object-oriented programming techniques has proven to be a good way of supporting this in our implementations.

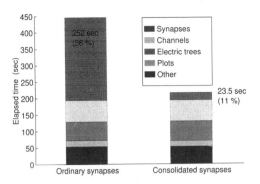

Figure 3. This execution profile of SPLIT illustrates the effect of Lytton's method [4] when simulating a cortical network with 2535 neurons and 150542 synapses (59 synapses/neuron).

These solutions are the basis for the multi-platform simulation library SPLIT, currently being used for large scale simulations.

REFERENCES

[1] Örjan Ekeberg, Per Hammarlund, Björn Levin, and Anders Lansner. SWIM — A simulation environment for realistic neural network modeling. In Josef Skrzypek, editor, *Neural Network Simulation Environments*. Kluwer, Hingham, MA, 1994.

[2] Per Hammarlund and Örjan Ekeberg. Large neural network simulations on multiple hardware platforms, 1996. (Submitted.).

[3] Anders Lansner and Erik Fransén. Improving the realism of attractor models by using cortical columns as functional units. In James M. Bower, editor, *The Neurobiology of Computation: Proceedings of the Third Annual Computation and Neural Systems Conference*, pages 251–256, Monterey, CA, July 21–26, 1994 1995. Kluwer, Boston, MA.

[4] William W. Lytton. Optimizing syaptic conductance calculation for network simulations. *neucomp*, 8(3):501–509, 1996.

[5] Tom Wadden, Jeanette Hellgren-Kotaleski, Anders Lansner, and Sten Grillner. Intersegmental coordination in the lamprey – simulations using a continuous network model, 1996. In Press.

143

OPTIMIZATION ANALYSIS OF COMPLEX NEUROANATOMICAL DATA

Claus C. Hilgetag,[*] Mark A. O'Neill, and Malcolm P. Young

Neural Systems Group
University of Newcastle upon Tyne
Department of Psychology
Ridley Building, Newcastle upon Tyne NE1 7RU, United Kingdom
cch@crunch.ncl.ac.uk
mao@crunch.ncl.ac.uk
malcolm@flash.ncl.ac.uk

ABSTRACT

Neuroanatomical data describing the numerous connections between brain structures contain valuable information about the organization of nervous systems. This information, however, cannot be assessed readily since the data are numerous, confusingly cross-referential, incomplete, contradictory, and of varying reliability. The classification of such data, moreover, allows vast numbers of different, equally possible interpretations that have to be evaluated. We have developed a computational approach that effectively deals with these difficulties by using stochastic optimization. We represented cortical connectivity data as 'black-box' objects that are linked with each other through a network of anatomical relations. This network can be arranged optimally according to suspected structuring principles. The approach makes it possible to analyze large amounts of complex anatomical data in a number of ways. We have successfully applied this technique to the analysis of processing clusters and hierarchies in cat and monkey cortical systems.

1. INTRODUCTION

Neuroscience historically has centered around anatomical studies of the brain. Ramón y Cajal (1990) began his revolutionary "New ideas on the Structure of the Nervous System in Man and Vertebrates" with the statement: "A thorough understanding of the

---

[*] Corresponding author

structure of the nervous system clearly provides an essential foundation for the disciplines of physiology and pathology." Since that time neuroscience has made enormous progress. Current functional studies often use highly sophisticated equipment that demands considerable experimental and scientific experience. Compared to these efforts, purely structural investigations may appear less important and unremarkable. Even so, the introduction of new staining and tracing techniques has caused the amount of available anatomical data to explode during the last two decades. This rapid accumulation of new data has, however, not been paralleled by an equal progress in understanding the organization of the brain. Consequently, many functional studies are carried out without considering such knowledge essential.

There are many questions, some dating back to Cajal's day, that are still answered unsatisfactorily. For example: What are the exact proportions of different cell types in the many different areas and regions of the brain, how and with what probability do these different cell types connect to each other? Is there designated cellular circuitry that mediates the connections between areas? What is the overall organization of the interconnected brain regions?

Given the complexity of neural systems, such questions can only be answered with the help of approaches that compile anatomical data according to rigorous standards, formalize the data, and attempt an objective analysis using mathematical and computational tools. Pursuing a 'systems level' approach, the detailed micro-structure of areas can be neglected, and the objective of the analysis is to interpret the organization of the complex network of connections between the areas.

Compilations of anatomical data for areas that correspond to central sensory systems often include hundreds of connections between several tens of areas In the monkey visual cortex, there exist more than 300 connections between more than 30 areas (Felleman & Van Essen 1991), and some 60 connections between at least 22 areas have been reported for the cat visio-limbic cortex (Scannell et al. 1995). There is disagreement on how these data should be interpreted, mainly because the amount of data is too large and is too strongly interlinked to reveal its structure by inspection alone. Moreover, the data are incomplete (not all the possible connections between proposed areas have been studied yet), some of the data are in contradiction to others, and some other data are not fully reliable. Attempts to classify these data require consideration of a vast number of different possibilities. The latter problem is due to the combinatorial explosion that has to be faced in many seemingly simple problems, like the Traveling Salesman task, where no obvious finite algorithm for finding the best solution exists.

2. METHODOLOGY

What approaches are available for the analysis of connection data? The unreliable and inconsistent nature of some of the data rules out straightforward categorization, and their complexity also makes exhaustive sorting impossible. Numerical computation, including artificial neural network approaches, does not seem to be feasible either, because the data are not usually characterized by numbers.

We developed a novel software system, CANTOR, that represents data as multi-relational objects which can be arranged by stochastic optimization according to given goals. The system consists of a multilevel dynamic database of objects which are linked to other objects, or indeed any computer entity (procedures, files, hardware-devices, etc.) with relational functions. The CANTOR environment provides routines which permit the arrangement of the objects within the database and which evaluate such modifications through cost functions. The costs are then optimized by a method based on simulated annealing (Laarhoven & Aarts 1987). Figure 1 illustrates the optimization algorithm that was used for the present analysis.

This architecture makes it possible to use high-level rules directly as input to the processor and to produce output in a form which is easy to understand. The data and the relations between them can be represented in a number of ways: symbolic (e.g., "A ≥ B"), graph-theoretical (e.g., "A -> B") or numerical (e.g., "distance(A, B) = 1.23"). The system is also able to categorize sets of data as meta-objects and to analyze output data statistically.

The software is written in ANSI-C and has been implemented on a wide variety of different UNIX computer systems. The system uses purpose-designed UNIX memory management and can handle computationally demanding tasks and large quantities of complex data by dividing tasks into sub-processes, which may run independently and in parallel on one or more computer processors.

3. RESULTS

Our optimization approach has been applied successfully to the analysis of cortical hierarchies in the cat and monkey brain. Connections between cortical visual areas can be classified as feedforward, lateral or feedback according to their pattern of origin and termination within the cortical layers (Rockland & Pandya 1979, Felleman & Van Essen 1991). This categorization yields a set of data consisting of pairwise hierarchical relations between cortical areas. These complex data have all the attributes outlined above. We used the CANTOR system to find hierarchical area arrangements that fitted the set of pairwise constraints optimally, that is, with the smallest possible number of constraint violations. We computed 200,000 such optimal hierarchies for the monkey visual system (Hilgetag et al. 1996) and 25,000 optimal hierarchies for the cat visual system (Hilgetag et al. 1995). These hierarchies were better hierarchical representations of the anatomical data than previously published schemes, and they suggested a number of hitherto unsuspected properties of the primate visual system. The system is likely to involve more hierarchical levels and a much more flexible way of processing than suspected previously. Certain cortical areas maintained fixed relations throughout all optimal solutions (for instance areas

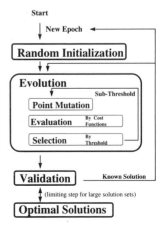

Figure 1. Evolutionary optimization algorithm used by the CANTOR system. Arbitrarily complex cost functions are possible. In the case of hierarchical analysis, the cost was simply the number of hierarchical constraint violations created by a particular area arrangement. For cluster analyses, the (component) cost function consisted of a repulsion term that counted non-existing area connections *within* the clusters and an attraction term counting existing connections *between* clusters. In both approaches the goal was to minimize the cost function. Point mutations were designed as the relocation of one area to another hierarchical level or cluster or swapping of two areas belonging to different levels/ clusters. The selection threshold was set at a level of 125% of parental cost for two successive offspring generations. These mutation and threshold parameters ensured that the solutions space was searched at the smallest possible resolution and that the search did not become trapped in local minima.

MT and V4t) whereas most others could be arranged flexibly. We also used the CANTOR system to explore hypotheses about sub-divisions of cortical areas and on the relevance of future experiments for resolving the indeterminacy of the primate visual hierarchy. This enabled us to make specific predictions for anatomical experiments (Hilgetag et al. 1996).

The number of levels in all optimal hierarchies ranged between 13 (for 11 hierarchies) and 24 (for 6 hierarchies), with the peak of the level distribution around 18 levels (13745 hierarchies) and 19 levels (13425 hierarchies). All of the optimal arrangements possessed 6 violations of the hierarchical constraints each. In all solutions the violations were FST ≤ MSTd (relative occurrence in all solutions 17%), FST < STPp (14%), LIP = PITv (12%), FST ≥ TF (2%), LIP ≤ MSTd (2%), MSTd < PITv (2%), FST ≥ PITd (<<1%), MSTd < PITd (<<1%), together with their corresponding symmetrical relations: e.g. MSTd ≥ FST. More details of the analysis and conclusions and predictions from these results can be found in Hilgetag et al. (1996) and at <http://www.psychology.ncl.ac.uk/www/hierarchy.html>.

The left-to-right arrangement of areas in this diagram into processing streams, namely a ventral and a dorsal stream, was inspired by the stream separation demonstrated in an previous multi-dimensional optimization analysis (Young 1992). This arrangement

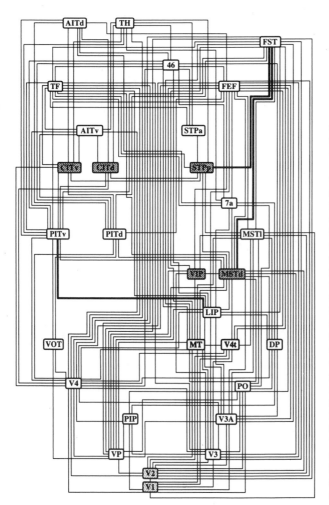

Figure 2. Optimal hierarchy of Macaque cortical visual areas based on the main area peaks in the distribution of all optimal visual hierarchies (Figure 1 in Hilgetag et al. 1996). The distribution took into account 150,000 optimal hierarchical arrangements and demonstrated that apart from early cortical areas V1 and V2, which where always found on the first and second hierarchical level respectively, the exact hierarchical position of cortical visual areas was indeterminate. Identically shaded areas had identical distributions in all optimal hierarchies, that is, their positions relative to each other in the hierarchies were fixed. Although the hierarchy depicted is only one out of a very large number of equally optimal arrangements, it has overlaps with many other solutions and therefore captures some essential features of all the optimal hierarchies. Established hierarchical connections between areas are represented as lines. Connections are generally reciprocal. The three pairs of constraints that could not be fitted for this hierarchy, notably FST ≤ MSTd, FST < STPp, LIP = PITv, and their respective counterparts, are shaded with dots.

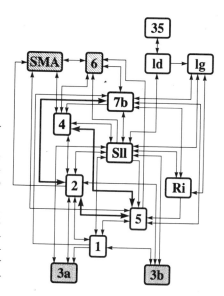

Figure 3. Optimal peak level hierarchy for the distribution of areas in 144 optimal hierarchies of Monkey somatosensory and motor areas. Arrowheads in the this hierarchy indicate whether connections are one- or two-way. The optimal hierarchies had between 9 (for 2 solutions) and 13 (for 24 solutions) levels, with the peak at 12 levels (56 hierarchies). All optimal hierarchies had three violations of the hierarchical constraints. The possible constraint violations were 7b<2 (33%), 4<5 (33%), 2<5 (25%) and 5<2 (8%), with the first three also being the violations in this peak hierarchy. The violated constraints are diagonally hatched. Areas shaded in the same color had fixed positions relative to each other in all optimal hierarchies.

has since been largely confirmed by a cluster analysis using CANTOR with a modified two-parameter cost function. The objective in this analysis was to minimize the number of connections existing *between* different clusters together with the number of non-existent connections *within* clusters. The only significant deviations from the depicted streams were V4 being assigned to the dorsal cluster and 7a being assigned to the ventral cluster. This figure, therefore, summarizes important aspects of the organization of the primate visual system.

The analysis used data from Felleman & Van Essen (1991) but a different classification scheme employing first- and second-order criteria to assign directions to the connections. Connections were generally considered to be feedforward if the terminated mainly in cortical layers 3 and 4 and feedback, if they avoided these layers. Other criteria were only in the absence of such evidence. This scheme allowed to classify a larger amount of the laminar data and reduced the number of resulting optimal hierarchies at the cost of a slightly increased number of violations. Future approaches will derive optimal hierarchical classification schemes.

4. DISCUSSION

The unsuspected results of our analyses reinforced the case for objective, computational analysis of neuroanatomical data (Young et al. 1995). Given the complex structure of the data, application of stochastic and optimization methods for data collection and analysis seems generally necessary. Work of this type also relies on the availability of high-quality compilations of formalized neuroanatomical data. A milestone example for this idea is NeuroBase (Burns *et al.* 1996), described elsewhere in this volume.

Our results raise further questions about the organization of cortical systems. Is the main hierarchical classification scheme used in this analysis (Felleman & Van Essen 1991) really applicable for all cortical areas in both Macaque and cat? The modified hierarchical categorization employed in the analysis of the somatosensory hierarchy already suggested that improvements on existing classification schemes are possible. Future approaches should attempt to select optimal schemes through automated classification. Recent results (C. Cusick, personal communication) also indicated that the laminar structure of input and output to cortical areas varies depending on the location of the area: the amount of lateral interconnections between areas appears increased for more frontal regions. Future work could attempt to separate cortical regions on the basis of these changes in connection patterns, or could attempt to delineate the borders of cortical hierarchy as such by including data for sub-cortical connections as well.

A related optimization approach will extend the analysis and visualization of neural connectivity by graph theory. Here the optimization of relational objects is required to assess strongly connected components in neural systems (incorporating the densities of the connections), to compute optimal path lengths of cyclic or acyclic pathways between anatomical complexes, and to determine critical paths in neural activation networks.

The CANTOR optimization system has proved sufficiently flexible and efficient that it has also been used in a number of different fields: in the determination of inter-species relationships from spatial biological databases, for the automatic optimal calibration of a camera sensor, or in the ecological modeling of different insect species. The system may also be suitable for further applications like structural modeling of biological macromolecules or the design of intelligent relational databases.

ACKNOWLEDGMENTS

The work has been supported by the MRC and the University of Newcastle upon Tyne. many thanks to G.A.P.C. Burns for numerous stimulating discussions

5. REFERENCES

G A P C Burns, M A O'Neill, M P Young: Calculating Finely-graded Ordinal Weights for Neural Connections from Neuroanatomical Data from Different Anatomical Studies. In *Computational Neuroscience '96* (ed. J. M. Bower). Boston: Plenum 1997.

S Ramón y Cajal: New ideas on the Structure of the Nervous System in Man and Vertebrates. Translated by N Swanson and L W Swanson. MIT Press: Cambrigde MA 1990.

D J Felleman, D C Van Essen: Distributed Hierarchical Processing in the Primate Cerebral Cortex, Cerebral Cortex 1 (1991), 1–47.

C C Hilgetag, M A O'Neill, J W Scannell, M P Young: A Novel Network Classifier and its Application: Optimal Hierarchical Orderings of the Cat Visual System from Anatomical Data, *Genetic Algorithms in Engineering Systems: Innovations and Applications*, IEE Publication No. 414, Sheffield 1995.

C C Hilgetag, M A O'Neill, M P Young: Indeterminate Organization of the Visual System, *Science* 271 (1996), 776–777.

P J M Van Laarhoven, E H L Aarts: *Simulated Annealing: Theory and Applications*, Kluwer: Dordrecht 1987.

K S Rockland, D N Pandya: Laminar Origins and Terminations of Cortical Connections of the Occipital Lobe in the Rhesus Monkey, *Brain Res* 179 (1979), 3–20.

J W Scannell, C Blakemore, M P Young: Analysis of Connectivity in the Cat Cerebral Cortex, *J Neurosci* 15 (1995), 1463–1483.

M P Young: Objective Analysis of the Topological Organization of the Primate Cortical Visual System, *Nature* 358 (1992), 152–155.

M P Young, J W Scannell, M A O'Neill, C C Hilgetag, G Burns, C Blakemore: Non-metric Multidimensional Scaling in the Analysis of Neuroanatomical Connection Data and the Organization of the Primate Cortical Visual System, *Phil Trans R Soc Lond B* 348 (1995), 281–308.

144

DETAILED SIMULATION OF LARGE SCALE NEURAL NETWORK MODELS

Anders Lansner,[*] Örjan Ekeberg, Erik Fransén, Per Hammarlund, and
Tomas Wilhelmsson

SANS — Studies of Artificial Neural Systems
Department of Numerical Analysis and Computing Science
Royal Institute of Technology
S-100 44 Stockholm, Sweden

E-mail: {ala,orjan,erikf,perham,towil}@nada.kth.se
http://www.nada.kth.se/sans/

INTRODUCTION

Mathematical modeling and computer simulation is rapidly gaining in use as tools in neuroscience. Models from the sub-cellular level, over single neurons, to networks and systems of interacting networks are formulated and simulated. The complexity of cell models differs considerably, from simplified population models, *e.g.* mathematical oscillators, to complex single cell models, *e.g.* of the Hodgkin-Huxley type, with many compartments, ionic conductances and biochemical processes.

Simulations are computationally quite demanding. Large networks with thousands of detailed model neurons have been out of reach so far. Simplifications can be introduced in terms of reduced cell models and limited cell numbers. However, real cells vary considerably in their properties. For each cell type, some aspects of the firing characteristics are often found to be crucial for network function. This makes it undesirable to reduce the cell model too much. Moreover, the size, complexity and intricate structure of biological neural networks makes it hard to simplify and sub-sample substantially at the network level. Population units are not always easy to relate to quantities of the type one can measure in experiments. Together this makes it important to be able to simulate large networks of detailed model neurons.

---

[*]To whom correspondence should be addressed.

Computational Neuroscience
edited by Bower, Plenum Press, New York, 1997

SIMULATION SOFTWARE

Computing power is continually increasing. Today it is possible to simulate biologically detailed cells comprised of hundreds of ionic conductances and thousands of compartments on a single processor workstation. GENESIS and NEURON are two examples of widely used simulator packages [1, 7]. We have also developed a number of simulators of this sort. They have been developed primarily for simulations of very large network of thousands of detailed neurons and millions of synapses rather than very detailed single cell models. The most well known program is SWIM [2, 3], which is also used outside the SANS group. SWIM uses a compartmental cell model and a numerical method originating from Hines. Voltage dependent ion channels (Na, K, Ca, $K_{(Ca)}$ and NMDA) are modeled using Hodgkin-Huxley-like equations. There are two calcium pools ([Ca_{AP}] and [Ca_{NMDA}]) with different time constants. Synaptic interaction include conventional chemical as well as voltage dependent NMDA receptor gated type synapses.

SWIM was primarily targeted towards single-processor workstations running UNIX. It was later ported to multiprocessor workstations and CRAY vector computers [4]. The simulation engine has also been implemented on the TMC Connection Machine [6]. Based on these experiences we have recently developed a new simulation library, SPLIT [5]. This library is designed to simulate large networks, providing the same functionality across multiple hardware platforms, and to allow for easy addition of new objects *e.g.* model extensions. One important design goal was to isolate the hardware idiosyncrasies from the user; the same software interface is used across different platforms. SPLIT is currently running on single and multiprocessor workstations, CRAY vector computers and distributed memory systems such as the IBM SP2.

In SWIM the network was specified at the level of individual neurons and synapses. This proved to be inefficient for large networks. In SPLIT the network is specified in terms of populations and projections containing multiple neurons and synapses. On parallel machines the populations and projections are automatically split up and mapped to individual processors. On distributed memory systems it is important to map the network in such a way that overhead from communication is minimized. Spike messages are buffered and axonal and synaptic delays are used to hide communication latencies. Neurons having many short-delayed synaptic interconnections are kept together on the same processor.

EXAMPLES OF LARGE NETWORK SIMULATIONS

The software developed has been used to simulate locomotor burst generation in a model of the lamprey spinal cord, and an attractor model of cortical associative memory, with Hebb's cell assembly hypothesis as a starting point.

Simulations of intersegmental coordination in lamprey

The SWIM simulator was used extensively to model the spinal rhythm generation in the lamprey at the single cell and local CPG level, as well as with populations of neurons and intersegmental coordination in a distributed CPG with two main cell types, excitatory interneurons (EIN) and contralaterally caudally projecting interneurons (CCIN). Both types of cells were modeled with five compartments, seven ionic conductances and two calcium pools. The largest models simulated to date have been comprised of 2400 cells and 700 000

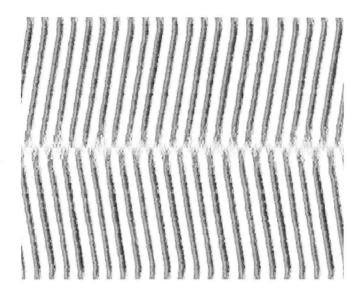

Figure 1. This figure shows the activity of the lamprey CPG network with 2400 cells. The network is driven by tonic bath excitation. Activity (soma membrane potential) is shown coded with high as dark and inactivity/inhibition as white. Activity in the rostral part is shown centrally and the caudal right and left parts upwards and downwards respectively. Time is along the horizontal dimension and is a 2 s simulation starting at time 1 s.

synapses. Simulations of 5 seconds on an 8 processor SparcServer took about 8 hours to perform [9]. This model produced an appropriate frequency range of 1:10 and reproduced several other aspects of lamprey spinal cord burst generation, for an example see figure 1.

Cortical associative memory simulations

The Columnar Model In the associative memory model the principal connectivity was derived from storing a number of binary patterns using a correlation-based learning rule. A columnar structure was added to obtain a more realistic connection structure [8]. Each column is composed of 12 regularly spiking pyramidal cells and three fast spiking inhibitory interneurons. The pyramidal cells have six compartments, ten ionic conductances, and three calcium pools while the interneurons have three compartments, six ionic conductances, and one calcium pool.

For a column network with 2535 cells and 127572 synapses a simulation of 500 ms with a time step of 50 μs took 169 minutes on a Sun SPARCstation-5 and 45 minutes on one IBM SP2 thin node. For an example of a simulation see figure 2. The table below shows timings and parallel speedup for multiprocessor execution on an IBM SP2 with thin nodes. For this size of the model, at least 16 nodes may be used efficiently. Larger models will scale up to larger number of processors.

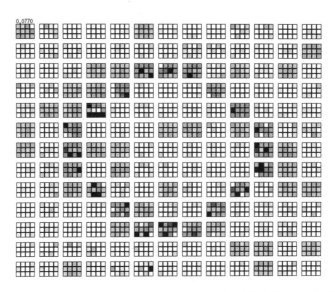

Figure 2. This figure shows one snapshot (77 ms after start of simulation) from a simulation of a 13x13 column network with a total of 2535 neurons and 127 572 synapses (kainate/AMPA and NMDA type). Only the 12 pyramidal cells in each column is shown. Twenty different memory patterns were stored. The pattern recalled looks like an "o". The activity is the result of a pattern completion process starting with a distorted input pattern similar to the memory recalled. Activity is coded with white for low activity to dark for high activity.

Nodes	Time (s)	Speedup
1	2708	1.0
2	1396	1.9
4	633	4.3
8	358	7.6
16	178	15.2
32	143	19.0

The Hypercolumn Model We are currently studying a cortical associative memory model composed of aggregates of columns, hypercolumns. With a higher order network of this type learning produces improved internal representations. The present network consists of 484 hypercolumns. Each such hypercolumn is a winner-take-all module comprised of eight of the columns mentioned above. The columns correspond to the eight different local features in the receptive field of the hypercolumn. This gives the network 58 080 cells and 4.4 million synapses. Simulating 100 ms required 4 hours on 8 wide SP2 nodes.

CONCLUSION

There is a need to simulate large neural networks while retaining details at the cellular level. Such simulations are indeed possible today using state-of-the-art computing resources. Simulators that take advantage of the parallelism inherent in neural network simulations may be implemented on powerful parallel computers. We have developed software which is used to simulate models with tens of thousands of cells and millions of synapses. Such simulations are valuable for establishing the relation between processes at the cellular and sub-cellular level on one hand and collective system phenomena on the other. We give two examples of large scale network models made possible by these tools: intersegmental coordination in the lamprey spinal network and associative memory in a cortical network with columnar organization.

REFERENCES

[1] James M. Bower and David Beeman. *The Book of GENESIS: Exploring Realistic Neural Models with the GEneral NEural SImulation System*. TELOS, Springer-Verlag, Santa Clara, CA, 1994.

[2] Örjan Ekeberg, Per Hammarlund, Björn Levin, and Anders Lansner. SWIM — A simulation environment for realistic neural network modeling. In Josef Skrzypek, editor, *Neural Network Simulation Environments*. Kluwer, Hingham, MA, 1994.

[3] Örjan Ekeberg, Peter Wallén, Anders Lansner, Hans Tråvén, Lennart Brodin, and Sten Grillner. A computer based model for realistic simulations of neural networks. I: The single neuron and synaptic interaction. *Biol. Cybern.*, 65(2):81–90, 1991.

[4] Per Hammarlund and Örjan Ekeberg. Large neural network simulations on multiple hardware platforms, 1996. (Submitted.).

[5] Per Hammarlund, Örjan Ekeberg, Tomas Wilhelmsson, and Anders Lansner. Large neural network simulations on multiple hardware platforms, 1997. (This volume).

[6] Per Hammarlund, Björn Levin, and Anders Lansner. BIOSIM — A program for biologically realistic neural network simulations on the Connection Machine. In Teuvo Kohonen, Kai Mäkisara, Olli Simula, and Jari Kangas, editors, *Artificial Neural Networks*, pages 1477–1480, Espoo, Finland, June 24–28 1991. Elsevier, Amsterdam. Proc. ICANN-91.

[7] Michael Hines. The NEURON simulation program. In Josef Skrzypek, editor, *Neural Network Simulation Environments*. Kluwer Academic Publishers, Norwell, MA, 1993.

[8] Anders Lansner and Erik Fransén. Improving the realism of attractor models by using cortical columns as functional units. In James M. Bower, editor, *The Neurobiology of Computation: Proceedings of the Third Annual Computation and Neural Systems Conference*, pages 251–256, Monterey, CA, July 21–26 1995. Kluwer, Boston, MA.

[9] Tom Wadden, Jeanette Hellgren-Kotaleski, Anders Lansner, and Sten Grillner. Simulations of intersegmental coordination using a continuous network model. In James M. Bower, editor, *The Neurobiology of Computation: Proceedings of the Third Annual Computation and Neural Systems Conference*, Monterey, CA, July 21–26 1995. Kluwer, Boston, MA.

145

TETRODES FOR MONKEYS[*]

J. S. Pezaris, M. Sahani, and R. A. Andersen

Computation and Neural Systems
Division of Biology
California Institute of Technology
Pasadena, California 91125

ABSTRACT

We have adapted the Recce-O'Keefe tetrode for use in monkey cortex. Eleven penetrations have been made over macaque area MT to evaluate the mechanical and electrical properties of our design. Experiments have been made with 15 and 25 μm insulated nichrome wire running through oil-filled 33 gauge stainless steel guide tubes. Electrically, the tetrodes compare favorably with traditional tungsten electrodes for single-unit isolation, and are superior for multi-unit recordings. Preliminary histology shows straight tracks at up to 12 mm tetrode extension.

1. INTRODUCTION

The primary contribution of this paper is to show that the Recce-O'Keefe tetrodes[1] can be used to successfully record from monkey cortex, and to detail the parameters and techniques used in that process. The design of our tetrodes is presented, including specific construction details, followed by the description of experiments performed to measure their characteristics. We close with a list of intended design modifications and follow-up experiments.

1.1. Tetrodes and Spheres of Sensitivity

A tetrode is a bundle of four individually insulated fine wire electrodes, whose tips lie closer together than their respective spheres of sensitivity. An exact analogy can be drawn between neural recordings obtained from such a device and quadraphonic musical recordings. A

---

[*] This paper was presented in poster form at CNS*95, but did not appear in that year's proceedings due to an administrative error. By agreement with the editor, it appears here in the proceedings for CNS*96.

Computational Neuroscience
edited by Bower, Plenum Press, New York, 1997

neuron that lies in the overlap of two or more of these spheres is detected by two or more electrodes. Assymetries, whether in electrode construction or relative geometry, insure that different electrodes record this neural signal through slightly different filters. The comparison of signal across channels allows the disambiguation of signals that appear identical to a single electrode. Later in the paper, a tetrode recording is presented which shows this effect.

1.2. Animal Preparation

Previous reports describing tetrodes have concerned themselves with rat[1,2] or cat[3] preparations. Two features of our awake monkey preparation constrained our design. First, while a recording chamber is chronically implanted, the surface of the brain is exposed and electrodes are inserted and removed during each recording session. Thus, the dura mater must remain intact and is subject to toughening. Second, the brain of the macaque is larger than those of either the rat or the cat and profoundly gyrated in comparison. Thus even cortical areas can lie quite far from the exposed surface. A mechanism is therefore required to penetrate the dura and deliver the delicate tetrode wires to relatively deep neural structures, while minimizing tissue insult. Our approach is described below.

We initially chose a method which uses a carrier tube to hold the tetrode wire within the brain (Mk I), combined with a normal guide tube to puncture the dura. A modified design under development combines the two functions (Mk II). These two designs are described in detail below.

2. METHODS

2.1. Tetrode Design, Mk I

The basic design is a four-conductor twisted electrode running in a carrier tube. 15 or 25 μm insulated nichrome (Reid or Sigmund Cohn, respectively) is used to construct the tetrode bundle. To penetrate dura, a 21 gauge stainless steel guide tube is used, inside of which a *carrier tube* assembly tapers to a 33 gauge oil-filled cannula holding the tetrode bundle. Conductive silver paint is used to electrically and mechanically connect the wires to a four-pin connector mounted on the carrier tube. Thus, the tetrode bundle runs from the recording end, up through the carrier tube, and exits in a loop which terminates on the connector. See Figure 1. Construction is described in the following paragraphs.

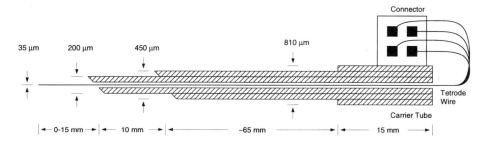

Figure 1. Tetrode Mk I, section. This carrier design is extended a few millimeters beyond the end of a guide tube (not shown) which has punctured the dura. The tetrode bundle, free to move inside the oil-filled center channel, is advanced into the area under investigation. Drawing not to scale.

The tetrode wires are twisted in a motorized jig. Once twisted, the four wires are glued together using cyanoacrylate glue, allowed to dry, carefully cut free of the jig, and set aside during the construction of the carrier tube.

The carrier tube is constructed by gluing three lengths of cannula, 9 cm of sharpened 33 gauge, 8 cm of sharpened 26 gauge, and 1.5 cm of 21 gauge, flush at the blunt ends with quick-setting epoxy. The four-pin connector is glued transversely to the outer cannula. The assembly is designed to withstand gentle handling and clamping in the microdrive while providing a small tip cross-section to minimize tissue insult. See Figure 1.

Once the glue has set, the carrier tube is reamed with music wire, cleaned with acetone, and filled with silicone oil. Oil-filling the carrier tube is important to increase construction yield and reduce the chances of kinking during use. The tetrode bundle is loaded into the carrier tube, and the free ends carefully attached to the four-pin connector using conductive silver paint (Silver Print, GP Electronics, Rockford, IL). No step is taken to remove the wire insulation: adequate electrical connection is made through the metal exposed at the end of the wire when it is cut. The recording end of the tetrode bundle is cut blunt with fine scissors, cleaned in isopropanol, and plated in 5% gold chloride ($AuCl_4$) solution for approximately 20 µA-s per tip. This yields a 1 kHz tip impedance of 0.1–0.2 MΩ at plating time, as measured on a BAK impedance tester.

2.2. Tetrode Design, Mk II

Our improved design, incompletely tested in the current experiment, simplifies use and increases reliability. Three nested lengths of cannula are again glued to form a carrier tube. The tetrode bundle is attached to a second sleeve of layered cannula which forms a cap that slides over the lower carrier. With the tetrode withdrawn, the carrier is advanced through the dura but short of the brain, and the tetrode extended under hydraulic control. This approach obviates the need for a separate guide tube, as required for Mk I. See Figure 2.

2.3. Equipment

Our experimental setup is as follows. A hydraulic microdrive (Fred Haer Corp, Brunswick, Maine) is used to position the tetrodes. Tetrode signals are amplified by a custom four-channel headstage amplifier ($A = 100$) feeding a custom four-channel variable-gain preamplifier ($A = 1$ to 5000, nominally 500). The preamplifier feeds anti-alias filters ($f_c = 6.4$ kHz, Tucker-Davis Technologies, Gainsville, Florida) and four-channel instru-

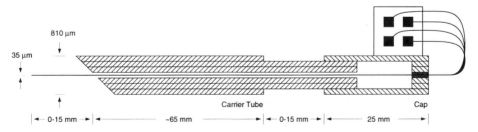

Figure 2. Tetrode Mk II, section. The carrier tube punctures the dura directly, but is not advanced into the brain. The cap is advanced over the carrier, forcing the tetrode wire, which has been fixed to it, into the brain. Drawing not to scale.

mentation-grade A/D (f_s = 51.2 kHz, decimated to 12.8 kHz, also TDT). Data are collected on an IBM-PC compatible and analyzed on Sun computers using custom software.

2.4. Experiment

In this experiment, we sought sought to verify that recordings could be made using this technology in semi-chronic monkey preparations. Subsequent to this verification, the following issues were explored: wire size, tip plating, penetration straightness, and tissue insult.

Eleven penetrations were made over the period of a week. Six penetrations were made with 15 μm wire tetrodes, three with 25 μm wire tetrodes, and two with traditional tungsten electrodes as controls. All penetrations were made in the left hemisphere of monkey 92–13 in an awake preparation. Penetrations were made to tetrode extensions of 4–12 mm, limited by variations in the mechanical designs of the four tetrodes used. Although recordings were made from all penetrations, electrical problems confounded the recordings from two. Some penetrations were made without leaving lesions, others had a pair of electrolytic lesions made 2 mm apart, and others had lesions made every 500 μm; all lesions were made during electrode withdrawal, after recording had finished. After the last day of recording, the brain was perfused using standard techniques.

3. RESULTS

An example tetrode recording is shown in figure 3. This is of particular interest because it clearly shows one unit (**a**) detected on channel 3, a second unit (**b**) on channel 2, and both on channel 1 at approximately equal amplitude, illustrating the monaural-versus-quadraphonic analogy suggested earlier. Two additional units (**c, d**) were detected in this recording. Notice how similar three of the four waveforms are on channel 1, making it very difficult to disambiguate between them given only the information from that trace.

3.1. Electrical Performance

Analysis of tetrode electrical performance was primarily comparative. Peak-of-signal-to-noise ratios of approximately 14:1 were found for single-unit isolation. *In situ* im-

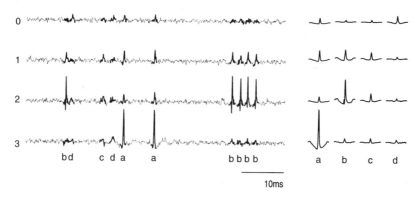

Figure 3. Example Tetrode Recording. On the left is a short stretch of the four channels of data. Heavier sections represent detected events. Letters signify events from four presumably different units. On the right are the averages of all events from each of the four identified cells.

pedance at 1 kHz was consistently 0.5–0.7 MΩ for plated electrodes, 1.0–2.1 MΩ for non-plated electrodes. This compares to 23:1 and 0.8 MΩ for the two control penetrations. Non-rigorous examination of recorded data reveals multiple distinct cells per stable recoding location for tetrodes.

3.2. Wire Diamemeter

No significant differences have been found between the various recordings or penetrations based on wire diameter. Further exploration may be necessary to clearly elucidate any distinctions.

3.3. Tip Plating

Plating the electrode tips is essential to producing good recordings. This is clearly depicted by comparing the spectral impedance of an unplated and plated tetrode tip, as seen in Figure 4. This is an area of active pursuit.

3.4. Penetration Straightness and Tissue Insult

We have histologically identified all marked penetrations and have determined that the tetrode wires ran straight and the tracks are not unusual. The deepest tetrode penetration, some 12 mm, included 12 lesions which lie in a straight line, even at the maximum extension. Preliminary comparison of tracks between the control penetrations and the two wire diameters show that all three are comparable. That is, the tetrode wire caused no additional tissue insult.

4. SUMMARY

We have described the design and construction of a tetrode design appropriate for use in the semi-chronic monkey preparation. Experiments were performed to verify the mechanical and electrical performance of the tetrodes which showed them to be functional and adequate for simultaneously detecting multiple neighboring cells in monkey cortex.

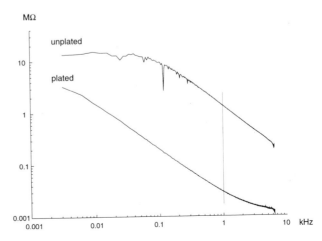

Figure 4. Impedance Curves. This graph shows a spectral analysis of tip impedance for a typical tetrode wire. The upper curve is from a freshly cut tip in saline (the spike at 120 Hz is due to power line interference). The lower curve is from the same tip after gold plating.

5. REFRENCES

1. Recce, M. L., and O'Keefe, J., "The tetrode: an improved technique for multi-unit extracellular recording," *Society of Neuroscience Abstracts*, **15**(2), p. 1250 (1989).
2. Wilson, M. A., and McNaughton, B. L., "Dynamics of the hippocampal ensemble code for space," *Science*, **261**(5124), pp. 1055–1058 (1993).
3. Gray, C. M., et al, "Tetrodes markedly improve the yield and reliability of multiple single unit isolation from multiunit recordings in cat striate cortex," *Society for Neuroscience Abstracts*, **20**(1), p. 625 (1994).

146

INTERNET BRAINS: COMBINING NEURONAL SIMULATION AND ROBOTS

Chris J. Roehrig

UBC Department of Computer Science
2366 Main Hall
Vancouver, BC
Canada V6T1Z4

ABSTRACT

The DSS protocol is a general-purpose mechanism for combining realistic neuronal simulation and robots without specialized robot programming. It permits a neuronal simulation to be distributed across multiple computers with minimal programming effort. Because it uses the actual time as a reference, it can be used to synchronize simulations in order to interact with robot sensors or tissue recordings in real time. In addition, because it allows each part of the simulation to run as an independent entity, it can be used to construct arbitrarily large simulations by using many networked computers. DSS is limited by network speeds and computer processing speeds and scales well as both increase.

1. INTRODUCTION

Robots are being used increasingly in computation neuroscience to provide behavioral grounding for simulations. Robot implementations of cockroach locomotion [1], [8] have provided insight that might not have been apparent from simulations alone.

We are developing a robot system for modelling the *C. elegans* tap withdrawal reflex [9]. *C. elegans* makes a good model system for robot neuroethology because it has well characterized behaviors [11], its nervous system has been completely mapped [10], and it has simple cells that are amenable to real-time modelling [12].

Current methods for combining robots with neuronal simulation require specialized expertise that most computational neuroscientists do not have. This paper describes a general-purpose method we are developing for doing robot neuroethology that doesn't require specialized robot programming.

Computational Neuroscience
edited by Bower, Plenum Press, New York, 1997

Figure 1. Robot System Overview

2. OVERVIEW

The basic system consists of a Lego robot tethered to a network of workstations (see Figure 1).

Most general-purpose robot controller chips are not as powerful as the typical workstations that are used by computational neuroscientists. Moreover, robot controller chips need to be programmed specifically for the problem at hand and require specialized programming skills; there is no general-purpose neuronal modelling software such as GENESIS [2] for robot controller chips. To make it accessible to neuroscience modellers, we therefore decided upon a tethered system. The robot uses an inexpensive microcontroller to govern all sensory and motor information. It is attached by a cable to a general-purpose UNIX workstation that performs the actual neuronal simulations. We chose the MIT Miniboard as an inexpensive and readily available controller and we have written software to operate it as a tethered system. This arrangement frees the modeller from the need to do any robot programming.

We constructed a general-purpose behavioral robot made of Lego based on a tank chassis. A tank chassis was chosen because it is simple to construct a reliable wheeled locomotion system that includes proprioceptive feedback with minimal components, and because its locomotion is "monotonic": it is capable of orienting to any direction whether moving or resting. This lends itself to simple biological behaviors better than other wheeled chassis which require three-point turns to reorient themselves.

In order for the system to scale to large circuits in real time, we decided on a distributed architecture. The system was designed to use inexpensive and readily available workstations such as PC's running Linux or FreeBSD, connected via TCP/IP (Internet), and was designed to scale well with future workstation and networking speeds. Networks based on TCP/IP are robust and scale well to vast proportions as the phenomenal growth of the Internet attests. There is a natural correspondence between networks of interconnected neurons and a distributed network of computers simulating those neurons and we are intrigued by the possibility of using computer networks to scale real-time neuronal simulation to brain-sized dimensions.

We have encapsulated the network signal transmission mechanism into a network protocol and associated programmer libraries we call DSS (Digital Signal Sockets). Fundamental to its design is the notion of a signal — the data that is communicated across the network. A signal is a generic representation of a physical quantity that changes over time, such as membrane potential or calcium concentration. A design goal was to make a generic, general-purpose network interface for a signal that is independent of the transmission and coding scheme — in effect, a programmatic interface to a continuous or analogue signal. The receiver can use the signal independently of knowing how it originated, whether it was generated by the solution to a system of ordinary differential equations (ODE's), or whether it is a live signal from a robot sensor or a tissue recording.

Because the system is designed for real-time operation while interacting with live external signals, the time reference is the actual "wall clock" time. This implies that the independent variable t in a system of ODE's is really the actual time. In order for multiple computers to exchange real-time signal data, they must all agree on the actual time. In our system, the internal clocks of all computers are synchronized by exchanging network packets containing timestamps. The variable network latencies inherent in Internet transmission are accounted for by statistical averaging of packet return times. Over time, computers can be synchronized to the accuracy of their own internal clocks using this technique. We are currently using NTP, a network protocol and set of programs that use this technique [6].

3. SAMPLING AND NETWORK TRANSMISSION

Sampling theory [7] states that a signal can be represented exactly by its discrete samples provided that the sampling frequency is at least twice that of the highest frequency component of the signal. Typically, passive membrane potentials have frequency components in the range of the reciprocal of their membrane time constant; spiking cells have higher frequency components. The sampling rate need not have any relationship to the integration step size of the system of differential equations. The integration step size is normally chosen to produce a suitably accurate and stable solution and may be at a much higher rate than needed to accurately represent the computed signal.

These samples are numerical data which are inserted in network packets addressed to the destination computer and sent on the computer network. Each signal packet contains a timestamp indicating the time at which the enclosed samples were taken. This allows the receiver to accurately reconstruct the signal independently of network transmission delays.

Computer networks have bandwidths many times that of typical neuronal signals, allowing thousands of low-bandwidth neuronal signals to be multiplexed on a single network cable. Each signal is transmitted over the network cable as a stream of individual packets interleaved with thousands of other packet streams, each with their own source and destination. Because the bandwidth of the network cable is finite, the number of signal streams on one cable is limited, and this limits the interconnectivity of the simulated neuronal network. However, it does not limit the total number of cells in the network and it should be possible to create arbitrarily large network simulations using this technique.

Network transmission latency affects the kinds of signals that can be communicated in real time and this will affect the way simulations can be distributed. Long network delays mean the signals must have low bandwidth or have long delays associated with them. Since neuronal signals travelling long distances have similar restrictions, this implies that the topology of a network simulation must in some sense reflect the anatomy of the neuronal system.

4. SIGNAL RECONSTRUCTION

There are two requirements for signal reconstruction: prediction and resampling. The receiving end of a signal is typically a device that interacts with the world (e.g. a robot actuator), or a numerical simulation that uses the signal as an input. Often, the value needed is the current value of the signal, to drive an actuator for instance. The typical state of a numerical simulation is that it has just finished computing results for time t_{n-1} which is a time now in the past, and is working on computing results for the next time step, t_n, which is a time in the immediate future. To solve systems, numerical algorithms often need to evaluate their equations at different time values between t_{n-1} and t_n in order to produce an accurate result for time t_n. This necessitates having values for the input signal for times between t_{n-1} and t_n. Because time t_n is still in the future, samples for time t_n cannot have arrived yet at the receiver. Therefore, some way of interpolating and predicting values for the signal is needed.

Frequently, the receiver of a signal needs to use the signal at different sampling intervals. For instance, our Miniboard robot acquires data from its sensors every 50 milliseconds. However, this data is used in a system of equations that are integrated with a 1 millisecond step size. Some method of interpolating (or "upsampling") the sensor data is required to avoid large discontinuities every 50 integration steps.

A continuous signal can be completely reconstructed from its samples provided the sampling rate is twice the signal bandwidth. The signal reconstruction formula is a doubly infinite sum:

$$x(t) = \sum_{n=-\infty}^{\infty} x_n \frac{\sin(\pi(t-nT)/T)}{\pi(t-nT)/T}, \tag{1}$$

where T is the sampling period. In practice, the infinite sum is truncated and there are many techniques for efficiently computing it using digital reconstruction filters [7]. Digital reconstruction filters provide the means to accurately interpolate the incoming signal and provide, in essence, an interface to an analogue signal.

A digital reconstruction filter imposes a delay in the signal. This delay has nothing to do with network transmission delays or computer processing times. It is inherent in the reconstruction process and is roughly one half the product of the "width" in samples of the symmetrically truncated sum in (1), and the sampling period of the signal.

There are several ways of dealing with this delay in order to obtain the current value of the signal. First, we can use this delay to model some physical aspect of the system, such as the synaptic delay between the change in presynaptic potential and the resulting postsynaptic current, or spike propagation delay along an axonal process. The spike generated at the axon hillock is a signal $u(t)$ which is propagated to the presynaptic site with some fixed delay δ. The signal at the presynaptic site could therefore be modelled as $v(t) = u(t-\delta)$. As long as δ is greater than the delay incurred by the signal reconstruction, there is no problem in reconstructing the signal $v(t)$ at the receiver.

Another way of dealing with the delay is to extrapolate the input signal to predict its current value. The prediction scheme therefore becomes part of the numerical solution technique and will introduce errors in the computed solution. By knowing the form of the signal and sampling it at higher rates than necessary ("oversampling"), it may be possible to do this prediction within error tolerances. For example, many biophysical processes are modelled as rate equations of the form

$$\tau \dot{h} = f(t) - h, \tag{2}$$

where τ is the rate constant for the process. Hodgkin-Huxley channel kinetics are an example [4]. We are currently investigating ways which $h(t)$ can be reconstructed and extrapolated at the receiving end from past samples of $f(t)$.

5.   DIGITAL SIGNAL SOCKETS

The DSS network protocol and programmer libraries perform all the necessary signal transmission and reconstruction.   The network protocol describes the format of the DSS packets and how they are used over network transport layers such as TCP/IP [3] and IEEE 1394 ("Firewire"), an emerging standard for a high-speed network that is particularly well suited for real-time applications [5].   The current version uses UDP [3] over IP for its low transmission overhead.

The DSS libraries provide an application programmer's interface ("API") for sending and receiving signals and managing connections. At the core of the API are the `dss_write` and the `dss_read` functions. The `dss_write` function takes a value and a timestamp and sends the data to its destination where it is collected in a buffer. The `dss_read` function is used by the simulation at the receiving end. It takes a timestamp parameter and returns the signal's value for the requested time. It performs any required interpolation or prediction and maintains an error estimate for the returned value. In addition, it can be configured to synchronize the simulation to real time in a simple manner by waiting until the appropriate time to return. Thus, existing neuronal simulations can easily be converted to distributed, real-time operation by using the DSS library.

6.   COMPUTER AND NETWORK SPEED ISSUES

The above discussion assumes that the network transmission and computer processing occurs sufficiently quickly; i.e. that the network can transmit a sample and the computer can use it to compute the simulation state for the next sampling interval, all before the next sampling interval actually arrives. For sufficiently low-bandwidth signals (such as potentials in *C. elegans* passive cell membranes), this will be the case. Faster computers and networks will permit higher bandwidth signals. If processing is too slow for real-time performance, the simulation can optionally slow to some percentage of real time. The DSS libraries monitor the load and provide an indication of real-time performance.

7.   CONNECTION MANAGEMENT

A large simulation may contain many signals and computers. The DSS protocol includes the means to identify and manage these connections. When a signal socket (input or output) is created, it is named and its name and network address is registered with a database so it can easily be found. Connections are established and managed via DSS control messages which are particular kind of DSS network packets. This allows for centralized connection management from a remote computer. We are developing a DSS object for the Java programming language to permit graphical connection management via a standard web browser.

8.   CONCLUSION

The DSS protocol is a general-purpose mechanism for carrying live, continuous (analogue) signals over digital packet networks such as the Internet. It permits a neuronal simulation to be distributed across multiple computers with minimal programming effort. Because it uses the actual time as a reference, it can be used to synchronize simulations to real time in order to interact with robot sensors or tissue recordings. In addition, because it allows each part of the simulation to run as an independent entity, it can be used to construct arbitrarily large simulations by using many networked computers. DSS is limited by network speeds and computer processing speeds and scales well as both increase. DSS can be found at `http://www.crispart.com/dss`.

REFERENCES

[1] R. D. Beer and H. J. Chiel. Simulations of cockroach locomotion and escape. In R. D. Beer, R. E. Ritzmann, and T. McKenna, editors, *Biological Neural Networks in Invertebrate Neuroethology and Robotics*, pages 267–285. Academic Press, 1993.

[2] J. M. Bower and D. Beeman. *The Book of Genesis*. Springer-Verlag, 1995.

[3] R. Braden. Requirements for Internet hosts – communication layers. RFC 1122, October 1989.

[4] A.L. Hodgkin.and A.F. Huxley. A quantitative description of membrane current and its applicaton to conduction and excitation in nerve. *J. Physiol. (London)*, 117:500–544, 1952.

[5] IEEE standard for a high performance serial bus, 1995. Institute for Electrical and Electronic Engineers, Standard No. 1394-1995.

[6] D. L. Mills. Network time protocol (version 3) specification, implementation and analysis. RFC 1305, March 1992.

[7] A. V. Oppenheim and R. W. Schafer. *Discrete-Time Signal Processing*. Prentice Hall, 1989.

[8] R. D. Quinn and K. S. Espenschied. Control of a hexapod robot using a biologically inspired neural network. In R. D. Beer, R. E. Ritzmann, and T. McKenna, editors, *Biological Neural Networks in Invertebrate Neuroethology and Robotics*, pages 365–381. Academic Press, 1993.

[9] C. H. Rankin, C. D. O. Beck, and C. M. Chiba. *Caenorhabditis elegans*: a new model system for the study of learning and memory. *Behavioral Brain Research*, 37:89–92, 1990.

[10] J. G. White, E. Southgate, J. N. Thomson, and S. Brenner. The structure of the nervous system of the nematode *Caenorhabditis elegans*. *Philos. Trans. R. Soc. Lond. (Biol.)*, 314:1–340, 1986.

[11] S. R. Wicks and C. H. Rankin. Integration of mechanosensory stimuli in *Caenorhabditis elegans*. *Journal of Neuroscience*, 15(3):2434–2444, 1995.

[12] S. R. Wicks, C. J. Roehrig, and C. H. Rankin. A dynamic network simulation of the nematode tap withdrawal circuit: Predictions concerning synaptic function using behavioral criteria. *Journal of Neuroscience*, 16(2):4017–4031, 1996.

147

MODELING BRAIN IMAGING DATA WITH NEURONAL ASSEMBLY DYNAMICS

M.-A. Tagamets[*] and Barry Horwitz

Laboratory of Neurosciences
National Institutes on Aging, National Institutes of Health
Bethesda, Maryland, 20982
mally@cs.umd.edu
horwitz@alw.nia.nih.gov

INTRODUCTION

Brain imaging methods, such as positron emission tomography (PET), are widely used to measure functional activity in human subjects. These methods provide data simultaneously about activity in many areas of the brain while a subject performs a cognitive task. Imaging studies in which regional cerebral blood flow (rCBF) is used as an index of neural activity typically are designed for identifying regions of brain that are involved in specific aspects of cognitive tasks, but the question of how these areas interact as a dynamic network to produce such data is poorly understood (Horwitz and Sporns, 1994). The goal of this work is to better understand how neuronal dynamics relate to rCBF imaging data.

A major problem in interpreting brain imaging data lies in the fact that these data do not directly reflect neuronal activity, but rather measure changes in blood flow, which are presumed to occur in reaction to changes in a brain area's electrical activity. It is generally thought that increased blood flow is primarily due to the energy demands of afferent synaptic activity (Sokoloff, 1993; Jueptner and Weiller, 1995) rather than the activities of neurons, such as action potentials. A consequence of this is that it is possible that both excitatory and inhibitory inputs can increase the measured rCBF in an area, even though excitation increases spiking activity of pyramidal neurons, while inhibition decreases it. An increase in glucose metabolism, which is known to reflect local energy consumption, has been shown to occur in the hippocampus after inhibition by prolonged stimulation of the fornix, for example [Ackermann, 1984]. It is likely that the changes in rCBF depend on relative proportions of excitatory and inhibitory connections, both within and afferent to

---

[*] to whom correpsondence should be addressed

an area. The question arises as to how the interaction between local and global connectivity affects the total synaptic activity in an area. If, for example, there were no local connectivity at all, then it would be expected that any afferent activity, excitatory or inhibitory, would increase local energy demands, and thus blood flow. In order to realistically model brain imaging data, these factors need to be taken into account.

METHODS

In order to study the potential effect of specific connectivity patterns on PET-measured rCBF, we have implemented a large-scale neural model composed of simulated neuronal assemblies that performs a simple object-matching task similar to those designed for PET studies. Here we describe the basic framework of the model and show how different patterns of afferent activities can affect simple circuits with known proportions of local excitation and inhibition. The basic unit of the model consists of one excitatory element and one inhibitory element, an example of a Wilson-Cowan unit, and is shown in Figure 1. Each unit represents a local population of neurons, such a cortical column.

It is known that most synapses in the cortex are excitatory [Braitenburg and Schòz, 1991]. Furthermore, recent evidence from cat visual cortex indicates that a majority of synapses originate locally and most are excitatory connections onto excitatory neurons [Douglas et al., 1995]. These results have given impetus to the idea of "amplification", in which relatively small inputs can result in high activation due to the local self-excitatory connectivity. From the data of Douglas et al. we can derive the expected proportions of different types of local and afferent connections for use in a model of Wilson-Cowan units. Excitatory-to-excitatory connections account for most of the local connectivity, and excitatory-to-inhibitory and vice versa together make up less than half the percentage of synapses. Afferent connections comprise about 10–20% of all synapses within an area. The exact proportions used are based on the data mentioned above and are shown in Figure 1.

To examine the changes due to the balance of afferent activity, simulations were performed on a small network as shown in Figure 2. There is one 9x9 group of each of the excitatory and inhibitory units, connected as shown in Figure 1. A fixed level of input excitation is shown by the connections coming from the input array on the left side of each area in the figure. This represents activity coming from some specific pathway, such as sensory input. Additional excitation or inhibition is provided by diffuse connections, shown on the right, that

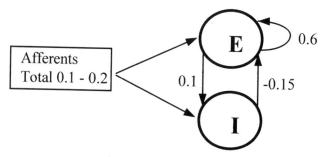

Figure 1. The basic unit used in the model is an example of a Wilson-Cowan unit and represents a local assembly, such as a cortical column. The weights used in our model reflect the estimate that about 60% of local connections are excitatory-to-excitatory, about 15% and excitatory-to-inhibitory and vice versa, and only about 10–20% of all connections come from other areas.

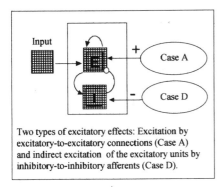

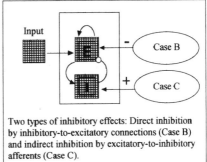

Two types of excitatory effects: Excitation by excitatory-to-excitatory connections (Case A) and indirect excitation of the excitatory units by inhibitory-to-inhibitory afferents (Case D).

Two types of inhibitory effects: Direct inhibition by inhibitory-to-excitatory connections (Case B) and indirect inhibition by excitatory-to-inhibitory afferents (Case C).

Figure 2. Network used for the simulation study. The single area consists of a 9x9 excitatory population (E) and a 9x9 inhibitory population (I) of elements. Pairs of E-I units at the same locations in the area are connected as shown in the basic unit of Figure 1. The input area is shown on the left of the E-I areas. Modulatory afferents are shown on the right in ovals. Connections from the input area are topographically arranged and are all excitatory. The afferent connections are weak and synapse onto all units equally. During one simulation run, only one of the four possible afferent types (i.e. A, B, C or D) is used. The input units are clamped in an L-shaped pattern to one of three levels and afferents are systematically increased from 0 to 9% of the total activity coming into the E-I area. Each possible combination is run for 200 iterations, during which both the neuronal activities and blood flow activity are computed.

project equally to all units in the area. This acts as a modulatory influence on the area. We investigated all four possible combinations of afferent modulatory activation, as shown in Figure 2: excitatory-to-excitatory (Case A), inhibitory-to-excitatory (Case B), excitatory-to-inhibitory (Case C), and inhibitory-to-inhibitory (Case D). For each case we examined both the blood flow and the mean electrical activity under different levels of modulatory activity. Simulations were run by presenting a pattern to the input area, while providing increasing levels of modulatory activity. The input activity was set to one of three possible levels of excitation: a low level, which is essentially a background level, and medium and high activations of an L shape on the input array. These simulate three different intensities of incoming activity via this pathway. 10 different levels of modulatory feedback are tested, ranging from 0 to 9% of the total afferent activity in the areas. Thus we examined a total of 30 different conditions for each type of afferent connectivity: 10 levels of afferent activity for each of the 3 levels of excitatory input activity.

Electrical activity of the neuronal elements is computed as a sigmoidal function of the weighted, summed inputs and characterizes a neuronal firing level of the assembly that each unit represents. Mean electrical activity is computed over the duration of each simulated trial by summing the activation of the units in the excitatory group. This corresponds to the pyramidal neurons whose activities are most frequently measured in single-unit studies of cortical encoding properties. Simulated blood flow is computed by summing the absolute values of all synaptic activity in both the excitatory and inhibitory groups. This corresponds to brain imaging data such as PET, which has a spatial resolution of about $1cm^2$ and includes both types of neurons.

RESULTS

The results of these simulations are shown in Figure 3. Solid lines are the blood flow activity, while dashed lines with O's show the mean electrical activity within the area. Af-

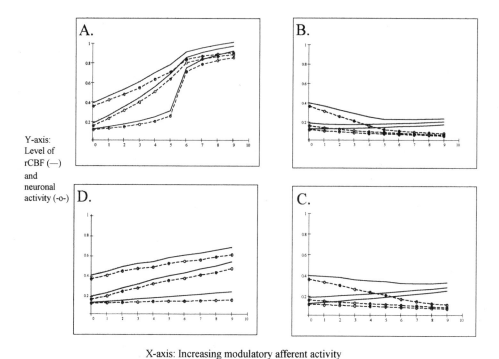

Y-axis:
Level of
rCBF (—)
and
neuronal
activity (-o-)

X-axis: Increasing modulatory afferent activity

Figure 3. The effects of different types of afferent activity on simulated blood flow and mean spiking activity in the area shown in Figure 2. Each part of the figure shows the effect of increasing one of the four types of afferent activity: excitatory-to-excitatory (A), inhibitory-to-excitatory (B), excitatory-to-inhibitory (C) and inhibitory-to-inhibitory (D). Solid lines: simulated blood flow. Dashed lines with O's: Mean neuronal electrical activity in the excitatory area. X axis: Level of afferent activity, ranging from 0 to 9 % of total activity within the units. The Y-axis indicates normalized levels of blood flow and electrical activity. Three different levels of activity from the input were tested, and the two curves depicting the blood flow and electrical activity at each level begin at approximately the same level on the Y-axis (low level inputs are shown by the curves that begin at about 0.1 on the Y-axis, the medium level begins at about 0.2 and the high level begins around 0.4).

ferent excitation causes a rise in both blood flow and electrical activity in this model, as expected. However, the effect of afferent inhibition depends on the existing level of excitation. With high levels of incoming activity from the input area, increasing inhibition causes a decrease in both the computed blood flow and the electrical activation. At lower levels of excitation, though, there is a dissociation between the blood flow, which tends to go up, and the electrical activity, which goes down. It is worth noting that inhibitory afferents result in much less pronounced changes in blood flow than do excitatory afferents.

SUMMARY AND DISCUSSION

This type of experiment is relevant for interpreting and modeling the results of a PET study, since frequently in these studies the sensory input is the same in two tasks but the cognitive tasks differ (e.g. Corbetta et al., et al., 1991). Presumably the differences in cognitive tasks are mediated by some form of feedback modulation of the sensory inputs. It is often the goal of an imaging study to identify such modulatory effects and their sources.

These simulations indicate that the local blood flow response depends on a number of factors, including: (1) the balance of afferent inputs; (2) the specific type of inhibition, i.e. direct inhibition of excitatory neurons or indirect inhibition by excitation of local interneurons populations; (3) the balance of connections in the local circuitry; and, (4) the existing level of activity in the area.

Both cases in which the modulation produced excitation of the excitatory population (Cases A and D) showed that neuronal activity and rCBF increased at about the same rate. This is consistent with the finding that local glucose metabolism increases approximately linearly with increasing spiking activity [Sokoloff, 1993]. It is interesting that in this model the rCBF and electrical activity change are similar, even though excitation is effected by different means in Cases A and D. On the other hand, the cases in which the net effect is inhibition (Cases B and C) both show a dissociation between blood flow and electrical activity, and in both cases the net change is much smaller than in the previous cases. This suggests that with the local connectivity balance that appears to exist in the cortex, inhibitory afferents tend to change local synaptic activity to a much lesser degree than excitatory afferents. This would be expected if the great majority of local connections are excitatory. Thus it is likely that areas subject to inhibitory modulation may not show appreciable changes in blood flow, even though there is substantial modulation of spiking activity in pyramidal neurons.

REFERENCES

Ackermann RF, Finch DM, Babb TL, Engel J Jr , (1984) Increased glucose metabolism during long-duration recurrent inhibition of hippocampal pyramidal cells. *J. Neurosci.* V. 4 (251–264).

Braitenburg V and and Schòz, (1991), *Anatomy of the Cortex. Statistics and Geometry.* Springer-Verlag, Berlin. 1991.

Corbetta M, Miezin FM Dobmeyer S, Shulman GL, and Petersen SE, (1991). Selective and divided attention during visual discriminations of shape, color, and speed: Functional. *J. Neuroscience*, V. 11 (2383–2402).

Douglas RJ , Koch C, Mahowald M, Martin KAC, and Suarez HH, (1995) Recurrent excitation in neocortical circuits. *Science*, V. 269, 18 August 1995 (981–984).

Horwitz B and Sporns O, (1994), Neural modeling and functional neuroimaging. *Human Brain Mapping*, V. 1 (269–283).

Jueptner M. and Weiller C, (1995) Review: Does Measurement of Regional Cerebral Blood Flow Reflect Synaptic Activity?—Implications for PET and fMRI. *Neuroimage*, V. 2 (148–156).

Sokoloff L, (1993) Sites and mechanisms of function-related changes in energy metabolism in the nervous system. *Dev. Neurosci.* V. 15 (194–206).

AUTHOR INDEX

SUBJECT INDEX